A Companion to the History of Science

BLACKWELL COMPANIONS TO WORLD HISTORY

This series provides sophisticated and authoritative overviews of the scholarship that has shaped our current understanding of the world's past. Each volume comprises between 25 and 40 essays written by individual scholars within their area of specialization. The aim of each contribution is to synthesize the current state of scholarship from a variety of historical perspectives and to provide a statement on where the field is heading. The essays are written in a clear, provocative, and lively manner, designed for an international audience of scholars, students, and general readers.

The *Blackwell Companions to World History* is a cornerstone of the overarching *Companions to History* series, covering British, American, and European History

WILEY BLACKWELL COMPANIONS TO BRITISH HISTORY

A Companion to Roman Britain
Edited by Malcolm Todd

A Companion to Britain in the Later Middle Ages
Edited by S. H. Rigby

A Companion to Tudor Britain
Edited by Robert Tittler and Norman Jones

A Companion to Stuart Britain
Edited by Barry Coward

A Companion to Eighteenth-Century Britain
Edited by H. T. Dickinson

A Companion to Nineteenth-Century Britain
Edited by Chris Williams

A Companion to Early Twentieth-Century Britain
Edited by Chris Wrigley

A Companion to Contemporary Britain
Edited by Paul Addison and Harriet Jones

A Companion to the Early Middle Ages: Britain and Ireland c.500–c.1100
Edited by Pauline Stafford

WILEY BLACKWELL COMPANIONS TO EUROPEAN HISTORY

A Companion to Europe 1900–1945
Edited by Gordon Martel

A Companion to Eighteenth-Century Europe
Edited by Peter H. Wilson

A Companion to Nineteenth-Century Europe
Edited by Stefan Berger

A Companion to the Worlds of the Renaissance
Edited by Guido Ruggiero

A Companion to the Reformation World
Edited by R. Po-chia Hsia

A Companion to Europe Since 1945
Edited by Klaus Larres

A Companion to the Medieval World
Edited by Carol Lansing and Edward D. English

A Companion to the French Revolution
Edited by Peter McPhee

A Companion to Mediterranean History
Edited by Peregrine Horden and Sharon Kinoshita

WILEY BLACKWELL COMPANIONS TO WORLD HISTORY

A Companion to Western Historical Thought
Edited by Lloyd Kramer and Sarah Maza

A Companion to Gender History
Edited by Teresa A. Meade and Merry E. Wiesner-Hanks

A Companion to the History of the Middle East
Edited by Youssef M. Choueiri

A Companion to Japanese History
Edited by William M. Tsutsui

A Companion to International History 1900–2001
Edited by Gordon Martel

A Companion to Latin American History
Edited by Thomas Holloway

A Companion to Russian History
Edited by Abbott Gleason

A Companion to World War I
Edited by John Horne

A Companion to Mexican History and Culture
Edited by William H. Beezley

A Companion to World History
Edited by Douglas Northrop

A Companion to Global Environmental History
Edited by J. R. McNeill and Erin Stewart Mauldin

A Companion to World War II
Edited by Thomas W. Zeiler, with Daniel M. DuBois

A Companion to the History of Science
Edited by Bernard Lightman

A Companion to Chinese History
Edited by Michael Szonyi

A Companion to Nazi Germany
Edited by Shelley Baranowski, Armin Nolzen, Claus-Christian W. Szejnmann

A Companion to Public History
Edited by David Dean

A COMPANION TO THE HISTORY OF SCIENCE

Edited by

Bernard Lightman

WILEY Blackwell

Registered Offices
John Wiley & Sons, Inc., 111 River Street, Hoboken, NJ 07030, USA
John Wiley & Sons Ltd, The Atrium, Southern Gate, Chichester, West Sussex, PO19 8SQ, UK

Editorial Office
111 River Street, Hoboken, NJ 07030, USA

For details of our global editorial offices, customer services, and more information about Wiley products visit us at www.wiley.com.

Wiley also publishes its books in a variety of electronic formats and by print-on-demand. Some content that appears in standard print versions of this book may not be available in other formats.

Library of Congress Cataloging-in-Publication Data

Name: Lightman, Bernard V., 1950– editor.
Title: A companion to the history of science / edited by Bernard Lightman.
Description: Chichester, UK ; Malden, MA : John Wiley & Sons, 2016. | Series:
 Blackwell companions to world history | Includes index.
Identifiers: LCCN 2015041597 (print) | LCCN 2015036870 (ebook) | ISBN
 9781118620779 (cloth) | ISBN 1118620771 (cloth) | ISBN 9781119121145 (pbk.) |
 ISBN 9781118620748 (Adobe PDF) | ISBN 9781118620755 (ePub)
Subjects: LCSH: Science–Historiography.
Classification: LCC Q124.7 .C65 2016 (ebook) | LCC Q124.7 (print) | DDC
 507.2/2–dc23
LC record available at https://lccn.loc.gov/2015041597

Cover Design: Wiley
Cover Image: Depiction of the Geo-Heliocentric Universe of Tycho Brahe, 17th century by
Andreas Cellarius © CM Dixon/Print Collector/Getty Images

Set in 10/12pt Galliard by SPi Global, Pondicherry, India
Printed and bound in Singapore by Markono Print Media Pte Ltd

10 9 8 7 6 5 4 3 2 1

Contents

About the Editor

Bernard Lightman is Professor of Humanities at York University and former editor of the journal *Isis* (2004–2014). Lightman's most recent publications include *Victorian Popularizers of Science, Victorian Scientific Naturalism* (co-edited with Gowan Dawson), *Evolution and Victorian Culture* (co-edited with Bennett Zon), and *The Age of Scientific Naturalism* (co-edited with Michael Reidy). He is currently working on a biography of John Tyndall and is one of the editors of the John Tyndall Correspondence Project, an international collaborative effort to obtain, digitalize, transcribe, and publish all surviving letters to and from Tyndall.

About the Contributors

Jim Bennett is Keeper Emeritus at the Science Museum, London. He was formerly Director of the Museum of the History of Science, University of Oxford. He has published on the history of instruments, of astronomy and of practical mathematics from the sixteenth to the nineteenth centuries.

Charlotte Bigg is a researcher at the CNRS, Centre Alexandre Koyré, Paris. She has published widely on scientific images and visual cultures in the nineteenth and twentieth centuries. She has co-edited *The Heavens on Earth: Observatories and Astronomy in the Nineteenth Century* (Duke University Press, 2010) and *Atombilder. Ikonografie des Atoms in Wissenschaft und Öffentlichkeit des 20. Jahrhunderts* (Wallstein Verlag, 2009). She is currently preparing *Astronomy and Photography*, to appear in Reaktion Books' *Exposure* series.

Jimena Canales holds the Thomas M. Siebel Chair in the History of Science at University of Illinois–UC and was previously Assistant and Associate Professor at Harvard University. She is the author of *The Physicist and the Philosopher: Einstein, Bergson and the Debate that Changed Our Understanding of Time* (Princeton University Press, 2015) and *A Tenth of a Second: A History* (University of Chicago Press, 2010) and has published widely on science, technology, art, and philosophy.

Valérie Chansigaud is a researcher associated with SPHERE laboratory (University Paris–Diderot–CNRS). She studies the relation between human beings and wild nature. She has published several books of the history of ornithology, naturalist illustration, and protection of nature. Her last book, *L'Homme et la Nature* (Delachaux et Niestlé, 2013), has been given the Prix Léon de Rosen of the Académie française.

Peter Dear teaches the history of science and science studies at Cornell University. He is the author of *Revolutionizing the Sciences: European Knowledge and Its Ambitions 1500–1700* (2nd ed., Princeton University Press, 2009).

Marwa Elshakry is Associate Professor in the History Department at Columbia University. She is the author of *Reading Darwin in Arabic, 1860–1950* (University of Chicago Press, 2013) and co-editor, with Sujit Sivasundaram, of *Science, Race and Imperialism*, volume 6 of *Victorian Literature and Science* (Pickering & Chatto, 2012).

Diarmid A. Finnegan is Senior Lecturer in Human Geography in the School of Geography, Archaeology, and Palaeoecology at Queen's University, Belfast, United Kingdom. His research interests center on the cultural geography of science and religion in the nineteenth century. His work includes the book *Natural History Societies and Civic Culture in Victorian Scotland* (Routledge, 2009) as well as several articles on scientific culture in Victorian Britain and Ireland.

Aileen Fyfe is Reader in Modern British History at the University of St. Andrews. Her research interests lie in the communication and popularization of the sciences. She is author of *Science and Salvation* (University of Chicago Press, 2004) and *Steam-Powered Knowledge* (University of Chicago Press, 2012), and co-editor of *Science in the Marketplace* (University of Chicago Press, 2007). She is currently leading a major research project "Publishing the *Philosophical Transactions*: The economic, social and cultural history of a learned journal 1665–2015."

Anita Guerrini is Horning Professor in the Humanities and Professor of History at Oregon State University. Trained in the history of science, she has written on the history of experimenting, animals, medicine, food, and the environment. Her books include *Experimenting with Humans and Animals: from Galen to Animal Rights* (John Hopkins University Press, 2003) and *The Courtiers' Anatomists: Animals and Humans in Louis XIV's Paris* (University of Chicago Press, 2015). She blogs at http://anitaguerrini.com/anatomia-animalia/.

Klaus Hentschel is professor for history of science and technology in Stuttgart. He has worked on relativity theory, quantum physics, spectroscopy, the interplay of instrumentation, experimentation and theory formation, on social networks and on invisible hands, on taxonomies in science, and on argumentation. For his studies on the physical sciences he has received five national and international prizes. cf. www.uni-stuttgart.de/hi/gnt/hentschel and …/gnt/dsi for his international "Database of Scientific Illustrators 1450-1950" with more than 10,000 entries.

Catherine Jackson is Assistant Professor of History of Science at University of Wisconsin, Madison. She previously held research fellowships at the University of Notre Dame and Chemical Heritage Foundation. Originally trained as a synthetic organic chemist, Jackson is completing a book on the origins of organic synthesis. She has published on Liebig, Hofmann, and the chemical laboratory, and co-edited (with Hasok Chang) *An Element of Controversy: The Life of Chlorine in Science, Medicine, Technology and War* (British Society for the History of Science, 2007).

Boris Jardine was the 2014-15 Munby Fellow in Bibliography, Cambridge University Library. He has published widely on scientific instrumentation, in particular the relation between instruments and texts. He has also worked at the Whipple Museum of the History of Science (Cambridge) and the Science Museum (London).

Kristin Johnson is Associate Professor of Science, Technology, and Society at the University of Puget Sound in Tacoma, Washington. She is the author of *Ordering Life: Karl Jordan and the Naturalist Tradition* (Johns Hopkins University Press, 2012).

Matthew L. Jones is the James R. Barker Professor of Contemporary Civilization at Columbia University. A Guggenheim Fellow, he is completing a book on the National

Security Agency, and is undertaking a historical and ethnographic account of "big data," its relation to statistics and machine learning, and its growth as a fundamental new form of technical expertise in business, political, and scientific research. His *Reckoning with Matter: Calculating Machines, Innovation, and Thinking About Thinking from Pascal to Babbage* is forthcoming from the University of Chicago Press.

Heike Jöns is Senior Lecturer in Human Geography at Loughborough University. She has widely published on the geographies of science and higher education with a focus on transnational academic mobility and knowledge production. Her books include the research monograph *Grenzüberschreitende Mobilität in den Wissenschaften* (Universität Heidelberg 2003) and the edited volume *Geographies of Science* (Springer, 2010). Her current research examines the history of European universities with an emphasis on Britain and Germany.

David A. Kirby is Senior Lecturer in Science Communication Studies at the University of Manchester. Several of his publications address the relationship between cinema and the cultural meanings of genomics. His book *Lab Coats in Hollywood: Science, Scientists, and Cinema* (MIT Press, 2011) examines collaborations between scientists and the entertainment industry. He is currently writing a book entitled "Indecent Science: Film Censorship and Science, 1930–1968" exploring how movies served as a battleground over science's role in influencing morality.

Robert E. Kohler is Professor Emeritus of History and Sociology of Science at the University of Pennsylvania. He has written extensively on the history of the field sciences.

Steven J. Livesey is Brian E. and Sandra O'Brien Presidential Professor of the History of Science at the University of Oklahoma. His research interests focus on medieval science, history of early scientific methodologies, science in medieval universities, and manuscript studies. His current research project, supported by the Fulbright Commission, investigates the medieval library of Saint-Bertin, and the project goal is to reconstruct the library by identifying modern survivors of a collection that has been dispersed since the French Revolution.

Paul Lucier is historian of the earth and environmental sciences and of their interconnectedness with the mining and energy industries. He is the author of *Scientists and Swindlers: Consulting on Coal and Oil in America, 1820-1890* (John Hopkins University Press, 2008) and is currently writing a history of science and capitalism in America.

Rory McEvoy is Curator of Horology at the Royal Observatory, part of Royal Museums, Greenwich. His research has a natural focus on the production, development, and use of precision horological instruments and associated technology as well as the broader history of the Royal Observatory, civil time, and its distribution.

Cyrus C. M. Mody is Professor and Chair in the History of Science, Technology, and Innovation in the Faculty of Arts and Social Sciences at Maastricht University. He is the author of *Instrumental Community: Probe Microscopy and the Path to Nanotechnology* (MIT Press, 2011). His current research focuses on the semiconductor industry's shaping of changes in US science and science policy since 1965, and on American physical and engineering scientists' creative responses to the dire conditions of the long 1970s.

Bruce T. Moran is Professor of History at the University of Nevada, Reno, and teaches courses in the history of science and medicine. He is the author of *Distilling Knowledge: Alchemy, Chemistry, and the Scientific Revolution* (Harvard University Press, 2005) and *Andreas Libavius and the Transformation of Alchemy: Separating Chemical Cultures with Polemical Fire* (Science History Publications, 2007). A co-edited volume, *Bridging Traditions: Alchemy, Chemistry, and Paracelsian Practices in the Early Modern Era*, appeared in 2015 from Truman State University Press. His current project examines the relationship between private sentiment and alchemical practice.

Iwan Rhys Morus is Professor of History at Aberystwyth University in Wales. He completed his PhD at Cambridge in 1989 and since then has worked largely in the area of Victorian science, with particular interests in the public culture of science and the backstage work of scientific performance. He is the author of *Frankenstein's Children: Electricity, Exhibition and Experiment in early Nineteenth-century London* (Princeton University Press, 1998), *When Physics became King* (University of Chicago Press, 2005), and *Shocking Bodies: Life, Death and Victorian Electricity* (History Press, 2011) as well as co-author of *Making Modern Science* (University of Chicago Press, 2005).

Joshua Nall is Curator of Modern Sciences at the Whipple Museum of the History of Science, in the Department of the History and Philosophy of Science at the University of Cambridge. His research focuses on mass media and material culture of the physical sciences after 1800. He is currently preparing a monograph on the role of mass media in *fin de siècle* debates over life on Mars.

Carla Nappi is Associate Professor of History and Canada Research Chair in Early Modern Studies at the University of British Columbia. Her first book was *The Monkey and the Inkpot: Natural History and its Transformations in Early Modern China* (Harvard University Press, 2009). She is currently working on the histories of translation, narrative, and embodiment in Ming and Qing China from the fifteenth through nineteenth centuries.

Tara Nummedal is Associate Professor of History at Brown University. She is the author of *Alchemy and Authority in the Holy Roman Empire* (University of Chicago Press, 2007) and is currently completing "The Lion's Blood: Alchemy, Gender, and Apocalypse in Reformation Germany."

Lynn K. Nyhart is Vilas-Bablitch-Kelch Distinguished Achievement Professor of the History of Science at the University of Wisconsin-Madison. The author of *Biology Takes Form* (University of Chicago Press, 1995) and *Modern Nature: The Rise of the Biological Perspective in Germany* (University of Chicago Press, 2009), she is currently a Senior Fellow at UW–Madison's Institute for Research in the Humanities, working on a history of ideas about biological individuals, parts, and wholes in the nineteenth century.

Brian Ogilvie is Associate Professor of History at the University of Massachusetts Amherst. He is the author of *The Science of Describing: Natural History in Renaissance*

Europe (University of Chicago Press, 2006). His current research focuses on insects in European art, science, and religion from the Renaissance to the Enlightenment. He is also writing a short book on the cultural history of the butterfly for the *Animal* series from Reaktion Books.

Donald L. Opitz is Associate Professor in the School for New Learning at DePaul University. His research concerns the role of science in Anglo-American Victorian culture, with an emphasis on gender, class, and sexuality. He is co-editor, with Annette Lykknes and Brigitte Van Tiggelen, of *For Better or For Worse? Collaborative Couples in the Sciences* (Birkhaüser, 2012) and principal editor, with Staffan Bergwik and Brigitte Van Tiggelen, of *Domesticity in the Making of Modern Science* (Palgrave Macmillan, 2015).

Katherine Pandora is Associate Professor in the Department of the History of Science at the University of Oklahoma. Her research focuses on questions of scientific authority, science and popular culture, and science communication, particularly in relation to nineteenth and twentieth-century natural history and social science. She is the author of *Rebels within the Ranks: Psychologists' Critique of Scientific Authority and Democratic Realities in New Deal America* (Cambridge University Press, 1997), and blogs at katherinpandora.net/petri_dish.

Denise Phillips is Associate Professor of History at the University of Tennessee, where she teaches German history and the history of science. She is the author of *Acolytes of Nature: Science and Public Culture in Germany* (University of Chicago Press, 2012) and the co-editor of *New Perspectives in the Life Sciences and Agriculture* (Springer 2015).

Kapil Raj is Directeur d'études (Research Professor) at the École des Hautes Études en Sciences Sociales in Paris. His research examines the global intercultural negotiations which have gone into its making, the subject of *Relocating Modern Science* (Palgrave Macmillan, 2007) which focuses on the role of circulation and encounter between South Asian and European skills and knowledges in the emergence of crucial parts of modern science. He has also co-edited *The Brokered World: Go-Betweens and Global Intelligence, 1770–1820* (Science History Publications, 2009) and has just co-edited another collective work on the history of knowledge and science in the long nineteenth century which will appear in French at the end of 2015. He is currently engaged in researching for his next book on the urban and knowledge dynamics of Calcutta in the eighteenth century.

Lukas Rieppel is David and Michelle Ebersman Assistant Professor of History at Brown University. He received his PhD at Harvard University in 2012 and is currently writing a book that explores what the history of dinosaur paleontology can tell us about the culture of capitalism in late nineteenth and early twentieth-century America. Under contract with Harvard University Press, the book is tentatively titled *Assembling the Dinosaur: Science, Museums, and American Capitalism, 1870–1930*. Together with Eugenia Lean and William Deringer, he is also editing the 2018 volume of *Osiris* on the theme of *Science and Capitalism: Entangled Histories*.

Nathan Sidoli received a BA in Liberal Arts from St. John's College, Santa Fe, and an MA and PhD from the University of Toronto in the History and Philosophy of Science and Technology, with a dissertation on the mathematics of Claudius Ptolemy. He was a principle-investigator postdoctoral fellow for the US National Science Foundation and the Japan Society for the Promotion of Science, before taking up a position at Waseda University (Tokyo, Japan), where he is currently Associate Professor of the History and Philosophy of Science. His current research focuses on foundations and practices in Greek mathematics and the transmission of Greek mathematical sciences in Arabic sources.

Josep Simon teaches the history of science, technology, and medicine at the Universidad del Rosario, Bogotá. He is the author of the award-winning *Communicating Physics: the Production, Circulation and Appropriation of Ganot's Textbooks in France and England (1851–1887)* (Routledge, 2011), the *Handbook* chapter "Physics Textbooks and Textbook Physics in the Nineteenth and Twentieth Centuries" (Oxford University Press, 2013), the *Encyclopedia of Science Education* entry "History of Science" (Springer, 2015) and a number of special issues, and articles on science, education, and their historiographical interfaces.

Robert Smith is Professor of History in the Department of History and Classics at the University of Alberta. In addition to books and numerous articles on twentieth-century astronomy and the history of large-scale science, he has also written on a range of topics in the history of nineteenth-century astronomy broadly conceived.

Mary Sunderland is a historian of science and technology at the University of California, Berkeley where she is affiliated with the Center for Science, Technology, Medicine, and Society and the Department of Nuclear Engineering. She is interested in the twentieth-century life sciences. At present, her research focuses on engineering education and translational research. Questions about how the societal roles of scientists and engineers are shaped by pedagogy motivate her work.

Liba Taub is Director and Curator of the Whipple Museum of the History of Science, and Professor of History and Philosophy of Science, at the University of Cambridge. Her research focuses on material culture of science, and Greco-Roman science. With Frances Willmoth, she co-edited *The Whipple Museum of the History of Science: Instruments and Interpretations, to Celebrate the 60th Anniversary of R.S. Whipple's Gift to the University of Cambridge* (Cambridge University Press, 2006). She retains fond memories of "Things of Science."

Joyce van Leeuwen is Postdoctoral Research Scholar at the Max Planck Institute for the History of Science in Berlin. She pursued graduate studies at the Humboldt University Berlin and Stanford University. Her research interests lie in Greek paleography, diagrammatic reasoning, history of mechanics, and early modern science. *The Aristotelian Mechanics: Text and Diagrams* will appear in 2015 in Springer's *Boston Studies in the Philosophy and History of Science*.

Hector Vera is a researcher at Instituto de Investigaciones sobre la Universidad y la Educación, at Mexico's National University (UNAM). He has a PhD in sociology and historical studies from The New School for Social Research. His doctoral dissertation, "The Social Life of Measures: Metrication in Mexico and the United States, 1789–1994," is a historical-comparative analysis on how diverse institutions and groups (state agencies, scientific societies, chambers of commerce and industry) appropriated and signified the decimal metric system. He is the author of *A peso el* *kilo. Historia del sistema métrico decimal en México* (Libros del Escarabajo, 2007), a monograph on the adoption of the metric system in Mexico. He is also co-editor, with V. García-Acosta, of a volume on the history of systems of measurement, *Metros, leguas y mecates. Historia de los sistemas de medición en México* (CIESAS, 2011).

Jeremy Vetter is Assistant Professor of History at the University of Arizona. He works at the intersection of history of science and technology, environmental history, and the history of the American West. He is author of *Field Life: Science in the American West during the Railroad Era* (University of Pittsburgh Press, forthcoming).

Paul White is an editor on the Darwin Correspondence Project and teaches in the Department of History and Philosophy of Science at the University of Cambridge. He is the author of *Thomas Huxley: Making the 'Man of Science'* (Cambridge University Press, 2003) and various articles on Victorian science, literature, and culture. He is working on a book on "Darwin and the Evolution of Emotion."

Nick Wilding is Associate Professor of History at Georgia State University. He works on early modern science and modern forgery. He is the author of several articles, reviews, and digital projects on Hooke, Wilkins, Galileo, Sagredo, and Kircher. *Galileo's Idol: Gianfrancesco Sagredo and the Politics of Knowledge*, which came out with the University of Chicago Press in 2014, has won the Aldo and Jeanna Scaglione Prize for Italian Studies from the Modern Language Association.

Acknowledgements

This was a particularly demanding project, both because of the size of the volume and the range of the topics. Since I began work on it in 2012 I have called upon many of my colleagues for help. I am indebted to those colleagues who suggested the names of possible chapter contributors, including Mario Biagioli, Dana Freiburger, Klaus Hentschel, Adrian Johns, Edward Jones-Imhotep, Daryn Lehoux, Lissa Roberts, Grace Shen, and Larry Stewart. Several colleagues helped not only to suggest possible contributors but also to conceptualize specific sections of the volume. They were David Livingstone, Alison Morrison-Low, Tacye Phillipson, and Klaus Staubermann. I am especially grateful to those who gave me advice on how to structure the entire volume, which was a particularly complicated task. They were Katey Anderson, Janet Browne, James Elwick, and Bob Westman.

Then there were a small number of scholars who I consider to be unofficial advisors to the project. I saved some of the most troublesome questions for them. Liba Taub was an invaluable help in working through the Tools section, the area covered by the volume about which I know the least. Rob Kohler and Lynn Nyhart gave me sound advice on the structure of the volume and many other difficult issues. When contributors were unsure what to read on developments in science that took place in Asia, South America, or Africa, I was fortunate to have Sonja Brentjes advise me on what to suggest. Finally, the basic structure of the volume was worked out one night over dinner with Anne and Jim Secord.

I have found the editors at Wiley to be well organized and efficient, as well as a pleasure to work with. I want to thank Sally Cooper, Tessa Harvey, Georgina Coleby, and Karen Shield for their guidance throughout the entire life of the project. Alec McAulay was a superb copy-editor and Shalini Sharma managed the production activities with great skill.

My greatest debt is to my wife, Merle, to whom I have been married for almost 40 years. Her love has sustained me through good times and bad. I dedicate this work to her.

Introduction

Bernard Lightman[1]

For those of us who populate the industrialized regions of the world, it is not very controversial to assert that our lives are profoundly shaped by science. In our everyday, mundane existence we are constantly encountering, using, and relying on specific technologies that are based on scientific discoveries. In addition, we see how science has transformed the physical world that forms the stage on which we go about our business day after day. Our relationship to nature, for better or for worse, is mediated through science. The very way we think is indebted to scientific ideas. The culture surrounding us is saturated with them. Popular films bring the lives of colorful scientists, such as Stephen Hawking, Alan Turing, and Albert Einstein, to the big screen. Controversies over scientific issues appear regularly in our media, whether it be the theory of evolution, the possibility of life on other planets, the dangers of climate change, or the authority of the modern scientist. But how and when did this come to be? Science was not always so central to human culture. And what is the larger significance of its centrality? These are among the questions tackled by historians of science.

Over the last 35 years the study of the history of science has been transformed by the gradual adoption of a new historiographical approach. Whereas the history of science previously stressed a big picture focusing on the theoretical progress made by great scientific heroes like Galileo and Newton, the field is now dominated by scholars offering rich, thickly descriptive, local studies. Rather than emphasizing the discovery of new scientific theories, historians of science became interested in how science was practiced in the laboratory as well as in other sites. A whole new cast of characters has been added to the story, most of them from outside the intellectual elite, including women, invisible assistants, popularizers, and members of the working class. Historians of science have integrated modes of scholarship from other fields into their work. They have looked to cultural studies, communication studies, women's studies, visual studies, and the scholarship on science and literature, to name just a few. The result has been the development of a dynamic field out of which has come some of the most exciting scholarship in the world of academe.

A Companion to the History of Science, First Edition. Edited by Bernard Lightman.
© 2016 John Wiley & Sons Ltd. Published 2020 by John Wiley & Sons Ltd.

Those of us who witnessed this seismic shift in the 1980s and 1990s, and who maybe even contributed to the upheaval, will have a particular book or article that inspired them to see the field in a different way, or that helped them understand just how much the ground had shifted underneath us. For me, and I suspect for many others, it was Paul Forman's stirring declaration "Independence, Not Transcendence, for the Historian of Science," published in *Isis* in 1991. The point of Foreman's article was to provide a "principled basis" for those historians of science who wanted independence from the sciences. He argued that the role of the scientist and the role of the historian of science were fundamentally different. While the scientist embraced transcendence, the historian of science cultivated independent moral judgment (Forman 1991, 71). Historians of science, then, had to supply their own agenda for their discipline rather than accepting that of the scientist. We could not be intellectually subservient like historians of science from earlier decades. Our business was not celebrating the past achievements of scientists. Nor was it studying those scientific theories that were considered correct by contemporary standards. If we had to understand the science of any period we sometimes had to look at scientific pursuits now seen as marginal or pseudoscientific. Whereas the defenders of the old scholarship would have considered an investigation of phrenology or mesmerism as a waste of time, those seeking independence had to be prepared to pursue the understanding of science in a particular period wherever it took them. Our job was to completely historicize "scientific knowledge—explaining possession of specific pieces or structures of it, not by appealing to a transcendent reality…, but by reference to mundane factors and human actors" (Forman 1991, 78). Forman believed that historians of science had been groping their way towards genuine intellectual autonomy; however, they had not fully grasped that the "new" history being developed was based on a renunciation of transcendence (Forman 1991, 85).

Forman's declaration of independence on behalf of historians of science was, for many, a revelation. It contributed to a reorientation of the discipline that was both exhilarating and daunting. Exhilarating because it opened up a whole new set of questions by casting a different light on some of the basic assumptions of the older scholarship. Was there a scientific revolution in the early modern period that led to the formation of what we think of as modern science—or not? Was there really such as thing as the "Darwinian revolution" in the nineteenth century? Could we really make the concept of progress the main feature of the story we told about science? But these big questions were daunting as well. They added up to one gigantic question: what, exactly, were historians of science studying? In other words, was there no essential "thing" that we could call "science" that began in ancient times and survived to the present? (Golinski 2012). In gaining our independence we had to reconstitute our discipline. The aim of the *Blackwell Companion to the History of Science* is not exactly to provide a single, unified "big picture"—something that many view as epistemologically suspect. Rather its object is to survey recent developments that have resulted from the effort to re-envision the field.

Deciding on a structure for this volume was anything but straightforward. The structure had to reflect the significant historiographical shift that took place since the 1980s. The chapters themselves had to be synthetic, midscale studies rather than microstudies (Kohler and Olesko 2012). But what topics should the chapters focus on, and how should the chapters be organized into parts? The initial temptation—almost irresistible for a historian—was to think along chronological lines. A chronological approach, starting with the ancient period and then moving through the middle ages, the early modern period, the eighteenth century, and the modern era, was fairly common for previous surveys of the history of science. Andrew Ede and Lesley Cormack's one-volume *A History of Science in Society* (2004) followed that format (Ede and Cormack 2004). So did the eight-volume Cambridge History of Science series, edited by Ronald Numbers and David Lindberg (Lindberg and Numbers 2003–). Moreover, there are many books that deal with specific periods in the history of science. I wanted to try something different. Perhaps a structure that combined chronological and thematic approaches would be best? In effect, this was the structure adopted by *The Routledge Companion to the History of Modern Science* (1990) (Olby, Cantor, Christie, and Hodge 1990). But there are 67 chapters in that book and I had fewer to work with. Trying to cover both key chronological periods and important themes would be impossible. I also wanted to have a tighter focus for the thematic chapters. After consulting widely with colleagues, I finally decided on a four-part thematic approach reflecting the broad analytical categories central to history of science today. Adopting this structure helps us to move the emphasis in the volume away from the discovery of abstract scientific theories, the theme of progress through the ages, and the contributions of specific elite scientists. There is a loose chronological order within some of the parts so that developments over time can be tracked. But the thematic structure has allowed contributors to cut across traditional chronological and geographic boundaries in exciting ways.

The first chapter is actually a prologue to the four parts. Here, Lynn Nyhart provides a much more detailed and nuanced discussion than the one in this introduction of the historiographical trends over the last 35 years that have made four analytical categories so important for historians of science. It is a complicated story, which illustrates how historians of science have borrowed from other disciplines as the ground beneath their feet began to shift. She examines the impact of social constructionism and feminist scholarship on the history of science, with their emphasis on how science has been constructed by a diverse group of individuals, by no means just male intellectuals, through a complex social process. Then she shows how this led historians to explore the nature of past scientific activity, or what has been called "scientific practice." Looking at the making of scientific knowledge opened up new doors. It led historians of science to investigate communicative practices, whether it be the movement of knowledge between scientists, or between scientists and the public, or even from the local or national context to the global context. The turn to practice also raised interesting questions about the material culture of science, the *stuff* that scientists worked with, from specimens to gigantic instruments. The four analytic categories, then, deal with the roles, places and spaces, communication, and tools of science.

The chapters in Part I, on "Roles," will explore the various roles of the "scientist" from ancient times to the present. The term is in quotation marks as the chapters will emphasize how the idea of the "scientist" has changed dramatically over time and to indicate that the chapters are more concerned with what could be called "roles in science." The term itself was not coined until 1834 by the English polymath William Whewell (Whewell 1834, 59). The article in which he introduced the term was actually a laudatory review of Mary Somerville's *On the Connexion of the Physical Sciences* (1834). Whewell did not have in mind the specialized, professional scientist that we are familiar with today. He was using the term to counter the tendency of his contemporaries to subdivide science into separate disciplines. Inventing the term "scientist" was part of Whewell's plea for unity in science and his rejection of specialization.

It is not likely that Whewell would have expanded the term "scientist" to include invisible technicians, instrument makers, artisans, or human experimental subjects. But given the diverse roles played by scientific figures in the past there is a good argument for including them. If we were to apply the term only to those who fit the current criteria for defining who is and who is not a scientist, the number of those who met the qualifications would diminish the further back we went into the past. Understanding how lines were drawn between who was considered to be in possession of natural knowledge and who was not is one of the goals of these chapters. The social role of individuals with special relationships to natural knowledge must be considered in various cultural settings located in different times and places.

The chapters in Part II, "Places and Spaces," all examine the situatedness of knowledge. All scientists, whatever role they assume, must perform that role in a specific place. Historical geographers of science like David Livingstone have emphasized that space is not a neutral "container" in which social life takes place. "Space," Livingstone asserts, "is not (to change the metaphor) simply the stage on which the real action takes place. Rather, it is itself constitutive of systems of human interaction" (Livingstone 2003, 7). When we are considering critical sites in the generation of knowledge, such as the university, the field, or the laboratory, we always need to ask, who manages that space? What are its boundaries? Who is allowed access? Paying attention to place, by contrast, means taking into account the local, regional, and national features of science. If we take Forman seriously then we will not think of science, as Livingstone puts it, "as some transcendent entity that bears no trace of the parochial or contingent." Rather we will cultivate a "geography of science" that reveals "how scientific knowledge bears the imprint of its location" (Livingstone 2003, 13).

Part II, then, will examine the sites from which scientific knowledge has emerged, and will concentrate more on the local rather than the regional or national scale. It is striking to see how sites of knowledge have varied from the ancient period to the present. Durable sites, such as the university and the observatory, have changed dramatically over time. But there is nothing analogous to some of the older sites, such as the European court of the sixteenth and seventeenth century, while new sites, such as the scientific society, did not exist prior to the early modern period. Some of these spaces, such as the laboratory and the museum, have long been recognized by historians as privileged places of power. But the importance of others, such as domestic and commercial spaces, have only recently been recognized. In this section we have only touched on a relatively small number of scientific sites. Studying the remarkable range of sites in which scientific work has been undertaken illustrates why space and place matter.

After examining how location figures into the generation of knowledge, Part III focuses on how ideas and images travel between sites. As they circulate, scientific ideas and images undergo translation and transformation, since people encounter representations differently in different circumstances (Livingstone 2003, 11). Jim Secord's widely cited article "Knowledge in transit" (2004) outlines the contours of this dimension of the historical mode of enquiry. Secord points out that focusing on how knowledge is generated locally, at times, produced an obstacle for historians.

"The more local and specific knowledge becomes," Secord declared, "the harder it is to see how it travels" (Secord 2004, 660). To counter this problem, Secord suggested that we understand science as a form of communication in which the processes of movement, translation, and transmission become central. "This means thinking always about every text, image, action, and object as the trace of an act of communication," Secord asserted, "with receivers, producers, and modes and conventions of transmission. It means eradicating the distinction between the making and the communicating of knowledge" (Secord 2004, 661).

In Part III, "Communication," the authors examine how knowledge was transferred between sites through a variety of media, including print, visual, and oral media. There are chapters on manuscripts, letters, periodicals, books, textbooks, lectures, film, radio, and television. But there is also some attention to the changing technologies of communication, in particular print forms of communication, as in the chapter on the printing press. The chapters deal both with how scientists communicated to each other, and how they communicated to the public. We could have included many more chapters on "Communication." It has been a topic of much scholarly interest since the turn of the century. Moreover, there are many more modes of communication central to science and connected with specific places that we could not cover due to space constraints, such as field notebooks, museum catalogues, and the experimental register. We have included, primarily, those modes of communication that have received the most attention from scholars.

Secord has pointed out that the key to creating a history of science as a form of communication "is our new understanding of scientific knowledge as practice. All evidence from the past is in the form of material things" (Secord 2004, 665). This is as true of periodicals, books, and notebooks as it is of experimental instruments, natural history specimens, and two-dimensional models. Studying the communication of science therefore leads us to the investigation of its material culture. Part IV deals with the tools of science, which also circulate between scientific sites. Chapters will cover important scientific instruments and material objects as a way to illuminate the changing practices of science. We will encounter chapters on timing, measuring, calculating, and recording devices; instruments, such as microscopes, telescopes, and spectroscopes, that enhance the senses; and material objects that have been used by scientists including specimens, collections, diagrams, and three-dimensional models.

Scientific objects are the things studied by scientists, whereas instruments are the tools by which those objects are studied. Instruments and objects have been the subject of investigation for several decades. Though they are treated by historians of science as part of material culture, this does not preclude attention to their epistemological dimensions. Daston's edited collection *Biographies of Scientific Objects* (2000) dealt with how "whole domains of phenomena—dreams, atoms, monsters, culture, mortality, centers of gravity, value, cytoplasmic particles, the self, tuberculosis—come into being and pass away as objects of scientific inquiry" (Daston 2000, 1). Daston was not just interested in objects as material. She wanted to understand how material objects contained a significant intellectual component. Though it is less obvious, instruments also have immaterial attributes. Liba Taub has noted that the turn toward scientific practice beginning in the 1990s brought with it attention to instruments. "At the same time," Taub affirms, "there was a growing fascination on the part of many scholars, in a range of disciplines, with 'materiality'" (Taub 2011, 690). However the fascination with materiality does not limit historians of science to the object *qua* thing. Taub argues that the work on instruments problematized them by forcing scholars to confront how they understood the term "instrument" itself (Taub 2011, 696), just as Daston asked how and when a scientific object came to be. "Object" and "instrument" both have material and immaterial attributes.

By focusing on the roles, places, communicative practices, and materials of science in the past we hope to capture what has made current scholarship in the field so vibrant and exciting. But the field continues to evolve. Undoubtedly, new analytical categories will be developed in the future by enterprising historians of science. These kinds of experiments in historical innovation are to be encouraged if the field is to retain its vitality and its relevance. Moreover, they are essential if historians of science hope to maintain their independence.

Endnote

1 I am indebted to Lynn Nyhart for her extremely helpful suggestions on how to strengthen this introduction.

References

Daston, Lorraine (ed). 2000. *Biographies of Scientific Objects*. Chicago: University of Chicago Press.

Ede, Andrew, and Lesley Cormack. 2004. *A History of Science in Society: From Philosophy to Utility*. Peterborough, Ontario: Broadview Press.

Forman, Paul. 1991. "Independence, not transcendence, for the Historian of Science." *Isis*, 82: 71–86.

Golinski, Jan. 2012. "Is it time to forget science? Reflections on singular science and its history." *Osiris*, 27: 19–36.

Kohler, Robert E., and Kathryn M. Olesko. 2012. "Introduction: Clio meets science." *Osiris*, 82: 1–16.

Lindberg, David, and Ronald L. Numbers (eds). 2003– . *The Cambridge History of Science*. 8 vols. Cambridge: Cambridge University Press.

Livingstone, David N. 2003. *Putting Science in Its Place: Geographies of Scientific Knowledge*. Chicago and London: University of Chicago Press.

Olby, R. C., G. N. Cantor, J. R. R. Christie, and M. J. S. Hodge (eds). 1990. *Companion to the History of Modern Science*. London: Routledge.

Secord, James A. 2004. "Knowledge in transit." *Isis*, 95 (December): 654–672.

Taub, Liba, 2011. "Introduction: Reengaging with instruments." *Isis*, 102: 689-696.

[Whewell, William.] 1834. "On the connexion of the physical sciences. By Mrs. Somerville." *Quarterly Review*, 51: 54–68.

CHAPTER ONE

Historiography of the History of Science

LYNN K. NYHART

Over the past 35 years or so, the subject matter, people, places, and processes associated with history of science have grown vastly. Exaggerating only slightly for effect, an older predominant history of science might be captured by the image of a tree of scientific ideas rooted in the base of Western culture (perhaps extending downward earlier to ancient Egypt and Babylonia); the task of the historian of science was to trace the tree's growth and branching. Today a more fitting image would be of the history of science as a densely tangled bank of people and material things teeming with social, cultural, economic, and religious life, that covers the globe. The historian's task now is to tease out how certain forms of knowledge and practice within this mass of activity came to be understood as "science;" what has sustained science socially, culturally, and materially; and who has benefitted and who has suffered in its formation. What happened in the past did not change: what we expect professional historians of science to know and care about has.

The four parts of this volume—Roles, Places and Spaces, Communication, and Tools of Science—reflect broad analytical categories central to today's history of science. They cut across historical periods, geographical locations, and sciences to provide a common vocabulary that helps tie our far-flung history together. Rather than reproduce these categories in the present essay, I sketch out some of the historiographic trends that made it possible—even commonsensical—to use them to thematize contemporary history of science scholarship written in English.

I focus first on the social constructionist turn of the late 1970s and early 1980s, and its consequences for how we think about the nature of scientific knowledge and who is involved in its making. I then turn to the subsequent (re-)formulation of approaches to answering two fundamental questions in our field. One focuses on *making* scientific knowledge, asking "How is scientific knowledge constructed in a given context?" Historians' answers to this question since the early 1990s have become increasingly attentive to scientific practice, its settings and material culture. A second question focuses on *moving* scientific knowledge. As James Secord (2004, 655) put it, "How

A Companion to the History of Science, First Edition. Edited by Bernard Lightman.
© 2016 John Wiley & Sons Ltd. Published 2020 by John Wiley & Sons Ltd.

and why does [scientific] knowledge circulate? How does it cease to be the exclusive property of a single individual or group and become part of the taken-for-granted understanding of much wider groups of people?" Scholars working on this question have highlighted the tropes of communication and circulation, and indeed often question the very distinction between making and moving.

Recent history of science has been profoundly shaped by its historians' interactions with scholars from other disciplines across and between the social sciences and humanities. In these exchanges, historians of science have both given and received, but they have often shied away from direct theoretical statements in favor of a more empiricist style that integrates analytical insights into narrative structures. Within the broad themes of this essay, I highlight works that articulate or exemplify analytical approaches and conceptual tools that might be applicable to different places and periods. While these often originate from individual authors, I have been particularly struck by the importance of thematic journal issues and that most maligned of genres, the multi-authored edited volume. Thematic volumes are notoriously hard to get published, yet they can raise the visibility of an approach or topic well above the level of the individual article or even book, and give a sense for the significant conversations in which our community participates. The liveliness of these conversations is evidenced by the large number of collective works cited in the present essay—and also, of course, by this volume, which as a whole attests to the community-based nature of the history we make.

Constructing Scientific Knowledge, Socially

Since the late 1970s, historians of science have gradually come to accept a predominantly social constructionist account that views the development of scientific knowledge as depending heavily on particulars of local circumstances, people, epistemes, and politics, and that doesn't necessarily drive ever closer toward a single truth. Although historians of science had long been interested in recovering earlier knowledge systems and the means by which they were transformed over time (e.g. Kuhn 2012), social constructionism offered new tools for doing so. The sociologists of the "Edinburgh School" and the "Bath School" developed many of these tools in the 1970s and early 1980s; despite differences in approach, they broadly articulated what was known as the "Strong Programme" of the social construction of scientific knowledge. (For retrospective analyses of the early situation, see Golinski 2005; Shapin and Schaffer 2011; Kim 2014; Soler et al. 2014).

The new sociologists of scientific knowledge participated in a broader postmodern rejection of our unmediated access to reality, often associated with other critiques of science's truth value. Michel Foucault (especially 1970, 1973) challenged historians to understand how the structures of knowledge, discourse, and institutions instantiated forms of power (the entire bundle called "epistemes") that were virtually invisible to those living inside their regimes. Since he offered no clues as to how one episteme turned into another, and little in the way of specific empirical evidence for his provocative claims, Foucault's work remained largely (if importantly) inspirational. From a different direction, feminist scientists would soon expand the purview of social constructionist criticism of science (Bleier 1984; Fausto-Sterling 1992). Uneasy with both the implications of radical social constructionism and the "all-seeing" stance represented in standard claims to objectivity, however, Sandra Harding (1986) and Donna Haraway (1988) developed, respectively, the crucial ideas of standpoint epistemology and "situated knowledges." Haraway (1988, 590) in particular advocated the "partial perspective," which lent the authority of agency to individuals previously without standing and demanded communal effort to arrive at shared reliable knowledge.

Such perspectives collectively challenged the received view of history of science in two fundamental ways. First, they demonstrated that scientific knowledge was *constructed* by human beings, not discovered in nature. Second, this process was not the work of individual minds but was ineluctably *social*. The implications for history were profound.

If knowledge of nature is made, not arrived at, then we should not expect that science will progress toward a pre-existing universal truth. One important implication is that the truth value of a claim in the past cannot be assessed by what we now believe to be true—an account of the success or failure of a scientific claim must be neutral with respect to that outcome. Evaluations of success must depend on other grounds—social, political, rhetorical—and both successes and failures must be treated similarly. In the 1980s cutting-edge historians of science adopted these principles of "neutrality" and "symmetry" (Bloor 1976), taking up the challenge of treating the outcomes of scientific controversies as determined not by the truth winning, but by social interactions.

The paradigmatic example of this sociological-historical approach is Steven Shapin and Simon Schaffer's *Leviathan and the Air-Pump* (1985). They interpreted the contest between Robert Boyle and Thomas Hobbes as not just over the existence and nature of the vacuum and its experimental proof, but over what sort of knowledge would be counted as scientific (or, more properly, "natural philosophical"), and what adjudged not. The very division between "science" and "non-science" was at stake, and the winner not only won the specific controversy but also the right to claim what kind of knowledge would be constituted as authoritative (experimental knowledge), who would be considered a natural philosopher in the future (Robert Boyle), and who would not (Thomas Hobbes).

Developing the commitment to neutrality with respect to the outcome of a controversy led Martin Rudwick to take a different tack. His *Great Devonian Controversy* (1985) experimented with a radically anti-teleological narrative of controversy, persuasion, and power that steadfastly resisted letting the reader know how this geological story came out until its end. It thereby called attention to the conventions of histories that anticipate the outcome, challenging readers to problematize the very structure of historical narrative and to recognize the contingency of the development of science.

Both books also forcefully showed the extent to which the construction of scientific knowledge was *social*, in the sense of involving many people (see also Smith 1998 on the collective "discovery" of the conservation of energy). The diversity of kinds of people included in this social reckoning has only expanded over time. If Michael Ruse was innovatively broad, in his 1979 *Darwinian Revolution: Science Red in Tooth and Claw*, for including over a dozen British male natural philosophers as the relevant community that helped to make the revolution in Darwin's name, its scope seems narrow today, when we see that revolution as preceding Darwin in many of its features (Desmond 1992; Secord 2000) and extending far into nineteenth-century British and European culture (e.g. Beer 1983; Glick and Engels 2008)—and indeed cultures worldwide (Pusey 1983; Elshakry 2013).

The key second claim of social constructionism, then, was that the development of science involved many people, doing many different kinds of things. As microsociological laboratory studies demonstrated the centrality of postdocs, graduate students, and technicians to making knowledge (Latour and Woolgar 1979), historians wondered, Who were the "invisible technicians" of the past (Shapin 1989; Hentschel 2007)? How were the social relations of knowledge production managed, and how did these change over time?

Feminist scholars observed that European women were in fact also involved in many aspects of making knowledge about nature, though only exceptionally afforded opportunities to "do science" in ways we easily recognize (Schiebinger 1989; Findlen 1993; Terrall 1995). Women participated in science as patrons and *salonnières*, as illustrators, as teachers of children, as popular writers (Shteir 1996), and as partners working with their scientific husbands (Pycior, Slack, and Abir-Am 1996) long before "careers" in science were generally available to women. As historians looked beyond European laboratories and the social structures that surrounded and sustained them, they found not only women but also men who helped make science in the field in these and other ways as well—as servants, collectors, and taxidermists; as translators, providers of local or indigenous knowledge, and other sorts of go-betweens; and as experimental subjects. (See Part I, "Roles," in this volume.) The peoplescape of contributors to science has grown accordingly.

As the kinds of people recognized as involved with science have diversified, the notion of the "scientist" itself has undergone new scrutiny, most prominently with the development of the idea of scientific *personae* (Daston and Sibum 2003). This concept simultaneously offers a theorized way to differentiate among kinds of scientists, describe certain collective patterns of scientific behavior, and offer an intermediate level of analysis between the individual and the institution. The "scientist as expert" has spawned a distinctive specialist literature as well (Lucier 2008; Broman 2012; Klein 2012). To be sure, more traditional biography has hardly disappeared from the history of science—indeed, four of the eleven winners of the History of Science Society's Pfizer Prize for best scholarly book between 2003 and 2013 were biographies (Terrall 2002; Browne 2003; Antognazza 2009; Schäfer 2013). Historians have also been inspired to revisit how scientific biographies themselves are constructed—by scientists (Otis 2007), by admirers (Rupke 2005), and by historians (Söderqvist 2007).

Doing Scientific Things with Scientific Things: Practice and Materiality

Historians of science today do not write only about scientists and others producing and supporting science. They write about the *stuff* of science: about glassware, computers, fruit flies, oceans, books, diagrams, maps, models, and particle accelerators. They write about theory, too—but their goal is less often to elucidate how scientists derived their theories than to present a broader historical web of scientific and cultural practices that in turn are solidly embedded in the physical world. This rich material tapestry has been woven together from diverse strands: the social-constructionism-inspired turn to experimental practice; the formerly distinct scientific instrument tradition; attention to natural history collections and fieldwork; and interdisciplinary studies of material culture.[1]

The central feature, which gained heft from the social constructionism of the 1980s, has been the turn toward practice (Soler et al. 2014). Literary postmodernists of the period might declare with Derrida that all thought is discourse, and thus all products of thought were forms of text, amenable to deconstruction. Not so analysts of science. Shapin and Schaffer (1985, 25), for instance, bent far backward to call written arguments "literary technologies," which along with material and social technologies established scientifically legitimate "matters of fact" in the Scientific Revolution. To them, seeing science as constructed meant focusing attention on the physical, material means of that construction. Since the 1980s, broader trends have helped to keep historians' attention on the materiality of science. The digitization and virtualization of our academic and social world has wrought renewed appreciation for physical things, while at the same time, ever-increasing awareness of our dependence on a rapidly degrading nature has lent new urgency to that appreciation. We can no longer afford to attend primarily to theory.

Attention to materiality is not new to the history of science. An older Marxist tradition insisted on the central role of material and economic needs in shaping science (Bernal 1971). Separately, a long tradition studied historical scientific instruments; with its valuation of object-connoisseurship connected to art history and museum work, this was often treated as a sideline in the field. Then in the mid-1990s, scholars of material culture—mostly working in museums—made new claims for their importance to the study of history of science and technology (Lubar and Kingery 1993; Kingery 1996). Combined with the history of science's new focus on practice, this helped push instruments and other materials toward the center of the field (van Helden and Hankins 1994).

Analyses of the material nature of scientific practice have looked different as they intervened in different historical subspecialties. In early modern studies, for instance, such analyses have carried forward the theme of the "scholar–craftsman" union (Zilsel et al. 2000; Roberts, Schaffer, and Dear 2007; Long 2011); a similar concern with the relationship between abstract knowledge and craft knowhow has animated recent work on ancient and non-Western understandings of nature (e.g. Robson 2008; Schäfer 2011). In the history of modern physics, the study of experimental practice challenged the historiographic dominance of theoretical physics. As Peter Galison (1997) has argued, developments in theoretical and experimental physics have not been yoked together; tracing the history of experimental physics, its instruments and material practices, yields new historical narratives that change our picture of "physics"—even challenging its unity as a science.

In the history of twentieth-century experimental life sciences, attention to practice and material culture led to new ways of thinking about the unique tools for investigating living processes (Clarke and Fujimura 1992). Robert Kohler's iconic *Lords of the Fly* (1994) analyzed the Morgan school of *Drosophila* geneticists, showing how the organisms themselves began to drive the systems of investigation (and indeed, the entire "moral economy" of the school) and analyzing how the scientists responded. Subsequent scholarship further refined analyses of knowledge-making systems involving people, model organisms and organic materials, and experimental set-ups in the life sciences (e.g., Rheinberger 1997; Creager 2002; Landecker 2007).

Historical studies of experimental practice, then, have shared a focus on the use of instruments and experimental systems that extend our senses and manipulate nature to tease out its processes, their underlying structures, and, ultimately, their laws. Historians of natural history have attended to quite different aspects of material practice, including not only the life and work of scientists in "the field" (Kuklick and Kohler 1996; Vetter 2011) as they searched for natural objects and materials, but also the practices of collection and preservation, and the organization of specimens into ordered collections (Heesen and Spary 2002; Endersby 2008; Johnson 2012). Here, the history of science has intersected with the history of museums and collections, and with the broader material culture perspective that museums have promulgated (Nyhart 2009; Alberti 2011; Poliquin 2012).

Such approaches have drawn attention to the spatial dimensions of scientific practice—another aspect of its materiality closely intertwined with social organization (Finnegan 2008). Modern scientific activity typically takes place in recognized kinds of venues: observatories, laboratories, museums, and "the field" are perhaps the four most prominent categories (see Part II, "Places and Spaces," this volume). Each of these has evolved over time and developed characteristic forms of social organization and practices, though historians have repeatedly noted how permeable and variable these sites are (e.g., Gooday 2008). This focus may be understood as part of a broader interdisciplinary "spatial turn" visible recently across the humanities and social sciences (e.g., Warf and Arias 2008). Geographers have offered taxonomies of scientific spaces and places that draw useful distinctions (such as that between particular locations in the world—Brazil, say—and kinds of places—such as "the tropics"), and have called attention to important differences in the scales at which spatial analysis of science may be undertaken (see esp. Livingstone and Withers 2011). Spatial and geographical language—referring to actual places, kinds of places, and metaphors of place and mapping—now provides a prominent vocabulary and mode of analysis among historians of science.

Moving Knowledge Around: Communication and Circulation

A long-accepted tenet of the social constructionist history of science is that scientific knowledge begins locally. If this is the case, then how does it spread? Over the last three decades historians have pursued this fundamental question in many directions, and the analysis of the ways in which people, ideas, and artifacts travel and communicate to move science around has yielded an especially rich set of intellectual tools.

The communicative practices within and surrounding science are central to its spread, and writing is the practice historians have studied longest and most deeply. For decades, if not centuries, historians of science have analyzed texts. In the 1980s rhetoricians joined them to examine anew both the persuasive strategies of scientists and the forms of scientific publication, especially the scientific article (e.g., Bazerman 1988; Dear 1991; Gross, Harmon, and Reidy 2002). Unpublished (if not always private) forms have also received scrutiny, especially as they reflect the broader social structures in which they were embedded, such as the correspondence network or the archive (Hunter 1998; van Miert 2013).

Beyond its rhetorical dimensions, the historical study of science communication has been transformed by the dramatic expansion and increasingly sophisticated historiography of "popular science" (often conflated with "science popularization"). An older, diffusionist model tended to treat popular science as a watered-down version of "real" science, popularizers as lesser lights who lacked the chops to do their own research, and readers as a passive audience. This has given way to a perspective in which both writers for the general public and that public itself are treated as active cultural interpreters and knowledge-makers worthy of study (Cooter and Pumfrey 1994). James Secord (2000) has shown just how far one can take this approach, with his classic study *Victorian Sensation*, which treats Robert Chambers' 1844 *Vestiges of the Natural History of Creation* as a remarkably fluid text: he shows how its many editions developed in conversation with its critics, while also illuminating localized styles and cultures of reading. More recently, Topham (2009) has suggested considering science popularization more seriously as an actor's category, while Daum (2009) has proposed a broader historiographic transformation that would consider popular science as part of a larger notion of public knowledge.

Daum has rightly criticized the existing historiography of popular science for its parochial focus on nineteenth-century Britain—a trend reinforced by the large number of literary scholars of Victorian culture who have reached out to meet historians of popular science, especially (though not exclusively) via a mutual interest in the genre of the general periodical (e.g., Cantor et al. 1994; Cantor and Shuttleworth 2004; Lightman 2007). It is refreshing, therefore, to see innovative analyses of popular science being developed for new contexts such as the twentieth-century Soviet Union and China, where the relationships among public science, the state, and forms of identity have been both fraught and different from British-inflected Western assumptions (Andrews 2003; Schmalzer 2008; Fan 2012a).

Communication has a material history, too, explored powerfully through its print culture. Historians of science have come to view books, atlases and encyclopedias, journals, and popular magazines not just as vehicles of scientific information but also as objects whose physical attributes offer important historical clues to the authors, artists, engravers, printers, publishers, and patrons who contributed to making the printed scientific work (and thus further expand the cast of characters involved in producing science). The material object also provides clues to which sorts of readers might have had access to it and where, how they might have read it, and indeed the broader culture of reading of which the work was a part. As the technologies and economics of printing and publication have changed, so, too, have the associated cultures of print (Johns 1998; Secord 2000; Apple, Downey, and Vaughan 2012).

Historical analysis of scientific communication extends beyond the study of writing. The history of "non-verbal communication in science" (Mazzolini 1993) has become increasingly broad and varied, and its analyses now often combine with those of other forms of scientific communication, analyzed within the overlapping interdisciplinary fields of visual, print, and material culture of science (Fyfe and Lightman 2007; Hopwood, Schaffer, and Secord 2010; Jardine and Fay 2013; Messbarger 2013; Hopwood 2015; cf. Topper and Holloway 1980). These non-verbal aspects have even become fully integrated into topics once judged exclusively philosophical, as demonstrated by Daston and Galison's *Objectivity* (2007). As the present volume illustrates, the study of science's communicative practices also encompasses in-person forms of transmission such as lectures and demonstrations, distance media like radio and television, and a host of visual and material forms that often blur the already soft lines among the technical, the didactic, and the popular.

Although the distinction between "making" and "moving" knowledge has some utility, a considerable body of literature demonstrates its superficiality. Historians and anthropologists have long recognized that scientific knowledge changes when moving from one place to another; thus, moving knowledge means, at the very least, re-making it in some ways. Older rubrics for this process included *knowledge transfer*, *reception*, and (following an older sociological tradition) *diffusion* (Dolby 1977). All these earlier terms placed the primary agency on a source understood to be scientific, which is then differentially adopted by recipient cultures. It is now appreciated how inadequate this perspective is: there is always more knowledge-making going on at the "receiving" end.

The analysis of linguistic translation is an obvious way in to understanding problems of cultural translation and transfer, tracking what remains more or less the same and what is transformed when ideas are brought into new cultural environments. Such analyses challenge the longstanding assumption that scientific knowledge is merely transposed in linguistic translation, and not transformed at all (Elshakry 2008). The nitty-gritty details of translation indicate some of the cultural challenges. What was the German professor–translator H. G. Bronn, Europe's highest paleontological authority, to make of Darwin's pigeon breeds, with their impossible names, and Darwin's easy assumption that these would help win over his audience to evolution (Gliboff 2008)? How much more was transformed beyond language in the centuries-long projects of translating Greek texts into Arabic (and commenting on them), which produced new documents that themselves served as the sources subsequently translated into Latin in medieval Europe and the Mediterranean! While later cast as the "rediscovery" of an ancient Classical heritage that was merely routed through the ancient Near East, scholars have shown how misleading this story is—how it ignores the power, autonomy, and creative contributions by the many cultures of western Asia and the Near East to what we call "science," and the many transformations accompanying translation (Montgomery 2000; Iqbal 2012). Textual translation was further complicated when the writing systems, visual culture, and technologies of text production differed (Fu 2012).

The complex relationship between moving and making scientific knowledge goes beyond the alterations undergone in transit. Analysts of science have argued in different ways that the movement of knowledge itself has been essential to making it scientific. One argument, focused especially on laboratory knowledge, goes roughly like this: for something to be true, it must be true in more than one place; hence the importance of replicating results. Drawing on this logic, historians and sociologists have examined how scientists have worked to recreate "the same" conditions and techniques in different places in order to render the laboratory a "placeless place" in which scientists might successfully replicate results and thus create empirically based assent (Gieryn 2002; Kohler 2002 and sources cited therein). Here, science is simultaneously made and moved by homogenizing and spreading its techniques and environments.

Another perspective has focused on how certain kinds of objects and information—in the sociologist Bruno Latour's (1987) term, "immutable and combinable mobiles"—have been extracted from "elsewhere" and brought together in specific "centers of calculation." At these centers—typically Western, metropolitan, and more powerful than the diffuse locations from which the objects come—scientists do the work that would yield the knowledge called "scientific." Such historical attention to the forging of scientific knowledge through the centralized accumulation, organizing, analysis, and classification of objects and information has increased along with attention to the natural-historical sciences, and more broadly, with what Lorraine Daston has called the "Sciences of the Archives" (see http://www.mpiwg-berlin.mpg.de/en/research/projects/DeptII_Daston-SciencesOfTheArchives).

A third, increasingly prominent, approach has participated in broader historiographic trends of studying the global movements of people, things, and ideas. Much of this work has gone under a general framework of (Western) science and (European) empire. It has highlighted the mutual accommodations made among Western scientists (especially naturalists), commercial interests, Christian missions, and expansionist states, as well as appropriations of indigenous materials and knowledge (e.g. Schiebinger 2004; Schiebinger and Swann 2004; Delbourgo and Dew 2007; Bleichmar et al. 2009; Mitman and Erickson 2010).

In one of the most ambitious of these accounts, Harold Cook (2005) has argued that the Scientific Revolution itself should be located in the constellation of values encouraged by the early modern Dutch commercial empire, which valorized an interest in detail and "matters of fact" that served both the global commerce undertaken by Dutch traders and, as it turns out, science. In Cook's picture, the knowledge that came to be considered scientific emerged from global interactions of people, organisms, and things that filtered back to Europe through circuitous and often contingent networks. In this view, "science" is not made in one place and then spread to another, nor is it located primarily in the organization of bits of information into complex systems at the metropole by leading knowledge producers. Rather, it is the historical product of many different people who themselves contributed, not always voluntarily, to a culture that valued things, their description, and the making of scientific meaning around them.

This sort of account has often been connected with the term "circulation" (e.g., Raj 2007; Terrall and Raj 2010; Lightman, McOuat, and Stewart 2013). This term has been used to emphasize the agency of those formerly considered merely passive instruments in the spread of scientific knowledge (either as receivers or as those whose local knowledge was appropriated), opening up analytical space to acknowledge their interests and their creative, knowledge-generating work. Such analyses have highlighted reciprocal interactions among historical actors, sometimes involving "go-betweens" (Schaffer et al. 2009), often at sites where "trading zones" (Galison 1997) existed or hybrid knowledge cultures persisted (Kohler 2002; Gómez 2013).

In conjunction with a global perspective, the circulation metaphor does important work: it displaces the unidirectionality of older center–periphery models centered on western Europe and the US, and flattens the status difference these models imply, elevating the status of non-Western contributors to Western knowledge and also the non-Western cultures and knowledge systems themselves. It also offers a new big-picture framework under which to unite a plethora of local studies. Because science has for so long been considered an exclusive product of the West, this is a salutary development.

Yet this vocabulary of global "circulation" and "flows" of knowledge has generated criticism from scholars such as Warwick Anderson, who has somewhat sardonically dubbed it the "hydraulic turn" (2014, 375). Fan (2012b, 252) has articulated the concern: "The image of circulation tends to impose too much unity, uniformity, and directionality on what was complex, multidirectional, and messy. . . . [It also] doesn't encourage a critical analysis of, say, power relations in science." Fan, Anderson, and others would prefer more attention to specific sites of resistance and stories of conflict, to remind us that, in historically specific situations, those "flows" may meet significant "blockages" worthy of our attention.

Scaling History of Science

The world covered by historians of science is bigger, more densely populated, and more complex than it once was. How shall we manage this multileveled intellectual terrain? How can we avoid getting lost in its lush vegetation? As we have seen, current high-visibility scholarship seeks to bind the local and the global through tropes of motion, bypassing well-worn social categories, such as the state and civic institutions, that operate at intermediate levels. Following people and objects around, as they travel the globe, allows the historian to collapse low and high levels of resolution into a single story, which is very appealing. Yet the broad range of intermediate levels of analysis should not be forgotten (Kohler and Olesko 2012). Attending to scales of analysis may in fact help us negotiate the tensions over global circulation mentioned above: a high-level focus on broad patterns tends to gloss over non-hegemonic voices, while lower levels of specificity bring them out. (See Misa 2009 for a similar analysis in history of technology.) Moreover, intermediate levels are crucial for tackling other leading questions not addressed in this essay, such as the comparative history of demarcation, which asks "How has science calved off historically from other activities into its own cultural field?" "How has such demarcation been supported socially and economically?" and "How has it been maintained (or not) in the face of contestation?"

As historians, we must attend to temporal scale as well. Localized stories often take place at the scale of a human lifespan or less, while questions about periodization remain a staple of mid-range temporal analysis. Scholarship on science and history extending temporal scales of analysis to yet broader expanses is emerging around us, drawing on archaeology, anthropology, and environmental history (Robin and Steffen 2007; Robson 2008; Safier 2010). It remains to be seen whether this scalar challenge is one historians of science are willing to take up, and if so, how.

The landscape of the history of science is one we simultaneously inhabit and cultivate: as both science and our broader cultural concerns continue to change, so, too, will the history of science. But the fundamental shift that has taken place since the late 1970s appears to be permanent. Historians of science now treat science as something that has been produced historically and contingently, not arrived at through an increasing recognition of truths. It has emerged instead through the cultivation of particular values that have sustained the investigation of the material world around us, in different directions at different times and places. People undertaking the activities we call "science" have created cultural space for themselves by advancing and taking advantage of new institutions and communicative forms; these in turn have been sustained by the commitments and livelihoods of many people who are not themselves "scientists."

Indeed, the science we depict is deeply embedded in its surrounding culture (even when scientists and spokesmen for science have argued otherwise)—yet that culture itself is typically not closed, but instead engaged in constant exchange with other cultures, feeding the wellsprings of scientific innovation, power, and conflict.

All of this makes the history of science a buzzing, dynamic field of action. Whether we examine it from close up, deep inside the tangle, from a mid-range that resolves certain actors and structures while leaving others fuzzy, or from a more distant view focused on large-scale patterns, our intellectual challenge is to explore diverse narrative and explanatory paths through this terrain. Our practical challenge is to illuminate these paths using all the tools we have available—academic monographs and articles, exhibitions, living history reconstructions and performances, films, podcasts, and the sweep of possibilities offered by new media—to invite others, not always historians of science, to come along with us.

Endnote

1   New attention to bodies in feminist and gender studies and the history of medicine forms a parallel topic that is unfortunately beyond the scope of this essay.

References

Alberti, Samuel J. M. M. 2011. *The Afterlives of Animals: A Museum Menagerie.* Charlottesville: University of Virginia Press.

Anderson, Warwick. 2014. "Making global health history: The postcolonial worldliness of biomedicine." *Social History of Medicine,* 27, No. 2: 372–84.

Andrews, James T. 2003. *Science for the Masses: The Bolshevik State, Public Science, and the Popular Imagination in Soviet Russia,* 1917–1934. College Station, TX: Texas A&M University Press.

Antognazza, Maria Rosa. 2009. *Leibniz: An Intellectual Biography.* Cambridge: Cambridge University Press.

Apple, Rima D., Gregory John Downey, and Stephen Vaughn. 2012. *Science in Print: Essays on the History of Science and the Culture of Print.* Madison, WI: University of Wisconsin Press.

Bazerman, Charles. 1988. *Shaping Written Knowledge: The Genre and Activity of the Experimental Article in Science.* Madison, WI: University of Wisconsin Press.

Beer, Gillian. 1983. *Darwin's Plots: Evolutionary Narrative in Darwin, George Eliot, and Nineteenth-Century Fiction.* London: Routledge & Kegan Paul.

Bernal, J. D. 1971. *Science in History.* 4 vols. Cambridge, MA: MIT Press.

Bleichmar, Daniela, Paula De Vos, Kristin Huffine, and Kevin Sheehan (eds.) 2009. *Science in the Spanish and Portuguese Empires,* 1500–1800. Stanford, CA: Stanford University Press.

Bleier, Ruth. 1984. *Science and Gender: A Critique of Biology and Its Theories on Women.* New York: Pergamon.

Bloor, David. 1976. *Knowledge and Social Imagery.* London: Routledge & Kegan Paul.

Broman, Thomas. 2012. "The semblance of transparency: Expertise as a social good and an ideology in enlightened societies." *Osiris,* 2nd series, 27: 188–208.

Browne, Janet. 2003. *Charles Darwin: The Power of Place.* Princeton, NJ: Princeton University Press.

Cantor, Geoffrey N., Gowan Dawson, Graeme J. N. Gooday, Richard Noakes, Sally Shuttleworth, and Jonathan R. Topham. 2004. *Science in the Nineteenth-Century Periodical: Reading the Magazine of Nature*. Cambridge: Cambridge University Press.

Cantor, Geoffrey N., and Sally Shuttleworth. 2004. *Science Serialized: Representation of the Sciences in Nineteenth-Century Periodicals*. Cambridge, MA: MIT Press.

Clarke, Adele, and Joan Fujimura (eds.) 1992. *The Right Tools for the Job: At Work in Twentieth-Century Life Sciences*. Princeton, NJ: Princeton University Press.

Cook, Harold John. 2007. *Matters of Exchange: Commerce, Medicine, and Science in the Dutch Golden Age*. New Haven, CT: Yale University Press.

Cooter, Roger, and Stephen Pumfrey. 1994. "Separate spheres and public places: Reflections on the history of science popularization and science in popular culture." *History of Science*, 32: 237–67.

Creager, Angela N. H. 2002. *The Life of a Virus: Tobacco Mosaic Virus as an Experimental Model, 1930–1965*. Chicago: University of Chicago Press.

Daston, Lorraine, and Peter Galison. 2007. Objectivity. New York: Zone Books.

Daston, Lorraine, and Heinz Otto Sibum (eds.) 2003. "Scientific Personae". Special Issue, *Science in Context*, 16, No. 1–2.

Daum, Andreas W. 2009. "Varieties of popular science and the transformations of public knowledge: Some historical reflections." *Isis*, 100: 319–32.

Dear, Peter Robert (ed.) 1991. *The Literary Structure of Scientific Argument: Historical Studies*. Philadelphia: University of Pennsylvania Press.

Delbourgo, James, and Nicholas Dew (eds.) 2007. *Science and Empire in the Atlantic World*. London: Routledge.

Desmond, Adrian. 1992. *The Politics of Evolution: Morphology, Medicine, and Reform in Radical London*. Chicago: University of Chicago Press.

Dolby, R. G. A. 1977. "The transmission of science." *History of Science*, 15: 1–43.

Elshakry, Marwa. 2008. "Knowledge in motion: The cultural politics of modern science translations in Arabic," *Isis*, 99: 701–730.

Elshakry, Marwa. 2013. *Reading Darwin in Arabic, 1860–1950*. Chicago: University of Chicago Press.

Endersby, Jim. 2008. *Imperial Nature: Joseph Hooker and the Practices of Victorian Science*. Chicago: University of Chicago Press.

Fan, Fa-ti. 2012a. "Science, state, and citizens: Notes from another shore." *Osiris*, 2nd series, 27: 227–49.

Fan, Fa-ti. 2012b. "The global turn in the history of science." *East Asian Science, Technology and Society*, 6, No. 2: 249–58.

Fausto-Sterling, Anne. 1992. *Myths of Gender: Biological Theories about Women and Men*. Revised edition. New York: Basic Books.

Findlen, Paula. 1993. "Science as a career in Enlightenment Italy: The strategies of Laura Bassi." *Isis*, 84: 440–69.

Finnegan, Diarmid A. 2008. "The spatial turn: Geographical approaches in the history of science." *Journal of the History of Biology*, 41: 369–88.

Foucault, Michel. 1971. *The Order of Things: An Archaeology of the Human Sciences*. New York: Pantheon Books.

Foucault, Michel. 1973. *The Birth of the Clinic: An Archaeology of Medical Perception*. London: Tavistock.

Fu, Liangyu. 2012. "Indigenizing visualized knowledge: Translating Western science illustrations in China, 1870–1910." *Translation Studies*, 6, No. 1: 78–102.

Fyfe, Aileen, and Bernard V. Lightman (eds.) 2007. *Science in the Marketplace: Nineteenth-Century Sites and Experiences*. Chicago: University of Chicago Press.

Galison, Peter. 1997. *Image and Logic: A Material Culture of Microphysics*. Chicago: University of Chicago Press.

Gieryn, Tom F. 2002. "Three truth-spots." *Journal of the History of the Behavioral Sciences*, 38, No. 2: 113–32.

Gliboff, Sander. 2008. *H. G. Bronn, Ernst Haeckel, and the Origins of German Darwinism: A Study in Translation and Transformation*. Cambridge, MA: MIT Press.

Glick, Thomas F., and Eve-Marie Engels (eds.) 2009. *The Reception of Charles Darwin in Europe*. London: Bloomsbury Academic.

Golinski, Jan. 2005. *Making Natural Knowledge: Constructivism and the History of Science*. Chicago: University of Chicago Press.

Gómez, Pablo F. 2013. "The circulation of bodily knowledge in the seventeenth-century Black Spanish Caribbean." *Social History of Medicine*, 26, No. 3: 383–402.

Gooday, Graeme. 2008. "Placing or replacing the laboratory in the history of science?" *Isis*, 99: 783–95.

Gross, Alan G., Joseph E. Harmon, and Michael Reidy. 2002. *Communicating Science: The Scientific Article from the 17th Century to the Present*. New York: Oxford University Press.

Haraway, Donna. 1988. "Situated knowledges: The science question in feminism and the privilege of partial perspective." *Feminist Studies*, 14, No. 3: 575–99.

Harding, Sandra. 1986. *The Science Question in Feminism*. Ithaca, NY: Cornell University Press.

Heesen, Anke te, and E. C. Spary (eds.) 2002. *Sammeln als Wissen: das Sammeln und seine wissenschaftsgeschichtliche Bedeutung*. Göttingen: Wallstein.

Hentschel, Klaus. 2008. *Unsichtbare Hände: Zur Rolle von Laborassistenten, Mechanikern, Zeichnern u. a. Amanuenses in der physikalischen Forschungs- und Entwicklungsarbeit*. Diepholz: GNT-Verlag.

Hopwood, Nick. 2015. *Haeckel's Embryos: Images, Evolution, and Fraud*. Chicago: University of Chicago Press.

Hopwood, Nick, Simon Schaffer, and James A. Secord (eds.) 2010. "Seriality and scientific objects in the nineteenth century." *History of Science*, 48, No. 3–4: 251–85.

Hunter, Michael (ed.) 1998. *Archives of the Scientific Revolution: The Formation and Exchange of Ideas in Seventeenth-Century Europe*. Woodbridge: Boydell Press.

Iqbal, Muzaffar (ed.) 2012. *Studies in the Making of Islamic Science: Knowledge in Motion*. Aldershot: Ashgate.

Jardine, Nicholas, and Isla Fay (eds.) 2013. *Observing the World through Images: Diagrams and Figures in the Early-Modern Arts and Sciences*. Leiden: Brill.

Johns, Adrian. 1998. *The Nature of the Book: Print and Knowledge in the Making*. Chicago: University of Chicago Press.

Johnson, Kristin. 2012. *Ordering Life: Karl Jordan and the Naturalist Tradition*. Baltimore: Johns Hopkins University Press.

Kim, Mi Gyung. 2014. "Archeology, genealogy, and geography of experimental philosophy." *Social Studies of Science*, 44, No. 1: 150–62.

Kingery, W. David (ed.) 1996. *Learning from Things: Method and Theory of Material Culture Studies*. Washington, DC: Smithsonian Institution.

Klein, Ursula (ed.) 2012. "Artisanal-scientific experts in eighteenth-century France and Germany." *Annals of Science*, 69, No. 3: 303–433.

Kohler, Robert E. 1994. *Lords of the Fly: Drosophila Genetics and the Experimental Life*. Chicago: University of Chicago Press.

Kohler, Robert E. 2002. "Labscapes: Naturalizing the lab." *History of Science*, 40, No. 4: 473–501.

Kohler, Robert E., and Kathryn M. Olesko. 2012. "Introduction: Clio meets science." *Osiris*, 2nd series, 27: 1–16.

Kuhn, Thomas S. 2012. *The Structure of Scientific Revolutions: 50th Anniversary Edition*. Chicago: University of Chicago Press.

Kuklick, Henrika, and Robert E. Kohler (eds.) 1996. "Science in the field." *Osiris*, 2nd series, 11: 1–265.

Landecker, Hannah. 2007. *Culturing Life: How Cells Became Technologies*. Cambridge, MA: Harvard University Press.

Latour, Bruno, and Steve Woolgar. 1979. *Laboratory Life: The Social Construction of Scientific Facts*. Beverly Hills, CA: Sage Publications.

Latour, Bruno. 1987. *Science in Action: How to Follow Scientists and Engineers through Society*. Cambridge, MA: Harvard University Press.

Lightman, Bernard V. 2007. *Victorian Popularizers of Science: Designing Nature for New Audiences*. Chicago: University of Chicago Press.

Lightman, Bernard V., Gordon McOuat, and Larry Stewart (eds.) 2013. *The Circulation of Knowledge between Britain, India and China*. Leiden: Brill.

Livingstone, David N., and Charles W. J. Withers (eds.) 2011. *Geographies of Nineteenth-Century Science*. Chicago: University of Chicago Press.

Long, Pamela O. 2011. *Artisan/Practitioners and the Rise of the New Sciences, 1400–1600*. Corvallis, OR: Oregon State University Press.

Lubar, Steven, and W. David Kingery (eds.) 1993. *History from Things: Essays on Material Culture*. Washington: Smithsonian Institution Press.

Lucier, Paul. 2008. *Scientists and Swindlers: Consulting on Coal and Oil in America, 1820–1890*. Baltimore: Johns Hopkins University Press.

Mazzolini, Renato G. (ed.) 1993. *Non-Verbal Communication in Science prior to 1900*. Firenze: L.S. Olschki.

Messbarger, Rebecca. 2013. "The re-birth of Venus in Florence's Royal Museum of Physics and Natural History." *Journal of the History of Collections*, 25: 195–215.

Misa, Thomas J. 2009. "Findings following framings: Navigating the empirical turn." *Synthese*, 168: 357–75.

Mitman, Gregg, and Paul Erickson. 2010. "Latex and blood: Science, markets, and American empire." *Radical History Review*, 107: 45–73.

Montgomery, Scott L. 2002. *Science in Translation: Movements of Knowledge through Cultures and Time*. Chicago: University of Chicago Press.

Nyhart, Lynn K. 2009. *Modern Nature: The Rise of the Biological Perspective in Germany*. Chicago: University of Chicago Press.

Otis, Laura. 2007. *Müller's Lab*. Oxford: Oxford University Press.

Poliquin, Rachel. 2012. *The Breathless Zoo: Taxidermy and the Cultures of Longing*. University Park, PA: Pennsylvania State University Press.

Pusey, James Reeve. 1983. *China and Charles Darwin*. Cambridge, MA: Harvard University Asia Center.

Pycior, Helena M., Nancy G. Slack, and Pnina G. Abir-Am (eds.) 1996. *Creative Couples in the Sciences*. New Brunswick, NJ: Rutgers University Press.

Raj, Kapil. 2007. *Relocating Modern Science: Circulation and the Construction of Scientific Knowledge in South Asia and Europe, 1650–1900*. Basingstoke: Palgrave Macmillan.

Rheinberger, Hans-Jörg. 1997. *Toward a History of Epistemic Things: Synthesizing Proteins in the Test Tube*. Stanford, CA: Stanford University Press.

Roberts, Lissa, Simon Schaffer, and Peter Dear (eds.) 2007. *The Mindful Hand: Inquiry and Invention from the Late Renaissance to Early Industrialisation*. Amsterdam: Koninklijke Nederlandse Akademie van Wetenschappen.

Robin, Libby, and Will Steffen. 2007. "History for the Anthropocene." *History Compass*, 5, No. 5: 1694–1719. doi:10.1111/j.1478-0542.2007.00459.x.

Robson, Eleanor. 2008. *Mathematics in Ancient Iraq: A Social History*. Princeton, NJ: Princeton University Press.

Rudwick, Martin John Spencer. 1985. *The Great Devonian Controversy: The Shaping of Scientific Knowledge among Gentlemanly Specialists*. Chicago: University of Chicago Press.

Rupke, Nicolaas A. 2005. *Alexander von Humboldt: A Metabiography*. New York: Peter Lang.

Ruse, Michael. 1979. *The Darwinian Revolution: Science Red in Tooth and Claw*. Chicago: University of Chicago Press.

Safier, Neil. 2010. "Global Knowledge on the Move: Itineraries, Amerindian Narratives, and Deep Histories of Science." *Isis*, 101: 133–145.

Schäfer, Dagmar. 2011. *The Crafting of the 10,000 Things: Knowledge and Technology in Seventeenth-Century China*. Chicago: University of Chicago Press.

Schaffer, Simon, Lissa Roberts, Kapil Raj, and James Delbourgo (eds.) 2009. *The Brokered World: Go-Betweens and Global Intelligence, 1770–1820*. Sagamore Beach, MA: Science History Publications.

Schiebinger, Londa. 1989. *The Mind Has No Sex? Women in the Origins of Modern Science*. Cambridge, MA: Harvard University Press.

Schiebinger, Londa. 2004. *Plants and Empire: Colonial Bioprospecting in the Atlantic World*. Cambridge, MA: Harvard University Press.

Schiebinger, Londa, and Claudia Swan (eds.) 2004. *Colonial Botany: Science, Commerce, and Politics in the Early Modern World*. Philadelphia: University of Pennsylvania Press.

Schmalzer, Sigrid. 2008. *The People's Peking Man: Popular Science and Human Identity in Twentieth-Century China*. Chicago: University of Chicago Press.

Secord, James A. 2000. *Victorian Sensation: The Extraordinary Publication, Reception and Secret Authorship of Vestiges of the Natural History of Creation*. Chicago: University of Chicago Press.

Secord, James A. 2004. "Knowledge in transit." *Isis*, 95, No. 4: 654–72.

Shapin, Steven. 1989. "The invisible technician." *American Scientist*, 77: 554–63.

Shapin, Steven, and Simon Schaffer. 1985. *Leviathan and the Air-Pump: Hobbes, Boyle, and the Experimental Life*. Princeton, NJ: Princeton University Press.

Shapin, Steven, and Simon Schaffer. 2011. *Leviathan and the Air-Pump: Hobbes, Boyle, and the Experimental Life. With a New Introduction by the Authors*. Princeton, NJ: Princeton University Press.

Shteir, Ann B. 1996. *Cultivating Women, Cultivating Science: Flora's Daughters and Botany in England, 1760–1860*. Baltimore: Johns Hopkins University Press.

Smith, Crosbie. 1998. *The Science of Energy: A Cultural History of Energy Physics in Victorian Britain*. Chicago: University of Chicago Press.

Söderqvist, Thomas. 2007. *The History and Poetics of Scientific Biography*. Aldershot: Ashgate.

Soler, Léna, Sjoerd Zwart, Michael Lynch, and Vincent Israel-Jost (eds.) 2014. *Science after the Practice Turn in the Philosophy, History, and Social Studies of Science*. London: Routledge.

Terrall, Mary. 1995. "Émilie du Châtelet and the gendering of science." *History of Science*, 33: 283–310.

Terrall, Mary. 2002. *The Man Who Flattened the Earth: Maupertuis and the Sciences in the Enlightenment*. Chicago: University of Chicago Press.

Terrall, Mary, and Kapil Raj (eds.) 2010. "Circulation and Locality in Early Modern Science." Special Issue, *British Journal for the History of Science*, 43, No. 4.

Topham, Jonathan R. 2009. "Introduction [to Focus Section: Historicizing popular science]." *Isis*, 100: 310–18.

Topper, David R., and John H. Holloway. 1980. "Interrelationships between the visual arts, science and technology: A bibliography." *Leonardo*, 13, No. 1: 29–33.

Van Helden, Albert, and Thomas L Hankins (eds.) 1994. "Instruments." *Osiris*, 2nd series, 9: 1–250.

Van Miert, Dirk (ed.) 2013. *Communicating Observations in Early Modern Letters (1500–1675): Epistolography and Epistemology in the Age of the Scientific Revolution*. London: Warburg Institute.

Vetter, Jeremy (ed.) 2011. *Knowing Global Environments: New Historical Perspectives on the Field Sciences*. New Brunswick, NJ: Rutgers University Press.

Warf, Barney, and Santa Arias (eds.) 2008. *The Spatial Turn: Interdisciplinary Perspectives*. London: Routledge.

Zilsel, Edgar, Diederick Raven, Wolfgang Krohn, and R. S. Cohen. 2000. *The Social Origins of Modern Science*. Dordrecht: Kluwer Academic Publishers.

PART I

Roles

CHAPTER TWO

Learned Man and Woman in Antiquity and the Middle Ages

Nathan Sidoli

In ancient and medieval societies, certain individuals were able to distinguish themselves *as learned* and to ensure that evidence, or narratives, of their learning survived to our time. In order to understand who these individuals were from a historical perspective, we should consider broadly the nature of the learning they controlled as well as give some description, where possible, of who these people actually were—how they lived, what kinds of stories circulated about them, and how they transmitted their learning.

What was meant by *learned* was different in different times and places. For example, the learned man might be seen as the goal of a general Buddhist monastic education, or as the successful candidate of civil examinations, who had mastered an appropriate interpretation of the Confucian classics (Elman 2000; Lee 2000, 111–70; Scharfe 2002, 158–9). Stories of the mathematician Archimedes (c. 250–212 BCE) were used to illustrate the learning that Roman conquerors could inherit from Greek scholars (Jaeger 2013). For Sanskrit grammarians, the pinnacle of learning was represented by the brahmans of Āryāvarta, who understood correct speech through an innate genius (Pollock 1985, 505). In all cases, however, the learned were those who had mastered something that we can call knowledge, or science, taken broadly. Often, however, the content of this theoretical knowledge was far removed from current forms. Lists of medieval Indian or Islamic sciences often include a number of transliterated terms—the implication being that these designate fields of knowledge are so alien to our current concepts that they cannot be fully conveyed by simple modern expressions.

The Sanskrit term *śāstra*—meaning rules, treatise, or knowledge—includes many concepts that are similar to what we mean by knowledge, or science, but others that are fairly divergent. *Śāstra* is divided, first, between *śruti*, heard texts, such as the Vedic hymns, mantras and various theological works, and *smṛti*, remembered texts, such as rules of behavior and other fields of knowledge. Lists of shastric teachings include subjects like the *Vedas* and *Upavedas*, history (*itihāsaveda*), statecraft (*arthaśāstra*),

A Companion to the History of Science, First Edition. Edited by Bernard Lightman.
© 2016 John Wiley & Sons Ltd. Published 2020 by John Wiley & Sons Ltd.

weapons and war (*dhanurveda*), music, medicine (*āyurveda*), logic or philosophy (*ānvīkṣikī*), Sanskrit grammar, metrics, astronomy/astrology, sacrificial procedures, economics (*vārttā*), cooking, erotology, law (*daṇḍanīti*), and so on. In general, the primary goal of *śāstra* was to regulate correct behavior. Just as some texts give us rules for solving mathematical problems, other texts give us rules for lovemaking. There is hardly a discernible difference between normative and descriptive knowledge (Pollock 1985, 2011; Scharfe 2002, 13–17).

Although the divisions and categorizations of the sciences were different for different individuals, in medieval Arabic discourse the most essential branch of knowledge (*ʿilm*) was that of the religious sciences, such as exegesis of the Qurʾān (*tafsīr*), study of the traditions of the prophet Muammad (*ḥadīth*), jurisprudence (*fiqh*), and rational theology (*kalām*). There were also fields that were sometimes described as educational, such as calculation, grammar and metrics, animal husbandry, and history. Finally, there were the sciences known as rational, ancient, and sometimes foreign, such as late Platonic or Aristotelian philosophy (*falsafa*), logic, arithmetic, geometry, astronomy, and medicine. Among these fields, however, we also find some disciplines that were produced within Islamic societies, as algebra (*ḥisāb al-jabr waʾl-muqābala*), timekeeping (*ʿilm al-mīqāt*), and cosmography or structural astronomy (*ʿilm al-hayʾa*). Even scholars who are best known to us as mathematicians or physicians, however, were often known in their own times as masters and scholars of the religious sciences (Rosenthal 1975, 54–70; Brentjes 2008; Brentjes 2014, 95).

Potentially more deceptive, however, are those fields that the Greeks designated by words that are the source of terms that we still use to name modern disciplines. Our physics has almost nothing to do with ancient studies of *phusis*—the essence of natural, and especially living, things. The word *mathēmatikē* was used to designate studies in harmonics, sundial construction, and astrology as well as number theory and geometry. *Philosophia* encompassed many things that we still understand as philosophy—such as ethics, political theory, and logic—but it also had broader meanings. It could designate a general inquiry into the life worth living, including spiritual practices, as well as more technical fields such as medicine or the mathematical sciences (Feke 2014; Tolsa 2014, 458).

All of these terms—*śāstra*, *ʿilm*, *philosophia*—must be understood first and foremost in the cultural context in which they arose. They also, however, meant something general, like knowledge, or *science*—when by "science" we mean an account of the things that we think we know.

For more recent historical periods, we generally regard a study of an individual's lived experience as essential to fully understanding the content of their thought. For pre-modern periods, however, our sources are often insufficiently detailed to elucidate this background. Even in the few cases where we have a rich source basis for discussing an individual's life, the authors of our sources were often motivated by other interests than that of conveying to us a complete picture of the lived experience of the subject at hand. Stories of learned men were sometimes related in order to develop a picture of an estimable life, or in order to provide moral lessons (Fancy 2013, 21–22; Jaeger 2013). We might be told about the learning of certain women so as to boost our estimation of the houses to which they belonged (Azad 2013, 81). In such cases, the roles of learned individuals can be analyzed both as narrative and as lived experience, although the two may in fact be different.

We also have cases of individuals who are known as learned but who left no texts, either intentionally, such as Pythagoras (late sixth–early fifth century BCE), or through the accidents of history, such as Hypatia (mid fourth–early fifth century CE). Here, all that remains to us are stories, which themselves change over time. The early Pythagoras was a sage who traveled to the east and brought back to the Greek colonies a love of wisdom and knowledge of an upright way of life (Zhmud 2012, 30–60). The later Pythagoras was a miracle-worker, who personified a righteous life and taught his followers the mathematical mysteries of the universe (O'Meara 1989). In the early sources, Hypatia appears as teacher of both pagans and Christians, who exemplified a broad-minded virtue, whereas in later sources she became a trope to depict conflicts between forms of knowledge that emphasize reason and others that emphasize faith (Dzielska 1995).

Finally, learned individuals were sometimes regarded as identical with the texts they left. A striking example is that of Euclid (early third century BCE), about whom we know almost nothing aside from the mathematical texts attributed to him. Whoever Euclid was, or whatever he thought, must be grasped through these texts and their connections to other texts. For ancient and medieval readers, however, a similar perspective was sometimes also taken towards authors whose life circumstances were recorded. For example, although accounts of the life of Aristotle (384–322 BCE) were known, these were rarely used as a means to analyze his thought. For most readers in the ancient and medieval periods, Aristotle was simply encountered *as his works*—but, of course, different readers had access to different texts, and read them in different ways. In this chapter, I will describe a number of examples of learned individuals in the ancient and medieval periods, sometimes with regard to their lived experience, sometimes through the stories that were told about them, and sometimes through the writings that they left.

Greek Mathematicians in the Hellenic Cities

Although we have no direct evidence for the activities of learned mathematicians of the classical Hellenic period (c. 500–c. 300 CE), it appears that they sought to distinguish their work both from practical traditions of mathematics that went back to Egypt and Babylonia, and from that of other groups of intellectuals, like philosophers and sophists. Nevertheless, it is clear that during this early period mathematicians never formed a professional group who earned their living through developing and teaching their mathematical skills, although some of them apparently did earn a living teaching mathematics (Asper 2003).

Indeed, although those mathematicians that we know anything about all came from privileged backgrounds, they appear to have performed diverse social roles. Archytas (mid-fifth–mid-fourth century BCE) was a statesman and a general; Hippocrates of Chios (mid-fifth century BCE) was a wealthy merchant; while Eudoxus (mid-fourth century BCE) was a legislator and a philosopher with many students. Nevertheless, the respect accorded to mathematicians in philosophical and literary texts indicates that they were able to secure a place for themselves in learned high culture, even if they had no institutionalized role in society (Netz 1997; Netz 2002).

We know almost nothing about the ways in which early mathematicians learned their discipline or how this might have been related to education in the philosophical schools. Nevertheless, it seems likely that Greek mathematicians most often worked alone, not in research groups or schools. Of course, there are some exceptions to this. In Athens, there were small groups of mathematicians who worked together, or at least on related problems. Some of these, such as Eudoxus, then returned to their homes and founded schools of mathematical and philosophical instruction. There were also peripheral schools, of which a striking example was the group at Cyzicus (Sedley 1976).

Families of Scholars in Persian and Hellenistic Uruk

Excavations of the ruins of the ancient Mesopotamian city of Uruk have given us direct evidence for the scholarly activity of a number of learned families, in the form of clay tablets written and owned by these scholars. While the city underwent major political transitions from Persian rule under the Achaemenids (c. 485–c. 330 BCE) and Greek rule under the Selucids (c. 330–c. 125 BCE), clans of scholars who could trace their lineages back for centuries continued the traditional scribal practices of their houses, integrating new methods in mathematical astronomy and predictive astrology with their performances of the ancient rites (Rochberg 2004, chapter 6; Beaulieu 2006; Robson 2008, chapter 8; Steele 2011).

They had a number of different titles, such as "scribe (*ṭupšar*) of *Enūma Anu Enlil* (a canonical omen text)," incantation priest (*āšipu*), or lamentation priest (*kalû*), but the evidence of the clay tablets that they wrote, or owned, indicates that the men of these families practiced a broad range of scholarly activities. Two families who lived successively in the same house during the Persian period owned tablets covering various omens, medical prescriptions and incantations, rituals and magic, hymns and literature, astronomy/astrology, mathematics, and the earliest-known tablet of predictive mathematical astronomy. During the Hellenistic period, we have a wealth of tablets mentioning four interrelated families. The tablets from these families deal with omens and rituals, incantations and lamentations, medical and magical astrology, horoscopic astrology, and predictive mathematical astronomy. As well as their responsibilities in preparing for and performing various rites, the men of these families also carried out other scribal functions, such as writing and witnessing legal and financial documents (Robson 2008, 229–50).

It appears from the evidence that these scholars formed a tight-knit, professional community. Their families intermarried, and by working together to train each successive generation they managed to keep their learning within the confines of these narrow circles for centuries. The fact that the colophons of a number of tablets describe the contents as secret, or exclusive, indicates that these scribal traditions were carefully guarded within certain family groups. Tablets that list correlations between astronomical signs and parts of rare animals may have been a kind of code for medicinal remedies based on astrological reasoning. Through such means, scribal and priestly families could insure their elite status as "learned" through many generations (Rochberg 2004, 211–13; Steele 2011, 335–8).

It was almost certainly members of these scholarly families, probably in Babylon and Uruk, who developed the sophisticated methods of predicting the behavior of the moon and the planets known as Babylonian mathematical astronomy (Neugebauer 1975, book 2; Ossendrijver 2012). The motivations for the production of this predictive apparatus probably came from the cult duties of these scholars and the needs of their temples. The priests of these families were charged with performing various rites in conjunction with meaningful celestial events, and with enacting rituals for warding off the ill tidings of certain omens. As the methods of mathematical astronomy developed, they were able to determine with considerable confidence when these events would occur and to predict beforehand when certain omens would appear. This secret knowledge allowed them, and them alone, to fulfill their sacred duties.

Scholars as Clients in Warring States to Early Han China

The transition from the Warring States to the formation of the first empires, Qin and Han, was a period of great social and political change. During this time, scholars both produced many of the concepts that would later assume a central place in Chinese intellectual activity, and they developed the idealized narratives of their own role in society.

Throughout the Warring States period (mid-fifth century–211 BCE), gentleman-scholars (*shi* 士) sought out patronage as clients, or guests (*ke* 客) at local courts. In this, they competed not only with each other, but also with a broad spectrum of probably more useful, and certainly more entertaining guests, such as a man who was an expert in the art of assassination, or one who could crow like a rooster. The primary role of scribes (*shi* 史) and scholarly retainers appears to have been that of determining the timing and format of ritual practices, and performing key roles in the rituals themselves. Some scholarly retainers, known as *ru* 儒, also educated the youth of noble houses in high culture, ethics and rituals (Cook 1995; Lee 2000, 108–10; Lloyd and Sivin 2002, 22–32).

Rulers accumulated retainers, however, not in order to promote scholarly research or education, but in order to increase their own prestige. Nevertheless, some rulers gathered together such large numbers of scholars that they created renowned centers of leaning. One of these was the Duke Huan of Qi, who is said to have appointed some from among his learned guests to serve as ministers in his government. Another famous example is that of Lu Buwei who, as chancellor of Qin, reportedly collected some 3000 guests. In this position, he oversaw the production of the encyclopedic *Springs and Autumns of Master Lu.*

This is the social background in which the ideas that later formed the basis of the Chinese cosmological synthesis were originally generated (Lloyd and Sivin 2002, 253–71). Although many different concepts were advanced and circulated during this period, some of these came to be understood as fundamental—such as *qi* 氣, an active principle of all matter; *yin–yang* 陰陽, used to explain polarities of opposites and complementaries; and the *five phases* (*wu-xing* 五行), a conceptual scheme used to describe interactions and relationships. It is important to recognize, however, that these concepts only became canonical in later re-readings of the ancient sources, and many related concepts that were introduced in the Warring States period were not further expounded in later texts (Cook 2013; Lo 2013).

In the Qin and early Han periods (211 BCE–c. 100 BCE), certain ritualists were able to convince the state to allow the teaching of only their preferred classics. Lineages of masters and disciples formed around the transmission of certain texts—usually, in the form of bundles of bamboo strips, which were treated as venerated material objects, sometimes gifted to rulers, or interred with the dead. During this period, the term *ru* came to designate those lineages that focused on transmitting the Confucian classics. In the Han, procedures were developed for selecting and qualifying individuals from these groups, on the basis of their mastery of the classical texts, to serve as salaried functionaries (*boshi* 博士) in the state bureaucracy—such as advisors, diplomats, historians, astronomers, and so on (Zufferey 1998; Lloyd and Sivin 2002, 32–42).

In these contexts, scholars generally addressed their works to rulers and sought to establish a place for themselves as minsters in the imperial bureaucracy.

A Roman Physician and a Roman Mathematician

Two scholars who became profoundly influential in their respective fields, Claudius Ptolemy (early to mid-second century CE) and Galen of Pergamum (129–ca 215), were contemporaries during the height of the Roman empire—a time of flourishing intellectual activity. Both were natives of Greek-speaking cities in the eastern part of the empire, most likely Roman citizens, and both were highly productive authors of original treatises. Although they worked in different areas—medicine and mathematics— their cultural outlook and philosophical approach was similar in many ways (Lehoux 2012, 6–8, 109–11). They were both members of a small, highly literate segment of society; they had been well educated in the mathematical and philosophical sciences; they appear to have become increasingly disillusioned with school-based philosophy, which they characterized as involving endless debate; and yet, nevertheless, throughout their lives, they continued to situate their own work within a broader conception of philosophy as the pursuit of a life worth living.

Because of Galen's habit of filling his writings with personal anecdotes, we are rather well-informed about the details of his life. A native of Pergamum, Galen came from a wealthy family of builders and architects. Following a private education in letters, mathematics, and philosophy, he pursued a medical education in Alexandria, the most important center of technical learning in the Greek-speaking empire. He then worked for a while as physician to a gladiatorial troupe in Pergamum and then went on to spend much of his career in Rome and the western empire, where he attended to the health of emperors and their heirs, and mixed with both the learned and the powerful (Nutton 2013, 222–35).

Galen practiced an erudite type of medicine. Although he certainly made careful observations, carried out some experiments, and dissected animals for audiences of the Roman elite, his primary approach was theoretical, and one of his most common modes of exposition was commentary on, and critique of, the work of his predecessors. He also attempted, however, to produce a kind of demonstration in medicine, modeled on mathematical proof. In particular, he often spoke of "analysis," by which he meant the use of theory to construct physical objects that performed certain functions, such as sundials or architectural features, that would be proven by actual use. On the whole, however, Galen's medical theorizing was not able to achieve the standard he set for it. Hence, he utilized narratives of his abilities in philosophy and the mathematical sciences to argue for his general scholarly competence as compared to that of his rivals.

In contrast, we know few of the details of Ptolemy's life. From later sources, we learn that he spent most of his time in Canopus, a suburb of Alexandria, perhaps in a temple of Osiris (Jones 2005, 64). Ptolemy, however, unlike Galen, does not use stories of his own life as part of his rhetorical strategy. In the rare cases where he speaks of himself in the first person, it is usually to refer to an observation that he claims to have actually made. For the most part, his authorial voice is that of the Greek mathematician, which uses the first person only as a stylistic trope.

Nevertheless, by examining the internal context of his writings, we can see how his attitude towards philosophy and his scholarship changed over time. In his early writings, his approach was almost purely theoretical and sometimes addressed only towards the philosophical concerns of his predecessors. As his career progressed, however, he began to place more emphasis on an empirical basis for knowledge and to relegate discussions of general philosophy to introductions and asides. Nevertheless, he continued to situate his work within the broader context of philosophy as theoretical knowledge and to argue that the exact sciences provided a template for true philosophy (Feke and Jones 2010; Tolsa 2014; Feke 2014).

This comparison exemplifies a tendency for the lived experience of the practitioner to play different roles in the rhetorical strategies of different disciplines. Galen created an image of his medical authority by referring to his education and actual experiences, which put him in position to develop the kinds of knowledge that he claimed. Ptolemy, on the other hand, created a sense of certainty by removing himself from the text and appealing to universally accepted principles and using mathematical methods to derive new knowledge from this foundation.

Contexts of Scholarship in Sanskrit Sciences

Just as we have seen that the relationship between the text and its context may be differently construed in the same time and culture, but in the different intellectual endeavors of medicine and mathematics, so, in different cultural arenas, lived experience may be addressed in a variety of ways. In Sanskrit scholarly writings, as in Greek mathematical texts, the individual author tends to disappear into the discourse, so that we have nearly "all text and no context" (Ganeri 2008, 553).

In order to develop a picture of how these scholars lived, we can study the various social settings in which learning was transmitted and developed, such as in monasteries and temples, Buddhist centers of general learning, training in caste occupations (*jāti*), or court positions and appointments. But we can rarely be certain of the biographical details of any particular author that we are reading (Scharfe 2002, 132–93; Plofker 2009, 178–81).

This lack of obvious context, however, also points us towards a fundamental preoccupation of the tradition. Authors of Sanskrit texts sought to associate themselves with a particular *śāstra*, which was conceived of not as a science maturing over time, but as a sort of "pre-existent, codified theoretical paradigm for activity" (Pollock 1985, 508). The process of producing treatises and commentaries was usually construed as one of rediscovery, not invention—even in cases where the rediscovered knowledge may strike us as novel. Little weight is given to the lived experience of the author as the locus of knowledge, whereas the texts themselves, as passed down from antiquity, are both the core field of inquiry and the primary source of knowledge (Pollock 1985; Ganeri 2008).

Translators and Other Scholars in Abbasid Baghdad

Under the Abbasid Caliphs—from al-Manṣūr (r. 714–775 CE) to al-Mutawakkil (r. 847–861 CE) and beyond—a cultural project of great significance was carried out in Baghdad. During this period, scholars were paid to study, translate, assimilate, and develop the technical knowledge found in Persian, Indian, Syriac, and, especially, Greek texts. Although the motivations for this movement—whether deriving from administrative needs of the Umayyad Caliphate or imperial ideologies of the Abbasid Caliphate—are not certain, it is clear that this cultural activity involved an unprecedented number of scholars in projects of synthesizing what they could find of the world's accumulated knowledge and using this as a basis on which to open up new avenues of approach (Gutas 1998, 28–120; Saliba 2007, 27–72; Dallal 2010, 13–16).

The Baghdadi translators and scholars came from diverse linguistic and religious backgrounds, and their work was patronized by a range of social groups in the higher strata of Abbasid society. Although the most culturally crucial support came from the Caliphs themselves and their families, other groups actually funded a larger number of translation projects and provided the vital intellectual motivations that made this work possible. Among these we should include the courtiers and companions of the Caliphs and other princes, the heads of warrior families that acted as military leaders and governors, state functionaries in the Abbasid administration, and leading physicians and scholars who commissioned translations of important works in their fields in order to further their own research and teaching agendas (Gutas 1998, 121–50).

The work of translating and studying these difficult texts was often carried out in loose groups of collaborators and competitors, most of whom were also experts in the fields transmitted by the treatises they studied. As this work progressed, the translations became more accurate and intelligible, so that the same work was often translated and corrected multiple times—sometimes by the same scholar.

Probably the most famous of these groups was that of Ḥunayn ibn Isḥāq (809–873 CE) and his students and colleagues, particularly Ḥubaysh (mid to late ninth century) and ʿĪsā ibn Yaḥyā (893–974), all of whom were both translators and learned physicians. Together they translated Greek medical works, including almost all of Galen's corpus, into Arabic or Syriac. Through this process, they became known for the development of a new translation style that involved fully understanding the intended meaning of the source language and then rendering that meaning with natural expressions in the target language, and in the process developing Arabic medical terminology on an intuitive basis (Rosenthal 1975, 20–21; Gutas 1998, 144–5).

Another important group was that which formed around al-Kindī (c. 800 – c. 870), sometimes known as "the philosopher (failasūf) of the Arabs." This group concentrated on the works of Plato, Aristotle, and the theological writings of the late ancient Platonic philosophers who had developed creative methods of reinterpreting authors like Plato, Aristotle, and Ptolemy in order to create a new synthesis of their often incompatible views. The motivation of al-Kindī and his associates in creating these translations and epitomes was, no doubt, a desire for new information and arguments in their project of creating a Neoplatonic philosophy that was compatible with what they took to be the tenets of Islam (Gutas 1998, 145–7).

The few detailed reports of this activity in our sources indicate that the translation of difficult technical works was often a complicated process—taking place over many years through repeated study of manuscript sources, involving a number of different individuals with various technical strengths.

Mandarins and Calendar Reform

One of the most conspicuous features of the landscape of scholarship in ancient and medieval China was the imperial bureaucracy, which drew its staff from a pool of the successful candidates of an elaborate system of civil service examinations (Elman 2000; Lee 2000, 111–70). Among the many ministries of these imperial systems were various astronomical bureaux, which were tasked with producing a state calendar that predicted the positions of the sun and moon, eclipses, certain planetary phenomena, and various aspects of divination, such as days that would be lucky or unlucky for certain undertakings—a sort of ephemeris or almanac. Despite the fact that these bureaux were generally staffed with fairly low-level functionaries, because the emperor was responsible for everything under the heavens, their activity was a matter of state importance (Lloyd 2002, 34–5; Sivin 2009, 35–8).

The astronomical bureaux oversaw activities in both mathematical calendrical studies (*lifa* 曆法) and in astronomy/astrology (*tianwen* 天文). Calendrical studies involved the use of observations and mathematical methods in the production of astronomical tables and canons of algorithms that reduced the determination of an annual calendrical almanac to a series of mechanical steps. Astrology involved finding various patterns in the heavens, making and recording observations of ominous phenomena and interpreting their import.

For a variety of different reasons—ranging from real or perceived technical deficiencies in the current system, through the need to legitimate regime or policy changes, to the personal goals of emperors or ministers—these calendrical systems were subject to numerous reforms (*gaili* 改曆). From the first century BCE to the middle of the eighteenth century, we have at least basic information about some hundred systems, of which around fifty were used throughout history to determine official ephemerides (Sivin 2009, 37–56). These reforms sometimes involved reshuffling the staff in the various ministries, or the development of a new directive in addition to the existing ones. In some cases, a new observatory was built to collect new data. In all cases, however, the motivations for these reforms went well beyond the practical needs of predicting astronomical events and regulating the calendar. These reforms helped to legitimize the imperial system and confirmed the role of the bureaucracy in assisting the emperor to carry out the mandate of heaven (Cullen 1993; Sivin 1995, 19–21; Sivin 2009).

Salaried Scholars in Damascus

Under the Ayyubids (mid-twelfth to mid-fourteenth century), Damascus experienced a resurgence as a political and intellectual center, and its cultural significance in scholarly circles continued even after the political capital of the region had moved to Cairo under Mamluk rule (mid-thirteenth to early sixteenth century). During this period, scholars from across the Islamic sphere came to Damascus to study, to teach for a while, or to settle into a life of salaried scholarship (Gilbert 1980, 107–11).

The mechanism underlying these changes was originally the use of the charitable foundation (*waqf*) to secure property for religious purposes. Many of these endowments established stipends (*manṣab*) for both students and teachers, often at a law college (*madrasa*), a house of instruction in the traditions of the Prophet (*dār al-ḥadīth*), or one of a variety of Sufi institutions. This gradual proliferation of stipends facilitated competition among the scholarly elite, and provided a mechanism for increasing numbers of men to earn a living as a result of their knowledge (Gilbert 1980, 113–26; Chamberlain 1994, 51–68; Hallaq 2009, 142–6).

Madrasas, however, were not purely colleges, and the stipends that they supported for lecturers and readers were not purely for teaching posts. Madrasas also served a range of extra-educational purposes, such as providing places of worship and burial, housing employees of the postal system (*barīd*), incarcerating prisoners, providing for the poor, and accumulating and securing household wealth. The terms of a stipend might dictate that certain subjects be taught, but this was often not enforceable, and the majority of stipends were free of formal constraints. Moreover, successful scholars often held stipends from a number of different institutions, including occasionally from an institution in a different city, such as Cairo. Madrasas had no corporate identity, granted no degrees, and had no fixed curriculum. Rather, they acted as an institutional nexus that brought together masters and students in the transmission of knowledge, which remained a fundamentally private process (Chamberlain 1994, chapter 2).

The core of medieval Damascene education was the relationship between the individual master (*shaykh*) and the student (Chamberlain 1994, chapter 4). Young people sought the companionship of a certain master, not enrollment at a particular school or institution. Teachers might be chosen as much for their manner of living and moral qualities as for their scholarly accomplishments, and the most important thing was that they be well-regarded as a link in a significant transmission of knowledge. Masters educated by reading texts with students—often accompanied by their own commentaries, lecturing—often in the form of an oral presentation of their own commentaries, and correcting their students' work in copying and memorizing these texts. Students who had sufficiently understood their master's ideas or texts could receive a written certificate (*ijāza*) to present legal opinions or to transmit certain texts. These practices established chains of transmissions of knowledge from one individual to another (*isnād*, *asnād*), which themselves became objects of study (Chamberlain 1994, 140–50; Brentjes 2002, 61).

In these circles, the religious sciences were always the core of scholarly activities, but some scholars also worked in, and transmitted, the rational or ancient sciences, such as logic, arithmetic and algebra, medicine, geometry, astronomy, and late Platonic and Aristotelian philosophy (Chamberlain 1994, 82–7; Brentjes 2002). The motivations for practicing and studying these fields were probably usually personal, based on interest and ability, but may sometimes have derived from career goals such as holding a position at a hospital or as timekeeper of a mosque (*muwaqqit*). For medieval Damascenes, however, the goal was mastery of many fields and even those who we regard as having contributed most significantly to medicine or the mathematical sciences are also praised in the biographical sources for their legal and religious scholarship.

Scholarly Women in the Ancient and Medieval Periods

Although our sources for the details of ancient and medieval scholars are often inadequate, a comparison of the numbers of men and women scholars in our sources show that very few women scholars left clear evidence of their activity (Netz 2002, 197; Plofker 2009, 180; Azad 2013, 57; Pomeroy 2013, 1). There are various reasons that could be advanced to explain this lack of information, and we cannot now be certain what the actual percentages of women scholars were at any given time and place. Nevertheless, these numbers are almost certainly indicative of real hurdles that women faced in leading a life of scholarship and making their work known.

With this general proviso, however, there were certain social settings that were more productive of woman scholars than others. For example, women were generally not active in areas that involved official, or semi-official, professionalization, such as the Chinese civil service examination system or the competition for stipendiary posts in medieval Islamic societies. In private settings, however, women of means could achieve a high level of learning. And religious orders and institutions often afforded women environments in which they could engage actively with scholarship.

Along with well-known examples, such as Rābiʿa al-ʿAdawiyya al-Qaysiyya (717–801 CE) or Hildegard of Bingen (1098–1179 CE), there is evidence that certain religious orders fostered women scholars. Some of the Daoist sects were said to have been founded by women, and some certainly included women in key roles. Women were ordained into the clergy, they taught, and they produced texts on the *way* (*dao* 道) and inner alchemy (Despeux 1990; Despeux 2000). In the Hellenic cities, the rites of female divinities were generally overseen by female officials, and in this capacity learned priestesses could attain civic recognition, and some political power (Connelly 2007). It seems that in these special surroundings women were more readily able to circumvent some of the obstacles of social and economic pressures and family obligations that otherwise often complicated their pursuit of learning.

Of course, there is also evidence for learned women outside of the context of religious orders. A number of famous literary authors were women. For example, two classics of Japanese literature were written at the beginning of the eleventh century by Sei Shōnagon and Murasaki Shikibu. Woman scholars are sometimes mentioned in Greco-Roman sources, without drawing particular attention to the fact that they are women. Porphyry (mid to late third century BCE) discusses the work of Ptolemaïs (third century BCE to first century CE), a harmonic theorist, and Pappus addresses himself to Pandrosion (both late third to early fourth century CE), a mathematical scholar (Levin 2009, 229–40; Bernard 2003). Biographical discussions of the prominent families of medieval Islamic cities often included accounts of learned women. They were usually scholars of the religious sciences, but in this regard they were no different from the men (Azad 2013).

Even in these cases, however, the sources often depict a close connection between woman scholars and a spiritual mode of life. One of the largest collections of learned writings by women from Greco-Roman sources come from the Pythagoreans (Pomeroy 2013). Although the Pythagoreans were not a religious order, they certainly advocated a spiritual mode of life, including detailed disciplines and rites. Perhaps the most famous woman scholar of Greco-Roman antiquity, Hypatia of Alexandria, also illustrates this tendency (Dzielska 1995). Hypatia was a philosopher, mathematical scholar and respected teacher of the youths of elite Alexandrian families. Her philosophical leanings, however, were those of late Platonism, a highly spiritual school, and her students wrote of her as a moral and spiritual guide. Along with her intellectual accomplishments, she was extolled for her virtues, especially that of temperance (*sōphrosunē*).

In many of these cases, women were able to mobilize narratives of traditionally feminine virtues—such as temperance, modesty, or chastity—in order to create a socially sanctioned space in which they could engage in serious scholarship.

Conclusion

As even this short survey of examples shows, very little of a general nature can be said about learned individuals of pre-modern periods. Who these people were, how they distinguished themselves as learned, and the roles they played within their societies was rather diverse even within the same times and cultures and quite different across larger geographic and temporal spans. Nevertheless, it is clear that an understanding of the role of learned individuals within their societies is inseparable from an articulation of the types of knowledge that they produced. By taking a narrow view it has sometimes appeared that an ancient or medieval author was the first to see some aspect of the world in essentially the same way that we do—that is, that they *discovered* it. For example, if we read only a few theorems of Euclid or a few passages of Ibn al-Nafīs, we might convince ourselves that Euclid's approach to geometry was the same as ours or that al-Nafīs gave an account of pulmonary transit that accords with our current understanding. But when we take a wider perspective and fit these fragments into a more coherent picture, as articulated in the works and practices for which we have evidence, it becomes clear that many aspects both of their knowledge claims and of their methods for producing knowledge do not accord with modern forms. In this way, we see that the discoveries that we attribute to them fit into a different network of beliefs and practices than any which we currently have, and hence they cannot be understood as straightforwardly equivalent to knowledge claims that we might make. In order to understand the learning, and the achievements, of past scholars, we must articulate their ideas in the various contexts that shaped them, and which they in turn shaped.

References

Asper, Markus. 2003. "Mathematik, milieu, text." *Sudhoffs Archiv*, 87: 1–31.

Azad, Arezou. 2013. "Female mystics in medieval Islam: The quiet legacy." *Journal for the Economic and Social History of the Orient*, 56: 53–88.

Beaulieu, Paul-Alain. 2006. "De l'Esagil au Mouseion: l'organisation de la recherche scientifique au IVe siècle avant J.-C." In *La transition entre l'empire achéménide et les royaumes hellénistique (vers 350-300 av. J.-C.)*, edited by Pierre Briant and Francis Joannès, 17–36. Paris: Éditions de Boccard.

Bernard, Alain. 2003. "Sophistic aspects of Pappus's *Collection*." *Archive for History of Exact Sciences*, 57: 93–150.

Brentjes, Sonja. 2002. "On the location of the ancient or 'rational' sciences in Muslim educational landscapes (AH 500–1100)." *Bulletin of the Royal Institute for Inter-Faith Studies*, 4: 47–71.

Brentjes, Sonja. 2008. "The study of geometry according to al-Sakhāwī (Cairo, 15th c) and al-Muḥubbī (Damascus, 17th c)." In *Mathematics Celestial and Terrestrial: Festschrift für Menso Folkerts zum 65*, edited by Joseph Dauben, Stefan Kirschner, Andreas Kühne, Paul Kunitzsch, and Richard Lorch, 323–41. Halle: Deutsche Akademie der Naturforscher Leopoldina.

Brentjes, Sonja. 2014. "Teaching the mathematical sciences in Islamic societies, eighth–seventeenth centuries." In *Handbook on the History of Mathematics Education*, edited by Alexander Karp and Gert Schubring, 85–107. Berlin: Springer.

Chamberlain, Michael. 1994. *Knowledge and Social Practice in Medieval Damascus, 1190–1350*. Cambridge: Cambridge University Press.

Connelly, Joan Breton. 2007. *Portrait of a Priestess: Women and Ritual in Ancient Greece*. Princeton, NJ: Princeton University Press.

Cook, Constance A. 1995. "Scribes, cooks, and artisans: Breaking Zhou tradition." *Early China*, 20: 241–71.

Cook, Constance A. 2013. "The Pre-Han period." In *Chinese Medicine and Healing*, edited by T. J. Hinrichs and Linda L. Barnes, 5–29. Cambridge, MA: Belknap Harvard University Press.

Cullen, Christopher. 1993. "Motivations for scientific change in ancient China: Emperor Wu and the Grand Inception astronomical reforms of 104 B.C." *Journal for History of Astronomy*, 24: 185–203.

Dallal, Ahmad. 2010. *Islam, Science, and the Challenge of History*. New Haven, CT: Yale University Press.

Despeux, Catherine. 1990. *Immortelles de la Chine anciénne: Taoïsm et alchimie fémine*. Puiseaux: Pardès.

Despeux, Catherine. 2000. "Women in daoism." In *Daoism Handbook*, edited by Livia Kohn, 384–412. Leiden: Brill.

Dzielska, Maria. 1995. *Hypatia of Alexandria*. Cambridge, MA: Harvard University Press.

Elman, Benjamin A. 2000. *A Cultural History of Civil Examinations in Late Imperial China*. Berkeley: University of California Press.

Fancy, Nahyan. 2013. *Science and Religion in Mamluk Egypt: Ibn al-Nafīs, Pulmonary Transit and Bodily Resurrection*. London: Routledge.

Feke, Jacqueline. 2014. "Meta-mathematical rhetoric: Hero and Ptolemy against the philosophers." *Historia Mathematica*, 41: 261–76.

Feke, Jacqueline, and Alexander Jones. 2010. "Ptolemy." In *The Cambridge History of Philosophy in Late Antiquity*, edited by Lloyd P. Gerson, 197–209. Cambridge: Cambridge University Press.

Ganeri, Jonardon. 2008. "Contextualism in the study of Indian intellectual cultures." *Journal of Indian Philosophy*, 36: 551–62.

Gilbert, Joan E. 1980. "Institutionalization of Muslim scholarship and professionalization of the 'Ulamā' in medieval Damascus." *Studia Islamica*, 52: 105–34.

Gutas, Dimitri. 1998. *Greek Thought, Arabic Culture*. London: Routledge.

Hallaq, Wael B. 2009. *Sharīayna: Theory, Practice, Transformations*. Cambridge: Cambridge University Press.

Jaeger, Mary. 2013. *Archimedes and the Roman Imagination*. Ann Arbor: University of Michigan Press.

Jones, Alexander. 2005. "Ptolemy's *Canobic Inscription* and Heliodorus' observation reports." *SCIAMVS*, 6: 53–98.

Lee, Thomas H. C. 2000. *Education in Traditional China: A History*. Leiden: Brill.

Lehoux, Daryn. 2012. *What Did the Romans Know?* Chicago: University of Chicago Press.

Levin, Flora. 2009. *Greek Reflections on the Nature of Music*. Cambridge: Cambridge University Press.

Lloyd, Geoffrey E. R. 2002. *The Ambitions of Curiosity*. Cambridge: Cambridge University Press.

Lloyd, Geoffrey E. R., and Nathan Sivin. 2002. *The Way and the Word*. New Haven, CT: Yale University Press.

Lo, Vivienne. 2013. "The Han period." In *Chinese Medicine and Healing*, edited by T. J. Hinrichs and Linda L. Barnes, 31–64. Cambridge, MA: Belknap Harvard University Press.

Netz, Reviel. 1997. "Classical mathematics in the Classical Mediterranean." *Mediterranean Historical Review*, 12: 1–24.

Netz, Reviel. 2002. "Greek mathematicians: A group picture." In *Science and Mathematics in Ancient Greek Culture*, edited by Christopher J. Tuplin and Tracey E. Rihill, 196–216. Oxford: Oxford University Press.

Neugebauer, Otto. 1975. *A History of Ancient Mathematical Astronomy*. Berlin: Springer.

Nutton, Vivian. 2013. *Ancient Medicine*, 2nd edition. London: Routledge.

O'Meara, Dominic J. 1989. *Pythagoras Revised: Mathematics and Philosophy in Late Antiquity*. Oxford: Clarendon Press.

Ossendrijver, Mathieu. 2012. *Babylonian Mathematical Astronomy: Procedure Texts*. Berlin: Springer.

Plofker, Kim. 2009. *Mathematics in India*. Princeton, NJ: Princeton University Press.

Pollock, Sheldon. 1985. "The theory of practice and the practice of theory in Indian intellectual history." *Journal of the American Oriental Society*, 105: 499–519.

Pollock, Sheldon. 2011. "The languages of science in early modern India." In *Forms of Knowledge in Early Modern Asia*, edited by Sheldon Pollock, 19–48. Durham, NC: Duke University Press.

Pomeroy, Sarah B. 2013. *Pythagorean Women*. Baltimore: Johns Hopkins University Press.

Robson, Eleanor. 2008. *Mathematics in Ancient Iraq: A Social History*. Princeton, NJ: Princeton University Press.

Rochberg, Francesca. 2004. *Heavenly Writing: Divination, Horoscopy, and Astronomy in Mesopotamian Culture*. Cambridge: Cambridge University Press.

Rosenthal, Franz. 1975. *The Classical Heritage in Islam*. London: Routledge.

Saliba, George. 2007. *Islamic Science and the Making of the European Renaissance*. Cambridge, MA: MIT Press.

Scharfe, Hartmut. 2002. *Education in Ancient India*. Leiden: Brill.

Sedley, David. 1976. "Epicurus and the mathematicians of Cyzicus." *Cronache Ercolanesi*, 6: 23–54.

Sivin, Nathan. 1995. "Shen Kua." In *Science in Ancient China*, by Nathan Sivin, part III. Aldershot: Variorum.

Sivin, Nathan. 2009. *Granting the Seasons: The Chinese Astronomical Reform of 1280*. Berlin: Springer.

Steele, John. 2011. "Astronomy and culture in late Babylonian Uruk." *Proceedings of the International Astronomical Union*, 7: 331–41.

Tolsa, Cristian. 2014. "Ptolemy and Plutarch's *On the Generation of the Soul in the Timaeus*: Three parallels." *Greek, Roman, and Byzantine Studies*, 54: 444–461.

Zhmud, Leonid. 2012. *Pythagoras and the Early Pythagoreans*. Oxford: Oxford University Press.

Zufferey, Nicolas. 1998. "Érudits et lettrés au debut de la dynastie Han." *Asiatische Studien*, 58: 915–65.

CHAPTER THREE

Go-Betweens, Travelers, and Cultural Translators

KAPIL RAJ[1]

Almost totally absent from the history and sociology of science, technology and medicine a decade ago, intermediation has recently begun to attract increasing scholarly attention in the field. The go-between has thus emerged as a significant new actor in science studies. However, if one can only be pleased with the appearance of new and, hopefully, powerful analytical concepts, there is also a need to explain why they were overlooked for the longest of periods.

It has until recently been widely assumed that the economic, political, social, and cultural interaction between the West and the Rest in the course of European expansion was a bipartite relationship between radically different cultural groups, linked at best through a one-way cultural and scientific diffusionism from European "center" to colonial "periphery." One of the main reasons adduced for this one-way flow is the purported superiority and methodological rigor of Western science (Basalla 1967; Law 1986), a strategy similar to that used by historians of science when treating the cases of non-European sciences in China, India, or the Islamic world (Huff 2003; Huff 2011; H. F. Cohen 2010). In spite of the vast difference in historical periods, contexts, and spaces considered, these studies resort to the common, classical methodology of comparativism, which constitutes its objects in contrasting, dualistic, terms, with clear boundaries between them. Unfortunately, this ontology is more widely accepted than one would want to believe, and even the works of great humanistic historians of science, such as Joseph Needham who can hardly be accused of Eurocentrism, are tarnished by this conception (Needham 1954–2005).

Although most historians of science have yet to openly question this dichotomous ontology (Hart 1999), social, economic and cultural historians have in recent decades revisited this all-too-often agonistic vision of the world on which traditional historiography is based. Scholarly attention has thus increasingly turned to global *inter*connections, intercultural encounter, and negotiation. In addition to a renewal of the traditional comparativist approach, new forms of "relational" history

A Companion to the History of Science, First Edition. Edited by Bernard Lightman.
© 2016 John Wiley & Sons Ltd. Published 2020 by John Wiley & Sons Ltd.

have emerged that offer more organic historiographical perspectives, such as history as "cross-roads" (*carrefour*), connected histories, *histoire croisée*, and circulation (Lombard 1990; Gruzinski 2001; Subrahmanyam 2001; Green 2004; Werner and Zimmermann 2006; Raj 2007; Raj 2013; Y. Cohen 2010). These new approaches have striven to bring out the inextricably enmeshed nature of cultures across the world, the commonalities on which intercultural contact is constructed and the ways people or groups of people cross cultural barriers.

Alongside these historiographical reflections, and sometimes relying on fairly conventional approaches, general historians have also produced concrete historical studies showing the organic nature of human societies. Drawing on anthropology, economics, and history, Philip Curtin, for example, studied trade and exchange across cultural lines in world history in his now-classic work (Curtin 1984). He has convincingly shown that trade diasporas have always transcended the barriers of locality and parochialism to link together widely separate parts of the globe. Starting from the premise that no human group could invent by itself more than a small part of its cultural and technical heritage, he adduces a plethora of examples of cultural transfer across human history to argue that the external stimulation provided by trade diasporas has been the most important single source of change and development in art, science, and technology. In so doing, he has shown the crucial need for mediation, brokerage, and cross-cultural communication in establishing sustained exchange between disparate and different cultures and the role that "cross-cultural brokers," or go-betweens, have played in this context, a subject of increasing academic interest over the last decades.

This chapter traces the variety of approaches on the subject adopted by cultural and social historians and anthropologists, as well as the theoretical reflections of classical sociologists. It then tackles the subject of go-betweens in the world of knowledge and science through the way they have been problematized in recent work in the history of science and the resulting change in the narrative from one of an agonistic "clash of cultures" to that of actively constructed connections. After presenting a concrete historical example from the late seventeenth century to show the decisive role of intermediation in the construction of knowledge in intercultural contexts, it concludes with some reflections on the continuing importance of go-betweens today for the construction of both intercultural and interdisciplinary knowledge.

Connecting Cultures in Recent Historiography

Go-betweens in themselves are, of course, not unfamiliar to historians. They have been widely studied in Romance literature as messengers and matchmakers in amorous encounters (Rouhi 1999; O'Hara 2000, chapter 3; Mieszkowski 2006). Their crucial role is emphasized even in the earliest manuals, such as the *Kamasutra*, on the conduct and management of amorous encounters (Vatsayana 2002, 115–21), but academic interest in the role of go-betweens for cross-cultural communication is more recent, gaining traction amongst historians only over the past decades. One of the first spaces to attract attention was the Indian Ocean world. This region has historically been characterized by trade, based mainly on spices, dyes, medicinal drugs, metals, saltpeter, rare timbers, cotton, silk, ceramics, and other manufactured and luxury goods, a part of which found their way into the Mediterranean and Europe, which trade built up during the first half of the second millennium CE. This trade was organized around peripatetic merchants on the one hand, and sedentary investors and brokers based in commercial, polyglot port cities like Aden, Calicut, and Melaka on the other. The whole network was in the hands of Arab, Hadrami, Iranian, Swahili, Chinese, Kashmiri, Armenian, Malay, Gujarati, Tamil, Kannada, and Telugu merchants, investors and bankers as well as a few from Venice and North Africa. The geographical spread of their origins was matched only by their religious diversity, comprising mainly Sunnis, Shias, Sufis, Bohras, Khojas, Chettiars, Baniyas, Jains, Jews, Zoroastrians, Theravada Buddhists, and Syrian Christians (Goitein 1987; Lombard and Aubin 1988; Ptak and Rothermund 1991; Middleton 2004).

It goes almost without saying that third parties, variously labeled "*passeurs culturels*," "culture brokers," or "go-betweens" were invariably at hand, "holding this disparate whole together," so to speak, ensuring the passage between the varied languages, customs and accounting techniques of the merchants and those of local communities of producers and suppliers, and influencing long-term interactions in new and fundamental ways. "It has been the ancient customes (sic) of the Indians to make all bargains by the mediation of Brokers, all Forreigners (sic) as well as the natives are compelled to submit to it: the Armenians, Turks, Persians, Jews, Europeans and Banyans," wrote Samuel Annesley, a senior merchant in the service of the English East India Company's factory at Surat in India at the turn of the eighteenth century (Chaudhuri 1978, 71). These intermediaries were designated by special appellations, such as *dallal* in Arabic, *terjuman* in Turkish (the origin of dragoman in English), *meturgeman* in Hebrew, and *dubash* (a term which refers to the multilingual faculty of go-betweens) in most languages of the Indian subcontinent, and constituted an obligatory passage point for all transactions. Already in the fourteenth century, Arab merchants were being advised to use the services of such factotums (ibn Ali al-Dimishqi [1318] 1955). In addition to translators, interpreters, moneychangers, bankers and moneylenders, the regional trade network was predicated upon specific maritime knowledge and skills. Pilots, navigators and theorists of navigation helped guide ships around maritime Asia and East Africa, thus forming yet another intermediary profession (ibn Majid al-Najdi [c.1490] 1971).

At the end of the fifteenth century, when the Portuguese first entered the highly organized and complex economic network of the Indian Ocean, they had to rely on the services of different local Muslim pilots to direct them up the Swahili coast and then to Calicut, their final destination (Subrahmanyam 1997, 115, 121–8). And when, after initially carrying out armed attacks on local powers, merchants and populations, they finally embarked on establishing an empire in the region, based on fortified littoral colonial settlements, private trade, and political and commercial treaties with regional polities, their interaction with the various communities and political authorities concerned was rendered possible only through the mediation of professional go-betweens with specific literary, technical, juridical, administrative, and financial skills (Boxer 1969; Couto 2003). The pattern set by the Portuguese in the sixteenth century was to continue into the following centuries and formed the basis of subsequent European interaction and maritime settlements in the Indian Ocean (Fischel 1956; Pearson 1988). In sum, we can distinguish at least four major functional types of go-betweens —the interpreter–translator, the merchant banker, the comprador, and the cultural broker who all also frequently carried and brokered specialized knowledges between communities. For instance, knowledge about the medicinal properties of plants and other natural products and their relative qualities depending on their provenance were commonly transmitted to medics and apothecaries by merchants and interpreters. In the South Asian context, each of these types of intermediaries could be Asian, North African, or European.

Of course, the Indian Ocean world was not unique. Many of the characteristics described here were shared with terrestrial Central Asia, the South China Sea and the Mediterranean (Szuppe 2004; Reid 1988–1993; Höfele and von Koppenfels 2005; Krstić 2012; Rothman 2012) and point to the fact that cultural brokerage is not limited to planetary connections, but is also a phenomenon to be studied for understanding intra-regional relations, not least through translation between languages (Burke 2005). Historians of the New World, too, have examined the borderlands of encounter between settlers and Native Americans in North America of the seventeenth and eighteenth centuries (White 1991). Other scholars have rightly stressed the performative character of these brokers' projects, and brought out the clear advantages of subjecting these agents to detailed and careful biographical scrutiny, since their performances offer unusually clear case-studies for the complexities of cultural negotiation and mediation (Hagedorn 1988; Karttunen 1994; Connell-Szasz 1994; Merrell 1999; Hinderaker 2002; Vaughan 2002).

Specialists of the early modern period of European navigation and the settlement of the Americas in regions such as Mexico and Brazil have found rich materials for the analysis of the work of such agents. Much of this analysis has focused on processes of hybridization and creolization in colonies of settlement. The studies of Serge Gruzinski and his colleagues have highlighted the ways in which these "*passeurs culturels*" involved the entanglement of very different forms of culture and society to forge new and robust systems of expression and knowledge of worldwide scope (Ares Queija and Gruzinski 1997; Bénat Tachot and Gruzinski 2001; Gruzinski [1999] 2002; O'Phelan and Salazar-Soler 2005).

Elsewhere, notably in the Portuguese encounters in Brazil, recent work has sought to classify the various processes of translation, settlement, mediation and intercourse that dominated new kinds of knowledge spaces (Metcalf 2005). They have argued that the go-betweens involved in these projects often adopted experimental and improvised strategies. Alida Metcalf gives a typology of these go-betweens in the Luso-Brazilian context, distinguishing between three types hierarchically arranged from the simplest to the highest form: physical/biological—those, such as sailors, passengers and slaves, who created material links between worlds by carrying flora and fauna, diseases and by bearing children of mixed race; transactional—principally translators, traders or cultural brokers, who made possible communication, exchange, trade, settlement, and conquest; and representational—chroniclers, priests, orators, map-makers, artists or writers, who represent the "other" culture through texts, images, or maps. This typology is, of course, purely formal: recent research convincingly shows that these three forms of activity are inextricably intertwined and are overwhelmingly practised by the same individuals. The fabled Doña Maria, also known as Malintzin or La Malinche, is a case in point. An Aztec woman whom Hernán Cortés took as part of a trophy during his conquest of Central America in 1519, she was to become his chief translator and negotiator while also being his mistress and mother of his first son (Lanyon 1999).

Travel and Translation

Mobility, travel, and translation literature is another focus of this new historiography. Until recently scholarship in this genre focused on accounts by European travelers and the ways in which they went out and reported the world through their representations of the "other" (Todorov [1982] 1984; Rubiés 1996; Rubiés 2000; Bourguet 1997; Elsner and Rubiés 1999). And even though there has been a burgeoning interest in non-European travel accounts, these narratives share the same general assumptions as their traditional European counterparts (Khan 1998; Fisher 2004; Chatterjee and Hawes 2008). However, the emphasis is now shifting to the interaction between mobile figures and the cultures they encounter, examining the negotiations, inflections, and reconfigurations that result from this process (Raveux 2004; Sheller and Urry 2006; Subrahmanyam 2011; Subrahmanyam 2012). Equally importantly, this approach also indicates a way of circumventing the prickly problem of changes of social, historical, and geographical scale that has so preoccupied the social sciences, in particular cultural anthropology and sociology. As one specialist of mobility studies notes, these mediators

> cross scales, places, territories, venture out into spaces with uncertain or moving boundaries, create or use networks. Nomadic experts are neither "local," nor "regional," nor for that matter are they "global." They cross "classical" territorial formations by juggling with possibilities and constraints, construct spaces tailored to their own activity, cultivate solutions of continuity [and] function through networks (Saunier 2005).

A focus on mobility thus allows us to study intercultural relationships and phenomena at all scales, from the local to the planetary and not only in the past, but also right up to the present day.

However, as we shall see more clearly through the example analyzed later in the chapter, "go-betweening" is not synonymous with peregrination; quite to the contrary, a crucial part of the process of intermediation, at least in knowledge-related domains, requires the central actors to be comparatively stationary.

Brokering in the Cultural and Social Sciences

If, in the recent past, social and cultural historians have expressed an increasing interest in these savvy, often elusive characters (whose paths are difficult to trace even in the best of circumstances, demanding an uncommon degree of perseverance), it has not been their prerogative alone. Cultural anthropologists, too, have been scrutinizing them for many decades, studying the crucial role of go-betweens in defining, objectifying, and maintaining the purported boundaries between cultures (Barth 1969). They have also looked at the processes of integration and change in contemporary post-colonial societies, focusing on the unique role played by go-betweens—notably teachers, preachers, curers, and community chiefs. They have examined in detail the eventual effects of mediation upon local and national cultures, as well as the relations between the two. New lines of inquiry have thus been opened concerning the mediator's quest for an identity that can accommodate the expectations of what anthropologists call the "great" tradition of national groups and the "little" tradition of local communities (Geertz 1959). One of these lines concerns the uneasy cultural or social identification, or "role ambiguity," crucial to the mobility and innovative behavior associated with the go-between (Press 1969). And, educational theorists who examine learning processes within and between groups of practitioners, have also thematized the notion of brokering in order for practices to move between different communities (Wenger 1998; Kubiak 2009).

As a whole, then, these studies point to the ubiquitous presence of go-betweens in all periods across the world, but, taken together, also show a keen sensitivity to the situated nature of intermediation, that is, to historical and spatial specificities. Nonetheless, there are a number of characteristics and strategies which go-betweens also share across space and time. The sociologist Georg Simmel's insightful reflections on what he called the "third element" are suggestive here. In addition to identifying the category of "impartial mediator" who reduces conflicts "to the objective spirit of each partial standpoint," thereby "produc[ing] the concord of two colliding parties," Simmel delineated other forms of intermediation which bestow power on the go-between: the arbitrator who, having been vested by both parties, "gains a special impressiveness and power over the antagonistic forces"; the *Tertius Gaudens* who, as the third party, draws advantage from the antagonism between the two sides; and the third element who "intentionally produces the conflict in order to gain a dominating position" (Simmel [1908] 1950, 145–69, on 146, 151, 162).

These contributions thus stress the historical contingency of these practices and mutations in knowledge and taxonomies introduced by movement itself. They also show that the itinerant subjects involved are not just passers-by or simple agents of cross-cultural diffusion; rather, they actively articulate relationships between disparate worlds or cultures by being able to translate between them. Simmel is once again inspiring with regard to this attribute. Seeking to make sense of their specificity by contrasting the wanderer "who comes today and goes tomorrow" with the stranger "who comes today and stays tomorrow,"—most commonly as a trader, as attested to by the history of economics—he sees the latter as someone whose

> position in [the host] group is determined, essentially, by the fact that he has not belonged to it from the beginning, that he imports qualities into it, which do not and cannot stem from the group itself. [...] He is not radically committed to the unique ingredients and peculiar tendencies of the group, and therefore approaches them with the specific attitude of "objectivity" [...] a particular structure composed of distance and nearness, indifference and involvement.

This distance can often bestow on the stranger the position of arbiter: "its most typical instance was the practice of those Italian cities to call their judges from the outside, because no native was free from entanglement in family and party interests" (Simmel [1908] 1950, 402–8). In a recent reflection on the stranger effect in early modern Asia, the historian Felipe Fernández-Armesto has differentiated between different configurations that strangers can find themselves in in a host society, from incorporating wisdom from afar to being king... or also a scapegoat, "the sump into which odium drains"—a predicament with strong evocations of that of millions of immigrants in today's world (Fernández-Armesto 2003).

Indeed, even those who were successful as intermediaries were often looked upon with contempt and suspicion, referred to variously as "arrogant, rebellious and audacious" or "shameless, bold, cunning, debauchees [and] liars" (Qaisar 1974, 221). Besides, attempting to articulate between disparate worlds and "holding the whole together" carried its own hazards; and Sanjay Subrahmanyam reminds us of the tragic fate of Sidi Yahya-u-Ta'fuft, a Berber notable and adventurer who acted as mediator between his compatriots and the Portuguese in early 16th-century Morocco and was assassinated by the former in the course of a mission on behalf of the latter—one of the probable, if unfortunate, ways of "fall[ing] deeply and irrevocably between two (or more) stools" (Subrahmanyam 2011, 1–10). Sidi Yahya was not the only one to pay dearly for straddling cultures. Quacy, a Guinean slave in Surinam, whose life straddled much of the eighteenth century, used his mastery over charms to become a healer and diviner for both colonials and his fellow-slaves. His skills in switching between the worlds of slaves and Europeans and being astutely able to connect them, sometimes also in playing one against the other, earned him worldwide fame, but also the distrust of his fellows, ultimately suffering the supreme humiliation of losing an ear for what the latter saw as his acts of betrayal (Parrish 2006, 1–6).

Go-Betweens Enter Science Studies

If historians of science have only recently begun to be interested in new approaches to global history and engage with this lively and fast-developing field of the social sciences, they have the decided advantage of all late-comers: having at hand masses of fine empirical research, much theoretical groundwork laid out and the cloth already cut out for them in this growing effort to re-map the emergence of the early modern and modern worlds as a global phenomenon, not one simply conceived of in western Europe and subsequently diffused to the rest of the world. However, the slowness of historians and sociologists of science to respond to and engage with this emerging field is not only due to the overwhelming hold of the dichotomous ontologies present since the inception of the field, but also to their insularity and reticence to fully incorporate the methods and tools of the anthropologist, a *sine qua non* for seriously engaging in this kind of research.

This is not to say that the problem of boundary-crossing and intermediation has been entirely absent in the fields of the sociology and history of science. On the contrary, and quite independently of the ambient dynamic in the neighboring social sciences, certain scholars in the field initiated, some decades ago, reflections on machines, objects, and mechanisms that either lie at the boundaries of many worlds or serve to mediate between them (Wise 1988; Wise 1993; Star and Griesemer 1989; Galison 1997, chapter 9; Rentetzi 2009, chapter 1; Star 2010). However, one of the first book-length works entirely dedicated to go-betweens from a history of science perspective appeared only in 2009 (Schaffer, Roberts, Raj, and Delbourgo 2009). This collection of 11 essays covers many of the themes described above across a number of world regions, but with a focus on the making, communication and reconfiguration of knowledge through the activities of knowledge go-betweens, and the global connections established thereby. Its most important feature was the historical conjuncture on which it focused, from 1770 to 1820. This period, which C. A. Bayly has labeled that of "proto-globalization" (Bayly 2002; Hopkins 2002, 5–6), is widely considered to be pivotal for the emergence of political, industrial and scientific modernity. This is, very importantly, the putative moment of the "rise of the West," the period when the Great Divergence between the West and Rest is said to have taken root, a period which has also been characterized as a "second Scientific Revolution" during which the new systems, institutions and professions of natural knowledge that undergirded the "industrial revolution" took shape (Kuhn 1977). Schaffer et al., on the other hand, focus on the crucial roles played by intermediaries, such as brokers, missionaries, spies, translators, entrepreneurs, and messengers, from all regions of the world during these decades in articulating the new racial, imperial, geographical, and disciplinary boundaries that mark the turn of the nineteenth century.

An exhibition, curated by Felix Driver and Lowri Jones, organized at the Royal Geographical Society in 2009, focused on the multifarious roles and agency of indigenous peoples and intermediaries in the history of exploration, especially guides, interpreters, and pilots. It sought to highlight what is obscured in standard narratives of exploration (Driver and Jones 2009). These exhibitions and publications have generated a fair amount of interest, and it is clear that they have also been convincing in showing the usefulness of using relational historiographies in order to understand the history of science in a non-Eurocentric, global perspective.

Amongst the more recent works that help bring out other aspects of the go-between in scientific travel literature is Marie-Christine Skuncke's recent book on the voyage of Linnaeus's pupil, Carl Peter Thunberg, to Japan in the late eighteenth century. It is an important re-evaluation of the means and methods of a naturalist voyager, not least his brokering talents, that breaks sharply with the common diffusionist perspective (Skuncke 2014). Mention must also be made here of Jordan Kellman's recent work on intermediaries in the Frenchman Tournefort's travels through the Levant in the early eighteenth century (Kellman 2014). And with a more gendered focus, albeit based primarily on textual and literary analysis, Narin Hassan has highlighted the role of European women and women doctors as go-betweens in the British empire who negotiated access to both the public world of medicine and the more intimate world of women, children, and private homes and significantly fashioned imperial medical culture (Hassan 2011).

In order to illustrate the usefulness and workings of intermediation in the production of knowledge, let us look briefly at the making of an early modern herbal in South Asia in the second half of the seventeenth century. This will also help highlight some of the main characteristics of go-betweens and dispel many misconceptions about them, notably the ubiquitous association between intermediation and mobility.

An Example

As West European trade with Asia was based largely on the spice, drug, and luxury-commodity trade, its sustained existence and development depended crucially on a detailed knowledge of natural things, notably the flora of the east. A knowledge of plants and their uses was important not only for introducing new commodities on the European markets but also to maintain the health of the thousands of sailors and traders who found themselves in the hostile climes of the tropics (Cook 1996; Raj 2007, 35–6). Thus, making inventories of the region's plants and drugs and their uses was an important part of European activity from the time of direct contact with

South and Southeast Asia (Cook 2007). Some of the best-known examples of such works are the Portuguese Garcia da Orta's *Colóquios dos simples e drogas da Índia* (1563) and Cristovão da Costa's *Tractado de las drogas y medicinas de las Indias Orientales* (1578), the Dutch medics Jacobus Bontius's posthumous *De medicina indorum* (1642) and Paul Hermann's *Paradisus Batavus* (1698), and the English surgeon, Samuel Browne's, multi-part contribution to the *Philosophical Transactions of the Royal Society* published under the title "An account of part of a collection of curious plants and drugs, lately given to the Royal Society by the East India Company," (1700–1701). Another English surgeon, Edward Bulkley (1651–1714), sent back to London at least five volumes of dried plants, fruits and drugs, with their local names sometimes transcribed in local characters.

In the 1670s, in response to a request for an inventory of local fauna from his superiors in the Dutch East India Company (VOC), the civil and military Commander of Dutch possessions on the Malabar coast in southwestern India, Hendrik Adriaan van Reede tot Drakenstein (1636–1691), had a gigantic work commissioned on the flora of this region. Its pen-and-ink-wash drawings of some 720 species were accompanied by a detailed description of each. The herbal was published in Latin, partly posthumously, under the title of *Hortus Indicus Malabaricus* in 12 folio volumes in Amsterdam between 1678 and 1693, and was soon to become the standard reference work for the flora of southwestern India. Indeed, van Reede's work, like that of Paul Hermann on Ceylon, was to form Linnaeus's main sources for the flora of Asia.

It is particularly instructive to see how van Reede's herbal was made. In the Preface to the third volume of the *Hortus Malabaricus* (1682), dedicated to the Raja of Cochin, he describes the construction of this work:

> By my orders, [...] Brahmin and other physicians made lists of the best known and most frequently occurring plants in their language. On this basis, others classified the plants according to the season in which they attracted notice for their leaves or flowers or fruit. This "seasonal" catalogue was then given to a certain number of experts, who were entrusted with the collection of the plants with their leaves, flowers, and fruit, for which they even climbed the highest tops of trees. These experts went out in groups of three to designated forests. Three or four draftsmen, who stayed with me in a convenient place, would accurately depict the living plants as the collectors brought them. To these pictures a description was added, nearly always in my presence (van Reede 1682, volume 3, viii).

However, in the first volume, published in 1678, van Reede includes three affidavits, the first (Figure 3.1), in Malayalam, in the *Aryaezuthu* script usually reserved for the higher castes, by the chief interpreter for the Dutch, a Luso-Indian named Emanuel Carneiro, followed by two others written by the physicians involved, in their respective languages and scripts, swearing to the veracity and exactitude of the information they had provided for the *Hortus Malabaricus*. The first is written in Malayalam, in the low-caste *Kolezuthu* script, by Itty Achudem, a well-known Malayali medic from a caste that climbed palm-trees and distilled alcohol, native of a village a few miles south of Cochin (Figure 3.2). The second, a collective certificate in the Konkani language written in the Nagari script, is signed by three Brahmin physicians: Ranga Bhatt, Vinayaka Pandit, and Appu Bhatt of Konkan origin, but settled in the Cochin region

Figure 3.1 The official interpreter, Emanuel Carneiro's affidavit in the first volume of *Hortus Indicus Malabaricus.* Hendrik Adriaan Van Reede. 1678. *Hortus Indicus Malabaricus.* Amsterdam, p. vii.

(Figure 3.3). Each is followed by a Latin translation of its contents made by Christian Herman van Donep, the civil secretary of the Dutch administration in Cochin.

Together, the translations make very interesting reading, for they open a window into the various stages of intermediation involved in the process of writing and communicating the plant descriptions and botanical and medicinal knowledge that the volumes contain. They tell us that they were first translated into Portuguese, the intermediary language between Indians and Europeans and between European nations as well in this part of India. From the affidavits, we learn that Itti Achudem's knowledge—partly textual, partly orally or tacitly appropriated, and partly empirically acquired—was put into writing in Portuguese by Emanuel Carneiro. Equally, Carneiro rendered his own testimony as well as Achudem's into Portuguese, and the Bhatts' and Vinayaka Pandit's collective testimony was translated by the latter as he had a sufficiently good command of the language, hailing from the Konkan coast, a region under Portuguese rule or influence for over 150 years at the time. The whole work was then translated into Portuguese by Carneiro and his team of interpreter-translators, in all probability assisted by Vinayaka Pandit. These letters seek to establish the credibility of the medics, the—diverse—sources and traditions of their knowledge, and the trust one could invest in them as well as in their testimony. Their account largely corroborates van Reede's, but with a few nuances: for instance it was the Konkani doctors who sent the men out to collect the flowers, fruits, and seeds of the desired plants, and it was they who compiled the descriptions from written sources as well from their own experience over a period of two years (and one can imagine that it was not always in

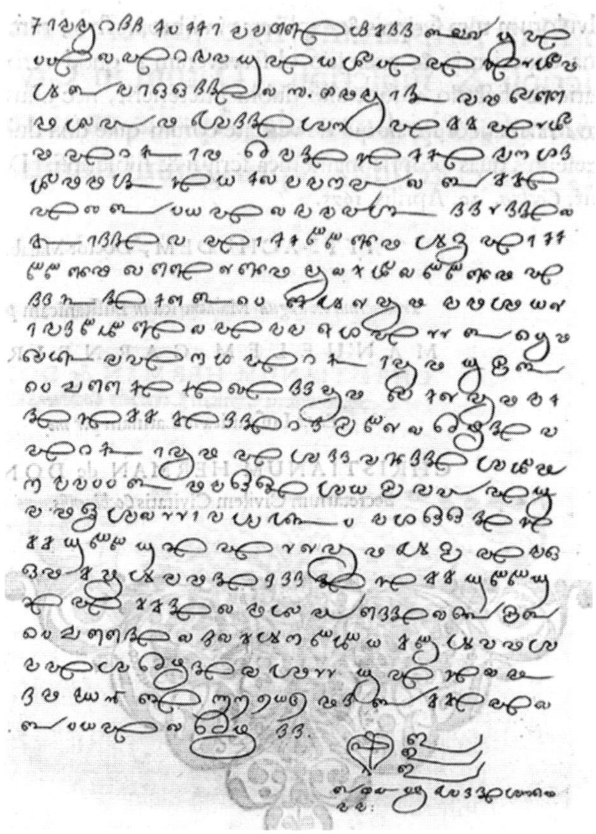

Figure 3.2 Itti Achudem's affidavit in Malayalam, written in the Kolezuthu script in the first volume of *Hortus Indicus Malabaricus*. Hendrik Adriaan Van Reede. 1678. *Hortus Indicus Malabaricus*. Amsterdam, p. ix.

van Reede's presence). These texts, and perhaps more written in different languages and scripts, were finally rendered via Portuguese and Dutch (since we know from van Reede's accounts that he himself made also made notes of many conversations with the medics) into Latin, the language in which the work was ultimately published. The latter translation was started by van Donep and his team in Cochin, then continued in Batavia (now Jakarta) and was finally completed in the Netherlands. The drawings were transformed into engravings in Amsterdam where the whole work was published in twelve folio volumes over a period of fifteen years.

Now, this account is radically different from the stories told so far about knowledge-making that involves indigenous peoples. In these accounts, the latter, commonly referred to as "informants," respond as adequately as they are able to questions determined by European investigators, who in turn are designated as "collectors" or "travelers." This information is then supposedly transformed into certified knowledge in the European metropole and can then be disseminated *urbi et orbi*.

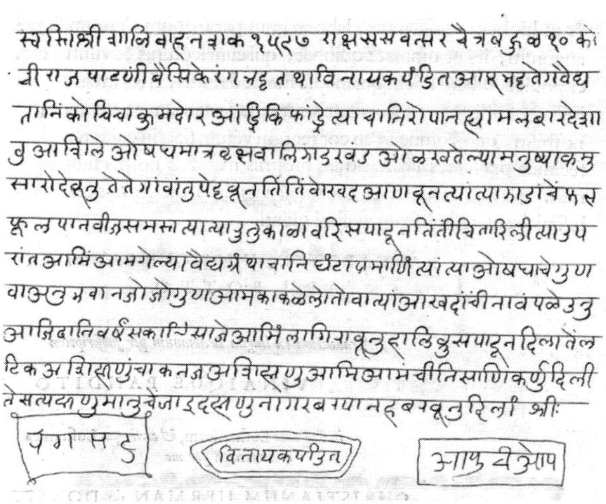

Figure 3.3 Joint certificate in Konkani in the Nagari script written by Ranga Bhat, Vinayaka Pandit, and Appu Bhat, in the first volume of *Hortus Indicus Malabaricus*. Hendrik Adriaan Van Reede. 1678. *Hortus Indicus Malabaricus*. Amsterdam, p. xi.

Focusing instead on relational processes in social interaction, the brief presentation above tells a very different story. In addition to identifying various go-betweens, it provides a very different glimpse into the uneven distribution of power among the protagonists and the complex nature of the process of intermediation which goes into the accomplishment of such projects. These intermediations involve persons—who are not merely vectors of information, but actually participate in the definition of the questions themselves—and languages (in this case Portuguese). They would also include other practical and material aspects of the project, for example, techniques, instruments, writing, etc., if the sources accessible exist to attest their existence. As for the people, it was for some a full-time, salaried function—for instance, Carneiro and the translators who worked under him—while for others it was a passing, opportunistic, activity, as for example the doctors and laborers involved in the enterprise, and, indeed, for van Reede himself—a senior functionary with no training in botany, who, sensing the potential advantages of this knowledge, brokered the execution of the work on behalf of his patrons to help expand their potential exports from India, as well as of Dutch naturalists and physicians.

To sum up then, although intermediation can sometimes be a professional activity, it is much more often an occasional, conjunctural, activity, not an essential part of the being of the persons considered. Notice also that the go-betweens in our story were not mobile people. Indeed, one of the preconditions for the success of the project lay in the maintenance of interpersonal relations based on sustained contact and trust between the interlocutors. The protagonists in this story are explicitly stationary, as is obvious for Carneiro and Itti Achudem, And even though the three Brahmin physicians were of Konkani origin and continued to use their own language, they or their ancestors had settled in Cochin where they practiced and also served as interpreters for diplomatic negotiations between Europeans and the Cochin rajas (Heniger 1986).

Even van Reede, who was furthest removed from the region, was comparatively sedentary during the years of the making of the *Hortus Malabaricus*. It was his investment
in building working relations with the local polity that was decisive in bringing to
fruition the whole project, the success of which crucially depended, as he himself
explicitly states, on being a *sedentary* resident commander in Cochin. Thus, and somewhat counterintuitively, intermediation is not synonymous with peregrination and is
not to be confused with mobility: our actors can only act as go-betweens because
they are comparatively stationary; if and when they do move, they do so as Simmel's
"stranger" rather than his "wanderer."

Mutatis mutandis, Serge Gruzinski's remarks about "borderline" persons in a given
society apply perfectly to go-betweens: their

> individual experiences enable the historian to reconstruct […] the sudden or progressive
> reorganization of a complex of cultural characteristics—[…] to analyse a dynamics that
> is generally obliterated by the static description of a symbolic system or of a mentality.
> These cases are also the only means of exploring concretely—to the extent, of course, that
> the sources permit—such crucial and subtle phenomena as […] the fascination exercised
> by an individual and the cultural features he manipulates, individual strategies of power
> and belief, and the modalities of putting into circulation original or modified cultural
> materials (Gruzinski [1985] 1989, 5).

Whither Go-Betweens Today?

In conclusion, I would like to point to some of the new directions that the concept
of go-betweens has opened, as well as its potential for the history and sociology of
science.

The first thing to stress is that, contrary to what many seem to think—an impression
that the body of research on the subject so far has unwittingly tended to reinforce—
intermediation is of interest not only for understanding pre-modern, or early modern
intercultural encounters in the light of European global expansion and imperial conquest. By taking on board the historical changes in the nature of the actors and institutions involved, it can be used—and indeed recently has been—by scholars within the
science studies tradition to argue, for instance, for the intermediary role of international organizations in instigating complex research involving disparate research communities (de Chadarevian 2015). It has also found increasing favor on the margins of
science studies, and maybe even independently of it, to study: the role of engineers
in connecting distinct practical traditions and in transporting diverse ways of doing
across different cultures of practice (Y. Cohen 2001; Downey and Lucena 2004); the
emergence of collaborative interdisciplinary research in the contemporary world, irre-
spective of the spatial scales involved—anywhere between the local and the global
(Beddoes 2011).

Science studies seems, however, largely to have ignored the analytical potential
of intermediation for the field, for instance in investigating communication across
the myriad quasi-autonomous subcultures that characterize much of science today
(Galison 1997, chapter 9). In his highly influential attempt to analyze the processes of
local coordination that exist to ensure cross-boundary transactions between disparate
scientific cultures, Galison coins the metaphor "trading zone," loosely inspired by the

analysis of trading practices between peasants and landowners in southwestern Colombia put forward by the Australian anthropologist, Michael Taussig (Taussig 1980). In these transactions, members of distinct cultures meet, negotiate a common understanding of procedures of exchange, transact and then go their own separate ways. They seem to live in a world of "pop-up store" culture, at best developing "pidgin" languages for "intercultural" transactions, but there is no attempt here to investigate the more substantial mutual transformations that sustained interactions induce in their respective practices (Pratt 1992, 6–7; Gruzinski [1999] 2002), nor for that matter the institutional, material, metrological, and procedural constraints imposed by the "market" (Callon 1998). And, crucially for the subject of this chapter, there is no mention of third parties that render encounters possible in the first place. Indeed, as at least one detailed sociological analysis has compellingly shown, the construction of markets themselves invariably requires the active presence of intermediaries in order to establish a stable and meaningful relationship between sellers and buyers (Garcia 1986). Ironically, and in spite of its limitations, "trading zones" has spawned a mass of literature which investigates communication across scientific sub-cultures in the universities of the "West," leaving "go-betweens" to connect the "Rest" (Gorman 2010), which brings us full circle back to where we started.

Independently of this schism, in order for it to be really insightful, intermediation needs to be further conceptualized in order to transform it from a largely descriptive category to an analytic one. This calls for a concerted, collective effort on the part of science studies and the history of science to renew the field once again—much as happened four decades ago to break with the positivist-idealist mode of classic history of science (Woolgar 1988; Pestre 1995)—and pick up the gauntlet of critical engagement with the other social sciences, notably translation theory, the history, sociology and anthropology of intercultural encounter, and gender studies—keeping in mind the porousness and power asymmetries between communities or cultures of practice. The resulting reflections could then be applied as much to long-distance encounters as to the very local encounters between different "subcultures" of research in the same region, country, university, or building. They just need to take into account the role of third parties that bring these disparate, unevenly matched, and *prima facie* incommensurable, worlds together and work toward creating inherently novel situations the results of which cannot be reduced to mere "hybrids," "pidgins," or "creoles." These notions need to be taken seriously, rather than as apparently self-explanatory metaphors. Finally, this approach would also help in forging a much-needed conceptual vocabulary to replace the traditional diffusionist one which inevitably thwarts any serious engagement with studies of encounter between knowledge cultures or specialities.

Endnote

1   Although invited to write this chapter in early 2013, a preliminary version was only presented at a panel entitled "Where are the Go-Betweens Going?" in the HSS Annual Meeting in Chicago, November 2014. I thank the organizers, Janet Browne and Jordan Goodman, for the honor of inviting me to comment on the papers, and the participants for their contributions and the lively discussion which helped me focus my thoughts on some of the key issues discussed here. I am also greatly indebted to Susan Gross

Solomon, Simon Schaffer, Maria Rentetzi, Jordan Kellman, Jordan Goodman, and the participants of my 2014–15 research seminar at the EHESS for their careful reading and insightful suggestions. In dealing with a tardy and absent-minded author, Bernie Lightman has been a model of patience and generosity.

References

Ares Queija, Berta and Serge Gruzinski (eds.) 1997. *Entre dos mundos. Fronteras culturales y agents mediadores.* Seville: EEHA.

Barth, Fredrik (ed.) 1969. *Ethnic Groups and Boundaries: The Social Organization of Culture Difference.* London: Allen & Unwin.

Basalla, George. 1967. "The spread of Western science." *Science*, 156, No. 3775: 611–22.

Bayly, C. A. 2002. "'Archaic' and 'Modern' globalization in the Eurasian and African arena, *c.* 1750–1850." In *Globalization in World History*, edited by A. G. Hopkins, 47–72. London: Pimlico.

Beddoes, Kacey. 2011. "Practices of brokering: Between STS and feminist engineering education research." Unpublished Ph.D. dissertation. Virginia Polytechnic Institute and State University.

Bénat Tachot, Louise, and Serge Gruzinski (eds.) 2001. *Passeurs culturels: mécanismes de métissage.* Paris: Presses universitaires de Marne-la-Vallée & Éditions de la MSH.

Bourguet, Marie-Noëlle. 1997. "La collecte du monde. Voyage et histoire naturelle (fin XVIIème–début XIXème siècle)." In *Le Muséum au premier siècle de son histoire*, edited by Claude Blanckaert, Claudine Cohen, Pietro Corsi and Jean-Louis Fischer, 163–96. Paris: Muséum d'histoire naturelle.

Boxer, Charles Robert. 1969. *The Portuguese Seaborne Empire, 1415–1825.* London: Hutchinson.

Burke, Peter. 2005. "The Renaissance translator as go-between." In *Renaissance Go-Betweens: Cultural Exchange in Early Modern Europe*, edited by Andreas Höfele and Werner von Koppenfels, 17–31. Berlin: Walter de Gruyter.

Callon, Michel. 1998. "Introduction: The embeddedness of economic markets in economics." In *The Laws of the Markets*, edited by Michel Callon, 1–57. Oxford: Blackwell/The Sociological Review.

Chatterjee, Kumkum, and Clement Hawes (eds.) 2008. *Europe Observed: Multiple Gazes in Early Modern Encounters.* Lewisburg, PA: Bucknell University Press.

Chaudhuri, Kirti Narayan. 1978. *The Trading World of Asia and the English East India Company, 1660–1760.* Cambridge: Cambridge University Press.

Cohen, H. Floris. 2010. *How Modern Science Came Into the World: Four Civilizations, One 17th-Century Breakthrough.* Amsterdam: Amsterdam University Press.

Cohen, Yves. 2001. *Organiser à l'aube du taylorisme. La pratique d'Ernest Mattern chez Peugeot, 1906–1919.* Besançon: Presses universitaires franc-comtoises.

Cohen, Yves. 2010. "Circulating localities: The example of Stalinism in the 1930s." *Kritika: Explorations in Russian and Eurasian History*, 11: 11–45.

Connell-Szasz, Margaret (ed.) 1994. *Between Indian and White Worlds: The Cultural Broker.* Norman: University of Oklahoma Press.

Cook, Harold J. 1996. "Physicians and natural history." In *Cultures of Natural History*, edited by Nicholas Jardine, James A. Secord, and Emma C. Spary, 91–105. Cambridge: Cambridge University Press.

Cook, Harold J. 2007. *Matters of Exchange: Commerce, Medicine, and Science in the Dutch Golden Age.* New Haven, CT: Yale University Press.

Couto, Dejanirah. 2003. "The role of interpreters, or *linguas*, in the Portuguese Empire during the 16th century." *E–Journal of Portuguese History*, 1: http://www.brown. edu/Departments/Portuguese_Brazilian_Studies/ejph/html/issue2/pdf/couto.pdf

Curtin, Philip D. 1984. *Cross-Cultural Trade in World History*. Cambridge: Cambridge University Press.

de Chadarevian, Soraya. 2015 (in press). "Human population studies and the World Health Organization." *Dynamis*, 35: 359–388.

Downey, G. L., and J. C. Lucena. 2004. "Knowledge and professional identity in engineering: Code-switching and the metrics of progress." *History and Technology*, 20: 393–420.

Driver, Felix, and Lowri Jones (eds.) 2009. *Hidden Histories of Exploration: Researching the RGS–IBG Collections*. London: Royal Holloway, University of London.

Elsner, Jaś, and Joan-Pau Rubiés (eds.) 1999. *Voyages and Visions: Towards a Cultural History of Travel*. London: Reaktion Books.

Fernández-Armesto, Felipe. 2003. "The stranger effect in early modern Asia." In *Shifting Communities and Identity Formation in Early Modern Asia*, edited by Leonard Blussé and Felipe Fernández-Armesto. Leiden: CNWS.

Fischel, Walter J. 1956. "Abraham Navarro: Jewish interpreter and diplomat in the service of the English East India Company (1682–1692)." *Proceedings of the American Academy for Jewish Research*, 25: 39–62.

Fisher, Michael H. 2004. *Counterflows to Colonialism: Indian Travellers and Settlers in Britain 1600–1857*. Delhi: Permanent Black.

Galison, Peter L. 1997. *Image and Logic: A Material Culture of Microphysics*. Chicago: University of Chicago Press.

Garcia, Marie-France. 1986. "La construction d'un marché parfait: le marché au cadran de Fontaines-en-Sologne." *Actes de la recherche en sciences sociales*, 65: 2–13.

Geertz, Clifford. 1959. "The Javanese Kijaji: The Changing Role of a Cultural Broker." *Comparative Studies in Society and History*, 2: 228–49.

Goitein, Shelomo Dov. 1987. "Portrait of a medieval Indian trader: Three letters from the Cairo Geniza." *Bulletin of the School of Oriental and African Studies*, 50: 449–64.

Gorman, Michael E. (ed.) 2010. *Trading Zones and Interactional Expertise: Creating New Kinds of Collaboration*. Cambridge, MA: MIT Press.

Green, Nancy L. 2004. "Forms of comparison." In *Comparison and History: Europe in Cross-Cultural Perspective*, edited by Deborah Cohen and Maura O'Connor, 41–56. London: Routledge.

Gruzinski, Serge. [1985] 1989. *Man-Gods in the Mexican Highlands: Indian Power and Colonial Society, 1520–1800*. Stanford: Stanford University Press.

Gruzinski, Serge. 2001. "Les mondes mêlés de la Monarchie catholique et autres 'connected histories'." *Annales HSS*, 56: 85–117.

Gruzinski, Serge. [1999] 2002. *The Mestizo Mind: The Intellectual Dynamics of Colonization and Globalization*. New York and London: Routledge.

Hagedorn, Nancy L. 1988. "'A friend to go between them': the interpreter as cultural broker during Anglo-Iroquois Councils, 1740–70." *Ethnohistory*, 35: 60–80.

Hart, Roger. 1999. "Beyond Science and Civilization: A post-Needham critique." *East Asian Science, Technology, and Medicine*, 16: 88–114.

Hassan, Narin. 2011. *Diagnosing Empire: Women, Medical Knowledge, and Colonial Mobility*. Farnham: Ashgate.

Heniger, Johannes. 1986. *Hendrik Adriaan van Reede Tot Drakenstein (1636–1691) and Hortus Malabaricus: A Contribution to the History of Dutch Colonial Botany*. Rotterdam: A. A. Balkema.

Hinderaker, Eric. 2002. "Translation and cultural brokerage." In *A Companion to American Indian History*, edited by Philip J. Deloria and Neal Salisbury, 357–75. Oxford: Blackwell.

Höfele, Andreas, and Werner von Koppenfels. 2005. *Renaissance Go-Betweens: Cultural Exchange in Early Modern Europe*. Berlin & New York: Walter de Gruyter.

Hopkins, A. G. (ed.) 2002. *Globalization in World History*. London: Pimlico

Huff, Toby E. 2003. *The Rise of Early Modern Science: Islam, China, and the West*. Cambridge: Cambridge University Press.

Huff, Toby E. 2011. *Intellectual Curiosity and the Scientific Revolution: A Global Perspective*. Cambridge: Cambridge University Press.

ibn Ali al-Dimishqi, Abu al-Fadi Ja'far. 1318 (in Arabic), 1955. *The Book of Knowledge of the Beauties of Commerce and of the Cognizance of Good and Bad Merchandise and of Falsifications*. In *English in Medieval Trade in the Mediterranean World: Illustrative Documents*, translated with introductions and notes by Robert S. Lopez and Irving W. Raymond, 23–7. New York: Columbia University Press, 1955.

ibn Majid al-Najdi, Ahmad. (c.1490) 1971. *Kitab al-Fawa'id fi usul wa'l qawa'id*, English translation by Gerald Randall Tibbetts, *Arab Navigation in the Indian Ocean before the Coming of the Portuguese*. London: Royal Asiatic Society of Great Britain and Ireland.

Karttunen, Frances. 1994. *Between Worlds: Interpreters, Guides, and Survivors*. New Brunswick, NJ: Rutgers University Press.

Kellman, Jordan. 2014. "Confident men and confidence men: Taxonomy and local agents in Joseph Pitton de Tournefort's Eastern Mediterranean voyage, 1700–1702." *Unpublished paper presented at the History of Science Society Annual Meeting*, Chicago, 6–9 November.

Khan, Gulfishan. 1998. *Indian Muslim Perceptions of the West during the Eighteenth Century*. Karachi: Oxford University Press.

Krstić, Tijana. 2012. "Of translation and empire: Sixteenth-century Ottoman imperial interpreters as Renaissance go-betweens." In *The Ottoman World*, edited by Christine Woodhead, 130–42. London and New York: Routledge.

Kubiak, Chris. 2009. "Working the interface: Brokerage and learning networks." *Educational Management Administration and Leadership*, 37: 239–56.

Kuhn, Thomas S. 1977. "The function of measurement in modern physical science." In *The Essential Tension: Selected Studies in Scientific Tradition and Change*, by Thomas Kuhn, 178–224. Chicago: University of Chicago Press.

Lanyon, Anna. 1999. *Malinche's Conquest*. St. Leonards, NSW: Allen & Unwin.

Law, John. 1986. "On the methods of long-distance control: Vessels, navigation and the Portuguese route to India." In *Power, Action and Belief: A New Sociology of Knowledge?* edited by John Law, 234–63. London: Routledge & Kegan Paul.

Lombard, Denys. 1990. *Le carrefour javanais. Essai d'histoire globale*, 3 vols. Paris: Éditions de l'EHESS.

Lombard, Denys, and Jean Aubin (eds.) 1988. *Marchands et hommes d'affaires asiatique dans l'Océan Indien et la Mer de Chine*. Paris: Éditions de l'Ecole des hautes études en sciences sociales.

Merrell, James H. 1999. *Into the American Woods: Negotiators on the Pennsylvania Frontier*. New York: W. W. Norton.

Metcalf, Alida C. 2005. *Go-Betweens and the Colonization of Brazil, 1500–1600*. Austin, TX: University of Texas Press.

Middleton, John. 2004. *African Merchants of the Indian Ocean: Swahili of the East African Coast*. Long Grove, IL: Waveland Press.

Mieszkowski, Gretchen. 2006. *Medieval Go-Betweens and Chaucer's Pandarus*. Basingstoke: Palgrave Macmillan.

Needham, Joseph. 1954–2005. *Science and Civilisation in China*, 7 vols. Cambridge: Cambridge University Press.

O'Hara, Diana. 2000. *Courtship and Constraint: Rethinking the Making of Marriage in Tudor England*. Manchester: Manchester University Press.

O'Phelan, Scarlett, and Carmen Salazar-Soler (eds.) 2005. *Passeurs, mediadores culturales y agents de la primera globalización en la Mundo Ibérico*. Lima: Pontificia Universidad Católica de Perú.

Parrish, Susan Scott. 2006. *American Curiosity: Cultures of Natural History in the Colonial British Atlantic World*. Chapel Hill, NC: University of North Carolina Press.

Pearson, Michael Naylor. 1988. "Brokers in western Indian port cities: Their role in servicing foreign merchants." *Modern Asian Studies*, 22: 455–72.

Pestre, Dominique. 1995. "Pour une histoire sociale et culturelle des sciences. Nouvelles définitions, nouveaux objets, nouvelles pratiques." *Annales HSS*, 50: 487–522.

Pratt, Mary Louise. 1992. *Imperial Eyes: Travel Writing and Transculturation*. London: Routledge.

Press, Irwin. 1969. "Ambiguity and innovation: Implications for the genesis of the culture broker." *American Anthropologist*, 71: 205–17.

Ptak, Roderich, and Dietmar Rothermund (eds.) 1991. *Emporia, Commodities and Entrepreneurs in Asian Maritime Trade, c.1400–1750*. Stuttgart: Franz Steiner Verlag.

Qaisar, Ahsan Jan. 1974. "The role of brokers in medieval India." *Indian Historical Review*, 1: 220–46.

Raj, Kapil. 2007. *Relocating Modern Science: Circulation and Construction of Knowledge in South Asia and Europe, 1650–1900*. Basingstoke: Palgrave Macmillan.

Raj, Kapil. 2013. "Beyond postcolonialism … and postpositivism: Circulation and the global history of science." *Isis*, 104: 337–47.

Raveux, Olivier. 2004. "Espaces et technologies dans la France méridionale d'Ancien Régime: l'exemple de l'indiennage marseillais." *Annales du Midi, revue archéologique, philologique et historique de la France méridionale*, 246: 155–70.

Reid, Anthony. 1988–93. *South East Asia in the Age of Commerce*, 2 vols. New Haven, CT: Yale University Press.

Rentetzi, Maria. 2009. *Trafficking Materials and Gendered Experimental Practices: Radium Research in Early 20th Century Vienna*. New York: Columbia University Press.

Rothman, E. Natalie. 2012. *Brokering Empire: Trans-Imperial Subjects between Venice and Istanbul*. Ithaca, NY: Cornell University Press.

Rouhi, Leyla, 1999. *Mediation and Love: A Study of the Medieval Go-Between in Key Romance and Near Eastern Texts*. Leiden: Brill.

Rubiés, Joan-Pau. 1996. "Instructions for travellers: Teaching the eye to see." *History and Anthropology*, 9: 139–90.

Rubiés, Joan-Pau. 2000. *Travel and Ethnology in the Renaissance: South India through European Eyes 1250–1625*. Cambridge: Cambridge University Press.

Saunier, Pierre-Yves. 2005. "Épilogue: À l'assaut de l'espace transnational de l'urbain, ou la piste des mobilités." *Géocarrefour*, 80: 249–253.

Schaffer, Simon, Lissa Roberts, Kapil Raj and James Delbourgo (eds.) 2009. *The Brokered World: Go-Betweens and Global Intelligence, 1770–1820*. Sagamore Beach, MA: Science History Publications.

Sheller, Mimi, and John Urry. 2006. "The new mobilities paradigm." *Environment and Planning A*, 38: 207–26.

Simmel, Georg. [1908] 1950. "The isolated individual and the dyad," "The triad," and "The stranger." In *The Sociology of Georg Simmel*, translated, edited and introduced by Kurt Heinrich Wolff, 118–44, 145–69 and 402–8 respectively. New York: Free Press, 1950.

Skuncke, Marie-Christine. 2014. Carl *Peter Thunberg, Botanist and Physician: Career Building across the Oceans in the Eighteenth Century*. Uppsala: Swedish Collegium for Advanced Study.

Star, Susan Leigh. 2010. "This is not a boundary object: Reflections on the Origin of a Concept." *Science, Technology, and Human Values*, 35: 601–17.

Star, Susan Leigh, and James R. Griesemer. 1989. "Institutional ecology, 'translations' and boundary objects: Amateurs and professionals in Berkeley's Museum of Vertebrate Zoology." *Social Studies of Science*, 19: 387–420.

Subrahmanyam, Sanjay. 1997. *The Career and Legend of Vasco Da Gama*. Cambridge: Cambridge University Press.

Subrahmanyam, Sanjay. 2001. "Du Tage au Gange au XVIe siècle. Une conjuncture millénaire à l'échelle eurasiatique." *Annales HSS*, 56: 51–84.

Subrahmanyam, Sanjay. 2011. *Three Ways to be Alien: Travails and Encounters in the Early Modern World*. Waltham, MA: Brandeis University Press.

Subrahmanyam, Sanjay. 2012. *Courtly Encounters: Translating Courtliness and Violence in Early Modern Eurasia*. Cambridge, MA: Harvard University Press.

Szuppe, Maria. 2004. "Cercles des lettrés et cercles littéraires. Entre Asie centrale, Iran et Inde du Nord (XVᵉ–XVIIIe siècle)." *Annales HSS*, 59: 997–1018.

Taussig, Michael T. 1980. *The Devil and Commodity Fetishism in South America*. Chapel Hill, NC: University of North Carolina Press.

Todorov, Tzvetan. [1982] 1984. *The Conquest of America: The Question of the Other*. New York: Harper & Row.

Van Reede tot Drakestein, Hendrik Adriaan. 1678–1693. *Hortus Indicus Malabaricus*, 12 vols. Amsterdam.

Vatsayana. 2002. *Kamasutra*, translated from the Sanskrit by Wendy Doninger and Sudhir Kakar. Oxford: Oxford University Press.

Vaughan, Alden T. 2002. "Sir Walter Ralegh's Indian interpreters, 1584–1618." *The William and Mary Quarterly*, 3rd series, 59: 341–76.

Wenger, Étienne. 1998. *Communities of Practice: Learning, Meaning and Identity*. Cambridge: Cambridge University Press.

Werner, Michael, and Bénédicte Zimmermann. 2006. "Beyond comparison: Histoire croisée and the challenge of reflexivity." *History and Theory*, 45: 30–50.

White, Richard. 1991. *The Middle Ground: Indians, Empires, and Republics in the Great Lakes Region, 1650–1815*. Cambridge: Cambridge University Press.

Wise, M. Norton. 1988. "Mediating machines." *Science in Context*, 2: 77–113.

Wise, M. Norton. 1993. "Mediations: Enlightenment balancing acts, or the technologies of rationalism." In *World Changes: Thomas Kuhn and the Nature of Science*, edited by Paul Horwich, 207–56. Cambridge, MA: MIT Press.

Woolgar, Steve. 1988. *Science: The Very Idea*. London: Tavistock Publications.

CHAPTER FOUR

The Alchemist

TARA NUMMEDAL

Most alchemists today have an ambivalent relationship to science. For example, while pursuing advanced graduate work in mathematics at the University of Vienna, the contemporary American alchemical practitioner, author, teacher, and entrepreneur Dennis William Hauck (2015) "became convinced that the alchemists had discovered universal principles of transformation that are part of the very fabric of the universe, but which modern science was only beginning to recognize." After a two-year "apprenticeship" in the late twentieth century with a practicing alchemist in Prague, Hauck came to believe that alchemy "represented a higher level of science." As Hauck's website explains, "The amazing demonstrations and experiments he witnessed convinced him that, just as the alchemists believed, consciousness was a force of nature that modern science had completely ignored." While academic historians would nuance and historicize Hauck's assessment of what "the alchemists believed," the point here is that for Hauck, science has its limits, in that it can explain only "our physical reality," not the deeper "hidden reality" from which the physical stems. He decided that alchemy offered one path into that hidden reality, and so, "Knowing that this path of discredited 'pseudoscience' might forever taint his career, Hauck decided it was worth the risk to discover solid evidence and challenge the dominant Newtonian paradigm." Hauck is now a prolific author, and has been at the center of efforts to institutionalize modern alchemy through organizations such as the International Alchemy Guild, the *Alchemy Journal*, the short-lived International Alchemy Conference, and Hauck's own extensive website, alchemylab.com. He and other contemporary alchemists frame their work as a continuation of an ancient alchemical tradition, reminding us that alchemy's history did not come to an end in the eighteenth century when parts of it were claimed for a newly differentiated practice and field called "chemistry." Alchemy today is as vibrant as it ever has been, but its practitioners tend to position their work very deliberately as something distinct from—and usually beyond or supplementary to—science, and most scientists would agree that alchemy no longer falls within their purview. These days, the only way to study alchemy at an accredited university, for

A Companion to the History of Science, First Edition. Edited by Bernard Lightman.
© 2016 John Wiley & Sons Ltd. Published 2020 by John Wiley & Sons Ltd.

example, or to obtain an NSF grant to pursue alchemical research, is to examine it as a historical phenomenon. Alchemy no longer is thought to have any bearing on the central questions of modern science.

If the scientist and the alchemist now occupy different intellectual, social, and cultural spaces, this separation is a relatively recent development. In fact, one of the central tasks of the "new historiography of alchemy," which contemporary alchemists generally discount in favor of their own origins stories, has been to understand precisely what role alchemy and alchemists have played in the history of science. The task is challenging, for both are moving targets. It is a matter of tracing out intersections between two complex entities, "alchemy" and "science," each of which has undergone multiple transformations of its own over the course of a very long history from late Antiquity to the present. Alchemists have taken on many different social, cultural, and intellectual roles over this long history, but the most enduring are artisan, natural philosopher, and prophet. Historians of science now appreciate alchemical theories of the nature of matter as central to any history of science. Likewise, they have begun to recognize the importance of artisanal epistemologies to the development of new ways of investigating nature that emerged in early modern Europe. Indeed, the fusion of theory and practice that was central to alchemy from its origins has made it emblematic of the "new science" that emerged in the sixteenth and seventeenth centuries (Moran 2005; Martinón-Torres 2011; Moran et al. 2011). Alchemists periodically have claimed to be able to access or incorporate other forms of knowledge as well, however: both divine, as in late Antiquity and the Middle Ages or, more recently, psychological, as Hauck's understanding of alchemy as a window to "consciousness" suggests (Jung 1937; Principe and Newman 2001; Tilton 2003, 18–21; Hanegraaff 2012, 277–95). Alchemists making these claims in the past two centuries have found themselves in an increasingly complicated position with respect to modern, institutionalized science, which has rejected "irrational" and individualized insights as foundations for generalizable, rational, knowledge of nature (Hanegraaff 2013). Alchemists' most important role in the history of science, therefore, lies in the way they have challenged and helped refine and reconfigure its boundaries, both deliberately and unintentionally, especially those between artisanal and philosophical knowledge, and between the natural and the supernatural, spiritual, or psychological.

One of the oldest and most enduring features of alchemy is a clear corpus of alchemical texts, as well as the scholarly practices associated with studying them; even the earliest extant alchemical texts refer back to earlier authors, creating a surprisingly cohesive intellectual tradition stretching from Greco-Roman Egypt, where alchemy seems to have originated, to the present (Principe 2012, 15–16, for example). This feature of alchemy positioned alchemists as scholars who located important knowledge in writings of their predecessors, and who saw value, therefore, in preserving and interpreting a long textual tradition. The alchemical canon was first assembled roughly between the seventh and the eleventh century by Byzantine scholars who created compendia of Greek alchemical texts originally composed centuries earlier, between the late first or second to the eighth century. This *Corpus alchemicum graecum* (Halleux 1981), as the Byzantine compendia are now known, is problematic. The compendia were assembled hundreds of years after the texts they contain were originally written, and were shaped by the editorial choices and priorities of the Byzantine compilers, who included only a fraction of the original Greco-Egyptian alchemical texts (most of which are now lost). Nevertheless, the Greek alchemical corpus is invaluable because it does contain some precious early texts from alchemy's origins in first to third-century Greco-Roman Egypt, as well as treatises from later Byzantine exegetes and commentators from the fourth to the thirteenth century. In assembling these authors in one place, the *Corpus alchemicum graecum* established a Greek canon of authors whose works have been considered central to the alchemical tradition ever since, including, for example, Maria Judea, Pseudo-Democritus, and Zosimos of Panopolis (Mertens 2006).

Some of the original Greco-Egyptian alchemical writings also travelled on another path parallel to the one they took through Byzantium. In the Islamic world, roughly between 750 and 1400, early Greco-Egyptian alchemical texts were translated into Arabic, digested, and developed in provocative and influential ways. Unfortunately the vast corpus of medieval Arabic alchemical works was never codified in the same way as the Greek alchemical corpus and perhaps in part for this reason it remains poorly understood. Even some of the most basic issues of authorship remain to be sorted out. For example, it remains unclear whether some of the Arabic texts (e.g. the famous *Emerald Tablet*, attributed to the fabled ancient Egyptian figure Hermes Trismegistus) are original Arabic compositions, or Arabic translations of earlier lost Greek originals completed as part of the broader Islamic engagement with Classical Greco-Roman culture and scholarship. Nevertheless, it is clear that scholars in the medieval Islamic world added considerably to the late Antique alchemical canon, including works by the eighth-century author Jābir ibn-Hayyān (or, more probably, a school of authors writing in his name), Abū Bakr Muhammed ibn-Zakarīyya al-Rāzī (c. 865–923/4 CE), ibn-Sīnā (or Avicenna, c. 980–1037 CE) and others, as well as the anonymous treatise usually known by its Latin title, *Turba philosophorum* (Principe 2012).

When Europeans first encountered alchemy, therefore, they faced a millennium's worth of alchemical texts. Alchemy first came to Europeans' attention in the twelfth century via Latin translations of Arabic texts from the Islamic world; only later, during the Renaissance, did they encounter the Greek corpus. In the thirteenth century, European scholars such as Roger Bacon, Paul of Taranto, who wrote under the pseudonym Geber (a Latinized version of the Arabic "Jābir"), and John of Rupescissa began to write original Latin alchemical treatises, sometimes attributing them to contemporary figures such as Thomas Aquinas, Arnald of Villanova, or Raymond Lull. By the sixteenth century, European numerous printers began to offer a new Latin and vernacular alchemical canon to eager readers in the form of printed compendia that included not only original Greco-Egyptian and medieval Arabic and Latin texts, but also new "modern" authors who continued to engage with and add to the alchemical tradition (*De alchimia opuscula* 1550; Manget 1702). These early modern anthologies revitalized the study of the alchemical corpus and made it available to new audiences, and they continue to circulate (albeit often in deeply problematic translations) via modern editions and, most recently, websites (for example McClean 2015).

The alchemical corpus has evolved over the centuries, therefore, but it also has a surprisingly stable core, suggesting that one way to define the art and its purview is as the preservation, study, and elaboration of a self-referential textual tradition stretching from Greco-Roman Egypt to the present. At the center of this tradition lies the fusion of theory and practice that has always marked alchemical work: desiderata such as the philosopher's stone, elixirs, and the "Great Work" of transmutation; the apparatus and technical operations required to create them; and the theoretical frameworks, such as the mercury–sulfur (and later salt) theory of metals, that explain their efficacy. Figures from antiquity such as Maria Judea, [pseudo-]Democritus, and Zosimos of Panopolis, texts by medieval authors such as Jābir ibn-Hayyān, the Latin Geber, Raymond Lull, Arnald of Villanova, and Albertus Magnus, as well as the early modern authors Paracelsus, Basilius Valentinus, or Eirenaeus Philalethes have become canonical alchemical authorities.[1]

Using this corpus as a point of entry into the history of alchemy focuses attention first and foremost on the practices of writing and reading, and frames alchemy primarily as a scholarly practice. From antiquity to the present, many alchemists have approached the art with a strong historical sensibility. They have located important alchemical expertise in the past, in the theoretical and technical knowledge of previous alchemists ("the ancient sages") and argued that past expertise could be recovered and accessed, at least in part, through texts. In this sense, then, alchemists must be understood in part as scholars who drew on the reigning scholarly techniques of their day to compile, copy, edit, translate, read, synthesize, and comment on the works of their forebears. The constitution of the Greek alchemical corpus, for instance, appears to have been part of a broad encyclopedic impetus in ninth and tenth-century Byzantium to constitute a range of "corpora," including the Hippocratic corpus and Hermetic corpus (Mertens 2006, 222–3). Likewise, scholars in the Islamic world and, later, in the Latin West first encountered alchemy as part of more general interest in translating ancient and more recent medical, mathematical, and astronomical literature, first from Greek to Arabic in places like eighth-century Baghdad, and later from Arabic to Latin in places like southern Italy and Iberia in the tenth, eleventh, and twelfth centuries. In their efforts to understand both individual texts and an increasingly complicated intellectual legacy, alchemists variously took up scholasticism, humanism, and, eventually, new exegetical techniques such as Christian cabala to translate, interpret, and synthesize alchemical texts (Moran 2007; Nummedal 2011b, especially 332–3; Forshaw 2013).

On one level, then, the history of alchemy is part of a broader story of the transmission, translation, and transformation of ideas—from both Greek natural philosophy and alchemy—in Byzantium, the Islamic world, and Europe (Lindberg 2007).[2] In all of these times and places, alchemical authors preserved and engaged with the writings of past alchemists, preserving, debating, and developing the theories and practices at the art's core. For example, one of the reasons we know anything about Maria Judea, whose writings have been lost, is that Zosimos of Panopolis wrote about her alchemical apparatus and techniques, including the method of heating a substance gently in a water bath still known today as a *bain-Marie* (Principe 2012, 15–16) (Figure 4.1). Alchemists also drew inspiration more broadly from classical authorities such as Aristotle, seeking to reconcile classical and alchemical ideas to craft an understanding of the metallic and mineral world that could account for matter's formation and transformation at the hands of art and nature alike. The eighth-century Arabic Jābirian corpus, for example, adapted the Aristotelian idea that metals originated in two exhalations in the earth, one smoky and the other moist, to formulate the influential mercury–sulfur theory of metals. Likewise, the Jābirian notion of an elixir that could perfect each metal and transmute it into gold was an alchemical reworking of Galenic medical ideas about the way in which medicines could bring the four humors into balance. Europeans would extend and further develop these and other ideas about the nature of metals

Figure 4.1 Bain-marie ("Mary's bath"), a water bath alchemists used to heat substances gently. The Greek Zosimos of Panopolis attributed its invention to the ancient Jewish alchemist Maria Judea. This image appeared in print in Philippus Ulstadius's compilation of distillation techniques, *Coelum philosophorum, seu De secretis naturae liber* (Ulstadius 2003, Woodcut Page XV, recto). This image encapsulates not only the longevity and transmission of alchemical knowledge from antiquity to early modern Europe (and beyond) and across cultures and linguistic communities, but also demonstrates the union of scholarly and artisanal practices at the heart of alchemists' work. Roy G. Neville Collection, Othmer Library of Chemical History, Chemical Heritage Foundation. Downloaded with permission from the Chemical Heritage Foundation, as part of the Wikipedian in Residence initiative.

and medicines. When medieval Latin alchemists encountered what they (erroneously) believed to be Aristotle's powerful argument against the transmutation of species, for example, they responded in kind with a vigorous defense of alchemy's potential (Newman 1989).[3] Eventually, alchemists would play a pivotal role in transforming medieval matter theory into modern experimental philosophy (Newman 2006).

Rarely have alchemists been interested in the subject solely as an intellectual activity, however; they have almost always sought to manipulate nature through hands-on work as well. Indeed, the alchemical corpus is as full of information about technical operations and the apparatus (particularly specialized stills and furnaces) required to carry them out as it is about theories of matter. Readers certainly mined the corpus, as well as a complementary recipe literature, for these insights into practice. In this sense alchemists have always been as much artisans as scholars. In fact, the earliest sources associated with the art document its roots in the artisanal manufacture of luxury materials in Greco-Roman Egypt. Two third-century papyri, the oldest extant documents related to alchemy, and now known as the Stockholm and Leyden Papyri, contain a typical range of recipes, all of which deal with coloring matter: processing base metals to resemble real silver or gold, for example, or the production of purple dyes or artificial gemstones to imitate more expensive or genuine products (Caley 1926, 1927; Halleux 1981). The first-century *Four Books* of Pseudo-Democritus, preserved in fragments and only in later copies, contain similar material, and, indeed, the association between alchemy and other coloring techniques persisted into the early modern period (Martelli 2013). The early medieval *Mappae clavicula* (*Little Key to the World*, Smith and Hawthorne 1974) and *Compositiones variae* (*Various Compositions,* Johnson 1939), for example, as well as Theophilus's c. 1125 *On Divers Arts* (Hawthorne and Smith 1979), combine recipes for glass, dyes, artist's pigments (not to mention incendiary materials) with processes for making and working with metals, including gold. The *Mappae clavicula*, in fact, even includes a direct translation of one of the recipes in the Leyden Papyrus X, suggesting a line of transmission from Greco-Roman Egypt to Latin Europe (Smith and Hawthorne 1974, 17). In the early modern period, the popular *Kunstbücher* (or "books of secrets") genre carried these technical recipe compendia into the world of print, often bearing the name "alchemy" as a catchall term for their contents (De alchemia 1541; Kertzenmacher 1570; Eamon 1994; Leong and Rankin 2011). To take only one example, 1400 years after the papyri were created, the title page of Giovanni Battista Birelli's 1603 *Alchimia nova* placed an expansive list of techniques under the title of alchemy, including, "how to make all kinds of alchemical and metal products, waters and oils, preparations of lime, the art of figuring, of making silver and gold, gemstones, glues, mixtures [*Mixtum*], and mirrors, salts, the arts of painting and drawing, as well as many other amusing and entertaining arts" (Birelli 1603).

The status of this sort of eclectic recipe collection with respect to a history of alchemy is contested. Some scholars do not see such recipes as "true" alchemy, for a range of reasons. First, the broad range of technical processes such collections contain do not seem focused enough to be distinguishable as alchemy, as opposed to, say, glassmaking or dyeing. Moreover, they record recipes, rather than an explicitly theoretical understanding of matter, and often propose only to *imitate* precious metals, rather than to make "real" gold and silver. Finally, they often lack some of what would become the signature goals of alchemical work, such as the philosopher's stone, which effected the transmutation of metals (Caley 1926, 1927; Principe 2012, 11, 13). Matteo Martelli (2013, S2, S57–63), however, has argued convincingly that any definition of "alchemy" must emerge from the sources themselves, even if it requires that scholars "reconsider how early 'alchemy' is to be defined." Quite simply, he counters the claim by some modern scholars that the *Four Books* of Pseudo-Democritus are a mere prelude to a "proper" alchemical tradition by underscoring "an ancient tradition that unanimously placed Democritus among the first and most important fathers of alchemy," concluding that we, too, "must accept that the topics covered in his books [i.e., the manufacture of gold, silver, purple dye, and gemstones] were considered to be key aspects of the discipline." This point can be extended to alchemy as a whole. In short, we need to use actors' categories, rather than imposing our own; if historical authors (or printers) included something in the category of alchemy, then historians, too, need to do the same.

Alchemy's earliest sources document a fairly broad and inclusive understanding of the art, as Martelli (2013, S59) has argued, but over time some alchemists came to focus more prominently on techniques for making real gold and silver (i.e., *chrysopoeia* and *agyropoeia*). This narrower definition of alchemy offered a point of focus for another technical thread running through the alchemical literature, namely discussions of how to prepare the philosopher's stone or elixir. Although the term "philosopher's stone" did not appear until sometime after the sixth century, the notion of a medicine or salve that could transform metals was already present in Pseudo-Democritus and Zosimos's works; the central question was how to create it and thereby accomplish the "Great Work" of transmutation (Principe 2012, 26). Several medieval Arabic texts, including the treatises attributed to Khālīd ibn-Yazīd, the *Emerald Tablet* attributed to Hermes, the Arabic classic now known by its Latin name, *Turba philosophorum*, and the Jābirian corpus, take up this question. The answers found in these texts are anything but straightforward; rather, they participated in a discourse of secrecy that lent them enough interpretive richness to sustain generations of readers. *Decknamen*, or "cover names" for common and alchemically processed substances, as well as enigmatic metaphors and intentional obfuscation (not to mention unintentional confusion due to copying or translation errors) not only made alchemical discussions of the philosopher's stone hermeneutically interesting, but also raise interesting questions about how alchemists attempted to translate texts into hands-on work.

To be sure, there were many points at which alchemists could go wrong, leaving them to decide whether failure was a result of misunderstanding one's sources or not executing the processes properly. And yet alchemists had several resources at their disposal when they ran into difficulties. The alchemical corpus itself, of course, offered one solution. According to a well-worn dictum, one book opens another (*liber enim librum aperit*), and certainly alchemists could turn to more books to help understand a particular process or apparatus before returning to the workshop to try again (Bonus 1702, Lib. III. Sect. I. Subsect. I, 33). Although sources documenting this in detail are rare, alchemical practice nearly always involved a complicated and repeated interplay of textual work, experimental practice, and writing (Newman and Principe 2002). Another solution might be to redirect the techniques used in allied fields, such as mining or medicine, towards more distinctly alchemical goals, such as the philosopher's stone. The sixteenth-century goldsmith, engraver, and painter Franz Brun, for example, demonstrates the fluid boundaries between various arts of metal. He undoubtedly drew on his expertise as a fine metalworker when he took up alchemy to make several alchemical "tinctures" for making silver and gold, repurposing his skills once again by turning to mining and assaying in the Harz mountains before returning to alchemy and decorative brass work at a princely court in Wolfenbüttel. While we may now see the decorative arts, mining, and alchemy as separate enterprises, Brun's trajectory suggests that experience in any one of these ways of working with metal could inform projects in another (Nummedal 2014).

Some texts also point towards another resource for alchemists who could find a way forward in neither books nor related crafts: personal initiation at the side of a more experienced adept. Alchemical lore is full of stories about aspiring alchemists and their teachers (Principe 2012, 117): the Umayaad prince Khālid ibn-Yazīd and the Christian monk/adept Marianos in seventh-century Egypt (Principe 2012, 28-30); the German noblewoman Anna Zieglerin and her lover/teacher Count Carl von Oettingen in sixteenth-century central Europe (Nummedal 2011a) and Robert Boyle and George Starkey in seventeenth-century England (Newman and Principe 2002). On one level, the master–disciple transmission of alchemical knowledge resembles craft knowledge more generally, which until at least the early modern period was typically transmitted in person and through demonstration, rather than through writing. Given alchemy's origins in the world of Egyptian artisans, it should not be surprising that the transfer of alchemical knowledge, too, would be figured as occurring through apprenticeship at the side of a master. The sense that alchemical techniques are secret, privileged knowledge is also to be expected in this context. A discourse of secrecy, understood in the context of "trade secrets" or proprietary knowledge, has often attended artisanal knowledge, and it has long permeated alchemy as well (Long 2001; Smith 2004; Principe 2012, 12).

Both secrecy and initiation have another set of meanings as well, and these aligned alchemy more closely with divine revelation and sometimes figured alchemists as prophets. Egyptian artisans in late Antiquity who specialized in precious metals and stones seem to have prepared ritual objects in temples, lending their expertise, and eventually alchemy, an aura as a "sacred and holy art," the knowledge of which was only for initiates (Martelli 2013, S66). Alchemy was also shaped by a late Antique syncretic tradition claiming that Greek philosophy was indebted to the divine wisdom acquired by Egyptian, Persian, and Babylonian sages. In this tradition, Platonic philosophy, for example, was understood to contain not only rational knowledge, but also a more profound understanding of the salvation of the soul; this wisdom was hidden, concealed in myths, architecture, and texts, detectable only to initiates (Hanegraaff 2012, 12–17, 191ff). Such currents were strong at alchemy's origins, and the art certainly absorbed some of these associations. For example, a legend claiming that the Greek atomist philosopher Democritus was initiated into the mysteries at the temple in Memphis, Egypt, and linking him to the Persian magus Ostanes became entangled with alchemy when the *Four Books* on alchemy appeared in Democritus's name (Martelli 2013, S63–73).

These sorts of syncretic legends linking alchemy, and Greek philosophy more generally, to Eastern wisdom percolated throughout the alchemical corpus, bubbling up occasionally in receptive climates to represent alchemy as a repository of eastern wisdom. Such legends account in part for the "initiatic style" characteristic of Jābirian alchemy and its imitators, as well as the appearance by the tenth century of numerous

Arabic treatises on alchemy attributed to the legendary Egyptian figure Hermes Trismegistus (Newman 1991, 90; Principe 2012, 44–45). In the European Renaissance, legends associating Greek philosophy with ancient revelations flourished again with the revival of Platonism and Hermeticism, which launched anew a project of recovering the divine wisdom of Hermes Trismegistus, Zoroaster, and Moses. According to Renaissance humanists, this wisdom had been lost through the ages, but was still hidden in the textual remains of Classical culture and could be recovered through careful exegesis (Hanegraaff 2012, 64).[4] Since alchemy had come to be associated with Hermes, it too took part in this Renaissance Hermetic revival; alchemy's longstanding tradition of hiding alchemical techniques and substances under "cover names," or *Decknamen*, meanwhile, made it perfect fodder for the growing interest in decoding secrets in early modern Europe (Newman 1996; Hanegraaff 2012, 205–7). In the Renaissance, in other words, alchemy became an important component of the search to restore ancient revelatory knowledge that scholars now refer to as the Western Esoteric Tradition and that continues to thrive today, albeit as a form of knowledge "rejected" in academic contexts (Hanegraaff 2012, 2013).

The discourse of revelation in alchemy was not limited to the recovery of *ancient* wisdom, however. A long tradition in both Islamic and, especially, Christian theology figured all knowledge as a "gift of God," or *donum dei*, and one can easily find the claim that alchemical insight, too, was in some sense a result of God's will (Principe 2012, 192–5). Some alchemists went a step further, moreover, and sought more direct divine assistance with their alchemical endeavors, either in Scripture or through prayer or angelic intermediaries (Crisciani 2008; Hedesan 2013). In the fourteenth century, Petrus Bonus argued in his *New Pearl of Great Price* (*Pretiosa margarita novella*) that reason and revelation worked hand in hand in alchemical work, and that the alchemist should seek out revelation when intellect failed to provide alchemical insights (Crisciani 1973). John Dee and Robert Boyle did just that in their pursuit of the philosopher's stone, and in doing so, they joined many other Europeans who believed that new secrets of nature, alchemical among them, would be revealed as the world hurtled towards its final moments (Nummedal 2014). Alchemical prophets such as John of Rupescissa (DeVun 2009) and the alchemist Anna Zieglerin (Nummedal 2011a), meanwhile, proposed that alchemical elixirs would have a special role to play in the End Times.

Alchemists' eschatological speculations serve as a reminder that many things in addition to intellectual curiosity motivate the study of nature. Alchemists have found connections between their art and everything from theology, poetry, and translation to dyeing, medicine, metallurgy, psychology, and natural philosophy. It is impossible, therefore, to confine the history of alchemy to the history of science. On the contrary, alchemy—always a rich and varied enough tradition to attract a range of practitioners from artisans and natural philosophers to those in search of revelatory wisdom—forces scholars to open up the history of science and see connections between science and craft, philosophy, religion, and culture.

And yet, alchemists have played a crucial role in the history of science. First, in joining the hands-on manipulation of matter with more theoretical speculations about its composition and transformation, alchemists (like physicians) modeled the extraordinary potential of the union of head and hand long before it became a hallmark of modern science in the sixteenth and seventeenth centuries. This was not without cost, as alchemists risked losing social and epistemological status because their art appeared overly artisanal. Many alchemists, therefore, sought ways to elevate their art and differentiate it from the work of other artisans (Martelli 2011), either by emphasizing distinctive apparatus (specialized stills, furnaces for sublimation, and the *kerotakis*) or substances (e.g. "divine water"), underscoring alchemy's philosophical credentials (Crisciani 1973), or highlighting alchemy's potential to participate in elite cultural discourses that might help it accrue social respectability (Principe 2012, 174–8). Eventually, of course, the productive entanglement of artisanal and scholarly cultures that had long been characteristic of alchemy would help remake science in early modern Europe (Smith 2004). Moreover, alchemists' longstanding claims not merely to imitate, but also create and even improve on, natural materials would stimulate some of the most ambitious claims about the power of technology in seventeenth-century Europe (Newman 2004). Alchemy's failure to find a secure footing in the institutions that have supported scientific knowledge, including the university, guilds, courts and academies, meanwhile, lays bare the agendas, priorities, and, ultimately, limits of these bodies as they sought to organize (and reorganize) the social production of knowledge (Moran 1991; Smith 1994; Nummedal 2007). Finally, the complete elimination of alchemy from modern scientific discourse (Newman and Principe 1998; Principe 2008; Hanegraaff 2012, 191–207; Powers 1998, 2012) offers a powerful example of the capacity of linguistic, social, and cultural forces to redraw maps of knowledge in ways that define which subjects are worthy of academic study at all. For alchemy to thrive in the twentieth century, as Dennis William Hauck noted, it would require rejecting modern science in turn.

Endnotes

1   It is important to underscore the pseudoepigraphic tradition in alchemy. Although Lull, Arnald, and Paracelsus, for example, all wrote genuine texts on other subjects, numerous alchemical texts were falsely attributed to them as well.

2   It is important to note that China, too, has a set of traditions that are often translated as "alchemy": *waidan* and *neidan*. The relationship between these and the alchemical tradition discussed in this essay, however, remains unclear. Translating *waidan* and *neidan* into English as "alchemy" emphasizes the similiaries between these two traditions, suggesting that "alchemy" is a nearly universal cultural and historical phenomenon. However, the differences between these two traditions—indeed the radically different cultural and intellectual contexts in which they emerged and thrived—are far more important, and argue against understanding *waidan* and *neidan* as part of the same tradition as the alchemy considered here. On *waidan* and *neidan*, see Sivin (1976) and Pregadio (2006 and 2011).

3   The actual author of this text, *De congelatione et conglutinatione lapidum*, was ibn-Sīnā/Avicenna (c. 980-1037), but the mistaken attribution to Aristotle made it impossible for Latin scholars to ignore.

4   As Wouter Hanegraaff has noted, "In other words: 'ancient wisdom' equaled 'hidden wisdom.' It may well be argued that the notions of secrecy and concealment are inherent in the very structure of the Renaissance narrative of ancient wisdom, for the simple reason that, in one way or another, its Christian adherents always needed to make the

argument that beneath the surface crust of pagan religion there lay a hidden core of Christian truth" (Hanegraaff 2012, 64).

References

Birelli, Giovanni Battista, and Peter Uffenbach. 1603. *Alchimia Nova, Das ist, Die Güldene Kunst Selbst, Oder Aller Künsten Mutter: Sampt dero heimlichen Secreten, unzehlichen verborgenen Kindern vnd Früchten; Von allerley Alchimistischen unnd Metallischen Geschäfften, Wässern vnnd Oelen, Bereitungen der Kälck, der Kunst ... Silber und Gold zumachen.* Franckfurt am Main: Palthenorus.

Bonus, Petrus. 1702. "Pretiosa margarita novella." In *Bibliotheca Chemica Curiosa*, edited by Jean-Jacques Manget. Geneva: Chouet, G. de Tournes, Cramer, Perachon, Ritter and S. de Tournes. http://archive.org/details/jojacobimangetim00mang.

Caley, Earl Radcliffe. 1926. "The Leyden Papyrus X: An English Translation with Brief Notes." *Journal of Chemical Education*, 3: 1149–1166.

Caley, Earl Radcliffe. 1927. "The Stockholm Papyrus: An English Translation with Brief Notes." *Journal of Chemical Education*, 4: 979–1002.

Crisciani, Chiara. 1973. "The Conception of Alchemy as Expressed in the Pretiosa margarita novella of Petrus Bonus of Ferrara." *Ambix*, 20, No. 3: 165–81.

Crisciani, Chiara. 2008. "*Opus* and *sermo*: The Relationship between Alchemy and Prophecy (12th–14th Centuries)." *Early Science and Medicine*, 13: 4–24.

De alchemia. 1541. *In hoc volumine De alchemia continentur haec.* Nuremberg: Johannes Petreius.

De alchimia opuscula. 1550. *De Alchimia Opuscula Complura Veterum Philosophorum.* Frankfurt: Cyriacus Jacob.

DeVun, Leah. 2009. *Prophecy, Alchemy, and the End of Time: John of Rupescissa in the Late Middle Ages.* New York: Columbia University Press.

Eamon, William. 1994. *Science and the Secrets of Nature: Books of Secrets in Medieval and Early Modern Culture.* Princeton, NJ: Princeton University Press.

Forshaw, Peter. 2013. "Cabala Chymica or Chemia Cabalistica—Early Modern Alchemists and Cabala." *Ambix*, 60, No. 4: 361–89. DOI: 10.1179/0002698013Z.0000000003

Halleux, Robert, ed. 1981–. *Les Alchimistes grecs.* Paris: Belles Lettres.

Hanegraaff, Wouter J. 2012. *Esotericism and the Academy: Rejected Knowledge in Western Culture.* Cambridge: Cambridge University Press.

Hanegraaff, Wouter J. 2013. *Western Esotericism: A Guide for the Perplexed.* New York: Continuum.

Hauck, Dennis William. 2015. "Biography." Accessed September 12, 2015. http://dwhauck.com/bio.htm.

Hawthorne, John G. and Cyril Stanley Smith. 1979. *On Divers Arts: The Foremost Medieval Treatise on Painting, Glassmaking, and Metalwork.* New York: Dover Publications.

Hedesan, Georgiana D. 2013. "Reproducing the Tree of Life: Radical Prolongation of Life and Biblical Interpretation in Seventeenth-Century Medical Alchemy." *Ambix*, 60, No. 4: 341–60. DOI: 10.1179/0002698013Z.00000000038

Johnson, Rozelle Parker. 1939. *Compositiones variae, from Codex 490, Biblioteca capitolare, Lucca, Italy, an introductory study, Illinois University, Illinois Studies in Language and Literature.* Volume xxiii, no 3. Urbana: University of Illinois Press.

Jung, Carl Gustav. 1937. "Die Erlösungsvorstellungen in der Alchemie." Eranos-Jahrbuch 1936. Zurich: Rhein-Verlag.

Kertzenmacher, Peter. 1570. *Alchimia das ist alle Farben, Wasser, olea, Salia, und Alumina.* Zu Franckfurt am Meyn: Bey C. Engenoffs Erben.

Leong, Elaine Yuen Tien, and Alisha Rankin (eds.) 2011. *Secrets and Knowledge in Medicine and Science*. Aldershot: Ashgate.

Lindberg, David C. 2007. *The Beginnings of Western Science: The European Scientific Tradition in Philosophical, Religious, and Institutional Context, Prehistory to A.D. 1450*. 2nd ed. Chicago: University of Chicago Press.

Long, Pamela O. 2001. *Openness, Secrecy, Authorship: Technical Arts and the Culture of Knowledge from Antiquity to the Renaissance*. Baltimore: Johns Hopkins University Press.

Manget, Jean-Jacques. 1702. *Bibliotheca Chemica Curiosa*. 2 vols. Geneva: Chouet, G. de Tournes, Cramer, Perachon, Ritter, & S. de Tournes.

Martelli, Matteo. 2011. "Greek Alchemists at Work: 'Alchemical Laboratory' in the Greco-Roman Egypt." *Nuncius*, 26: 271–84.

Martelli, Matteo. 2013. "The Four Books of Pseudo-Democritus (Sources of Alchemy and Chemistry: Sir Robert Mond Studies in the History of Early Chemistry)." *Ambix*, 60 (Supplement 1.

Martinón-Torres, Marcos. 2011. "Some Recent Developments in the Historiography of Alchemy." *Ambix*, 58, No. 3: 215–37. DOI: 10.1179/174582311X13129418299063

McClean, Adam. 2015. "The Alchemy Web Site." Accessed September 12, 2015. http://www.alchemywebsite.com.

Mertens, Michèle. 2006. "Graeco-Egyptian Alchemy in Byzantium." In *The Occult Sciences in Byzantium*, edited by Paul Magdalino and Maria Mavroudi, 205–30. Geneva: La Pomme d'Or.

Moran, Bruce T. 1991. *The Alchemical World of the German Court: Occult Philosophy and Chemical Medicine in the Circle of Moritz of Hessen (1572–1632)*. Stuttgart: Franz Steiner Verlag.

Moran, Bruce T. 2005. *Distilling Knowledge: Alchemy, Chemistry, and the Scientific Revolution*. Cambridge, MA: Harvard University Press.

Moran, Bruce T. 2007. *Andreas Libavius and the Transformation of Alchemy Separating Chemical Cultures with Polemical Fire*. Sagamore Beach: Science History Sublications.

Moran, Bruce, Lawrence M. Principe, William R. Newman, Tara E. Nummedal, and Ku-ming (Kevin) Chang. 2011. "Forum: Alchemy and the History of Science." *Isis*, 102, No. 2: 300–337.

Newman, William R. 1989. "Technology and Alchemical Debate in the Late Middle Ages." *Isis*, 80: 423–45.

Newman, William R., ed. 1991. *The Summa Perfectionis of Pseudo-Geber: A Critical Edition, Translation and Study*. Leiden: Brill.

Newman, William R. 1996. "'Decknamen' or Pseudochemical Language? Eirenaeus Philalethes and Carl Jung." *Revue d'Histoire des Sciences et de leurs Applications*, 49: 159–88.

Newman, William R. 2004. *Promethean Ambitions: Alchemy and the Quest to Perfect Nature*. Chicago: University of Chicago Press.

Newman, William R. 2006. *Atoms and Alchemy: Chymistry and the Experimental Origins of the Scientific Revolution*. Chicago: University of Chicago Press.

Newman, William R., and Lawrence M. Principe. 1998. "Alchemy vs. Chemistry: The Etymological Origins of a Historiographic Mistake." *Early Science and Medicine*, 3, No. 1: 32–65.

Newman, William R., and Lawrence M. Principe. 2002. *Alchemy Tried in the Fire: Starkey, Boyle, and the Fate of Helmontian Chymistry*. Chicago: University of Chicago Press.

Nummedal, Tara. 2007. *Alchemy and Authority in the Holy Roman Empire*. Chicago: University of Chicago Press.

Nummedal, Tara. 2011a. "Anna Zieglerin's Alchemical Revelations." In *Secrets and Knowledge in Medicine and Science*, edited by Elaine Yuen Tien Leong and Alisha Rankin, 125–41. Aldershot: Ashgate.

Nummedal, Tara. 2011b. "Words and Works in the History of Alchemy." *Isis*, 102, No. 2: 330–37. Article DOI: 10.1086/660142

Nummedal, Tara. 2014. "The Alchemist in His Laboratory." In *Goldenes Wissen. Die Alchemie—Substanzen, Syntheses, Symbolik*, edited by Petra Feuerstein-Herz and Stefan Laube, 121–8. Wolfenbüttel: Herzog August Bibliothek.

Powers, John C. 1998. "'Ars sine arte': Nicholas Lemery and the End of alchemy in Eighteenth-century France." *Ambix*, 45, No. 3: 163–89.

Powers, John C. 2012. *Inventing Chemistry: Herman Boerhaave and the Reform of the Chemical Arts*. Chicago: University of Chicago Press.

Pregadio, Fabrizio. 2006. *Great Clarity: Daoism and Alchemy in Early Medieval China*. Stanford, CA: Stanford University Press.

Pregadio, Fabrizio. 2011. *The Seal of the Unity of the Three*. Mountain View, CA: Golden Elixir Press.

Principe, Lawrence M. 2008. "Transmuting Chymistry into Chemistry: Eighteenth-century Chrysopoeia and its Repudiation." In *Neighbours and Territories: The Evolving Identity of Chemistry*, edited by José Ramón Bertomeu-Sánchez, Duncan Thorburn Burns, and Brigitte Van Tiggelen, 21–34. Louvain-la-Neuve: Memosciences.

Principe, Lawrence M. 2012. *The Secrets of Alchemy*. Chicago: University of Chicago Press.

Principe, Lawrence M., and William R. Newman. 2001. "Some problems with the historiography of alchemy." In *Secrets of Nature: Astrology and Alchemy in Early Modern Europe*, edited by William R. Newman and Anthony Grafton, 385–432. Cambridge, MA: MIT Press.

Sivin, Nathan. 1976. "Chinese Alchemy and the Manipulation of Time." *Isis*, 67: 513–27.

Smith, Cyril Stanley, and John G. Hawthorne. 1974. "Mappae Clavicula; A Little Key to the World of Medieval Techniques." *Transactions of the American Philosophical Society* (New Series, 64, No. 4):1–128. DOI: 10.2307/1006317

Smith, Pamela. 1994. *The Business of Alchemy: Science and Culture in the Holy Roman Empire*. Princeton, NJ: Princeton University Press.

Smith, Pamela H. 2004. *The Body of the Artisan: Art and Experience in the Scientific Revolution*. Chicago: University of Chicago Press.

Tilton, Hereward. 2003. *The Quest for the Phoenix: Spiritual Alchemy and Rosicrucianism in the Work of Count Michael Maier (1569–1622)*. Berlin: Walter de Gruyter.

Ulstadius, Philippus. 1528. *Coelum philosophorum, seu De secretis naturae liber*. Argentorati: Arte et impensa Joannis Grienynger.

CHAPTER FIVE

The Natural Philosopher

PETER DEAR

The Natural Philosopher in the University; Mathematics and Physics

The natural philosopher in the Latin West of the high middle ages had, typically, been an inhabitant of a university community of teachers whose role was not vocational, but temporary and occasional. "Natural philosophy," *philosophia naturalis* (sometimes *scientia naturalis*, and sometimes simply *physica*) was a label attached to a subject of the medieval university's arts curriculum, and its teachers were commonly students in one of the higher faculties, whether theology, medicine, or law. As such, these teachers did not regard their role, that of a lecturer in natural philosophy, as vocational; one did not usually make one's career as a natural philosopher. By the end of the sixteenth century, this situation remained, in important ways, unchanged, but now universities sometimes also contained teachers whose primary role was that of natural philosopher rather than that of theologian- or physician-in-training: there were now *professors* of natural philosophy (often a position held along with another, more important chair). In the year 1600, natural philosophy as a recognized endeavor, and the natural philosopher as the occupant of a recognized social role, resided centrally within the setting of the university (Blair 2006; Gaukroger 2006). In that setting, the texts of Aristotle were the usual reference points: natural philosophy was a knowledge-enterprise that took its form from established traditions of interpretation of Aristotelian texts. This is not to say, however, that there were no other knowledge enterprises concerning knowledge of the natural world, or other claims to natural philosophy than those of scholars found in institutions of higher learning.

One of the difficulties in understanding knowledge enterprises in this period concerns appropriate categories. Natural philosophy, natural history, mathematics, and medicine (one could add others) were all distinct endeavors, recognized as such and corresponding to similarly distinct, although also overlapping, personas or social roles (Shapin 2003, 2006; cf. Gaukroger 2006, 2010). The "natural philosopher" at the beginning of the period was different from the more encompassing figure characteristic of the eighteenth century. The academic definition that gives us our most

A Companion to the History of Science, First Edition. Edited by Bernard Lightman.
© 2016 John Wiley & Sons Ltd. Published 2020 by John Wiley & Sons Ltd.

solid orientation in 1600 became increasingly challenged in the course of the seventeenth century; by the end of that century other institutional markers become more apposite.

The persona of the natural philosopher was one that various sorts of people in the seventeenth century adopted for themselves, sometimes tagged by use of the label "physics" or the prefix "physico." One of the more important areas in which new claims to being a natural philosopher collided with longstanding Aristotelian understandings of that role was that of the mathematical sciences. Mathematicians, who had not previously been generally regarded as engaged in the same domain of knowledge as natural philosophers, began to identify areas of natural philosophy ("physics") in which their own expertise should be relevant, potentially making a difference in the framing of problems by Aristotelian natural philosophers themselves.

Mathematical disciplines had previously, within the dominant Thomistic–Aristotelian epistemological orthodoxy that held general academic sway by the sixteenth century, been sharply demarcated from natural philosophy. The most important institutional locus for this development was the Jesuit college, of which a widespread network had become established throughout Catholic Europe by the early seventeenth century. With its centralized curricular structure and official adherence to Thomistic Aristotelianism, the Jesuit college became a crucible in which controversies about the nature and conduct of the so-called "mixed" mathematical sciences (astronomy, optics, mechanics, and others modelled on them) drove some Jesuit mathematicians to develop ways of managing their sciences so as to legitimate them as means of learning about the natural world on a par with natural philosophy (Dear 2011). This endeavor involved countering the claims of philosophers that mathematical demonstrations were not truly scientific because not causal (a reference to one of Aristotle's criteria for a genuine science), as well as achieving the establishment of universal knowledge claims on the basis of a few specific observations (a technique imported into experimental work from observational astronomy). The artificiality of experimental practices needed less explicit defense, since these practices were carried out in the context of the mixed mathematical sciences, and thus made no pretense to the identification of final causes. Quantity supposedly had no direct relevance to essence, substance, or natural causation, and dealt only with magnitudes and their relationships. Nonetheless, both Galileo and Kepler began to assert, in the first two decades of the seventeenth century, that their own work amounted to a "philosophical" astronomy that used the subject matter of a classical mathematical science, astronomy (which concerned the apparent motions of celestial bodies), and derived from it knowledge that impinged directly upon the concerns of natural philosophers. Kepler wanted to make inferences about the physical causes of the motions of the planets in concert with a much more precise characterization of the mathematical form of those motions themselves (Voelkel 2001). Galileo used his mathematical, geometrical inferences about the proximity to the surface of the sun of the newly discovered sunspots—something that had direct relevance to Aristotelian physical ideas about the substantial immutability of the heavens—to claim, much like Kepler, that his work amounted to "philosophical" astronomy (Dear 1995, chapter 4).

Of a piece with these boundary-crossing claims by Kepler and Galileo, some Jesuit mathematicians began to speak of aspects of their own work as "physico-mathematics," similarly folding together the natural-philosophical and the mathematical to create a new role for the mathematician that could also turn him into a new kind of "natural philosopher." The process is conveniently encapsulated, if one moves ahead in time to the end of the century, in Isaac Newton's famous book title *Philosophiae naturalis principia mathematica* (1687), or "Mathematical Principles of Natural Philosophy," in which the role of the mathematician as a kind of natural philosopher is roundly proclaimed: whether Newton meant to suggest that natural philosophy was properly founded on mathematical principles alone or whether he wanted (more probably) to indicate that *this work* was about natural philosophy's mathematical principles, without denying that it might also rest upon others in addition, he clearly signaled a central place for mathematical skills in the armamentarium of the natural philosopher. Indeed, he complained about natural philosophers who thought (in his view) that mathematics was merely an ornamentation to their own qualitative work; for Newton, mathematics was central. In 1686 he complained in a private letter to Edmund Halley, the publisher of the *Principia*, about the presumption of Robert Hooke in claiming some credit for the idea of inverse-square-law gravitational attraction as the explanation for planetary orbits:

> Now is not this very fine? Mathematicians that find out, settle and do all the business must content themselves with being nothing but dry calculators & drudges and another that does nothing but pretend & grasp at all things must carry away all the invention as well of those that were to follow him as of those that went before. (As quoted in Koyré 1965, 235.)

There had by this time come into wide usage a new term to designate the kind of "physics" that used mathematical demonstrations: "physico-mathematics." The term is first known from the Dutch schoolmaster and philosopher—and friend of Descartes—Isaac Beeckman. He used it in a private journal entry in 1618, referring to his new acquaintance René Descartes. "This Descartes has been educated with many Jesuits and other studious people and learned men. He says however that he has never come across anyone anywhere, apart from me, who uses accurately this way of studying that I advocate, with mathematics connected to physics." The novel terminology appears in Beeckman's marginal note: "Very few physico-mathematicians" (Dear 1995, 171). But the term, and self-identification, evidently spread very quickly, whether through Beeckman's acquaintances Descartes and Mersenne (the latter an indefatigable and influential correspondent with many people across Europe), or through rapidly efflorescing use among Jesuit mathematicians over the succeeding two decades. It is noteworthy that Beeckman introduced the term by using a version that focused on kinds of *people*: "physico-mathematicians."

Isaac Newton's mentor and predecessor as Lucasian Professor at Cambridge, Isaac Barrow, in a published version of his mathematical lectures originally given in the 1660s, explained the division of mathematics into "pure" and "mixed," and noted that studies falling under the latter category were sometimes dubbed, in Latin, *physico-mathematicas*; by Newton's time the term had become a commonplace (Dear 1995, 178). In short, in the course of the seventeenth century the expert in mathematical sciences had become one of the available personae of the natural philosopher. Those mathematical endeavors were also (as another hallmark of "physico-mathematics") increasingly associated with experimental methodologies.

Experimenters

The development of cultural practices centered on artificial experimentation was a characteristic feature of the sciences in the seventeenth century, whether or not associated with mathematical conceptualizations. The social loci of these nascent experimental cultures were varied, but all involved cognitive and epistemological assumptions that took the specificity of witnessed events to be foundational for any philosophy of nature. The networks of association that sustained experimental production linked together such sites as the private laboratories of alchemists, princely courts, and learned academies in addition to classrooms and scholars' closets; more generally, they connected individuals who often regarded themselves and their correspondents as participating in a "Republic of Letters" that transcended national or even confessional divides.

Endeavors that employed contrivance to uncover natural behaviors had certainly formed part of European cultural practices prior to the seventeenth century: chemical/alchemical texts from the high and late middle ages often detail such procedures, sometimes taking the form of accounts of specific events (Newman 1997). Such knowledge enterprises resemble in this respect the instrumentally facilitated procedures of astronomy, where specific observational data provided grist to the mill of astronomical generalization and modeling. Aristotle had characterized the mathematical sciences as having no regard for physical causes; they were instead concerned with accurate descriptive generalizations, as with the astronomical modelling of celestial motions through the use of circles (Dear 2011).

Chemical experimentalism in the sixteenth century attempted to capture craft techniques in formal recipes, or to record the achievement of some long-sought material transformation (Eamon 1996). The recipes provided in sixteenth-century "books of secrets" (a genre especially prevalent in, but not restricted to, Germany and Italy) lacked the investigative aspect of the seventeenth-century event-experiment, being instructions on how to produce an already-known phenomenon, property, or substance. Alternatively, and especially when the alchemist worked within the textual tradition of Paracelsus, an aspect of spiritual experience typically transfigured chemical work, and made the apparent "experiment" into a unique communion between man and God's Creation (Principe 2013). The experimental outcome was less about determining a typical (and hence generalizable) piece of natural behavior, and more resembled the routine miracle of the transubstantiation of the Eucharist (and cf. Weeks 1997). Its epistemic function, that is, was very different from the academically rooted scholastic-Aristotelian conception of philosophical experience.

Ideal Types and Natural Philosophers

It is tempting to construct an "ideal type" of the new (non-university) natural philosopher at various moments in the period under consideration. For the first three or four decades of the seventeenth century, such a person might look like this: an academically trained educated man, either independently wealthy or (more likely) a clergyman and/or a family retainer, often a tutor, sustained by the patronage of a noble family or as part of a royal court. Such figures as Descartes, Mersenne, Gassendi, Harriot, as well as Kepler, serve as exemplars, as does Galileo after his move to the Tuscan court in 1610. Such people can usefully be seen as linked together through the vague ideology of the "Republic of Letters." Around the middle of the century, and increasingly thereafter, the natural philosopher was more likely to be associated with a voluntary society of some kind, whether an arm of the state, as in the case of the Académie Royale des Sciences in Paris, an extension of aristocratic patronage (Accademia del Cimento), or a voluntary association (such as the Montmor academy, or the Académie de Caen); many such groups arose starting in the middle of the seventeenth century and continued to effloresce in the eighteenth. The identity of the natural philosopher was now typically attached to these groups, at the least in the form of a corresponding membership or publication in a growing array of widely distributed associated periodicals (the *Philosophical Transactions*, or the *Journal des Sçavans*, for example). The natural philosopher might well still approximate to the earlier social profile, but the label had acquired a more concrete sense relating to the *activities* that such a person might engage in. Thus by the middle of the eighteenth century figures such as John Michell, a rural clergyman in Yorkshire, and Henry Cavendish, scion of a famous aristocratic family, each figured as natural philosophers who published in the Royal Society's *Philosophical Transactions* (Jungnickel and McCormmach 1999; McCormmach 2012).

One reason to take such an "ideal type" seriously is that some historical figures generally regarded as significant producers of natural knowledge in this period do not always qualify as "natural philosophers" nonetheless. A celebrated example is that of Thomas Sydenham, an important medical writer in Restoration England, who, unlike a good few other physicians at the time, such as Thomas Willis, never became a fellow of the Royal Society: he was a physician first and foremost. Such examples warn us against too easy a subsuming of contemporary categories to the anachronistic one of "science" in its modern sense: "natural philosopher" was not a social role similar to that of "scientist" (*contra* Ben-David 1971).

Role: Public or Private?

The role of the natural philosopher crucially involved an understanding of its social character. The central issue concerned whether knowledge-making should be a private, closed enterprise accessible only to the chosen few, or open and public, available in principle to all. The issue was further complicated by the ambiguity of the category "public": was the state to be understood as comprised of public institutions, or were its workings closed to all but the licensed few? All these concerns were in play and debated extensively in the formation of the experimental philosophy that rose to prominence in the seventeenth century and that came to characterize the natural philosopher (Shapin 1992, 1994).

A criticism of alchemical practitioners frequently made in the early seventeenth century held that their knowledge was suspect because it was fabricated in private, secretly, and not for the benefit of all. The German schoolmaster and chemist Andreas Libavius vociferously maintained such a position against rival Paracelsian alchemists, arguing that the proper way to develop chemistry as a science was to make it a publicly accessible enterprise: in short, to make it *teachable* (Hannaway 1975, 1986; Moran 2007; the theme was also pursued in Christie and Golinski 1982). For him, the chemical enterprise was operational, practical, and experimental, but also, and crucially, it was a *public* good, and should not be hidden away in the shadows.

Around the same time, in England, Francis Bacon assailed the alchemists in a similar way, saying that their secrecy amounted to the sins of pride and greed; they wanted the fruits of their discoveries only for themselves (Rossi 1968). The American alchemist George Starkey's alchemical notebooks from the mid-seventeenth century perhaps exemplify to some degree the kind of private experimentalism that Libavius and Bacon had in mind (Starkey 2005). Bacon's famous promotion of a kind of experimental philosophy purported to open up natural investigation to all; the "levelling of men's wits" aimed at rendering everyone competent to contribute in some way to his grand enterprise. That was the moral of Bacon's fable of Solomon's House in the *New Atlantis* (1626), in which the grand research enterprise was conducted by an elaborate hierarchical social structure that included twelve "merchants of light," who roamed the world collecting information on innovations and discoveries so as to return them home, where they would be incorporated into an edifice of experimental research performed by various classes of workers directed from the top by three "interpreters of nature." The resultant deep knowledge of natural processes (called "axioms") constituted the final product of the endeavor. Bacon's talk elsewhere of the public utility of this knowledge is called into question, however, by the following remark made by the "father" of Solomon's House:

> we have consultations, which of the inventions and experiences which we have discovered shall be published, and which not; and take all an oath of secrecy for the concealing of those which we think fit to keep secret; though some of those we do reveal sometime to the State, and some not. (Bacon 1635; punctuation etc. modernized.)

Here, the State is treated as a distinct entity, separate at least from the administration of Solomon's House; whether it can be identified with the common good (or "the people") is unclear, although the implication is that it cannot. Bacon's view of the natural philosopher, on this model, conceives of him as necessarily disciplined by an authoritarian hierarchical social structure that guarantees the truth of philosophical "axioms" through the rigid application of Bacon's own "New Organon," or logic of discovery.

René Descartes considered problems surrounding the social organization of experimental knowledge-production, and expressed a firm opinion, in his *Discours de la méthode* of 1637, that collective or shared investigations were inefficient and unreliable. Only the hiring of artisans, "or such persons as he could pay, and for whom the hope of gain, which is a very effective means, would make them do exactly all the things that he ordered of them," could help. Real collaboration was impractical, because voluntary assistants "would want without fail to be paid by the explanation of various difficulties, or at least by compliments and useless conversations, so that they wouldn't be able to save him as much time as he lost"[1] (Descartes 1637, Pt.6; see also Garber 2001, chapter 5). For Descartes, all knowledge was created inside a single mind; understanding followed from the cognitive activities of a fully integrated knower. Experimental work, and attendant experimental results, were useful to that knower only to the extent that the competence of the work, and the reliability of the results, were certain. Consequently, Descartes preferred that would-be collaborators simply send him money to hire assistants. A more totalitarian cognitive politics can scarcely be imagined.

Rigid authoritarian control was well suited to a political regime with absolutist pretensions, like many European regimes of the period. It is unsurprising that the chemical projector Johann Joachim Becher outlined a system for the production of alchemical (transmutational) knowledge around 1680 that shared with Solomon's House the strict control and segregation of its channels of information: the left hand not knowing the actions of the right hand. Workers were to be kept in ignorance of one another, so as not to have a complete knowledge of the various parts of the processes being conducted, in some cases illiteracy was to be prized, and only the "counselors" at the head of the business, reporting to the prince, would properly know the meaning of what was occurring; only they could be real natural philosophers (Smith 1994, 237–9).

Societies and Experiment

The establishment of societies of natural inquirers, typically under aristocratic patronage, became common in the second half of the seventeenth century; membership of them soon came to define the typical character of the natural philosopher. The virtues of collectivity seem in most cases to have been assumed rather than theorized, with the most obvious available model being the academic college. "Fellows" of such groups collaborated to various degrees, and experimental activities formed an important part of their ideologies of knowledge. In England, such groups often invoked Bacon as their chief spokeman (Webster 1976). This Baconian enthusiasm culminated in the establishment in the early 1660s of the Royal Society of London. Its self-avowed brief was the promotion of "physico-mathematicall experimental learning," and the specification of experiment occurred with an explicitly Baconian reference. The details of Bacon's methodological precepts may not have been followed to the letter, but the ethos of experimentation dominated the group's self-presentation and much of its communal activity (Hunter 1994 details the early fellowship). The form of social organization underlying its knowledge enterprise was predicated on the moral force of common practices rooted in specificity and trust. Much as in English legal practice, and in the ideal comportment of English gentlemen, so in this community of gentlemanly philosophers was trust placed in another's credit concerning reports of matters of fact. Belief in experimentally and observationally accredited facts about the world grounded a shared and active knowledge enterprise in which a social collective could know things beyond the reach of the single Cartesian knower (Shapin and Schaffer 1985; Shapiro, B. 1983, 2000; Shapin 1994).

The ideology of "Baconianism" also existed at this time elsewhere in Europe. Christiaan Huygens, one of the most prominent natural philosophers brought in by Colbert to populate Louis XIV's prestigious new *Académie des Sciences*, founded at the end of 1666, himself sometimes exalted Bacon's experimental project; the "physical" division of the early Académie incorporated much experimental investigation, while its "mathematical" division incorporated contrived demonstrations of the sort usual in the mathematical sciences. The political organization of knowledge in the Académie, however, resembled less the free association of gentlemen celebrated by the Royal Society's historian, Thomas Sprat, and more the King's private philosophical and technical school. The King's patronage shaped the knowledge-economy of the Académie, such that its members were always understood to be contributing to the absolutist project of enhancing the crown's prestige and glory (Biagioli 1992, 1995; Licoppe 1994; Licoppe 1996). The Académie's practice prior to its new constitution of 1699 was to publish its monographs collectively rather than under the names of individual authors; this policy applied to work done under the rubric of "physics," meaning such endeavors as natural history and descriptive, "factual" sciences. Anything smacking of the hypothetical, however, usually retained the name of its individual author (Hahn 1971; Sturdy 1995). Collective publication was intended to represent the imprimatur of the institution itself: the results presented were not claims made by a specific person, but findings vouched for by the Académie, itself an arm of the state, speaking, that is, for the king. By contrast, the specificity of authorship of the work published by Fellows of the Royal Society was essential to its quite distinct politics of knowledge: only by knowing *who* supported the assertions made in a report could members of the experimental community assess the credibility of its contents.

The other prominent group of experimenters in these years around 1660 was the Florentine Accademia del Cimento. This was a closed group of natural philosophers brought together under the patronage of Prince Leopoldo of Tuscany, younger brother of the Grand Duke and later a Cardinal. Beginning in 1657, its work continued sporadically until the publication of its collected *Saggi* in 1667 (Biagioli 1992, 1995; Licoppe 1994, 1996; Boschiero 2007; Beretta, Clericuzio, and Principe 2009). Like many publications of the early Académie, the *Saggi* were presented collectively rather than having particular experiments and their descriptions attributed to specific named authors. Much like the accounts produced by Fellows of the Royal Society, however, those found in the *Saggi* are historically narrated, and with much circumstantial detail. The impersonal authority of the aristocratically accredited assembly was thus supplemented by the actuality of specific and plausible events (cf. Shapin and Schaffer 1985, chapter 2, on virtual witnessing; Dear 1985).

These variously disciplined knowledge economies interacted successfully (for the most part) due to a widespread network of individuals who could act as nodes connecting such groups together (Lux and Cook 1998; Harris 1996, 2006). In the second half of the seventeenth century, besides the publications already noted, individual correspondents, most notably Henry Oldenburg in London, who both edited the *Philosophical Transactions* and mediated between the activities of the Royal Society's members and others, chiefly in continental Europe, could provide news of their own activities as well as those of groups and societies of which they were themselves members. Such interactions sometimes required translation and negotiation between different knowledge economies, but enabled a broader sense of a common natural-philosophical enterprise (Shapin 1994, chapter 5). In the first half of the century, similar roles had been played by Marin Mersenne and Nicolas-Claude Fabri de Peiresc, each the center of a vast body of correspondence across Europe and beyond that often crossed confessional as well as continental lines. Their personal endeavors mirror the large correspondence archives created by members of the Jesuit order, including those of such scientific luminaries as Christopher Clavius and Athanasius Kircher, and the later correspondences of the Paris Académie des Sciences. Correspondence networks like these enabled individuals to connect with and mediate between institutions. The nascent ideology consolidating this social network was that of the "republic of letters," something more widely accepted in the eighteenth century, when it can be seen as one of the tropes of "Enlightenment." (On the "republic of letters" in the seventeenth century, see Daston 1991.)

Natural philosophy thus formed part of local as well as geographically distributed communities, not all of which shared completely the same norms of cognitive and material practice. By the early eighteenth century a particular form of so-called "experimental philosophy" had grown up around the Royal Society, particularly associated with the writings of Robert Boyle. The term was then adopted by Isaac Newton, first being used by him at the beginning of the eighteenth century and then by others in connection with "Newtonianism" in the following decades (Shapiro, A. 2004; Anstey 2005). French Enlightenment thinkers in particular related this Newtonian experimental philosophy to the empirical, associationist epistemology of John Locke. It can be seen as a kind of phenomenalist approach to natural philosophy, for which Newton's own publicly asserted nescience regarding the causes of gravitational attraction acted as an exemplar (Gaukroger 2010).

Experimental work outside England also prospered in the eighteenth century, sometimes associated with Newtonianism (as in the Netherlands) and often not. Cartesian natural philosophy of a more hypothetical kind had been promoted in the latter part of the previous century by Jacques Rohault and others; it proved a flexible resource for work in chemistry (Nicolas Lémery, Guillaume Homberg) as well as in the nascent study of electricity (Charles François de Cisternay du Fay [Dufay]), with its talk of variously configured interlocking corpuscles as well as force-exerting vortices in a subtle aetherian matter (Holmes 1989; Kim 2003; on Dufay, e.g., Heilbron 1979). This experimental work was thus guided and interpreted using non-Newtonian theories of matter and causation; the Cartesian approach allowed only mechanical contact-action transfers of force, unlike the action-at-a-distance forces permitted by Newtonians. However, the phenomenalist language of experimental outcomes, divorced from such theoretical interpretations, permitted much emulation and interchange among experimental natural philosophers across Europe, no matter their natural-philosophical inclinations. By the second half of the century, the explicitly phenomenological language associated with Newton tended to identify many kinds of experimental natural philosophy as "Newtonian," and its practitioners as typical natural philosophers, independently of their precise forms of matter theory (Licoppe 1996; Hellyer 2005, on Jesuit colleges and their cabinets of instruments).

Natural philosophers in the eighteenth century were thus increasingly typified by their self-proclaimed empiricism, as well as by their membership of appropriate societies separate from those of traditional academic culture (McClellan 1985). At the same time, that academic culture made space for some people, usually in the established areas of mathematics and medicine (which allowed for work in astronomy and chemistry respectively), who possessed the aura of the new kind of natural philosopher associated with a simultaneous involvement with scientific societies and associated international networks. Notable figures in the chemistry of the eighteenth century, such as Georg Ernst Stahl in Berlin, Hermann Boerhaave in the Netherlands, or Joseph Black in Scotland, were academic physicians by training and profession; one should remember also the Savilian Professorship in astronomy at Oxford University, occupied by such notables as John Keill and James Bradley, a chair that Christopher Wren had also occupied in the seventeenth century. At the end of the century, educational reforms in post-Revolutionary France brought major figures, especially from the mathematical-physical sciences, into the educational staff of the new Ecole Polytechnique (Fox 1974; Gillispie 2004, 520-40). Nonetheless, natural philosophers remained into the nineteenth century generally only adventitiously members of universities; natural philosophy as a research area remained an avocation for the university professor.

Natural Philosophy and Natural History

There were aspects of what counted as natural philosophy in the eighteenth century that (from a modern perspective) seem oddly continuous with its Aristotelian antecedents. For Aristotle, natural philosophy—"physics"—dealt with all aspects of the natural world: all those things that contained within themselves a principle of change. This meant that living things were as much a part of "physics" as were rocks and elements, although categorical distinctions between the living and the non-living were still observed, associated as they generally were with distinct Aristotelian texts, and had even found their place within the divisions set up within the Paris Académie des Sciences in the later seventeenth century. Natural history, by contrast, had since antiquity been regarded as a quite separate endeavor from that of natural philosophy (Pliny the Elder's *Historia naturalis* being the *locus classicus*). Natural history, unlike natural philosophy, was understood to be about *description*, not causal *explanation*. The physicist explained, whereas the historian (natural or civil) simply described. But in the eighteenth century the descriptive work of the naturalist had begun to bleed over into the explanatory work of the natural philosopher, a movement that had begun in the seventeenth century with the ever-closer association of experimental natural philosophy with the interests of the (historical, descriptive) antiquarian (Ashworth 1990). The natural philosopher, thanks to his association with experiment, evidence, and the telling *fact*, became ever more the expert on nature who knew things, multifarious things that begged to be reduced to some order. Natural history, the repository of facts about nature, became an arm of the natural philosopher's trade.

Georges-Louis Leclerc, comte de Buffon, is in some ways the type (and in others, not) of the natural historian in the eighteenth century, yet he began his illustrious career as a self-styled Newtonian and mathematician; the natural historian was not, by mid-century, clearly distinct from the natural philosopher. A characteristic feature of Buffon's kind of natural history was his insistence on concrete, causal concepts in natural history, rather than what he saw as abstract, metaphysical forms of taxonomy. In England, the collapsing together of the two roles was promoted by natural theology, the use of apparent designfulness in the natural world to evidence the hand of a benevolent Creator (Brooke 1991). The natural theologian deployed materials from both the organic and inorganic worlds to accomplish his end; a good instance is William Derham, the author of two very successful works of natural theology, *Physico-Theology* (1713), which especially focused on examples from the organic world, and *Astrotheology* (1714). Such theological, or quasi-clerical, roles for the natural philosopher grew naturally out of the representation of the natural philosopher as a "priest of nature," as it was promoted in the later seventeenth century by Robert Boyle (Harrison 2005).

By the end of the eighteenth century, however, the natural philosopher's social role had begun to move towards more familiar, and nineteenth-century, contours: those of the "man of science" who framed and facilitated practical projects of interest to the state (Gillispie 1980; Shapin 2003). Such a role had been promoted by the Royal Society and the Académie des Sciences, among other groups, since the seventeenth century, in connection with Baconian visions; only now did the reality begin to correspond to those dreams.

Endnote

1   "[O]u telles gens qu'il pourrait payer, et à qui l'espérance du gain, qui est un moyen très efficace, ferait faire exactement toutes les choses qu'il leur prescriroit"; "voudraient infailliblement être payés par l'explication de quelques difficultés, ou du moins, par des compliments et des entretiens inutiles, qui ne lui sauroient coûter si peu de son temps qu'il n'y perdît."

References

Anstey, Peter R. 2005. "Experimental versus speculative natural philosophy." In *The Science of Nature in the Seventeenth Century: Patterns of Change in Early Modern Natural Philosophy*, edited by Peter R. Anstey and John A. Schuster, 215–42. Dordrecht: Springer.

Ashworth, William B. 1990. "Natural history and the emblematic world view." In *Reappraisals of the Scientific Revolution*, edited by David C. Lindberg and Robert Westman, 303–32. Cambridge: Cambridge University Press.

Bacon, Francis. 1635. *The New Atlantis*. London.

Ben-David, Joseph. 1971. *The Scientist's Role in Society: A Comparative Study*. Englewood Cliffs, NJ: Prentice-Hall.

Beretta, Marco, Antonio Clericuzio and Lawrence M. Principe (eds.) 2009. *The Accademia del Cimento and Its European context*. Sagamore Beach, MA: Watson Publishing.

Biagioli, Mario. 1992. "Scientific revolution, social bricolage, and etiquette." In *The Scientific Revolution in National Context*, edited by Roy Porter and Mikulás Teich, 11–54. Cambridge: Cambridge University Press.

Biagioli, Mario. 1995. "Le prince et les savants: La civilité scientifique au 17e siècle." *Annales: Économies, Sociétés, Civilisations*, 50: 1417–53.

Blair, Anne. 2006. "Natural philosophy." In *The Cambridge History of Science* vol. 3: *Early Modern Science*, edited by Katherine Park and Lorraine Daston, 365–406. Cambridge: Cambridge University Press.

Boschiero, Luciano. 2007. *Experiment and Natural Philosophy: The History of the Accademia del Cimento*. Dordrecht: Springer.

Brooke, John Hedley. 1991. *Science and Religion: Some Historical Perspectives*. Cambridge: Cambridge University Press.

Christie, John R. R. and J. V. Golinski. 1982. "The spreading of the word: New directions in the historiography of chemistry, 1600–1800." *History of Science*, 20: 235–66.

Daston, Lorraine J. 1991. "The ideal and reality of the republic of letters in the Enlightenment." *Science in Context*, 4: 367–86.

Dear, Peter. 1985. "*Totius in verba*: Rhetoric and authority in the early Royal Society." *Isis*, 76: 145–61.

Dear, Peter. 1995. *Discipline and Experience: The Mathematical Way in the Scientific Revolution*. Chicago: University of Chicago Press.

Dear, Peter. 2011. "Mixed mathematics." In *Wrestling With Nature: From Omens to Science*, edited by Peter Harrison, Michael Shank, and Ronald Numbers, 149–72. Chicago: University of Chicago Press.

Descartes, René. 1637. *Discours de la méthode*. Leiden.

Eamon, William. 1996. *Science and the Secrets of Nature*. Princeton: Princeton University Press.

Garber, Daniel. 2001. *Descartes Embodied: Reading Cartesian Philosophy through Cartesian Science*. Cambridge: Cambridge University Press.

Fox, Robert. 1974. "The rise and fall of Laplacian Physics." *Historical Studies in the Physical Sciences* 4: 89–136.

Gaukroger, Stephen. 2006. *The Emergence of a Scientific Culture: Science and the Shaping of Modernity 1210–1685*. Oxford: Clarendon Press.

Gaukroger, Stephen. 2010. *The Collapse of Mechanism and the Rise of Sensibility: Science and the Shaping of Modernity, 1680–1760*. Oxford: Clarendon Press.

Gillispie, Charles C. 1980. *Science and Polity in France at the End of the Old Regime*. Princeton, NJ: Princeton University Press.

Gillispie, Charles C. 2004. *Science and Polity in France: The Revolutionary and Napoleonic Years*. Princeton, NJ: Princeton University Press.

Hahn, Roger. 1971. *The Anatomy of a Scientific Institution: The Paris Academy of Sciences 1666–1803*. Berkeley: University of California Press.

Hannaway, Owen. 1975. *The Chemists and the Word: The Didactic Origins of Chemistry*. Baltimore: Johns Hopkins University Press.

Hannaway, Owen. 1986. "Laboratory design and the aim of science: Andreas Libavius versus Tycho Brahe." *Isis*, 77: 585–610.

Harris, Steven J. 1996. "Confession-building, long-distance networks, and the organization of Jesuit Science." *Early Science and Medicine*, 1: 287–318.

Harris, Steven J. 2006. "Networks of travel, correspondence, and exchange." In *The Cambridge History of Science* vol. 3: *Early Modern Science*, edited by Katherine Park and Lorraine Daston, 341–360. Cambridge: Cambridge University Press.

Harrison, Peter. 2005. "Physico-theology and the mixed sciences: The role of theology in early modern natural philosophy." In *The Science of Nature in the Seventeenth Century: Patterns of Change in Early Modern Natural Philosophy*, edited by Peter R. Anstey and John A. Schuster, 165–84. Dordrecht: Springer.

Heilbron, John L. 1979. *Electricity in the Seventeenth and Eighteenth Centuries: A Study in Early Modern Physics*. Berkeley, California: University of California Press.

Hellyer, Marcus. 2005. *Catholic Physics: Jesuit Natural Philosophy in Early Modern Germany*. Notre Dame, IN: University of Notre Dame Press.

Holmes, Frederic L. 1989. *Eighteenth-Century Chemistry as an Investigative Enterprise*. Berkeley: Office for History of Science and Technology, University of California at Berkeley.

Hunter, Michael. 1994. *The Royal Society and Its Fellows, 1660–1700*. British Society for the History of Science Monograph 4. Chalfont St. Giles: British Society for the History of Science.

Jungnickel, Christa, and Russell McCormmach. 1999. *Cavendish: The Experimental Life*. Lewisburg, PA: Bucknell University Press.

Kim, Mi Gyung. 2003. *Affinity, That Elusive Dream: A Genealogy of the Chemical Revolution*. Cambridge, MA: MIT Press.

Knowles Middleton, W. E. 1971. *The Experimenters: A Study of the Accademia Del Cimento*. Baltimore: Johns Hopkins University Press.

Koyré, Alexandre. 1965. *Newtonian Studies*. Chicago: University of Chicago Press.

Licoppe, Christian. 1994. "The crystallization of a new narrative form in experimental reports (1660–1690): Experimental evidence as a transaction between philosophical knowledge and aristocratic power." *Science in Context*, 7: 205–44.

Licoppe, Christian. 1996. *La formation de la pratique scientifique: Le discours de l'expérience en France et en Angleterre, 1630–1820*. Paris: Éditions La Découverte.

Lux, David S. and Harold J. Cook. 1998. "Closed circles or open networks?: Communicating at a distance during the scientific revolution." *History of Science*, 36: 179–211.

McClellan, James E. 1985. *Science Reorganized: Scientific Societies in the Eighteenth Century.* New York: Columbia University Press.

McCormmach, Russell. 2012. *Weighing the World: The Reverend John Michell of Thornhill.* Dordrecht: Springer.

Moran, Bruce T. 2007. *Andreas Libavius and the Transformation of Alchemy: Separating Chemical Cultures with Polemical Fire.* Sagamore Beach, MA: Science History Publications.

Newman, William R. 1997. "Art, nature, and experiment among some Aristotelian alchemists." In *Texts and Contexts in Ancient and Medieval Science: Studies on the Occasion of John E. Murdoch's Seventieth Birthday*, edited by Edith Sylla and Michael McVaugh, 305–17. Leiden: Brill.

Principe, Lawrence M. 2013. *The Secrets of Alchemy.* Chicago: University of Chicago Press.

Rossi, Paolo. 1968. *Francis Bacon: From Magic to Science.* Chicago: University of Chicago Press.

Shapin, Steven. 1992. "'The mind is its own place': Science and solitude in seventeenth-century England." *Science in Context*, 4: 191–218.

Shapin, Steven. 1994. *A Social History of Truth: Civility and Science in Seventeenth-Century England.* Chicago: University of Chicago Press.

Shapin, Steven. 2003. "The image of the man of science." In *The Cambridge History of Science* vol.4: *Eighteenth-Century Science*, edited by Roy Porter, 159–83. Cambridge: Cambridge University Press.

Shapin, Steven. 2006. "The man of science." In *The Cambridge History of Science* vol. 3: Early *Modern Science*, edited by Katherine Park and Lorraine Daston, 179–91. Cambridge: Cambridge University Press.

Shapin, Steven and Simon Schaffer. 1985. *Leviathan and the Air-Pump: Hobbes, Boyle, and the Experimental Life.* Princeton, NJ: Princeton University Press.

Shapiro, Alan E. 2004. "Newton's 'Experimental Philosophy'." *Early Science and Medicine*, 9: 185–217.

Shapiro, Barbara J. 1983. *Probability and Certainty in Seventeenth-Century England: A Study of the Relationships between Natural Science, Religion, History, Law, and Literature.* Princeton, NJ: Princeton University Press.

Shapiro, Barbara J. 2000. *A Culture of Fact: England, 1550–1720.* Ithaca, NY: Cornell University Press.

Smith, Pamela H. 1994. *The Business of Alchemy: Science and Culture in the Holy Roman Empire.* Princeton, NJ: Princeton University Press.

Starkey, George. 2005. *Alchemical Laboratory Notebooks and Correspondence.* Edited by William R. Newman and Lawrence M. Principe. Chicago: University of Chicago Press.

Sturdy, David J. 1995. *Science and Social Status: The Members of the Académie des Sciences, 1666–1750.* Woodbridge: Boydell Press.

Voelkel, James R. 2001. *The Composition of Kepler's Astronomia Nova.* Princeton, NJ: Princeton University Press.

Webster, Charles. 1976. *The Great Instauration: Science, Medicine and Reform, 1626–1660.* London: Duckworth.

Weeks, Andrew. 1997. *Paracelsus: Speculative Theory and the Crisis of the Early Reformation.* Albany: State University of New York Press.

CHAPTER SIX

The Natural Historian

KRISTIN JOHNSON

Our understanding of the history of natural history and its practitioners has benefited greatly from the new historiographical approaches adopted within the history of science over the past thirty years. Within the "big picture" narratives of theory change, Charles Darwin (1809–1882) made an appearance, of course. But the huge efforts to collect, catalog, and describe the natural world by his fellow naturalists often remained hidden. This was despite the fact Darwin himself captured that effort quite eloquently in his introduction to *On the Origin of Species* (1859) when he wrote: "In considering the Origin of Species, it is quite conceivable that a naturalist, reflecting on the mutual affinities of organic beings, on their embryological relations, their geographical distribution, geological succession, and other such facts, might come to the conclusion that each species had not been independently created, but had descended, like varieties, from other species" (Darwin 1985, 66).

Understanding how and why the patterns upon which Darwin's imaginary naturalist reflected became known in the first place requires attention to a wonderfully varied pantheon of characters. Yet when the history of science became a professional discipline, those characters—here titled "Natural Historians," or Naturalists—had already been demoted to, at best, science's unglamorous fact-gatherers. They were useful in their own way, of course, but hardly respectable subjects for serious students of the history of science. Naturalists themselves documented the history of their tradition with loyalty and attention, but their results remained largely outside the primary interests and narratives of professional historians of science (Allen 1976). Natural history remained largely outside traditional accounts of the Scientific Revolution focused on physics and astronomy. This was despite the fact Charles Raven provocatively noted that for historians to define the origins of science in such limited terms was profoundly unhistorical. "In the sixteenth and seventeenth centuries," he wrote, "it is obvious that the scientific revolution owed more to the botanists and zoologists and to the doctors and explorers than to the astronomers" (Raven 1942, as quoted in Hoeniger 1985, 147).

A Companion to the History of Science, First Edition. Edited by Bernard Lightman.
© 2016 John Wiley & Sons Ltd. Published 2020 by John Wiley & Sons Ltd.

The turn toward a broader concept of appropriate and exciting subjects for historians of science in the final quarter of the twentieth century opened up rich justifications for closer attention to both naturalists and the continuation of the natural history tradition into the twentieth century. Calls for examining the science of particular times and places on its own terms rather than our own, the study of "scientific commoners" rather than just the "greats," and the examination of scientific practice rather than just theory change, have each created an ideal framework within which to study natural historians. Indeed, some of the most important manifestos for expanding historians' gaze into the past have appeared in studies of natural history. Paul Farber, for example, noted the problems that arise when historians unduly emphasize theory for understanding the nature and history of science by analyzing the growth of natural history collections, the day-to-day workload of organizing and naming specimens, reworking classification systems, and institutional frameworks. In doing so Farber demonstrated the strong continuity of practice and problems often overlooked in the search for historical discontinuity on the theoretical level (Farber 1985). Attention to new sites of the production of natural knowledge like natural history museums has taught us more about the individuals working, and forming identities, within those institutions (Kohlstedt 1995). Reexaminations of traditional "Big Picture" narratives of the Scientific Revolution have uncovered a new and central place for natural history in the rise of modern science, as Raven predicted (Levine 1983; Hooykaas 1987; Gascoigne 2013). And an expansion of who must be included in the history of science—including women, "amateurs," laymen, taxidermists, specimen dealers, and schoolteachers—has opened a rich field for examining the diverse meanings and practices that fall under the grand rubric of the Naturalist Tradition (Keeney 1992; Shteir 1996; Barrow 2000; Nyhart 2009).

Nearly twenty years ago, the fruits of these broadened perspectives culminated in the now classic volume *Cultures of Natural History*. There, in marked contrast to the twentieth-century stereotypes of natural history as "stamp collecting" (Johnson 2007), one learns of "the importance of the roles assigned to natural history in the commonwealth of learning: as a universal discipline, prior to political, social and moral order; as the partner with civil and sacred history in the revelation of the workings of divine providence; as the universal and stable foundation for the transitory and speculative systems of natural philosophy; as the basis for the agricultural, commercial and colonial improvement of the human estate" (Jardine, Secord, and Spary 1996, 3). Of course, these various roles raise numerous challenges for the historian attempting to characterize naturalists' role. Synthesizing current scholarship on the role of the "natural historian" is further complicated by the fact that such literature calls a wide range of "subject headings" home, a legacy of the fact that nineteenth-century specializations based on the organism of interest (ornithologists, entomologists, botanists, to take a few) gave way in the twentieth century to categories based on biological problems (ecologists, evolutionary biologists, conservation biologists, and ethologists). Thus the history of these fields, and fields such as taxonomy and systematics, all become relevant to an examination of the history of the naturalist tradition. But facing these challenges is worth it. For there are fascinating stories to uncover as naturalists reinvented and reclassified themselves, gave themselves new names, and adapted to changing social and scientific values and rules.

This chapter will focus on how natural historians have defined themselves and been defined by others. Such an approach allows us both to understand naturalists on their own terms, but also highlights recent scholarship focused on the varied contexts within which naturalists have studied nature. To begin, a word about terminology. As the ecologist Paul Sears (1891–1990) noted in 1944, "if we choose to go by the lexicon, a naturalist is any student of natural sciences," yet Sears knew that astronomers, physicians, chemists, and even most geologists and biologists, did not consider themselves to be naturalists. "So, regardless of the dictionary," he noted, "something is happening to the word 'naturalist'" (Sears 1944). Something had, in fact, been happening to the word for some time. The first appearance of the term "natural historian" in the pages of the *Philosophical Transactions of the Royal Society* occurs in 1720, to describe James Petiver (c. 1665–1718) (Blaire et al. 1720, 32). It appears once in the 1740s, and then a dozen times or so in the 1770s and 1780s. Naturalist, by contrast, is used almost four hundred times since the journal's founding. But the terms seem to be used interchangeably, and thus both will be used below.

The Early Modern Naturalist

To speak of the "natural historian" in the early modern period raises a number of interesting problems if one pays close attention to historical figures' own categories. For many twentieth-century observers, a "natural historian" was someone with the virtue or vice (according to one's concept of good science) of making "painstaking and prolonged observation" the primary method of studying the natural world, in contrast to the experimental methods of the physiologist (Southern 1945), or the generalizing ambitions of the natural philosopher (Sears 1944). Yet historians have noted that during the latter half of the seventeenth century "natural history and natural philosophy became deeply entangled... to the point that many regarded the major occupation of the natural philosopher to be the practice of natural history" (Anstey 2012, 13). Paying attention to how thinkers used such terms in the past has even recovered the important role natural history played in the thought and work of those traditionally examined as though they were loyal to later categorizations. Setting aside the usual focus on Robert Boyle's (1627–1691) corpuscular hypothesis and his role in the rise of the mechanical philosophy, for example, unveils Boyle's strong belief that natural history must first be compiled "in Order to" do natural philosophy. Indeed, Anstey and Hunter (2008, 96) argue that "it is the dovetailing of speculative natural philosophy with the compilation of natural histories which provides the key to an integrated understanding of Boyle's natural philosophical endeavours." The good natural philosopher was, in other words, also a careful and observant natural historian.

Parsing how natural historians eventually distinguished themselves from other students of nature requires close attention to how naturalists understood and justified their work at different times, an endeavor that in turn helpfully embeds naturalists firmly within the shifting values of their society and culture. Studies of how naturalists navigated the transition from Renaissance natural history to the programs of the Royal Society, for example, has improved our understanding of how natural historians' roles changed over time. In the earlier period, natural history served "emblematic" purposes in which naturalists viewed animals as symbols conferring moral knowledge.

Thus natural historians considered fables, proverbs, allegories, and morals appropriate topics to include in accounts of animals (Ashworth 1990). The natural historian was, in turn, a self-cultivated individual who, in learning the ways of God and of man by studying animals, improved himself. But at the beginning of the seventeenth century, Francis Bacon (1561–1626) helped transform the role of the naturalist from humanists' emphasis on the use of natural knowledge for individual improvement, to self-betterment as a means of Christian charity and social betterment (Lancaster 2012). Bacon's vision meant the naturalist could claim an important role in efforts to improve society. The committed natural historian could thus see something greater at stake in the careful descriptions of the microscopic anatomy of a louse, or the accumulation of hundreds of bird specimens, than simply self-improvement or curiosity. So, too, could his courtly patrons, as naturalists developed varied justifications to those willing to fund museums and the network of collectors required for such a painstakingly detailed and time-consuming endeavor (Findlen 1994).

Toward the century's end John Ray (1627–1705), in his *The Wisdom of God* of 1691, made the case for natural historians' important role in countering atheism through their detailed demonstrations of God's wisdom and beneficence. Gillespie's examination of the religious and social context of such arguments demonstrates how the merger of natural history and natural theology in Britain was rooted in the religious and social turmoil of the late seventeenth century. Ray's efforts must be understood within the context of his contemporaries' desperate need to reconcile the English Civil War, the Restoration, and the Glorious Revolution, with ideas about God's providence. Ray's combination of natural history and natural theology popularized a strong apologetics that would not only be convincing to all but provide a pious use of one's time. Meanwhile, criticisms of the use of final cause in explanations by Baconians and Cartesians only strengthened the empiricism of the tradition. As Gillespie notes, "The temptation to fall into sterile and lazy speculation, which gave merely verbal explanations or which confused theological and physical causes, was always present. The best guard against this was careful research" (Gillespie 1987, 28). Furthermore, Ray's manifestos inspired the natural historian to become increasingly devoted to *living* nature (even if that entailed the examination of the purposeful parts of dead specimens), establishing a tradition of natural historians as interpreters of God's works that would eventually, of course, influence Darwin.

Ray's works indicate that even in the seventeenth century naturalists had to spend time defending their role within science and society. The target of Ray's defenses of natural history were those who contented themselves with "the knowledge of tongues, and a little skill in philology, or history perhaps, and antiquity," while ignoring natural history and the works of Creation. That natural history was in need of a defense is hinted at in his lament that "I wish that this might be brought in fashion among us; I wish men would be so equal and civil, as not to disparage, deride, and vilify those studies which themselves skill not of, or are not conversant in" (169). The knowledge of animals and plants was pleasant, Ray wrote, and it satisfied and fed the soul. Indeed, Ray seemed to think heaven would be full of men and women piously studying God's creations—in other words, Heaven would be full of naturalists!—with the important difference that we would then be able to see "to our great satisfaction and admiration, the ends and uses of these things, which here were either too subtle for us to penetrate and discover" (171). Still, in the meantime, Ray consoled his mortal readers with a promise that the contemplation of natural history "will afford matter enough to exercise and employ our minds" (172). One could obtain a glimpse of the meaning in some small section of nature, if one observed long and well. And if that did not convince, Ray tried an appeal to conscience: "Some reproach methinks it is to learned men, that there should be so many animals still in the world, whose outward shape is not yet taken notice of, or described, much less their way of generation, food, manners, uses, observed" (Ray, 1691, 178). It is an appeal that sounds familiar, though the reasons we should feel guilty for failing in the task have certainly changed.

The Enlightenment Naturalist

By the dawn of the eighteenth century a broad consensus that natural history consisted of observation, collection, and description of flora and fauna does seem to have emerged, though precisely how this should be done, and why it should be done, changed. One senses a unity to natural history that allowed naturalists to see themselves as part of an important tradition. Indeed, naturalists' success in establishing this tradition led one historian to describe natural history as the "big science" of the late seventeenth and eighteenth centuries (Findlen 2006, 436). Certainly, the values to which subsequent, self-styled naturalists would often appeal had been established: emphasis on observation and experience as a prerequisite to and corrective aid to reason; an emphasis on the particular rather than the abstract; broad participation by a range of sectors of society; a tight relationship with geographical exploration; and beliefs regarding the utility of knowledge about animals, plants and minerals that drew on both religious and economic justifications.

Through the manifestos and work of Buffon (1707–1788) and Linnaeus (1707–1778), natural history reached its modern, familiar form, with part of that familiarity being rather heated debates over what precisely that form should be (Farber 2000). Indeed, close analysis reveals divergent traditions of what being a naturalist entailed, each captured by a different hero. Those focused on naming and ordering life, or taxonomists, looked to Linnaeus as guide, while those amassing detailed descriptions of each species (including habits, distribution, and life history), with the aim of establishing general laws, appealed to the legacy of Buffon. For those studying life histories yet uninterested in Buffon's demand for discovering the laws of life, there was always Gilbert White (1720–1793). While those doing careful comparative anatomy and morphology could call on the spirit of Louis-Jean-Marie Daubenton (1716–1799). Even experimental naturalists could find a hero in Lazzaro Spallanzani's (1729–1799) effort to understand basic vital functions. In view of twentieth-century narratives, this diversity of heroes deserves attention (Farber 1982).

In the context of eighteenth-century thought and philosophy, Linnaean naturalists who focused on description and classification did provide a central endeavor through which many thought the study of nature should proceed. This belief provided one of the main intellectual defenses of what became a diverse network of individuals engaged in amassing natural history specimens for both public museums and private collections. But, as is well known, classification did have its detractors. Buffon soundly criticized Linnaean taxonomy and hoped to use the great collections of the natural history museum to establish causal laws (Roger 1997). During a time in which naturalists tried to classify themselves with as much rigor as they did their tiny objects of study,

Buffon's manifestos created some confusion among his contemporaries regarding what natural history was and how it related to natural philosophy. If one began speculating, was one still doing "natural history" or something else? Furthermore, for Buffon, if one simply described anatomy, and did not include life history and habits, one was not in fact including anything *historical* (Hoquet 2010).

Of course, not everyone worried about such debates. Naturalists came in many guises, from those in charge of the great collections like the Jardin du Roi, to the enormous network of collectors and correspondents sending specimens back home (Spary 2000). The steady stream of careful descriptions filling naturalists' journals demonstrate that, while the giants argued, a general consensus regarding natural historians' empiricist virtues justified a huge amount of work. Careful description was needed, no matter what grand naturalists like Cuvier, Buffon, and Goethe were up to. Indeed, when making the case that corals were animals rather than plant-animals, John Ellis (c. 1710–1776) appealed to this ethos as a counter to speculation: "Hence it appears, that this metamorphosis of a plant to an animal is a flowery expression, and in my opinion, better suited to the poetical fancy of an OVID, than to that precise method of describing which we so much admire in a natural historian" (Ellis 1776). And the model offered by Gilbert White provided plenty of room for generations of naturalists to carry on careful description—of form, behavior, and distribution—while considering themselves part of a venerable and useful tradition.

The Nineteenth-Century Naturalist

The opening of the nineteenth century witnessed yet another contest over the meaning and goals of the naturalist. Jean-Baptiste Lamarck (1744–1829) famously delivered the opening salvo by titling his 1809 excursion into explaining the patterns discovered by naturalists, *Zoological Philosophy*. Georges Cuvier's (1769–1832) famous "Memoir of Lamarck" was a clear counter-attack to Lamarck and Buffon's broadened definitions of what a good naturalist should be up to (Burkhardt 1977; Appel 1987). Cuvier's vision of (at least the average) naturalist as disciplined describer triumphed for a time, especially within nations retrenching into strict Baconianism after the speculative excesses of the French Revolution. In contrast to those proposing theories of the formation and movement of glaciers, for example, De Saussure was "the great natural historian of the Alps" (Murchison 1844, clii). This reestablished consensus proved an ideal stage for the creation of natural history museums and journals focused on amassing careful descriptions. Indeed, during much of the nineteenth century the study of natural history as a science was generally the province of naturalists at large museums, and both the curators of national collections and wealthy owners of private collections provided the expertise and authority that determined the direction of scientific (as opposed to popular) natural history (Farber 1980).

Empire served as an additional crucial context within which naturalists flourished (Browne 1983). By the time Darwin returned from his voyage in 1836, the Zoological Society's museum overflowed with specimens, and John Henslow (1796–1861) famously warned him that the zoologists would think a number of undescribed creatures a nuisance. France, Belgium, and, eventually, Germany joined in the competition for these furred, scaled, and feathered demonstrations of imperial power. Meanwhile, naturalists on the other side of the Atlantic accompanied American expansion West with their guns and butterfly nets, joining exploring expeditions and following the railroads, eventually as part of large scale, government-funded natural history surveys (Kohler 2006). Bolstered by a methodological framework and institutional imperatives that supported such endeavors, the description of new species occupied museum-based naturalists while the publication of novelties filled natural history journals. Eventually, division of labor developed in the face of overwhelming diversity, as naturalists aligned into disciplines based on the taxa of study. Collections of specimens anchored the descriptive work published in entomology, ornithology, and botanical journals. In Britain, the Zoological Society of London had its own collection, while the work of private collection-owners dominated the Entomological Society of London and the British Ornithologists' Union. Though not excluding accounts of live animals (the Zoological Society of London also had its own menagerie), the journals produced by these societies focused on the accumulation of specimen-based facts. All of this work was in turn bolstered by a vision of naturalists' virtues as their careful attention to detailed observation unguided by any pet hypothesis or theory.

We now have wonderful, contextual biographies that place Darwin firmly within the naturalist tradition, in all its varied social, political, and theological grandeur (Browne 1995; Desmond and Moore 1997). Secretly, of course, Darwin was breaking down traditional definitions of what a good respectable, British naturalist should be spending his time on. As Burkhardt (1974) notes, part of what made *Origin of Species* so shocking to Darwin's contemporaries (and why it was largely ignored in naturalists' journals), was that he departed from a methodological consensus that good natural historians described rather than theorized about nature. Worse, Darwin's natural explanation of purposeful parts, not to mention his fabulously understated wish to examine "the moral sense or conscience" "exclusively from the side of natural history" in *The Descent of Man,* seemingly destroyed naturalists' pretentions to be central planks in the effort to uphold a God-given moral and social order. Of course, later naturalists like William Ritter (1856–1944) would not only find much to praise in Darwin's inclusion of the human mind as "a proper subject for the natural historian to study" (Ritter 1938, 322), some would even try to reestablish their role as guides to God's ways and intentions well into the twentieth century (Bowler 2001). Darwin's work may thus have convinced a new generation of naturalists of their great importance, had not grand new banners been raised by the likes of Thomas Henry Huxley (1825–1895). For under the rubric of "the new biology," a rising generation of professionals marched self-consciously away from museums, natural theology, specimens, and taxonomy (Caron 1988).

The various reorientations of what constituted good science inevitably influenced natural historians' sense of their role and the naturalist tradition in general. As Allen (1998, 361) notes, "suddenly, they found themselves being told that what they had all along been accustomed to think of as useful and even in some cases valuable scientific work was no longer of very much moment and, worse, ought for preference to be abandoned and a quite different approach adopted in its stead." At the end of the century, C. Hart Merriam (1855–1942) lamented that "when it became fashionable to study physiology, histology and embryology, the study of systematic natural history was not only neglected, but disappeared from the college curriculum, and the race of naturalists became nearly extinct" (Merriam 1893). In defending the naturalists' approach to nature, Merriam walked along the edge of a precipice that had often threatened to expunge naturalists from the realm of science. For he criticized college biology curriculums for deadening students' interest in the natural world, writing that "to reconstruct a general naturalist at the present day, I would rather have the farmer's boy who knows the plants and animals of his own home than the highest graduate in biology of our leading university." The boy, Merriam argued, brought an enthusiasm and love for nature that set the true naturalist apart from his more "scientific" fellows. Others echoed Merriam's ideal of the natural historian as one with "enthusiasm for the outdoor world, (with) his eyes, ears, and nose attuned to the pungency of field and hedgerow" (Hare 1954, 323). Of course, given the demand for natural explanations, the source of that enthusiasm had been secularized over the centuries. Merriam did not justify attention to natural history on the grounds that it brought one closer to God, though he did think it made men and women better. But the appeal to the naturalist's love of nature as a defining characteristic could prove problematic when the "man of science" was defined as distant, objective, and, ideally, in possession of a lab coat and experimental apparatus. Henry Fairfield Osborn's (1857–1935) insistence (1924, 3) that he preferred the naturalist to the "scientist" "because there is less of the ego in him" appealed to the traditional associations of natural history with a strict empiricism. But the fact Osborn included John Burroughs, Louis Pasteur, William Wordsworth, and Huxley in his list of "great naturalists" must have provided a more confusing than helpful impression of what it was, exactly, a "great naturalist" did.

The Modern Naturalist

By the twentieth century, the theory of evolution had transformed the lens through which many naturalists explained the facts and patterns they discovered, though the effect on their day-to-day practice proved less marked. In contrast to the naturalists who, in the 1860s, saw Darwin's theory as outside the appropriate purview of the working naturalist, by 1938, the natural historian could be described as one interested in "illuminating the evolutionary processes which have occurred." (Epling 1938, 560). A decade later, another described how "anatomical studies of the present period appropriately adopt the dynamic viewpoint of the natural historian and analyze the anatomy of an animal with reference to its living conditions and activities," including the role of natural selection in determining morphology (Reed 1951, 513). Though the primary virtues of the natural historian, evolutionary or no, remained a cautious, detailed empiricism, hours in the field or museum, and an appreciation of the whole organism, these reformations in what a good naturalist did reflect the fact natural historians with an eye on institutional space, government funds, and hiring patterns had to adapt to a changing landscape, especially after the world wars. A new generation of editors of taxa-defined journals transformed what counted as appropriate papers (including increased space for theoretical excursions), while new journals appeared with titles like *Ecology* and *Evolution*. Meanwhile, in the face of naturalists' (ironic) role in undermining traditional versions of natural theology, some reworked naturalists' role as upholders of the moral and social order on more secular grounds (Pauly 2002).

The self-conscious shifts toward more theoretical and experimental papers in naturalists' journals and in new biology curricula betimes inspired ardent defenses of naturalists' traditional virtues. In the US, William Morton Wheeler (1865–1937) was particularly adept at defending the natural historian's approach, updating Merriam's laments into cutting criticisms of the rising stars of science—the geneticists. William Ritter reminded readers that Darwin had done his great work because he remained "true to his qualities as a typical, unspoiled naturalist." He had observed, described, named, and classified the objects of nature "from a perfectly naive acceptance of the sense data he collects and reflects upon" (Ritter 1938, 319). As one ecologist defending natural historians' approach noted, sciences had abandoned the term naturalist as they became more mathematical, technical, and experimental, until the term "naturalist" was used as a pejorative description for apparently superficial, uncoordinated study of nature "that is out-of-doors and does not involve the use of apparatus" (Sears 1944, 44). Like Ritter, Sears turned to Darwin, "an authentic, confessed and conceded, naturalist," to defend naturalists' approach. Their attention to the complexity of living nature—the inseparability of life and environment—countered any claim that the laboratory was "a sufficient source of scientific truth." But Sears acknowledged that genetics had by proclamation gained the advantage, while natural history "slipped into the temple of learning through a side door... mumblingly christened 'oecology'" (48).

Some questioned whether such laments reflected reality. "Only the most bigoted scientists would assert," one chemist wrote, "that the natural historian and the field naturalist do not carry out perfectly *scientific* activities" even though they do not use quantitative methods (Hutchinson 1964, 41). Others rebelliously donned the naturalist cape, old-fashioned or no. In reaction to the heady speculations of previous generations, the anthropologist A. L. Kroeber proudly called himself a "natural historian of culture" whose first interest had always has been in phenomena and their ordering. It was, he explained, "akin to an aesthetic proclivity, presumably congenital" (Kroeber 1952, 3). But by mid-century it was clear to some, at least, that the "bigoted scientists" were winning the ear of hiring committees. E. O. Wilson (b. 1929) recalled James Watson, his "brilliant enemy" in the halls of Harvard's science buildings, wondering why anyone in their right mind would hire an ecologist (Wilson 1994). Some naturalists called on history as part of their response to such dismissals. Most famously, Ernst Mayr (1904–2005) recovered the critical place of naturalists in the history of evolution theory in order to counter narratives that gave sole credit for the evolution synthesis to geneticists and experimentalists (Mayr and Provine 1980).

Mayr characterized the response of naturalists to both nature and evolution theory in distinctive ways, as opposed to experimentalists, characterizations that in turn mapped on to stories of the synthesis as a unification of naturalist and experimentalist-thinking (Allen 1975, 1979). But, as in all classification, the categories proved tricky when faced with a diverse reality, and historians have since questioned not only naturalists' supposed demise, but even the supposed boundary between naturalists and experimentalists (Rainger, Benson, and Maienschein 1988; Benson, Maienschein, and Rainger 1991; Kingsland 1991). In contrast to the implication that the museum tradition died out during the move to laboratories, museums experienced a heyday in both the United States and Germany at precisely the point in time that the experimentalists supposedly triumphed (Nyhart 1996). Naturalists (sometimes under the new disciplinary identities of ethologists or ecologists) worked to maintain and carve out new places in twentieth century science (Hagen 1984; Hagen 1986; Hagen 1999; Vernon 1993; Burkhardt 1999). Historians taking up the call to carry out detailed, local, and contextual examinations of naturalists and their institutions have demonstrated how the success of naturalists' efforts varied considerably even within particular countries (Kraft and Alberti 2003). They have also expanded the basis of our narratives outside the sphere of those working in Europe and the United States to the work of naturalists in Latin America (Lopes and Podgorny 2000), China (Fan 2004) and elsewhere (Sheets-Pyenson 1988).

Meanwhile, the legacy of reforming, defining, and defending the natural historian's role continues, both for naturalists themselves and for historians telling stories about their past. Most recently, the biodiversity crisis has provided new justifications for natural historians' work. It has also been the context for new heroes, such as Wilson, who proudly titled his 2006 autobiography *Naturalist.* As Secord (1996, 458) so aptly noted, as a result "the history of natural history needs to become part of environmental history." Even more provocatively, recent work on those who cross boundaries between field and lab (Kohler 2002) and scholarship that examines what exactly scientists do rather than the titles they claim (naturalist or experimentalist) (Strasser 2012) is expanding the potential realm of naturalists' jurisdiction over the natural world. There may, indeed, be more natural historians out there than we thought.

References

Allen, David. 1976. *The Naturalist in Britain: a Social History.* London: Allen Lane.

Allen, David. 1998. "On parallel lines: Natural history and biology from the late Victorian period." *Archives of Natural History*, 25: 361–71.

Allen, Garland E. 1975. *Life Science in the 20th Century.* New York: John Wiley & Sons.

Allen, Garland E. 1979. "Naturalists and experimentalists: The genotype and the phenotype." *Studies in History of Biology*, 3: 179–209.

Anstey, Peter, and Michael Hunter, 2008. "Robert Boyle's 'Designe about Natural History.'" *Early Science and Medicine*, 13: 81–126.

Anstey, Peter. 2012. "Francis Bacon and the classification of natural history." *Early Science and Medicine*, 17: 11–31.

Appel, Toby. 1987. *The Cuvier–Geoffroy Debate: French Biology in the Decades before Darwin.* Oxford: Oxford University Press.

Ashworth, William B., Jr. 1990. "Natural history and the emblematic world view." In *Reappraisals of the Scientific Revolution*, edited by David C. Lindberg and Robert S. Westman, 303–32. Cambridge: Cambridge University Press.

Barrow, Mark V. 2000. "The specimen dealer: Entrepreneurial natural history in America's Gilded Age." *Journal of the History of Biology*, 33: 493–534.

Benson, Keith R., Jane Maienschein, and Ronald Rainger (eds.) 1991. *The Expansion of American Biology.* New Brunswick, NJ: Rutgers University Press.

Blaire, Patrick, James Jackson, Elizabeth Bell, Charles Browne, and Gilbert Anthone. 1720—1721. "A discourse concerning a method of discovering the virtues of plants by their external structure." *Philosophical Transactions (1683–1775)*, 31: 30–38.

Bowler, Peter. 2001. *Reconciling Science and Religion: The Debate in Early Twentieth-Century Britain.* Chicago: University of Chicago Press.

Browne, Janet. 1983. *The Secular Ark: Studies in the History of Biogeography.* New Haven, CT: Yale University Press.

Browne, Janet. 1995. *Charles Darwin: Voyaging.* New York: Knopf.

Burkhardt, Frederick. 1974. "England and Scotland: The learned societies." In *The Comparative Reception of Darwinism*, edited by T. F. Glick, 32–74. Austin: University of Texas Press.

Burkhardt, Richard W. 1977. *The Spirit of System: Lamarck and Evolutionary Biology*. Cambridge: Cambridge University Press, 1977.

Burkhardt, Richard W. 1999. "Ethology, natural history, the life sciences, and the problem of place." *Journal of the History of Biology*, 32: 489–508.

Caron, Joseph A. 1988. "Biology in the life sciences: A historiographical contribution." *History of Science*, 26: 223–68.

Darwin, Charles. 1985. *The Origin of Species*. London: Penguin Classics.

Desmond, Adrian, and James Moore. 1992. *Darwin*. New York: Warner Books.

Ellis, John. 1776. "On the nature of the Gorgonia; That it is a Real marine animal, and not of a mixed nature, between animals and vegetable. By John Ellis, Esq. F.R.S. in a Letter to Daniel Solander, M.D. F.R.S." *Philosophical Transactions of the Royal Society of London*, 66: 1–17.

Epling, Carl. 1938. "Scylla, Charybdis and Darwin." *The American Naturalist*, 72: 547–61.

Fan, Fa-ti. 2004. *British Naturalists in Qing China*. Cambridge, MA: Harvard University Press.

Farber, Paul Lawrence. 1980. "The development of ornithological collections on the late eighteenth century and early nineteenth centuries and their relationships to the emergence of ornithology as a scientific discipline." *Journal of the Society for the Bibliography of Natural History*, 9: 391–4.

Farber, Paul Lawrence. 1982. "Research traditions in eighteenth-century natural history." In *Lazzaro Spallanzani e la biologia del settecento: Teorie, esperimenti, istituzioni scientifiche*, edited by Walter Bernardi and Antonello La Vergata, 397–403. Florence: Leo Olschki.

Farber, Paul Lawrence. 1985. "Theories for the birds: An inquiry into the significance of the theory of evolution for the history of systematics." In *Religion, Science, and Worldview: Essays in Honor of Richard S. Westfall*, edited by Margaret Osler and Paul Lawrence Farber, 321–39. Cambridge: Cambridge University Press.

Farber, Paul Lawrence. 2000. *Finding Order in Nature: The Naturalist Tradition from Linnaeus to E. O. Wilson*. Baltimore, MD: John Hopkins University Press.

Findlen, Paula. 1994. *Possessing Nature: Museums, Collecting, and Scientific Culture in Early Modern* Italy, Berkeley: University of California Press.

Findlen, Paula. 2006. "*Natural history*." In The Cambridge History of Science. Volume 3. Early Modern Science, edited by Katherine Park and Lorraine Daston, 435–68. Cambridge: Cambridge University Press.

Gascoigne, John. 2013. "Crossing the Pillars of Hercules: Francis Bacon, the scientific revolution and the New World." In *Science in the Age of Baroque*, edited by O. Gal and R. Chen-Morris, 217–237. Dordrecht: Springer.

Gillespie, Neal C. 1987. "Natural history, natural theology, and social order: John Ray and the 'Newtonian Ideology'." *Journal of the History of Biology*, 20: 1–49.

Hagen, Joel B. 1984. "Experimentalists and naturalists in twentieth-century botany: Experimental taxonomy, 1920–1950." *Journal of the History of Biology*, 17: 249–70.

Hagen, Joel B. 1986. "Ecologists and taxonomists: Divergent traditions in twentieth-century plant geography." *Journal of the History of Biology*, 19: 197–214.

Hagen, Joel B. 1999. "Naturalists, molecular biologists, and the challenges of molecular evolution." *Journal of the History of Biology*, 32: 321–41.

Hare, F. Kenneth. 1954. "[Review of] *Climate and the British Scene*." *Geographical Review*, 44: 323–4.

Hoeniger, F. David. 1985. "How plants and animals were studied in the mid-sixteenth century." In *Science and Arts in the Renaissance*, edited by John W. Shirley and F. David Hoeniger, 130–48. Washington, DC: Folger Shakespeare Library.

Hooykaas, R. 1987. "The rise of modern science: When and why?" *British Journal for the History of Science*, 20: 453–73.

Hoquet, Thierry, 2010. "History without time: Buffon's natural history as a nonmathematical physique." *Isis*, 101: 30–61.

Hutchinson, Eric. 1964. "Science and responsibility." *American Scientist*, 52: 40–50.

Jardine, N. J., J. A. Secord, and E. C. Spary (eds.) 1996. *Cultures of Natural History*. Cambridge: Cambridge University Press.

Johnson, Kristin. 2007. "Natural history as stamp collecting: A brief history." *Archives of Natural History*, 34: 244–58.

Keeney, Elizabeth B. 1992. *Botanizers: Amateur Scientists in Nineteenth-Century America*. Chapel Hill: University of North Carolina Press.

Kingsland, Sharon. 1991. "The battling botanist: Daniel Trembly MacDougal, mutation theory, and the rise of experimental evolutionary biology in America, 1900–1912." *Isis*, 82: 479–509.

Kohler, Robert E. 2002. *Landscapes and Labscapes: Exploring the Lab–Field Border in Biology*. Chicago: University of Chicago Press.

Kohler, Robert E. 2006. *All Creatures: Naturalists, Collectors, and Biodiversity, 1850–1950*. Princeton, NJ: Princeton University Press.

Kohlstedt, Sally Gregory. 1995. "Essay Review: Museums: revisiting sites in the history of the natural sciences." *Journal of the History of Biology*, 28: 151–66.

Kraft, Alison, and Samuel J. M. M. Alberti. 2003. "'Equal though different': Laboratories, museums and the institutional development of biology in late-Victorian northern England." *Studies in History and Philosophy of Science Part C*, 34: 203–36.

Kroeber, A. L. 1952. *The Nature of Culture*. Chicago: University of Chicago Press.

Lancaster, James A. T. 2012. "Natural knowledge as a propaedeutic to self-betterment." *Early Science and Medicine*, 17: 131–96.

Levine, Joseph, 1983. "Natural history and the history of the scientific revolution." *Clio*, 13: 57–73.

Lopes, Maria Margaret, and Irina Podgorny. 2000. "The shaping of Latin American museums of natural history, 1850–1890." *Osiris*, 15: 108–34.

Mayr, Ernst, and William Provine (eds.) 1980. *The Evolutionary Synthesis: Perspectives on the Unification of Biology*. Cambridge, MA: Harvard University Press.

Merriam, C. Hart. 1893. "Biology in our colleges: A plea for a broader and more liberal biology." *Science*, 21: 352–5.

Murchison, Roderick Impey. 1844. "Address to the Royal Geographical Society of London." *Journal of the Royal Geographical Society of London*, 14: xlv–cxxviii.

Nyhart, Lynn K. 1996. "Natural history and the new biology." In *Cultures of Natural History*, edited by N. J. Jardine, J. A. Secord, and E. C. Spary, 426–43. Cambridge: Cambridge University Press.

Nyhart, Lynn K. 2009. *Modern Nature: The Rise of the Biological Perspective in Germany*. Chicago: University of Chicago Press, 2009.

Osborn, Henry Fairfield. 1924. *Impressions of Great Naturalists*. New York: Charles Scribner's Sons.

Pauly, Philip J. 2002. *Biologists and the Promise of American Life: from Meriwether Lewis to Alfred Kinsey*. Princeton, NJ: Princeton University Press.

Rainger, Ronald, Keith Benson, and Jane Maienschein (eds.) 1988. *The American Development of Biology*. Philadelphia: University of Pennsylvania Press.

Raven, Charles. 1942. *John Ray: Naturalist*. Cambridge University Press.

Ray, John. 1691. *The Wisdom of God Manifested in the Works of Creation*. London: Smith.

Reed, Charles A. 1951. "Locomotion and appendicular anatomy in three soricoid insectivores." *American Midland Naturalist*, 45: 513–671.

Ritter, William. 1938. "Mechanical ideas in the last hundred years of biology." *The American Naturalist*, 72: 315–23.

Roger, Jacques. 1997. *Buffon: A Life in Natural History.* Ithaca, NY: Cornell University Press.

Sears, Paul B. 1944. "The future of the naturalist." *The American Naturalist,* 78: 43–53.

Secord, James A. 1996. "The crisis of nature." In *Cultures of Natural History,* edited by N. J. Jardine, J. A. Secord, and E. C. Spary, 447-59. Cambridge: Cambridge University Press.

Sheets-Pyenson, Susan. 1988. *Cathedrals of Science: the Development of Colonial Natural History Museums during the Late Nineteenth Century.* Kingston and Montreal: McGill-Queen's University Press.

Shteir, Ann B. 1996. *Cultivating Women, Cultivating Science: Flora's Daughters and Botany in England, 1760–1860.* Baltimore: Johns Hopkins University Press.

Southern, H. N. 1945. "[Review of] *Studies in the Life History of the Song Sparrow* by Margaret Morse Nice." *Journal of Animal Ecology,* 14: 54.

Spary, E. C. 2000. *Utopia's Garden: French Natural History from Old Regime to Revolution.* Chicago: University of Chicago Press.

Strasser, Bruno. 2012. "Collecting nature: Practices, styles, and narratives." *Osiris,* 27: 303–40.

Vernon, Keith. 1993. "Desperately seeking status: Evolutionary Systematics and the taxonomists' search for respectability, 1940–1960." *British Journal for the History of Science,* 26: 207–227.

Wilson, E. O. 1994. *Naturalist* Washington, DC, Island Press.

CHAPTER SEVEN

Invisible Technicians, Instrument-makers and Artisans

IWAN RHYS MORUS

To man the fleet that lay idle at Brest would call for twenty thousand men. The seamen – what seamen there were – would have to march hundreds of miles from the merchant ports of Le Havre and Marseille if they were not sent round by sea. Twenty thousand men needed food and clothing, and highly specialised food and clothing moreover. The flour to make biscuits, the cattle and pigs and the salt to salt them down, and the barrel staves in which to store them – where were they to come from?

C. S. Forester, *Hornblower and the Hotspur* (1962)

Who makes knowledge and who does it belong to? The problem of scientific authority has always been closely bound up with the identity of the knowledge-maker. One common image of the scientist is of the lonely, isolated thinker, absorbed in their own mind. Another is of the white-coated experimenter, wrestling with nature's secrets in their laboratory. What both images have in common is that they are of individuals in isolation—and that is usually how we think of the scientist. Historians of science have long argued how misleading these sorts of images of scientists and their work really are. The process of making knowledge is rarely, if ever, the sole concern of a single individual. Making knowledge is a collective process that usually involves large numbers of people. Modern scientists, whether they are experimenters or theoreticians, work in groups rather than as individuals in isolation. Behind the individual scientists lie whole armies of support workers, just as Forester had his hero Hornblower realize the collective labor that lay behind Napoleon's genius. Their contribution to the process of making knowledge usually goes unacknowledged by historians. In a seminal discussion of the emergence of modern notions of scientific authority and identity, the historian Steven Shapin (half) jokingly suggested that the names of individual laboratory technicians usually only appear in the historical record when

A Companion to the History of Science, First Edition. Edited by Bernard Lightman.
© 2016 John Wiley & Sons Ltd. Published 2020 by John Wiley & Sons Ltd.

something has gone wrong in the performance of an experiment. They are otherwise historically invisible (Shapin 1989).

This historical invisibility of technicians, instrument-makers and others involved behind the scenes in the process of making scientific knowledge highlights that some of the ways in which we routinely talk about the nature of scientific activity carry particular connotations about the relative value of different ways of knowing. The distinction in status between work that is done with the head and work that is done with the hands is one that has a long history in Western culture. Ancient Greek culture accorded even quite highly skilled handwork a considerably lower status than was given to intellectual work, for example. Philosophy mattered more than pottery (Applebaum 1992). There is, nevertheless, also historical evidence to support the view that ancient Greek craftsmen themselves took a different view of the status of their manual labor and the ways in which it contributed to their sense of personal identity. By the late eighteenth and early nineteenth centuries, artisans' sense of their own personal and cultural identity was tightly bound with their sense of possessing a property of skill that set them apart from unskilled laborers. In the volatile context of the early industrial revolution this often meant that arguments about the place and possession of knowledge carried an explicitly political edge (Rule 1989).

This is nicely illustrated in the dispute that flared up during the 1820s between the English natural philosopher Charles Babbage and his technician Joseph Clement around their work on the calculating engine (Schaffer 1994). Babbage employed Clement at the recommendation of Marc Isambard Brunel to engineer the calculating engine's components to a very high degree of precision. Despite an initially productive working relationship, the two soon fell out over precisely these sorts of problems regarding the ownership of knowledge and self-identity. They clashed over the question of who owned the products of their collaboration (as Clement saw it). In particular, Clement took the view, as was traditional among such skilled workers, that the tools he had designed and made in order to produce the calculating engine's components were his own property. He was also unwilling to accede to Babbage's demand that he would not make further copies of the calculating engine without Babbage's express permission. These were disputes not just about the ownership of the engine and the tools needed to make it, but about the ownership of the knowledge involved in its production and how credit for that knowledge should be distributed. Babbage was operating within an epistemological tradition that suggested the technician's contribution should be tacit. Clement, clearly, was not.

Arguments about the visibility of scientific authorship were often conducted in the language of discovery and invention. When Charles Wheatstone and William Fothergill Cooke—co-patentees of the electromagnetic telegraph—fell out over their respective contributions that was exactly the language they and their arbitrators adopted. The dispute between Cooke and Wheatstone was only one of a number of quarrels between inventors over the proprieties of dividing the spoils during the 1840s and beyond (Morus 1998). For some at least of these protagonists and their cohorts, these were political battles with important consequences for demarcation disputes. Deciding priority in discovery and invention—and deciding what was to be labeled as discovery or invention—was often a matter of deciding what was and was not appropriately labeled scientific as well. The *Mechanic's Magazine*'s combative editor was forthright in his view that efforts to reallocate scientific credit away from artisans and towards gentlemen of science were no less than inequitable attempts to deprive them of their proper status—to put them outside the boundaries of science. Arguments about what was discovery and what was (mere) invention were understood to be concerned with the visibility of scientific work and its proper attribution.

In this chapter I want to trace some of the key ways in which the place of instrument-makers, technicians, and other manually skilled workers in the production of scientific knowledge has shifted since the Scientific Revolution. It should be clear that these sorts of questions about attribution and authority go to the heart of our understanding of what scientific knowledge is, how it is made, and who gets the credit for its production. Looking at the ways in which the contributions of such individuals to the production of knowledge—and the varying extents to which they might themselves be credited as autonomous makers of scientific knowledge in their own right—provides historians with a way of approaching how understandings of the nature of science itself are embedded in specific historical contexts. It also draws to our attention the extent to which epistemological discussions invariably have cultural and political consequences. Historical questions about how knowledge is made cannot be adequately answered without drawing attention to the identity of the maker.

Instrumental Knowledge

In his magisterial *Scientific Instruments of the 17th & 18th Centuries and their Makers* (1972) the French chemist and historian Maurice Daumas argued that instrument-makers had played a crucial role in the Scientific Revolution and the emergence of modern science. He suggested that "scientists of the seventeenth century, most of whom were also craftsmen, could not have created their apparatus without the collaboration of the professional craftsman." Men of science might be responsible for designing new instruments, "but it needed the regular work of the instrument maker to reproduce and improve the model and to make of it an instrument of everyday use" (Daumas 1972, 2). Daumas's argument was that instrument-makers were not just to be regarded as implementing innovations developed by the proponents of the New Science, but that their distinctive craft skills—the tacit knowledge they embodied— were a vital element of the intellectual and practical ecology within which the New Science flourished. This was a view of the origins of the Scientific Revolution that ran directly counter to the accounts offered by some of its pioneering historians such as Alexandre Koyré (Koyré 1968). Koyré was unequivocal that science owed nothing to craftsmen and technicians. He was responding, in part at least, to Marxist accounts of the origins of the Scientific Revolution which placed practical technical know-how at the center of the picture (Zilsel 1942).

Instrument-making as a skilled trade was already well-established by the beginning of the seventeenth century. It had its origins within the various trades associated with metalworking, such as engraving and turning. Scientific instrument-making also had strong links to clock-making. Much of this trade was in the production of instruments for surveying and navigation such as compasses or various kinds of sighting apparatus. Instruments for astronomical purposes such as astrolabes were also popular. During the seventeenth century, skills associated with glass-making such as glass-blowing and grinding came to be increasingly prominent as new instruments such as the telescope were developed. As well as instruments for practical use, there was an increasingly lucrative trade in instruments designed for collection and display. The men who made these instruments were skilled craftsmen who had typically entered the trade through apprenticeship and membership of a guild. In terms of social status they were unmistakably tradesmen rather than gentlemen during a period when natural philosophy was increasingly associated with gentility. By the end of the seventeenth century nonetheless, some instrument-makers were making names for themselves as natural philosophers. The former draper's apprentice Francis Hauksbee progressed from being one of the Royal Society's instrument-makers to being one of its Fellows, elected in 1706 thanks to the patronage of its president, Isaac Newton.

Throughout the eighteenth century, instrument-making continued to provide a route to status as a natural philosopher for its most elite practitioners such as Hauksbee. Instrument-makers such as John Dollond, celebrated for his optical instruments, or his son-in-law Jesse Ramsden, equally celebrated for his astronomical instruments, were both Fellows of the Royal Society, for example. The older Francis Hauksbee's nephew and namesake followed a career that usefully illustrates the ways in which a promising instrument-maker could use his family background, his talents as a skilled craftsman as well as elite patronage to make a name for himself. The younger Hauksbee was apprenticed to an optician before setting up as an instrument-maker on his own account. As well as making instruments, he offered lectures in natural philosophy and published a number of books. He worked occasionally for the Royal Society and in 1723 was appointed the Society's clerk and housekeeper. Unlike his uncle, he never achieved the status of a Fellow, but he attended the Fellows' dining club and seems to have been treated, at times at least, as a social equal. Entry into the trade was still usually by apprenticeship. Instrument-making became increasingly centralized in larger cities such as London or Paris during the course of the century. Larger concerns often farmed out some of the processes involved to smaller craftsmen who specialized in particular aspects of the industry (Sorrenson 2013).

By the nineteenth century, however, instrument-making was becoming a less promising route to membership of elite institutions such as the Royal Society. When Sir Humphry Davy retorted that the electrician and instrument-maker William Sturgeon, who had originally been apprenticed as a boot-maker, should "stick to his last" he was articulating an increasingly pervasive view that mere instrument-makers should not aspire to be recognized as philosophers (as quoted in Morus 1998, 68). The poisonous, long-running dispute between the astronomer James South and his erstwhile telescope-maker Edward Troughton is another good, if extreme, example of the ways in which powerful new scientific constituencies were seeking to marginalize instrument-makers' claims to be recognized as knowledge-makers. After legal proceedings forced South to pay Troughton for the unsatisfactory telescope mountings Troughton had made for him, South expressed his contempt with a final calculated insult, breaking up the instrument (which had cost him well over a thousand pounds) and selling it for scrap by public auction. In 1839 he advertised a "quantity of Mahogany, other Wood, and Iron, being the Polar Axis of the Great Equatorial Instrument made for the Kensington Observatory, by Messrs. Troughton and Simms." The rest was sold off a few years later, the posters for this sale fulminating about the "Mechanical Incapacity of English Astronomical Instrument Makers of the present day." It demonstrates how the ownership of particular kinds of practical hands-on skill was being displaced as a guarantor of philosophical credentials in favor of more abstract notions of expertise (as quoted in Hoskin 1989, 198).

In many ways, the dispute was symptomatic of the emergence of new notions of how scientific knowledge was constituted that celebrated abstraction. John Herschel's *Preliminary Discourse on the Study of Natural Philosophy*, published in 1830, advocated just this sort of view. In the *Preliminary Discourse*, Herschel tried to set out a systematic account of what was—and was not—distinctively scientific. One distinction that Herschel was particularly keen to draw was between science and craft. True natural philosophy was open and transparent, craft was closed and secretive. Accessibility was a defining feature of what it meant to be scientific. Proper natural philosophy was "divested, as far as possible, of artificial difficulties, and stripped of all such technicalities as tend to place it in the light of a craft and a mystery, inaccessible without a kind of apprenticeship." Craftsmanship on the other hand had a tendency to "bury itself in technicalities, and to place pride in particular short cuts and mysteries known only to adepts; to surprise and astonish by results, but conceal processes" (Herschel 1830, 71–2). Instrument-making might be an essential part of making scientific knowledge, but instrument-makers themselves could not be knowledge-makers, in other words. The view of how knowledge was properly made and by whom that became dominant during the Victorian period did not accommodate easily the view that knowledge was embodied in its instruments.

By the end of the nineteenth century, scientific instrument-making was an important and significant industry. The rise of new industries—and the telegraph and electric power industries in particular—led to the development of new markets for precision instrumentation that might previously only have been needed by relatively small numbers of users. Scientific instrument-makers developed close links with new university departments and laboratories on the one hand, and the new technical industries on the other. The Cambridge Scientific Instrument Company, founded by Charles Darwin's son Horace, offers an instructive example. Horace Darwin graduated in mathematics from Trinity College Cambridge in 1874 before entering into an engineering apprenticeship with Easton and Anderson. In 1878, shortly after completing his apprenticeship he established the Cambridge Scientific Instrument company along with Albert George Dew-Smith, who worked as a lens grinder at the university observatory. As a Cambridge graduate, Darwin had useful connections to key figures such as his near-contemporary at Trinity, Richard Glazebrook, soon to be the Cavendish Laboratory's demonstrator. "Horace Darwin's Shop," as the company was often known, soon became one of the Cavendish Laboratory's main suppliers of scientific instruments (Cattermole and Wolfe 1987). In Germany, the electrical engineering firm of Siemens and Halske fulfilled a similar role for the Physikalische Technische Reichsanstalt, which Werner von Siemens had played such a key role in establishing (Cahan 1989).

At the turn of the twentieth century, scientific instrument-making firms were sometimes crucial repositories of key skills that would be essential for establishing the new physics. Ernest Rutherford's researches into radioactivity at Manchester, for example, relied heavily on the delicate instruments provided by a local instrument-maker and glass-blower, Otto Baumbach. Baumbach had established himself as an instrument-maker in Manchester during the first few years of the century, working first as the university's glass-blower and then running his own instrument-making business, though on the university's premises and on the understanding that university requirements would be given priority. Producing the glass tubes for Rutherford's alpha particle experiments required a particularly delicate hand, since as Rutherford noted, they had to be "sufficiently thin to allow alpha particles from the emanation and its products to escape, but sufficiently strong to withstand atmospheric pressure." Baumbach's ability to blow "such fine tubes very uniform in thickness" was, as Rutherford acknowledged, essential to Rutherford's own success as an experimenter in this instance (Rutherford and Royds 1909, 283; Hughes 2008). Making the experimental apparatus that underpinned the laboratory regimes and revolutionary physics of the early twentieth century required new and complex interactions between laboratory managers, their co-workers, technicians and instrument-makers. Only relatively rarely, as in the case of Rutherford's acknowledgement of the key role Baumbach's instruments had played in the success of his experiments, was the existence of such invisible networks made visible.

Behind the Scenes

Much of the recent literature on the role of technicians and their comparative neglect in the history of science takes as its starting point a number of Steven Shapin's publications of the early 1990s, in which he drew attention to the sources of authority and trust in early modern experimental natural philosophy (Shapin 1994). Shapin was particularly interested in looking at the cultural origins of trust in natural philosophers and the ways in which authority was represented, arguing that early modern natural philosophers, such as the founding Fellows of the Royal Society, helped themselves to already existing norms of trust in disinterested gentility in their attempts to define their community. One reason that technicians were so often invisible, he noted, was that they were assumed to be lacking the necessary prerequisites for trust. Whilst gentlemen's independence underlined their trustworthiness, the technician's subservience meant that their capacity for truth-telling was compromised. It is striking how often, in early modern visual representations of scientific work for example, that work is seen to be performed by disembodied hands, rather than by a human laborer. (Figure 7.1) Removing the technician from the picture was a way of showing where scientific authority really resided. Following Shapin's lead, historians of science have taken the task of making invisible technicians visible again to be central to the business of disassembling scientific knowledge-making and displaying its collective nature.

The usual term employed during the early modern period for a person performing the sorts of activities that would now be carried out by a laboratory technician was "laborant" or, more rarely, "laborator." They were specific individuals employed and paid to carry out specific tasks associated with experiment. By and large, such individuals bore the character of servants, along with the cultural connotations that came with that characterization. As employees, they were extensions of their master, rather than autonomous agents, regardless of the amount of individually acquired skills they brought to their duties. They were not regarded as reliable witnesses to experimental observations, and it seems likely that they were required to sign a declaration that they would "not knowingly discover to any person whatsoever, whether directly or indirectly, any process, medicine, or other experiment, which he shall enjoin me to keep secret." Their names rarely appear in contemporary experimental accounts, and, as Shapin points out, where their names are known, as in the case of Denis Papin,

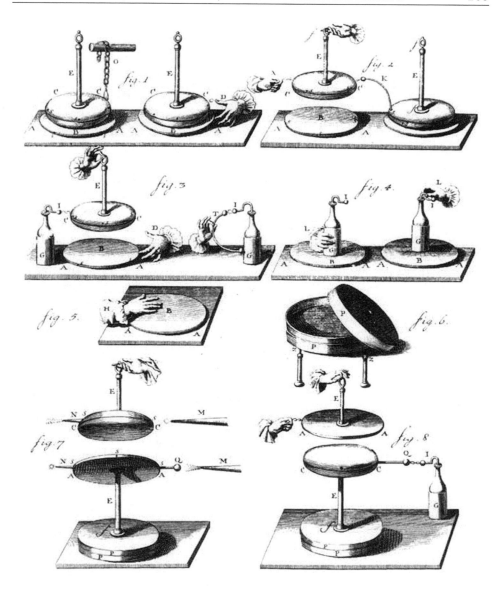

Figure 7.1 An illustration demonstrating the use of Volta's electrophorus. The disembodied hands represent the otherwise invisible technician. From Alessandro Volta. 1918–1929. *Le Opere,* 7 Volumes. Milan: Hoepli, volume 3, 101.

one of Robert Boyle's technicians, it seems likely that they occupied a more elevated and trusted role than that of most ordinary technicians, who remained entirely anonymous. In Papin's case, for example, it seems clear that Boyle allowed him significant freedom to experiment independently (Shapin 1994, 403).

Papin's background and previous associations with other natural philosophers suggest that he was not a typical technician too. He had a degree in medicine from the

University of Angers, and seems to have worked previously with both Christiaan Huygens and Gottfried Leibniz before Boyle employed him as an assistant. He also worked as an assistant to another experimental natural philosopher who had previously acted as Boyle's technician—Robert Hooke. Hooke worked for Robert Boyle at Oxford during the 1650s, when he assisted him with his early experiments on the spring of the air. In 1662 he was appointed the Royal Society's curator of experiments—a position that whilst it offered Hooke rather more status than that of an ordinary laborant, also confirmed his subservient status as an employee. As curator of experiments Hooke was there to carry out instructions given to him by the Fellows, rather than to act as an autonomous experimental investigator in his own right. Hooke's ambivalent position at the Royal Society underscores the status of early modern laboratory technicians. On the one hand, their experimental and instrumental skills were recognized and rewarded by their philosophical employers, on the other, they were servants to be instructed to carry out specific tasks (and complained about when they failed to fulfil them adequately) rather than to be themselves the authors of their experiments (Iliffe 1992).

Exactly the same kind of ambivalence is evident a century and a half later during the course of Michael Faraday's transition from being Humphry Davy's laboratory assistant at the Royal Institution to being himself a Fellow of the Royal Society and director of the Royal Institution's laboratory in his own right (Morus 1998). Faraday had completed his apprenticeship as a bookbinder before being employed by Davy as his assistant in 1813, when he replaced the previous assistant William Payne, who had been dismissed for brawling with the Royal Institution's instrument-maker John Newman. The list of Faraday's duties as laboratory assistant specified that he was expected to "attend and assist the lecturers and professors in preparing for and during lectures. Where any instruments or apparatus may be required, to attend to their careful removal from the model-room and laboratory to the lecture-room, and to clean and replace them after being used, reporting to the managers such accidents as shall require repair, a constant diary being kept by him for that purpose" (as quoted in Morus 1998, 17). It is also clear, however, both that Faraday had ambitions beyond such menial occupations and that Davy was willing to indulge those ambitions. Nevertheless, there was an ambiguity about Faraday's status that made it possible for Davy to ask him to act as his valet during their continental tour, and that allowed Davy's wife to treat him as a servant—much to Faraday's indignation.

More generally, it does appear that there were circumstances in which laboratory technicians during the nineteenth century could aspire towards an autonomous identity as men of science. William Robert Grove's assistant at the London Institution was George Thomas Fisher, for example. Fisher had been a student at King's College London with ambitions to enter the medical profession. As Grove's assistant he was expected to maintain the laboratory and its apparatus as well as keep a laboratory diary. During his tenure there, Fisher also published a number of books, including *Photogenic Manipulations* (1843), *A Practical Treatise on Medical Electricity* (1845) and *Microscopic Manipulations* (1846) as well as a handful of articles in the *Westminster Review*. He seems to have been trying to make a name for himself as an author of popular scientific texts (Morus 1998). William Barrett, one of John Tyndall's laboratory assistants at the Royal Institution went on to carve out a considerable reputation for himself as a physicist and psychical researcher. Barrett worked with Tyndall on his experiments on sensitive flames at the Royal Institution. In fact it was Barrett who had first noticed the phenomenon. As he described it:

> while preparing the experiments for one of the Christmas lectures at the Royal Institution, I noticed that the higher harmonics of a brass plate (which I was sounding with a violin-bow in order to obtain Chladni's figures) had a remarkable effect on a tall and slender gas-flame that happened to be burning near. At the sound of any shrill note the flame shrank down several inches, at the same time spreading out sideways into a flat flame, which gave an increased amount of light from the more perfect combustion of the gas. (Barrett 1867, 216–17)

It seems clear that there were contexts in which nineteenth-century laboratory technicians could represent themselves as autonomous experimenters, or that work as a technician could in some circumstances lead to such autonomy. Laboratory assistants at places such as the Royal Polytechnic Institution, men like Thomas Tobin or Thomas Hepworth, appear to have regarded themselves as men of science in their own right. Hepworth, for one, certainly took part in scientific lecture tours of the provinces during his time at the Polytechnic (Weedon 2008). Whilst technicians were only rarely named as collaborators in published accounts of experiments there were other opportunities for claiming status as a man of science. Benjamin Davies, Oliver Lodge's laboratory assistant, wrote scientific articles for a Welsh-language Liverpool newspaper, for example. In fact some of Lodge's publications during the 1890s did acknowledge that he had been "assisted by Mr. Benjamin Davies." Davies went on to work for the Eastern Telegraph Company at Lodge's recommendation, eventually becoming head of the company's research departments and filing a number of patents under his and the company's name (Rowlands 1990). There were clearly ways around invisibility. What this suggests, though, is that the epistemological status of technicians (just like the epistemological status of science and scientists for that matter) was never stable. It shifted with context and according to the various constituencies to whom claims for authority were being made.

One group of scientific assistants whose epistemological status remained particularly murky throughout the early modern and Victorian periods was women. Early modern commentators supposed that there were good reasons why women should not be considered as reliable knowledge-makers. They lacked the manly independence of the gentleman scholar. It is clear nevertheless that women did sometimes act as assistants for male practitioners—often their fathers, brothers, or husbands. Marie-Anne Paulze often acted as her husband Antoine Lavoisier's laboratory assistant, for example. It is clear from surviving notebooks that she actively participated in laboratory experiments, kept notes and helped with managing the laboratory, as well as translating scientific works into French for her husband. Caroline Herschel, William Herschel's sister and John Herschel's aunt, similarly assisted actively in her brother's astronomical work and helped make some of his telescopes. Following her brother's death in 1822 she participated in her nephew's work on cataloguing nebulae. Caroline Herschel did in fact gain some visibility for her astronomical activities, being awarded the Gold Medal of the Royal Astronomical Society in 1828. In that respect, at least, she was unusual. Most women scientific workers remained invisible—at least in public. Even at the end of the nineteenth century as women started to enter scientific laboratories in their own right as knowledge-makers, the role of female family members in scientific work remained largely hidden.

Assembly-line Knowledge

As Shapin points out, his decision to highlight the contemporary and historical invisibility of technicians in particular was in part strategic and pragmatic: "the network of support personnel about which one could (and in principle should) speak is potentially infinite" (Shapin 1994, 368). Modern science does not just depend on those who make, operate, and maintain its instruments. Its routines require the participation of regiments of workers carrying out a wide range of mundane and often tedious operations. Experiments require routine observation as well as active manipulation. Much of the work of calculation was also often routine and repetitive until human calculators were replaced by mechanical, and then by electronic varieties. Outside laboratories and observatories, field sciences like geology, botany, or zoology depend on legions of local informants, specimen hunters, and support personnel of all kinds in order to operate successfully. The huge industrial-scientific institutions of big science that grew up during and after the Second World War depend for their existence on complex networks and infrastructures. The networks that sustain modern science extend well beyond the kinds of spaces and people that we would conventionally recognize as being scientific. Even within spaces more conventionally understood as scientific, much of the labor that underpins practice is often invisible (Agar 2012).

When George Bidell Airy took up his appointment as Astronomer Royal in 1835, and the management of the Royal Observatory at Greenwich which came with that position, he instituted a new style of observational working that owed a great deal to Adam Smith's discussions of the division of labor. Airy instituted a strict hierarchy and a division of labor in which routine and repetitive calculations were carried out by cadres of clerks before being passed on for checking, just as would be done in an accountant's office. Greenwich under Airy was an astronomical factory. Much of what went on there depended on the activities of invisible laborers who were rarely, if ever, credited as scientific workers (Smith 1991). Similarly, the kinds of large-scale astronomical expeditions that took place throughout the century, such as the magnetic crusades spearheaded by Edward Sabine, depended on the deployment of significant amounts of largely invisible labor. Apart from the shipboard labor essential to getting the expeditions to where they wanted to be, many of the routine observations were carried out by naval officers, for example. Expeditions later in the century such as those to observe the transit of Venus, were just as dependent on the contributions of a variety of trained personnel to carry out a range of mundane and sometimes highly skilled tasks (Cawood 1979).

The activities of botanists, geologists, natural historians and other practitioners of field sciences was (and is) similarly underpinned by invisible labor and local knowledge. Early Victorian geologists needed the services of local guides who could show them how to find their way around the terrain. Fossil hunters like Mary Anning were a vital and usually unacknowledged source of specimens for gentlemen of science. Anning is herself a good example of how class and gender sometimes determined whether contributions to knowledge were admitted or not by elite practitioners (Torrens 1995). Some practitioners who eventually became significant figures in their own right, such as Alfred Russel Wallace, started their scientific careers in this kind of capacity. Before his famous letter to Darwin in 1858 cast him into the scientific limelight, he eked out a living for himself as a scientific collector. Between 1848 and 1862 Wallace spent most of his time abroad, first in the Amazon basin and then in the Malay Archipelago, collecting specimens to be sold in London. He was one of several such individuals whose activities were essential in providing the raw materials for gentlemanly theorizing back in the metropolis (Raby 2001). Local informants were an important resource for the collecting activities of these explorers as well. They depended on people who could lead them to where likely specimens could be found and to help transport them back to their markets, for example. Race, as well as class and gender, made these individuals' activities even more invisible than those of the collectors they worked for. Nineteenth-century British naturalists in China, for example, depended for information and specimens on the army officers, civil servants, and missionaries who spread inland from the British enclaves—and in turn they depended on the cooperation and knowledge of local informants (Fan 2004).

Twentieth-century big science depended on the marshalling of huge armies of workers carrying out a range of routinized tasks. But even before the Manhattan project inaugurated this way of doing science on a massive scale much scientific activity was already carried out in routine fashion by workers skilled at specific tasks. Thomas Alva Edison at his Menlo Park laboratory employed dozens of assistants to carry out routine experiments such as testing. A substantial office staff was needed to keep records of the laboratory's experimental activities. This kind of routinized and often repetitive approach to intensive experiment depended on the subdivision of tasks and on bureaucratic oversight to collate the experimental output. Edison's reputation as genius inventor depended on being able to make himself the sole author of the intensive and collaborative work that went on in the laboratory (Israel 2000). Early experiments in radioactivity similarly depended on difficult and repetitive observation carried out of cadres of observers trained for a specific task. The physicist Ernest Rutherford's experiments in radioactivity at Manchester between 1907 and 1919 needed the services of numbers of assistants (often female) whose task it was to count the number of flashes on a scintillation screen. This was difficult and tedious work that required sustained concentration so that observers worked in shifts to ensure that they continued to work with the greatest efficiency and accuracy (Hughes 2008).

The Manhattan project to develop and manufacture an atomic bomb depended on the massed labor of hitherto unprecedented numbers of workers. The Oak Ridge installation where large cyclotrons were used to separate out the fissile material from uranium employed about 22,000 operators to run the machinery. Another equally massive and labor-intensive site was developed at Hanford to purify plutonium. The Los Alamos laboratory where the bomb itself would be designed started its operations on a comparatively small scale with about a hundred workers. The number of personnel employed there then doubled in size approximately every nine months for the remainder of the war (Hughes 2002). This was scientific activity on a gargantuan scale and was crucially dependent not just on scientists and trained technicians but on whole bodies of workers needed to keep the whole operation ticking over. The large-scale scientific projects that emerged during the Cold War and its aftermath (primarily in the physical sciences but more recently in the life sciences as well) also depended on the deployment of a whole range of workers carrying out an array of tasks without which the more obviously scientific work of institutions such as CERN, Fermilab or NASA could not take place. This was science on an industrial scale and that was managed and overseen in much the same way that a factory might be managed—just as Airy had modelled the Greenwich Observatory on early Victorian factories (Agar 2012).

Conclusions

It is not a historian's task to judge how demarcations between scientific and non-scientific work should be made. It is a historian's task to enquire into how and why such discriminations are historically constructed. Defending the heroes of the Scientific Revolution, Alexandre Koyré insisted that "science is not made by engineers or craftsmen, but by men who seldom built or made anything more real than a theory. The new ballistics was made not by artificers or gunners but against them" (Koyré 1968, 17). Representing the opposite camp, Edgar Zilsel argued that "beneath both the university scholars and the humanistic literati the artisans, the mariners, shipbuilders, carpenters, foundrymen, and miners worked in silence on the advance of technology and modern society" (Zilsel 1942, 551) They were the "the real pioneers of empirical observation, experimentation, and causal research." Both perspectives are polemic. Their authors wanted to locate the origins of modern science not just in a specific period, but in a specific group of people. One of the reasons that Boris Hessen's notorious essay on "The social and economic roots of Newton's *Principia*" aroused such opprobrium was because of its argument that the origins of modern science lay in collective labor rather than individual genius (Hessen 1931). Fundamentally, these were arguments about just where the boundaries around scientific knowledge and its origins should be drawn.

Artisans, instrument-makers and technicians have tended to be invisible actors in the production of knowledge because, generally speaking, our modern culture tends not to credit the sorts of activities in which they engage as knowledge-making. There is ample evidence that many such individuals themselves saw things differently. Nineteenth-century artisan autodidacts developed their own ways of acquiring and engaging with scientific knowledge, for example, outside the prevailing gentlemanly culture of science. They carved out spaces where they could engage with scientific activities on their own terms and construct alternative representations of scientific authority. One way of understanding organizations such as the short-lived Electrical Society of London, founded by William Sturgeon in 1837, is as an attempt by instrument-makers and artisans to articulate an alternative account of scientific authority that did give credit to instrumental knowledge. The relative invisibility of such instrumental knowledge is a historical construct. There have been times, as we have seen, when claims to scientific authority based on skilled practice have carried more traction than they have had at other historical junctures. Such claims are often (always arguably) politically loaded too, carrying as they do implications about the social distribution of knowledge.

Paying close attention to the activities of artisans, instrument-makers and technicians in the making of scientific knowledge offers historians ways, therefore, of looking more closely at how scientific authority is constructed and how. Two aspects are worth emphasizing in conclusion. Looking at invisible technicians offers a strategy for foregrounding the collective nature of scientific knowledge. Since the late eighteenth century at least, we have tended to ascribe knowledge to individuals—and very particular kinds of individuals at that. Science is generated by genius. Going behind the individual focus provides a way of re-emphasizing the role of the collective instead. It gives historians an opportunity to demonstrate that the notion of science as the product of individual genius is itself a construct. As well as collectivity, putting technicians back into the picture gives us a way of displaying the tacit dimension of science. One way of understanding Herschel's remarks about the attributes of proper science being openness and transparency is as a plea to banish tacit knowledge (and its owners) from claims to authority. Historians of science following the sociological turn of the 1970s know that such an attempt is hopeless. Tacit knowledge is integral to science, and unless we pay due attention to the bodies in which that knowledge resides we will end up leaving the third dimension out of our historical images of science in the making.

References

Agar, Jon. 2012. *Science in the 20th Century and Beyond*. Cambridge: Polity Press.

Applebaum, Herbert. 1992. *The Concept of Work*. Albany: State University of New York Press.

Barrett, William. 1867. "Note on sensitive flames." *Philosophical Magazine*, 33: 216–22.

Cahan, David. 1989. *An Institute for an Empire: The Physikalisch-Technische Reichsanstalt 1871–1918*. Cambridge: Cambridge University Press.

Cattermole, M. J. G., and A. F. Wolfe. 1987. *Horace Darwin's Shop: A History of the Cambridge Scientific Instrument Company*. London: Adam Hilger.

Cawood, John. 1979. "The Magnetic Crusade: Science and politics in early Victorian Britain." *Isis*, 70: 492–518.

Daumas, Maurice. 1972. *Scientific Instruments of the 17th & 18th Centuries and their Makers*. London: Portman Books.

Fan, Fa-ti. 2004. *British Scientists in Qing China: Science, Empire and Cultural Encounter*. Cambridge, MA: Harvard University Press.

Herschel, John. 1831. *Preliminary Discourse on the Study of Natural Philosophy*. London: Longman, Reese, Orme, Browne, Greene and Taylor.

Hessen, Boris. 1931. "The social and economic roots of Newton's *Principia*." In *Science at the Crossroads*, edited by Nikolai Bukharin, 149–212. London: Kniga Ltd.

Hoskin, Michael. 1989. "Astronomers at war: South vs. Sheepshanks." *Journal for the History of Astronomy*, 20: 175–212.

Hughes, Jeff. 2002. *The Manhattan Project: Big Science and the Atom Bomb*. London: Icon Books.

Hughes, Jeff. 2008. "William Kay, Samuel Devons and memories of practice in Rutherford's Manchester Laboratory." *Notes & Records of the Royal Society*, 62: 97–121.

Iliffe, Rob. 1992. "In the warehouse: Privacy, property and priority in the early Royal Society." *History of Science*, 30: 29–68.

Israel, Paul. 1998. *Edison: A Life of Invention*. New York: John Wiley.

Koyré, Alexandre. 1968. *Metaphysics and Measurement*. London: Gordon Breach.

Morus, Iwan Rhys. 1998. *Frankenstein's Children: Electricity, Exhibition and Experiment in early Nineteenth-century London*. Princeton, NJ: Princeton University Press.

Raby, Peter. 2001. *Alfred Russel Wallace: A Life*. Princeton, NJ: Princeton University Press.

Rowlands, Peter. 1990. *Oliver Lodge and the Liverpool Physical Society*. Liverpool: Liverpool University Press.

Rule, John. 1989. "The property of skill in the period of manufacture." In *The Historical Meanings of Work*, edited by Patrick Joyce, 99–118. Cambridge: Cambridge University Press.

Rutherford, Ernest, and Thomas Royds. 1909. "The nature of the alpha particle from radioactive substances." *Philosophical Magazine*, 17: 281–6.

Schaffer, Simon 1994. "Babbage's intelligence: Calculating engines and the factory system." *Critical Inquiry*, 21: 203–27.

Shapin, Steven. 1989. "The invisible technician." *American Scientist*, 77: 554–63.

Shapin, Steven. 1994. *A Social History of Truth*. Chicago: University of Chicago Press.

Smith, Robert. 1991. "A national observatory transformed: Greenwich in the nineteenth century." *Journal for the History of Astronomy*, 22: 5–20.

Sorrenson, Richard. 2013. *Perfect Mechanics: Instrument Makers at the Royal Society of London in the Eighteenth Century*. Boston: Docent Press.

Torrens, Hugh. 1995. "Mary Anning of Lyme: The greatest fossilist the world ever knew." *British Journal for the History of Science*, 28: 257–84.

Weedon, Brenda. 2008. *The Education of the Eye: A History of the Royal Polytechnic Institution 1864–1881*. Cambridge: Granta Editions.

Zilsel, Edgar. 1942. "The sociological roots of science." *American Journal of Sociology*, 47: 544–62.

Chapter Eight

Scientific Illustrators

Valérie Chansigaud[1]

"Illustration" is defined as "a picture, drawing, or photograph used for decorating a book or explaining something" (*Macmillan Dictionary* 2014). "Illustrator" is accordingly defined as "a person who draws or creates pictures for magazines, books advertising etc." (*Oxford Dictionary of English*), thus a person who constructs pictures for printed documents, whatever technique is used: drawing, engraving or photography. That dual definition is useful since it places the illustrator in the context of the line of production aimed at the creation of a printed document. The definition, however, remains theoretical, since it does not consider the diversity of the roles played by the science illustrator, or changes in those roles due to the evolution of the techniques used to reproduce illustrations. Indeed, the creation of a scientific illustration requires several intervening persons, from the "expert" who conceives the picture to the people who practically construct it, so that it can be printed. The conventional definition of a scientific illustrator meets thus with two difficulties: first, the role and the function of the various participants in the creation of an image vary with time, disciplines, and editorial projects; second, it is difficult to evaluate the artistic, technical, and intellectual contributions of the illustrator, since his or her role could as well be that of a collaborator to the scientist as that of a mere technical assistant.

Debates between Darwin and Illustrators of his Books

Some of the reasons why the position of an illustrator is difficult to define are well shown by the debates around the illustrations in Charles Darwin's book *The Expression of the Emotions in Man and Animals*. The book was published in 1872; it contained 21 woodcuts and 8 photos reproduced by heliotypy, a newly introduced procedure. Darwin began the assembly of graphic elements for his book in 1871; their origins are diverse and they each deserve a different graphic approach. Darwin "mentioned to Mr. Bartlett (Abraham Dee Bartlett (1812–1897), superintendent of London Zoological Garden) his wish to have some work done at The Gardens which required unusual care" (Palmer 1895, 192). Bartlett suggested Joseph Wolf (1820–1899), one

A Companion to the History of Science, First Edition. Edited by Bernard Lightman.
© 2016 John Wiley & Sons Ltd. Published 2020 by John Wiley & Sons Ltd.

of the most prominent scientific illustrators of the nineteenth century, who worked for the Zoological Society of London, and whose capabilities he highly praised: "Mr. Wolf has got an eye like photographic paper, it will seize on anything!" (Palmer 1895, 194). Darwin wished to reproduce an observation by Clarke Abel (1780–1824) cited by William Charles Linnaeus Martin (Martin 1841, 405). The observation concerned the reaction of monkeys faced with a living tortoise. Darwin asked to have a tortoise placed in the cage of several monkeys kept at the London Zoo, so that the expressions shown on the faces of the monkeys could be observed. According to Darwin, monkeys would express fear, astonishment, pleasure or even laugh. Wolf attended the experiments and, at the request of Darwin, drew sketches of the facial expression of the Celebes crested macaque (*Macaca nigra*), a difficult exercise because of the expressiveness of those monkeys. Darwin was satisfied with the drawings but asked, at the moment of engraving, for changes in the shape of the monkeys' ears.

Most of the illustrations in the book were made from life, although some, also signed by Wolf, are pure inventive creations. This is the case for a set of drawings that Darwin asked Wolf to do to illustrate affection or fear in cats and dogs. Darwin was not satisfied with the result, regretting, for example, that the dogs had a collar, or that the threatening dog had bristling fur. Darwin preferred to approach another illustrator, Briton Rivière (1840–1920), but even in that case, archives show frequent exchanges between Darwin and Rivière (Prodger 2009, 151–3). The final illustration of the affectionate dog, rather different from Rivière's original drawings, is a kind of combination of Darwin's requirements and the illustrators' remarks, particularly concerning the attitude of the dog. In a different sketch, Wolf chose to show an affectionate cat rubbing against a chair. Darwin did not use the drawing as supplied; he asked Thomas William Wood (fl. 1855–1872) to alter the illustration rather significantly: the chair was replaced by the leg of a man, in order to strengthen the impression of affection. All of these examples show the difficulties faced by illustrators who were asked to respond to the requests of the author while trying to impose their own expertise. Some illustrators, even though they finally complied with author's requests, expressed their disagreement. Joseph Wolf, who had a wide knowledge of animals due to years of observation (Schulze-Hagen and Geus 2000), did not share the opinion of Darwin on the point that monkeys could laugh: "I never believed that that fellow Monkey was laughing, although Darwin said he was. I am not one of those who place absolute belief in all 'authority' " (as quoted in Palmer 1895, 195–6).

The photos published in *The Expression of the Emotions in Man and Animals* give a different example of the relation between illustrators and scientists. This was not a case of Darwin imagining a picture to be made by others; rather, he selected his illustrations from a collection of already existing pictures. His approach was similar to the present-day use of a picture database. Just at the time he was preparing *The Expression of the Emotions* for publication, Darwin met Oscar Gustave Rejlander (1813–1875), a photographer of Swedish origin, famous for his photographs of the expression of feelings by human beings (fear, indignation, surprise etc.) and for his montages (Prodger 2009). Darwin was interested in Rejlander's work because it was very close to that of a painter or a draughtsman: he actually constructed his photographs (the subjects posed according to the desired effect, or he combined several photographs in a montage). The result was not an exact reproduction of any reality (always assuming that photography can approach reality (Freund 1974)). It is worth noting that Darwin explicitly asked for the photographs to be reproduced using the new technique of heliotypy (a lithographic stone is directly prepared from a negative, thus obviating the need for hand engraving), a request which caused problems for the publisher and the printer, both in keeping to schedules and with achieving a consistent quality of print.

Thus, a published illustration is the final product of the intellectual dialogue between an author and "his" illustrator, whose respective roles are often difficult to separate. Faced with such intertwined tracks it seems necessary to simplify the study of "the illustrator" by using a chronological approach, corresponding to the sequential introduction of new techniques in the reproduction of images. We consider here only natural sciences (biology, chemistry, physics, astronomy and earth sciences), and exclude medicine.

Early Stages

The first printed scientific books showed a great diversity both in the scientific project and in the technical making of illustrations. Some of the latter were very schematic, as in *Herbarius*, published in Mainz around 1484, which showed plants with a perfect bilateral symmetry, a feature rarely found in nature. Images were for the most part copied from earlier images rather than from life, a process resulting in a loss of accuracy in detail and in shape of leaves, flowers etc. Two-thirds of the prints of *Hortus sanitatis*, published in 1491, are copied from a plant collection published in 1485 in Mainz: defects are imputable to the work of the copyist-illustrator who failed to respect the intention of the original artist (Arber 1912). However, during this first period in the history of scientific illustration, one can notice an improvement in technique of illustration. In *Das Buch der Natur* (1475) woodcuts have been carved for that very book, not copied from preexisting models. For the first time, images were not intended merely to decorate the work, but to accompany and explicate the text (Arber 1912).

A genuine revolution in the design of images took place in 1530 with the publication of *Herbarum vivae eicones* by Otto Brunfels (1488–1534). The illustrations are of a quality surpassing that of all earlier images. It is believed that Hans Weiditz (c. 1500–c. 1536), a disciple of Dürer, made the woodcuts. They obviously are based on observations from nature, as indicated in the title ("living images of plants"): the entire plant is shown and the drawing has a magnificent precision. It is worth noting that the printer/publisher, Johann Schott of Strasbourg, rather than the author, hired the illustrator and supervised the work himself. The book was a success and it was re-published many times with increases in the number of species described, from 135 to 260.

The Illustrator as an Interpreter of Nature

In 1542, Leonard Fuchs (1501–1566) published *De Historia Stirpium Commentarii Insignes* in Basel, a landmark in natural history iconography: the illustrator is clearly the interpreter of a scientific discourse and expresses himself as an important player in the chain of scientific communication.

The difference between the illustrations in the books of Brunfels and Fuchs lies not in the technical quality of the woodcuts (they show similar artistic command), but in the manner that plants are depicted. In Fuchs, the idea is not to describe individual specimens with associated anomalies (such as leaves eaten by insects. or broken stalks) but to display an archetypical plant. The illustrator constructs a chimeric scientific image derived from the features of several individuals, combined to produce a perfect example of the species. This new kind of image took its place in a larger scientific project aimed at defining a taxonomy of the vegetable kingdom. The illustrator was thus in the position of an interpreter, translating reality as it is seen into a scientific description, visually transcribing the content of textual scientific analysis— often more efficiently than the author. This example can be defined as the moment scientific illustration became determined by spoken or written scientific discourse: it therefore lost its proper and specific meaning. This explains why the work of illustrators was thereafter placed under such tight control. Fuchs wrote: "Insofar as concerns the illustrations themselves... we have taken particular care that they should be most perfect... and not allowed the craftsmen to indulge their fancy in such a way that the drawing does not correspond to the truth" (as quoted by Marcus 1944, 379). Fuchs also makes it clear that the illustrator should not be allowed to practice as an artist: "We have purposely and deliberately avoided the hiding of the natural forms of the plants by shadows and unnecessary things by which artists sometimes wish to win praise" (as quoted by Marcus 1944, 380–81).

De Historia Stirpium Commentarii Insignes features a rare event in the history of scientific illustration: not only did Fuchs name the illustrators, but their portraits appeared in the book. They are Heinrich Fullmaurer, who painted the plants from life, Albrecht Meyer, who translated the paintings into a drawing on wood, and Veit Rudolf Speckle, who made the engraving. This indeed constituted a genuine recognition of illustrators and their role. Moreover, the succession of events in the production of an illustrated book was clearly defined. It can be summarized as: the author supervises the creation of the illustration (a work sometimes devolved to the printer/publisher), the painter produces an original image, most often colored, and an engraver translates that image on to a printable support, wood, metal or stone, most often for a black and white printing. This scheme of production process remained more or less unchanged until the nineteenth century, before techniques of photoengraving replaced the engraver and photography replaced painting.

From the sixteenth century, the names of the individuals involved were often given at the bottom of an illustration, followed by abbreviations describing their contribution, such as: the painter (*pinx., pinxt., pinxit...*), the draughtsman (*del., delt., delin...*), the engraver (*sc., sculp., sculpt...*) without any prejudice to the engraving technique used (with some exceptions, such as *lith., litho., lithog...* for a lithoengraver). The abbreviation *f., fec., fectit...* indicated the person who prepared the engraving and perhaps the printer as well, whereas *imp., impressit,* were for the printer only. A combination of abbreviations, such as *del.et lith.,* meant that the same person made the original drawing and transcribed it into lithography.

Some images were not signed—for no simple or unique reason. Anonymity was the usual case for diagrams, graphs, and descriptions of apparatus: the illustrator was considered as merely carrying out orders, reproducing the model prepared by the scientist; the expertise of the illustrator added nothing to the value of the image. Since the Renaissance, the role of the illustrator has thus oscillated between that of a simple, anonymous technician to that of a recognized expert playing a crucial, creative role in the production of an image, genuinely participating in the scientific discourse.

The System of Conventions in the Illustrator's Profession

From the Renaissance up to the present day, many books claim that illustrations have been made "from life." Fuchs is an example of "from life" images not strictly reflecting reality but rather some ideal representation connected to a scientific definition or project. The expression "from life" is an assertion of the original and accurate feature of an illustration, something which is not a mere copy or a compilation of previously published images. It is to be understood that the illustrator has respected a particular idea—a convention—of the image (for example, chimeric, archetypal illustrations of plants in Fuchs) and he or she did not seek accurately to reproduce the reality of a unique, isolated, sample.

In this context, the relationship between the scientist and the illustrator can be tracked through the codes and unwritten conventions which apply to illustrations, some of them guided by the scientific demands specific to each discipline. Natural history offers several significant examples of such implicit rules that govern the illustrator's work.

Images of plants are a good example. From the eighteenth century onwards, the illustrations of botany books detailed the anatomy of the flower because it is the key element in the determination of species. This leads to illustrated books in which each flower is dissected and each elementary floral component drawn separately, as after a necropsy, such as in *Illustratio systematis sexualis Linnaeani* (1789) by John Miller and Friedrich Wilhelm Weiss. By contrast, the plates of books relevant to pharmacy showed the details of plants with particular attention paid to the parts used in the preparation of drugs (roots and bark principally)—details usually ignored in illustrations in strictly botanical books. The *Herbier de la France* (1780) by Pierre Bulliard, and the *Medicinisch-pharmaceutische botanik* (1841) by Karl Stupper, offer striking examples of these codes. Finally, illustrations of horticulture books did not pay much attention to anatomical details, instead focusing on the terminal, flowered, parts of plants. This convention is particularly visible in the first books on garden flowers, such as *Paradisi in sole paradisus terrestris* (1629) by John Parkinson. These three types of images of plants did not differ in their artistic quality, in their technical production, nor even in the species illustrated (the same plant can appear in each type of book). They primarily differed in the author's intended field of science (botany, pharmacy or horticulture), and the illustrator had to adapt his or her expertise to the type of illustration that was demanded.

Another example of a convention operating during the Renaissance period is the presentation of plants and animals in botany and zoology books as isolated individuals, without any information on their environment, or, if so, in merely anecdotal form. The background of the plates in Buffon's *Histoire naturelle* are oversimplified and, at best, merely hint at geographical details, or a particular trait or habit of an animal, or even evoke legends about it. At the opposite extreme, many illustrations of geography books (from the seventeenth century) or books on hunting (from the eighteenth century) placed the animals in more or less natural, but often imaginary, environments. Attempts to illustrate the natural environment of a plant or animal first appeared in zoology books in the middle of the nineteenth century: this evolution in the figuration of animals allowed some illustrators to become genuine experts since the accurate rendition of the environment of a species depended on the artist's knowledge of natural history.

A final example of rules affecting representations concerns the manner animals and plants were dealt with: in general, illustrators had at their disposal dead samples kept in collections (preserved animals or their skin, dried plants in herbaria etc.). The drawing of minerals, fossils, or shells obeyed classical rules of representation. By contrast, the drawing of mammals, birds and plants implied the re-creation of the latter in a living form: this process often resulted in mistakes (rigidity and lack of realism in the attitude of the animals, errors in the orientation of the flowers of plants or of the legs of invertebrates). The illustrator, therefore, needed to possess an extended knowledge in order to prepare images that respected a reality of which they did not always have direct experience. The point is that they needed not only mastery of graphical techniques, but also knowledge of the anatomy and behavior of the organisms to be drawn. This explains why illustrators often specialized in a discipline, and sometimes in a subdiscipline, much in the same way as scientists themselves.

Improvement in the quality of the depiction of animals and plants, along with the requirement that images appear as natural as possible, is noticeable from the beginning of the nineteenth century, and was directly associated with the increasing number of zoological gardens (Kisling 2000) and botanical gardens, places where illustrators could observe living specimens. The pressure for a better observation of nature explains the participation of illustrators in numerous scientific expeditions. They could make sketches of plants and animals when still alive, or at least, soon after they were collected or caught (for example, colors of fish fade rapidly after they are taken out of the water). The collection of specimens for the production of illustrations on return was no longer acceptable (Stafford 1984). The illustration was considered a scientific tool, as important as the collection of specimens. Moreover, the presence of illustrators in expeditions allowed the preservation of the memory of the explored landscapes, before the use of photography had spread (Figure 8.1).

The Illustrator and the Economy of Scientific Books

Illustration has always been a major factor in the cost of a publication. A study of the profits and losses of the archives of Plantin-Moretus, an Antwerp publisher, shows that during the seventeenth century the cost of illustration could amount up to 75 per cent of the total cost of a publication (Voet 1863–1972). The publishers did their best to minimize that cost, often in such a way that the initial link between the illustrator and the scientist was weakened or broken. For example, previously published engravings were re-used, or engravings owned by other publishers were acquired: in 1659 Plantin-Moretus acquired about 3180 engraved blocks of plants from the publisher Pieter van Waesberghe. An inventory of the Plantin-Moretus assets showed that drawings *per se* did not have a significant value, but that engraved metal and wood plates did, particularly since original drawings were often destroyed during the process of engraving (they were used as a tracing guide). Plantin-Moretus therefore assembled a base of images dissociated from their original association with a text. Gradually,

Figure 8.1 The illustrator fully participates in scientific work, much in the same way as other technicians such as collectors of specimens, laboratory technicians, taxidermists, and so on. In the present case the illustrator's role is to draw animals and freshly collected plants, particularly if their shape or their colors do not easily persist. Sophie Wörishöffer. 1888. *Das Naturforscherschiff, oder fahrt der jungen Hamburger mit der Hammonia nach den besitzungen ihres vaters in der Südsee*. Bielefeld: Velhagen & Klasing, frontispiece by H. Merle (?).

original images came to be routinely preserved, particularly because collectors sought them, which gave them a monetary value.

Copperplates, the use of which dominated the seventeenth century, followed later on by lithography, which appeared at the beginning of the nineteenth century, permitted sharper and more accurate engravings, but, in contrast with woodcutting, required separate printings of the image and the text, resulting in additional costs. An interesting consequence of this technical point was the development of a connoisseur market, distant from scientific interests, in which the illustrator took the top place, since the value of the book primarily rested on his work. From the eighteenth century onwards, a market specializing in splendid editions of scientific books developed, largely for collectors. Those books shared several characteristics: they were often in color, the plates being printed in black then painted using several different techniques (some of these books were published in two versions, black and color, as for the volumes on birds of the *Histoire naturelle* by Buffon); print runs were small, in general 200 to 350 copies; formats were often large (*The Birds of America* by John James Audubon reached the record dimension of 99 by 66 cm); and the book was published in parts and by subscription. Some scientists, illustrators, and entrepreneurs, such as John Gould (1804–1881) and Philip Lutley Sclater (1829–1913), specialized in producing books for collectors, and were the authors of remarkable books on birds and mammals. Another example was the three George Brettingham Sowerbys (I, II and III), who published well-illustrated books on shells.

Here, too, the place of the illustrator was variable: John Gould produced numerous paintings or sketches for his publications, but their conversion into lithographs was carried out by other artists, particularly his spouse Elizabeth Gould (1804–1841), the previously mentioned Joseph Wolf, and the poet and illustrator Edward Lear (1812–1888). Many monographs on naturalist iconography (Lambourne 2001 for instance) are restricted to the study of this kind of illustration—splendid indeed—but one should not ignore the vast field of black and white pictures, often of small size and inexpensive: these plain images circulated widely in educated society and contributed to the appearance of a scientific culture.

The Industry of Illustrated Scientific Books

The development of wood engraving by Thomas Bewick (1753–1828), artist and naturalist, was a milestone in the history of illustration. Wood engraving consisted in engraving the end grain of hard boxwood, thus perpendicular to the usual way of cutting wood: the work was more difficult and required the use of steel tools, but the engraved block was more resistant and permitted larger print runs. Moreover, since the part that receives ink is higher than on a metal or stone plate, it became possible to mix text and illustration on the same page. This innovation was a major cause of illustrations being granted much more space in general and scientific publications of the nineteenth century. Supply and demand for books and journals increased, inducing in turn the emergence of a genuine market for popular science journals, most often illustrated. Moreover, the beginning of the nineteenth century saw technical improvement in the production of paper and in printing methods enabling a decrease in the cost of printed matter and an even greater increase in the use of illustration.

The Penny Magazine illustrates this novel enthusiasm: between 1832 and 1845 the magazine published 7,000 pages and 2,900 illustrations; illustrations covered about 20 percent of the printed area and required a budget of £20,000 (Bennett 1984). The *Penny Magazine* met with an immense success with a distribution of 200,000 copies in 1832, a year after it had been launched (Austin 2010). Zoology was the second most popular topic, after travel. For scientific articles (excluding papers on geography and anthropology) the proportion of space devoted to illustrations was as high as 22 percent. The amount of space devoted to by zoology was not only due to the magazine's support for the diffusion of scientific knowledge, but also because images of animals were very attractive to the readers.

The editorial market for science which developed during the second half of the nineteenth century covered a wide range of productions with, at one end, deluxe books and at the other, less costly publications with much higher print-runs. The impressive sales of illustrated books resulted in an increase in the number of illustrators and in their specialization. While the illustrators of luxury books are well known, those employed by popular science editors are largely anonymous and poorly studied, partly because documents about them are rare and fragmentary (Houfe 1978). As for so many other jobs in the field of book production, the world of illustrators was almost exclusively masculine, women illustrators of science operating only within the context of luxury books or in certain highly specific scientific domains (pteridology, algology, and nature education). There is no simple way to investigate the professional status and wages of scientific illustrators of the nineteenth century due to the diversity of their status, ranging from that of general illustrators directly employed by scientific editors to that of scientists acting as illustrators, passing through the body of specialized and independent illustrators and, obviously, those employed by scientific institutions. The Muséum national d'histoire naturelle de Paris was most probably one of the first institutions to employ a full-time illustrator. A professorship for icononography, or the art of drawing and painting the features of nature, was established in 1793 for the painter Gérard van Spaendonck (1746–1822). The position was cancelled upon his death. The function of institutional illustrators was not only to illustrate publications but also to build reference collections of images: the Royal Horticultural Society recruited in 1896 a young artist called Nellie Roberts to paint prize-winning orchids and retain a record of cultivars and hybrids whose long-term growth is unstable (Elliott 2010).

The multiplication of publications gave certain illustrators the opportunity to escape anonymity, although that appears to have been far from being the rule. Pierre-Joseph Redouté (1759–1840), John James Audubon (1785–1851), and Joseph Wolf (1820–1899) were among those illustrators who acquired a certain fame. It is worth noting that a biography of Wolf was written when he was still alive, an exceptional event indeed (Palmer 1895). While Redouté and Audubon can be considered as businessmen, Wolf was attached to a scientific institution, the Zoological Society of London, where he was the celebrity illustrator (Schulze-Hagen and Geus 2000).

Many illustrators earned their living by simultaneously working on genuinely scientific books and on more popular ones, the frontier between the two being difficult to identify. That was the case for Joseph Wolf who, in addition to his plates for the Zoological Society of London and several scientific monographs, also illustrated travel reports by naturalists (*The Naturalist on the River Amazon* by Henry Walter Bates, 1863), folk tales (*Reynard the Fox* adapted by Goethe, 1855) and poems (*Poems* by Eliza Cook, 1861).

If many illustrators were highly specialized—Wolf only worked on zoological illustrations of great mammals and birds—others were less so. In that respect, Édouard Riou (1833–1900), who worked for Hachette publishers of Paris, created engravings for books on zoology (Figuier 1869), geology (Figuier 1863), geography (for the magazine *Le Tour du Monde*), travel (Bouyer 1867) as well as literary works (Verne 1863). Such versatility permitted the illustrator to obtain more work than scientific works alone could supply. For the same reasons, illustrators in astronomy created images for works of fiction, such as Chesley Bonestell (1888–1986) (Miller 1996). The illustrators working for scientists or popular science writers (including novelists) contributed to the diffusion of precise knowledge into society.

Women Illustrators

Though a minority, a few women have worked as illustrators (Gates 2002; Shteir and Lightman 2006). One assumes that illustration was one of the few scientific domains (with pedagogy and popular sciences) accessible to women even though that must be tempered by several considerations: women illustrators constituted a small minority and they were not considered to be professionals.

Some of them, however, earned enough on their own to work exclusively in the field of illustration. This was the case with Maria Sibylla Merian (1647–1717) (Wettengl 1997), the author of an important work on butterflies and other insects, and with Marianne North (1830–1890), who painted a rich collection of botanical watercolors now kept in a dedicated pavilion in Kew Gardens. Merian and North shared the distinction of being much traveled (Surinam for the former and Brazil for the latter) at a time when women rarely traveled alone.

Most women illustrators did not have a genuine scientific activity of their own but assisted their husbands in their scientific work. Such was the case with Elizabeth Gould (1804–1841), one of the main illustrators of the books published by her husband. Only at the end of the nineteenth century did women appear who were both illustrators and scientists. Anna Botsford Comstock (1854–1930) was first known through the illustrations published in books by her husband, the entomologist John Henry Comstock (1849–1931). Remarkably enough, Comstock was eager that the name of his wife appeared as the co-author of *A Manual for the Study of Insects* (1895). After these publications, Anna Comstock worked on pedagogy and actively participated in the expansion of the Nature Study movement. In 1897, she became the first woman to teach at Cornell University, but received a full professorship only in 1929.

The extremely small number of women among pioneers of scientific photography is worthy of comment. The situation in photography was similar to that noted in other examples of the diffusion of a new technique—electricity, motor cars, or computers— access was restricted and obstructed. Similarly, few women were known in the field of animal photography probably because the latter is considered as an adjunct of hunting, a barely feminine activity (Dunaway 2000), but also because it required physical strength, due to the weight of the early photographic machines.

Construction of a Visual Scientific Culture

Illustrators have largely contributed to the emergence of a vernacular scientific culture, particularly through popular science journals. The following three examples will explain their role in those publications.

The first example concerns the spread of knowledge of paleontology. The book *La Terre avant le déluge* by Louis Figuier (1819–1894), first published in 1863, was an immense success, with a fourth printing in 1864 and a total print run of 25,000 copies. Aware of new scientific discoveries, Figuier modified some plates to keep pace with progress in science: for instance, a plate depicting a landscape of the Carboniferous period as a marsh was replaced by a dense wood. To construct that image, Riou used the geological plates of Josepf Kuwasseg (1802–1877) published in *Die Urwelt in iren verschiedenen Bildungsperioden* (1851) by Franz Unger (1800–1870). These books largely contributed to the spreading awareness of the latest discoveries in paleontology (O'Connor 2007), as did Benjamin Waterhouse Hawkins's (1807–1889) reconstruction of prehistoric animals presented at the Crystal Palace.

The second example deals with zoological pictures and the notion of environment. *Illustrites Tierlebens* by Alfred Edmund Brehm (1829–1884) began publication in 1864. It was one of greatest successes in scientific illustration of the nineteenth century: the six volumes of the 1864–1869 edition were expanded to ten volumes for the second edition (1890–1893), marked by the introduction of color. Considered by Charles Darwin to be one of the landmarks of naturalist illustration, its success was largely determined by its images rather than the text written by Brehm, which was gradually replaced in subsequent editions. Not only was the book translated and adapted in most European languages, but its images were re-used in many other books; also, illustrators discovered through their participation in *Illustrites Tierlebens,* were asked to contribute. This publication marked the existence of a genuine German school of illustrators that included Robert Kretschmer (1812–1872), Gustav Mützel (1839–1893), Eduard Oskar Schmidt (1823–1886) and Wilhelm Kuhnert (1865–1926). The great innovation of the *Tierlebens* was the replacement of a static posed form of animal representation (as if animals were stuffed and kept in a museum) by attitudes that were as natural as possible, and set in their natural environment (appropriate vegetation always present). The great plates were not only wonders of naturalist iconography, they also introduced a new manner of perceiving the living world: an animal is not a dead individual, detached from its environment and studied in the context of a laboratory, but rather a living creature interacting with other animals, as well as with the surrounding world. The authors were conscious of that objective, actually asserted by Brehm in the preface to the first volume of *Illustrites Tierlebens,* a credo shared by many zoologists of the time (Nyhart 2009).

Another example deals with the construction of extraterrestrial landscapes. *Hector Servadac* (1877), a novel by Jules Verne, presented the first examples of Space Art with an engraving by Henri Félix Emmanuel Philippoteaux (1815–1884) showing the rings of Saturn viewed from the surface of the planet (Miller 1996). The idea was not to show the reality of Saturn as seen through an earth-based telescope, but rather as we could see it from the surface of Saturn, if the journey could be made. This approach was not unique to anticipation stories, since it was also found in books of astronomy such as *The Moon* by James Hall Nasmyth (1808–1890) and James Carpenter

(1840–1899): high-quality photographs of the surface of the moon were shown beside artist's views of moon landscapes. The idea was to illustrate the relief of the surface of the moon by changing the perspective of the viewer.

In these three examples the role played by the illustrator was to make perceptible a reality that a textual description, or even a photograph, was unable to do by itself. The knowhow of the different illustrators permitted the perception of an artificial reality, though respecting the knowledge of a discipline at that time. Such a creative work very often contributes to the adoption of a scientific representation by the collective imagination, much more effectively than scientific texts can achieve.

The Photography Revolution

The search for an alternative to the couple "illustrator–engraver" was one of the prime motivations of the inventors of photography: "Every man his own printer & publisher," William Henry Fox Talbot (1800–1877) wrote to Sir John Frederick William Herschel (1792–1871) in 1839. The dream that the scientist can produce, print, and release his images by himself may never truly be attained, but photographic equipment rapidly became an instrument of optics and enabled the production of images difficult to create through classical illustration, astronomy being the discipline in which that need is the most evident (Norman 1938). From the last quarter of the nineteenth century onwards, a massive devotion of scientists to photography took place: a study of French dissertations in the sciences between 1880 and 1909 shows the growing importance of illustration: 86 percent were illustrated, 12 percent contained photographs (Fieschi 2000). The success of scientific photography was largely due to technical innovations which enabled a larger public to take photographs (particularly because of the introduction of the gelatin–silver process), and facilitated the printing of photographs. One of the key landmarks in the evolution of techniques was the publication in 1895 of the first book entirely illustrated by photographs of animals in their natural environment: *British Birds' Nest* by Richard Kearton with photographs by his brother, Cherry. Within a short space of time, photography had become as important a tool for the production of scientific images as painting and drawing (Figure 8.2).

Of course, photography had peculiarities of its own. For example, pioneers of animal photography presented themselves as a kind of modern hero and took much care to detail their inventiveness in taking photographs, and the risks they faced. Everything had to be invented: methods for shooting, careful approach to the animals, mastering of light, etc. The photography of large fauna was very risky since the animals had to be approached closely enough (early lenses did not allow photographs from distance) and the animals could thus attack the photographer. For the first time, the practical activities of a scientific illustrator became a central theme of books (Kearton 1898).

The study of scientific illustrators is often difficult because only vague indications are known concerning their life, training, and career. The renowned illustrators (Redouté, Audubon, Riou, etc.) are rare exceptions to the rule that illustrators remain anonymous, a characteristic they share with several other participants in science, such as technicians and collectors of specimens. The study is made even more complex because the place and the role of the illustrator changes with time, disciplines, and editorial projects, and also with techniques used. The freely accessible website Stuttgart

DESCENDING AN OVERHANGING CLIFF.

Figure 8.2 Photographers of animals are the first illustrators to show how their images are made. Pioneers in photography are genuine sportsmen and often please themselves by showing the risks they take. Richard Kearton. 1898. *Wild Life at Home: How to Study and Photograph It.* London: Cassell and Company, 27.

Database of Scientific Illustrators 1450–1950 (DSI) allows a search among more than 10,000 entries of draughtsmen, engravers, lithographers, photographers, model-makers etc. active in natural history, medicine, the biological, physical, technical and geo-sciences, etc. in more than 100 countries.[2] However, the activities of illustrators are rich in questions that deserve further study: How are images built up? In what type of economy is the activity of the illustrator placed? How is the elaboration of illustrations influenced by the technical and editorial environment? Illustration thus offers an excellent subject to investigate the ways in which science is constructed, how it is communicated, and how it participates in a shared scientific culture.

Endnotes

1   I thank Gabriel Gachelin for his assistance.
2   http://www.uni-stuttgart.de/hi/gnt/dsi The database is searchable in 20 search fields, including alternative name spellings, preferred techniques, patronage and client names for whom these illustrators worked. In the subfields websites and sources, hundreds of further sources and reference works are listed.

References

Arber, Agnes. 1912. *Herbals, their Origin and Evolution. A Chapter in the History of Botany, 1470–1670.* Cambridge: Cambridge University Press.

Austin, April Louise. 2010. "Illustrating animals for the working classes: *The Penny Magazine* (1832–1845)." *Anthrozoös*, 23, No. 4: 365–82.

Bennett, Scott. 1984. "The editorial character and readership of *The Penny Magazine*: An analysis." *Victorian Periodicals Review*, 17, No. 4: 127–41.

Bouyer, Frédéric. 1867. *La Guyane française.* Paris: Librairie de L. Hachette et Cie.

Dunaway, Finis. 2000. "Hunting with the camera: Nature photography, manliness, and modern memory, 1890–1930." *Journal of American Studies*, 34: 207–30.

Elliott, Brent. 2010. "The Royal Horticultural Society and its orchids: A social history." *Occasional Papers from the RHS Lindley Library*, 2: 3–53.

Fieschi, Caroline. 2000. "L'Illustration photographique des thèses de science en France (1880–1909)." *Bibliothèque de l'École des chartes*, 158: 223–45.

Figuier, Louis. 1863. *La Terre avant le déluge.* Paris: Hachette.

Figuier, Louis. 1869. *Les Poissons, les Reptiles et les Oiseaux.* Paris: Hachette.

Freund, Gisèle. 1974. *Photographie et Société.* Paris: Seuil.

Gates, Barbara Timm. 2002. *Kindred nature: Victorian and Edwardian women embrace the living world.* Chicago: University of Chicago Press.

Houfe, Simon. 1978. *Dictionary of British Book Illustrators and Caricaturists, 1800–1914.* Woodbridge: Baron Publishing.

Kearton, Richard. 1898. *With Nature and a Camera: Being the Adventures and Observations of a Field Naturalist and an Animal Photographer.* London: Cassell.

Kisling, Vernon (ed.) 2000. *Zoo and Aquarium History: Ancient Animal Collections to Zoological Gardens.* Boca Raton, FL: CRC Press.

Lambourne, Maureen. 2001. *The Art of Bird Illustration: A Visual Tribute to the Lives and Achievements of the Classic Bird Illustrations.* Royston: Eagle Editions.

Macmillan Dictionary. 2014. "illustration—definition." Accessed September 14, 2015 http://www.macmillandictionary.com/us/dictionary/american/illustration.

Marcus, Margaret Fairbanks. 1944. "The Herbal as Art." *Bulletin of the Medical Library Association*, 32, No. 3: 376–84.

Martin, William Charles Linnaeus. 1841. *A General Introduction to the Natural History of Mammiferous Animals, with a Particular View of the Physical History of Man, and the More Closely Allied Genera of the Order Quadrumana, or Monkeys.* London: Wright and Co.

Miller, Ron. 1996. "The archaeology of Space Art." *Leonardo*, 29, No. 2: 139–43.

Norman, Daniel. 1938. "The development of astronomical photography." *Osiris*, 5: 560–94.

Nyhart, Lynn K. 2009. *Modern Nature: The Rise of the Biological Perspective in Germany.* Chicago: University of Chicago Press.

O'Connor, Ralph. 2007. *The Earth on Show: Fossils and the Poetics of Popular Science, 1802–1856.* Chicago: University of Chicago Press.

Palmer, Alfred Herbert. 1895. *The Life of Joseph Wolf, Animal Painter.* London: Longman, Green.

Prodger, Phillip. 2009. *Darwin's Camera: Art and Photography in the Theory of Evolution.* Oxford: Oxford University Press.

Schulze-Hagen, Karl, and Armin Geus. 2000. *Joseph Wolf (1820–1899): Tiermaler.* Marburg an der Lahn: Basilisken-Presse.

Shteir, Ann B., and Bernard V. Lightman (eds.) 2006. *Figuring It Out: Science, Gender, and Visual Culture.* Hanover, NH: Dartmouth College Press.

Stafford, Barbara Maria. 1984. *Voyage into Substance: Art, Science, Nature and the Illustrated Travel Account, 1760–1840.* Cambridge: Cambridge University Press.

Verne, Jules. 1863. *Cinq semaines en ballon.* Paris: J. Hetzel.

Voet, Leon. 1969–1972. *The Golden Compasses: The History of the House of Plantin-Moretus.* Amsterdam: Vangendt.

Wettengl, Kurt (ed.) 1997. *Maria Sibylla Merian. Künstlerin und Naturforscherin 1647–1717.* Frankfurt: Hatje Cantz Verlag.

CHAPTER NINE

The Human Experimental Subject

ANITA GUERRINI

Humans have been experimental subjects in Western science for almost as long as experimentation has taken place. As Western-style biomedicine has come to dominate most of the world in the past century, so Western-style experimentation has also taken root. Medical anthropologist Margaret Lock notes, "medical knowledge and practices in all societies are inevitably associated with moral judgments and with ideas about what is normal and abnormal" (Selin 2003, 155). Experimenting on humans has come to be considered normal. The definition of experimentation varies widely over time and place, and includes external observation and manipulation; the testing of toxins in poisons, foods, or diseases; and vivisection, the surgical cutting open of the body, among other interventions in a body's normal functioning. But experimental subjects over the centuries have generally been drawn from a few classes of people, what historian Grégoire Chamayou has referred to as "vile bodies." These include prisoners, orphans, prostitutes, people with disabilities, the mentally ill, hospital patients, slaves, and the colonized (Chamayou 2008, 7). To these might be added certain racial and economic groups, and soldiers. But it is not inevitable that these people would become experimental subjects, and how one became a "vile body" has shifted with time and circumstance. The critical relationship is one of power: those who possess political or medical authority have deemed that certain individuals occupy the boundaries of health, or the law, or the state and that they are therefore expendable as experimental subjects for a greater good. Yet self-experimentation among researchers constitutes a significant category of experimentation that these definitions do not cover. Human experimental subjects have only recently become subjects of historical scrutiny, as knowledge-making has come to be viewed more broadly.

Only in the twentieth century did ethical theories and accompanying statutes emerge in Western nations to limit, mitigate, and in some cases forbid human experimentation. Not coincidentally, that century also witnessed the most flagrant violations of human bodily integrity, at times despite legal restrictions. Historians have argued that the concept of consent is the most important legacy of twentieth-century

A Companion to the History of Science, First Edition. Edited by Bernard Lightman.
© 2016 John Wiley & Sons Ltd. Published 2020 by John Wiley & Sons Ltd.

bioethics, and that revelations of Nazi atrocities during the Nuremberg trials in 1946 led to the codification of consent of the subject as a prerequisite for experimenting. But the story is more complicated: both Chamayou (2008) and Lederer (1995) have shown that researchers viewed consent as necessary long before 1946. Moreover, numerous studies beginning with the revelations of Maurice Pappworth and Henry Beecher in the 1960s have shown that the criterion of consent continues to be inconsistently applied.

"Informed consent," the notion that the human experimental subject must be advised of potential risks and benefits, also has a long history before Nuremberg. But its definition continues to be imprecise, and legal and ethical obligations do not entirely coincide. Factors such as fear, hope, respect for authority, and financial incentive may make consent less than free and informed, particularly among the populations most likely to become experimental subjects.

This account offers a selective look at some of the categories of human subjects over history: prisoners, slaves, children, patients, and self-experimenters. For much of history, the greater public has viewed human experimentation with horror and disgust; this has made it an easy shorthand to vilify political or ideological enemies. Sometimes this vilification has been justified. My story ends in 1993 with the revelation of secret human experiments conducted by the US government, which led to further regulations. New arenas for experimentation and abuse emerged with the biotechnology revolution of the late 1990s and 2000s.

Prisoners

Although there are surgical traditions in many parts of the world, there are few instances of dissection outside Greco-Roman medical culture (Selin 2003). During the eras in Western science when human dissection was allowed for research, the bodies of executed criminals were most commonly used. Dissection therefore entailed collusion between researchers and the executing authorities. The British Murder Act of 1752 codified what was usually a local transaction, specifying that the corpses of those executed for murder would be available for dissection. Whether criminals or prisoners of war, the incarcerated have historically also been the primary "vile bodies" employed as experimental subjects. Imprisonment for whatever reason places humans completely under another's control. Even more than slavery, it effaces individual identities and rights.

Alexander the Great founded the city of Alexandria in Egypt in 331 BCE as a cosmopolitan center of learning. Its combination of willing physicians, ambitious rulers, and what historian Heinrich von Staden has called "scientific frontiersmanship" (von Staden 1989, 141) allowed the dismissal of old taboos, including Greek prohibitions of the mutilation of the human body, living or dead. There is good evidence that, around 280 BCE, Herophilus of Chalcedon and his younger contemporary Erasistratus dissected humans. In addition, the king granted them condemned criminals to dissect alive to investigate further the workings of the human body. But the Alexandrians' moment was short, and sanctioned human dissection did not reappear for over 1500 years.

When human dissection began again around 1300, executed criminals were the most common subjects. Although some anatomists dreamt longingly of the days of Herophilus, human vivisection remained unthinkable; but as dissection became increasingly practiced after 1500, rumors abounded that the most prominent anatomists, including Jacopo Berengario da Carpi and Andreas Vesalius, did not always wait for their anatomical subjects to expire before cutting them open. In Oxford in 1650, Anne Greene, convicted and hung for infanticide, was not the first to revive at the touch of the anatomist's knife.

The use of prisoners to test poisons, drugs, and remedies also has a long history. King Attalus III of Pergamon around 150 BCE tested poisons and their antidotes on criminals condemned to death. Shortly after, the notorious King Mithradates of Pontus also tested poisons on prisoners, as well as on members of his court. Between antiquity and the Renaissance, information about experimentation is scanty. A hostile commentator attributed to the thirteenth-century Holy Roman Emperor Frederick II a number of experiments involving prisoners, but modern historians have dismissed these accounts as exaggerated. However, medieval prisoners of war were treated harshly, and the line between torture and experimenting might have been thin.

The first modern use of prisoners as test subjects came in the context of smallpox variolation (deliberately infecting with smallpox to induce immunity) in London in the early 1720s. In the wake of Lady Mary Wortley Montagu's successful variolation of her own children, royal physician Hans Sloane arranged for an experimental trial with six prisoners from London's Newgate prison, three men and three women. Once inoculated, five came down with smallpox and survived; it turned out one had already had the disease. The prisoners gained their release, but one of them, Elizabeth Harrison, aged 19, further proved her immunity by nursing smallpox victims.

Prison itself could be an experiment. The Pennsylvania System of incarceration based on isolation, introduced in 1829 at the Eastern State Penitentiary in Philadelphia, has subsequently been viewed as a wide-scale (and long-term) human experiment in the effects of prolonged solitary confinement. Intended to induce penitence, solitary confinement often led to insanity. Charles Dickens wrote of it in 1842, "I hold this slow and daily tampering with the mysteries of the brain to be immeasurably worse than any torture of the body; and because its ghastly signs and tokens are not so palpable to the eye,... and it extorts few cries that human ears can hear" (Dickens 1842). This system endured until the early twentieth century. But by that time, scientists and prison administrators began to realize that prison populations provided ample human material for a variety of experiments.

Amnesty or a reduced sentence for participation in experiments proved to be a powerful incentive. The 1915 pellagra experiments of Joseph Goldberger in Mississippi provided a model for subsequent use of prisoners in experiments. Goldberger, a physician with the US Public Health Service, believed that the cause of pellagra, which was endemic in the South, was not a germ but a dietary deficiency. He had tested this hypothesis in two Mississippi orphanages and a Georgia mental asylum. With the agreement of the state's governor, Goldberger then embarked on a controlled dietary experiment with prison volunteers. All of the chosen volunteers where white; Goldberger argued that because pellagra was less common among whites, this would offer a more convincing demonstration, and the telltale rash was more easily discernible on lighter skin. But historian Jon Harkness has argued that the predominance of white subjects in this and other prison experiments reflected social and racial divisions in prison culture, where participation in experiments (and its rewards) was viewed as a privilege (Harkness 1996). The experiment successfully demonstrated the relation between pellagra and diet, but critics remained and Goldberger resorted to self-experimentation to make his case. There was little criticism, however, of his use of prisoners.

Between 1915 and the early 1970s, many experiments were conducted on American prison populations, ranging from testicular implants to research on tuberculosis, cancer, and malaria. The peak of prison experimenting occurred between 1951 and 1974 at Holmesburg Prison in Philadelphia, which journalist Allen Hornblum described in his book *Acres of Skin* (1999). Directing these experiments was dermatologist Albert Kligman. In exchange for payments of between $10 and $300, inmates tested consumer products such as detergents and hair dyes as well as radioactive, hallucinogenic, infectious, and toxic materials for pharmaceutical companies and government agencies. Kligman developed the popular acne medication Retin-A from experiments at Holmesburg. Following the Tuskegee revelations in 1972, experimenting stopped at Holmesburg and Federal laws enacted in 1978 restricted prison experimenting to studies with minimal risk to inmates. Some have argued that this law goes too far and that therapeutic experiments on prisoners should be allowed.

Slaves

Next to prisoners, slaves undoubtedly had the least agency and were therefore quintessential "vile bodies." Roman *praegustatores* or food tasters, who were slaves or freedmen, continued earlier traditions of servants in noble households who tasted food to ensure its safety. Historians have recently argued that, far from being unwitting victims of poisoners, the *praegustators* were skilled toxicologists who prided themselves on their ability to detect poisons, employing a variety of methods (Johnston 2013).

Slaves were of increasing importance in the global economy in the eighteenth century, and they were often too valuable to use in experiments. Eighteenth-century physicians such as Zabdiel Boylston, who practiced variolation in Boston, or Thomas Fowler, who published "trials" of the therapeutic uses of tobacco and arsenic, preferred to use their patients. Historian Londa Schiebinger cites the colonial physician John Quier's variolation of black slaves in Jamaica in the 1760s as an example of experimentation on slaves. But Quier's variolation did not differ from standard general practices and the slaves were also his patients (Quier 1780; Schiebinger 2004, 401–2). In the United States, however, historian Todd Savitt argues that in the antebellum south, "Blacks were considered more available and more accessible" to medical research, owing particularly to their legal invisibility (Savitt 1982, 332). In the nineteenth century, a developing scientific medicine based on clinical observation demanded human subjects, alive and dead. In the south the majority of hospital patients were black, solicited by offers to slave owners of free medical care, and most student dissections were performed on black bodies.

Several historians cite two more sinister uses of slaves for the advancement of medical knowledge. In Georgia in the 1820s and 30s, Dr. Thomas Hamilton subjected a slave named Fed to experiments on heat stroke. On several occasions, Fed sat in a pit surrounded by hot embers to test various remedies. In the 1840s, Dr. James Marion Sims tested surgical techniques in the repair of vesico-vaginal fistula on three slave women in Alabama. Each woman experienced around thirty operations, which in this pre-anesthetic, pre-antiseptic age were both painful and risky. Sims finally succeeded, and found that his white patients on whom he subsequently operated seemed much less tolerant of pain than the slave women, perpetuating a myth of racial difference in pain perception. Other examples of surgical trials on slaves included tests of ovariotomy, anaesthesia, and caesarian section (Savitt 1982, 346–7).

World War II victims of medical experimenting by the Axis powers may more properly be defined as slaves than as prisoners, since their incarceration was a function of who they were rather than the result of a crime. The experiments at Nazi concentration camps and the Manchurian site known as Unit 731 have been thoroughly documented. In Germany, medical scientists had elaborated long before World War II the distinctively German strain of eugenics known as *Rassenhygiene* or racial hygiene, which classified certain ethnic groups, particularly Jews and Roma, along with homosexuals and the mentally ill, as threats to the nation's health. These groups became the "vile bodies" of Nazi experimentation. Bioethicist Arthur Caplan has argued that "mainstream biomedicine in Germany boarded the Nazi bandwagon early, stayed for the duration of the Nazi regime, and suffered few public second thoughts or doubts about the association even after the collapse of the Reich" (Caplan 1992, 57).

Although Nazi physicians at the "doctors' trial" at Nuremberg in 1946–47 claimed that they were merely following orders in wartime, the ideological basis of Nazi science made such arguments unconvincing, and revelations of mass euthanasia in concentration camps made any claims for the scientific value of Nazi experiments equally dubious. However, debate has continued on the ethics of using Nazi experimental data. Does the nature of the experimental subjects make the data invalid? In the 1980s, some scientists argued that the use of Nazi data on hypothermia would recognize the sacrifice of the experimental subjects, but in 1990 physician Robert Berger showed that the experiments were deeply flawed (Berger 1990). On the other hand, Nazi physician Eduard Pernkopf compiled what is widely regarded as one of the most accurate anatomical atlases ever produced. It has become increasingly clear that Pernkopf used victims of Nazi violence to compile his atlas, and researchers continue to disagree about the ethics of its use (Pringle 2010).

Although the Japanese physicians who experimented with chemical and biological warfare at Unit 731 believed, like the Nazis, that their victims—mostly Chinese and Russian—were racially inferior, they were prisoners of war, political prisoners, and others identified as subversive; in other words, their identity as "vile bodies" was political rather than ideological. Of course, these motivations made little difference to the victims themselves. However, unlike with Nazi doctors, the occupying Americans did not try Japanese physicians for war crimes, instead appropriating their research on biological weapons. In 1949, the USSR tried 12 Japanese officers in connection with experiments on Russians, but full revelation of Japanese experiments only occurred from the 1980s onward (Harris 1994).

Patients

By the seventeenth century, human dissection had become increasingly common not only for medical instruction, but for research into human structure and function. The English physician William Harvey tied tourniquets on arms in order better to observe

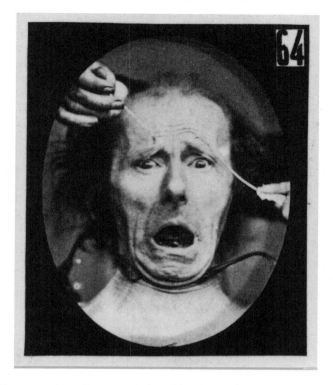

Figure 9.1 The expression of terror as induced by electrical stimulation of facial muscles. Duchenne used his patients for his experiments on the electrical stimulation of muscles. G.-B. Duchenne. 1862. *Mécanisme de la physionomie humaine.* Paris: Renouard, Plate 7, Image 64. Wellcome Library, London.

the veins and arteries, but that was the extent of his research on live humans. His concept of the circulation of the blood (1628) led to a number of experiments on blood transfusion in the 1660s, especially in England and France. Transfusion experiments had begun with animals, but the therapeutic potential of exchanging old, sick blood for fresh, young blood seemed too promising to pass up. With little understanding of the composition of blood, experimenters believed that the blood of young animals such as calves or lambs could be particularly beneficial for humans. The first transfusions, of seriously ill people, were inconclusive. Trials were then made on mentally ill (but otherwise healthy) subjects. In London, Arthur Coga survived two transfusions of lambs' blood and his "too warm" brain calmed; in Paris, however, Antoine Mauroy died, either from the effects of the transfusions or from some other cause. (Figure 9.1)

Human transfusion then ceased until 1818, when James Blundell, a London obstetrician, suggested blood transfusion as a way to mitigate the consequences of uterine hemorrhage after birth. After numerous experiments with dogs (during which he determined that injecting blood from other species was fatal), Blundell transfused blood from one human into another in 1825, and several times subsequently. Not all of his patients survived, but some did. A few others adopted his procedures with mixed success. Until greater knowledge of blood types and anticoagulants emerged in the twentieth century, blood transfusion remained a hit-or-miss procedure. Little is known about Blundell's subjects, but it is likely that they were his hospital patients rather than from his private practice, and therefore were poor women.

I have argued elsewhere (Guerrini 2003) that smallpox inoculation in the eighteenth century was the first large scale human experiment, conducted largely by doctors on their patients. While early trials of the procedure were made on the "vile bodies" of prisoners, variolation soon became widespread and eagerly sought by populations across Europe. Smallpox was common and greatly feared: contagious, disfiguring, and in up to a third of cases, fatal. Survivors were unlikely to contract it again, but the nature of immunity was not understood. In parts of Asia and Africa, children had long been variolated. The most common method involved inserting some dried scabs into a scratch on the arm, although powdered scabs were sometimes inhaled.

An account of this practice in Constantinople appeared in the *Philosophical Transactions* of the Royal Society of London in 1714; a few years later, Lady Mary Wortley Montagu, the wife of the British ambassador to the Ottomans, had her young son inoculated by the embassy's surgeon. Back in London when a smallpox epidemic broke out in 1721, her daughter was also inoculated. The girl contracted a mild case of smallpox and survived, and the medical world took notice. Following the experimental trial with six prisoners, variolation spread among the British population.

Meanwhile, an epidemic broke out in New England. The clergyman and fellow of the Royal Society Cotton Mather had learned of variolation from the *Philosophical Transactions* as well as from his African slave. He persuaded only one Boston physician, Zabdiel Boylston, to try it, who inoculated his sons, his slaves, and several patients. Boston physicians debated the efficacy of variolation, but statistics—a very new science—was on Boylston's side; only 12 of the 400 he inoculated died, versus 500 out of 3600 natural cases. In the late 1720s, the British physician James Jurin made a larger statistical case for variolation, determining that only 1 in 60 died of it in Britain, versus 1 in 8 of natural smallpox.

Widely practiced, variolation remained experimental. It did not always work: induced cases of smallpox were not always mild, and the inoculated could spread the disease to others. The statistical methods of Jurin and later Daniel Bernoulli (1700–1782) in France also turned individual patients into numbers, a step toward the depersonalization of the experimental subject. Variolation and later vaccination introduced statistics and an early version of cost-benefit analysis to human experimentation. But it also amounted to doctors experimenting on their patients. Because smallpox was such a feared disease, it could even be argued that the consent of those who volunteered for variolation was not entirely uncoerced. These issues would arise again.

Hospital patients became experimental subjects by the eighteenth century, and some argue they remain so. Until the end of the nineteenth century, those who could afford it received care from a physician in their homes. Those who could not went to hospitals. Historical opinion differs on the level of care provided in these institutions, but they were fruitful ground for the observation of diseases and for research into remedies and their effectiveness. Most therapies were experimental in that their actions and effectiveness were poorly understood, and therefore any patient, in or out of hospital, might be considered an experimental subject. Hospital patients provided a captive and largely powerless pool of subjects. Slave patients in antebellum hospitals were therefore doubly captive.

By the early nineteenth century, the practice of therapeutic experimentation had become standardized; in 1814 French physician Auguste Chomel summarized the rules that medical students should follow in such experiments. He distinguished two varieties of trials, one to determine the action of a therapeutic agent against a known illness, the other to find the specific body function that the therapy affected. One must therefore know the remedy, the experimental subject, and the illness (Chamayou 2008, 206). Experiments, added Chomel, should aim to cure a patient, not to make him ill. Over a century later, Beecher and Pappworth detailed dozens of harmful experiments that took place in hospitals in the 1950s and 60s.

In the heyday of colonialism in the eighteenth and nineteenth centuries, colonial subjects were deemed excellent experimental subjects; in status they fell between slaves and their European rulers. For example, in the 1840s, Scottish surgeon James Esdaile experimented in Calcutta on Indian subjects on the efficacy of mesmerism as an anesthetic (Winter 1998). The discovery of the mosquito vector for malaria in the 1890s included experiments on British colonial subjects and patients in China and India.

The most famous human experimental subject in the nineteenth century was neither a slave nor a hospital patient, and the experiments had no connection to therapy. In 1822, a French-Canadian voyageur in upper Michigan named Alexis St. Martin (1794–1880) received a gunshot wound in his side. The wound did not close, leaving an opening into his stomach. This gastric fistula provided a window into still mysterious digestive processes, and his doctor, Army surgeon William Beaumont (1785–1853), decided to investigate these. St. Martin could not continue fur-trapping and served in Beaumont's household as a handyman. As historian Alexa Green has pointed out, St. Martin's agreement with Beaumont was an employment contract rather than one protecting his rights as an experimental subject (Green 2010).

Beaumont poked bits of food into St. Martin's stomach through the fistula. He also took samples of gastric fluid and inserted a long thermometer, which was quite painful. Between 1825 and 1833 Beaumont performed hundreds of experiments on St. Martin, and published several articles and a book that made them both famous. St. Martin disliked the experiments and periodically ran away. He left Beaumont's household permanently in 1833, and although he lived in poverty, he resisted the surgeon's pleas to return for more experiments. St. Martin was hardly an equal in the relationship with Beaumont; like hospital patients, his poverty and low social status made him particularly susceptible to exploitation as a "vile body."

The noted French physiologist Claude Bernard greatly admired Beaumont's work. In the preface to his influential *Introduction to the Study of Experimental Medicine* (1865), Bernard echoed Chomel in justifying human experimentation in certain circumstances; yet none of these applied to St. Martin:

> It is our duty and our right to perform an experiment on man whenever it can save his life, cure him or gain him some personal benefit. The principle of medical and surgical morality, therefore, consists in never performing on man an experiment which might be harmful to him to any extent, even though the result might be highly advantageous to science. ... So, among the experiments that may be tried on man, those that can only

harm are forbidden, those that are innocent are permissible, and those that may do good are obligatory (Bernard 1957, 101–2).

A dozen years after these lines were written, the emergence of the germ theory of disease afforded a new set of experimental imperatives for medicine, in the development of the clinical trial. The modern clinical trial tests a medical treatment or a prevention strategy (Collier 2009). In the mid-eighteenth century, Scottish physician James Lind tested the efficacy of citrus juice in the prevention and treatment of scurvy on 12 sailors; only the two fed citrus recovered from the disease. While notions of placebo and control group developed in the nineteenth century, the additional concepts of blinding and randomization only emerged in the late 1940s. In a blind and randomized trial, neither the researcher nor the experimental subject knows who is receiving the treatment being tested and who receives a placebo; furthermore, assignment to each group is random. These measures minimize researcher bias.

This design was not fully realized when researchers from the United States Public Health Service began a project in 1932 to examine the effects of syphilis. This study was not a clinical trial, but an observational study of the effects of untreated syphilis in black men, although the 600 subjects, poor southern men, included 201 who did not have the disease and served as controls. They received free medical examinations and burial insurance. They did not, however, receive information about the nature of the study when they consented to participate. It continued for 40 years, and even when penicillin, highly effective in the early stages of the disease, became available in the 1940s, the men were not offered it. The study was finally terminated in 1972 after its revelation in the press led to public outcry. On the heels of the whistle-blowing of Beecher and Pappworth, the Tuskegee case led to a reexamination of human subjects protection in the US, culminating in the National Research Act of 1974, which codified written consent and established Institutional Review Boards to evaluate human subjects research. The act also set up a National Commission on the Protection of Human Subjects to identify the basic ethical principles underlying human research. Building on the 1964 Helsinki Declaration of the World Medical Association and subsequent amendments, the commission's 1979 Belmont Report named three key principles: respect for persons, beneficence, and justice.

Children

A thirteenth-century chronicler claimed that Emperor Frederick II caused several orphaned children to be raised in complete silence to determine if they would spontaneously speak Hebrew, thought to be the original language of the world and therefore the language of God. Orphans and other institutionalized children, like prisoners, had no one to speak for them other than institutional authorities. After Sloane conducted his smallpox trial at Newgate Prison, a group of orphans received inoculation. Only then did it gain royal approval.

Not all children were as vulnerable as these. But the threat of a deadly disease could compel parents to offer their children as guinea pigs. Vaccination is one example. Edward Jenner, an English country physician, knew of many cases of milkmaids and dairymen who had contracted cowpox—a bovine disease with minor effects in humans—and then seemed resistant to smallpox. He tested this hypothesis in 1796 on an eight-year-old boy, James Phipps, inoculating him with cowpox matter from a milkmaid. James fell ill with cowpox but recovered, and Jenner then inoculated him with smallpox with no effect. Vaccination (from the Latin *vacca*, cow) quickly supplanted variolation.

We don't know what the parents of James Phipps (who were poor farm laborers) thought when Edward Jenner injected him with cowpox. But the mother of nine-year-old Joseph Meister took him to Paris in the summer of 1885 to see the celebrated scientist Louis Pasteur. Joseph had been bitten by what appeared to be a rabid dog, and Pasteur had recently announced that he had developed a rabies vaccine that was effective in dogs. It had not, however, been tried in humans. Pasteur agreed to treat Joseph, who survived.

The development of the polio vaccine in the mid-twentieth century affords many examples of children as experimental subjects. Poliomyelitis, also known as "infantile paralysis," was a virus that particularly attacked children. A common and largely unrecognized disease before 1900, new sanitation and public health measures ironically made it less common and more virulent. Most cases of polio did not result in paralysis, but those that did made it a much-feared disease in the first half of the twentieth century, particularly in the US (Oshinsky 2006).

Public pressure to develop a vaccine was enormous. Two researchers in the mid-1930s announced they had developed a vaccine. Maurice Brodie first tested his killed-virus vaccine on himself and some of his lab assistants before inoculating a dozen children who had been volunteered by their parents. Brodie eventually inoculated 9000 children. Several developed allergic reactions to the vaccine, which turned out to be ineffective. Around the same time, John Kolmer developed a live (but weakened) virus vaccine. He too tested it on himself and on his own children and then on a sample of 23 children with parental consent before inoculating some 10,000 children. A dozen or more became ill, and nine died.

After World War II, many researchers sought a polio vaccine, and children continued to be experimental subjects. In this post-Nuremburg era, consent of parents or guardians was required. In 1950, Hilary Koprowski tested a live-virus vaccine on 20 children in a New York state facility for intellectually disabled children. Two years later, Jonas Salk tested his killed-virus vaccine on two groups of institutionalized children, including both the physically and the intellectually disabled. Neither of these were clinical trials; the purpose was to determine whether the vaccines produced antibodies. Many of the children in these institutions were under the guardianship of the state, which gave consent. Koprowski had already tested his vaccine on himself; Salk injected himself and his family after these first human trials.

The success of Salk's trials led to a massive field trial of his polio vaccine on 600,000 children, the first large randomized double-blind clinical trial. A second parallel trial of 1,000,000 children did not administer a placebo, but simply observed the non-vaccine subjects. Parents overwhelmingly agreed to allow their children to become "Polio Pioneers." Salk's field trial concluded in June 1954; six months later, Albert Sabin tested his live-virus vaccine on 12 prison volunteers. Koprowski and Sabin later conducted wide-scale field trials, mainly on children, in what were then the Belgian Congo and the USSR. Salk's vaccine was licensed for use in 1955, Sabin's in 1961.

Goldberger in 1915 manipulated the diets of Mississippi orphans before his prison experiments. Although there was some criticism of Koprowski and Salk for their use of institutionalized children for their early trials, the extent of the use of such children did not become clear until much later. Between 1946 and 1953, more than 100 boys at the Fernald State School in Massachusetts were fed oatmeal laced with radioactive iodine and calcium to trace the absorption of nutrients. At the Willowbrook School on Staten Island, experiments with Hepatitis A between 1956 and 1970 included deliberately infecting children. The Willowbrook experiments were revealed in the early 1970s and became part of the greater revulsion over Tuskegee. But the Fernald experiments only came to light, with other radiation experiments, in the 1990s.

Self-experiment

Researchers have experimented on themselves for centuries. Who could be a better research subject? The ancient king Mithradates also tried out poisons on himself. He discovered that by giving himself tiny doses of certain poisons he became over time immune to their noxious effects (Mayor 2009, 237–46).

Journalist Lawrence Altman's *Who Goes First?* (1998) remains the classic study of self-experimentation in medicine. In the early seventeenth century, Italian physician Santorio Santorio spent the better part of 30 years sitting for several hours a day on a large balance of his own design. By measuring his bodily intake and discharges, Santorio sought to understand metabolism. He concluded that the body lost a quantity of fluid each day in what he called "insensible perspiration." A century later, his countryman Lazzaro Spallanzani studied digestion by swallowing cloth bags and wooden tubes filed with food and then vomiting them up. He also swallowed and retrieved a sponge to obtain a sample of gastric fluid. Around the same time, the renowned London surgeon John Hunter reported injecting pus from a patient with gonorrhea into the penis of an unnamed experimental subject. Most historians agree that he was the subject, and that he probably contracted both gonorrhea and syphilis from the experiment.

Before chemical assays were developed (and even after), smelling, tasting, and ingesting foreign substances were standard methods of identification. In the early nineteenth century, inhaling was added to this list. The English chemist Humphry Davy discovered the anesthetic (and pleasurable) properties of nitrous oxide by trying it on himself. In the 1840s, American and British doctors tested a number of potential anesthetics on themselves. Some, like acetone and benzene, had no effect. Ether, like nitrous oxide, was a recreational drug and its medicinal uses were recognized almost by accident. Chloroform, also effective, was less flammable than ether. But it was also addictive, as the dentist Horace Wells discovered. Later in the century, first Sigmund Freud and then the Americans Richard Hall and William Halsted experimented with the use of cocaine as a local anesthetic. Hall and Halsted became addicted, and Halsted weaned himself off cocaine with morphine, to which he remained addicted for the rest of his illustrious surgical career.

While the germ theory of disease illuminated the causes of many diseases, it obscured the causes of others. Goldberger's prison experiments on pellagra had convinced him that its cause was dietary, but his contemporaries remained skeptical. He therefore convened what he called a "filth party." He and his assistant injected blood from a pellagra victim into each other. In addition, they swabbed out secretions from the nose and throat of another victim and applied them to their own noses and throats, and even gave each other capsules they had made of the scabs from pellagra rashes. Goldberger's wife later joined the party. None of them got pellagra, but their critics remained unconvinced (Harkness 1996).

Throughout the twentieth century, doctors infected themselves or took drugs they hoped would kill microbes or offer immunity. In 1900, the four physicians of the US Yellow Fever Commission in Cuba—Walter Reed, Aristides Agramonte, Jesse Lazear, and James Carroll—decided to test on themselves the new theory that a mosquito was the disease vector before testing soldier volunteers. Lazear and Carroll allowed themselves to be bitten by mosquitoes, and both fell ill with yellow fever. Carroll survived, but Lazear died. Half a century later, many early polio researchers tested their vaccines on themselves and their families before proceeding to human trials. And in the 1970s, David F. Clyde allowed himself to be bitten thousands of times by malarial mosquitoes in order to test an experimental vaccine before trials on prison volunteers.

Conclusion

The year 1993 proved to be a turning point in human experimentation in the US. The revelation of long-term secret human experiments on the part of the US government, particularly those conducted under the auspices of the US Atomic Energy Commission between 1947 and 1974, led to an outcry even greater than over the Tuskegee experiments twenty years earlier. The Tuskegee case was one of neglect; experiments revealed in 1993 included injection of radioactive substances, administration of drugs, and other examples of deliberate exposure across the spectrum of human subjects: patients, prisoners, children, the handicapped, the poor: all had been secretly used as guinea pigs in the two decades that followed the Nuremberg trials. A presidential Advisory Committee on Human Radiation Experiments, appointed in 1994, further emphasized the necessity of informed consent that the Belmont Report had established. While the regulatory framework is stronger than it has ever been in history, one may not assume that abuses, errors, or failed experiments no longer occur. Moreover, stronger regulation in the US and Europe has led to the export of clinical trials to poorer nations with more lax oversight (Shah 2006).

The biotechnology revolution that began in the late 1990s has complicated the ethical and regulatory picture. Older laws do not cover new techniques such as cloning, gene therapy, genetic engineering, and stem cell therapy. New reproductive and therapeutic technologies call into question the very definition of human. The human experimental subject of the future may indeed be a cell.

References

Altman, Lawrence. 1998. *Who Goes First? The Story of Self-Experimentation in Medicine.* Berkeley: University of California Press.

Beecher, Henry K. 1966. "Ethics and clinical research." *New England Journal of Medicine*, 274: 1354–60.

Berger, Robert L. 1990. "Nazi science—The Dachau hypothermia experiments." *New England Journal of Medicine*, 322: 1435–1440.

Bernard, Claude. 1957. *An Introduction to the Study of Experimental Medicine*. Translated by Henry Copley Greene. New York: Dover.

Caplan, Arthur (ed.) 1992. *When Medicine Went Mad: Bioethics and the Holocaust*. Totowa, NJ: Humana Press.

Chamayou, Grégoire. 2008. *Les corps vils. Expérimenter sur les êtres humains aux XVIIIe et XIXe siècles*. Paris: La Découverte.

Collier, Roger. 2009. "Legumes, lemons and streptomycin: A short history of the clinical trial." *Canadian Medical Association Journal*, 180: 23–4.

Dickens, Charles. 1842. *American Notes for General Circulation*. New York: Harper & Brothers.

Green, Alexa. 2010. "Working Ethics: William Beaumont, Alexis St. Martin, and medical research in antebellum America." *Bulletin of the History of Medicine*, 84: 193–216.

Guerrini, Anita. 2003. *Experimenting with Humans and Animals. From Galen to Animal Rights*. Baltimore: Johns Hopkins University Press.

Harkness, Jon M. 1996. "Prisoners and pellagra." *Public Health Reports*, 111: 463–7.

Harris, Sheldon H. 1994. *Factories of Death. Japanese Biological Warfare, 1932–45, and the American Cover-up*. London: Routledge.

Hornblum, Allen M. 1999. *Acres of Skin: Human Experiments at Holmesburg Prison*. New York: Routledge.

Johnston, Chelsea. 2013. *Beware of that Cup! The Role of Food-tasters in Ancient Society*. M.A. thesis, University of Otago.

Lederer, Susan. 1995. *Subjected to Science: Human Experimentation in America before the Second World War*. Baltimore: Johns Hopkins University Press.

Mayor, Adrienne. 2009. *The Poison King: The Life and Legend of Mithradates, Rome's Deadliest Enemy*. Princeton, NJ: Princeton University Press.

Oshinsky, David M. 2006. *Polio: An American Story*. New York: Oxford University Press.

Pappworth, Maurice. 1967. *Human Guinea Pigs: Experimentation on Man*. London: Routledge and Kegan Paul.

Pringle, Heather. 2010. "Confronting anatomy's Nazi past." *Science*, 329: 274–5.

Quier, John. 1780. *Letters from Mr John Quier, Practitioner of Physic in the Island of Jamaica, to Dr D. Monro, Jermyn-street, London, on the Small-pox and Inoculation, Measles, &c.* In *Medical commentaries. … Collected and published by Andrew Duncan*. London: printed for Charles Dilly.

Savitt, Todd. 1982. "The use of Blacks for medical experimentation and demonstration in the Old South." *Journal of Southern History*, 48: 331–48.

Schiebinger, Londa. 2004. "Human experimentation in the 18th century: Natural boundaries and valid testing." In *The Moral Authority of Nature*, edited by Lorraine Daston and Ferdinando Vidal, 384–408. Chicago: University of Chicago Press.

Selin, Helaine (ed.) 2003. *Medicine across Cultures: History and Practice of Medicine in Non-Western Cultures*. Dordrecht: Kluwer.

Shah, Sonia. 2006. *The Body Hunters. Testing New Drugs on the World's Poorest Patients*. New York: New Press.

Von Staden, Heinrich. 1989. *Herophilus. The Art of Medicine in Early Alexandria*. Cambridge: Cambridge University Press.

Winter, Alison. 1998. *Mesmerized. Powers of Mind in Victorian Britain*. Chicago: University of Chicago Press.

CHAPTER TEN

Amateurs

KATHERINE PANDORA

In identifying and assessing the roles played by amateurs in the pursuit of scientific knowledge, one immediately encounters core assumptions about what properly constitutes "science"—not only as a form of life, but as such a preeminent form of life that its existence is cast as coextensive with the emergence, development, and viability of modern society. These foundational assumptions are embedded in the work of mid-twentieth-century scholars who found the origins of modern science within a European "Scientific Revolution" of the seventeenth century, and who sought to legitimize history of science as a professional academic enterprise in the post-World War II period. They began with assumptions that the origins of modern science were the origins of the modern mind, and they worked at illuminating the theoretical breakthroughs of those who had wielded intellectual expertise in ways that evoked the status of modern professional scientists, even as they sought to understand such thought as embedded within historical place and time. Discussions of scientific amateurism occurred as a by-product of interest in the highest levels of what were seen as professional-grade achievements, and developed in tandem with the standard narrative of great men and great discoveries, both reflecting and reinforcing this historiographic imperative.

Defining Modern Science: The Amateur as Foil

Judged by historians as less skilled and theoretically oriented than their more celebrated peers, these participant–observers were at the same time more well-informed and directly engaged in the new ventures than the general populace; the "amateur" designation offered acknowledgment of values they shared with elite colleagues despite differences in intellectual stature. As members of the Royal Society of London, for example, men of affairs such as naval administrator Samuel Pepys expended means, time, and effort to witness developments and converse about what they had seen, oftentimes putting a hand in themselves, despite the press of worldly obligations and the lure of other diversions. A. R. Hall described the emergent roles

A Companion to the History of Science, First Edition. Edited by Bernard Lightman.
© 2016 John Wiley & Sons Ltd. Published 2020 by John Wiley & Sons Ltd.

of expert and enthusiast being invented at the Royal Society with a simple formulation: "Newton was as much a self-taught professional as Pepys was a self-taught amateur." The companionate amateurs were "interested in natural phenomena, especially curiosities, and made an effort to understand"; their differences with the professionals "lay in the latter's penetrating to the roots of understanding in mathematical analysis, dissection, systematic observation, or experiment" (Hall 1983, 384). Amateurs provided patronage and publicity, and were able to serve as contemporary conduits for mainstreaming ideas, forms of discourse, and behavioral patterns among wider social groupings. In their self-fashioned roles they joined with the major figures in making the revolution viable.

Defining Modern Science: The Amateur as Proto-Professional

Although historians of science typically avoid the terms "professional" and "amateur" as anachronistic in regard to natural philosophy in the early modern and enlightenment eras, they have been confident nonetheless that they could distinguish between those who possessed the intellectual expertise to speak for nature as legitimate authorities and those whose interest in natural knowledge was merely recreational, utilitarian, or idiosyncratic. This care for precision even extends into the nineteenth century, as exemplified in Martin Rudwick's characterization of the British "leaders of science" who "dominated" the highest levels of geological investigation and debate as "gentlemanly specialists"—the term "specialist" indicating an elite amateur who held to professional standards of work, even if the pursuit of scientific knowledge itself was not yet a profession (1985, 21). Rudwick's gentlemanly specialists were analogous to those christened as "Grand Amateurs" by Allan Chapman (1998) in his survey of Victorian amateur astronomy—acknowledged experts such as Sir John Herschel who had the self-funding and the leisure to pursue cutting-edge research which set the terms of inquiry in their fields. Beyond these eminent amateurs were enthusiastic groups of non-experts—"lesser amateurs"—who enjoyed pastimes related to scientific pursuits or who were engaged in scientific activities as paid employment. Such lesser amateurs could play critical roles in the research nexus through the contribution of observations and specimens, as did Mary Anning, a commercial fossil collector. The value of Anning's state-of-the-art excavations of such then-extraordinary phenomena as plesiosaur, ichthyosaur, and pterosaur skeletons increased after they left her hands through their use in developing the reputations of the eminent amateurs whose conceptual property they became. (Figure 10.1) Some sense of the variety of social groupings and the forms of interchange that could exist among members of these various amateur communities and expert practitioners have been detailed in updated disciplinary histories and occasionally assayed more broadly (Keeney 1992; Goldstein 1994; Barrow 1998; Chapman 1998; Welch 1998). That the pursuit of scientific knowledge had not yet taken on the professionalized form that postgraduate certification, publication in disciplinary journals, and paid positions would bring did not negate the fact that this more loosely organized and amorphously populated amateur world of thought and action had a hierarchical structure in which some amateurs were more equal than others, both in the past and historiographically. As a working-class figure with a rudimentary education, a woman, a tradesperson, and an inhabitant of a region outside the metropolis, Anning's opportunities to participate as a learned

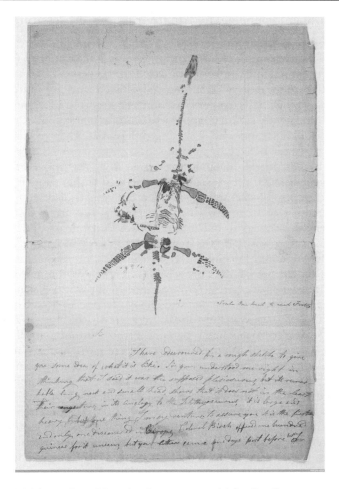

Figure 10.1 1823 letter from Mary Anning, commercial fossil collector, concerning the discovery of a plesiosaurus fossil. Archives and Manuscripts, Wellcome Library, London. Copyrighted work available under Creative Commons Attribution only licence CC BY 4.0 http://creativecommons.org/choose/results-one?license_code=by&jurisdiction=&version= 4.0&lang=en.

colleague were negligible: the concept of a "working-class specialist" needing historical elaboration rather than existing as a footnote was a fraught one at the time she lived, and for later scholars (Torrens 1995). Although historians recognized that a remarkable enthusiasm for science was evident cross-culturally and across sectors of society in the nineteenth century, the existence of amateur and public attention to science served more as a convenient historical marker testifying to the burgeoning influence of the concerns and goals of the specialist experts than as an area itself in need of critical reconnaissance, despite the dynamic emergence in the historical discipline of the new social history in the 1960s and 1970s, which gave priority to exploring how to research and to write histories of ordinary people, "from the bottom up."

Defining Modern Science: The Amateur as the Professional's Inverse

That the figure of the amateur was the closest most historians of science came to thinking about what science from the bottom up might look like accounts for some of the disconnect that the discipline displayed in regard to the new social history, especially when the negative connotations of the term are taken into account. As a descriptor, the term "amateur" had acquired yet another unstable meaning for advanced thinkers in the nineteenth century who had committed themselves to science as a vocation (paid or unpaid), and who wished to differentiate their pursuits from "amateurish" efforts that they believed would lessen, by association, the prestige of those working at the highest levels. In hindsight, a trajectory that moved from a state of incipient professionalism for a few who were embedded within a much larger, diffuse, and dispersed collection of scientifically minded individuals displaying varying levels of engagement to that of a circumscribed and rationalized professional structure in the twentieth century seemed all but inevitable. An integral component of the logic of the standard narrative was that "the growing complexity of science demanded formal scientific education and full-time professional work, not the casual, intermittent attention of self-taught amateurs" (Bruce 1987, 4). It was the declension of amateurs as legitimate participants in the coming-into-being world of true professionals that stood out as an historical certainty to many, despite their substantial cultural presence. Speaking of the extant historical literature in the mid-1970s, Sally Gregory Kohlstedt observed that when amateurs were mentioned at all by historians, "they are noted in passing as important supporters of particular scientific activity, as beginning professionals in a temporary situation or as time-consuming nuisances to the researchers" (1976, 174). When placed beside an ever-burgeoning historiography that featured the standard narrative of notable names and famous discoveries, the thin nature of what was presumed to be nonprofessionals' "merely personal" interests reinforced an impression of the insubstantiality of amateur scientific experience. Nor was this circumstance simply a result of the obscurity of amateurs compared to knowledge elites. Even famed thinkers such as Henry David Thoreau would be given little attention by historians of science; as an individual engaged in nature study without an official affiliation, his scientific status was categorized as being more of an amateur sort than not, and his authorial output presumed to be better-suited as a subject for scholarly analysis in other academic fields, such as literature or American Studies.

Re-mapping the Amateur Sphere: Historiographic Challenges to the Standard Narrative

It would not be until the 1980s and 1990s that interpretive space would open up regarding the foundational assumptions of the history of science discipline and the role played by nonprofessionals, fostered by three key historiographic shifts: research into the history of women in science; the construction of histories of natural history and science in the field; and the incorporation of perspectives and methods from cultural history, including the study of popular culture and the popularization of science. This unsettling mix of new approaches, topics, and methods brought histories of what existed *behind* conventional understandings of amateurs and professionals into play.

While a focus on companionate amateurs as informal enthusiasts would continue to exist as a topic, the merging of aspects of what had been considered to be the amateur sphere into a larger, more heterogeneous and complicated domain designated by the term "popular culture" elicited different kinds of questions. Thinking in terms of popular culture foregrounded a world in which the laity predominated, and while it was a domain within which exchanges of ideas, practices, values, and relationships contributed to shaping the professional world, it was neither coextensive with it or subsumable to it (Topham 2009). As a more detailed sense of what science within popular culture looked like was developed case by case, two possible outcomes were foregrounded: that these new histories would be read supplementally as evidence of how those "outside" of professional science "responded" to events generated by those on the "inside," or, alternatively, that they could result in a decentering of the standard narrative. The conservative nature of the discipline's historiography tended to favor the first outcome, at least initially. That the second would become a possibility by the present day has resulted in an unanticipated set of potential ramifications and increased the odds against business as usual.

Challenges Posed by the History of Women in Science

Inattention to the presence of scientific women in the historical record was reversed as the interdisciplinary field of women's studies emerged in the 1970s and 1980s. Difficult questions were pursued about whether or not qualified women had been actively excluded from professional opportunities in the nineteenth century, especially when there were more women who were highly educated in science by the 1870s and 1880s than at any time previously (Rossiter 1982). Fuller and more specific accounts of the obstacles placed in the way of women as aspirants to a scientific vocation demonstrated that the only category of scientific practice open to most of them was that of "amateur," that is one who might collect specimens, make detailed observations of natural phenomena, produce scientific illustrations, or author books and articles for children and the general public. In the latter half of the nineteenth century a small but growing phalanx of individual women sought out graduate-level education with the intention that their training would admit them into the specialist world occupied by men of science. While a handful of women managed to find a workable arrangement such as a professorship at an undergraduate-only women's college, these gains came with restrictions (such as the requirement to remain unmarried) and marginal funding for outfitting laboratories and conducting research. As male scientific leaders began to experience the era's influx of educated women into scientific societies membership requirements were often scrutinized, raised, and even set doubly high to bar women's full participation, placing them at a further disadvantage. These women existed in a nebulous state in relation to the scientific community, in which cohorts of the "most qualified women who had the most to contribute" to specific fields were kept from "fellowship with suitable colleagues" (Rossiter 1982, 74). Categorized as inadequate professionals, scientifically educated women experienced a kind of demotion to a "gendered amateurism" in which they were directed to work at lower levels than that those for which they had been trained, as a way to maintain the gender hierarchy of organized science—a tactic that had the result of creating a form of "obligatory amateurism" (Ogilvie 2000).

This attention to the difficulties of women in moving beyond the amateur sphere is one of the factors that sensitized historians of science to sociological boundary-keeping mechanisms in the nineteenth and twentieth centuries (Gieryn 1983). Boundary-work was used to restrict the definition of who was or was not doing legitimate science in relation to different groups at different times and for different reasons (the clergy, artisans, inventors, engineers, unorthodox theorists, indigenous peoples, and so on), but as the contingent characteristics of professionalization became more widely acknowledged, recognition that it had been designed to reproduce discriminatory sociopolitical structures became harder to ignore. As with women as a group, for those located outside the boundaries of the authorized scientific community, a kind of *de facto* amateurization occurred in how they were treated—individually, and in regard to practices, settings, and agendas—when the professionalizing community either recommended or acquiesced in downgrading their credentials as worthy of peer status. In the United States, for example, from the later nineteenth through the mid-twentieth century, even those African-Americans who had managed to secure educations at leading universities were compelled to work as scientists outside the (white) academy after earning their degrees, as they were unwelcome to serve as faculty at the institutions that had trained them. The case of experimental biologist Ernest Everett Just—a Dartmouth undergraduate and 1916 University of Chicago PhD who faced a constant struggle to acquire research facilities consonant with his career expectations throughout his Howard University professorship—is perhaps the best-known example in history of science (Manning 1983). The existence of African-American scientists behind the color line at historically black colleges and universities, and their research, has been largely unacknowledged by historians of science, as with articles published by black social scientists in segregated journals, and even for figures who conducted high-profile sociological research studies, such as W. E. B. Dubois and Charles S. Johnson (Holloway and Keppel 2007). In terms of amateur science and social class, the nature of the sustained working-class interest in science has been expanded upon in research that works deeply with primary sources in relation to artisan naturalists in the early nineteenth century (A. Secord 1996), political radicalism and evolutionary theory (Desmond 1992), and the pursuit of scientific self-education among late-century Victorian laborers (McLaughlin-Jenkins 2003). These studies have provided evidence for the latest cohorts of historians of science of what the historical profession in general had already learned from social and labor historians a generation before: that it is erroneous to assume that the thoughts and efforts of less-privileged members of society are irrelevant in constituting shared cultural forms.

The problematic status of women in relation to professional science also resulted in serious consideration as to how women investigated, studied, interpreted, and intervened in nature from their disadvantaged positions. This research effort took scholars in interdisciplinary directions, including literary studies, history of education, philosophy, childhood studies, environmental studies and more, as they brought such figures as Arabella Buckley, Anna Botsford Comstock, Jane Marcet, Mary Kingsley, and Rachel Carson to the fore. The most expansive vein was that of women's literary efforts as popularizers, and the heterogeneity of genres in which they worked (Norwood 1993; Gates and Shteir 1997; Gates 1998; Baym 2001; Gianquitto 2007). Who *was* authorized to speak for nature, and on what basis was this authorization granted? Barbara Gates put this question at the center of her study of Victorian and Edwardian women who pursued active interventions into the scientific cultures of their times through their writing, in which they "interrupted revised, ignored and some-times disrupted masculine discourse as they participated in conceptualizing, describing, representing, and preserving the natural world" (1998, 7; 2002).

Challenges Posed by the History of Natural History

Feminist scholars argued for an historical model of scientific communit(ies) that incorporated a wider body of members than historians of science had typically recognized. In the case of work such as "nature writing," this was an area that had been willingly ceded to literary scholars; and yet it had returned and was being used to query whether or not the discipline was leaning too heavily on an idealized vision of the dynamics of scientific professionalization with subsequent inattention to amateur practices. This challenge might have remained at a stalemate if it had not been for its convergence with a related historiographic shift that historians of science could not ignore as being a mere side issue: the rise to prominence of the history of natural history. Both the publication of the edited survey *Cultures of Natural History* (Jardine, J. Secord, and Spary 1996) and the 1996 *Osiris* volume on "Science in the field," edited by Henrika Kuklick and Robert Kohler, revealed the diversity of approaches and the vitality of questions that had been kept off-stage for far too long, overshadowed as natural history and the field sciences had been by the insistence that the birth of modern science was a result of the pre-eminence of physics, mathematics, and laboratory experimentalism. The disciplinary weight that was now extended to natural history, field, and observational sciences also highlighted the amateur sphere precisely because it was these scientific areas—archaeology, astronomy, botany, entomology, geology, natural history, ornithology, paleontology—that had seen the largest number of nonprofessional participants and enthusiasts, for both the nineteenth and twentieth centuries. These sciences were particularly suited for non-specialists, as the minimal threshold for informal involvement in collecting beetles, observing the behavior of birds, recording the paths of comets, and similar tasks could be done with little in the way of expensive equipment, elaborate training, and intensive time commitments compared to work at the expert level. The importance of natural history as a source of scientific engagement among nonprofessionals and as a representation of "the people's science" is an important one, for "unlike the laboratories of physics or chemistry the natural world appears to belong to everybody" (Drouin and Bensaude-Vincent 1996, 417). Natural history was such a vital source of reflection that it remained as a generative intellectual commons in the popular realm even after it lost its disciplinary power within academic science in the twentieth century, as media icons such as Jane Goodall and Jacques-Yves Cousteau attest.

Challenges Posed by the New Cultural History

The widespread popularity of natural history across society and the practical, emotional, rational, and imaginative investments that readers, lecture audiences, visitors to zoos, aquaria, and exhibitions, and museum-goers made in it brought historians of science face-to-face with interdisciplinary work characteristic of "the new cultural history" in the 1980s. Drawing on anthropological sensibilities in approaching the dynamics of cultural meaning-making, the analytical challenge lay not in apprehending historical actors assigned within "an array of natural kinds, fixed types divided by sharp qualitative differences," but instead in assessing a field of endeavors that was "at once fluid, plural, uncentered, and ineradicably untidy" (Geertz 1980, 166). In subsequent cultural reconstructions of the history of science, how to situate nonprofessionals in relation to professional scientists in terms of differential power, status, or impact would prove to be not only matters of sociological import but cognitive as well. The intellectual agency and efficacy of those who did not possess elite privileges was open for study rather than being presumed to be negligible at the outset – it also included within its framework a wide array of differently positioned members of the culture as a matter of course, suggesting the salience of new geographies of knowledge for understanding "the scientific." Drawing on examples and methods from European, British, and American cultural history, this quest for a decentralized historical understanding was one that relied on close examination of local experiences "and on open-mindedness to what those examinations will reveal rather than on elaboration of new master narratives or social theories." From the standpoint of conventional frameworks, explorations in cultural history could seem perplexingly akin to an indiscriminate "doubling back over territory already presumably covered" (Hunt 1989, 22, 21).

In history of science, new cultural histories brought the so-called amateur sphere back into consideration. Provocative revisitations of the "birth" of modern science such as Pamela Smith's *The Body of the Artisan: Art and Experience in the Scientific Revolution* (2004) and Deborah Harkness's *The Jewel House: Elizabethan London and the Scientific Revolution* (2007) examine knowledge-making practices among a fluid, plural, untidy, and dispersed network of practitioners conceived broadly and diversely. Smith argues that experiential authority legitimated the "vernacular" epistemology (142) that goldsmiths, painters, carpenters, and other artisans possessed, an authority that is historically opaque to scholars when they unreflectively "replicate the hierarchy of cognition and knowledge that emerged from nineteenth century positivist views of science and colonialist views of primitive societies" (148) in their research. Harkness likewise shifts the location of the revolution away from elite precincts to the spaces occupied by lowlier members of society such as "empirics, alchemists, old women, compilers of recipe books, manual laborers, artists, and craftsmen" (247), detailing vibrant communities of inquiry conducted through face-to-face interactions by urban inhabitants who displayed "a shared desire to exchange books, specimens, techniques, and tools" (8), resulting in a "vernacular science" at odds with gentlemanly specialist sensibilities.

Appropriating Scientific Authority from within Amateur Realms: Nineteenth- and Twentieth-Century Ecologies of Knowledge

Rather than normalizing intellectual advance by presenting professional science as a necessary and sufficient natural kind—instead of as an abridgement of a larger set of dynamics—these revisionist histories portray the scientific revolution as existing within environments in which there existed "a state of heady confusion when it came to natural knowledge and questions of science" (10), underscoring a challenge to the standard narrative itself. This reconfiguration of the amateur applies as well to subsequent eras, even those contemporary to us when the heavily organized pursuit of science has brought a strongly realized regime of order and control into being: new information, practices, and unanticipated events nevertheless generate disruptive consequences, especially given the temporal rapidity and the global scale at which they occur. A potent case in point again is the nineteenth century, a similarly tumultuous period of intellectual and cultural upheaval in the study of nature that intrigued people from all walks of life, enticing them to engage with intimates and strangers in handling, speculating, and improvising pathways through a superabundance of natural curiosities and technological innovations that challenged experiences of space, time, and meaning. The overwhelming profusion of information about the natural world that was at hand continued to multiply, along with political pressures to expand the public sphere and to educate coming generations as citizens rather than as subjects, opening up new opportunities for a widely construed community of scientific practitioners. The explosion of popular venues, publications, and practices in the modern era worked against the considerable demarcationist efforts of elites to authorize an exclusive scientific sphere. Indeed, there is growing evidence that the success of these elites in consolidating a constricted professionalized zone of activity did *not*, however, also eliminate vernacular epistemologies of scientific knowledge and vernacular networks that nurtured the generation, expression, interpretation, and dissemination of science, renewing a need to rethink the place of the amateur and ecologies of knowledge within popular culture (Pandora 2001).

Dynamic appropriations of "the scientific" in the nineteenth century from within amateur realms are well-displayed by scholars who have taken us deeply into the literary expressions of this era, demonstrating the wide-ranging, inventive, and spirited reconfiguration, interpretation, speculation, and reconsideration given to new scientific ideas and to the questions and answers offered by specialists. In James Secord's *Victorian Sensation* (2000), a multiscaped world is reanimated by tumbling down the rabbit-hole of one nineteenth-century book, the production and reception of Robert Chambers's evolutionary-themed treatise, *Vestiges of the Natural History of Creation* (1844). Secord invokes a culture-wide set of conversations in which materials, voices, images, objects, and conceptual leaps occur across a panoramic view of British life. In both probing the book's celebrity as a prompt for re-imagining nature's realities for a startlingly diverse conglomeration of readers and as a node in a complex industrial-strength socioeconomic system powered by steam, the idea of a passive public is rendered dubious, despite the longevity of the diffusion model of scientific communication—itself a relic of the Victorian period—in positing one. A second case in point is the wide-open framework adopted for an intensive scan of earth science in this period in Ralph O'Connor's *The Earth on Show* (2007), with the tremendous vitality of the geological imaginary conveyed by taking account of hundreds of diversely interrelated primary sources. Here the first half of the nineteenth century is shown to have been introduced, seduced, and invited to conceptualize the drama of "deep time" by means of a variety of literary formats, visualizations, and performative textual renderings that evoked museum displays and exhibition hall pageantry on a cosmic scale—forms of epistemic validity that defy a reductive approach to explaining changing theoretical commitments. Another powerful demonstration of scientific knowledge created across an extensive cultural field of participants comes via Bernard Lightman's *Victorian Popularizers of Science* (2007) which utilizes the work and lives of more than thirty figures who ranged across multiple domains and endeavors as they experimented with modes of scientific popularization that both drew from a broad, deep culture of curiosity and contributed to its further expansion in intellectual, social, and cultural terms. The point is not that these densely aggregated textual, theatrical, and interpersonal cultural performances were merely additive in effect, but that they were maximally consequential in that they were experienced in relation to each other. Those of Lightman's figures who were authors of best-selling books that were as prominent (if not more so) than canonical works such as *Origin of Species* did not simply relay the thoughts of those "above" them in the intellectual hierarchy, but were designed to convey meanings of their own "while subverting the agenda of the would-be scientific professional" (xi). The most pointed challenge these works embodied was the assertion that the answer to the question of "who speaks for nature?" was that many did, amateurs and nonprofessionals included.

As various as the largely natural history/evolutionary-based cultural discussions described in these three texts were, they represent but one dimension of vernacular science during the nineteenth century, as supported within an intellectual commons. As a further example, electrical theatricalism was a site of knowledge-claiming, as explored in Iwan Rhys Morus's (1998) treatment of a world beyond the genteel theorists' sphere, following instrument makers, artisans, mechanics, and commercial entrepreneurs as they engaged in the public experimentation of electricity to metropolitan audiences in London's Galleries of Practical Science; they, too, were making modern science, through an "alternative craft-based articulation of experimental practice" (1998, xii). It is important to recognize that these epistemic touchpoints were not simply temporally proximate ones, because there were long lines of influence that fed into these cultural experiences, which they in turn would feed as well. Electricity as a public form of scientific entertainment and investigative inquiry extended back to the eighteenth century, and it would extend into the twentieth as well (de la Peña 2003; Delbourgo 2006). These public sites could be multiplied yet further, as with museums, zoos, aquaria, world's exhibitions, expeditions, films, and other media events where spectacle, entertainment and education were produced amidst competing visions of what it meant to represent nature as well as how to conceive of the constituencies that were being served (Rydell 1993; Mitman 1999; Alberti 2001; Fyfe and Lightman 2007; Nyhart 2009; Rader and Cain 2014). The ecologies of knowledge made visible in these high-profile venues are likely to have parallel knowledge-making spaces within the amateurized personal domains of private life, whether it is natural history collections and freshwater aquariums and illustrated periodicals in the parlor in the nineteenth century or home photography labs and radio hobbyists in the basement, experimental biology in residential gardens, and microcomputing hacking in the garage in the twentieth (Douglas 1989; Curry 2014; Maines 2013; Gotkin 2014). Until very recently, the stigma of the word "hobby" as signifying superficial relaxation in the twentieth century has left unrecognized the proliferation of scientific and technological activities pursued within domestic spaces, driven by individuals' own rationales, and on their own time. In a similar vein, the extracurricular experiences of children—the ultimate amateurs—with nature and science has been barely touched on by historians, but is an important aspect of science in the home as well. The nineteenth-century saw "philosophical toys" in the form of optical instruments such as phenakistoscopes and zoetropes brought into the drawing room for play, pleasure and learning at the same time that scientific specialists were using optical illusions to study how the eye and the brain function in understanding the world; in the twentieth century exploration of the solar system was being debated among adults while science fiction on radio and in comic books along with toy robots and backyard rocketry inspired imaginative "what ifs?" among youngsters (Turner 1987). A related avenue to pursue are histories of what have been broadly termed "citizen science" projects from the cold war era until the present-day, given the enormous amplification of lay assistance that is now possible via internet-powered networks (Ferris 2003; McCray 2007). The citizen science tradition is in fact under some competition from peer-to-peer grassroots scientific projects that are emerging outside of academic and commercial scientific settings (Wohlsen 2005; Delfanti 2012). Indeed, the coalescence of the open access ethic, the proliferation of interactive social media, the rapid development of digital humanities tools, and experimentation with new forms of open online scientific education and communication may result in historians of science needing to rethink our own work in light of the possibilities that now exist for amateurs outside the domain of professional history of science to make their voices heard and to design vernacular histories that speak to questions we have yet to conceive, for audiences we have yet to envision.

History of Science's Amateurs: The Question of Where the Center Lies

The study of amateurs, nonprofessionals, and science and popular culture is not simply an exploration of science in an informal public guise. Embedded within these explorations of scientific life as conceived capaciously are alternative understandings of the development of modern science as a consequence of members of a culture thinking out loud together about the natural world, not in a necessary relation of leaders and followers, but moving in and out of communities of discourse, critiquing, questioning, and negotiating matters of intellectual meaning. At times, prominent scholars have scored such histories as indulging in a careless relativism about the discipline's obligations to scientific knowledge, positing that the history of science as a research field is in danger of "losing its science" (Broad 1980; Voosen 2014). It is true that, when read collectively, these studies of an amateur realm reveal an historical density and cultural reach at odds with binary assumptions of a professional world that is complex and a lay world that is necessarily superficial. This is a challenge to assumptions of disciplinary self-definition that are likely to be exacerbated in pursuing future directions for research: the next steps to be developed require not only ranging further temporally, geographically, and topically, but in becoming cognizant of histories beyond those that are primarily middle-class, white, and Anglo-European in provenance. Toward this end, more sociologically diverse and transnational and global frameworks that include indigenous epistemologies and forms of expertise need to be generated, not simply to have a more complete picture, but because the individuals and groups involved in these histories, along with their practices and values, have traveled and circulated, and were appropriated and transformed and recirculated again, existing in tension with authorized scientific norms (Wohlforth 2005; O'Neill 2007; Tilley 2010). Bringing new figures, groups, activities and events to prominence is likely to result in recalculating historical periodizations beyond the standard narrative as well as understandings about properly constituted "science, with theoretical consequences that will continue to raise the question of whether the history of science is losing its science, or advancing its history.

References

Alberti, Samuel J. M. M. 2001. "Amateurs and professionals in one county: Biology and natural history in late Victorian Yorkshire." *Journal of the History of Biology*, 34: 115–47.

Barrow, Mark V. 1998. *A Passion for Birds: American Ornithology after Audubon*. Princeton, NJ: Princeton University Press.

Baym, Nina. 2001. *American Women of Letters and the Nineteenth-Century Sciences: Styles of Affiliation*. New Brunswick, NJ: Rutgers University Press.

Broad, William. 1980. "History of science losing its science." *Science*, 207: 389.

Bruce, Robert V. 1987. *The Launching of Modern American Science, 1846–1876*. New York: Knopf.

Chapman, Allan. 1998. *The Victorian Amateur Astronomer*. Chichester: John Wiley.

Curry, Helen Anne. 2014. "From garden biotech to garage biotech: Amateur experimental biology in historical perspective." *The British Journal for the History of Science*, 47: 539–65.

de la Peña, Carolyn Thomas. 2003. *The Body Electric: How Strange Machines Built the Modern American*. New York: New York University Press.

Delbourgo, James. 2006. *A Most Amazing Scene of Wonders: Electricity and Enlightenment in Early America*. Cambridge, MA: Harvard University Press.

Delfanti, Alessandro. 2012. "Users and peers: From citizen science to P2P science." *Journal of Science Communication*, 9: 1–5.

Desmond, Adrian. 1992. *The Politics of Evolution: Morphology, Medicine, and Reform in Radical London*. Chicago: University of Chicago Press.

Douglas, Susan. 1989. *Inventing American Broadcasting, 1899–1922*. Baltimore: Johns Hopkins University Press.

Drouin, Jean-Marc and Bernadette Bensaude-Vincent. 2006. "Nature for the people." In *Cultures of Natural History*, edited by Nicholas Jardine, James A. Secord, and Emma C. Spary, 408–25. Cambridge: Cambridge University Press

Ferris, Timothy. 2003. *Seeing in the Dark: How Amateur Astronomers are Discovering the Wonders of the Universe*. New York: Simon and Schuster.

Fyfe, Aileen and Bernard Lightman (eds.) 2007. *Science in the Marketplace: Nineteenth-Century Sites and Experiences*. Chicago: University of Chicago Press.

Gates, Barbara. 1998. *Kindred Spirits: Victorian and Edwardian Women Embrace the Living World*. Chicago: University of Chicago Press.

Gates, Barbara (ed.) 2002. *In Nature's Name: An Anthology of Women's Writing and Illustration, 1780–1930*. Chicago: University of Chicago Press.

Gates, Barbara, and Ann B. Shteir (eds.) 1997. *Natural Eloquence: Women Reinscribe Science*. Madison: University of Wisconsin Press.

Geertz, Clifford. 1980. "Blurred genres: The refiguration of social thought." *The American Scholar*, 49: 165–79.

Gianquitto, Tina. 2007. "*Good Observers of Nature*": *American Women and the Scientific Study of the Natural World, 1820–1885*. Athens: University of Georgia Press.

Gieryn, Thomas F. 1983. "Boundary-work and the demarcation of science from non-science: Strains and interests in professional ideologies of scientists." *American Sociological Review*, 48: 781–95.

Goldstein, Daniel. 1994. "'Yours for Science': The Smithsonian Institution's correspondents and the shape of scientific community in nineteenth-century America." *Isis*, 85: 573–99.

Gotkin, Kevin. 2014. "When computers were amateur." *Annals of the History of Computing, IEEE.* 36: 4–14.

Hall, A. R. 1983. "Science." In *The Diary of Samuel Pepys.* Volume X: *Companion,* edited by Robert Latham and William Matthews, 381–90. Berkeley: University of California Press.

Harkness, Deborah. 2007. *The Jewel House: Elizabethan London and the Scientific Revolution.* New Haven, CT: Yale University Press.

Holloway, Jonathan Scott and Ben Keppel (eds.) 2007. *Black Scholars on the Line: Race, Social Science, and American Thought in the Twentieth Century.* South Bend, IN: University of Notre Dame Press.

Hunt, Lynn (ed.) 1989. *The New Cultural History.* Berkeley: University of California Press.

Jardine, Nicholas, James A. Secord, and Emma C. Spary (eds.) 1996. *Cultures of Natural History.* Cambridge: Cambridge University Press.

Keeney, Elizabeth. 1992. *The Botanizers: Amateur Scientists in Nineteenth-Century America.* Chapel Hill: University of North Carolina Press.

Kohlstedt, Sally Gregory. 1976. "The nineteenth-century amateur tradition: The case of the Boston Society of Natural History." In *Science and its Public: The Changing Relationship,* edited by Gerald Holton and William A. Blanpied, 173–90. Dordrecht: D. Reidel.

Kuklick, Henrika and Robert E. Kohler (eds.) 1996. *Science in the Field.* Special Issue, *Osiris,* 11.

Lightman, Bernard. 2007. *Victorian Popularizers of Science: Designing Nature for New Audiences.* Chicago: University of Chicago Press.

Maines, Rachel. 2013. "'Stinks and bangs': Amateur science and gender in twentieth-century living spaces." *Icon,* 19: 33–51.

Manning, Kenneth R. 1983. *Black Apollo of Science: The Life of Ernest Everett Just.* Oxford: Oxford University Press.

McCray, W. Patrick. 2007. *Keep Watching the Skies! The Story of Operation Moonwatch & the Dawn of the Space Age.* Princeton, NJ: Princeton University Press.

McLaughlin-Jenkins, Erin. 2003. "Walking the low road: The pursuit of scientific knowledge in late Victorian working-class communities." *Public Understanding of Science,* 12: 147–66.

Mitman, Gregg. 1999. *Reel Nature: America's Romance with Wildlife on Film.* Cambridge, MA: Harvard University Press.

Morus, Iwan Rhys. 1998. *Frankenstein's Children: Electricity, Exhibition, and Experiment in Early-Nineteenth-Century London.* Princeton, NJ: Princeton University Press.

Norwood, Vera. 1993. *Made from This Earth: American Women and Nature.* Chapel Hill: University of North Carolina Press.

Nyhart, Lynn. 2009. *Modern Nature: The Rise of the Biological Perspective in Germany.* Chicago: University of Chicago Press.

O'Connor, Ralph. 2007. *The Earth on Show: Fossils and the Poetics of Popular Science, 1802–1856.* Chicago: University of Chicago Press.

O'Neill, Daniel. 2007. *The Firecracker Boys: H-Bombs, Inupiat Eskimos and the Roots of the Environmental Movement.* New York: Basic Books.

Ogilvie, Marilyn Bailey. 2000. "Obligatory amateurs: Annie Maunder (1868–1947) and British women astronomers at the dawn of professional astronomy." *The British Journal for the History of Science,* 33: 67–84.

Pandora, Katherine. 2001. "Knowledge held in common: Tales of Luther Burbank and science in the American vernacular." *Isis,* 92: 484–516.

Rader, Karen, and Victoria Cain. 2014. *Life on Display: Education, Exhibition and Museums in the Twentieth-Century United States.* Chicago: University of Chicago Press.

Rossiter, Margaret. 1982. *Women Scientists in America: Struggles and Strategies to 1940.* Baltimore: Johns Hopkins University Press.

Rudwick, Martin. 1985. *The Great Devonian Controversy: The Shaping of Scientific Knowledge among Gentlemanly Specialists*. Chicago: University of Chicago Press.

Rydell, Robert W. 1993. *World of Fairs: The Century-of-Progress Expositions*. Chicago: University of Chicago Press.

Secord, Anne. 1996. "Artisan botany." In *Cultures of Natural History*, edited by Nicholas Jardine, James A. Secord, and Emma C. Spary, 378–93. Cambridge: Cambridge University Press.

Secord, James A. 2000. *Victorian Sensation: The Extraordinary Publication, Reception, and Secret Authorship of* Vestiges of the Natural History of Creation. Chicago: University of Chicago Press.

Smith, Pamela. 2004. *The Body of the Artisan: Art and Experience in the Scientific Revolution*. Chicago: University of Chicago Press.

Tilley, Helen. 2010. "Global histories, vernacular science, and African genealogies; Or, is the history of science ready for the world?" *Isis*, 101: 110–19.

Topham, Jonathan. 2009. "Introduction: Focus Section on historicizing 'popular science.'" *Isis*, 100: 310–18. (Plus the accompanying articles of this special Focus Section.)

Torrens, Hugh. 1995. "Presidential Address: Mary Anning (1799–1847) of Lyme; 'The greatest fossilist the world ever knew.'" *British Journal for the History of Science*, 28: 257–84.

Turner, Gerard L'E. 1987. "Scientific toys." *The British Journal for the History of Science*, 20: 377–98.

Voosen, Paul. 2014. "Historians of science seek détente with their subject." *Chronicle of Higher Education*, Research, May 27, 2014. Accessed September 14, 2015 http://chronicle.com/article/Historians-of-Science-Seek/146759/.

Welch, Margaret. 1998. *The Book of Nature: Natural History in the United States, 1825–1875*. Boston: Northeastern University Press.

Wohlforth, Charles. 2005. *The Whale and the Supercomputer: On the Northern Front of Climate Change*. New York: Macmillan.

Wohlsen, Marcus. 2011. *Biopunk: DIY Scientists Hack the Software of Life*. New York: Current.

CHAPTER ELEVEN

The Man of Science

PAUL WHITE

The expression "man of science" is scarcely used today: it sounds old fashioned or indeed, with its explicit gendering of scientific persona, highly suspect. Yet it was a distinctly modern invention, becoming the most common generic term for a scientific practitioner in Britain and North America from the middle of the nineteenth century through the 1920s. It appears as early as the seventeenth century, but was used simply to designate a man of knowledge or learning rather than a collective pursuit or vocation. It gains currency during a period of profound change in the social structure and disciplinary formation of knowledge that some have likened to the second scientific revolution. Most of these developments have received considerable scholarly attention: the founding of specialist societies and journals, the rise of disciplines, the creation of new, broader audiences and criteria of participation, and the forging of institutional structures of science in government, industry, and education. It is perhaps not surprising that a new scientific identity was called forth by, and helped to usher in, these transformations. But questions of identity or personae are relatively recent in the historiography of science, and combine considerations of public image, rhetoric and ideology on the one hand, and personal character, ethos, and selfhood on the other (Daston and Sibum 2003; Outram 1984). The form that scientific identity took in the nineteenth century is comparatively little known and often misunderstood. It is widely assumed that the new type of practitioner who emerged was the scientist, that familiar figure we now associate with the laboratory and the white coat. Many are aware that the Cambridge philosopher William Whewell coined the word "scientist" in 1834, but for a variety of reasons it was not taken up until nearly a hundred years later (Ross 1962). Why then did the "man of science" emerge in its stead? What significance did the term hold for nineteenth-century practitioners and what purposes did it serve? How does it compare with scientific identities in other language cultures, where social conditions were often quite different?

At the beginning of the nineteenth century, most persons engaged in serious scientific work would have been called naturalists or natural philosophers, depending on the

A Companion to the History of Science, First Edition. Edited by Bernard Lightman.
© 2016 John Wiley & Sons Ltd. Published 2020 by John Wiley & Sons Ltd.

orientation of their research, or they would have been identified by their special branch of study. Many of the modern terms that designate specialty were in common use in the eighteenth century or before, thus zoologist, geologist, and botanist (but not yet physicist or biologist). The older generic terms persisted, but were increasingly problematic. "Naturalist" for some connoted a lack of expertise, especially as laboratory approaches grew more authoritative in medicine and the life sciences (Cunningham and Williams 1992). "Natural philosopher" became something of an oxymoron as the methods and results of experimental physics, chemistry, and applied mathematics grew remote from anything resembling traditional philosophy (Yeo 1993). This was already an issue in the 1830s. Whewell's remarks on coining the word scientist suggest an overriding concern with fragmentation: "the mathematician turns away from the chemist; the chemist turns away from the naturalist...." Yet he addressed the problem with a joke, introducing "scientist" in a derogatory fashion as a word akin to atheist, sciolist (one who merely pretends to knowledge), and even tobacconist (Whewell 1834, 59; a more sober plea appears in Whewell 1840, 1: cxiii). In other writings, Whewell hinted that his worry was less about over-specialization, and more about the hubris of science. Influential works of the period, such as John Herschel's *Preliminary Discourse on Natural Philosophy* (1831), presented science as the most authoritative or exclusive form of knowledge, a truth akin to revelation, a utopian force of progress (Secord 2014). Reviewing Herschel's *Discourse*, Whewell drew a sharp contrast between philosophy as the pursuit of general laws and causes, and science as the working out of technical details: "the two characters differ like that of a great general and of a good engineer" (Whewell 1831, 65). A scientist, for Whewell, thus denoted a narrow specialist, one who lacked authority outside of a particular domain, one who surrendered hard-gained facts to the superior mind of the theorist.

It was precisely this implied subordination of science to other forms of knowledge that Whewell's contemporaries rejected. Many agreed with the elevation of the heroic discoverer or generalizer above the humble gatherer of facts, but they asserted this distinction *within* science itself. In his widely discussed *Reflections on the Decline of Science in England* (1830), Charles Babbage proposed that knowledge-making be organized like a factory, with much of its production subject to mechanization, or like an army, with a mass of foot-soldiers and a coterie of commanding officers, or indeed like a parliamentary government. This kind of rank-and-file structure was one of the aims behind the founding of the British Association for the Advancement of Science in 1831, as well as many specialist societies around the same time: a hierarchy of scientific observers and collectors arranged under an elite group of practitioners, many of whom were based at metropolitan centers (Morrell and Thackray 1981). Such a neat division of labor was rarely achieved in practice, and was resisted in a variety of ways; however, the community of scientific practitioners as it evolved over the first half-century was an increasingly heterogeneous body. The expression "man of science" came to prevail in part because it could designate this mixed community: persons of technical expertise as well as broad interests and, so it was claimed, superior powers of mind.

Significantly, a man of science was not a paid professional. Science did not become a career in the modern sense until the end of the nineteenth century. France and the German lands, where a structure of salaried positions was established in major universities by the early decades, present possible exceptions, but these were just a handful of elite posts (Fox and Weisz 1980; Cahan 2003; Gillispie 2004). In Britain and North America, no systematic course of training or career path would emerge until the last quarter of the century, when the pattern of state-funded, academic research science became more dominant. Paid work was scarce and often lowly in status: instrument-making, specimen-selling, curating, illustrating, translating, editing. Even professorships might provide insufficient means to support a household, and so had to be supplemented by other work, such as private lecturing and tutoring, examining, or review writing. In the United States, a substantial income could be gained through scientific consulting in industry and agriculture, a career pioneered in the 1830s by Benjamin Silliman Sr. while professor of chemistry and natural history at Yale (Lucier 1995). In the absence of remunerative posts, science was undertaken by persons in a variety of other occupations or of independent means, such as landed property. In Britain, the role of the clergyman-naturalist appealed to Darwin until the good fortune of the *Beagle* voyage opened other avenues, and the prospect of inheriting substantial wealth made a paid career unnecessary. Darwin's neighbor, John Lubbock, also the heir to a large fortune, was first an entomologist, then a banker, then an anthropologist and Member of Parliament. Voyaging and collecting had long provided entry into scientific society, but again there was no established career pattern. Some travelled as independent gentlemen, others as ship's surgeons, others as specimen hunters for hire. Thomas Huxley voyaged to the South Pacific as an assistant surgeon, collecting and dissecting marine invertebrates in between his ship's duties. Ambitious for a research career in zoology, he quit the Navy in hopes of gaining a university lectureship, surviving for four years on journalism and translating until a position opened at the Museum of Practical Geology in London. But the income from his post was insufficient for a growing family, and he spent the rest of his life campaigning to improve the place of science in British education and to increase its status in general culture (White 2003).

For even the most ardent reformers, however, the ideal of scientific work that prevailed in this period was not one of paid profession, but of vocation. Science was conceived as a calling, pursued for its intrinsic merits and with a sense of higher purpose that was quite apart from any material or financial gain. The usefulness of knowledge, and the relationship between science, commerce, and industry, were vexed and controversial. The role of scientific expertise in government, manufacturing, and the military increased substantially over the course of the century. Yet positions in industrial chemistry, electricity, and so forth were often not regarded as truly scientific. Engineering, one of the new technical professions to emerge in the first part of the century, and one of the most celebrated in the literature on national progress, remained ambiguous in relation to science (Marsden and Smith 2005; Marsden 2013). In Britain, the state established a new teaching institution, the Royal School of Mines, to train experts in the industrial arts, but its numbers remained small, and most businesses were reluctant to hire its graduates. The universities that later arose in manufacturing towns like Birmingham and Manchester, typically founded by local industrialists, developed quickly into institutions of liberal education in imitation of Oxford and Cambridge. The ethos of scientific life thus remained one in which practical considerations were firmly secondary. Links to utility were asserted indirectly through a discourse of future benefits and by-products, of civic-mindedness and public spiritedness, so that claims to disinterestedness could be preserved. This was another important feature of the 'man of science': he was purportedly a moral figure standing outside politics and above private interests, and therefore able to provide knowledge that was truly beneficial to all of society. This equation was epitomized by the Royal Institution and its professors, from Humphry Davy to John Tyndall, all of whom did extensive work in relation to agriculture, mining, industrial chemistry, and navigation, while maintaining an image of selflessness and public service (Berman 1978).

The virtues of men of science could assume heroic proportions in biographical writing. Tyndall's life of Michael Faraday extolled the sacrifices made for truth, the purity of purpose, and the special powers of mind required for scientific discovery (Tyndall 1868). Such portraits drew on a romantic tradition of genius, an intellectual force that was innate like nobility, but that often resided in those of humble birth. Its characteristics—intuition, mental suppleness, refined discrimination—marked out a few individuals as destined to be leaders of men and spirits of the age (Schaffer 1990). In the nineteenth century, genius was usually coupled with the virtues of industriousness and hard work, becoming an endowment of the self-made man who had to labor and struggle for truth. Such qualities were drawn from more general models of middle-class manhood in a society still partly governed by inherited titles and noble birth. Even Darwin, a gentleman of leisure, continually characterized his scientific activity as hard work, tallying up days, months and years on which he spent writing each of his books (De Beer 1959). Francis Galton's 1874 survey, *English Men of Science: Their Nature and Nurture*, compiled over a hundred comparable accounts: men who walked 50 miles a day without fatigue in search of specimens, men who worked habitually until two or three in the morning. Using autobiographical testimony, he documented their leading characteristics as perseverance, steadiness, determination, and "the secretion of nervous force": "many have laboured as earnest amateurs in extra professional hours long into the night ... they have climbed the long and steep ascent from the lower to the upper ranks of life" (Galton 1874, 75). Possessing a rare combination of virtues in the highest degree, men of science were thus like an intellectual peerage or a new priesthood.

Scholars have noted parallels between the reorganization of knowledge in scientific institutions and political reforms during the first half-century that challenged traditional webs of power and patronage toward the landed aristocracy and the crown (MacLeod 1983). Men of science often aligned themselves with social causes that fell under the banner of progress. They positioned themselves as critics, educators, and agents of improvement. In so doing, they sought to gain the confidence not only of government elites, but of newly enfranchised classes, a diverse body of publics from educated readers of highbrow newspapers and periodicals, to artisans and workers who sought knowledge as a means of social improvement. If men of science no longer openly canvassed aristocratic patrons, they did court large audiences: in public lectures, extension courses, works of popularization, and the vastly expanded periodical press. Scientific controversies over evolution, matter, and energy were aired in literary magazines, where the moral, religious, or political implications of technical research were drawn out and debated with theologians, philosophers, economists, and statesmen (Cantor et al. 2004).

Such heated exchanges have been interpreted by some scholars as a contest of authority between a new scientific profession and an old entrenched one (the Anglican clergy), or as part of a more general class struggle between the industrial middle classes and the traditional landed elite (Young 1985; Turner 1993; Desmond 1998). Such interpretations are hard to sustain if one looks closely at the religious convictions or the social station of most men of science. The point is well-illustrated by the career of Joseph Dalton Hooker who, as director of the Royal Botanic Gardens at Kew, examiner for the University of London, and employee of the Department of Public Works, stood at the center of an expanding network of science in the service of agricultural improvement, technical education, and colonial administration. Hooker was a leading reformer and a new type of salaried, institution-based practitioner, and yet he conducted himself as a landed aristocrat with Kew gardens as his titled estate, drawing scientific authority from his gentlemanly manners, and maintaining allegiance to the Anglican church (Endersby 2008). He regarded many of the centralizing, democratizing, and secularizing reforms that in retrospect seem fundamental to modern science as demeaning and disgraceful impediments to the pursuit of knowledge.

A vast body of scholarship now questions the "conflict" model of science and religion, epitomized by public controversies between clergymen and men of science, such as the 1860 Oxford debate between Samuel Wilberforce and Thomas Huxley (Brooke 1991; Fyfe 2004). Yet underlying assumptions about the secularization of scientific life in the nineteenth century persist (Shapin 2008). Certainly, natural theology as a framework for scientific research and writing waned after mid-century, but then so did its importance for theologians. Clerical membership in the most prestigious scientific societies declined appreciably also, but the authority of clergymen as commentators did not. British Association meetings, if they did not always feature an Anglican bishop squaring off with a professor of zoology, were still accompanied by sermons in local churches, their remarks on leading addresses covered in detail by the press. Both publically and privately, Christian culture endured, and respect for religion was crucial for full participation in social and political life. Some of the most protracted debates between men of science and members of the clergy were focused on educational institutions and curricula. The state, which assumed more control over education at all levels, was increasingly non-sectarian, but not secular. Scientific subjects were introduced to schools and universities through appeals to liberal education, mental and moral discipline, and as complementary to traditional classical and religious learning. Such reforms established a division of scientific and religious authority, not a triumph of one over the other. As they gained prominence in education, government committees, and the periodical press, men of science became part of a broad community of learned elites who saw themselves as the collective bearers of natural knowledge and Christian culture (White 2005).

How did such an elite conception of scientific identity apply to the greatly enlarged community of practitioners that had come to exist by the middle decades of the century, the museum curators, geological surveyors, and artisan botanists, the gentlemen and women who microscopized and dissected in their homes, the public lecturers and demonstrators, the popular authors and periodical writers? Were all of these practitioners men of science? The expression was often used inclusively. It might refer to the fellows of the Royal Society, or to the much larger and more heterogeneous membership of the British Association (Barton 2003). The man of science functioned as a collective persona, evoked in circumstances where distinctions between science and other forms of work or culture needed to be enunciated, or when the interests of a larger community were being defended. Thus, 54 members of the British Association

wrote as "men of science" to the head of the Treasury on the inadequate provision
for natural history by the present board of trustees of the British Museum, dominated
as it was by "men of rank and general attainments" (*Parliamentary Papers*, 1847, 34:
253–6). In moments of crisis or controversy, the term could convey a strong sense of
solidarity between practitioners. The zoologist St George Mivart wrote to Darwin in
defense of a highly critical review of *Descent of Man,* claiming his place in the community to which they both belonged: "as a man of science I have no choice but to pursue
'truth' to the best of my ability" (Burkhardt and Smith 2013, 13).

In a period when participation in the sciences was increasing dramatically, and where
the boundaries of the scientific community were highly permeable, the man of science
could move between the elite and the popular, connoting eminence and distinction
on the one hand, and on the other hand a pursuit open to almost anyone with the
requisite skills and devotion. Much of the literature of scientific popularization, as
well as more specialist educational works, promoted this view of science as a form of
mental discipline and body of knowledge ostensibly open to all (Lightman 2007). As
late as 1887, *Nature* magazine declared: "Science has gathered around it a crowd of
workers ... they take part [in] the investigation of truth for truth's sake alone. They
may be professors or manufacturers, soldiers or physicians. If only they are imbued
with the desire to penetrate a little further into the mysteries which surround them,
if only they are willing to add something to the sum of human knowledge, they are
scientific men" (Anon. 1887, 217). But the man of science could also designate an
elite within a larger community, or an ideal to which the many should aspire but
could not hope to obtain. Galton asserted that there were no more than 300 men of
science in all of Britain, a number that would not even extend to all fellows of the
Royal Society. The criteria that he used were largely self-validating, such as holding a
professorial chair, a position of leadership, or a medal for scientific work, all of which
required election by elite committee (Galton 1874, 2–7). Such standards could make
the scientific community seem like a private gentlemen's club.

The exclusivity of the title is most clear with regard to gender. There was no
monopoly of knowledge in the nineteenth century by the male sex. But the "man of
science" was a form of masculinity: a cultural type composed of highly gendered characteristics and virtues. Women's minds and bodies were considered softer and weaker;
their intellects could be sensitive and refined, but not penetrating or rigorous. If they
devoted themselves exclusively to science or intellectual work of any kind, they might
be criticized for abandoning their proper sphere and even losing their woman's nature
(Russett 1989; Richards 1997). As is well known, women participated in science at
a variety of levels, as observers and experimenters, illustrators, assistants, and indeed
at what was traditionally the highest level, that of patron (Shteir 1996; Gianquitto
2007). Yet at the same time, new spaces of work and exchange opened in the early
decades from which they were strictly excluded. Women could not enter universities
or belong to the leading scientific societies. They could attend meetings of the British
Association, but only as members of the audience, and if a paper by a woman was
accepted, it was read by a man. Women formed their own groups to read and discuss
science, usually under the polite banner of Literary and Philosophical Societies. Correspondence was a more egalitarian space, and many women were able to participate
in science through letter-writing; however, public recognition for their contributions
was rare. The eventual admittance of women to institutions of higher education and

the opening of careers in medicine and science, coincides with the gradual eclipse of the "man of science" by the gender-neutral "scientist" (though, of course, gender prejudices and inequities have persisted).

The "man of science" was an Anglo-American construct and had no exact counterpart in other language cultures. The proliferation of specialties and disciplines was a general feature of nineteenth-century science, but it did not give rise to the same identity problems that prompted Whewell and his contemporaries to search for a new collective persona. In France, the generic term for learned person, "savant", continued to designate scientific practitioners until the early twentieth century. This may have reflected the more established place that the sciences held among other branches of learning within state institutions. From the days of the French Revolution and Napoleon, there were faculties of science within the universities, science degrees and concentrations at secondary schools, and prestigious technical colleges, all state-funded and under the direction of a ministry of public instruction (Fox 2012). In the German universities also, the sciences were well represented. They were not set in opposition to other forms of learning as they were in Britain; instead all forms of knowledge were conjoined under the heading *Wissenschaft*. But the career of the research professor and the state-funded research institute were largely German innovations, and German laboratories drew many students from Britain and America in the physical and life sciences. In German culture, a new scientific identity did emerge in roughly the same period as the man of science and with many of the same connotations: *Naturforscher*. The reasons for this may lie in the special role played by private societies in challenging and extending the boundaries of scientific activity outside of the universities. The British Association had been modelled in part on the Gesellschaft Deutscher Naturforscher und Ärtz founded in Leipzig in 1822. A large number of "second tier" societies were established from the 1830s onwards, especially on the provincial level, where the study of nature became part of the self-fashioning of new middle-class professionals: schoolteachers, medical men, chemists, and engineers (Phillips 2012).

A final context to consider in the formation of scientific identity is the European colonies. There was a long tradition of naturalists on survey expeditions, as well as naval officers and medical personnel whose duties extended to scientific observation and collecting. In the early nineteenth century, a specifically romantic ideal of discovery emerged in connection with voyages to the "New World" (Leask 2002). This was epitomized in the work of the Prussian naturalist Alexander von Humboldt, whose lavish descriptions of Spanish and Portuguese America provided Europeans with a vision of exotic nature that needed to be mapped, cataloged, and measured with elaborate scientific precision. Through an ambitious series of publications, Humboldt helped to fashion a heroic model of expeditionary science that combined a style of personal narrative and aesthetic appreciation with mechanical exactitude and philosophical vision (Dettlebach 1996). But over the course of the century, as exploration and commercial prospecting gave way to settlement and eventually, government, a pattern that had been followed since the first European voyages to the West Indies, new forms of identity were needed for those who would make their homes in the colonies.

It would be an easy matter if the Anglo-American 'man of science,' the French *savant*, or the German *naturforscher* were merely exported to the places where Europeans settled. But this view of science and its personnel spreading to different parts of the world by a simple process of diffusion is in fact a legacy of nineteenth century imperialism, and a part of its expansionist ideology. The role that the sciences played in European empires is evident, for example, in theories of racial superiority and oriental-ism, which exoticized and reclassified native peoples and traditions in Western terms, or in material transformations, most visible in parts of the West Indies and British India where whole landscapes were converted into plantations, with botanical gardens serv-ing as commercial outposts and centers of exchange (Drayton 2000). But science was not just a tool of empire. The mapping, classifying, and collection of "exotic" nature relied upon substantial local knowledge and labor, and was instrumental in the reshap-ing of European perspectives, products, and taste. The rhetoric of improvement may have been one-sided, but in practice colonization was a trans-cultural process affecting European settlers as well as native peoples (Pratt 2008, Sivasundaram 2010).

The identities that emerged for science in colonial settings were thus extremely diverse and often hybrid in nature. In Latin America, Humboldt's Eurocentric writ-ings extolling progress and civilization could be appropriated by creole elites to fash-ion themselves as liberal "white" nationals. The statesman Simon Bolivar followed in Humboldt's footsteps in attempting to ascend the highest peak in the Andes, and seized the Prussian's rhetoric of sublime nature to describe his own vision of a conti-nent free from European dominion (Pratt 2008, 177–8). In China, where Protestant missionaries promoted science as an instrument of religious conversion, empirical or experimental approaches to knowledge were re-presented as textual and philosophical traditions in order to align them with Confucianism and Daoism (Elshakry 2010). Where European rule was more established, men of science in London or savants in Paris might be nominally in control of networks of observation and collection in the colonies, overseeing local diplomats, military men, engineers, doctors, or geographers, acquiring specimens, herbaria, or artifacts, and communicating papers from abroad to learned societies back home. But in practice, such networks were more tangled, and authority within them was more widely distributed. Colonial regimes were often built on class systems and structures of patronage that were already in place. Local gov-ernors might set themselves up like native kings, taking up residence on the site of an ancient palace and gardens (Sivasundaram 2013). A British ethnographer might impersonate a Muslim pilgrim, or ghost author an epic poem of Sufi wisdom. Richard Francis Burton began a career in the foreign service as a captain in the army of East India Company, later working as a cartographer for the Royal Geographical Society, and a British consul in Fernando Po, Damascus, and Santos (Brazil). He was a founder of the London Anthropological Society, a fount of scientific racism in Britain, yet he was notorious for his knowledge of foreign languages and customs, often travelling in disguise to observe native practices or to visit places such as Mecca from which Euro-peans were excluded. He produced a series of controversial translations, such as *Kama Sutra*, *A Thousand and One Nights*, and *Kasidah*, an epic poem by a Sufi mystic, who was in fact an invention of Burton himself (Kennedy 2005). An orientalist and spy, a white supremacist who was dubbed a "White nigger" by his fellow officers, Burton was only an extreme case of the kind of mediator and go-between who emerged in the "contact zone" between European and native cultures.

It is striking that, in spite of these national and global variations, the terms now used to designate a scientific practitioner all gain currency about the same time. *Der wissenschaftler*, *le/la scientifique*, like "scientist," came into widespread use from the 1920s onwards as the career progression from undergraduate coursework and advanced degree to university teaching and research was standardized internationally. There are important continuities between the man of science and the modern scientist, especially the heroic depiction of great discoverers or theorists, the public image of purity, and the more private sense of a vocation whose main motive is truth. In the late nineteenth century, however, this ethos began to be detached from the individual "man" and identified instead with a highly formalized and impersonal system of training, and relocated within new research institutions that defined themselves as havens from commercial and political interests. Speaking at the new Johns Hopkins University, the professor of physics, Henry Rowland, condemned the common practice of private consulting as disreputable and corrupting for the true scientist who must aspire to "something higher and nobler" (Rowland 1883, 243; Lucier 2009). Such pleas for "pure science" were typically made from institutions financed by wealthy donors or lucrative government contracts. Possessing the job security and institutional prestige that men of science had lacked, modern scientists became a more insular community, with closed meetings and peer review, and a highly technical language comprehensible only to the specialist. Scientists might occasionally address the public, but they have lost the sense of civic duty and public mission, together with the gentlemanliness and masculinity, and the links with Christian and classical culture, that were defining features of their predecessors (Porter 2014). The scientist does not, as a matter of vocation, campaign for social and political reform, or engage in moral or religious debate, unless perhaps to defend the very purity of science itself. The connotations of the term that Whewell intended, and that his contemporaries rejected as derogatory, are now apt.

References

Anon. 1887. "Professor Tyndall and the scientific movement," *Nature*, 36, No. 923 (July 7): 217–18.

Babbage, Charles. 1830. *Reflections on the Decline of Science in England, and on Some of Its Causes.* London: Fellowes.

Barton, Ruth. 2003. "'Men of Science': Language, identity and professionalization in the mid-Victorian scientific community." *History of Science*, 41: 73–119.

Berman, Morris. 1978. *Social Change and Scientific Organization: The Royal Institution, 1799–1844.* Ithaca, NY: Cornell University Press.

Brooke, John. 1991. *Science and Religion: Some Historical Perspectives.* Cambridge: Cambridge University Press.

Burkhardt, Frederick, and Sidney Smith (eds.) 2013. *The Correspondence of Charles Darwin*, vol. 20. Cambridge: Cambridge University Press.

Cahan, David (ed.) 2003. *From Natural Philosophy to the Sciences: Writing the History of Nineteenth–Century Science.* Chicago: University of Chicago Press.

Cantor, Geoffrey, Gowan Dawson, Graeme Gooday, Richard Noakes, Sally Shuttleworth, and Jonathan Topham (eds.) 2004. *Science in the Nineteenth-Century Periodical: Reading the Magazine of Nature.* Cambridge: Cambridge University Press.

Cunningham, Andrew, and Perry Williams (eds.) 1992. *The Laboratory Revolution in Medicine.* Cambridge: Cambridge University Press.

Daston, Lorraine, and Otto Sibum. 2003. "Scientific personae and their histories." *Science in Context*, 16: 1–8.

De Beer, Gavin (ed.) 1959. "Darwin's journal." *Bulletin of the British History (Natural History). Historical Series*, 2: 3–21.

Desmond, Adrian. 1998. *Huxley: From Devil's Disciple to Evolution's High Priest*. Harmondsworth: Penguin.

Dettlebach, Michael. 1996. "Global physics and aesthetic empire: Humboldt's physical portrait of the tropics." In *Visions of Empire: Voyages, Botany, and Representations of Nature*, edited by David Philip Miller and Peter Hans Reill, 258–92. Cambridge: Cambridge University Press.

Drayton, Richard. 2000. *Nature's Government: Science, Imperial Britain and the "Improvement" of the World*. New Haven, CT: Yale University Press.

Elshakry, Marwa. 2010. "When science became Western: Historiographic reflections." *Isis* 101: 98–109.

Endersby, Jim. 2008. *Imperial Nature: Joseph Hooker and the Practices of Victorian Science*. Chicago: University of Chicago Press.

Fox, Robert. 2012. *The Savant and the State: Science and Cultural Politics in Nineteenth-Century France*. Baltimore: Johns Hopkins University Press.

Fox, Robert, and George Weisz (eds.) 1980. *The Organization of Science and Technology in France, 1808–1914*. Cambridge: Cambridge University Press.

Fyfe, Aileen. 2004. *Science and Salvation: Evangelical Popular Science Publishing in Victorian Britain*. Chicago: University of Chicago Press.

Galton, Francis. 1874. *English Men of Science: Their Nature and Nurture*. London: Macmillan.

Gianquitto, Tina. 2007. *"Good Observers of Nature": American Women and the Scientific Study of the Natural World, 1820–1885*. Athens, GA: University of Georgia Press.

Gillispie, Charles Coulton. 2004. *Science and Polity in France: The Revolutionary and Napoleonic Years*. Princeton, NJ: Princeton University Press.

Herschel, John. 1830. *Preliminary Discourse on the Study of Natural Philosophy*. London: Longman.

Kennedy, Dane Keith. 2005. *The Highly Civilized Man: Richard Burton and the Victorian World*. Cambridge, MA: Harvard University Press.

Leask, Nigel. 2002. *Curiosity and the Aesthetics of Travel Writing, 1770–1840*. Oxford: Oxford University Press.

Lightman, Bernard. 2007. *Victorian Popularizers of Science: Designing Nature for New Audiences*. Chicago: University of Chicago Press.

Lucier, Paul. 1995. "Commercial interests and scientific disinterestedness: Consulting geologists in antebellum America." *Isis*, 86: 245–67.

Lucier, Paul. 2009. "The professional and the scientist in nineteenth-century America." *Isis*, 100: 699–732.

MacLeod, Roy. 1983. "Whigs and savants: Reflections on the reform movement in the Royal Society, 1830–48." In *Metropolis and Province: Science in British Culture, 1780–1850*, edited by Ian Inkster and Jack Morrell, 55–90. Philadelphia: University of Pennsylvania Press.

Marsden, Ben. 2013. "Re-reading Isambard Kingdom Brunel: Engineering literature in the early nineteenth century." In *Uncommon Contexts: Encounters Between Science and Literature, 1800–1914*, edited by Ben Marsden, Hazel Hutchinson, and Ralph O'Connor, 83–109. London: Pickering and Chatto.

Marsden, Ben, and Crosbie Smith. 2005. *Engineering Empires: A Cultural History of Technology in Nineteenth-Century Britain*. Basingstoke: Palgrave Macmillan.

Morrell, Jack, and Arnold Thackray. 1981. *Gentlemen of Science: Early Years of the British Association for the Advancement of Science*. Oxford: Oxford University Press.

Outram, Dorinda. 1984. *Georges Cuvier: Vocation, Science and Authority in Post-Revolutionary France*. Manchester: University of Manchester Press.

Phillips, Denise. 2012. *Acolytes of Nature: Defining Natural Science in Germany, 1770–1850.* Chicago: University of Chicago Press.

Porter, Theodore. 2014. "The fate of scientific naturalism: From public sphere to professional exclusivity." In *Victorian Scientific Naturalism: Community, Identity, Continuity,* edited by Bernard Lightman and Gowan Dawson, 265–87. Chicago: University of Chicago Press.

Pratt, Mary Louis. *Imperial Eyes: Travel Writing and Transculturation,* 2nd edition. London: Routledge, 2008.

Richards, Evelleen. 1997. "Redrawing the boundaries: Darwinian science and Victorian women intellectuals." In *Victorian Science in Context,* edited by Bernard Lightman, 119–42. Chicago: University of Chicago Press, 1997.

Ross, Sydney. 1962. "'Scientist': The story of a word." *Annals of Science,* 18: 65–85.

Rowland, Henry. 1883. "A plea for pure science." *Science,* 29: 242–50.

Russett, Cynthia. 1989. *Sexual Science: The Victorian Construction of Womanhood.* Cambridge: Cambridge University Press.

Schaffer, Simon. 1990. "Genius in romantic natural philosophy." In *Romanticism and the Sciences,* edited by Andrew Cunningham and Nicolas Jardine, 82–98. Cambridge: Cambridge University Press.

Secord, James A. 2014. *Visions of Science: Books and Readers at the Dawn of the Victorian Age.* Chicago: University of Chicago Press.

Shapin, Steven. 2008. *The Scientific Life: A Moral History of a Late Modern Vocation.* Chicago: University of Chicago Press.

Shteir, Ann. 2006. *Cultivating Women, Cultivating Science: Flora's Daughters and Botany in England, 1760–1860.* Baltimore: Johns Hopkins University Press.

Sivasundaram, Sujit. 2010. "Sciences and the global: On methods, questions, and theory." *Isis,* 101: 146–58.

Sivasundaram, Sujit. 2013. *Islanded: Britain, Sri Lanka, and the Bounds of an Indian Ocean Colony.* Chicago: University of Chicago Press.

Turner, Frank. 1993. *Contesting Cultural Authority: Essays in Victorian Intellectual Life.* Cambridge: Cambridge University Press.

Tyndall, John. 1868. *Faraday as a Discover.* London: Longman, Green, and Co.

[Whewell, William]. 1831. "Modern science—inductive philosophy." *Quarterly Review,* 45: 374–407.

Whewell, William. 1834. "Review of *On the Connexion of the Physical Sciences.*" *Quarterly Review,* 51: 54–68.

Whewell, William. 1840. *The Philosophy of the Inductive Sciences,* 2 volumes. London: John Parker.

White, Paul. 2003. *Thomas Huxley: Making the "Man of Science."* Cambridge: Cambridge University Press.

White, Paul. 2005. "Ministers of Culture: Arnold, Huxley, and the Liberal Anglican reform of learning." *History of Science,* 43: 115–38.

Yeo, Richard. 1993. *Defining Science: William Whewell, Natural Knowledge, and Public Debate in Early Victorian Britain.* Cambridge: Cambridge University Press.

Young, Robert M. 1985. *Darwin's Metaphor: Nature's Place in Victorian Culture.* Cambridge: Cambridge University Press.

CHAPTER TWELVE

The Professional Scientist

CYRUS C. M. MODY

Before 1800, there were hardly any people professionally engaged in the study of nature. A few people conducted scientific research as their paid occupation, but were rare enough that they had difficulty sustaining quasi-professional organizations, as exemplified by the decline of Britain's Royal Society and turbulence in the Paris Academy around 1800. Across many scientific disciplines, though, workaday scientists became common enough over the nineteenth century to build and maintain the institutions of professional science in the twentieth: organizations and means of communication (societies, journals, etc.); codes of conduct and jurisdictions of expertise; codified training and standards; specialized tools, clothes, and workplaces. Those institutions secured substantial, though incomplete, intellectual autonomy. Rigorous training and the judgment of peers gave scientists an objective-enough expertise both to pursue esoteric topics and to pronounce on public matters without seeming beholden to vested interests. By 1900, scientists had manufactured such a reputation for objectivity and autonomy that other nascent professions—especially medicine (Starr 1982) and engineering (Kline 1995)—clamored to be accepted as "applied sciences."

As Paul White (Chapter 11 of this volume) shows, the pace of professionalization of science was uneven in the nineteenth century. In Western Europe it accelerated after 1870, and in the United States it attained feverish intensity between 1890 and 1900 (Kohler 1990). Professional societies proliferated to patrol disciplinary boundaries and ratify individual identification with a scientific community. The PhD was largely confirmed as the standard highest degree. Countries such as Russia, Japan, and the United States with rudimentary PhD-granting machinery routinely sent students abroad (especially to Germany) for training, even as those countries built up their own higher education systems (Ito 2005; Hall 2008). That pattern held true throughout the twentieth century, though the countries that students came from and went to shifted continually (Bound, Turner, and Walsh 2009; Wang 2010). Finally, one early twentieth-century innovation was to formalize post-PhD scientists' spending time at senior colleagues' institutions via the postdoctoral fellowship.

A Companion to the History of Science, First Edition. Edited by Bernard Lightman.
© 2016 John Wiley & Sons Ltd. Published 2020 by John Wiley & Sons Ltd.

The great transformations of the nineteenth century produced entirely new types of professional scientist. Industrialization, in particular, created a class of scientists clad in business suits and lab coats, working for the era's high-tech firms (Hoddeson 1981; Reich 1985; Wise 1985; Hounshell and Smith 1988). Industrial scientists could invent, refine, and monitor the quality of their employers' products, enhance firms' reputations, preserve monopoly power, and provide legal leverage over bothersome independent inventors.

Scientists employed by public and private universities and government bureaus were less scarce in 1800 than industrial scientists, but important features of their practice emerged in the nineteenth century. In particular, the management of administratively complex units became a prominent part of professional science. For instance, the German research university, much copied in the late nineteenth century, was characterized not just by integrated research and teaching, but by the management of an institute or large lab group by professors, populated by students, technicians, and assistant faculty of varying seniority and expertise, with a division of labor apportioned across a research portfolio (Johnson 1985). Though small, disorganized lab groups never vanished, the trend in the twentieth century in many disciplines was for research groups to grow larger and more complex. In universities, managerial faculty increasingly spend their time securing resources and building networks, leaving technical work to students, postdocs, technicians, and various "academic marginals" (Hackett 1990).

Elite scientific managers were in such demand in the twentieth century that they migrated effortlessly among academic institutes, industrial firms, and government committees. Perhaps the best example of this recurring type was Vannevar Bush, the civilian overseer of American science's contribution to World War Two (Zachary 1999). As a faculty member at MIT, Bush worked his way up through the academic administration while consulting for electrical utilities and co-founding a company, Raytheon. Then, in the 1930s, he took the reins of a philanthropic foundation, the Carnegie Institution, as well as the quasi-governmental National Advisory Committee on Aeronautics. From there, Bush had a beachhead and the administrative expertise to begin moving pawns and presidents to integrate science seamlessly into the American national security state.

While Bush was in a class of his own, throughout the war he relied on similar scientist-administrators such as Warren Weaver, Mina Rees, James Conant, Mervin Kelly, and others whose professional expertise consisted largely in overseeing research bureaucracies and lubricating frictions among universities, agencies, firms, and non-governmental organizations. Figures similar to Bush appeared all over the world in the twentieth century, ceaselessly crossing administrative boundaries while tying together the organizations separated by those boundaries. Often, like Bush, such figures put their expertise in managing scientific networks at the disposal of national security establishments: Fritz Haber, Homi Bhabha, Igor Kurchatov, A.Q. Khan, and countless less famous peers.

Scientists supplied states with managerial expertise; in return, states supplied resources for ever larger organizations for scientists to manage. "Big Science" was already apparent in some nineteenth century disciplines. For instance, the staff of the Royal Observatory at Greenwich expanded more than twenty-five-fold between 1811 and 1900 (Hughes 2002). Indeed, the organizational complexity of nineteenth-century observatories gave rise to social sciences such as sociology, psychology, and statistics (Schaffer 1988). In the twentieth century, Big Science spread relentlessly across disciplines and nations (Galison and Hevly 1992).

In some cases, justifications for organizational growth gestured to the scientific state of the art: forefront measurements could only be made with instruments so large, expensive, and complicated that they required well-managed, usually state-sponsored organizations to build and run them. The technical logic of instrumental and organizational growth held even in the social sciences (e.g., large teams needed for econometric and sociological surveys), but was especially evident in physical science experiments: particle accelerators, gravitational radiation flux detectors, nuclear weapons, and optical telescopes (Collins 2004; McCray 2004). By the 1960s, accelerator experiments were so organizationally complex that veterans of the Manhattan Project such as Luis Alvarez adopted military models of administration and labor discipline to safely boost production of collision events (Galison 1997). For later generations of accelerator physicists, the demands of overseeing multinational hundred-author collaborations required going to business school to get an MBA.

Of course, small science still made important contributions. Even professional astronomers continued to rely on dispersed networks of amateurs for observations of phenomena such as variable stars (Williams and Saladyga 2011). And some professional scientists were disenchanted with bureaucratized professionalism. For instance, Donald Glaser, the physicist who invented the bubble chamber, originally hoped it could facilitate tabletop research. Instead, he saw his invention engulfed by the militarized bureaucracies that Luis Alvarez and others constructed around enormous (and enormously dangerous) hydrogen bubble chambers. Thus, Glaser departed high-energy research for biophysics and eventually co-founded Cetus, one of the first and most famous academic biotechnology start-ups (Vettel 2006).

More ominous for Big Science, though, was that its logic became increasingly tenuous after 1970. The boundless gigantism of forefront experiments in fields such as high energy physics—combined with strains on state support for research in many NATO and Warsaw Pact countries—meant fewer forefront experiments could be built, and then only by complicated multinational, multiorganizational consortia (Hoddeson, Kolb, and Westfall 2008). As the Owl of Minerva leaves its perch on the Cold War, we can perhaps say that complex, hierarchically organized, state-sponsored Big Science ceded its dominance around 1990 to network forms of organization, in which science is conducted by small, simple, locally administered units where connections among those units endow an emergent organizational form to the enterprise as a whole (Kevles 1997; Owen-Smith and Powell 2004). If that is the case, science may become less profession- and organization-centric, and scientists may become post-professional "venture labor" (Neff 2012) navigating networks rather than organizations.

A Middling Sort of Science

When science professionalized in the nineteenth century, it did so as a bourgeois occupation. In the twentieth century, wherever the middle class grew, so too did the corps of professional scientists. In turn-of-the-century Germany and Austria, for instance, the shared middle-class background of scientists such as Hermann von Helmholtz and Max Planck allowed them to ground physical arguments in the piano-playing and concert-going expected of any good bourgeois (Hui 2013). In the United States, many research organizations were dominated by scientists and engineers from society's middle strata: white, middle-class, middlebrow men, often from Middle West states (Hughes 1989; Kaiser 2004; Gertner 2012).

Similarly, the Bolsheviks learned quickly during the Russian Revolution that they risked losing technical experts precisely because scientists lacked proletarian skills of food and fuel acquisition needed to survive cold, hungry post-World War I winters. Over the opposition of the Proletarian Culture movement (which wanted to populate Soviet science with working-class cadres), Lenin permitted special grants of supplies to scientists and other intellectuals so that they would remain alive long enough to aid Bolshevik "government construction" (Josephson 1991).

As middle-class professionals, scientists disavowed natural philosophy's patrician amateur tradition. Occasionally, aristocratic gentlemen of leisure still maintained private laboratories on their estates, such as the Wall Street millionaire Alfred Loomis, whose personally-funded research contributed to the interwar development of radar and the atomic bomb (Conant 2002). Yet Loomis and his few peers were nearly mythical exceptions to the rule.

At the same time, the professional scientist demarcated his expertise from craft, artisanal, and working class knowledge. By and large *his* because professionalization in science (as in many fields) co-evolved with the marginalization of women, ethnic/sectarian minorities, and people from the Global South (Abir-Am and Outram 1987; Harding 1993; Slaton 2010). The process of marginalization was particularly evident in the twentieth century's new fields, such as computer science, which often started as havens for women, minorities, and other subalterns, only to see their representation diminish as those fields "matured" (Ensmenger 2010; Abbate 2012). Of course, members of such groups remained essential to even the most mature fields (Rossiter 1982). Yet as professionals, male, middle-class (and, in Western countries, white) scientists could disproportionately move into positions of managing, while women, workers, and other subalterns were disproportionately tracked into less visible positions of being managed (Barley and Orr 1997).

Professionalization's occlusion of women, ethnosectarian minorities, and scientists from Europe's current or former colonial possessions made for their invisibility in professional histories of science as well (Yruma 2008). As Amit Prasad (2014) has shown, for example, the thick literature by historians and science studies scholars on the development of magnetic resonance imaging (MRI) has taken for granted that the technique was invented by Raymond Damadian and Paul Lauterbur in the United States and Peter Mansfield in the United Kingdom. Yet scientists in India—especially G. Suryan and A. K. Saha, in a line extending back to C. V. Raman—were deeply connected to, and at the forefront of, the global research community working on nuclear magnetic resonance techniques that gave rise to MRI. To say that MRI was invented "in" the US or UK is to isolate just one part of an interdependent transnational network.

That isolation happens all the time, though, partly as a consequence of the formation of professional bodies in Europe, North America, and to a lesser extent the Pacific Rim that represented their members as the sole sources of scientific expertise and discovery. In the case of MRI, even Indian scientists have come to believe that the technique originated elsewhere and belatedly came to South Asia. The category of professional science contains historical asymmetries of geography, race, class, and gender that disrupt fully visible participation in professional practice by actors on the wrong side of those asymmetries.

To upend those asymmetries, revolutionary movements throughout the twentieth century attempted to open science to non-Western, subaltern, and/or working-class cadres and expertise. In the People's Republic of China, for instance, the Cultural Revolution ensnared many thousands of scientists who were demoted, beaten, conscripted into hard labor, and occasionally killed. As Peter Neushul and Zuoyue Wang (2000) show in their study of marine biologist C. K. Tseng, a peasant upbringing was one of the few things that could save a scientist from the Cultural Revolution. Conversely, an interest in basic research topics, connections to scientific communities in the West, attempts to model Chinese scientific institutions on templates from the West, or any scientific analysis of communist ideology or state projects that arrived at unpalatable conclusions could put scientists in jeopardy. Tseng himself—despite heroic service developing mariculture of kelp—was "beaten, starved, placed in solitary confinement," forced into hard labor, and forbidden from discussing science (Neushul and Wang, 83).

Curiously, many aspects of experimental and field research could be mistaken for proletarian labor. Such research requires hands-on, often dirty, sometimes unpleasant work. Some fields, such as pre-digital observational astronomy or field geology, involve hard, uncomfortable physical tasks. Many experimental fields are accompanied by occupational hazards (e.g., exposure to dangerous levels of electricity, radiation, or toxic chemicals) more commonly associated with factory and agricultural labor than bourgeois professions (Herzig 2005; Sims 2005).

Professional science also involves a great deal of reading, writing, thinking, listening, conversing, and even—as George Gamow once joked—sleeping on the job (Kaiser 2005). Critics on the left and right have taken these characteristics as grounds to attack scientists as effete intellectuals, dangerous to their nations because of inauthenticity to some norm of male manual labor. As both Kaiser and Charles Thorpe (2002) have shown, for instance, postwar anti-communist witch hunts directed at American physicists often played on anti-Semitic tropes of an untrustworthy, effete physique. Theorists, in particular—and especially Robert Oppenheimer—were described with suspicion as ethereal ectomorphs, not quite red-blooded enough to withstand the Red Menace.

Supporting and Subverting their Sponsors

Scientists' perceived utility to publics and patrons (especially states) was usually sufficient to fend off such attacks. Some forms of scientific utility were obvious: new bombs and means of delivering them; means of surveillance and communication; administrative tools to make populations, resources, and economies more intelligible, calculable, controllable; pharmaceuticals, fertilizers, and consumer technologies to make populations healthier, happier, longer-lived, more productive, less restive. Patrons' desires for direct benefits from research incentivized scientists to ask some questions while ignoring others (Forman 1987; Proctor and Schiebinger 2008). Yet, scientists were skilled at arguing that even questions of primarily professional relevance could indirectly benefit their patrons. High-energy physicists, for instance, could tell funders they were searching for the "God particle," but to a significant extent American President

Lyndon Johnson saw Fermilab as a horse to trade for civil rights legislation (Ploeger 2002), and European states have long sponsored CERN partly out of a belief that the facility aids continental political integration (Krige 1996).

Scientists' professionalism also routinely led them to upend patrons' expectations. For instance, Western colonial regimes cultivated a technical autochthonous class in their overseas possessions in order to supply middle managers for overstretched European and American imperial bureaucrats, secure the loyalty of colonial elites, and impress metropolitan values on colonized populations (Headrick 1988; Anderson and Pols 2012). What imperial administrators presumably did not foresee was that the colonial scientific class would be rife with nationalist advocates for independence (Bassett 2009a; Chakrabarti 2009).

Warwick Anderson and Hans Pols (2012) show that doctors and life scientists were the leading voices of decolonization in the Philippines, Indonesia, and Taiwan. For instance, José Rizal—primarily remembered in the Philippines as a novelist, martyr, and revolutionary—was also an ophthalmologist trained by the liberal German medical scientist Rudolf Virchow. In Indonesia, the founders of the leading nationalist parties and newspapers, such as Cipto Mangunkusumo, Abdul Rasyid, Tirto Adi Suryo, and Sutomo, often had some medical training, usually obtained partly in the metropole. Indeed, organizations of medical students from Indonesia in the Netherlands—students who increasingly saw themselves as "Indonesian"—produced many leaders of the independence movement. Anderson and Pols argue that these medico-nationalists described decolonization in starkly medical terms—analogies to healing an ailing body politic by cutting out a foreign cancers, or curing an invasive infection, abounded in their rhetoric.

In the Cold War, both blocs followed the old colonial powers by supporting their allies' (and potential allies') scientific communities, again with sometimes ironic outcomes. Immediately after the war, for instance, American policymakers made rebuilding scientific institutions a central part of the Marshall Plan (Krige 2006)—only to discover, to their horror, that some prominent Western European scientists (e.g., the Joliot-Curie clan) were card-carrying communists!

In the 1960s, non-aligned India played the blocs against each other by instigating a university-building race among the United States, the United Kingdom, West Germany, and the Soviet Union. As a result, India gained four of the first five campuses of the now-famous Indian Institutes of Technology. But Indian hopes that Cold War competition would stock an indigenous, nationalist technical elite fell short. Students were not so mindful of Gandhian and Nehruvian ideals; having been trained at universities modeled on foreign exemplars, they tended to pursue careers abroad. As a stock gag put it, "when a student enters the IIT, his soul ascends to America; when he graduates his body follows" (Leslie and Kargon 2006; Bassett 2009b).

Professional scientists proved hard for the superpowers to control domestically, as well. In the Soviet Union, the regime pressed scientists—in physics, linguistics, biology, etc.—to conform ideologically and administratively. Scientists worked to Five Year Plans, and Stalin arbitrated technical debates as the "coryphaeus of science" (Pollock 2006). Yet the most successful Soviet scientist-administrators, such as Igor Kurchatov, head of the atom bomb program, were also skilled advocates for scientific autonomy from party ideology and patrons of colleagues who had run afoul of Stalin.

In both authoritarian and liberal democratic regimes, political leaders struggled against truly independent scientific expertise, just as many scientists struggled with their *dependence* on state-supplied resources, aims, and even data. For instance, during World War II the new field of operations research gained popularity among British officers—so long as operations researchers made recommendations officers wanted to hear (Rau 2005). When left-leaning operations researchers such as P. M. S. Blackett concluded that strategic bombing of civilian homes was unlikely to break the will of the German populace, they were denied access to data about military operations—the lifeblood of *operations* research.

Famously, in the early 1950s United States, Robert Oppenheimer ran afoul of a similar logic. As long as his policy recommendations aligned with the US nuclear establishment, he remained part of that establishment. But when Oppenheimer expressed doubts about the hydrogen bomb and nuclear airplane, his views were dismissed as "political" rather than "technical" and his security clearance—the passport for a scientist to influence nuclear policy—was revoked (Thorpe 2002).

Professionals Unbound

In fact, states often relied on scientists themselves to proclaim their colleagues' political unreliability. Stalin's purges were facilitated by Soviet scientists' willingness (sometimes eagerness) to denounce their peers as idealists and cosmopolitans. In the US, the Atomic Energy Commission and Air Force needed Kenneth Pitzer and Edward Teller to raise doubts about Oppenheimer's sympathies so that he could be classified as a security risk.

In other words, scientists are people, given to infighting as much as to in-group camaraderie. Scientists' disagreements are good for historians—they help us tell a compelling story, and to find proxies for any argument or counterargument. Disputes posed a thorny paradox for science as a profession, though. Scientists' corporate authority derived in large part from their reputation for objective rationality. The laws of logic and nature imply that any reasonable scientist should come to the same conclusions, regardless of their particular interests. Yet in the actual conduct of science, reasonable individuals regularly contradict each other, sometimes on questions of enormous public import.

Such disputes did not simply "spill out" into public view; scientists appropriated public attention as a resource to pursue aims relative to their professional communities, while ordinary citizens and a variety of mediators (e.g., journal editors) appropriated scientific attention as a resource to pursue societal aims. Scientists in the twentieth century routinely insinuated themselves into the public sphere through participation in social movements, museum work, political lobbying, and mass-market books, films, and radio and television shows.

Thus, science was never a monastic profession, quarantined from external influence. Scientists benefited from representing their profession as a discrete community, separated from the public by a distinctive rationality and expertise, yet they gained as much from the porosity of the profession to its publics. Divisions within the scientific community were mirrored in—made common cause with—divisions within the wider public sphere. In examining postwar American science, Kelly Moore (2008) describes this as the "unbounding" of the profession—noting, of course, that science's boundaries are continually made, remade, and unmade. Moore's study examines three modes in which more-or-less pacifist and left-leaning American scientists opened their profession to new audiences and critiques: a Quaker-inflected discourse of individual responsibility to do ethical science; a liberal discourse of providing objective information so citizens could act and vote with greater understanding; and a revolutionary discourse of transforming science along lines of race, class, gender, and international unity.

Liberal, pacifist, social democratic, feminist, postcolonial, and Marxist scientists were never alone in rending the boundaries of their profession, though. Many of the left-leaning actors in Kelly Moore's study consciously followed their adversaries' playbook: scientific eugenicists, fascists, defense intellectuals, etc., all of whom helped stitch their profession into extraprofessional social movements and political institutions through advocacy in the public sphere. Each of these alliances brought science resources and patrons, if sometimes also disrepute. But competition among these alliances also undermined the image of impartiality that long seemed professional science's distinctive characteristic. If some scientists allied with those who claimed the intellectual inferiority of certain races, while others disputed the reality of race (Reardon 2005); if some stoked fears of fallout from peacetime nuclear testing while others asserted that an all-out nuclear war would be a walk in the park (Hamblin 2013); if some argued vehemently for, and others vehemently against, unchecked population growth (Sabin 2013)—well, then weren't scientists the same as lawyers, politicians, even clerics, with their irresolvable, unending, and flexible arguments? If anyone could claim to be a scientist, and if the professional trappings of science—journals, societies, departments, etc.—are easily mimicked (Gordin 2012), then how to decide which scientists merit public trust?

In general, in the first half of the century, attempts to identify the real experts sought objective criteria to demarcate true science and scientists from their less reliable alter-egos. Some demarcations were informal—a scientist looked, dressed, talked, and behaved in a recognizable way. Other demarcations were formal, if never entirely successful even on their own terms. Even so, the search for demarcation criteria was enormously influential in philosophy, law, and science itself.

It is perhaps fair to say, though, that the second half of the twentieth century was characterized less by the unity of science than by a "nice derangement of epistemes" (Zammito 2004). Informally, it became more difficult to tell scientists apart from ordinary people, since they now came in a greater variety of shapes and colors and from more parts of the world and laid claim to more kinds of expertise. Formally, the ideas of demarcation and unity were undermined by the counterculture's and the academy's postmodern turn.

In place of philosophical demarcation of science from non-science, the second half of the century saw many experiments in what we might call distributed and postprofessional demarcationism—where the expert and his/her expertise are evaluated not by formally accredited peers but by some emergent collectivity. Perhaps the earliest influential forays in this direction came from cybernetics. As Andrew Pickering (2010) has shown, the basic cybernetic idea—that environments shape and are shaped by the entities within them (i.e., that "environments" are composed of many mutually shaping entities)—led some cyberneticians to conclude that their expertise could only be constructed ethically and robustly in symmetric dialogue with other thinking, evolving minds. Gregory Bateson and R. D. Laing for instance, proposed a form of psychiatric therapy in which the demarcation between patient and therapist dissolved—replaced by an environment of dialogue in which participants acted symmetrically upon each other's psyches.

Cybernetics' moment was fleeting, but Fred Turner (2006) argues that it influenced an emblematic form of distributed demarcation in the current moment—the social network or electronic village. In the late 1960s and early 1970s, cybernetic principles underwrote new media forms which freely incorporated reader response in the published text. In the 1980s the impresarios of such collectively authored texts, such as Kevin Kelly and Stewart Brand, became champions and hosts of electronic forums where participants performed their expertise for each other to gain attention, readers, and jobs. The ideology of such forums was profoundly anti-professional; influence accrued not from formal qualifications or recognition by professional societies, but by winning over other members of the network.

Finally, the economic dislocations experienced by many industrialized countries in the 1970s contributed to what might be termed neoliberal demarcationism. Neoliberal economists believe that knowledge has a value determined by free markets and thus the invisible hand decides who is a real expert, usually on the basis of whose discoveries get embedded in lucrative products. For neoliberals, industry representatives, by dint of their proximity to the marketplace, are more expert than accredited academic or government scientists (Vinsel 2012). Of course, the market is a better system for divining value than truth (Mirowski 2011). Denialism of climate science, oncology, and biochemistry contribute to the corporate bottom line, even if they turn facts on their heads (Oreskes and Conway 2010).

End of an Epoch?

The growing influence of neoliberal economists on science policy in the United States and in global institutions such as the International Monetary Fund has led to some hand-wringing by historians, sociologists, and philosophers of science. Much recent work (Gibbons et al. 1994; Ziman 2000; Forman 2007; Nordmann, Radder, and Schiemann 2011) claims an "epochal break" took place (usually located in or around 1980) in which science transitioned from "Mode 1" (disciplinary, knowledge-oriented) to "Mode 2" (interdisciplinary, application-oriented), became "post-academic," and was transformed into the adjunct of technology (where before technology had been the auxiliary of science). Lorraine Daston and Peter Galison (2007) have argued that the years since 1980 produced a new form of objectivity based not on automatic, mechanical "reading" of nature, nor on professional judgments about those readings accrued through long experience, but on the ability to fabricate hybrid natural–technological devices which exhibit scientifically and commercially interesting behaviors.

Certainly, some important things happened around 1980. In the United States alone in that year, Ronald Reagan was elected president, the Bayh–Dole Act permitted universities to patent innovations arising from federally funded research, and the Supreme Court opened the door to patenting living organisms—all events with a significant effect on the conduct and ideology of science in the United States and around the world. Perhaps the most important event of 1980 was the initial public stock offering of Genentech, one of the first biotechnology start-up companies co-founded by a professor (Herbert Boyer of University of California San Francisco) and a venture capitalist (Robert Swanson). Boyer became an instant millionaire and, almost as quickly, his academic colleagues went from criticizing his supposed violations of professional norms to emulating him by founding start-ups of their own (Jones 2009; Smith Hughes 2011). Since then, the Genentech model has become a powerful "sociotechnical imaginary" (Jasanoff and Kim 2013), as nations and regions scramble to incubate clusters of high-tech start-ups and entrepreneurial faculty, usually with poor to mixed results.

Still, the epochal break literature—whether celebratory or elegiac—misses some important points. Much science is still "Mode 1;" much was always "Mode 2." Scientists have long made hybrid "devices" with a view to their industrial or commercial application. Conversely, accusations of unprofessional commercialism still carry weight in scientific controversies. When Stanley Pons and Martin Fleischman claimed they had discovered/invented a cold fusion reaction in 1989, their finding was treated as suspect in part because they announced it in a press conference not a peer-reviewed journal. Many scientists participate in market activities less out of a post-professional craving for profit than out of a sense that commercial science is an intellectual challenge and that firms are, in some respects, more tolerant of free inquiry than the academy (Rabinow 1996; Shapin 2008). When commercial interests are perceived as threatening the pursuit of knowledge, some (though not all) scientific communities have pushed back and/or found accommodations that preserve their ability to ask professionally-relevant questions (Murray 2010; Nelson 2015). In any case, fears that the market is everywhere displacing the profession as the arbiter of scientific expertise seem premature, though certainly not groundless.

So, yes, the end of the Cold War brought significant changes to professional science. Some distinctive institutions of twentieth-century professional science—such as the corporate basic research laboratory and scholarly journal—have struggled to maintain relevance and credibility in the post-Cold War era (Reich 2009). Others, such as professional societies, funding agencies, and the increasingly market-oriented research university, have been on a slow but steady path of change since the late 1960s—a path defined by the need to demonstrate more clearly their benefit, and responsiveness, to a variety of stakeholders (Berman 2012). Yet pressures on professional scientists to seem less exclusively beholden to their professional peers are nothing new, and do not, on their own, spell the end of professional science as such. Rather, the course of the twentieth century shows that the form and function of professional science has continually co-evolved with an ever-changing global political economy. Professional science is indeed unbounded enough that what scientists know and do—and who they are—has been given form by the environments of which science is a part. And yet, professional ties, status, and institutions have given scientists the cohesiveness to shape those environments as much as those environments shape them.

References

Abbate, Janet. 2012. *Recoding Gender: Women's Changing Participation in Computing.* Cambridge, MA: MIT Press.

Abir-Am, Pnina, and Dorinda Outram (eds.) 1987. *Intimate Lives: Women in Science, 1789–1979*. New Brunswick, NJ: Rutgers University Press.

Anderson, Warwick, and Hans Pols. 2012. "Scientific patriotism: Medical science and national self-fashioning in southeast Asia." *Comparative Studies in Society and History*, 54: 93–113.

Barley, Stephen R., and Julian E. Orr (eds.) 1997. *Between Craft and Science: Technical Work in U.S. Settings*. Ithaca, NY: IRL Press.

Bassett, Ross. 2009a. "MIT-trained Swadeshis: MIT and Indian nationalism, 1880–1947." *Osiris*, 24: 212–30.

Bassett, Ross. 2009b. "Aligning India in the Cold War era: Indian technical elites, the Indian Institute of Technology at Kanpur, and computing in India and the United States." *Technology and Culture*, 50: 783–801.

Berman, Elizabeth Popp. 2012. *Creating the Market University: How Academic Science Became an Economic Engine*. Princeton, NJ: Princeton University Press.

Bound, John, Sarah Turner, and Patrick Walsh. 2009. "Internationalization of U.S. doctorate education." In *Science and Engineering Careers in the United States: An Analysis of Markets and Employment*, edited by Richard B. Freeman and Daniel L. Goroff, 59–98. Chicago: University of Chicago Press.

Chakrabarti, Pratik. 2009. "'Signs of the times:' Medicine and nationhood in British India." *Osiris*, 24: 188–211.

Collins, H.M. 2004. *Gravity's Shadow: The Search for Gravitational Waves*. Chicago: University of Chicago Press.

Conant, Jennet. 2002. *Tuxedo Park: A Wall Street Tycoon and the Secret Palace of Science that Changed the Course of World War II*. New York: Simon & Schuster.

Daston, Lorraine, and Peter Galison. 2007. *Objectivity*. New York: Zone Books.

Ensmenger, Nathan. 2010. *The Computer Boys Take Over: Computers, Programmers, and the Politics of Technical Expertise*. Cambridge, MA: MIT Press.

Forman, Paul. 1987. "Behind quantum electronics: National Security as basis for physical research in the United States, 1940–1960." *Historical Studies in the Physical and Biological Sciences*, 18: 149–229.

Forman, Paul. 2007. "The primacy of science in modernity, of technology in postmodernity, and of ideology in the history of technology." *History and Technology*, 23: 1–152

Galison, Peter, and Bruce Hevly (eds.) 1992. *Big Science: The Growth of Large-Scale Research*. Stanford, CA: Stanford University Press.

Galison, Peter. 1997. *Image and Logic: A Material Culture of Microphysics*. Chicago: University of Chicago Press.

Gertner, Jon. 2012. *The Idea Factory: Bell Labs and the Great Age of American Innovation*. New York: Penguin.

Gibbons, Michael, Camille Limoges, Helga Nowotny, Simon Schwartzman, Peter Scott, and Martin Trow. 1994. *The New Production of Knowledge: The Dynamics of Science and Research in Contemporary Societies*. London: Sage.

Gordin, Michael D. 2012. *Pseudoscience Wars: Immanuel Velikovsky and the Birth of the Modern Fringe*. Chicago: University of Chicago Press.

Hackett, Edward J. 1990. "Science as a vocation in the 1990s: The changing organizational culture of academic science." *Journal of Higher Education*, 61: 241–79.

Hall, Karl. 2008. "The schooling of Lev Landau: The European context of postrevolutionary Soviet theoretical physics." *Osiris*, 23: 230–59.

Hamblin, Jacob Darwin. 2013. *Arming Mother Nature: The Birth of Catastrophic Environmentalism*. New York: Oxford University Press.

Harding, Sandra. 1993. *The "Racial" Economy of Science: Toward a Democratic Future*. Bloomington: Indiana University Press.

Headrick, Daniel R. 1988. *The Tentacles of Progress: Technology Transfer in the Age of Colonialism*. New York: Oxford University Press.

Herzig, Rebecca M. 2005. *Suffering for Science: Reason and Sacrifice in Modern America*. New Brunswick, NJ: Rutgers University Press.

Hoddeson, Lillian, Adrienne W. Kolb, and Catherine Westfall. 2008. *Fermilab: Physics, the Frontier, and Megascience*. Chicago: University of Chicago Press.

Hoddeson, Lillian. 1981. "The emergence of basic research in the Bell Telephone System, 1875–1915." *Technology and Culture*, 22: 512–44.

Hounshell, David A., and John Kenly Smith. 1988. *Science and Corporate Strategy: Du Pont R&D 1902–1980*. Cambridge: Cambridge University Press.

Hughes, Jeff A. 2002. *The Manhattan Project: Big Science and the Atom Bomb*. New York: Columbia University Press.

Hughes, Thomas P. 1989. *American Genesis: A Century of Invention and Technological Enthusiasm, 1870–1970*. New York: Viking.

Hui, Alexandra. 2013. *The Psychophysical Ear: Musical Experiments, Experimental Sounds, 1840–1910*. Cambridge, MA: MIT Press.

Ito, Kenji. 2005. "The Geist in the institute: The production of quantum physicists in 1930s Japan." In *Pedagogy and the Practice of Science: Historical and Contemporary Perspectives*, edited by David Kaiser, 151–84. Cambridge, MA: MIT Press.

Jasanoff, Sheila, and Sang-Hyun Kim. 2013. "Sociotechnical imaginaries and national energy policies." *Science as Culture*, 22: 189–96.

Johnson, Jeffrey A. 1985. "Academic chemistry in Imperial Germany." *Isis*, 76: 500–524.

Jones, Mark Peter. 2009. "Entrepreneurial science: The rules of the game." *Social Studies of Science*, 39: 821–51.

Josephson, Paul R. 1991. *Physics and Politics in Revolutionary Russia*. Berkeley: University of California Press.

Kaiser, David. 2004. "The postwar suburbanization of American physics." *American Quarterly*, 56: 851–88.

Kaiser, David. 2005. "The atomic secret in Red hands? American suspicions of theoretical physicists during the early Cold War." *Representations*, 90: 28–60.

Kevles, Daniel J. 1997. "Big science and big politics in the United States: Reflections on the death of the SSC and the life of the Human Genome Project." *Historical Studies in the Physical and Biological Sciences*, 27: 269–97.

Kline, Ronald R. 1995. "Construing 'technology' as 'applied science:' Public rhetoric of scientists and engineers in the United States, 1880–1945." *Isis*, 86: 194–221.

Kohler, Robert E. 1990. "The Ph.D. machine: Building on the collegiate base." *Isis*, 81: 638–62.

Krige, John (ed.) 1996. *History of CERN, vol. III*. Amsterdam: Elsevier Science.

Krige, John. 2006. *American Hegemony and the Postwar Reconstruction of Science in Europe*. Cambridge, MA: MIT Press.

Leslie, Stuart W., and Robert Kargon. 2006. "Exporting MIT: Science, technology, and nation-building in India and Iran." *Osiris*, 21: 110–30.

McCray, W. Patrick. 2004. *Giant Telescopes: Astronomical Ambition and the Promise of Technology*. Cambridge, MA: Harvard University Press.

Mirowski, Philip. 2011. *Science-Mart: Privatizing American Science*. Cambridge, MA: Harvard University Press.

Moore, Kelly. 2008. *Disrupting Science: Social Movements, American Scientists, and the Politics of the Military, 1945–1975*. Princeton, NJ: Princeton University Press.

Murray, Fiona. 2010. "The oncomouse that roared: Hybrid exchange strategies as a source of distinction at the boundary of overlapping institutions." *American Journal of Sociology*, 116: 341–88.

Neff, Gina. 2012. *Venture Labor: Work and the Burden of Risk in Innovative Industries*. Cambridge, MA: MIT Press.

Nelson, Andrew J. 2015. *The Sound of Innovation: Stanford and the Computer Music Revolution*. Cambridge, MA: MIT Press.

Neushul, Peter, and Zuoyue Wang. 2000. "Between the devil and the deep blue sea: C. K Tseng, mariculture, and the politics of science in modern China." *Isis*, 91: 59–88.

Nordmann, Alfred, Hans Radder, and Gregor Schiemann (eds.) 2011. *Science Transformed? Debating Claims of an Epochal Break*. Pittsburgh, PA: University of Pittsburgh Press.

Oreskes, Naomi and Erik M. Conway. 2010. *Merchants of Doubt: How a Handful of Scientists Obscured the Truth on Issues from Tobacco Smoke to Global Warming*. London: Bloomsbury Press.

Owen-Smith, Jason, and Walter W. Powell. 2004. "Knowledge networks as channels and conduits: The effects of spillovers in the Boston biotechnology community." *Organization Science*, 15: 5–21.

Pickering, Andrew. 2010. *The Cybernetic Brain: Sketches of Another Future*. Chicago: University of Chicago Press.

Ploeger, Joanna S. 2002. "The art of science at Fermi National Accelerator Laboratory: The rhetoric of aesthetics and humanism in the national laboratory system in the late 1960s." *History and Technology*, 18: 23–49.

Pollock, Ethan. 2006. *Stalin and the Soviet Science Wars*. Princeton, NJ: Princeton University Press.

Prasad, Amit. 2014. *Imperial Technoscience: Transnational Histories of MRI in the United States, Britain, and India*. Cambridge, MA: MIT Press.

Proctor, Robert N., and Londa Schiebinger (eds.) 2008. *Agnotology: The Making and Unmaking of Ignorance*. Stanford, CA: Stanford University Press.

Rabinow, Paul. 1996. *Making PCR: A Story of Biotechnology*. Chicago: University of Chicago Press.

Rau, Erik. 2005. "Combat science: The emergence of Operational Research in World War II." *Endeavour*, 29: 156–61.

Reardon, Jenny. 2005. *Race to the Finish: Identity and Governance in the Age of Genomics*. Princeton, NJ: Princeton University Press.

Reich, Eugenie Samuel. 2009. *Plastic Fantastic: How the Biggest Fraud in Physics Shook the Scientific World*. Basingstoke: Palgrave Macmillan.

Reich, Leonard S. 1985. *The Making of American Industrial Research: Science and Business at GE and Bell, 1876–1926*. Cambridge: Cambridge University Press.

Rossiter, Margaret. 1982. *Women Scientists in America: Struggles and Strategies to 1940*. Baltimore: Johns Hopkins University Press.

Sabin, Paul. 2013. *The Bet: Paul Ehrlich, Julian Simon, and Our Gamble over Earth's Future*. New Haven, CT: Yale University Press.

Schaffer, Simon. 1988. "Astronomers mark time: Discipline and the personal equation." *Science in Context*, 2: 115–45.

Shapin, Steven. 2008. *The Scientific Life: A Moral History of a Late Modern Vocation*. Chicago: University of Chicago Press.

Sims, Benjamin. 2005. "Safe science: Material and social order in laboratory work." *Social Studies of Science*, 35: 333–66.

Slaton, Amy E. 2010. *Race, Rigor, and Selectivity in U.S. Engineering: The History of an Occupational Color Line*. Cambridge, MA: Harvard University Press.

Smith Hughes, Sally. 2011. *Genentech: The Beginnings of Biotech*. Chicago: University of Chicago Press.

Starr, Paul. 1982. *The Social Transformation of American Medicine*. New York: Basic Books.

Thorpe, Charles. 2002. "Disciplining experts: Scientific Authority and liberal democracy in the Oppenheimer case." *Social Studies of Science*, 32: 525–62.

Turner, Fred. 2006. *From Counterculture to Cyberculture: Stewart Brand, the Whole Earth Network, and the Rise of Digital Utopianism*. Chicago: University of Chicago Press.

Vettel, Eric James. 2006. *Biotech: The Countercultural Origins of an Industry*. Philadelphia: University of Pennsylvania Press.

Vinsel, Lee Jared. 2012. "The crusade for credible energy information and analysis in the United States, 1973–1982." *History and Technology*, 28: 149–76.

Wang, Zuoyue. 2010. "Transnational science during the Cold War: The case of Chinese/American scientists." *Isis*, 101: 367–77.

Williams, Thomas R., and Michael Saladyga. 2011. *Advancing Variable Star Astronomy: The Centennial History of the American Association of Variable Star Observers*. Cambridge: Cambridge University Press.

Wise, George. 1985. *Willis R. Whitney, General Electric, and the Origins of US Industrial Research*. New York: Columbia University Press.

Yruma, Jeris Stueland. 2008. "How experiments are remembered: The discovery of nuclear fission, 1938–1968." PhD dissertation, Princeton University.

Zachary, G. Pascal. 1999. *Endless Frontier: Vannevar Bush, Engineer of the American Century*. Cambridge, MA: MIT Press.

Zammito, John H. 2004. *A Nice Derangement of Epistemes: Post-Positivism in the Study of Science from Quine to Latour*. Chicago: University of Chicago Press.

Ziman, John. 2000. *Real Science: What It Is and What It Means*. Cambridge: Cambridge University Press.

PART II

Places and Spaces

CHAPTER THIRTEEN

The Medieval University[1]

STEVEN J. LIVESEY

After surveying the revival of Latin texts, jurisprudence, historical writing, the translation of Greek and Arabic (and we might add Hebrew) texts into Latin, and the recovery of ancient science and philosophy, Charles Homer Haskins turned to a completely new element of *The Renaissance of the Twelfth Century* (Haskins 1927), "The Beginnings of Universities." Although in 1927 Haskins was not the first to do so, he recognized that all of these elements of transformative culture in the long twelfth century were embraced in an institution that itself was embedded in its social, economic, political, and intellectual context and that spawned an incipient professional class of learned elites. To the extent that part of Haskins' plan was to show that modern European culture extended back beyond the more famous Italian Renaissance, he also noted that the university, perhaps uniquely among the legacies of the medieval world, retained its significance while the other elements of medieval culture fell into obscurity.

Traditionally, medieval society was subsumed under the concise formulation *sacerdotum, regnum, studium* (Grundmann 1951). Ecclesiastical monopoly did not extend beyond the Reformation, and monarchy suffered a similar blow in the eighteenth century. But while the university has experienced several modifications since the twelfth century, many of its core medieval hallmarks are recognizable in the modern institution, and, rather than shrinking in significance, modern universities are a worldwide institution.

While the university was certainly not the only venue for the study of science, it was an especially important one, both for the inception and evolution of medieval scientific ideas and for their transmission during the middle ages and subsequently. In this chapter, we will survey the institutions of learning available to scholars prior to the twelfth century as well as the preparatory schools in place after the formation of universities; the cultural and intellectual foundations of universities; the evolution of the institution from a handful of universal entities, drawing on a wide catchment for students, to national or regional foundations by the end of the middle ages; the

A Companion to the History of Science, First Edition. Edited by Bernard Lightman.
© 2016 John Wiley & Sons Ltd. Published 2020 by John Wiley & Sons Ltd.

curriculum, especially as it concerns scientific issues; techniques of teaching, learning, and transmission of knowledge; the place of the university within the larger society; and educational shifts at the end of the middle ages.

Pre-university Education

Before students arrived at university—indeed, before the evolution of university structures in the twelfth century—elementary education was dispensed in a variety of forms and venues. Among a small fraction of the aristocracy, the home might be the setting, with family or a private tutor as the instructor, but elementary education was generally achieved in a grammar school, where Latin vocabulary and pronunciation were taught either by chanting liturgical texts or through memorization of texts in primers. At the next level students studied grammatical forms with Donatus' *Ars grammatica*, or the more recent *Doctrinale* of Alexander de Villa Dei, supplemented by exercises drawn from simple classical texts of Cato or Aesop. Beyond this, students advanced to studies of logic, the classics, and the sciences of the *quadrivium*, derived from texts like the handbooks of Boethius. Together, this early training in letters might consume 10 or 12 years, after which most entered into careers as notaries, scribes, or chancellery secretaries, who did not need university educations (Verger 2000, 40–5; Orme 2006).

The instructors in these elementary schools were seldom university graduates themselves. In general, attrition in all levels of medieval education was sufficiently high that so long as the teacher's proficiency exceeded that of the students, the heads of schools made little attempt to check educational credentials. And while urban centers were more likely to have several schools, rural communities were not consigned to illiteracy. According to Nicholas Orme, rural counties in England possessed several schools, a situation replicated elsewhere in Europe (Orme 2006, 346–72).

Prior to the early eleventh century, monastic houses possessed a virtual monopoly on education in Europe. Families of means whose younger sons and daughters faced an uncertain future frequently gave them (and a "gift" of support) to religious houses, in hopes that the child would be educated and subsequently rise to a prominent position within the order (Johnson 1991, 18–27; Costambeys 2011, 138–48). While these internal schools educated children entering the religious life, monasteries frequently became proprietors for *scholae exteriores*, the grammar or song schools already mentioned.

The internal school's purpose was the development of future monks, which included a very particular perspective on learning: reading, and for that matter all learning, was practiced individually, contemplatively, and often silently. In the external schools, and especially in the growing number of schools attached to cathedrals, the purpose was entirely different, focused on the vocational needs of the secular clergy: preaching, performance of the liturgy, knowledge of canon law, and dispensation of the sacraments. Under decrees that were part of Charlemagne's reforms, each cathedral was required to establish a school to train the next generation of diocesan clergy. As urban centers began to grow and monastic orders effected reforms emphasizing spirituality over external responsibilities, cathedral schools eclipsed the monopoly in the educational world once enjoyed by the regular orders (Jaeger 1994; Contreni 1995).

While there was considerable variation among cathedral schools, most were rather limited, consisting of a single master who taught across the curriculum. To the extent that science entered the curriculum, it was limited to the handbooks of the late ancient or early medieval world, and directed to the practical functions of the Church. By the early twelfth century, however, some schools, like those in Paris, Laon, or Chartres, attained prominence for scholarship under a plurality of masters and incorporated new texts, including translations of ancient natural philosophical, mathematical, and medical texts (Gabriel 1969; Southern 1970; Ferruolo 1985).

As schools became larger, with multiple masters and more students drawn from broader catchment areas, how was the system funded? Students traditionally paid fees for instruction directly to masters, but in the interest of promoting the liberal arts and theology over the more lucrative disciplines (law and medicine), Alexander III and Innocent III at the third (1179) and fourth (1215) Lateran Councils required each cathedral church to provide a master to teach poor students without charge. This was to be funded by setting aside the income of one prebend in the cathedral for the master. While this unfunded mandate was neither cheerfully nor universally adopted by bishops, it opened the door for future initiatives—like Honorius III's bull *Super speculam* (1219) that discharged holders of benefices from the obligations of residence while they taught elsewhere—and eventually, in the fourteenth century, to widespread requests for and grants of benefices to clerics at universities (Post 1932; Courtenay and Goddard 2002–2013).

Universities and Scholastic Culture

Over the past generation, scholars have reaffirmed the continuity of the twelfth-century schools and the newer universities of the thirteenth century. However distinctive the university eventually became, the institutions did not spring *ex nihilo* and spontaneously, in the words of Jacques Verger, because they reflected the views of masters, students, civil and ecclesiastical authorities about the value and limits of schools as they knew them (Verger 1995). Among modern historians, the principal motivation for the noticeable shifts toward university organization has been more contested. An older view, particularly focused on the University of Paris, has emphasized the quest for legal autonomy by the early masters and students. Under this conflict interpretation, the struggle for power waged between masters and the chancellor, between local and papal authorities, between town and gown, or indeed between different groups of scholars eventually secured the privileges, security, and legitimacy of the university and its members (Rashdall 1895, 1936, vol. 1, chapter 5). Another perspective holds that universities emerged from changed intellectual circumstances, chiefly the influx of Greek and Arabic works in Latin translation, which in turn demanded new pedagogical techniques and institutions (Grundmann 1964). Diametrically opposed is the view that the university was the creation of economic, political, and social elites, who sought talented and sophisticated scholars in the service of the ruling class (de Ridder-Symoens 1992, 10). But for the most part, none of these extreme positions holds much allegiance today. Rather, universities are seen as microcosms of the societies in which they arose and evolved, neither hovering above the society in which they appeared nor crudely determined by social pressure (Classen and Fried 1983).

As they evolved in the twelfth and thirteenth centuries, universities acquired four characteristics that ancient schools, or indeed prior medieval schools, never possessed. First, they were highly utilitarian: although neither the universities themselves nor the masters who taught in them explicitly considered their mission vocational training, students at universities acquired specific skills that would make them useful for emerging careers in medieval society (de Ridder-Symoens 1992, chapter 8). Second, they became corporate bodies, fully recognized under the law as fictitious persons with rights and responsibilities detailed in charters and documents and represented more graphically in the iconography of the institutions, in seals or historiated initials in manuscripts. Third, they rapidly created fixed curricula that had to be followed to obtain the fourth distinguishing characteristic, the academic degree (de Ridder-Symoens 1992, chapters 10–13; Weijers and Holtz 1997).

As a corporation, the *universitas* designated a plurality, the whole group of scholars, masters, and students residing in a place for a single purpose, that is, training apprentices in letters. By the end of the thirteenth century, universities were distinguished from earlier monastic and cathedral schools by the fact that they drew students from a much broader geographical territory, rather than the local diocese or county. They featured a plurality of masters who came to be recognized as specialists in particular disciplines, not just a single master who was jack-of-all-disciplines. Universities also created higher faculties and more advanced degrees, and by conferring the license to teach, the *ius ubique docendi*, asserted that their alumni were qualified to teach in any other school or university (Post 1929).

The customary age of matriculation at universities was frequently 14 or 15, reflecting the more rudimentary pre-university preparation. There were no entrance requirements for students wishing to attend the university; in many universities, the only criterion was legitimate birth, and even those who were not sure about that probably assumed that they qualified. As one might expect, the student body had tremendous variation, and attrition was frequently high. Many stayed only a year; others—generally better prepared at matriculation—were set on the course of a long and arduous training. From matriculation to achievement of the pinnacle of university achievement (a doctorate in theology) could take as much as 20 years (de Ridder-Symoens 1992, chapter 6).

Most universities were subdivided into faculties focused on a single area of study, with the usual division being arts, theology, law, and medicine, though several southern universities failed to evolve plural faculties in part because of the concentration on professional disciplines like law and medicine and because the arts were seen as propaedeutic to medicine in a combined faculty of arts and medicine. Particularly in northern universities following the Paris model, matriculating students were obliged to attach themselves to a master who both protected and controlled them. On entrance, students swore an oath to preserve the statutes and promote the welfare of the university and to maintain the peace; once duly inducted, the student could expect to receive the protection of not just his own master, but of the entire authority of the university. This was, in its most basic form, the substance of the symbiotic relationship in the guild of scholars (de Ridder-Symoens 1992, chapter 6).

Reflecting the far-flung geographical origins of university students and masters, the two oldest universities—Bologna and Paris—formed "nations," that is, subsidiary entities through which administrative business was transacted (de Ridder-Symoens 1992,

114ff). A third administrative unit of European universities was the college, the first of which seem to have appeared in Paris toward the end of the twelfth century. Initially, these were houses supplied with an income endowed as a charitable benefaction by a prominent founder. Originally largely residential in nature, colleges took on greater pedagogical responsibilities by the end of the middle ages. In a period during which universities did not possess libraries, the colleges—thanks to the benefactions of former fellows (*socii*)—provided both intellectual resources and a comfortable life for scholars (Cobban 1975, 122–59).

As George Makdisi suggested, the college as a charitable trust can be seen in both Islamicate and Western European Christian societies, but the university as a corporation arose exclusively in the West (Makdisi 1981, chapter 4). Aside from the fact that the university represented a legal fictitious rather than physical personality, scholars in the West acquired privileges and protections while attached to the institution, while in the Muslim world such protections were accorded to people as citizens rather than scholars.

The character and function of the medieval university are only intelligible within a discussion of scholastic culture, since scholasticism refers to the pedagogical technique of the schools (Livesey 2005). The recovery of Roman legal texts, the translation of Greek and Arabic medical and philosophical works, and the creation and consolidation of canon law and systematic theology in the long twelfth century coincided with, and in many respects instigated, the educational revolution described above. The central focus of scholastic education was the authoritative text, be it legal, philosophical, medical, or theological, a prominence that can be seen in each of the two main pedagogical techniques of medieval education, the lecture and the disputation (Grabmann 1909–1911).

As students proceeded through the university curriculum, they were introduced to the authoritative text in the lecture (*lectura*, literally a reading). While each institution developed its own sequence of texts required for degrees, in general, in the arts faculty these included additional work in grammar beyond whatever the student had acquired before matriculation, plus Aristotle's logical works and the *Physics*. The remaining works of Aristotle—including *De anima*, *Metaphysics*, *Ethics*, *Politics*, and the *Parva naturalia*—were usually required, though not always with the same degree of emphasis. Finally, depending on the particular qualifications of the university faculty, additional works in mathematics, astronomy, and music could be required, though often these courses were considerably shorter than those devoted to the core texts of Aristotle. Given the underlying goal of inculcation of a canon of philosophical material, it was not uncommon for students to have "heard" these books more than once during the prescribed period leading up to the degree.

Lectures were distinguished by both the content and the time of day in which they were given. In the morning, fully qualified masters gave their detailed and comprehensive "ordinary" lectures on the core texts of the curriculum. These were followed in the afternoon by "extraordinary" or "cursory" lectures delivered by bachelors—that is, apprentice scholars whose lectures were part of their training for the degree—over the same books (essentially providing the medieval equivalent to the modern review session) or over secondary books in the curriculum. The lecture itself followed a formal pattern. First, the act of reading the base text sometimes provided a copy of the work itself; the frequent complaints of students that the reading proceeded too quickly and the countervailing injunctions of university authorities against reading too slowly suggest that transmission of the text *viva voce* did occur (Piltz 1977, 28 n. 92). Second, the lecturer "established the text," that is, provided corrections to errors within circulating copies, thereby ensuring that all students in the class were using the same text. Third, he noted the hierarchical divisions of the text, which in surviving student copies often appear as gibbet-like symbols. Fourth, the master explained linguistic, terminological difficulties, as well as the positions adopted by previous authoritative commentators on the text, both as preliminaries to his own more extended analysis. Finally, important questions or issues within the section of text under discussion that day were analyzed in greater detail. Particularly during the first century of the university's existence, the lecture provided an economical method of assimilating relatively new material in Europe (Glorieux 1968; O'Boyle 1998, 192–201).

The second pedagogical technique, the disputation, assumed the assimilation of this textual tradition and encouraged the creative juxtaposition of elements from the texts to resolve specific problems. Disputations were central to the university scholar's formation: part of the bachelor's training involved attendance at his master's disputations, and in time he was obliged to "respond" in a private mock dispute with his master or other students. Magisterial careers included the expectation of engaging regularly in disputations, either the ordinary kind, in which positions were carefully proposed and prepared in advance, or *de quolibet*, in which the master would debate any question with any person. Such exercises had multiple purposes, including the demonstration of both competence and creativity, the ability to "think on one's feet," but they also were a means of expanding and extending the tradition of the text (Lawn 1993; Weijers 2002).

Although the lecture and disputation addressed different aspects of scholastic education, one can also see how the disputed question evolved from the lecture. As masters prepared their lectures, certain parts of the text proved problematic and necessitated prolonged resolution. By themselves, these questions raised within the context of the lecture did not constitute disputations, but it appears that by the opening years of the thirteenth century, a repertoire of such questions had been detached from the lecture and formed autonomous exercises in their own right. This process was aided by the masters' growing recognition that education involved active engagement of the text, the creation of several compendia of "sentences"—the opinions of authoritative authors—and the growing assimilation of the new logic of Aristotle, especially the two *Analytics*, the *Topics*, and the *Sophistical Refutations*.

While the precise formulation of the disputed question varied across European universities and evolved through the high middle ages, a central format can be seen within the genre. First, the question, appropriately formulated and answerable either in the affirmative or the negative, is enunciated. Following this, in support of one response—generally the one that is ultimately rejected—the author presents several "principal arguments." Next, in the *Sed contra*, the author observes the contrary position, generally supported by an authoritative quotation. Following this, the author presents his own extended discussion of the issue (the *responsio*) in a format that displayed considerable variation throughout the middle ages. By the fourteenth century, for example, it was not uncommon for authors to present multiple opinions expressed by previous scholars and arguments against those opinions as well as subsidiary conclusions and doubts that serve as preliminaries to the author's ultimate resolution of the question.

Finally, the author returns to the "principal arguments" and replies to each, often drawing upon the distinctions and conclusions developed in the *responsio*.

Comparing this method to similar formulas in Arabic texts, George Makdisi sought to establish that Western scholasticism owed its very existence and its definition to Islamic institutions and intellectual practices. Makdisi (1974, 649) argued that

> it was in the very nature of Islam to develop the *sic-et-non* method. In other words, the development of this method in Christianity could very well not have happened at all, whereas without it, Islam could not have remained Islamic.

Because it had no councils, synods, or authoritative ecclesiastical hierarchy—no pope—Islam had to depend on consensus to define orthodoxy. As it developed in Islam, the scholastic *sic-et-non* method served as the mechanism for expressing different opinions about doctrinal positions; each generation debated ideas about the previous generation's positions, until dissent dissolved into consensus. By contrast, the Christian West, with its ecclesiastical hierarchy, councils, and synods for the purpose of defining orthodoxy, had no need of the *sic-et-non* method, and therefore could have developed without it. Scholasticism was thus an intrusive element within Christian Europe, and therefore must have been imported rather than indigenous.

This position has found little support among Western medievalists. In particular, it represents a misguided perspective on the way councils, synods, and the hierarchy functioned in Western Europe. In most instances, actions by these bodies followed, rather than preceded, the vigorous debate on both sides of the issue. Moreover, Makdisi noted quite correctly that the scholastic technique originated in legal contexts in both Arabic and Latin literatures, but as Charles Burnett (1984) noted, "it was precisely the Islamic koranic and legal literature that was *not* translated into Latin in the twelfth and thirteenth centuries." Finally, Makdisi proposed that the technique passed through the agency of medical scholars in both cultures, but such a transmission would have necessitated passage first from law to medicine in Islamicate society, then back from medicine to law in the West while apparently having no effect on the medical literature, for according to Brian Lawn (1993, 67), "purely medical *quaestiones disputatae* ... are very rare in the first half of the thirteenth century."

The centrality of both the lecture and the disputation can be seen by their inclusion in the ceremonial inception or investiture of the new master. Having satisfied all the requirements of the curriculum, the student demonstrated his proficiency in the basic duties of the university master, first in the *vesperie*—the afternoon disputation at the conclusion of which the supervising master recommended the candidate—and then in the *principium*—the inaugural lecture before all the regent masters of the faculty. The candidate now became the fully-fledged master, adorned with biretta and entitled to occupy the magisterial chair (Weijers 1987, 407–22).

On some occasions, both lecture and disputation were recorded in *reportationes*, that is, a transcription prepared during the oral session by someone else, a *reportator*, though the extent to which these texts represent what actually took place in the classroom is not at all clear (Courtenay 1994; Flüeler 1999). Scholastic materials survive in a complex array of formats, from the notes used by masters in the oral sessions, to private student notes, *reportationes* and *ordinationes* (the revised, edited versions), and finally derivatives of these materials, which sometimes were themselves used in classroom settings. A collection of ancillary literatures grew up to help scholastic authors in the preparation of lectures, disputations, and the literary products of university instruction. Chief among these was the *florilegium*, a collection of extracts taken from authoritative authors. *Florilegia* that focused on the bible or the Fathers were extremely popular among sermon writers and theologians, but philosophical *florilegia*, like the *Auctoritates Aristotelis* or the *Propositiones Aristotelis*, were mined for the commentary literature, both in the arts and in theology (Hamesse 1974; 1994). Union lists of books and catalogues of libraries, arranged alphabetically and thematically, appeared in the thirteenth century and proved to be enormously valuable in the search for materials on which to base lectures and commentaries (Rouse, Rouse, and Mynors 1991). And finally, in the service of those preparing for examinations, compendia of questions and responses like those found in Barcelona, Archivio de la Corona de Aragón, Ripoll 109, served as convenient (if frequently misleading) study aids to overburdened students of the scholastic curriculum (Lafleur and Carrier 1997; Kneepkens 2007).

One of the most significant developments of university science concerned quantifications of nature, stretching across natural philosophy (including motion, heat, and light), medicine and pharmacology, and fundamentally the methodology of science itself. Because much of the discussion centered on what came to be known as the "intension or remission of forms," early discussions of quantification required investigation of formal change. Well into the fourteenth century, scholars insisted that forms were ontologically static, yet this was inconsistent with common observation that heat or light or motion changes, temporally or spatially, in the object in which it inheres. Late medieval scholastics responded variously, some preferring a succession, others a coalescence of forms that explained change in things, and still others a part-by-part addition or reduction to explain augmentation and diminution of qualities (Kaye 1998).

Many of the principals in these discussions were primarily theologians rather than natural philosophers, and in part this movement reflects the doctrinal issues inherent in form and quality. Thirteenth-century scholastics observed that divine grace in the soul of the believer is a form, one that both grew and reduced, and these discussions drew upon the very same language and terminology used to describe intension and remission of qualities in nature. Beginning with Anneliese Maier, historians have suggested that fourteenth-century tendencies to quantify qualities drew upon these links between the theological and natural realms (Courtenay 1987, chapter 9).

Other scholars have pointed to analytical or measure languages as a bridge between the natural and the logico-mathematical. Thirteenth-century theories of supposition (the function of the word in a proposition, and thus a theory of reference) became a foundation for quantification discussions, often in solving sophisms, and frequently involving a metalinguistic treatment of problems. Under this technique, propositions speaking in natural philosophical terms—for example, "instant," "point," "line," "continuous magnitude," "begin," "cease," and the like—were translated into others in which these terms do not occur, manipulated under rules of propositional logic, and then retranslated back into natural philosophical terms that express meaning in the natural world (Murdoch 1984).

It is also clear that institutional structures and environments precipitated quantified natural philosophy. At Oxford, Robert Grosseteste's (d. 1253) dedication to Neo-Platonic light metaphysics and scholastic medicine, especially under the influence of Islamicate theories of compounded drugs, emphasized degrees of hot, cold, wet, and dry. Scholastic pedagogy is also seen in the work of the so-called Oxford Calculators—especially Thomas Bradwardine (d. 1349), William Heytesbury (d. 1372/73), and Richard Swineshead (fl. 1340–1354)—who often embedded their positions in disputations *de sophismatibus*, simultaneously instructing undergraduates in logic while practicing the oral exchanges of medieval universities. So prominent were these texts and so closely associated with medieval education, both in fourteenth-century Oxford and fifteenth-century Italian circles, that when the curriculum and pedagogical techniques changed in the Renaissance, the Calculators and their works suffered wilting criticism from humanists and declining attention in the schools (Sylla 1973).

New Directions

By 1378, at the start of the Great Schism, some 43 universities had been founded, though in that year only some 28 to 30 were still functioning. After an initial flurry of growth in the first half of the thirteenth century, no universities were founded between 1256 and 1290, when the second wave of foundations commenced. While the original universities—Bologna, Paris, Oxford, and a few others that emerged in the thirteenth century—were universal institutions, in the sense that they drew students and masters from across Europe, this began to change in the mid-fourteenth century, when regional universities offered a new avenue for learning (see Figure 13.1 and Table 13.1).

Prior to the middle of the fourteenth century, university development was confined to Southern Europe, probably because of the fragmented political organization of Northern Europe and longer legal traditions in the South. Although universities were established in Prague (1347), Kraków (1364), and Vienna (1365), after 1378 the pace of northern university foundations accelerated: by 1500, 18 universities were founded north or east of the Rhine. As in many areas of medieval society, the Great Schism altered the presumed universality of institutions. While the University of Paris and many of its French masters and students were loyal to the Avignonese pope, the Germans and Eastern Europeans at the university supported the Roman pope, and many left Paris after 1378. A growing sense of national or cultural (though not necessarily political) identity, combined with the princes' determination to control the clergy within their domains and provide training for functionaries closer to home, encouraged this new wave of university foundations. These new universities generally adopted the organizational structure of Paris, in part because many were influenced by the experience of expatriate masters of Paris who returned home when disputes over papal allegiances at Paris made collegial life difficult. Several embraced new currents of thought—astronomy at Kraków and Vienna, humanism at Freiburg-im-Breisgau, Ingolstadt, Tübingen, and Vienna, and nominalism at Heidelberg, Freiburg, and Tübingen (Heath 1971; Rosińska 1975; Hoenen 2003).

Seen from this perspective, the fifteenth century can hardly be seen as a period of university decline. While Paris was experiencing challenges to its supremacy, the regional northern French universities (Bourges, Caen, Poitiers, Nantes) and especially

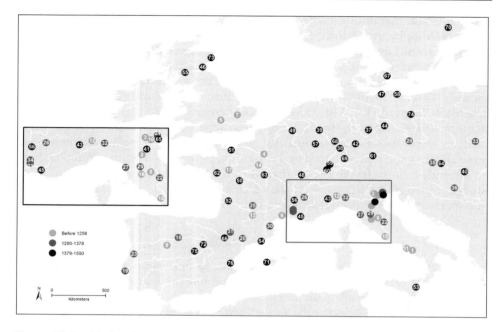

Figure 13.1 Medieval university foundations. While early thirteenth-century foundations were largely confined to Southern Europe and drew students universally, universities of the fourteenth and fifteenth centuries expanded northward and served a more regional clientele. Cartography by Jeffrey M. Widener, data collection by Steven J. Livesey, and data from the GIS of the European Commission.

the German universities were growing substantially during the late fourteenth and fifteenth centuries, which is all the more striking when viewed in the context of the aftermath of plagues (1347–1349 and 1361–1363) and the Hundred Years War and other instabilities. In fact, by the end of the fifteenth century, German universities saw an overabundance of graduates who could not be absorbed by church or state (Schwinges 1998; Schwinges 2000).

Beyond the growth in number, size, and geographic spread of universities, there is also evidence of qualitative improvements in old and new institutions. Colleges, which heretofore had largely been residential only, now assumed a pedagogical mission. The model for this innovation may have been a still earlier presence at universities, the mendicant *studia*. Shortly after their foundation, the Franciscans and Dominicans established networks of schools in proximity to universities, in some cases—notably at Bologna—providing services that augmented the teaching of secular masters (de Ridder-Symoens 1992, 414–17). Moreover, the mendicant orders established a network of *studia*, in provincial towns as well as university centers, and students moved through them as they progressed disciplinarily and proficiently. At the end of the middle ages, another form of religious reform also contributed to pedagogical innovation. The Brethren of the Common Life and the canons of Windesheim, who emerged in the Low Countries during the Great Schism and sought to respond to a yearning for greater spiritual and contemplative life, adopted several pedagogical reforms in their own schools. Chief among these was the designation of educational levels

Table 13.1 Medieval University Foundations. Acceleration of university foundations after the Great Schism (1378) addressed regions north and east of the Rhine, previously unserved in the early university movement, and provided regional alternatives to the older universal universities in the south. See Figure 1

	University	Foundation		University	Foundation		University	Foundation
1	Salerno	before 1200	27	Pisa	1343	53	Catania	1444
2	Bologna	before 1200	28	Prague	1347	54	Barcelona	1450
3	Vicenza	1204	29	Florence	1349	55	Glasgow	1451
4	Paris	early 13th century	30	Perpignan	1350	56	Valence	1452
5	Oxford	early 13th century	31	Huesca	1354	57	Trier	1454
6	Montpellier	early 13th century	32	Pavia	1361	58	Greifswald	1456
7	Cambridge	1209–1225	33	Kraków	1364; re-established 1397	59	Freiburg im Breisgau	1457
8	Arezzo	1215; refounded 1355	34	Orange	1365	60	Basel	1459
9	Salamanca	before 1218/19	35	Vienna	1365	61	Ingolstadt	1459
10	Padua	1222	36	Pécs	1367	62	Nantes	1460
11	Naples	1224	37	Erfurt	1379	63	Bourges	1464
12	Vercelli	1228	38	Heidelberg	1385	64	Pozsony	1465
13	Toulouse	1229	39	Cologne	1388	65	Venice	1470
14	Orléans	c. 1235	40	Buda	1389	66	Saragossa	1474
15	Rome	1245 (Curia); 1303 (University)	41	Ferrara	1391	67	Copenhagen	1475
16	Siena	1246; refounded 1357	42	Würzburg	1402	68	Mainz	1476
17	Angers	c. 1250	43	Turin	1404	69	Tübingen	1476
18	Valladolid	end of 13th century	44	Leipzig	1409	70	Uppsala	1477
19	Lisbon	1290	45	Aix-en-Provence	1409	71	Palma de Mallorca	1483
20	Lérida	1300	46	St Andrews	1411	72	Sigüenza	1489
21	Avignon	1303	47	Rostock	1419	73	Aberdeen	1495
22	Perugia	1308	48	Dole	1422	74	Frankfurt-on-Oder	1498
23	Coimbra	1308	49	Louvain	1425	75	Alcalá	1499
24	Treviso	1318	50	Poitiers	1431	76	Valencia	1500
25	Cahors	1332	51	Caen	1432			
26	Grenoble	1339	52	Bordeaux	1441			

within the curriculum, which replaced the medieval scholastic practice of repetitive exposure to the same material. The center of this movement was the Brethren's school at Deventer, whose alumni included Erasmus, Nicholas of Cusa, Rudolf Agricola, and Jan Standonck, who subsequently transplanted the techniques first in the Collège de Montaigu at Paris (1490) and then in the Collège de Standonck at the University of Louvain (Verger 2000, 61–3; Bakker 2007, 92–3, 110).

In sum, the notion that universities at the end of the middle ages were institutionally weak, pedagogically static and inflexible, and intellectually ossified is more mythological than real. In many respects, this modern storyline has arisen because we have taken Renaissance humanist criticisms at face value, although, as John Murdoch observed in his final essay, these criticisms presented a "skewed view" (Murdoch 2011). As the university renovated or restocked its pedagogical toolkit, many of the techniques that once were the heart of medieval education—*disputationes de sophismatibus*—were now preserved in texts whose content appeared to be very silly indeed (Sylla 1986), and it is largely these outmoded and ephemeral medieval residues that humanists despised. But while some of the content or the technique ceased to inhabit the core of the institution, universities themselves continued to play important roles in the Renaissance and Reformation movements, spawning religious reform and educating the elites of European society. The university proved itself to be a remarkably resilient institution in the face of war, plague, famine, and intellectual and personal disputes for some 400 years (Gascoigne 1990; Grendler 2004).

Endnote

1   The author would like to express his gratitude to Taylor and Francis Group LLC Books for allowing him to use material from a previously published article, "Scholasticism," in *Medieval Science, Technology, and Medicine: An Encyclopedia*, edited by Thomas F. Glick, Steven J. Livesey, and Faith Wallis, 453–455. London: Routledge, 2005.

References

Bakker, Paul J. J. M. 2007. "The statutes of the Collège de Montaigu: Prelude to a future edition." *History of Universities*, 22, No. 2: 76–111.

Burnett, Charles. 1984. "Review of Makdisi, *The Rise of Colleges*." *History of Universities*, 4: 187–88.

Classen, Peter, and Johannes Fried. 1983. *Studium und Gesellschaft im Mittelalter*. Stuttgart: A. Hiersemann.

Cobban, Alan B. 1975. *The Medieval Universities: Their Development and Organization*. London: Methuen.

Contreni, John J. 1995. "The Carolingian renaissance: Education and literary culture." In *The New Cambridge Medieval History*, edited by Rosamond McKitterick, volume 2, 709–57. Cambridge: Cambridge University Press.

Costambeys, Marios. 2011. *Power and Patronage in Early Medieval Italy: Local Society, Italian Politics and the Abbey of Farfa, c.700–900*. Cambridge: Cambridge University Press.

Courtenay, William J. 1987. *Schools and Scholars in Fourteenth-Century England*. Princeton, NJ: Princeton University Press.

Courtenay, William J. 1994. "Programs of study and genres of scholastic theological production in the fourteenth century." In *Manuels, programmes de cours et techniques d'enseignement dans les universités médiévales*, edited by Jacqueline Hamesse, 325–50. Louvain-la-Neuve: Institut d'Etudes Médiévales de l'Université Catholique de Louvain.

Courtenay, William J., and Eric D. Goddard. 2002–2013. *Rotuli Parisienses: Supplications to the Pope from the University of Paris, 1316–1394.* 3 volumes in 4. Leiden: Brill.

de Ridder-Symoens, Hilde (ed.) 1992. *Universities in the Middle Ages.* Cambridge: Cambridge University Press.

Ferruolo, Stephen C. 1985. *The Origins of the University: The Schools of Paris and Their Critics, 1100–1215.* Stanford, CA: Stanford University Press.

Flüeler, Christoph. 1999. "From oral lecture to written commentaries: John Buridan's *Commentaries* on Aristotle's *Metaphysics.*" In *Medieval Analyses in Language and Cognition: Acts of the Symposium, the Copenhagen School of Medieval Philosophy, January 10–13, 1996*, edited by Sten Ebbesen and Russell L. Friedman, 497–521. Copenhagen: Royal Danish Academy of Sciences and Letters.

Gabriel, Astrik. 1969. "The Cathedral Schools of Notre-Dame and the beginning of the University of Paris." In *Garlandia: Studies in the History of the Medieval University*, 39–64. Frankfurt am Main: Josef Knecht.

Gascoigne, John. 1990. "A Reappraisal of the role of universities in the Scientific Revolution." In *Reappraisals of the Scientific Revolution*, edited by David C. Lindberg and Robert S. Westman, 207–60. Cambridge: Cambridge University Press.

Glorieux, Palémon. 1968. "L'enseignement au moyen âge, Techniques et méthodes en usage à la Faculté de Théologie de Paris, au XIIIe siècle." *Archives d'histoire doctrinale et littéraire du moyen âge*, 35: 65–186.

Grabmann, Martin. 1909–1911. *Die Geschichte der scholastischen Methode nach den gedruckten und ungedruckten Quellen*, 2 volumes. Freiburg im Breisgau: Herdersche Verlagshandlung.

Grendler, Paul F. 2004. "The Universities of the Renaissance and Reformation." *Renaissance Quarterly*, 57: 1–42.

Grundmann, Herbert. 1951. "Sacerdotum, regnum, stadium." *Archiv für Kulturgeschichte*, 34: 5–27.

Grundmann, Herbert. 1964. *Vom Ursprung der Universität im Mittelalter.* Berlin: Akademie-Verlag.

Hamesse, Jacqueline. 1974. *Les Auctoritates Aristotelis: un florilège médiéval: étude historique et édition critique.* Louvain: Publications universitaires.

Hamesse, Jacqueline. 1994. "Les florilèges philosophiques, instruments de travail des intellectuels à la fin du Moyen Âge et à la Renaissance." In *Filosofia e teologia nel Trecento: studi in ricordo di Eugenio Randi*, edited by Luca Bianchi, 479–508. Louvain-la-Neuve: Fédération internationale des instituts d'études médiévales.

Haskins, Charles Homer. 1927. *The Renaissance of the Twelfth Century.* Cambridge, MA: Harvard University Press.

Heath, Terrence. 1971. "Logical grammar, grammatical logic, and humanism in three German universities." *Studies in the Renaissance*, 18: 9–64.

Hoenen, Maarten J. F. M. 2003. "'Via antiqua' and 'via moderna' in the fifteenth century: doctrinal, institutional and political factors in the 'Wegestreit'." In *The Medieval Heritage in Early Modern Metaphysics and Modal Theory, 1400–1700*, edited by Russell L. Friedman and Lauge Olaf Nielsen, 9–36. Dordrecht: Kluwer Academic.

Jaeger, C. Stephen. 1994. *The Envy of Angels: Cathedral Schools and Social Ideals in Medieval Europe, 950–1200.* Philadelphia, PA: University of Pennsylvania Press.

Johnson, Penelope D. 1991. *Equal in Monastic Profession: Religious Women in Medieval France.* Chicago: University of Chicago Press.

Kaye, Joel. 1998. *Economy and Nature in the Fourteenth Century: Money, Market Exchange, and the Emergence of Scientific Thought.* Cambridge: Cambridge University Press.

Kneepkens, Corneille H. 2007. "How to prepare for a BA in the late middle ages: Reparationes or study aids for logic." In *University, Council, City: Intellectual Culture on the Rhine (1300–1550): Acts of the XIIth International Colloquium of the Société Internationale pour l'Étude de la Philosophie Médiévale, Freiburg im Breisgau, 27–29 October 2004*, edited by L. Cesalli, Nadja Germann, and M. J. F. M. Hoenen, 63–95. Turnhout: Brepols.

Lafleur, Claude, and Joanne Carrier. 1997. *L'enseignement de la philosophie au XIIIᵉ siècle: autour du "Guide de l'étudiant" du ms. Ripoll 109: actes du colloque international*. Turnhout: Brepols.

Lawn, Brian. 1993. *The Rise and Decline of the Scholastic "Quaestio Disputata" With Special Emphasis on its Use in the Teaching of Medicine and Science*. Leiden: Brill.

Livesey, Steven J. 2005. "Scholasticism." In *Medieval Science, Technology, and Medicine: An Encyclopedia*, edited by Thomas F. Glick, Steven J. Livesey, and Faith Wallis, 453–5. London: Routledge.

Makdisi, George. 1974. "The scholastic method in medieval education." *Speculum*, 49: 640–61.

Makdisi, George. 1981. *The Rise of Colleges. Institutions of Learning in Islam and the West*. Edinburgh: Edinburgh University Press.

Murdoch, John E. 1984. "The analytic character of late medieval learning: Natural philosophy without nature." In *Approaches to Nature in the Middle Ages*, edited by Lawrence Roberts, 171–213. Binghamton, NY: Center for Medieval & Early Renaissance Studies.

Murdoch, John E. 2011. "A skewed view: The achievement of late medieval science and philosophy as seen from the Renaissance." In *Crossing Boundaries at Medieval Universities*, edited by Spencer E. Young, 177–88. Leiden: Brill.

O'Boyle, Cornelius. 1998. *The Art of Medicine: Medical Teaching at the University of Paris, 1250–1400*. Leiden: Brill.

Orme, Nicholas. 2006. *Medieval Schools: From Roman Britain to Renaissance England*. New Haven, CT: Yale University Press.

Piltz, Anders. 1977. *Studium Upsalense: Specimens of the Oldest Lecture Notes taken in the Mediaeval University of Uppsala*. Uppsala: Institutionen för klassiska språk.

Post, Gaines. 1929. "Alexander III, the *licentia docendi* and the rise of the universities." In *Anniversary Essays in Mediaeval History Presented on His Completion of Forty Years of Teaching*, edited by Charles Holt Taylor and John Life La Monte, 255–77. Boston: Houghton Mifflin.

Post, Gaines. 1932. "Master's salaries and student-fees in Mediaeval universities." *Speculum* 7: 181–98.

Rashdall, Hastings. 1895, 1936. *The Universities of Europe in the Middle Ages*. Oxford: Oxford University Press.

Rosińska, G. 1975. "L'école astronomique de Cracovie et la révolution copernicienne." In *Avant, avec, après Copernic: La représentation de l'univers et ses conséquences épistémologiques: 1–7 Juin 1973*, 89–92. Paris: A. Blanchard.

Rouse, Richard H., Mary A. Rouse, and R. A. B. Mynors. 1991. *Registrum Anglie de libris doctorum et auctorum veterum*, London: British Library in association with the British Academy.

Schwinges, Rainer Christoph. 1998. "The Medieval German university: Transformation and innovation." *Paedagogica Historica*, 34: 374–88.

Schwinges, Rainer Christoph. 2000. "On recruitment in German universities from the fourteenth to sixteenth centuries." In *Universities and Schooling in Medieval Society*, edited by William J. Courtenay, Jürgen Miethke, and David B. Priest, 32–48. Leiden: Brill.

Southern, Richard W. 1970. "Humanism and the School of Chartres." In *Medieval Humanism and Other Studies*, 61–85. Oxford: B. Blackwell.

Sylla, Edith D. 1973. "Medieval concepts of the latitude of forms: The Oxford calculators." *Archives d'histoire doctrinale et littéraire du moyen âge*, 40: 223–83.

Sylla, Edith D. 1986. "The fate of the Oxford calculatory tradition." In *L'homme et son univers au moyen âge: actes du septième congrés international de philosophie médiévale (30 août–4 septembre 1982)*, edited by Christian Wenin, 692–8. Louvain-la-Neuve: Editions de l'Institut supérieur de philosophie.

Verger, Jacques. 1995. "*Nova* et *vetera* dans le vocabulaire des premiers statuts et privilèges universitaires français." In *Les universités françaises au moyen âge*, 37–52. Leiden: Brill.

Verger, Jacques. 2000. *Men of Learning in Europe at the End of the Middle Ages*. Notre Dame: University of Notre Dame Press.

Weijers, Olga. 1987. *Terminologie des universitaires au XIIIe siècle*. Rome: Edizioni dell'Ateneo.

Weijers, Olga. 2002. *La 'disputatio dans les Facultés des arts au moyen âge*. Turnhout: Brepols.

Weijers, Olga, and Louis Holtz, ed. 1997. *L'enseignement des disciplines* à la Facultè *des arts: Paris et Oxford, XIIIe–XIVe siècles: actes du colloque international*. Turnhout: Brepols.

CHAPTER FOURTEEN

The Observatory

ROBERT W. SMITH

Early Observatories

The observatory as a research institution reached its maturity in the Islamic World. However, if we take an observatory to be defined as (following the *Oxford English Dictionary*) a "building or place set apart for, and furnished with instruments for making, observations of natural phenomena; especially for astronomical, meteorological or magnetic observations," then there were certainly observatories before the emergence of the Islamic world. If we stretch this definition to include buildings or places set apart for making observations of astronomical phenomena with no concern for instruments, then there are substantially more examples.

Various students of the megalithic monuments of Britain, Ireland, and north west France have long argued for their astronomical significance. The eminent astrophysicist of the late nineteenth and early twentieth century as well as long time editor of *Nature*, Norman Lockyer, claimed early in the twentieth century that, based on accurate measurements of numerous sites, these megalithic monuments had been built to observe and mark the rising and setting places of the heavenly bodies. Alexander Thom was one of the leading figures in the recent development of what is now generally known as archaeoastronomy. He published his *Megalithic Lunar Observatories* in 1971, for example. Here and in other works Thom claimed a technical sophistication for prehistoric peoples, including, for instance, a knowledge of the motions of the moon, that later astronomers would not obtain until the seventeenth or eighteenth centuries. Thom's studies, however, drew much criticism, especially from archaeologists. Thom, it is now generally agreed, read too much into his evidence, but it is also now widely accepted that there are astronomical features 'built in' to many megalithic sites and the debates have often shifted to the interpretations of these features rather than their existence.

These debates have to proceed in the absence of written evidence, but in the case of, for example, the sky watchers of Mesoamerica there are, despite the tragic obliteration of so many pre-Columbian texts by the Spanish invaders, a few original works as well as

A Companion to the History of Science, First Edition. Edited by Bernard Lightman.
© 2016 John Wiley & Sons Ltd. Published 2020 by John Wiley & Sons Ltd.

documents written by Spanish missionaries that have survived. These, in conjunction with the evidence gleaned from different sites, make clear that there were numerous buildings that were involved with various sorts of astronomical observations. But as the leading student of Mesoamerican astronomy, Anthony Aveni, has argued, these 'observatories' served a "divinatory and ritualistic function more than an astronomical one, and it would be entirely erroneous to regard such an institution as we do its modern counterpart" (Aveni 2001, 38).

China

The study of celestial phenomena was regarded as extremely important in ancient and medieval China, as such phenomena were reckoned to be very closely related to the emperor's sovereign power. The Emperor was regarded as the Son of Heaven, and although himself mortal, he was empowered by Heaven to rule human beings. Heaven made known its judgment of the Emperor's performance as the Son of Heaven through celestial events. In this political cosmology, then, a sacred obligation was placed on the Emperor to monitor the heavens. The study of heavenly phenomena therefore became a state function and the 'Imperial Astronomical Bureau' played a major role in all the ruling houses after the Han dynasty (which ran from 206 BCE to 220 CE). The staff were generally of low status and the observing instruments they employed had to be authorized and supplied by edicts from the Emperor. There was, however, no one study of astronomy and both *lifa*, the investigation of calendrical methods, and *tianwen*, the examination and analysis of celestial patterns, were reckoned to be very important for the effective functioning of the state and almost all their practitioners were members of the Imperial civil service. The task of students of *tianwen* was both to note unusual and unpredicted celestial phenomena and to explain their import for the human world.

The scrutiny of the skies—a watch was kept day and night—for celestial anomalies meant each astronomical bureau was provided with an observatory or "observatory platform." Chinese observatories sometimes operated on a large scale. The Yuan observatory, completed in 1286, shared a compound with the Astrological Commission's headquarters in Beijing. There were 70 scholars divided into three bureaus: computation, observation, and water clocks (Sivin 2008, 171–5).

Thatcher Deane has made an extensive study of the Imperial Astronomical Bureau in the Ming Dynasty (1368–1644). He has found that the bureau's staff came from a hereditary pool, and when the main capital was transferred to Beijing in 1402, both an Astronomical Bureau and a new observatory were established there. The observatory was located on the top of a section of the Beijing city wall. The two main instruments employed in the Ming were an armillary sphere and what was known as a 'simplified instrument,' neither of which matched the best Islamic instruments in terms of accuracy. As part of their policy of maintaining duplicates of most central government offices in Nanjing, Ming emperors also established an auxiliary astronomical bureau and observatory (Deane 1994, 131).

The Islamic World

A number of large-scale observatories were built by Islamic astronomers, often, because of a strong interest in astrology, with support from rulers and powerful figures.[1] The last years of the reign of Al-Ma'mūn (813–833) saw the beginnings of what has become known as the translation movement as well as a major program of observations undertaken from Baghdad and Damascus in order to improve on Ptolemaic observations of the sun and moon. Although the initial observations from Baghdad were not regarded as satisfactory, they were followed up by a team of astronomers at Damascus where observations were made daily for at least one year and then less regularly for an additional two or three years. There was, then, a coordinated and collective effort over an extended period by astronomers addressing a specific research program using instruments at special locations. As Dallal has argued, this research project

> endowed astronomical activity in the Islamic world with formal prestige. It also set a precedent for future support of scientific activity by other rulers and established patronage as one of the modes of supporting scientific activity (Dallal 2010, 23; see also Sayili 1960)

But were there observatories at Baghdad and Damascus or were they what might be termed observation posts? There is no such doubt about a number of later institutions.

Malikshāh (1073–1092) was the third and most powerful of the Seljūq sultans and his court at Nishapur drew students of nature as well as writers from across the Islamic world including the poet Omar Khayyam. He also established an observatory, probably at Isfahān. It was initially planned that its astronomers would follow Saturn through one revolution (Saturn was a very important planet for astrological reasons). However, with the death of its founder, the observatory was closed after 18 years (it would have taken nearly 30 years to have followed Saturn once around the sky).

The most renowned and influential Islamic observatory was built in the thirteenth century at Marāghā (modern day Marageh in Iran) for the famous astronomer Nasīr al-Dīn al-Tūsī. Its patron was the Mongol ruler of Persia, Hulagu, a devotee of astrology. Construction began in 1259 and the observatory, as it was funded by *waqf* revenues (that is, the observatory and its assets were held by a trust), outlived for a time the death of its patron and al-Tūsī's move to Baghdad in 1274, so that observations were made at Marāghā into the fourteenth century. It was equipped with the latest observational instruments, including a mural quadrant 14 feet in radius.

Marāghā, moreover, was a mix of observatory and research institute as it boasted an extensive library and drew astronomers and students from across the Islamic world. The Marāghā astronomers produced planetary and trigonometric tables (which have not been particularly well regarded by historians) and performed advanced research in planetary theory (the fruits of which have been very highly regarded by historians). The astronomers who did this work are now generally known as the Marāghā school.

Marāghā was also a model for the later Muslim observatories at Samarkand (established by Ulugh Beg in 1420), Istanbul (construction began in 1575), and later in Jaipur in India (construction ran from 1727 to 1734). Ulugh Beg was himself an enthusiastic astronomer and under his guidance astronomical tables were produced at Samarkand. These included a compilation of more than 1000 stars, which has been described as the one important star catalogue of the middle ages.

Tycho Brahe

The construction of the Istanbul Observatory was finished just in time for observations of the very bright comet of 1577. A young Johannes Kepler was taken to a hillside to watch the comet by his mother, and in Denmark Tycho Brahe paid it very close attention. The year before, Tycho had been granted the small island of Hven just off Copenhagen by King Frederick II of Denmark in part as a reward for the fame he had gained through his observations of the famous nova of 1572. Tycho's aim was to fashion an observatory that would serve his goal of reforming astronomy by means of unprecedentedly accurate observations. The result was Uraniborg, the finest observatory the world had ever seen, which was in operation from 1580 to 1597 (Thoren 1990; Christianson 2000; Mosley 2007).

Uraniborg was equipped not just with an impressive range of astronomical instruments, but also observing rooms, a printing press, a paper mill, a library, an alchemical laboratory, a workshop for building instruments, as well as accommodations for the many assistants. Although many of the instruments sprang from what has been called the neo-Ptolemaic tradition, a number also exhibited novel features and Tycho developed new designs for quadrants and sextants.

In his pursuit of accuracy, Tycho not only lavished care on his instrument designs but he employed teams of assistants so that simultaneous observations could be made with different instruments in an efficient and coordinated manner. Around 1584, Tycho added a satellite observatory, Stjerneborg, with instruments located below ground level in order to escape the buffeting of the wind.

The Seventeenth Century

The observatories that were built in Europe in the first decades after Tycho's death had strong Tychonic associations and influences. The first university observatory was established at Leiden, and its first instrument was provided by two of Tycho's former assistants in 1632. A "Royal Stjerneborg" was built at the University of Copenhagen between 1637 and 1642, with Longomontanus, second only to Kepler among the assistants who had served under Tycho, playing a significant role.

Before the seventeenth century, observatories tended to last for relatively short periods. The death of a patron (as was the case for Tycho following the death of King Frederick II in 1588) or of the astronomer in charge often brought the lives of observatories to an end as these people had provided the resources and energy to bring the institutions into being and to keep them operating.

A number of the official observatories established in the seventeenth century, however, were able to outlive their initial patrons and initial staff. Perhaps the most important of all seventeenth century observatories was that founded in Paris in 1667. Richly backed by the Sun King, Louis XIV, and managed by the French Academy of Sciences (itself established in 1665), most of the construction was completed by 1672 though work continued on the interior of the building until 1683. By 1670, Paris had become the foremost center for astronomy in Europe.

The initial instrumentation was made up of two main types. The first was for positional astronomy and consisted of mural sectors and quadrants, but, unlike Tycho's versions of these instruments, they were equipped with telescopic sights and micrometers. The second type of instrument was made up of long-focus refractors with objective lenses fashioned by the best makers of the period, Campani and Divini. While the management by the Academy of Sciences meant the observatory, very unusually, did not possess an official director, four generations of the Cassini family were effectively in charge between 1671 and 1793. The first of the Cassinis, J. D., helped put the observatory on the map with the discovery of four moons of Saturn. These finds go to underline that research at Paris was somewhat more wide-ranging than that at the Royal Observatory at Greenwich, founded in 1675.

King Charles II of England was persuaded in 1674 to issue a warrant for a Royal Commission to examine a proposal to solve the problem of determining longitude at sea. The proposal centered on the use of the position of the moon with respect to the stars as a sort of celestial clock and, while reasonable in principle, was unworkable due to the limited state of knowledge of star positions. John Flamsteed was already familiar with a better method that also employed the moon. Later it came to be known as the method of lunar distances.

In 1674, the Royal Society had begun to lay plans for an observatory. Sir Jonas Moore offered to meet all the expenses and proposed that Flamsteed be the observer. The upshot of these various developments was that the King signed a royal warrant that appointed Flamsteed as the observer for a Royal Observatory, instructing that he should forthwith:

apply himself with the most exact care and diligence to rectifying the tables of the motions of the heavens, and the places of the fixed stars, so as to find the so-much-desired longitude of places for the perfecting the art of navigation (Howse 1980, 28)

Charles II (a great grandson of King Frederick II of Denmark who had patronized Tycho) funded the building of the observatory and paid an annual salary to Flamsteed. He did not, however, pay for the instruments (still in many ways patterned on those employed at Hven) or assistants. In these circumstances, Flamsteed regarded the observations he secured as his own, a belief that led to controversy later. Flamsteed bought some instruments for himself and received others as gifts. As well as long refractors, he used a mural arc and a 7-foot equatorial sextant divided to new levels of accuracy and these, in combination with pendulum clocks, meant that angular measurements made at Greenwich were the most accurate in the world by 1690.

When Flamsteed died his widow removed the instruments from the Observatory, and the Astronomer Royal who followed Flamsteed, Edmond Halley, had to start over. It was only in 1764, after the records of Halley's successor, James Bradley, had proven extremely difficult to recover that the Royal Society composed regulations that laid down the rights and responsibilities of the Astronomer Royal, a move that in itself was a recognition of the permanence of the Royal Observatory even though Astronomers Royal would come and go.

The private observatory in Danzig of Johannes Hevelius was the last of its kind to make important contributions to fundamental research in positional astronomy. After Hevelius's death in 1687, official observatories very much took the lead in these investigations.

The Eighteenth Century

There were two crucial changes to practical astronomy in the eighteenth century. The first concerned the rapid rise in the number of official observatories in Europe, that is, observatories supported by states, universities, scientific societies, and religious foundations. In a period when there were relatively few professional scientists, the staff of these observatories likely outnumbered those in any other branch of science. The second key development was that these astronomers reached a consensus that the proper pursuit of astronomy involved securing angular measurements, most particularly of the positions of stars (Bennett 1988; 1992). Given the great emphasis astronomers placed on stability and accuracy, the observations had in practice to be limited to the meridian, that is, to observations employing an instrument pointing directly south, and so having a fixed mounting.

Also key for the eighteenth century developments was the emergence of instrument makers who could equip the observatories with large, state-of-the-art instruments. Before this, the instruments had generally been made locally. The leading figure in this shift was the London scientific instrument maker George Graham. His designs of instruments for the new sort of official observatory lasted for most of the eighteenth century and in some places beyond that. Instruments designed by Graham, as well as those by John Bird and Jonathan Sisson, whose designs stemmed from Graham's, were shipped from London across Europe and set the standard for mural quadrants, transit instruments, zenith sectors, astronomical regulators (very accurate timekeepers), and equatorial sectors. When, for example, in 1768 J. J. L. Lalande established the new observatory at the École Militaire in Paris he equipped it with an 8-foot mural quadrant by Bird.

There thus emerged a common set of goals for astronomers across Europe that was shaped fundamentally by instrument makers. Indeed, the instruments housed in the new sorts of official observatories established an agreement about the correct conduct and practices of an observatory, and so a community of active researchers with a unified program emerged (Bennett 1992).

The eighteenth century was also a crucial period for the emergence of new forms of scientific organization, with astronomy very much in the lead. Widmalm (2010) has summarized these developments: "Networks of scientific correspondence centered on metropolitan academies spread across Europe as astronomers eagerly collected observational data, information about instruments, their makers and about each other." The result was that astronomers' traditional roles expanded beyond observing and interpreting the skies. Now

> they became information specialists, experts in collecting and analyzing large amounts of data ... Astronomical practices were supported and developed in close relation with economic ... projects undertaken by the State, such as the longitude problem, land and coastal surveys, almanac production, etc. (Widmalm 2010, 174)

From their observatories astronomers observed the heavens, but their main job was to map the world. Indeed, it has been calculated that around 50 percent of all active astronomers at the turn of the nineteenth century were involved in one way or another with measuring geographical positions.

Giant Reflectors

By the late 1780s the most famous astronomer in Europe was William Herschel, the discoverer of the planet Uranus in 1781. But his interests and approaches to astronomy were also profoundly different from the positional astronomy of the mathematical astronomers. Rather, Herschel was a self-styled natural historian of the heavens and his great goal was what he termed "Construction of the Heavens." To this end he strove to build the biggest reflecting telescopes ever seen. These were deemed too large for their own structures. Herschel's telescopes sat in the open air and were located next to his home rather than housed in their own building.

Lord Rosse followed the same approach with the giant reflectors that he built a few decades later at the family estate at Birr Castle in the middle of Ireland. Rosse's most famous telescope was the celebrated "Leviathan of Parsonstown" with a primary mirror 72-inches in diameter slung between two walls of masonry and kept in the open air. The observatories built by the Liverpool brewer and amateur astronomy William Lassell, however, represented the wave of the future. He built 9-inch and 24-inch reflectors and housed these in buildings with rotating domes, a section of which could be opened to view the sky. There was, however, no such structure for Lassell's biggest telescope, a 48-inch reflector, presumably because housing it in such a similar building would have been difficult as well as prohibitively expensive.

The First Half of the Nineteenth Century

Herschel's astronomical practices were far out of the astronomical mainstream, which until the 1860s was centered on positional astronomy, as reflected in terms of observatories' practices and goals, the instruments they housed, and the staff employed. During the nineteenth century the number of astronomical observatories rose dramatically, from less than 36 to more than 200.

Many astronomers in the middle of the nineteenth century regarded the Pulkovo Observatory as the leading such institution and as the most complete and significant astronomical observatory in the world (Werrett 2010). In 1830, F. G. W. Struve was appointed director of the newly established observatory at Pulkovo, a small village to the south of St. Petersburg. It was officially opened in 1839. In Struve's hands both its instrumentation and its approach to research were German. Williams has also linked Pulkovo's architectural design with its astronomical functions (Williams 1989). Equipped with living facilities and an extensive library as well as rooms that housed both permanent and portable instruments and spaces for the astronomers to perform calculations, Pulkovo, Williams emphasizes, set new standards for careful design and for the rigor of its administration.

The undertakings of some other observatories were more lax. The Paris Observatory, for example, was notorious for the sloppiness of its operations, and unreduced observations piled up for years. This situation only changed with the appointment in 1854 of the autocratic and tough-minded U. J. J. Leverrier as director. In contrast to Paris before Leverrier, the operations at the Royal Observatory at Greenwich as run by George Biddell Airy displayed mid-nineteenth century British views of business efficiency. Airy incorporated German methods, but also extended them. Some historians have judged Greenwich to have been run on factory-like lines. There was a strict hierarchy of staff together with a rigid division of labor. Observation, Schaffer has emphasized, was mechanized and observers themselves were now subjected to scrutiny just as were the stars themselves (Schaffer 1988; Smith 2003, 159).

As well as establishing positions on land, astronomy was also crucial for seafarers to determine their position at sea, a task which was intimately related to timekeeping. For a number of observatories work with chronometers was therefore extremely important. Here the most notable example is Airy's Greenwich. When Airy became the Astronomer Royal in 1835 many responsibilities had fallen on Greenwich beyond positional astronomy narrowly defined. Airy did not reject the Observatory's efforts to improve chronometers but he did object to the routine rating and maintenance of chronometers. He viewed his staff as astronomical observers and calculators and not clerks. In 1840, about a third of the staff time was devoted to chronometers, and in 1865, thirty years after he was appointed as Astronomer Royal, the fraction was still one quarter.

The chronometer work was not really associated with the Observatory's scientific charge, at least not by the 1830s. Instead, it was related to what's been termed the Observatory's social function (Bennett 1980), what, through custom and practice as well as directions from the British government, it was required to do. Under Airy, this social function grew in scope as he thought it entirely appropriate to mesh the observatory with the practical world (Meadows 1975, 63). For him, this meant the improvement of chronometers, the distribution of accurate time, and, more broadly, work that tended to the advantage of geography and navigation (Smith 1991, 9).

Magnetism

The determination of the nature of the Earth's magnetic field was often seen as among the most important problems, if not the most important problem, in the physical sciences in the early nineteenth century. The wide interest in magnetism led to links between astronomical practice and magnetic practice at observatories. In the early nineteenth century, the leading center of magnetic research was France, and in particular the Paris Observatory, which even became for a time the point of reference for other observatories that pursued magnetic studies (Cawood 1979, 495). Often the astronomical and magnetic practitioners were the same people. By the early nineteenth century magnetic houses or observatories were common features of what are often regarded as solely astronomical observatories.

When the planning was underway in the 1840s for what would become the US Naval Observatory, James Gillis was sent to Europe for ten weeks to better understand the state of the art in observatory and instrument design. As a result of this trip, the new institution's magnetic observatory was placed underground as Gillis had seen was the case at Munich; it was connected by a tunnel to the main Observatory buildings (Dick 2003, 65–8).

The push to map the Earth's magnetic field spawned a number of large-scale enterprises during the nineteenth century, perhaps the most well known of which was the "Magnetic Crusade," led by British scientists and in imitation of earlier continental efforts. The Magnetic Crusade was termed by William Whewell as "by far the greatest scientific undertaking the world has ever seen" (Whewell 1857, 55) and by 1840 a range of observatories were working together on the enterprise (Cawood 1979). As an offspring of the Magnetic Crusade, several of the leading British scientists pressed for a new experimental observatory to follow on the already established observatories that Humboldt and others had been involved with at Berlin in the 1820s as well as the observatory at Gottingen in the 1830s. At the new observatory the staff would verify the performance of instruments as well as develop new sorts of astronomical and magnetic instruments and experiments. It became a kind of depot for accrediting the work of instrument makers, and, all in all, as Schaffer has argued, became a site at which the new branch of "experimental astronomy" could be pursued (Schaffer 1995). In an important recent work, Aubin, Bigg, and Sibum have further argued that nineteenth-century observatories should be understood in terms of what they call the observatory sciences, a group that as well as astronomy includes cartography, meteorology, geodesy, as well as to some degree physics and statistics (Aubin, Bigg, and Sibum 2010).

The Observatory as Laboratory

In writings on the rise of astrophysics, it has been widely agreed that the observatory first began to resemble a laboratory in the 1860s in the wake of the fundamental and path-breaking research on spectrum analysis by Gustaf Robert Kirchhoff and Robert Bunsen. Their research gave investigators new powers to interpret the spectral lines that could be observed in the light of the sun and stars when passed through a prism. This founding claim is well expressed in an account of "The New Astronomy" by the pioneering English astrophysicist William Huggins. In 1897, Huggins looked back to the 1860s and the origins of astrophysics. He had established his own observatory at Tulse Hill in London in the mid-1850s, but soon, he recalled, became dissatisfied with the routine character of ordinary astronomical work. Fortunately for Huggins, in 1862 he attended a lecture by his friend, neighbor, and Professor of Chemistry at King's College, Dr. William Allen Miller, on "The new method of spectrum analysis." In Huggins's retrospective account, Miller's lecture was his personal epiphany. Huggins abandoned positional astronomy in favor of the exciting new study of celestial spectroscopy. "Then it was," he reminisced,

> that an astronomical observatory began, for the first time, to take on the appearance of a laboratory. Primary batteries, giving forth noxious gases, were arranged outside one of the windows; a large induction coil stood mounted on a stand on wheels so as to follow the positions of the eye end of the telescope, together with a battery of several Leyden jars; shelves with Bunsen burners, vacuum tubes and bottles of chemicals, especially of specimens of pure metals, lined its walls (Huggins 1897, 913; also Becker 2011, 46–63).

The great goal now was to pursue, via spectrum analysis, the chemical and physical composition of the heavenly bodies. In 1897, Huggins was writing very much as one of the members of what might be called the spectroscopic vanguard and one of the winners in the disputes and debates of the last four decades of the nineteenth century over what constituted appropriate astronomical practices. The spectroscope, for Huggins, had remade the observatory (Figure 14.1).

As we have seen in the discussion of magnetism, as well as in the notion of the observatory sciences advanced by Aubin, Bigg, and Sibum, some observatories were already

Figure 14.1 The interior of William Huggins's observatory in the mid-1860s. William Huggins and Margaret Huggins. 1899. *An Atlas of Representative Stellar Spectra from λ 4870 to λ 3300.* London: William Wesley and Son, 4. From author's collection.

like laboratories by 1860 and had been so for some time. Huggins's portrait of observatories before 1860 as devoted solely to positional astronomy is misleading. There is, however, no denying that the observatories devoted principally to astrophysics were quite different from the observatories of the traditional positional astronomy. Further, in 1874, Potsdam, with backing from the Kaiser, became the first observatory to be founded by a state for the pursuit of astrophysics. Soon other astrophysical observatories followed, and astrophysics was also often incorporated into the activities of existing institutions.

The adoption of photographic methods and their linkage to the spectroscope in the late nineteenth century also made possible very extensive observing programs on stellar spectra. Here we see that technical changes and social changes in observatories were intertwined. Most notably, E. C. Pickering at the Harvard College Observatory in Cambridge, Massachusetts, took Airy's factory-like approach to new levels in terms of a division of labor and a strict hierarchy in the pursuit of the collection and analysis of the light from hundreds of thousands of stars. But at the bottom of Pickering's hierarchy now were women.

The Rise of American Observatories

In the late nineteenth and early twentieth century the success of American astronomers in securing funds from wealthy patrons led to the reform of the Harvard College

Observatory as well as new state-of-the-art observatories. Of these astronomers, George Ellery Hale was much the most persuasive and successful in extracting money from patrons. His deft handling of philanthropic foundations and wealthy individuals produced new observatories for the University of Chicago in the shape of the Yerkes Observatory, in 1897, and the Mount Wilson Solar Observatory, in 1904 (in 1920 it was to become the Mount Wilson Observatory). He later laid the foundations for the Mount Palomar Observatory and was involved in its initial planning, though he did not live to see its great 200-inch reflecting telescope come into operation in the late 1940s.

The Mount Wilson Observatory in California was the most important of Hale's creations. Not only did it house big telescopes (when its 100-inch telescope went into operation in 1919 it was easily the most powerful in the world) but under

> Hale's directorship, Mount Wilson became in many ways a concrete manifestation of the cooperative research (what would later be termed "interdisciplinary" research) in which he, and many American scientists of his generation, believed. He brought physicists into the observatory by establishing a physical laboratory, as well as machine shops, in nearby Pasadena (Smith 2003, 164).

Astronomers had traditionally been plagued by the problems of securing funds to keep observatories running effectively with adequate staffing, the upkeep of instruments, and adding new instruments. Hale largely solved these problems with the help of the Observatory's main patron, the Carnegie Institution of Washington, founded by the steel baron Andrew Carnegie in 1901. Through Hale's leadership and Carnegie's money, Mount Wilson became the leading astrophysical observatory in the world (Van Helden 1984, 138).

By the time Mount Wilson was established, astronomers were eager to escape the limits of observing sites in or close to cities or at cloudy locations that limited the operations of their instruments. This led to observatories at high sites that enabled them to get above a substantial portion of the obscuring layers of the Earth's atmosphere. Thus developed a common separation between an observatory on a mountain (like Mount Wilson) and the staff offices and associated laboratories (housed in Pasadena).

The first major mountaintop observatory was in fact established in 1882 at Pic-du-Midi in the French Pyrenees (which is at an altitude of nearly 3,000 meters), although it was initially employed principally for meteorology. Six years later the Lick Observatory went into operation on Mount Hamilton in California and became what has been termed as one example of a "factory observatory" dedicated to the routine production of astrophysical data (Lankford 1997). In 1889, Harvard's Arequipa station in Peru was opened, although it was transferred to South Africa in 1927. The move to mountaintop observatories took off in the second half of the twentieth century and it became usual to put observatories at altitudes of 2,000 meters or more.

The massive influx of government monies after World War II transformed many areas of science, including astronomy. But in the US, private support was still often very important. Perhaps the leading example here is the founding of the W. M. Keck Observatory at an altitude of more than 4,000 meters on the dormant volcano Mauna Kea in Hawaii. By 1996, the Keck Observatory boasted two 10-meter telescopes.

Space Observatories

Observatories were also sent into space in the second half of the twentieth century. The development of space astronomy was enabled by an outpouring of government monies that was astonishingly high by earlier standards, especially in the US where it was driven by the Cold War competition with the Soviet Union. After World War II, astronomers launched balloons and rockets to carry telescopes and instruments high into, or above, the atmosphere. In time these developments led to the launch of astronomical satellites, most importantly in 1990 the launch of the single most costly instrument in the history of science, the Hubble Space Telescope, which by the early 2010s had cost over $20 billion to build, operate, and maintain. While the telescope orbits hundreds of miles above the Earth its operations are planned and controlled by astronomers and engineers at the Space Telescope Science Institute in Baltimore and the nearby Goddard Space Flight Centre. It is the experts at the Space Telescope Science Institute who make the actual observations based on observing proposals from astronomers (often working in sizable teams) who have competed for observing time with Hubble. Many more observing proposals are rejected than accepted due to the limited observing time that is available and so astronomers have developed elaborate processes to decide which observations should be pursued. Also, unlike the situation in, say, the early twentieth century where astronomers were very largely restricted to making observations at their home observatories, government support for the establishment and operation of observatories has opened up access for astronomers to a wider range of telescopes and observatories.

Similarly, astronomers at ground-based telescopes now often 'observe' with telescopes even though far from the physical location of the observatory, with much of the work done by staff experts at the observatory. With the extension of astronomy beyond observations in optical wavelengths to the full range of the electromagnetic spectrum, new kinds of instruments such as radio telescopes came into being with new sorts of observatories to house them. In these ways the traditional notion of the observatory has been extended.

Endnote

1 Here we are following Ragep's use of the term "Islamic" as one that is "taken in the sense of the civilization, rather than the religion, because much of the astronomy was secular, and also in acknowledgement of the many non-Muslims who worked within these traditions" (Ragep 1997, 17).

References

Aubin, David, Charlotte Bigg, and H. Otto Sibum (eds.) 2010. *The Heavens on Earth: Observatories and Astronomy in Nineteenth-Century Science and Culture.* London: Duke University Press.

Aveni, Anthony. 2001. *Skywatchers: A Revised and Updated Version of Skywatchers of Ancient Mexico.* Austin: University of Texas Press.

Becker, Barbara. 2011. *Unravelling Starlight. William and Margaret Huggins and the Rise of the New Astronomy.* Cambridge: Cambridge University Press.

Bennett, J. A. 1980. "Airy and horology." *Annals of Science*, 27: 269–85.

Bennett, J. A. 1988. *Divided Circle: A History of Instruments for Astronomy, Navigation and Surveying*. London: Phaidon.

Bennett, J. A. 1992. "The English quadrant in Europe – instruments and the growth of consensus in practical astronomy." *Journal for the History of Astronomy*, 23: 1–14.

Cawood, John. 1979. "The Magnetic Crusade: Science and politics in early Victorian Britain." *Isis*, 70: 493–518.

Christianson, John Robert. 2000. *On Tycho's Island. Tycho Brahe and His Assistants, 1570–1601*. New York: Cambridge University Press.

Dallal, Ahmad. 2010. *Islam, Science, and the Challenge of History*. New Haven, CT: Yale University Press.

Deane, Thatcher. 1994. "Instruments and observation at the Imperial Astronomical Bureau during the Ming Dynasty." *Osiris*, 9: 127–40.

Dick, Steven J. 2003. *Sky and Ocean Joined: The U. S. Naval Observatory, 1830–2000*. New York: Cambridge University Press.

Howse, Derek. 1980. *Greenwich Time and the Discovery of Longitude*. Oxford: Oxford University Press.

Huggins, William. 1897. "The new astronomy: A personal retrospective." *Nineteenth Century: A Monthly Review*, 41: 907–29.

Lankford, John. 1997. *American Astronomy: Community, Careers, and Power, 1859–1940*. Chicago: University of Chicago Press.

Meadows, A. J. 1975. *Greenwich Observatory: The Royal Observatory at Greenwich and Herstmonceux, 1675–1975, vol. 2: Recent History (1936–1975)*. London: Taylor & Francis.

Mosley, Adam. 2007. *Bearing the Heavens: Tycho Brahe and the Astronomical Community of the Late Sixteenth Century*. Cambridge: Cambridge University Press.

Ragep, Jamil. 1997. "Arabic/Islamic astronomy." In *History of Astronomy: An Encyclopedia*, edited by John Lankford, 17–21. New York: Garland.

Sayili, Aydin. 1960. *The Observatory in Islam and Its Place in the General History of the Observatory*. Ankara: Türk Tarih Kurumu Basimevi.

Schaffer, Simon. 1988. "Astronomers mark time: Discipline and the personal equation." *Science in Context*, 2: 115–45.

Schaffer, Simon. 1995. "Where experiments end: Tabletop trials in Victorian astronomy." In *Scientific Practice. Theories and Stories of Doing Physics*, edited by Jed Z. Buchwald, 257–97. Chicago: University of Chicago Press.

Sivin, Nathan. 2008. *Granting the Seasons: The Chinese Astronomical Reform of 1280, With a Study of Its Many Dimensions and a Translation*. New York: Springer.

Smith, Robert W. 1991. "A national observatory transformed: Greenwich in the nineteenth century." *Journal for the History of Astronomy*, 22: 5–20.

Smith, Robert W. 2003. "Remaking astronomy: Instruments and practice in the nineteenth and twentieth centuries." In *The Cambridge History of Science. Volume 5: The Modern Physical and Mathematical Sciences*, edited by Mary Jo Nye, 154–73. New York: Cambridge University Press.

Thoren, Victor (with contributions by J. R. Christianson). 1990. *The Lord of Uraniborg: A Biography of Tycho Brahe*. Cambridge: Cambridge University Press.

Van Helden, Albert. 1984. "Building telescopes, 1900–1950." In *Astrophysics and Twentieth-Century Astronomy to 1950*, edited by Owen Gingerich, 134–52. New York: Cambridge University Press.

Werrett, Simon, 2010. "The astronomical capital of the world: Pulkovo Observatory in the Russia of Tsar Nicholas I." In *The Heavens on Earth: Observatories and Astronomy in Nineteenth-Century Science and Culture*, edited by David Aubin, Charlotte Bigg, and H. Otto Sibum, 33–57. Durham, NC: Duke University Press.

Whewell, William. 1857. *A History of the Inductive Sciences*, 3rd ed., 3 vols., Vol. 1II. London: John W. Parker.

Williams, M. E. W. 1989. "Astronomical observatories as practical space: The case of Pulkowa." In *The Development of the Laboratory: Essays on the Place of Experiment in Industrial Civilization*, edited by Frank A. J. L. James and J. V. Field, 118–36. New York: American Institute of Physics.

Widmalm, Sven. 2010. "Astronomy as military science. The case of Sweden, ca. 1800–1850." In *The Heavens on Earth: Observatories and Astronomy in Nineteenth-Century Science and Culture*, edited by David Aubin, Charlotte Bigg, and H. Otto Sibum, 174–198. Durham, NC: Duke University Press.

CHAPTER FIFTEEN

The Court

BRUCE T. MORAN

Courts in early modern Europe were more than buildings and households. At a time when political patronage and networks of clientage served as instruments of government administration, courts comprised locations that were social as well as spatial, and evolved as sites of contact between a ruler and those seeking to influence the direction of royal or princely power (Asch and Birke 1991; Asch 2003; Kolsky 2003; Gosman, MacDonald, and Vanderjagteds 2003–2005; Knecht 2008). Members of court might be in immediate attendance upon the ruler, or they might serve from a distance as part of a courtly circle. Whether large or small, courts reflected the personal tastes, desires, and interests of their sovereigns, and, as a result, each defined an "ethos" as well as an institution, displaying specific cultures and employing an array of customs, rituals, and evaluative systems for directing the behavior of clients seeking to gain or maintain its beneficence (Evans 1991).

Outside Europe courtly patrons employed techniques of gift giving and social/bureaucratic advancement to promote interests related to the continuance of dynastic rule (Fuess and Hartung 2011). Attention to *materia medica* during the Song dynasty aided the systematization of medicine in China (Goldschmidt 2008) and the early Ming promotion of craft production helped to draw craftsmanship into the scholarly realm (Schäfer 2011). Early mathematical writings in Sanskrit served political and commercial purposes and were collected later as part of the cultural ambitions of colonial rule (Bretelle-Establet 2010). In the Middle East, observational and mathematical requirements for accurate astrological predictions, and the composition of reliable calendars, turned the attention of rulers in Islamic states, and within the Ottoman Empire, to the construction of accurate measuring instruments, schools, and observatories (Sayili 1960; Saliba 2007). Statecraft and the study of nature merged in China in the person of the Kangxi emperor (1654–1722), whose study of mathematics and astronomy, written observations of natural phenomena, and expertise in surveying, cartography, and other forms of practical mathematics not only helped to increase the

A Companion to the History of Science, First Edition. Edited by Bernard Lightman.
© 2016 John Wiley & Sons Ltd. Published 2020 by John Wiley & Sons Ltd.

social and intellectual status of native mathematicians, but became an essential part of a novel conception of state management (Elman 2005; Jami 2012).

Those who sought court appointments or who pursued social and economic advantage through princely patronage represented a wide range of talents and motives. Descriptions of the Renaissance courtier range from portrayals of honorable advisors, who sought to win "the mind and favor of the prince" so as to guide his character toward virtue, to characterizations as "base sycophants" and "crumb-catching parasites," constantly anxious about image, social standing, and prestige. Self-fashioning within a courtly context was a competitive endeavor, a contest played for status, title, and reputation as well as for wealth, and one in which all could vanish with a loss of the prince's esteem and favor (Ducci 1607; Angelo 1977; Asch 2003). Patronage within this context reflected a particular sort of sociability and took its place among other kinds of social relationship including friendship networks, family ties, and confessional loyalties. Nevertheless, when aligned with specific princely interests, patronage and the dynamics of court reputation fostered individual artistic and literary talents and promoted skills, discoveries, and inventions related to the expansion of natural knowledge as well as to the development of practical mathematics and technology. By the later seventeenth century, however, styles of patronage, reflecting new state bureaucracies and the reorganization of learning, began to change, and the private sponsorship of princes gave way to the formation of public societies and academies focused upon experimental, philosophical, and utilitarian pursuits.

Engineering, Instruments, and Practical Mathematics

Princes sought economic, commercial, and military advantage over rival states, and those seeking the patronage of princes needed to keep in mind the practical needs of ruling families as well as specific princely curiosities. Applying for patronage from the Sforza Duke Ludovico, Leonardo da Vinci (1452–1519) made a list of what he could offer in terms of what he knew the prince most urgently desired. Under nine headings he listed attainments and inventions in military engineering. In a tenth he advanced his qualifications as a civil engineer and architect. Only at the very end did Leonardo make reference to what he could do in painting and sculpture (Paoletti and Radke 1997, 318). The same emphasis upon military and civil engineering also characterized his contracts with later patrons, Cesare Borgia (1475–1507) in Florence and Francis I (1494–1547) in France.

In Urbino, the support of military engineering and mathematical practice by the princes Federico da Montefeltro (1422–1482) and, later, Francesco Maria II della Rovere (1549–1631), led to the emergence of the city as a center of instrument making and applied mathematics. There artisanal skills and mathematical expertise combined in the city's well known *botteghe* and produced for the court an array of wondrous mechanical objects, including automata and clocks, many given as gifts to other courts. The career of the practical mathematician, Mutio Oddi (1569–1639), combined the mathematical cultures of both city and court. While losing the favor of one prince, the Urbinate Duke Francesco Maria, Oddi gained the support of another, the Milanese Cardinal Federico Borromeo (1564–1631), whose patronage in Milan generated useful friendships for Oddi and helped launch a lucrative business. Much of that business was oriented toward aristocratic consumption, as mathematical instruments, mathematical literacy, and knowledge of *designo* increasingly became part of the image of the proficient courtier (Marr 2011).

Part of Oddi's patronage strategy was to provide mathematical instruments, sometimes as gifts, to the social elite and then to teach his aristocratic supporters and other purchasers how to use them through publications and direct contact. Galileo (1564–1642) did the same, designing a mathematical compass, or sector, and providing purchasers with personal instruction in its civil and military uses. Among those to benefit from hands-on demonstration and "by word of mouth" were Johann Frederick, Prince of Holstein, the Archduke Ferdinand of Austria, Phillip, the Landgrave of Hesse, as well as the Duke of Mantua, Vincenzo Gonzaga, to whom Galileo sent the instrument with a personalized inscription in his pursuit of a court appointment (Galileo 1978). In these cases mathematical literacy and the realm of craft crossed paths within an aristocratic milieu.

Clockwork-driven machines and automata were popular items of prestige and conspicuous display at many European courts, and represented a major part of the material culture of mathematics in the sixteenth century. Along with other instruments, clockwork devices found places at court in exhibitions of curiosities and luxury goods that reinforced claims to courtly status. Such instruments served also as tools in crafting political ambitions, exchanged as gifts in the pursuit of princely alliance. At some courts, several of the most important in German-speaking lands, design, mathematics, and precision clockwork merged within a competitive technical culture, bringing together mathematicians and clockmakers in the pursuit of mechanical proficiency and, in certain instances, scientific inquiry.

In the Palatinate, the Tübingen mathematician Philipp Immser and the Heidelberg clockmaker Gerhard Emmoser collaborated in the tedious pursuit of an astronomical clock capable of the precision demanded by the Elector Ott-Heinrich (Moran 1977). At Hesse-Kassel, the Landgrave Wilhelm IV (1532–1592) promoted the technical skills of Eberhard Baldewein, a court attendant whose mechanical talents came to the notice of the prince who thereafter directed Baldewein's courtly service toward projects in fine mechanical engineering. The influential humanist and educational reformer Peter Ramus (1515–1572) noted the Landgrave's part in reassigning professional roles:

> Gellius tells us that Protagoras, who was a porter, was made into a philosopher by Democritus. The Germans say that Eberhart [Baldewein] who was a tailor [*sartore*] was made into a singularly fine artificer of astronomical instruments by the Landgrave of Hesse who is endowed with Democritean cleverness. (Ramus 1569, 67)

Wilhelm was a prominent prince-practitioner, not only supporting court projects but designing and taking part in them (Moran 1981). Of special importance were astronomical projects of stellar observation that involved the construction of observational instruments capable of greater precision and that utilized a procedure of time keeping by means of an innovative cross-beat escapement for purposes of astronomical measurement. The device emerged from the designs of the court instrument maker and mathematician Jost Bürgi (1552–1632), who utilized court connections to develop innovations in both practical and theoretical mathematics. At Kassel, and later in the service of Holy Roman Emperor Rudolf II (1552–1612) at Prague, Bürgi designed and constructed imaginative clockwork models of the heavens and developed logarithms to help simplify calculations (Mackensen, von Bertele, and Leopold 1982).

Cosmology and Philosophical Speculation

The Kassel court also provided an indulgent and intellectually open environment for the court's mathematician–astronomer Christoph Rothmann (1550?–1608?) and for visitors like the Danish astronomer Tycho Brahe (1546–1601) with regard to the examination, debate, and expression of Copernican views (Goldstein and Barker 1995; Granada 1999). Elsewhere, too, courts afforded tolerant cultural alternatives to the Aristotelian constraints of university learning in regard to exploring the secrets of nature. At Prague, traditions of astronomical innovation and diverse philosophical opinion were already well established by the time Tycho Brahe and Johannes Kepler (1571–1630) joined the court as imperial mathematicians.

Tycho possessed aristocratic status from birth and had already made use of his noble rank to advance his astronomical interests. From the Danish King Frederick II (1534–1588) he received an estate on the small island of Hven in the Öresund strait near Copenhagen and funding to build an observatory. From his castle-observatory, called Uraniborg, and from a later observatory built largely underground, called Stjerneborg, Tycho carried out programs of detailed celestial observation using large-scale observational instruments with innovative calibrations for purposes of precise measurement. He also maintained a paper manufactory and a printing press from which he published his observational results and his own interpretation of planetary movement. After leaving Hven in 1597, the result of a falling out with the new Danish King Christian IV (1577–1648), Tycho settled finally at the Rudolphine court in 1599, where he was appointed imperial mathematician. In that capacity he built another castle-observatory at Benátky nad Jizerou, assisted there from 1600 to his death a year later by Johannes Kepler. While succeeding Tycho as imperial mathematician, Kepler's initial interest in coming to Prague was less focused on obtaining a court appointment than in acquiring the records of Tycho's systematic observations. Those observations required years of effort to collect, and their publication, with the title *Tabulae Rudolphinae* (Rudolphine Tables) in honor of their patron, did not occur until 1627 (long after Rudolf's death), a delay due in part to a dispute with Tycho's heirs who demanded a share in the profits as well as the right of censorship.

The connection between Tycho and Kepler was, then, based both in court and professional relationships. To have access to Tycho's observational records Kepler agreed to synchronize, as much as possible, the new observations with Tycho's own geoheliocentric theory of the planets and to defend Tycho's claims to theoretical innovation. Once named court *mathematicus*, however, Kepler pursued his own course. The result was the publication of some of the most significant, and revolutionary ideas in the history of astronomy, introducing elliptical and non-uniform motions to the planets and, later, revealing a new kind of celestial harmony. The Rudolphine court supported the intellectual autonomy made manifest in Kepler's texts, and in a particular case became involved in directly financing the publication of Kepler's conceptual and mathematical insights. Both Kepler's *Astronomiae pars optica* (Optical Part of Astronomy, 1604) and his *Astronomia nova* (New Astronomy, 1609)—the result of a ten-year long study of the motion of Mars that he described as his "warfare with Mars"—were dedicated to the Emperor.

"New Things" and the Medici

In the support of observational and theoretical innovation, the Medici court in Florence stands out prominently. Medici patronage had long encouraged a variety of arts and influenced the circulation of knowledge during the Renaissance. The elder Cosimo de' Medici (1389–1464) was well aware of the significance of supporting humanist pursuits in the self-presentation of a cultured patron and gained prestige by founding libraries, encouraging the collection of classical manuscripts, and sponsoring the Platonic Academy, an organized fellowship surrounding Plato's translator, Marsiglio Ficino (1433–1499). Humanist programs continued to have a prominent place at the Medici court and were expanded by Cosimo's grandson, Lorenzo (1449–1492). Cosimo I (1519–1574) went further, founding the Florentine Academy as a means not only of encouraging the arts, but circulating knowledge through public lectures. It was, however, in its patronage of Galileo that the Medici court found itself at the center of theoretical revolution.

When Antonio de' Medici (1576–1621), brother of Grand Duke Cosimo II (1590–1621), made known to Galileo his interest in learning about "new things" observed through "the looking glass discovered by you," Galileo responded with a detailed letter describing and illustrating his observations of the moon, fixed stars, Jupiter, and Jupiter's moons, three of which were "totally invisible by virtue of their smallness" (as quoted in Westman 2011, 442–7). Galileo sought a court appointment and merged a desire for social status with reports of celestial novelties. However, the extent to which his commitment to Copernicanism evolved as part of a strategy of self-promotion remains controversial. The report of his discoveries fit Medici dynastic rhetoric. Galileo named the four moons that he observed orbiting Jupiter the "Medicean stars" to honor Cosimo and his three brothers, and his published report of observations made by means of an *occhiale* (spyglass), the *Sidereus Nuncius* (1610), was dedicated to Cosimo II. He also used the opportunity to give astrological significance to Jupiter's place in the ducal horoscope (Galileo 1989). According to one historical argument, Galileo's ambitions at court extended beyond rhetoric to coloring theoretical allegiance as well. In this case, Galileo not only fashioned his reports about the heavens to correspond with Medici mythology, but defended the theoretical novelty of the Copernican system as a form of spectacle that demonstrated to court patrons his mathematical and natural philosophical cunning. Another historical view differentiates between intellectual commitments and different kinds of "instrumental sociabilities" in Galileo's study of the heavens, emphasizing friendships and pedagogical practices as the primary features of his social world and decentering the pursuit of patronage as a compelling motive in fashioning theory choice. In both views, however, the court prevails as a crucial site of science (Biagioli 1993; Westman 2011, 434–54).

In the transformation from university mathematician to court, philosopher Galileo navigated between various markets and institutional domains (Biagioli 2006). Regardless of the social emphasis, the Medici court itself advanced Galileo's discoveries by playing a role in disseminating the book and the instrument that recorded its own elite status in the heavens. To advertise the "Medicean stars" at locations outside Italy, Tuscan ambassadors in Prague, Paris, London, and Madrid were promised copies of Galileo's *Sidereus Nuncius* and looked forward to receiving telescopes constructed by Galileo and paid for from the Medici court treasury.

Collections, Wonder, and the Order of Nature

Mathematical instruments, automata, and clockwork driven machines served scientific and mechanical interests at court and were also objects of art. As such they were often collected together with other artistic marvels and curious natural wonders within princely *Kunst* and *Wunderkammern*. Before the Renaissance, collections of precious objects tended to be associated with items related to specific dynasties and religious institutions. The fourteenth and fifteenth centuries, however, began to see collections of a more natural sort brought together by members of the higher nobility and urban elite. An emphasis upon civility among the social practices of court collecting established Italian courts among the "sites of knowledge" of early modern scientific culture (Findlen 1994). The most celebrated were to be found at the palaces of the Medici and Gonzaga, but at other sites as well, especially in Naples, Rome, and the duchy of Milan, naturalists and collectors brought practices relevant to the study and questioning of nature into elite society (Figure 15.1).

The culture of collecting flourished outside Italy as well, especially at the Hapsburg courts of Archduke Ferdinand II at Ambras, and at the courts of his imperial relatives, the emperors Maximillian II in Vienna and Rudolf II at Prague. Among smaller German courts, the *Kunstkammer* of Albrecht V of Bavaria in Munich inspired the first treatise to focus upon an ordered system of *Kunstkammer* collecting, a text that promoted pragmatic curiosity and visual knowledge within aristocratic circles (Quiccheberg 2014). At Dresden, the confluence of artisanal expertise, mathematical instruments, and uncommon material objects in the collection of the Elector August of Saxony gave shape to a site of particular value to the study of optics. While some objects in the collection engaged courtiers and visitors by playful interaction, others became objects of "experiment." Johannes Kepler, who visited the *Kunstkammer* sometime before 1604, was intrigued and entertained by the court's *camera obscura*, an experience, among others, that influenced the construction of his own theory of optical images. Kepler made no secret of the value of such courtly visits in turning his attention to mathematics. His curiosity concerning the symmetry of the snowflake was moved, he admitted, by having seen a dodecahedron within an architectural ornament at the Dresden *Stallhof* (Watanabe-O'Kelly 2002; Dupré and Korey 2009).

Of the inventoried groups of objects in the Dresden *Kunstkammera*, a quarter comprised tools from a range of trades and crafts (Bäumel 2004). While the degree of such a focus was not reflected in other court collections, it does suggest the potential role of the court as a point of interaction between scholarly traditions on the one hand and practical or craft traditions on the other. Similar practices of joining workshops and collections in building a demonstrative repository of *naturalia, artificialia,* and artisanal know-how became also a prominent feature of the Medici court. At the Casino di San Marco, Francesco I de Medici (1574–1587) first set up a *Fonderia*, a laboratory/workshop for things made by fire. Thereafter, laboratories and artist workshops were erected also at the Uffizi, and were administratively brought together by

Figure 15.1 Natural History Museum and *Wunderkammer*, from Ferrante Imperato. 1672. *Historia naturale di Ferrante Imperato Napolitano.* Venetia: Presso Combi, & la Noù. Wellcome Library, London.

Francesco's successor, Ferdinand I (1587–1609), as the *Galleria dei lavori*. The *Fonderia* produced a wide variety of metallurgic, chemical, and pharmaceutical products that combined craft techniques with court desires. Making precious stones, fireworks, explosives, imitation Chinese porcelain, poisons, and antidotes combined there with projects in painting, sculpture, and goldsmithing. Collaboration was often the result as alchemists and painters helped one another create new techniques that imaginatively turned works of nature into works of art (Kieffer 2014). In 1588 Ferdinand appointed the Roman nobleman Emelio de' Cavalieri as superintendent of artistic productions, requiring him to oversee all "jewelers, carvers of any type, cosmographers, goldsmiths, the makers of miniatures, gardeners of the gallery, turners, confectioners, clockmakers, artisans of porcelain, distillers, sculptors, painters, and makers of artificial gems" (Butters 2000, 144). The relationship between metalworkers and alchemists fit into the symbolic expression of power at the Medici court as alchemists developed recipes for hardening metallic cutting tools that made possible working on *pietre dure* like porphyry, a technique that proclaimed the power of a dynasty to shape and control even the most resistant natures. Francesco made frequent appearances within court workshops and was fascinated by technical expertise. The French essayist Michel de Montaigne (1533–1592) noted when recording his travels

The same day we saw the palace of the duke, where he himself takes pleasure in working at counterfeiting oriental stones and cutting crystals: for he is a prince somewhat interested in alchemy and the mechanical arts (as quoted in Rossi 1998, 64).

Curiosities and Visual Knowledge

Collecting often brought together natural oddities that challenged received opinions about the order to nature. The collection of Ferdinand II (1529–1595), a son of the Emperor Ferdinand I, at Ambras focused especially upon natural deviations and anomalies as well as upon the display of artisanal virtuosity in regard to working precious and rare materials (Irblich 1996). That same interest in exploring the imperfect or "borderline" parts of nature became the focus of Federico Cesi (1585–1630), the scion of an aristocratic family, who briefly became Lord of Acquasparta. In 1603, Cesi, with three friends, founded an informal fellowship called the *Academia dei Lincei* (Academy of the Lynx) aimed at the study of nature by means of observation, experiment, and inductive reasoning. The Academy, while never numbering more than 20 members, nevertheless supported the publication of Galileo's *Letters on Sunspots* (1613) and his *Assayer* (1623), and also defended Galileo in his controversies with church authorities. Cesi himself pursued projects that redirected attention away from categorically defined wonders and marvels and toward the study of "middle natures" that exposed the ambiguities of nature and demonstrated that "oddity might be related to regularity" (Freedberg 2002, 376).

To systematize irregularity, pictures were helpful, and Cesi emphasized the value of visually depicting "middle natures." Patronage, at other locations, also encouraged visual representations of plants and animals. Austrian Archduke Ferdinand II collected natural anomalies and, in 1562, commissioned Giorgio Liberale (1527–1597/80) to paint all the fish of the Adriatic. The endeavor would require 17 years to complete, and the images, thereafter, became part of the princely *Kunstkammer* (Cronberg 1971). Picturing nature also became part of a project that linked the physician and naturalist Pietro Andrea Mattioli (1501–1577) to the imperial court of Ferdinand I at Prague. Although Mattioli's *Commentaries on Dioscorides* had appeared years earlier, in 1544, Ferdinand and the Bohemian estates enlisted Mattioli to construct a new herbal, based on the earlier work. Mattioli, who was appointed personal physician to Ferdinand, moved to Prague in 1554 and remained there for the next 11 years, in part organizing the Herbarium that appeared first in Czech in 1562 accompanied by exquisite illustrations of species drawn from living plants in great part by Giorgio Liberale and Wolfgang Meyerpeck. Further editions appeared in German, Latin, and Italian with added illustrations. The Czech edition, however, was truly an aristocratic production that enlisted the support of the Emperor, the Bohemian estates, and the city of Prague. It contained a dedication to 14 Bohemian lords and 14 Bohemian knights, replete with illustrations of their coats of arms. Archduke Ferdinand of Tyrol became personally involved in collecting illustrations for the herbal, requesting, in special letters, that the cities of Augsburg, Nürnberg, and Strassburg help prepare woodcuts for printing 50 plant sketches. The text in Czech was prepared by Tadeáš Hájek von Hájek (after 1525–1600), a nobleman and well-known astronomer, mathematician, and physician with important connections to courtly circles (Bohatcová 1985).

Various styles of collecting continued well into the seventeenth and eighteenth centuries. The best-known collections were the *Kunst und Naturalienkammer* of the Brandenburg Electors Friedrich-Wilhelm and Friedrich III in Berlin and the *Cabinet du Roi* inaugurated by the French King Louis XIII. Occasionally princely cabinets came ready made, assembled earlier by private collectors. In this way the collection of *naturalia* and *artificialia* of the Danish physician, university professor, and court advisor Olaus Worm (1588–1654) found its way into the Copenhagen *Kunstkammer* of the Danish King Frederick III in 1650. Sometimes potentates just liked what they saw, and bought it. The Russian Czar Peter the Great paid 30,000 Dutch guilders for the cabinet of anatomical parts, especially infant parts, prepared and assembled by the chief anatomist of the Amsterdam surgeon's guild, Frederik Ruysch (1638–1731). Ruysch's collection caught also the attention of the popular author, natural philosopher, and secretary of the *Académie des Science*, Bernard de Fontenelle (1657–1757), who aligned it more with the pursuits of nobility than those of a private person.

> His museum, or repository of curiosities, contained such rich and magnificent variety, that one would have rather taken it for the collection of king than the property of a private man (as quoted in Hansen 1996, 670).

Gender, "Human Empire," and Global Knowledge

Except for the male dominated papal court and the courts of ecclesiastical officials, women also played a role in establishing and participating in patterns of patronage and clientelism. Noble networks served aristocratic women who sought answers to questions in natural philosophy, mathematics, and medicine. Questions posed to Galileo's student and supporter Benedetto Castelli (1578–1643) by the Grand Duchess Christina of Tuscany (1565–1637) concerning the compatibility of Copernican cosmology and Scripture prompted Galileo's formal response, *Letter to the Grand Duchess Christina*, an essay written in 1615 that circulated widely in manuscript before being published in 1636. The patronage of figures like Christina of Sweden (1626–1689), Anna of Denmark (1574–1619), and Margaret Cavendish, Duchess of Newcastle (1623–1673), merged natural philosophy with literary cultures and brought aristocratic women within reach of scientific communities and debates (Åkermann 1991; Barroll 2002; Sarasohn 2010). Other noblewomen, in charge of kitchens and domestic affairs within court households, became well known for their pharmaceutical expertise, aligning patronage, empirical knowledge, and court cultures within traditions of practical medicine (Rankin 2013).

Domestic practices as well as foreign ambitions gave rise also to projects aimed at advancing monetary interests and economic security. The type of "utilitarian patronage" that resulted produced an avalanche of maps, charts, and local surveys, and, while stamping the preoccupations of most princes, stands out most prominently at courts in Elizabethan and Jacobean England. The patronage of William Cecil (Lord Burghley, 1520–1598) promoted projects in agriculture and the mechanical arts and supported alchemical endeavors that sponsored the art of making copper and quicksilver by means of transmutation. Another favorite of the Elizabethan court, Robert Dudley (Earl of Leicester, 1532–1588) supported the talents in military engineering and surveying of the well-known astronomer Thomas Digges (ca. 1545–1595)

(Pumfrey and Dawbarn 2004). On occasion, pragmatic policies merged also with esoteric ideologies. Occult philosophy especially attracted Elizabeth I (1533–1603) and members of her court, and the combination of utilitarian and esoteric interests nurtured within court circles a thorny political relationship between English courtiers and the Queen's consultant, the practical magus, mathematician, and natural philosopher, John Dee (1527–1609) (Parry 2011). Elizabeth's successor, James I (1566–1625), directed his own learned endeavors toward literature and political philosophy. Nevertheless, his Attorney General and Lord Chancellor, Francis Bacon (1561–1626), pursued a grand vision of the expansion of learning by means of cooperative inquiry and royal patronage. By means of an organized learned society, "the noblest foundation … that ever was upon earth," members would collect and compare knowledge of "the sciences, arts, manufactures, and inventions of all the world" and find out "the true nature of all things whereby God might have more glory in the workmanship of them, and men the more fruit in their use of them," enlarging thus "the bounds of human empire" and effecting "all things possible" (Bacon 1627).

Bacon sought the acquisition of global knowledge, and the pursuit of useful knowledge created commercial empires in the early modern era (Cook 2007). The desire for empirical learning in regard to New World colonies particularly linked patronage, exploration, and empire building at the Spanish court of Philip II (r. 1556–1598) and helped define the contours of cosmographical learning. Within the imperial bureaucracy, the *Casa de la Contratación* and the Council of the Indies assembled physical information concerning natural resources and the secrets of nature, holding most of the accounts secret in order to protect territorial claims. Geographic and ethnographic descriptions collected by imperial administrators by means of questionnaires produced a rich cache of natural and mathematical knowledge that converged in a collection of empirical data known as the *relaciones-geográficas de Indias*. To aid navigation imperial mathematicians devised instruments and organized a global program of lunar observations for the purpose of determining longitude. Political and economic interests encouraged colonial expansion, motivated novel methods of gathering botanical knowledge, and generated imaginative ways for verifying reports of discoveries and claims to innovative technologies. Through vice royal courts in New Spain the crown organized practices determining the validity of claims concerning medicines and brought together local experts, merchants, and artisans to examine artifacts and potential commodities (Barrera-Osorio 2006; Portuondo 2009; Bleichmar, de Vos, Huffine, and Sheehan 2009; López Terrada 2011). Prompted by commercial interests, Philip II sent the physician Francisco Hernández (ca. 1515–1587) to the New World to "gather information generally about herbs, trees, and medicinal plants." Hernández was ordered specifically to:

> find out how the above mentioned things are applied, what their uses are in practice, their powers, and in what quantities the said medicines are given, as well as the places in which they grow and their manner of cultivation (as quoted in Barrera-Osorio 2009, 231).

Natural curiosities from the New World possessed both commercial and ornamental value. Both were of value to royal patronage and Saville became a collecting point for the distribution of rare plants to the gardens of the Spanish king.

Medicine, Pharmacy, and Society

In some instances court figures sought to persuade patrons to support new projects or attempted to influence intellectual habits (Nummedal 2007). At the Kassel court of Moritz of Hesse (1572–1632), the court physician Jacob Mosanus (1564–1616) asserted the value of preparing chemical medicines over the prince's interests in chrysopoeia (gold making). The change of direction altered patterns outside the court as well, as Moritz created a new professorship of chemical medicine (*chymiatria*) at the Hessian University of Marburg. As the first professor of the new discipline Moritz appointed Johannes Hartmann (1568–1631), who altered his own career path, turning from mathematics to chemistry and medicine so as to align himself with the prince's interests (Moran 1991; Krafft 2009). A similar attempt at patronage refashioning relates to the court physician and mathematician Joachim Becher (1635–1682) who sought to shift the attention of the Bavarian prince away from the uncertainties of alchemical ventures toward more predictable commercial and technical projects (Smith 1994).

Interests in practical alchemy often coincided with interests in magic, occult traditions, and non-traditional medical philosophies and practices. Paracelsian physicians in particular relied upon court appointments to establish credibility for their medical theories and procedures. The French Paracelsian physician Joseph Duchesne (Quercetanus) (ca. 1544–1609) assembled a list of princes who sponsored chemical medicine. These included the emperor and the king of Poland, as well as the archbishop of Cologne, the Duke of Saxony, the Landgrave of Hesse, the Margrave of Brandenburg, the Dukes of Braunschweig and Bavaria, and the Princes of Anhalt. The list missed the Duke of Neuburg, who later became Elector of the Palatinate, Ott-Heinrich, whose physician, Adam von Bodenstein, was one of the first publishers of Paracelsus's works. With the financial support of the Calvinist prince Christian I of Anhalt-Bernburg, the court *medicus ordinarius* (chief medical representative) Oswald Croll (ca. 1560–1608) constructed one of the most important expositions of Paracelsian remedies, a book dedicated to his prince, the *Basilica chymica* (1609). In Denmark, the professor of medicine and royal physician, Petrus Severinus (1542–1602), brought Paracelsian ideas into line with academically acceptable philosophical traditions that also pleased the court of the Danish king (Shackelford 2004). At Paris, Paracelsian physicians at the court of Henry IV (Jean Ribit, sieur de la Rivière [ca. 1571–1605], Duchesne, and Theodore de Mayerne [1573–1655]) contended with the university faculty of medicine concerning chemical medicines and their uses (Kahn 2007).

Courtly sites and the patronage of princes sheltered speculation, encouraged curiosity, fashioned collaborations, and stimulated debate in the early modern era. Court projects also oriented professional roles toward the demonstration of artisanal know-how and technical expertise, skills that altered the instruments and procedures of discovery and linked the pursuit of novelty and wonder to circles of collector–naturalists, mathematicians, artisans, and mechanical practitioners. Yet, courtly sites did not stand apart from other sites of knowledge, and intermingled with schools, cities, guilds, and academies. As scientific societies emerged in the mid seventeenth century, courtly etiquette still influenced codes of behavior and the language of civil discourse. But now, the interest of the court in scientific and technical endeavors was more often aroused in response to the popularization of "new learning" by academy members or by government ministers who identified projects of economic usefulness to the state (Hunter 1995; Biagioli 1996). In the new spaces of seventeenth-century experimental science, representatives of court remained as moral witnesses to the debates and projects of academies; and, in so doing, might still bestow indirectly intellectual prestige and social legitimacy upon scientific claims.

References

Åkerman, Susanna. 1991. *Queen Christina of Sweden and her Circle: The Transformation of a Seventeenth-Century Philosophical Libertine*. Leiden: Brill.

Angelo, Sydney. 1977. "The Courtier: The Renaissance and Changing Ideals." In *The Courts of Europe: Politics, Patronage, and Royalty, 1400–1800*, edited by A. G. Dickens, 33–54. New York: McGraw-Hill.

Asch, Ronald. 2003. *Nobilities in Transition 1550–1700: Courtiers and Rebels in Britain and Europe*. London: Arnold.

Asch, Ronald, and Adolf Birke (eds.) 1991. *Princes, Patronage and the Nobility: The Court at the Beginning of the Modern Age*. Oxford: Oxford University Press.

Barroll, Leeds. 2002. *Anna of Denmark, Queen of England: A Cultural Biography*. Philadelphia: University of Pennsylvania Press.

Bacon, Francis. 1627. *New Atlantis*. London: J. H. for William Lee.

Barrera-Osorio, Antonio. 2006. *Experiencing Nature: The Spanish American Empire and the Early Scientific Revolution*. Austin: University of Texas Press.

Barrera-Osorio, Antonio. 2009. "Knowledge and Empiricism in the Sixteenth Century Spanish Atlantic World." In *Science in the Spanish and Portuguese Empires, 1500–1800*, edited by Daniela Bleichmar, Paula de Vos, Kristin Huffine, and Kevin Sheehan, 219–32. Stanford: Stanford University Press.

Bäumel, Jutta. 2004. "Electoral Tools and Gardening Implements." In *Princely Splendor: The Dresden Court 1580–1620*, edited by Dirk Syndram and Antje Schemer, 160–75. Milan: Electa.

Biagioli, Mario. 1993. *Galileo Courtier: The Culture of Science in the Culture of Absolutism*. Chicago: University of Chicago Press.

Biagioli, Mario. 1996. "Ettiquette, Interdependence, and Sociability in Seventeenth-Century Science." *Critical Inquiry*, 22: 193–238.

Biagioli, Mario. 2006. *Galileo's Instruments of Credit: Telescopes, Images, Secrecy*. Chicago: University of Chicago Press.

Bleichmar, Daniela, Paula de Vos, Kristin Huffine, and Kevin Sheehan (eds.) 2009. *Science and the Spanish and Portuguese Empires, 1500–1800*. Stanford: Stanford University Press.

Bohatcová, Mirjam. 1985. "Prager Drucke der Werke Pierandrea Mattiolis aus den Jahren 1558–1602." *Gutenberg-Jahrbuch*, 60: 167–85.

Bretelle-Establet, Florence (ed.) 2010. *Looking at it from Asia: The Processes that Shaped the Sources of History of Science*. Dordrecht: Springer.

Butters, Suzanne. 2000. "'Una Pietra Eppure non una Pietra Pietre Dure e le Botteghe Medicee nella Firenza del Cinquecento." In *Arti Fiorentine: La Grande Storia dell' Artigianato*, vol. 3, edited by Franco Franceschi and Gloria Fossi, 144–63. Florence: Giunti Gruppo Editoriale.

Cook, Harold. 2007. *Matters of Exchange: Commerce, Medicine, and Science in the Dutch Golden Age*. New Haven, CT: Yale University Press.

Cronberg, Guglielmo Coronini. 1971. "Giorgio Liberale e i suoi Fratelli." In *Studi di Storiadell'Arte in Onore di Antonio Morassi*, edited by Arte Veneta, 85–96. Venice: Alfieri.

Ducci, Lorenzo. 1607. *Ars Aulica; or, the Courtiers Arte,* translated by Edward Blount. London: Melchior Bradwood for Edward Blount.

Dupré, Sven, and Michael Korey. 2009. "Inside the Kunstkammer: The Circulation of Optical Knowledge and Instruments at the Dresden Court." *Studies in History and Philosophy of Science,* 40: 405–20.

Elman, Benjamin. 2005. *On Their Own Terms: Science in China, 1550–1900.* Cambridge, MA: Harvard University Press.

Evans, R. J. W. 1991. "The Court: A Protean Institution and an Elusive Subject." In *Princes, Patronage, and the Nobility: The Court at the Beginning of the Modern Age,* edited by Ronald Asch and Adolf Birke, 481–91. Oxford: Oxford University Press.

Findlen, Paula. 1994. *Possessing Nature: Museums, Collecting, and Scientific Culture in Early Modern Italy.* Berkeley, CA: University of California Press.

Freedberg, David. 2002. *The Eye of the Lynx: Galileo, his Friends, and the Beginnings of Modern Natural History.* Chicago: University of Chicago Press.

Fuess, Albrecht, and Jan-Peter Hartung (eds.) 2011. *Court Cultures in the Muslim World Seventh to Nineteenth Centuries.* London: Routledge.

Galileo Galilei. 1978. *Operations of the Geometric and Military Compass 1606,* translated by Stillman Drake. Washington: Smithsonian Institution Press.

Galileo Galilei. 1989. *Sidereus Nuncius or The Sidereal Messenger,* translated by Albert van Helden. Chicago: University of Chicago Press.

Goldschmidt, Asaf. 2008. *The Evolution of Chinese Medicine: Northern Song Dynasty, 960–1127.* Abingdon: Routledge.

Goldstein, Bernard, and Peter Barker. 1995. "The Role of Rothmann in the Dissolution of the Celestial Spheres." *British Journal for the History of Science,* 28: 385–403.

Gosman, Martin, Alasdair MacDonald, and Arjo Vanderjagteds. (eds.) 2003–2005. *Princes and Princely Culture, 1450–1650,* 2 vols. Leiden: Brill.

Granada, Miguel. 1999. "Christoph Rothmann und die Auflösung der himmelschen Sphären. Die Briefen an den Landgrafen von Hessen-Kassel 1585." *Acta Historica Astronomiae,* 5: 35–57.

Hansen, Julie van. 1996. "Resurrecting Death: Anatomical Art in the Cabinet of Dr. Frederik Ruysch." *Art Bulletin,* 78: 663–79.

Hunter, Michael. 1995. "The Crown, the Public, and the New Science, 1698–1702." In *Science and the Shape of Orthodoxy: Intellectual Change in Late Seventeenth-Century Britain,* edited by Michael Hunter, 151–66. Rochester: Boydell Press.

Irblich, Eva. 1996. "Naturstudien Erzherzog Ferdinand II: zur Kunstkammer auf Schloß Ambrasbei Innsbruck." In *Thesaurus Austriacus: Europas Glanzim Spiegel der Buchkunst, Handschriften und Kunstalben von 800 bis 1600,* edited by Eva Irblich, 209–25. Vienna: Das Nationalbibliothek.

Jami, Catherine. 2012. *The Emperor's New Mathematics: Western Learning and Imperial Authority during the Kangxi Reign.* Oxford: Oxford University Press.

Kahn, Didier. 2007. *Alchimie et Paracelsisme en France à la fin de la Renaissance (1567–1625).* Geneva: Librairie Droz.

Kieffer, Fanny. 2014. "The Laboratories of Art and Alchemy at the Uffizi Gallery in Renaissance Florence." In *Laboratories of Art: Alchemy, and Art Technology from Antiquity to the Eighteenth Century,* edited by Sven Dupré, 105–28. Heidelberg: Springer.

Knecht, Robert. 2008. *The French Renaissance Court.* New Haven: Yale University Press.

Kolsky, Stephen. 2003. *Courts and Courtiers in Renaissance Northern Italy.* Aldershot: Ashgate.

Krafft, Fritz. 2009. "Das Zauberwort Chymiatria – und die Attraktivität der Marburger Medizin-Ausbildung, 1608–1620. Eine etwas andere Frequenzbetrachtung." *Medizinhistorisches Journal,* 44: 130–78.

López Terrada, and María Luz. 2011. "Flora and the Hapsburg Crown: Clusius, Spain, and American natural history." In *Silent Messengers: The Circulation of Material Objects of Knowledge in the Early Modern Low Countries,* edited by Sven Dupré and Christoph Lüthy, 43–68. Münster: Lit Verlag.

Mackensen, Ludolf von, Hans von Bertele, and John Leopold (eds.) 1982. *Die Erste Sternwarte Europas, mit ihren Urhen und Instrumenten.* Munich: Callway Verlag.

Marr, Alexander. 2011. *Between Raphael and Galileo: Mutio Oddi and the Mathematical Culture of Late Renaissance Italy.* Chicago: Chicago University Press.

Moran, Bruce. 1977. "Princes, Machines, and the Valuation of Precision in the 16th Century." *Sudhoffs Archiv,* 61: 209–28.

Moran, Bruce. 1981. "German Prince-Practitioners: Aspects in the Development of Courtly Science, Technology, and Procedures in the Renaissance." *Technology and Culture,* 22: 253–74.

Moran, Bruce. 1991. *The Alchemical World of the German Court: Occult Philosophy and Chemical Medicine in the Circle of Moritz of Hessen (1572–1632).* Stuttgart: Franz Steiner.

Nummedal, Tara. 2007. *Alchemy and Authority in the Holy Roman Empire.* Chicago: University of Chicago Press.

Paoletti, John, and Gary Radke. 1997. *Art in Renaissance Italy.* New York: Harry N. Abrams.

Parry, Glyn. 2011. *The Arch-Conjuror of England, John Dee.* New Haven, CT: Yale University Press.

Portuondo, Maria. 2009. *Secret Science: Spanish Cosmography and the New World.* Chicago: University of Chicago Press.

Pumfrey, Stephen, and Frances Dawbarn. 2004. "Science and Patronage in England, 1570–1625: A Preliminary Study." *History of Science,* 42: 137–88.

Quiccheberg, Samuel. 2014. *The First Treatise on Museums: Samuel Quiccheberg's Inscriptiones, 1565,* translated by Mark Meadow and Bruce Robertson. Los Angeles: Getty Research Institute.

Ramus, Peter. 1569. *P. Rami Scholarum Mathematicarum Liber Unuset Trigenta.* Basiliae: Per Eusebium Episcopium et Nicolai Fratris Haeredes.

Rankin, Alisha. 2013. *Panaceia's Daughters: Noblewomen as Healers in Early Modern Germany.* Chicago: University of Chicago Press.

Rossi, Paolo. 1998. "Sprezzatura, Patronage, and Fate: Benvenuto Cellini and the World of Words." In *Vasari's Florence: Artists and Literati at the Medicean Court,* edited by Philip Jacks, 55–69. Cambridge: Cambridge University Press.

Saliba, George. 2007. *Islamic Science and the Making of the European Renaissance.* Cambridge, MA: MIT Press.

Sarasohn, Lisa. 2010. *The Natural Philosophy of Margaret Cavendish: Reason and Fancy during the Scientific Revolution.* Baltimore: Johns Hopkins University Press.

Sayili, Aydin. 1960. *The Observatory in Islam.* Ankara: Türk Tarih Kurumu.

Schäfer, Dagmar. 2011. *The Crafting of 10,000 Things: Knowledge and Technology in 17th Century China.* Chicago. University of Chicago Press.

Shackelford, Jole. 2004. *A Philosophical Path for Paracelsian Medicine: The Ideas, Intellectual Context, and Influence of Petrus Severinus: 1540–1602.* Copenhagen: Museum Tusculanum Press.

Smith, Pamela. 1994. *The Business of Alchemy: Science and Culture in the Holy Roman Empire.* Princeton: Princeton University Press.

Watanabe-O'Kelly, Helen. 2002. *Court Culture in Dresden: From Renaissance to Baroque.* New York: Palgrave.

Westman, Robert. 2011. *The Copernican Question: Prognostication, Skepticism, and Celestial Order.* Berkeley, CA: University of California Press.

CHAPTER SIXTEEN

Academies and Societies

DENISE PHILLIPS

In 1998, the American Association for the Advancement of Science (AAAS) celebrated its 150th anniversary. To commemorate this event, the academy did what many learned societies before it had done on similar occasions: it commissioned a history of itself. This scholarly volume, assertively entitled *The Establishment of Science in America,* stands in a long tradition (Kohlstedt, Sokal, and Lewenstein 1999). Since Thomas Spratt's 1667 *History of the Royal Society of London,* scientific societies have been eager celebrators of their own achievements. Such publications have not been ancillary to the main purpose of scientific societies; in fact, they offer an important clue as to why this particular kind of intellectual organization has proven so popular and durable for the past four centuries. Learned societies have helped scientists stake a public claim to the importance of their work. These groups have been vehicles through which scientists have celebrated and promoted themselves, and in the process secured patronage, prestige, and attention. Learned societies have done more than that, however. As other chapters in this book explain (e.g., Chapter 11), categories like "science" and "scientist" only became stable cultural reference points over the course of the nineteenth century. During the period when the enterprise we call "science" was in formation, learned societies and academies played a key role in shaping the boundaries and structure of this enterprise.

The collective pursuit of knowledge has taken a variety of different forms in different settings, but the particular organizational tradition discussed in this chapter has fairly distinct beginnings in seventeenth-century Europe. By the twentieth century, however, scientific societies had become global in their reach, an integral part of knowledge making in Asia, Africa, and the Americas, as well as in Europe.

In the centuries since they first appeared in early modern Europe, these kinds of organizations have undertaken many different types of activities. They have built museums, lobbied for new government policies, sponsored prize questions, held public lectures, and provided audiences for novel experiments. They have given their

A Companion to the History of Science, First Edition. Edited by Bernard Lightman.
© 2016 John Wiley & Sons Ltd. Published 2020 by John Wiley & Sons Ltd.

members a forum for presenting research, and often published the results of their meetings in periodical form. In the most general terms, learned societies have served to formalize intellectual and cultural allegiances, making these allegiances more powerful and more visible by giving them collective voice. They have also played an important role in organizing scientific credit and exchange. Alongside the university, the learned society has arguably been modern science's most important institutional home.

This history of academies and societies provides insight into two interlocking aspects of the production of modern scientific knowledge. On the one hand, the history of these groups can be written as a history of internal consolidation—of like-minded men and women coming together to create networks that sustain new kinds of research practices, new disciplines, or new intellectual causes. On the other hand, the practices of learned societies also point outwards, towards society at large, and are imbedded in the broader history of modern state formation and the creation of modern civil society.

In the seventeenth century, scientific societies were intertwined with the emergence of new forms of experimental and empirical philosophy, and these groups helped the new philosophers secure legitimacy and authority. In the eighteenth century, learned societies provided a site where natural knowledge forged an alliance with the broader ethical and utilitarian impulses of the Enlightenment. In the nineteenth century, societies and academies were important vehicles for the spread of scientific practices and the popularization of scientific knowledge; conversely, this organizational tradition continued to be valuable to elite researchers who created what eventually became modern professional science. Societies continued to multiply in the twentieth century, serving as guardians of more localized forms of knowledge and as key sites in national and international scientific networks. The following chapter traces the development of scientific academies and societies through these four centuries, showing the ways in which this evolving organizational tradition helped constitute the enterprise we now call science.

New Philosophies and New Forums

From the earliest scholarship on the subject, academies and societies have been seen as integral to the organizational consolidation of modern science. In particular, early academies and societies played an important role in older accounts of the so-called "scientific revolution." Scholars who believed that modern science first appeared in full flower in the seventeenth century understood the Parisian *Académie des Sciences* and London's Royal Society as forums designed to nurture this new kind of knowledge (Hahn 1971; Ornstein 1975; for an account that continues this kind of argument into the eighteenth century, see McClellan 1985).

More recent research has refined this picture in important ways, but without severing the link between the intellectual innovations of the early modern period and the emergence of societies and academies. Other developments within the history of science have made "the scientific revolution" no longer such a self-evident category, and several key analytical assumptions of the older literature have been called into question. "Science" was once taken as a relatively self-evident and coherent category for the seventeenth century. Supposedly, when this new kind of pursuit emerged, men of science sought to create a supportive social space for themselves by founding academies and societies. More recent scholarship has emphasized the many things that differentiated seventeenth-century natural philosophy from modern science and has also paid closer attention to the diversity of intellectual and methodological commitments held by early modern learned men. As our view of the seventeenth century has become more thoroughly historicized, scholars have also developed a more sophisticated understanding of the relationship between social and epistemic practices; that is, they have paid more attention to how practices of scientific sociability and communication shaped the making of knowledge.

In this newer historiography, the emergence of academies and societies is still understood as connected to the intellectual novelties of the seventeenth century. But instead of simply assuming the existence of "science" as a unitary and coherent pursuit, historians have analyzed the specific epistemic commitments embodied in the practices of scientific societies. For example, older accounts of the scientific revolution depicted empirical science replacing a medieval scholastic tradition that had ignored the evidence of the senses. In actuality, both scholastic philosophers and the new experimental philosophers cared about experience; they just cared about different *kinds* of experience and evidence. Medieval philosophers had privileged experiences that were familiar and commonly recognized to be true. In contrast, the new natural philosophers of the seventeenth century were interested in unusual events, experiments such as Robert Boyle's efforts to create a vacuum with an air pump. The Royal Society's publications provided a new literary genre that allowed for the presentation of this new kind of empirical evidence. The research reports that appeared in the society's *Proceedings* were short descriptions of specific events, very different from the lengthy and erudite commentaries that had been the staples of earlier learned communication (Dear 1985). The collective approval of London's Royal Society, Italy's Academy of Linceans, and France's Academy of Sciences helped bolster the cultural authority of new kinds of singular, exceptional evidence. In the process, early modern learned societies stabilized a new brand of scientific authorship, one that drew on the social status of noble patrons or elite practitioners to produce new forms of intellectual credit (Biagioli 1996).

Through the creation of societies and academies, the practitioners of the new philosophies were able to draw support from some of the most powerful political and social forces in seventeenth-century Europe. The French Academy of Sciences was founded in 1666 as part of Jean-Baptiste Colbert's plans to consolidate and expand the royal power of his patron, the Sun King Louis XIV (Hahn 1971). The academicians enjoyed the rare privilege of being paid to pursue their philosophical work, and their accomplishments were supposed to reflect back glory on their royal patron. The King's support, in turn, imbued the activities of the academy with special status (Biagioli 1996). In contrast, the Royal Society of London (f. 1660) constituted a new identity for English natural philosophers by drawing on more widely shared cultural ideals of gentlemanliness. In seventeenth-century English society, the word of a gentleman could supposedly be trusted because gentlemen (unlike, say, women, children, or servants) were independent and hence capable of speaking honestly and forthrightly. The Royal Society appropriated these broader habits of social trust and used them to stabilize the authority of its natural philosophical investigations (Shapin 1994).

Enlightened Sociability and the Growth of Civil Society

In 1700, Europe had only a handful of learned academies and societies. By the early nineteenth century, there were hundreds. Casting a wide net, we can estimate that there were perhaps roughly 650 relevant groups founded in the eighteenth century.[1] Creating a definitive list of the "scientific" societies of this time period is difficult, however, for reasons that illuminate something fundamental about the nature of intellectual sociability in the Age of Enlightenment. The spread of learned societies and academies happened as part of a broader expansion of voluntary associations, and one could find an interest in natural history, chemistry, and natural philosophy in many different kinds of groups. While some historians have understood the proliferation of societies in the eighteenth century as a further attempt to fix the intellectual accomplishments of the scientific revolution in institutional form (McClellan 1985; Lowood 1991), such a formulation mischaracterizes the intellectual geography of the eighteenth century. In this period the study of nature could have many allies and bedfellows, and eighteenth-century groups typically did not understand themselves to be furthering a unified enterprise called "science."

For analytical purposes, one can distinguish three motivational strains that supported the growth of eighteenth-century societies and academies. First of all, societies and academies were important organs within the Republic of Letters, the pan-European community of intellectually active people. Prominent learned men (and very occasionally women) built their public reputations in part through their membership in learned societies, and the correspondence networks that these societies helped sustain were important communicative structures within European learned life (Goodman 1994). Berlin's Society for Nature-Researching Friends, for example, exchanged letters and specimens with its corresponding members across Europe. The flow of naturalia generated through their correspondence allowed the group to build up a sizable natural history collection, and the society's core members in Berlin devoted considerable collective effort to the identification, care, and preservation of their naturalia. The Society of Nature-Researching Friends, as its name suggests, focused only on nature, but many other learned groups cultivated a broader range of subjects. The French provincial academies, for example, had antiquarian as well as natural historical interests (Roche 1996). Like the French academies, many other groups in this period sought to cultivate all the sciences and arts together, rather than merely focusing on the study of nature.

Secondly, the public attention that such groups generated was also useful in the broader field of cultural politics. It was a mark of distinction for a city or state to possess a significant society or academy, and an expanding number of eighteenth-century elites were interested in calling attention to themselves and their *patria* in this way. We have already seen how France's Academy of Sciences magnified the symbolic power of the French monarchy. Similarly, Frederick the Great reformed the Royal Academy of Sciences in Berlin in part to promote the cosmopolitan, enlightened image he hoped to cultivate (Terrall 1990). The Dutch Society of Sciences had heavy support from Holland's elite class of regents (Roberts 1999). Less powerful, more provincial elites also used societies to raise their cultural profile. The Literary and Philosophical Societies of late-eighteenth-century Britain tried to reproduce an echo of the cultural offerings of London in smaller, provincial cities (Porter 1980). A similar dynamic was at work in the Americas, where cities like Philadelphia and Boston wanted to show that the new fledging American republic was robust enough to support the arts and sciences (Dupree 1976).

Finally, an interest in practical economic improvement also helped fuel the growth of new associations. By one estimate, there were 562 groups founded in eighteenth-century Europe that had the cultivation of useful knowledge as their aim (Stapelbroek and Marjanen 2012). In these associations, the study of nature was linked with the practical improvement of handicraft, manufacturing, and agriculture. Groups like Russia's Free Economic Society were founded with utility as their primary aim, but even in groups that identified themselves primarily as learned societies, useful knowledge was often an important item on their agenda. The aforementioned Society of Nature-Researching Friends published essays on agriculture, and the venerable Royal Society of London, which had always had useful knowledge in its sights, continued to cultivate this interest, particularly in the second half of the century (Miller 1999). The Dutch Society of Sciences pursued improvements to navigation and shipbuilding, topics near and dear to the hearts of Holland's mercantile elite (Roberts 1999).

The balance of these three motives—learned communication, cultural aggrandizement, and economic improvement—varied from society to society, but most eighteenth-century groups were sustained through a mix of all three, with philosophical curiosity, cultural ambition, and practical concerns existing in varying measure side by side. The famous Lunar Society of Birmingham, for example, brought together men like James Watt, Josiah Wedgwood, Joseph Priestley, and Erasmus Darwin, and their collective interests ranged from natural philosophical experiment to practical invention (Uglow 2002). Though eighteenth-century intellectuals talked a great deal about the relationship between theory and practice, they did not usually feel the need to segregate "pure" science from its more practical relatives within organized cultural life, and as a result, Europe's economic societies form an important part of the history of scientific sociability in the Enlightenment.

Taken collectively, these groups made important contributions to the constitution of eighteenth-century civil society. Along with other kinds of voluntary associations, they were part of a new public sphere that came into being between the formal organs of the state and the private lives of individuals (Nyhart and Broman 2002). As numerous historians have argued, voluntary associations provided new spaces for debate and communication, helping build up a public whose voice could begin to offer a counterpoint to the voice of the state. Voluntary associations also introduced new forms of sociability. They brought together members from different social groups who might not have otherwise rubbed shoulders in Old Regime society, and they created spaces governed by new standards of civility and politeness (Melton 2001; Hoffmann 2006).

One can see both of these broader trends at work in the history of scientific societies. In the comparatively oppressive environment of the Russian Empire, for example, learned societies played an important role in creating a sphere of civil society that could serve as a conversation partner for the state (Bradley 2009). In 1765 Catherine the Great allowed a group of officials, academics, and courtiers to found Russia's earliest voluntary association, the Free Economic Society. The group was an imitation of predecessors in the West, and it was allowed a unique degree of independence from the state (hence its status as a "free" society) because of the perceived importance of its practical economic mission. The society attracted influential members of the Russian nobility, and over the second half of the eighteenth century it developed a new image of the noble landowner as an effective, enlightened manager of his estates. Though the brand of enlightenment the society crafted was limited to a narrow elite, it nonetheless led some members of the nobility to understand their relationship to the polity in new ways. The society and its members also exerted significant pull at Catherine's court (Leckey 2011). In the more liberal environment of London, one can find the Royal Society taking on a similar role in relation to the British state. Its members, too, were well-connected men, often of noble birth. Over the course of the eighteenth century, the Royal Society acted as an advisor to the state on various technical and scientific matters (Gascoigne 1999). More diffusely, many eighteenth-century groups described themselves as "patriotic" endeavors devoted to the promotion of the common good. Enlightened definitions of patriotism had a dual cast. To be a patriot was to contribute to the prosperity of one's commonwealth or dynastic state, but it was also to be a friend of humanity in general, a contributor to a cosmopolitan project of moral and material improvement (Lindemann 1990).

Scientific societies also provided a meeting ground for people from different social estates. Zurich's Physical Society was a cooperative effort between the city's patrician families and local members of the learned professions, and a number of eighteenth-century groups fit this basic profile. At times, the intended leveling effect of society membership was made explicit. Berlin's Society for Nature-Researching Friends, like the Freemasons, wanted to create a space free from signs of external marks of rank, where noblemen and commoners encountered each other as (temporary) equals (Böhme-Kassler 2005). In the Enlightenment, sociability was also seen as a way to civilize and reform humanity's baser nature, and some groups were founded with this aim as well. When Peter the Great created a new Academy of Sciences in St. Petersburg, for example, he hoped that the discussions of the Academy would serve to introduce Western forms of civility into his court and capital city. The Academy, in other words, was to be one more tool in his efforts to cure his nobility of their supposedly crude and non-European habits (Gordin 2000).

High and Low Science

Early nineteenth-century Lancashire was home to several local botanical clubs. The artisans who belonged to these societies were a far cry from the powerful and wealthy men whose names filled the rolls of London's Royal Society or Russia's Free Economic Society. The Lancashire artisans met together in small local pubs to compare their plant specimens. The brighter lights of British botany tended to dismiss them as mere collectors of plants rather than true naturalists, but the artisans saw things differently. They staked a claim to the higher title of "botanist," developing their own criteria of skill and expertise (Secord 1994).

Eighteenth-century societies, in both their learned and practical guises, are best understand as part of a single, loosely organized cultural network. In the nineteenth century, in contrast, one begins to see a stronger split emerging within associational life. By the end of the century, a clear gap had opened up between elite and more popular kinds of science, a gap that appeared along the fault line that one can already see emerging in the case of Lancashire's artisan botanists. Scientific societies played an important role on both sides of this divide. They were the primary home of Europe and America's expanding "low" scientific culture; they also continued to be essential to the world of high science as well.

Some historians have argued that the "Age of Societies" ended around 1800, when the universities and other research institutes took over as the key sites of scientific research. According to this view, nineteenth- and twentieth-century learned societies played an ancillary role in the process of scientific professionalization and disciplinary specialization, but they were no longer significant incubators for scientific research (McClellan 1985). This way of framing the chronology, however, obscures important continuities between the eighteenth and the nineteenth century. Because it tends to cut off the world of elite "professional" science from other branches of scientific associational life, it also makes it harder to analyze the persisting interrelationships between high and low scientific circles.

The nineteenth century was a period of increased state support for science, and scientific associations fit within this period's new institutional landscape in different ways in different national cases. In Britain, where the universities were reformed relatively late, voluntary associations remained at the heart of elite science throughout the century (Morrell and Thackray 1981, 26–7; Desmond 2001). In France, in contrast, the post-Napoleonic educational system was heavily centralized, its lines of authority radiating out from the powerful center of Paris. The members of provincial academies and societies became, in Robert Fox's phrase, "voices from the periphery." A fairly sharp divide emerged between the state-funded academic scientists supported through the Ministry for Public Instruction and provincial scientific devotees (Fox 2012, 52–73). In other places, however, learned societies developed complementary relationships with universities and remained elite spaces throughout the nineteenth century. In 1895, for example, the physics professor Wilhelm Conrad Röntgen published his first preliminary paper on X-rays in the proceedings of his local scientific society, Wurzburg's Physical-Medical Society (Röntgen 1895). Like Wurzburg, most German university towns had learned societies that served as important intellectual forums for their faculty (Phillips 2012). When the Russian Empire reformed its universities in 1863, this German model was the one they attempted to reproduce. As part of the reforms, Russian officials created university-affiliated learned societies that were supposed to enhance the research productivity of the universities (Loskutova and Fedotova 2015, 17). In the United States, too, colleges like Yale and Harvard had affiliated learned societies (Dupree 1976, 29).

The nineteenth century also saw the emergence of prominent national associations for the promotion of science. Forums like the Society of German Natural Researchers and Doctors and the British Association for the Advancement of Science provided a public platform for leading researchers to advocate for the social and political importance of science. Now, unlike in earlier periods, science was conceptualized as a unitary and coherent cultural cause (Morrell and Thackray 1981; Phillips 2012). These national organizations played a significant role in defining and managing science's public image throughout the nineteenth century (Kohlstedt, Sokal, and Lewenstein 1999).

In the increasingly dense associational landscape of nineteenth-century science, however, prominent national organizations and elite learned societies were not the only voices that defined what counted as science. Scientific societies continued to proliferate at a rapid clip throughout the nineteenth century, and historians of science have only begun to explore this rich field of activity. Local and regional groups became not just more numerous, but also more specialized and differentiated over the course of the century. To take one typical example, in the eighteenth century the city of Lyon had been the home to an Academy of the Sciences, Literature and the Arts, a group that provided a broad umbrella for all areas of learning. By the end of the nineteenth century, Lyon had ten different groups whose interests overlapped with those of the venerable old academy, among them societies for botany, meteorology, and antiquities (Fox 2012, 59–60). Other major European cities had dozens of similar groups by the 1880s and 1890s. The list of provincial cities and towns with scientific societies also grew longer. Scotland alone had 60 different natural history societies founded in the nineteenth century (Finnegan 2009). Natural scientific societies proliferated rapidly throughout the German states as well (Daum 1998).

This massive expansion in the number of scientific societies came along with a corresponding change in their predominant intellectual orientation. The new, less elite groups that emerged in the nineteenth century often turned their attention primarily to the study of regional natural history (and often regional human history, too). This regional focus reflected the skills and interests of many society members; it also reflected an attempt by non-academic naturalists to differentiate their own forms of expertise from the expertise of elite scientists. Many local scientific societies claimed that they were happy to leave the study of grand generalities to metropolitan experts. Their field of endeavor was more constrained (and also more reasonable, they sometimes hinted). They wanted to conduct thorough studies of regional landscapes. For example, Scotland's natural history societies were imbued with a strong sense of civic pride and often engaged in collective excursions into the countryside to collect plant, animal, and mineral specimens (Finnegan 2009). One can see similar developments in the German states, where many of the new societies founded from the mid nineteenth century onward focused primarily on regional natural history. The Napoleonic Wars had radically simplified the German political landscape, greatly reducing the number of independent states, and natural history societies were often part of the reinvention of regional identities within this new, more centralized political landscape. For example, the Pollichia Society of the Palatinate was founded by residents of a largely Protestant region who wanted to set themselves apart from Bavaria, the dynastic state to which they had been assigned in the post-Napoleonic settlement (Phillips 2012, 177–201).

Late-nineteenth-century biologists often dismissed natural history as an uninteresting popular enthusiasm, something less than fully "scientific." Earlier historians of eighteenth and nineteenth-century societies have sometimes agreed. These scholars tended to measure scientific societies with a single yardstick. They were confident that they knew what serious science looked like, and felt comfortable passing judgment on groups that seemed to be doing something else. A common target of criticism, for example, was the so-called "naive Baconianism" of many groups. Already in the eighteenth century, natural historical collecting and observation made up an important component of many societies' activity, and to historians who took their model of science from physics, that kind of work seemed little better than stamp collecting. Daniel Roche (1996), a historian of France's eighteenth-century provincial academies, wrote that these groups had been able to further science *despite* their attachment to natural history. Similarly, Henry Lowood (1991) expressed surprise that the learned societies of eighteenth-century Germany had not realized that all of their fact-gathering was not bringing them any closer to discovering universal laws.

This very distinction—between supposedly mindless empiricism and law-generating science—was one that was produced in the cultural negotiations of the nineteenth century. Many elite life scientists in this period saw regional natural historical activity as a clearly second-tier form of science, or perhaps not real science at all (Alberti 2001). The proliferation and success of scientific voluntary associations testify to the limited impact that such views had within the broader public sphere. Natural history remained a pursuit with widely recognized civic and intellectual value. Local natural history groups devoted enormous energy to excursions and collecting, and even elite scientists, despite their dismissive rhetoric, sometimes continued to interact with a variety of scientific associations (Olesko 1994; Alberti 2001). Many societies continued their efforts well into the twentieth century, their activities interrupted but not eclipsed by the upheavals of political revolution and two World Wars. The regional scientific society of Iaroslavl, for example, survived the Russian Revolution and remained an important focus of local civic pride through the first decades of communism, adapting relatively well to a regime that valued science as a counterweight to conservative religious ideologies. It took the more aggressive intrusion of the Stalinist state to finally dissolve this space of independent provincial civic initiative (Andrews 2001).

Regional natural history was one frequent vehicle for collective pride, but not the only possible one. A good example of another kind of cultural strategy is the Senckenberg Society, an association founded in 1817 in the independent city-state of Frankfurt. The group was a joint enterprise of Frankfurt's learned and mercantile elites, who pooled their resources to create one of the best natural history museums in nineteenth-century Europe. The society's impressive collection did not focus on local naturalia. Its ambitions were much broader, and the initial core of the collection came from Eduard Ruppell, a Frankfurt native who had won acclaim through his extensive travels in Africa. The Frankfurt society illustrates how scientific associations could be used to help a small polity display its dynamism on the European stage. While Frankfurt had no hopes of making its presence felt in great-power politics, the Senckenberg Museum served to advertise the small city-state's continued cultural relevance in the landscape of post-Napoleonic Europe (Sakurai 2013).

In short, the history of scientific societies has been intricately bound up with the politics of place, particularly in the broader negotiations between political and cultural centers and their peripheries (Figure 16.1). Perhaps the most trenchant examples of this dynamic come from outside Europe, where scientific societies could sometimes become a tool used to negotiate with and potentially short-circuit European cultural imperialism. The Geological Society of China is an excellent case in point. In the early twentieth century, the Geological Society was founded in part as a reaction to European scientists' tendency to ignore the work of Chinese geologists. The association gave Chinese geologists a forum in which they could make their own research more visible to the international scientific community. It also allowed them to assert a symbolic claim to Chinese mineral resources, merging geological research with the process of Chinese nation building (Shen 2013).

Indeed, by the twentieth century scientific societies could be found around the globe. Settler colonists of European descent founded some of these organizations,

THE

JOURNAL

OF THE

ROYAL ASIATIC SOCIETY

OF

GREAT BRITAIN AND IRELAND.

QUOT RAMI TOT
ARBORES

A.D. MDCCCXXIII. INST.

VOLUME THE SEVENTH.

LONDON:
JOHN W. PARKER, WEST STRAND.

M.DCCC.XLIII.

Figure 16.1 The Royal Asiatic Society of Great Britain and Ireland (f. 1824) was devoted to the study of science, literature, and the arts as they related to Asia. The group offers one particularly specialized example of how academies and societies organized and mediated the knowledge that Europeans acquired as they expanded their influence commercially, militarily, and politically around the globe.

which dated, in some cases, all the way back to the eighteenth century. In India, the Asiatic Society of Bengal served as a platform for British scholarship and science from 1784 onward (Baber 1996, 153–60; Arnold 2000, 28–9), and the Royal Society of South Africa (f. 1877) and the South African Association for the Advancement of Science (f. 1902) had similar cultural profiles. In many cases, however, the spread of scientific societies occurred when non-Western elites' creatively adapted this Western tradition to support new forms of nationalist autonomy. The Japanese were early adopters of Western-style societies, for example. Scientific groups began appearing in Japan in the 1870s, and by the early twentieth century the country had at least 28 different learned societies devoted to the study of science, technology, or medicine (Takata 1997). The Chinese nationalist government founded the Academia Sinica in 1928 to coordinate science and technology in the service of national modernization; later, when the People's Republic of China was founded in 1949, the new Communist government replaced this earlier organization with a new Chinese Academy of Science (Yao 1989).

Conclusion

Following the history of academies and societies allows us to observe the expansion of European scientific networks in several directions. Through the study of organized scientific sociability, one can watch the modern professional scientific community slowly coalescing out of its early modern counterpart. One can also see how an interest in the organized study of nature expanded downward to other, less prestigious social groups, and outward to other kinds of elites. Finally, it allows us to map the geographic expansion of Western science as it paralleled the expansion of European imperial power around the globe.

To talk only about expansion, however, can be misleading. In watching scientific societies and academies proliferate over the centuries, we are not watching the simple replication of the same thing over and over again. The founders and members of new societies always adapted this organizational tool to their own purposes. The local allegiances that supported natural knowledge in any given place could be diverse, as could the cultural meaning of the activity. In one location, a society might be a private supplement to official bureaucratic activity, in another, an act of opposition to encroaching state authority. These groups could glorify royal power or celebrate the prestige of independent citizens. Their members might be illustrious or obscure—university professors, noblemen, and wealthy merchants, or apothecaries, schoolteachers, and artisans.

Many popular natural history associations founded in the nineteenth century are still around today, supporting the work of local natural history museums or sharing news of local bird sightings. The Nature-Researching Society of Bamberg, for example, was first founded in 1834 and is still active, as are many other of its ilk. The society still publishes a yearly volume of proceedings, the contents of which mostly contain articles devoted to regional natural history. Its members still take excursions together out into the local countryside. In addition, twenty-first century elite science and scholarship is still full of societies and academies. The American Council of Learned Societies, an umbrella group to which most major national societies in the humanities and social sciences belong, has 73 member organizations. Most academics go to multiple conferences a year, attending meetings sponsored by both very general disciplinary groups (the American Chemical Society) or more specialized ones (the American Society for Cell Biology).

William Clark has pointed out that the persistence of these meetings signals the continued importance of embodied, face-to-face contact in the conduct of modern science and scholarship (Clark 2006). Two hundred years ago, German Romantic philosophers saw this facet of learned societies as their most valuable characteristic. Friedrich Jacobi thought that regular social contact among the members of a society made intellectual connections possible that simply could not be achieved through more geographically distant forms of communication. "Sciences that seemed foreign to one another realize their close and increasing relation; narrowness of perspective disappears," Jacobi wrote in 1807 (Jacobi 1807, 17). From their beginnings, scientific societies have been, in one way or another, about managing both face-to-face relationships and global, "virtual" networks of contacts. There is every reason to think that they will remain important even in a world where new virtual forms of communication are proliferating. Since the seventeenth century, this basic organizational form has proven to be remarkably adaptable, both as a mechanism for managing credit and recognition among researchers, and as a strategy for gaining attention and acclaim from the broader social and political order.

Endnote

1 Stapelbroek and Marjanen (2012) estimate that there were about 562 economic societies founded in the eighteenth century; McClellan (1985) counts about 110 learned societies. My estimate, which is admittedly imprecise, is a combination of these two numbers, with some allowance for the possibility that some groups might have been included on both authors' lists.

References

Alberti, Samuel J. M. M. 2001. "Amateurs and professionals in one county: Biology and natural history in late Victorian Yorkshire." *Journal of the History of Biology*, 34: 115–47.

Andrews, James T. 2001. "Local science and public enlightenment. Iaroslavl naturalists and the Soviet State, 1917–31." In *Provincial Landscapes: Local Dimensions of Soviet Power, 1917–1953*, edited by Donald J. Raleigh, 105–24. Pittsburgh, PA: University of Pittsburgh Press.

Arnold, David. 2000. *Science, Technology and Medicine in Colonial India*. Cambridge: Cambridge University Press.

Baber, Zaheer. 1996. *The Science of Empire: Scientific Knowledge, Civilization, and Colonial Rule in India*. Albany, NY: State University of New York Press.

Biagioli, Mario. 1996. "Etiquette, interdependence, and sociability in seventeenth-century science." *Critical Inquiry*, 22: 193–238.

Böhme-Kassler, Katrin. 2005. *Gemeinschaftsunternehmen Naturforschung: Modifikation und Tradition in der Gesellschaft Naturforschender Freunde zu Berlin 1773–1906*. Stuttgart: Franz Steiner.

Bradley, Joseph. 2009. *Voluntary Associations in Tsarist Russia: Science, Patriotism and Civil Society*. Cambridge, MA: Harvard University Press.

Clark, William. 2006. *Academic Charisma and the Origins of the Research University*. Chicago: University of Chicago Press.

Daum, Andreas. 1998. *Wissenschaftspopularisierung im 19. Jahrhundert: Bürgerliche Kultur, naturwissenschaftliche Bildung und die deutsche Öffentlichkeit, 1848–1914.* Munich: R. Oldenbourg Verlag.

Dear, Peter. 1985. "Totius in verba: Rhetoric and authority in the early Royal Society," *Isis*, 76: 145–61.

Desmond, Adrian. 2001. "Redefining the X-axis: 'Professionals,' 'amateurs' and the making of mid-Victorian biology." *Journal of the History of Biology*, 34: 3–50.

Dupree, A. Hunter. 1976. "The national pattern of American learned societies." In *The Pursuit of Knowledge in the Early American Republic: American Scientific and Learned Societies from Colonial Times to the Civil War*, edited by Alexandra Oleson and Sanborn C. Brown, 21–32. Baltimore: Johns Hopkins University Press.

Finnegan, Diarmid A. 2009. *Natural History Societies and Civic Culture in Victorian Scotland.* London: Pickering & Chatto.

Fox, Robert. 2012. *The Savant and the State: Science and Cultural Politics in Nineteenth-Century France.* Baltimore: Johns Hopkins University Press.

Gascoigne, John. 1999. "The Royal Society and the emergence of science as an instrument of state policy." *British Journal for the History of Science*, 32: 171–84.

Gordin, Michael. 2000. "The importation of being earnest: The early St. Petersburg Academy of Sciences." *Isis*, 91: 1–31.

Goodman, Dena. 1994. *The Republic of Letters: A Cultural History of the French Enlightenment.* Ithaca, NY: Cornell University Press.

Hahn, Roger. 1971. *Anatomy of a Scientific Institution: The Paris Academy of Sciences, 1666–1803.* Berkeley, CA: University of California Press.

Hoffmann, Stefan-Ludwig. 2006. *Civil Society.* Basingstoke: Palgrave Macmillan.

Jacobi, Friedrich Heinrich. 1807. *Ueber gelehrte Gesellschaften, ihren Geist und Zweck.* Munich: E. A. Fleischman.

Kohlstedt, Sally Gregory, Michael M. Sokal, and Bruce V. Lewenstein. 1999. *The Establishment of Science in America: 150 Years of the American Association for the Advancement of Science.* New Brunswick, NJ: Rutgers University Press.

Leckey, Colum. 2011. *Patrons of Enlightenment: The Free Economic Society in Eighteenth-Century Russia.* Newark, NJ: University of Delaware Press.

Lindemann, Mary. 1990. *Patriots and Paupers: Hamburg, 1712–1830.* Oxford: Oxford University Press.

Loskutova, Marina V. and Anastasia A. Fedotova. 2015. "The rise of applied entomology in the Russian Empire: Governmental, public and academic responses to insect pest outbreaks from 1840 to 1894." In *New Perspectives on Life Sciences and Agriculture*, edited by Denise Phillips and Sharon Kingsland, 139–62. Berlin: Springer.

Lowood, Henry E. 1991. *Patriotism, Profit, and the Promotion of Science in the German Enlightenment: The Economic and Scientific Societies, 1760–1815.* New York: Garland.

McClellan, James E. 1985. *Science Reorganized: Scientific Societies in the Eighteenth Century.* New York: Columbia University Press.

Melton, James Van Horn. 2001. *The Rise of the Public in Enlightenment Europe.* Cambridge: Cambridge University Press.

Miller, David Philip. 1999. "The usefulness of natural philosophy: The Royal Society and the culture of practical utility in the later eighteenth century." *British Journal for the History of Science*, 32: 185–201.

Morrell, Jack and Arnold Thackray. 1981. *Gentlemen of Science: Early Years of the British Association of Science.* Oxford: Clarendon.

Nyhart, Lynn and Thomas Broman (eds.) 2002. *Science and Civil Society: Osiris 17.* Chicago: University of Chicago Press.

Olesko, Kathryn Mary. 1994. "Civic culture and calling in the Königsberg period." In *Universalgenie Helmholtz: Rückblick nach 100 Jahren*, edited by Lorenz Krüger, 22–42. Berlin: Akademie Verlag.

Ornstein, Martha. 1975 [Reprint of 1913 edition]. *The Role of Scientific Societies in the Seventeenth Century*. New York: Arno Press.

Phillips, Denise. 2012. *Acolytes of Nature: Defining Natural Science in Germany, 1770–1850*. Chicago: University of Chicago Press.

Porter, Roy. 1980. "Science, provincial culture and public opinion in enlightenment England." *British Journal for Eighteenth-Century Studies*, 3: 20–46.

Roberts, Lissa. 1999. "Going Dutch: Situating science in the Dutch enlightenment." In *The Sciences in Enlightened Europe*, edited by William Clark, Jan Golinski, and Simon Schaffer, 350–88. Chicago: University of Chicago Press.

Roche, Daniel. 1996. "Natural history in the academies." In *Cultures of Natural History*, edited by N. Jardine, J. A. Secord, and E. C. Spary, 127–44. Cambridge: Cambridge University Press.

Röntgen, Wilhelm Conrad. 1895. "Über eine neue Art von Strahlen. Vorläufige Mitteilung." In *Aus den Sitzungsberichten der Würzburger Physik.-medic. Gesellschaft Würzburg*, 137–47.

Sakurai, Ayaku. 2013. *Science and Societies in Frankfurt am Main*. London: Pickering & Chatto.

Secord, Anne. 1994. "Science in the pub: Artisan botanists in early nineteenth-century Lancashire." *History of Science*, 32: 269–315.

Shapin, Steven. 1994. *A Social History of Truth: Civility and Science in Seventeenth-Century England*. Chicago: University of Chicago Press.

Shen, Grace Yen. 2013. *Unearthing the Nation: Modern Geology and Nationalism in Republican China*. Chicago: University of Chicago Press.

Stapelbroek, Koen and Jani Marjanen (eds.) 2012. *The Rise of Economic Societies in the Eighteenth Century*. New York: Palgrave Macmillan.

Takata, Seiji. 1997. "Activity of Japanese physicists in the learned societies from 1877 to 1926." *Historia scientiarum*, 7: 81–91.

Terrall, Mary. 1990. "The culture of science in Friedrich the Great's Berlin." *History of Science*, 28: 333–64.

Uglow, Jenny. 2002. *The Lunar Men: Five Friends Whose Curiosity Changed the World*. New York: Farrar, Straus and Giroux.

Yao, Shuping. 1989. "Chinese intellectuals and science: A history of the Chinese Academy of Sciences." *Science in Context*, 3: 447–73.

CHAPTER SEVENTEEN

Museums and Botanical Gardens

LUKAS RIEPPEL

Museums and botanical gardens are physical spaces dedicated to the accumulation, storage, and display of knowledge. They are distinct from the library as well as the archive, which serve a related scholarly function, in that they tend to focus on the collection and exhibition of specimens; that is, material fragments of the natural world. In addition, they often feature products of human ingenuity, including technological innovations, artworks, and scientific instruments. Because they gather so many different kinds of things in one place, museums and gardens have sometimes been seen as a kind of microcosm or world in miniature. Often, they also double as spaces in which learned naturalists conduct scientific research. As such, they do not only *represent* and *display* knowledge. Museums and gardens have also played a crucial role in shaping the way knowledge is generated.

This essay traces the history of modern museums and botanical gardens from their origins in the European Renaissance to the early twentieth century, paying particular attention to how the development of these sites reflect broader changes in political economy. I will suggest that a coherent, if somewhat schematic, account can be had by attending to the question of who underwrote the production and dissemination of knowledge in these spaces and to what ends. Museums and gardens, I argue, have long served as a way to demonstrate the social, cultural, and political power of their patrons and benefactors, and they continue to do so today. The question of who underwrote these institutions and to what ends therefore reveals deep connections between the practice of gathering knowledge about nature and the accumulation of social, cultural, as well as financial capital. Thus, although they are often seen to be quaint and old-fashioned, this essay shows that museums and gardens have long been inextricably tied to the development of our modern, knowledge-based economy.

Renaissance and Early Modern Collections

The history of museums stretches back to medieval and early modern Europe. Although there were museums (places dedicated to the muses) in classical antiquity, it

A Companion to the History of Science, First Edition. Edited by Bernard Lightman.
© 2016 John Wiley & Sons Ltd. Published 2020 by John Wiley & Sons Ltd.

was during the middle ages that the modern enthusiasm for collecting truly emerged. Made up of rare, valuable, and wonderful objects, medieval collections tended to be kept in monasteries and houses of worship, where they were prized as both a material and spiritual treasure. In addition to bringing together illuminated manuscripts, artworks, and religious relics, they often included valuable objects of nature, including exotic things that were brought back from faraway lands by travelers and crusaders. For example, in addition to gemstones, medieval collections showcased spectacular and awe-inspiring items like unicorn horns, elephant tusks, and ostrich eggs.

During the high middle ages, collecting became increasingly associated with a bourgeoning court culture. Princes, dukes, and other nobility went to great lengths to surround themselves with marvelous possessions as evidence of their own exalted and chosen status. As the historians Katharine Park and Lorraine Daston have put it, collections of objects that evoked a strong sense of wonder and awe were seen to reflect "the wealth and power of those who owned them," enveloping "those around them with an aura of nobility and might" (Daston and Park 1998, 68). By the fifteenth and sixteenth century, an emerging class of wealthy merchants and bankers became avid collectors as well. Prominent families, such as the Medici in northern Italy and the Fuggers in southern Germany, invested sizable parts of their fortunes to amass impressive collections of natural and man-made objects. Because they were so closely associated with the landed aristocracy, impressive collections served as a display of their material riches and personal refinement. The practice of collecting therefore inserted wealthy families into the cultural nexus and intellectual milieu of the ruling elite, helping to cement their social status while simultaneously advancing their business goals. In addition, extensive collections of valuable objects also doubled as a convenient means of hoarding one's wealth, a kind of treasury that could be converted to cash should one find oneself facing a liquidity crisis (Findlen 1994; Meadow 2001).

These early collections were called *Wunderkammern* or Cabinets of Curiosity. As their name suggests, their significance in Renaissance culture did not just derive from the rare, strange, surprising, and opulent things they contained, but also the emotions they were understood to elicit: a sense of wonder, awe, and, above all, curiosity. Etymologically derived from the Latin word *cura* for "care," curiosity signified the mental disposition of being careful, assiduous, and inquisitive. When faced with a wonderful object it did not fully understand, the curious mind sought to know more by meticulous, detailed examination. Not only that, but curious persons were also expected to register their findings and philosophical reflections in meticulous yet clear and plainspoken reports (Whitaker 1996). Curiosity therefore came to be associated with detailed, factual, and empirical knowledge. This was a useful and practical kind of knowledge—connoisseurship or *kennen* rather than *wissen* or *savoir*—and it exemplified the epistemic virtues associated with a new set of practices dedicated to generating reliable knowledge about nature: science.

Curiosity and connoisseurship were characteristic of commerce as well as the new science. As Harold Cook has argued, the ability to inspect material goods and make fine-grained distinctions on issues of quality was crucial for financial success in an increasingly mercantile economy. For this reason, the cultivation of refined tastes, the amassing of detailed practical knowledge, and the skill required to make reliable judgments were all held in high esteem among merchants, traders, and bankers in seventeenth-century Europe. The knowledge it took to construct and appreciate an impressive cabinet of curiosities was precisely the same kind of knowledge that ensured success in the world of trade and commerce. This not only helps to explain why collecting became such a popular means of performing one's social distinction. It also sheds light on the enormous enthusiasm that wealthy elites harbored for the science of natural history (H. J. Cook 2007, 1–42).

Natural history originated with classical authors such as Aristotle, Theophrastus, and Pliny the Elder, but experienced a revival during the Renaissance. Initially, early modern humanists were primarily interested in the textual scholarship of translating ancient encyclopedic works, but with time they increasingly began to analyze the contents of these authoritative works as well, standardizing terminology and ironing out disagreements between them. This involved not only comparing one text to another, but also consulting real plants and animals. Althought it was rooted in the bookish culture of medieval scholarship, natural history therefore become an increasingly empirical body of knowledge. Moreover, the discovery of the New World led to a flood of reports about exotic new plants and animals entering Europe. To accommodate this abundance of new factual matter, naturalists increasingly supplemented traditional practices of textual scholarship with new techniques of handling, preserving, and studying material remnants of the natural world (Ogilvie 2006).

A powerful motive force driving the accumulation of empirical knowledge about natural history in early modern Europe was its practical applications. This was especially true of knowledge about plants, which were often found to have medicinal properties. In order to study the usefulness of different plants as *materia medica* (literally "medical substances" or pharmaceuticals), and to maintain a living inventory of God's botanical creation, several universities began cultivating botanical or *physic gardens*. Often, these were located in close proximity to the anatomy theater, where professors provided learned lectures on the makeup of the human body as an assistant dismembered a corpse. Most of the earliest botanical gardens were located in Italy, including at the Universities of Padua and in Bologna, but the University of Leiden and others in northern Europe (including Basel and Montpellier) also boasted impressive collections of live plants. Often superintended by a professor of medicine, these *physic gardens* tended to lay out their plots in elaborately conceived geometrical patterns, cleverly designed such that the spatial relationship among different plants reflected current knowledge about their botanical and medicinal relationships (Cunningham 1996).

By the late sixteenth and early seventeenth century, several universities began to build up extensive collections of natural history specimens to complement their botanical gardens. An account of Padua's medical school from 1591, for example, describes a number of purpose-built structures lining the botanical garden that were used to conduct medicinal and alchemical experiments and to house extensive collections of minerals, birds, and other kinds of terrestrial and marine animals. As the historian Paula Findlen has noted, botanical gardens became "part of a research and teaching complex that housed an anatomy theater and various scientific collections accumulated by the medical faculty" (Findlen 2006, 285). In addition to university professors and physicians, apothecaries (whose business was in the trade of *materia medica*) also amassed impressive collections of natural history specimens. By the early seventeenth century, these kinds of collections became increasingly bound up with official civic and university life. Not long before his death in 1605, for example, Ulisse Aldrovandi, the University Physician, bequeathed his entire collection to the city of Bologna. Similarly, nearly a century later, an English alchemist and noted apothecary donated an impressive collection to Oxford University to found the Ashmolean Museum.

These scholarly developments only further cemented the social connection between cabinets of curiosity and the culture of learning. Just as wealthy merchants came to value the kind of practical knowledge represented by the formal study of nature, the latter was becoming increasingly tied to physical collections of live plants and exotic museum collections from all over the world. The culture of university scholarship and that of high society therefore came into direct contact over their shared enthusiasm for natural history. Indeed, aristocratic noblemen and wealthy merchants often hired learned naturalists to curate their own collections and tend to their pleasure gardens.

An established feature of court culture, pleasure gardens were outdoor spaces modeled on the biblical paradise, designed to instill a sense of peace, calm, and well-being. If the wilderness represented all that was fearsome, threatening, and dangerous about the outside world, pleasure gardens were precisely the opposite: exclusive, tightly controlled, and exquisitely civilized. Indeed, medieval and Renaissance pleasure gardens were so far removed from the wilderness that they tended to place more emphasis on artistic and architectural than on organic features. However, with the growing popularity of natural history, wealthy elites increasingly began to fill their gardens with opulent though well manicured plants. Not only that, but they also developed an enthusiasm for well-stocked menageries full of exotic animals, such as lions and elephants, captured in faraway lands.

Museums, Gardens, and the Modern State

By the late seventeenth century, then, large numbers of natural history specimens were being exported from faraway lands and brought to Europe, where they were housed in purpose-built museums and gardens. As one might expect, the task of bringing exotic plants and animals from a tropical climate to a temperate one was daunting to say the least, and success hinged on a process called acclimatization. Acclimatization began as soon as these objects were extracted from their original context, as they had to be prepared and packaged to survive the arduous conditions of a long ocean voyage. Different procedures were developed for different organisms, but in each case it was essential to keep rats and other destructive animals at bay, to prevent their exposure to salty seawater, and to keep them supplied with fresh water throughout the voyage. Once they arrived in Europe, exotic plants and animals were often transferred to elaborate architectural structures designed to maximize their chances of survival. For example, great pleasure gardens often included buildings called orangeries that featured large windows to maximize exposure to light, and stoves that dispersed heat via a complex system of underground pipes. Despite being inordinately expensive to construct and maintain, several university *physic gardens* (including those at Amsterdam and at Leiden) also had botanical hothouses of this kind (H. J. Cook 2007, 304–39).

As powerful European states such as England, France, Holland, and Spain competed with one another to expand the scope of their colonial possessions, the exchange of natural history specimens continued to grow. Indeed, the historian Londa Schiebinger has argued that European colonialism was only made possible "through the fecund coupling of naval prowess to natural history" (Schiebinger 2004, 8). An early example of this coupling was the celebrated British physician and naturalist Sir Hans Sloane. In 1687, Sloane was invited to become the personal assistant to Christopher Monck, the second Duke of Albemarle, who had recently been appointed Lieutenant Governor of Jamaica. Sloane readily accepted the position and used his stay in the Caribbean to identify some 800 new species of plants, which he eventually catalogued in an ambitious encyclopedia publication (Sloane 1725). When the Duke of Albemarle died unexpectedly less than two years after his arrival in Jamaica, Sloane returned to London and quickly established himself as a central node in an extended network of corresponding naturalists. Leveraging his position to acquire additional materials via purchase and exchange, Sloane built up one of the largest collections of his time, amassing over 70,000 objects in all. When he died in 1753, an official Act of Parliament resulted in funds being raised to create a new institution—The British Museum—that was built around the nucleus of Sloane's extensive collection (De Beer 1953).

In a provocative turn of phrase, the historian and anthropologist Bruno Latour has described eighteenth-century botanical gardens and natural history museums as "centers of calculation" (Latour 1987). Latour's description has found favor for the evocative way it contextualizes natural history within the European colonial enterprise. Museums in the eighteenth century served as official repositories for valuable materials painstakingly gathered up at the periphery of far-flung empires and transported back to their metropolitan centers. However, Latour's language should not be taken too literally, and the flows of biological knowledge and specimens were far more complex than a simple center–periphery model suggests. This is especially true for botanical gardens, which developed into a vast, interconnected network that crisscrossed all parts of the globe.

The science of botany played an absolutely central role in Europe's colonial enterprise. The latter was motivated by a theory of statecraft Adam Smith derisively labeled as mercantilism, whose primary objective was to augment a state's wealth and power by effecting a favorable balance of trade (Smith 1776). Colonial possessions were highly valued because their imports and exports could be tightly controlled, with the circulation of raw materials directed to maximize the empire's ability to profit from the production of finished goods. During the eighteenth century, mercantilist policy increasingly came to regard the science of natural history as an indispensable tool in the service of imperial self-sufficiency. By deploying the expertise of its naturalists to cultivate exotic resources at home, the state could prevent hemorrhaging bullion to rivals. For example, when the British textile industry experienced a crisis due to the growing popularity of soft, Merino wool from Spain, an elaborate scheme was launched to breed Spanish sheep in the British colony of New South Wales (Gascoigne 1998; Drayton 2000).

India furnishes a particularly compelling site in which to examine how natural history was yoked into the service of political economy. When the British East India Company began contemplating the construction of a botanical garden in Calcutta during the 1780s, the influential and politically well-connected director of England's Royal Botanical Gardens at Kew, Joseph Banks, urged swift action on the grounds that doing so would serve "the purpose of cultivating plants likely to become useful to commerce" (Gascoigne 1998, 137). Banks envisioned a complex network of interconnected colonial gardens, with Calcutta serving as an important node for commercially lucrative research. This vision coincided perfectly with that of the Calcutta Garden's first director, Lt. Col. Robert Kyd, who insisted that such institutions were "not for the purpose of collecting rare plants as things of curiosity or furnishing articles for the gratification of luxury." Rather, Kyd explained, botanical gardens served an important economic purpose for the imperial state, "establishing a stock for disseminating such articles as may prove beneficial to … Great Britain, and which ultimately may tend to the extension of the national commerce and riches" (Brockway 1979, 75).

The importance of utilitarian or economic botany, as it came to be called, only increased during the nineteenth century. As Britain's empire continued to grow, her citizens developed increasingly cosmopolitan tastes. Perhaps chief among these was a penchant for tea, the most sought-after varieties of which were imported from China. In order to overcome this imbalance in trade, Britain began to experiment with the cultivation of tea in India. At first success proved elusive, but a weakening of the ruling Qing Dynasty during the 1830s and 40s allowed English botanists to succeed in exporting not only Chinese tea plants but also local experts to India. Around the same time, it was discovered that a wild variety of tea that grew in the Indian region of Assam could be made to yield a pleasing beverage if it was assiduously pruned and tended. By the late 1840s, Britain had succeeded in setting up large-scale tea plantations in India, the most famous of which were located in Assam and Darjeeling.

Useful knowledge about the natural world was not simply gathered in the periphery and brought home to be analyzed in the metropolitan center. Rather, it was a product of circulation. Knowledge, expertise, and, perhaps most importantly, people, flowed in every direction at once, suggesting that circulation was a constitutive element of the knowledge-making enterprise, not just an afterthought (Raj 2007). For this reason, the development of natural history and the economic growth it made possible must be understood as a global, not merely a European, phenomenon. An especially compelling example is the elaborate data-processing system devised by the Spanish crown to collect, organize, and ultimately exploit the local knowledge generated by indigenous populations in South America. These institutions—what one historian has described as a set of "intensely scrutinized and increasingly standardized mechanisms for gathering, producing, and distributing useful knowledge about the New World"— rendered the curative properties of substances initially discovered by native inhabitants into modern, scientific facts that could travel across the entire globe (Barrera 2002, 165).

Although the importance of botanical gardens for success in a mercantile economy can hardly be overstated, European societies valued natural history as more than just a commercial pursuit during the eighteenth and nineteenth centuries. Just as it had during the sixteenth and seventeenth centuries, a thorough knowledge of nature continued to serve as a badge of learning and a display of one's intellectual cultivation. In eighteenth-century Paris, for example, the celebrated savant Georges-Louis Leclerc, Comte de Buffon ranked among the most desirable guests at fashionable salons. Buffon served as the intendant (or director) of the prestigious botanical garden in Paris, the *Jardin du Roi*, from 1739-1788. Initially founded in 1626 by Guy de la Brosse, the personal physician to Louis XII, its original purpose was to house, study, and disseminate *materia medica*. During his tenure, however, Buffon transformed the *Jardin* from a relatively small medical teaching institute into an internationally recognized center of scientific research. He succeeded in doing so by deftly navigating the highly complex patronage system of eighteenth-century France, ingratiating himself to Parisian high society via his command of a vast storehouse of natural facts and skillful use of poetic language. This not only made him a sought-after conversationalist and dining companion, but Buffon's monumental, eighteen-volume *Histoire Naturelle* was universally celebrated as a masterpiece of both learning and high literary style (Spary 2000).

The *Jardin du Roi* changed dramatically during the French Revolution in the last decade of the eighteenth century. Within a remarkably short time, it was transformed from a potent symbol of royal power and privilege to a visible display of republican virtues. Changing its name to the *Jardin des Plantes*, the Paris botanical garden managed to thrive and even expand by refashioning itself into a public institution dedicated to the spread of universal reason. Celebrating the capacity of science to improve the collective lot of mankind, naturalists such as Georges Cuvier, Jean-Baptiste de Lamarck, and Geoffroy Saint-Hilaire offered public lectures in the garden's *Muséum d'Histoire naturelle* to curious visitors. In addition, the *Muséum* began issuing its own print publications to disseminate news of the research conducted in its collections. For much of the nineteenth century, the Paris *Muséum* was regarded as the world's leading center for natural history research. Before long, the model of a state-sponsored but scientifically administered museum spread throughout Europe, with similar institutions established in Berlin, Leiden, Vienna, London, and elsewhere (Nyhart 2009). Art museums too underwent a similar transformation, after the King's royal collection at the Louvre Palace was turned into a public museum in 1793.

Science, Education, and Popular Culture

During the long nineteenth century (roughly the period between the French Revolution and the outbreak of the First World War), natural history became an immensely popular leisure pursuit. In fact, something similar is true for the rest of modern science as well, and ordinary people flocked to attend public lectures and spectacular demonstrations of the latest advancements in knowledge. The sciences were caught up in the development of commercial popular culture, and natural history was no exception. Thus, while they remained spaces dedicated to the display and cultivation of learning, museums and gardens increasingly became places of public amusement as well (see Figure 17.1).

The long nineteenth century was a tumultuous time, characterized by a significant upheaval in social, political, and cultural life. Among other things, large parts of Europe and the fledgling United States underwent a rapid process of industrialization. This helped to bring about enormous prosperity, and it contributed to the establishment of an increasingly liberal political economy. Together, these developments helped to drive a tremendous growth in what Karl Marx and Friedrich Engels described as the bourgeoisie, a class of people whose wealth and status derived from their access to financial capital rather than noble birthright or traditional kinship ties (Marx and Engels 1848).

Members of the nineteenth century bourgeoisie celebrated the growth of modern science with great enthusiasm. In part, this was due to the decisive role that it played in supplying the technological innovations that fueled so much of the period's economic expansion. Perhaps nowhere was this more evident than in the popular expositions that

Figure 17.1 Interior of Bullock's Museum, a proprietary institution on Piccadilly Road in central London. Aquatint by Thomas Shepherd, 1810, published in John B. Papworth. 1816. *Select Views of London*. London: R. Ackermann, 89. Courtesy of the Wellcome Library, London.

grew increasingly large and elaborate as the century wore on. Although these expos were initially confined to showing off the productions of a single region or country, the organizers of an 1851 exhibition in London decided to invite other countries to participate as well, thereby creating the first world's fair. The event's organizers hoped that a friendly competition with other industrialized nations would help to encourage further technological progress and lead to an expansion in the market for British manufactured goods. Housed in a single, vast building constructed entirely out of metal and glass called the Crystal Palace, the resulting extravaganza was a huge popular success, attracting around six and a half million visitors in all (Greenhalgh 2011). In time, the world's fair became one of the most recognizable features of late nineteenth and early twentieth century culture. London's Crystal Palace Exposition set the tone for the genre in many respects, especially with its obsession to classify, to compare, and most especially, to rank the cultural, technological, industrial, and scientific achievements of various nations. Above all else, the genre emphasized size, wealth, and progress. Around the turn of the twentieth century, new technological wonders such as electricity, the phonograph, the telephone, artificial light, as well as the X-ray all dazzled large audiences.

Museums and gardens resembled world's fairs in that they, too, developed into a mass popular spectacle, especially in nineteenth-century America. During the period after the Civil War, two different kinds of natural history museums proliferated in the United States. The first of these were learned institutions dedicated to higher education and scientific research. Among the earliest examples of its kind was a museum founded in Philadelphia during the 1780s by Charles Willson Peale. Others included the museums of the Philadelphia Academy of Natural Sciences, the Boston Society for Natural History and the New York Lyceum, as well as those at Amherst College, Harvard University, and the Smithsonian Institution. Far more numerous, however, were the countless popular entertainment venues that would eventually come to be called "dime museums." These did not primarily cater to men of science, but to a much larger, paying public. The most famous was P. T. Barnum's American Museum in New York, but other, smaller, dime museums became a staple of America's cultural landscape.

What distinguished dime museums from more learned institutions was their status as commercial amusement venues. For this reason, they tended to emphasize the strange, the rare, and the surprising over the typical and the instructive. In some ways, they resembled Renaissance cabinets of curiosity more than nineteenth-century academic museums. For example, Barnum famously made a small fortune from the exhibition of the "Feejee Mermaid," playfully taunting audiences by strategically suggesting the possibility that it was a fraud or a hoax rather than a genuine scientific specimen. However, that does not mean these museums were necessarily seen as disreputable or lowbrow. On the contrary, Barnum deliberately catered to a bourgeois audience whose conventional notions of civility prevented them from patronizing most of the city's other amusement venues, such as the theater, the concert hall, and the saloon. Dime museums shrewdly offered a form of "rational recreation" that was immensely popular among the respectable classes, consistently playing up the moral and educational benefits of their exhibits. In addition to portraits of well-known celebrities and panoramic views of exotic and faraway places, Barnum's museum included taxidermic as well as live animals, moral dramas, variety shows that featured monstrous births and other aberrations of natural order, as well as popular lectures by respected authorities on natural history such as Louis Agassiz (J. Cook 2001).

Just as nineteenth-century cities witnessed a proliferation of proprietary museums, so, too, did their inhabitants patronize a number of commercial pleasure gardens. Originating during the seventeenth century but thriving especially during the eighteenth and the nineteenth centuries, commercial pleasure gardens resembled dime museums more than courtly gardens in that both were for-profit amusement venues emphasizing ornament, variety, and eclecticism. Among the oldest and most famous was London's Vauxhall Gardens, located not far from the city center on the south bank of the Thames, but imitations soon sprang up all over England as well as abroad. Pleasure gardens proliferated across Europe, but they became especially popular in the United States, so much so that even small and remote cities like Butte, Montana or Charleston, South Carolina had one. In fact, New Orleans could boast of having over fourteen in all! During the evenings in the warm summer months, large crowds of visitors flocked to these places, where they enjoyed a multimedia experience that ranged from sweet-smelling flowers and other plants to beautiful paintings and statuary, musical performances, and spectacular firework displays. By the mid to late nineteenth century, many American pleasure gardens also included a space for theatrical performances. Indeed, *The Black Crook,* widely credited as the first Broadway musical, was originally performed in the theater of Niblo's Garden, which was located on the corner of Broadway and Prince Street in modern day SoHo (Conlin 2013).

Not all nineteenth-century gardens and museums were commercial in nature, though. For one thing, government-sponsored institutions continued to thrive. These ranged from existing establishments such as Kew Gardens or the *Muséum d'Histoire naturelle* to more recent additions like Berlin's *Museum für Naturkunde*. In addition, countless schools, colleges, and universities amassed natural history collections for instructional purposes. Finally, this period also saw the creation of many new museums and gardens sponsored by philanthropic organizations whose aim was to popularize modern science. Initially, the popular science movement was especially strong in England, where it was aimed at spreading or "diffusing" the latest discoveries among a broad and economically diverse audience, in the hopes that doing so would help edify and uplift the working classes. During the first several decades of the nineteenth century, these efforts primarily centered on print culture, and charitable organizations issued a flood of cheap publications aimed at a mass audience, including the Bridgewater Treatises and the *Penny Cyclopedia*. Present-day historians and nineteenth-century readers alike have been struck by the politics of popular science, whose emphasis on reasoned discourse was seen as a counterweight to radical movements such as Chartism. As one working-class reader reportedly put it, the principal aim of popular science was "to stop our mouths with *kangaroos*" (Secord 2000, 48).

Popularization movements help to illustrate a second reason why the nineteenth-century bourgeoisie was so enthusiastic in its promotion of modern science. Besides supplying the technical innovations on which industrial factories thrived, the practice of science was seen as a good way to cultivate personal attributes such as objectivity and disinterestedness that meshed well with other characteristically bourgeois values. The epistemic virtues of modern science were tailor made for a class of people who valorized personal discipline and moral restraint in addition to learning and education. Of all the modern sciences, natural history was especially highly regarded. In part, this was because it consisted of drawing conclusions based on the careful, detailed, and sustained inspection of material objects. In this way, a visit to the museum or garden was understood to cultivate the faculty of attention, a core goal of pedagogical practice at the time. More importantly, though, it was also expected to teach valuable object lessons in Christian theology, illustrating the benevolence and omnipotence of a supernatural creator whose presence was immanent in the natural world.

Because they meshed so well with the aims of popular science, the mid to late nineteenth century saw the creation of numerous new museums and gardens aimed at a broad and diverse audience. In Europe these efforts tended to be spearheaded by voluntary organizations and local governments. For example, the British Parliament passed an act in 1845 that authorized municipalities to levy a tax for the creation and maintenance of museums and libraries, in the hopes that doing so would lead to the diffusion of useful knowledge (Alberti 2009). By contrast, the largest and most ambitious museums created in nineteenth-century America were financed by private individuals working alone or in concert. For example, J. P. Morgan, Andrew Carnegie, and Marshall Field—all of whom had amassed an immense personal fortune during a period in American History that Mark Twain famously satirized as the "Gilded Age"— donated huge sums to pay for the creation of natural history museums in New York, Pittsburgh, and Chicago, respectively (Twain and Warner 1873).

Prominent nineteenth-century philanthropists had complex motivations that ranged from a genuine sense of moral responsibility to the desire to justify their own wealth. The American Gilded Age was an immensely prosperous period, but one that also witnessed a sharp increase in inequality and a number of violent confrontations with workers and labor unions. The highly public act of giving back to the community by building libraries, museums, parks, and public gardens therefore did not only exemplify traditionally bourgeois virtues. It also allowed wealthy industrialists to argue that, despite its shortcomings, modern capitalism produced genuine public goods (Carnegie 1900). In addition, the philanthropic support of popular science also provided a visible stage for the performance of social distinction. Many industrialists who rose to power during this period hailed from a relatively humble, artisanal background and therefore lacked the high social status to match their immense fortunes. Adopting the terminology of the French sociologist Pierre Bourdieu, philanthropy can be described as a way to convert economic wealth into cultural capital (Bourdieu 1986). By associating themselves with the institutions of elite culture, wealthy industrialists could show off their refined tastes, epistemic virtues, and elite sensibilities.

Adopting the trappings of an older, European aristocracy, including the enthusiasm for natural history, America's *nouveau riche* sought to acquire, cement, and legitimize their newfound elite status. However, whereas the collections and gardens of early modern kings, princes, and wealthy merchants were personal and private affairs, it was essential for their late nineteenth-century philanthropic counterparts to be understood as public spaces. This is because the ability to draw in a large and diverse audience was crucial to their institutional mission. Only this way could they meet the demands of popular science and succeed as a visible display of their wealthy benefactors' civic munificence at a time of considerable labor unrest.

Conclusion: the hybrid spaces of popular science

There is a kind of irony to the history presented here, in that we have ended up in much the same place where we began. It is striking how much late nineteenth- and early twentieth-century philanthropic institutions such as the Carnegie Museum in Pittsburgh and the Bronx Zoo in New York resemble Renaissance princely collections. Both were a province of the wealthy and politically powerful. Then as now, the patronage of scientific research—and displays of the knowledge thereby produced—have served as an expedient means to exhibit one's learning and philosophical high-mindedness. As a result, naturalists in both periods gained a great deal of epistemic prestige from their association with persons and institutions of high social, cultural, and political rank. At the same time, something of the reverse holds true as well. Just as the collection and display of masterful artworks allowed wealthy elites to distinguish themselves from the rest of society, so too did they find it expedient to associate themselves with the generation and display of natural knowledge. Thus, it may fairly be said, the credibility and status of natural knowledge has always been bound up with that of the sites in which it is produced and displayed.

At the same time, there is a danger of overstating historical continuities and losing sight of what is unique and distinctive about each of the spaces examined above.

One especially noteworthy difference is the status of modern museums and gardens as public rather than private institutions. Of course, this does not mean they are equally accessible to all, and various barriers to entry continue to exist. Among others, these include the price of admission and that fact that they tend to be geographically located in well-to-do neighborhoods. Still, because they are explicitly designed to popularize science among a large and diverse audience, modern museums and gardens work hard to attract visitors from all walks of life. Moreover, their self-conception as being beholden to an educational mission has had dramatic consequences for the way they are organized and how they are run. By way of conclusion, it is worth spelling out the most important of these developments in brief.

During the heyday of popular science in the mid to late nineteenth century, natural history museums underwent a major institutional reorganization. Often referred to as the "new museum movement," its early articulation is usually credited to the Keeper of Zoology at the British Museum in London, John Edward Gray. In an influential essay from 1864, Gray distinguished between what he saw as the two major functions of modern museums. First, they were responsible for "the diffusion of instruction and rational amusement among the mass of people." At the same time, they were also dedicated to scientific research. Gray charged that museums had failed to distinguish between these two missions, with the consequence that they did not do a very good job of accomplishing either. Chief among the many problems that Gray identified was the tendency to cram every specimen in the museum's collection into its exhibition halls, with predictable results. Whereas "the general visitor perceives little else than a chaos of specimens," learned naturalists found it difficult to gain much useful knowledge or insight from objects that had been mounted to attract the public's attention. For this reason, Gray counseled, museums ought to divide their collections in two: a study collection for the use of learned naturalists and an exhibition collection for public consumption (Gray 1864).

By the time wealthy capitalists in America endowed philanthropic institutions like the American Museum in New York or the Field Museum in Chicago, the new museum movement had gone mainstream. In addition, by the turn of the twentieth century, cutting-edge botanical gardens also understood themselves as having a complex, hybrid mission that combined public education with popular amusement and scientific research. In official reports, these three motives were often described as mutually complementing and reinforcing each other. However, from time to time tensions did come to the fore. For example, in 1907 the anthropologist George A. Dorsey excoriated the American Museum's ethnographic exhibit for abdicating its educational responsibilities by catering too much to popular tastes. In a lengthy response, the museum's curator of anthropology, Franz Boas, argued that education and entertainment ought to go hand in hand. As the previous generation of American museum proprietors, most notoriously P.T. Barnum among them, well understood, ordinary New Yorkers had to be coaxed into attending an educational exhibit. Harkening back to the well-worn notion of using sugar to induce a child to consume its wholesome bread, Boas pointed out that "people who seek rest and recreation resent an attempt at systematic instruction while they are looking for some emotional excitement." For this reason, he argued, if they were to instruct, museums must, "first of all, be entertaining" (Boas 1907).

References

Alberti, Samuel. 2009. *Nature and Culture: Objects, Disciplines and the Manchester Museum.* Manchester: Manchester University Press.

Barrera, Antonio. 2002. "Local herbs, global medicines: Commerce, knowledge, and commodities in Spanish America." In *Merchants & Marvels: Commerce, Science, and Art in Early Modern Europe,* edited by Pamela H. Smith and Paula Findlen, 162–81. New York: Routledge.

Boas, Franz. 1907. "Some principles of museum administration." *Science,* New Series, 25, No. 650: 921–33.

Bourdieu, Pierre. 1986. *Distinction: A Social Critique of the Judgment of Taste.* London: Routledge & Kegan Paul.

Brockway, Lucile. 1979. *Science and Colonial Expansion: The Role of the British Royal Botanic Gardens.* New York: Academic Press.

Carnegie, Andrew. 1900. *The Gospel of Wealth, and Other Timely Essays.* New York: Century.

Conlin, Jonathan (ed.) 2013. *The Pleasure Garden: From Vauxhall to Coney Island.* Philadelphia: University of Pennsylvania Press.

Cook, Harold John. 2007. *Matters of Exchange: Commerce, Medicine, and Science in the Dutch Golden Age.* New Haven, CT: Yale University Press.

Cook, James. 2001. *The Arts of Deception: Playing with Fraud in the Age of Barnum.* Cambridge, MA: Harvard University Press.

Cunningham, Andrew. 1996. "The culture of gardens." In *Cultures of Natural History,* edited by Nicholas Jardine, James A. Secord, and E. C. Spary, 38–56. Cambridge: Cambridge University Press.

Daston, Lorraine, and Katharine Park. 1998. *Wonders and the Order of Nature, 1150–1750.* Cambridge: Zone Books.

De Beer, Gavin. 1953. *Sir Hans Sloane and the British Museum.* London: Oxford University Press.

Drayton, Richard. 2000. *Nature's Government: Science, Imperial Britain, and the "Improvement" of the World.* New Haven, CT: Yale University Press.

Findlen, Paula. 1994. *Possessing Nature: Museums, Collecting, and Scientific Culture in Early Modern Italy.* Berkeley: University of California Press.

Findlen, Paula. 2006. "Anatomy theaters, botanical gardens, and natural history collections." In *Early Modern Science,* edited by Katharine Park and Lorraine Daston, *272–89.* Cambridge: Cambridge University Press.

Gascoigne, John. 1998. *Science in the Service of Empire: Joseph Banks, the British State and the Uses of Science in the Age of Revolution.* Cambridge: Cambridge University Press.

Gray, James Edward. 1864. "On museums, their use and improvement, and on the acclimatization of animals." *Annals and Magazine of Natural History,* 14: 283–97.

Greenhalgh, Paul. 2011. *Fair World: A History of World's Fairs and Expositions, from London to Shanghai, 1851-2010.* Winterbourne: Papadakis.

Latour, Bruno. 1987. *Science in Action.* Cambridge, MA: Harvard University Press.

Nyhart, Lynn. 2009. *Modern Nature.* Chicago: University of Chicago Press.

Marx, Karl, and Friedrich Engels. 1848. *Manifest Der Kommunistischen Partei.* London: Gedruckt in der Office der "Bildungs-Gesellschaft für Arbeiter" von J.E. Burghard.

Meadow, Mark. 2001. "Merchants and marvels: Hans Jacob Fugger and the origins of the *Wunderkammer.*" In *Merchants and Marvels,* edited by Pamela Smith and Paula Findlen, 182–200. London: Routledge.

Ogilvie, Brian W. 2006. *The Science of Describing: Natural History in Renaissance Europe.* Chicago: University of Chicago Press.

Raj, Kapil. 2007. *Relocating Modern Science: Circulation and the Construction of Knowledge in South Asia and Europe, 1650–1900.* Basingstoke: Palgrave Macmillan.

Schiebinger, Londa. 2004. *Plants and Empire: Colonial Bioprospecting in the Atlantic World.* Cambridge, MA: Harvard University Press.

Secord, James A. 2000. *Victorian Sensation: The Extraordinary Publication, Reception, and Secret Authorship of* Vestiges of the Natural History of Creation. Chicago: University of Chicago Press.

Sloane, Hans. 1725. *A Voyage to the Islands Madera, Barbados, Nieves, S. Christophers and Jamaica.* London: Printed by B. M. for the author.

Smith, Adam. 1776. *An Inquiry into the Nature and Causes of the Wealth of Nations.* London: W. Strahan and T. Cadell.

Spary, Emma. 2000. *Utopia's Garden: French Natural History from Old Regime to Revolution.* Chicago: University of Chicago Press.

Twain, Mark and Charles Dudley Warner. 1873. *The Gilded Age: A Tale of To-Day.* Hartford, CT: American Pub. Co.

Whitaker, Katie. 1996. "The culture of curiosity." In *Cultures of Natural History,* edited by Nicholas Jardine, James A. Secord, and E. C. Spary, 75–90. Cambridge: Cambridge University Press.

CHAPTER EIGHTEEN

Domestic Space

DONALD L. OPITZ

Among a rich variety of sites where science was practiced, residences, or research facilities coextensive with residential spaces, have been among the most prevalent and significant. After a period of neglect in the historiography, domestic space—especially its private and public character and entanglement with gender—has attracted historical scrutiny, particularly with the ascendancy of approaches informed by sociology and geography. This chapter reviews the developments in the historiography of science concerned with domestic space, highlights the most important issues and themes, and identifies new directions for further research.

Situating the Domestic Production of Scientific Knowledge

Encouraged by the rise of the sociology of scientific knowledge (SSK) during the 1980s, scholars began elevating the importance of private and public spaces, and their interrelation, within the history of science. Adherents to SSK emphasized the situatedness of knowledge and its transformation from private experiences into public matters, or as Steven Shapin (1995, 305) explained, SSK illuminated "the textual and informal means by which scientists labor to persuade others, to extend experience from private to public domains." In this vein, Shapin and Simon Schaffer argued in their highly influential *Leviathan and the Air Pump* (1985) that the private pneumatic trials made by Irish chemist Robert Boyle and his assistants in England during the late sixteenth century became credible knowledge through techniques—material, literary, and social—of persuading public audiences. Building upon SSK's emphases, historical geographers advanced a literature concerned with private and public "geographies of science," offering rich, empirical analyses of the practices and performances of science as "spatially distributed" (Smith and Agar 1998; Livingstone 2003; Livingstone and Withers 2011). Meanwhile, a growing body of studies concerned with popularizations of science, popular science, vernacular science, and variations on these themes, added further perspectives on science as "public" knowledge, permeating culture both in and

A Companion to the History of Science, First Edition. Edited by Bernard Lightman.
© 2016 John Wiley & Sons Ltd. Published 2020 by John Wiley & Sons Ltd.

out of domestic doors (Cooter and Pumphrey 1991; Pandora 2001; Lightman 2007, 1–38).

Decidedly, the landmark study focusing on domestic spaces—and, particularly, the domestic threshold as a regulator of private and public access to knowledge—is Shapin's (1988) examination of the usage of gentlemen's residences for experimental research, especially by Boyle and other members of the young Royal Society of London. Shapin argued that among a range of venues in seventeenth-century England, including the shops of apothecaries and instrument makers, coffeehouses, royal palaces, and college rooms, private residences of gentlemen were "by far the most significant," with the "overwhelming majority of experimental trials, displays, and discussions that we know about" having occurred within them (378). Despite others' recognition of the wider applicability of this assessment well beyond this context, Alix Cooper (2006) noted in her survey of scientific homes and households in the early modern period, "Few historians of science have paid attention to these kinds of 'private' spaces" (224). A recent upsurge in studies focusing on families, domestic contexts and domesticity has to some degree ameliorated this neglect (Coen 2007; Winterburn 2011; Terrall 2014; Opitz, Bergwik, and van Tiggelen 2015).

Historians have noted how divisions between public and private knowledge-making often distinguished ideological forms of knowledge, with public knowledge characterized as open, communal, and useful, versus private as secretive, insular, and obscure. This opposition is canonical in the literature on the scientific revolution, illustrated, for example, in analyses of the founding of the Royal Society of London, an embodiment of efforts to make science more public against a backdrop of keeping "secrets of nature" (Eamon 1994, 319–50). Natural philosophers' ideological placement of scientific practice within recessed workshops, like alchemists' cellars, or, alternatively, within publicly accessible rooms thus influenced the designs of early modern residential research sites (Hannaway 1986). In advocating for state-sponsored endowments for scientific research, spokespersons made related distinctions, for example between public "applied" research, in service of the common good, and private "fundamental" research, in service of creativity and discovery, unfettered from utilitarian ends (Alter 1987, 214–45). Public and private mappings of domestic science have thus carried distinctive moral judgments that have shifted from context to context. Captured within popular images, stereotypes have ranged from the early nineteenth-century Gothic portrayal of Victor Frankenstein's madness, fermented privately "in a solitary chamber, or rather cell, at the top of the house" where he kept his "workshop of filthy creation" (Shelley 1992, 38), to the late twentieth-century American crowdsourcing experience, engaging a citizenry of amateur "garage scientists" in collective, altruistic pursuits of solutions to synthetic biological problems (Hitt 2012).

Gender and Domestic Productions

Value-laden conceptualizations of domestic research have been strongly shaped by gender conventions. According to a rich cultural history literature, the modern demarcation of things private from things public emerged with the post-Enlightenment privatization of domestic life apart from politics, civics and commercial affairs (Perrot 1990). Within a range of European contexts, this cultural development was entangled with the rise of the middle classes (Habermas 1962). Historians of science,

particularly those focusing on women's roles, have analyzed the bearing of this separation of spheres on the spatial and social redistribution of scientific production between
private households and public institutions, even as they questioned the degree of separation between these spheres and their stability as analytical categories.

Representative of this body of scholarship, Londa Schiebinger (1989) contrasted
the status of women in the professional sciences of the late nineteenth century with
their status in early modern European fields closely aligned with private craft traditions.
In those traditions, homespun industry provided a structure for collaborative scientific
work among families, especially in observational fields like entomology and astronomy. Amid the proliferation of academic laboratories by the early twentieth century,
and given the tenacious exclusion of women from those new spaces, homes increasingly became retreats where men of science still carried out private researches and yet
received the "hidden" assistance of wives and other family members. Deborah Harkness (1997) illuminated a stark example of wifely assistance in "managing an experimental household" within John Dee's sixteenth-century home alchemical and medical
studies. Debra Lindsay (1998) added more recent, American cases, but the broader
pattern could include devoted mothers and sisters (Hunter and Hutton 1997). Antic-
ipating Schiebinger's argument, a pioneering collection of studies examined within
various contexts the entanglement of women's scientific pursuits with their "intimate
lives" and, during the progressive professionalization of the sciences, the sidelining
of domestic realms and, concomitantly, those women amateurs who operated within
them (Abir-Am and Outram 1987, 1–16). Others, however, have challenged the presumed hidden and private character of women's domestic scientific pursuits, citing
cases of European aristocratic women who ran salons or country estates connected
with civic affairs. To this end, Elena Serrano (2012) detailed the civic natural history
pursuits of the Duchess of Osuna at her estate, *El Carpicho*, in late eighteen-century
southern Spain.

Although usually barred from the professions, women with scientific interests nevertheless often took advantage of the resources and opportunities available at home,
resulting, for some, in exceptional achievement and recognition. Arguably, collaboration with husbands and other family members proved the most common route for
women to participate in private research, and as Cooper (2006, 225) noted, much
of what is known about domestic-based science comes from the robust literature
attending to the familial context of women's contributions (Pycior, Slack, and Abir-
Am 1996). More recent studies have emphasized the complementary value of wives'
scientific illustrations as well as the variability of men's and women's roles within familial collaborations (Sheffield 2006; Lykknes, Opitz, and van Tiggelen 2012). Nevertheless, women's independent home-based research also thrived. Experimenting on
surface films during the decades around 1900 in her kitchen, German spinster Agnes
Pockels realized the possibilities yet experienced the constraints of working at home.
Despite conveniences like experimental space, household appliances, and family assistants, the burdensome demands of domestic responsibilities—in Pockels's case, elder
care—more often than not precluded her and other women's full-time devotion to
long-term research confined to domestic spaces (Giles and Forrester 1971).

Whether conducted independently or collaboratively, women's domestic work typically received little or contingent public recognition. The astronomer Lady Margaret
Huggins received a Civil List Pension from the British government in 1910 for her

"services to Science in collaboration with her husband" (Sir William) at their private observatory at Tulse Hill, south of London. Intended to help sustain her as a widow, the grant given so late in life originally belonged to her husband, though his achievements relied upon her assistance for 35 years (Becker 2011, 204). The more typical experience was that of Huggins's contemporary, English mathematician Grace Chisholm Young, who helped her husband William privately without due professional recognition, despite his frank acknowledgment that she deserved joint authorship on his publications (Jones 2009, 109–10). The reigning post-Enlightenment paradigm defined professional, public institutions (and paid positions within them) as masculine preserves; and, consistent with a range of cultural mandates—especially ones based in Christian teachings—the domestic, private sphere, inclusive of home-based occupations and family life, continued to be cast as a feminized domain (Davidoff and Hall 1987). Within this paradigm, women's domestic work was valued for being complementary to, and supportive of, men's paid, professional work. In the twentieth century, this gendered division of labor became institutionalized through a variety of means, such as American universities' anti-nepotism rules barring credentialed women from holding positions in the same scientific departments or even institutions as their husbands (Rossiter 1982, 194–7). Given the gender prescriptions associated with the separation of spheres, women's domestic research typically failed to achieve a standing on par with men's laboratory work—even that performed in private settings (see Rossiter 1993; Jones 2009, 117–42).

During the women's movements of the late nineteenth century, conservative reformers strategically drew upon popular associations between femininity and domesticity to elevate "domestic science" (or, alternatively, "home economics") as an academic discipline, thereby creating a significant "niche" for women to publicly pursue the sciences within acceptable social conventions (Rossiter 1982; Leavitt 2002). In parallel, scientific and technological innovations reshaped the practice and standards of housekeeping, impacting familial relations and gendered divisions of labor. In Ruth Schwartz Cowan's (1983) succinct summary of the overarching pattern, new domestic technologies enabled greater efficiencies that, in turn, fueled higher standards of cleanliness that, ironically, created "more work for mother." As scholars reexamined the historical relationship between domestic science and women's advancement, the quotidian familial experience of science and technology also attracted renewed scholarly interest, with studies on East Asian households among the field's path-breakers (Stage and Vincenti 1997; Bray 2008).

Domesticity, Social Class, and Professionalization

Despite the gendered undervaluation of the domestic sphere, scientists' efforts to advance new fields through the voluntary assistance of amateurs nevertheless could elevate the importance of domestic productions, whether performed by male or female practitioners. So, British geneticist William Bateson assembled a volunteer corps of private breeders to help amass observations for his studies on variation and inheritance in the decades around 1900 (Richmond 2006; Opitz 2011). This form of distributed, yet coordinated, research reinvented gentlemen-amateur practices, especially that of Charles Darwin—for Bateson, the archetypal field naturalist—who worked in the seclusion of his country retreat, Down House, while engaging in a wide correspondence circle (de Chadarevian 1996; Browne 2002). During the gradual professionalization of the sciences in the nineteenth and twentieth centuries, such networking among amateurs situated at distant domestic sites constituted a fairly commonplace enterprise supported by an infrastructure of local scientific societies, field clubs, and museums. As Lynn Nyhart (1996) pointed out in the case of Germany, the rise of the "new" biology alongside natural history at the end of the nineteenth century involved an expansion of research sites, with new academic laboratories coexisting (and often competing) with older museums where taxonomic studies continued to flourish, in part owing to the abundance of field-based collections assembled by amateurs working from home. The theme of cooperation between amateur and professional naturalists based in a diversity of institutions recurs in the historiography touching on France (Drouin and Bensaude-Vincent 1996), England (Alberti 2001), and the United States (Kohlstedt 1976).

In addition to professional norms, social class intermingled with domesticity in ways that differentiated forms of home-based research according to class membership. Across the centuries, the European aristocratic intelligentsia—exemplified by scientific personages like the Dane Tycho Brahe, Irishman Robert Boyle, and Frenchman René Réaumur—conferred gentlemanly honor on the scientific fields to which they made contributions based on their researches conducted in domestic or quasi-domestic spaces (Thoren 1990; Shapin 1994; Terrall 2014). Interwoven with intellectual merit, gentlemanly savants won cultural prestige through class position and social connections, as well as skillful execution of polite manners (Nye 1997). Gentlemen's country and town houses served as venues for both private research and intellectual interchanges. Annual local meetings of the national scientific societies often included visits at members' grand estates. When the British Association for the Advancement of Science met in Dublin in 1857, the Earl of Rosse received members on an excursion to his Birr Castle in Parsonstown. Guests enjoyed both the astronomical wonders of Rosse's celebrated "Leviathan" telescope as well as the gastronomical delights of a dinner banquet. Inspired by his visit, amateur American astronomer Henry Draper returned home to New York to establish a private observatory on his family's estate, where he received his father's assistance and, upon marriage, that of his wife Mary Palmer (Chapman 1998, 98–9; Jones and Boyd 1971, 211–45). Journalists enamored by the pomp and circumstance of organized country-house excursions often rendered them as curiosities, but as Victoria Carroll (2004) underlined in her analysis of Walton Hall, in Warwickshire, England, replete with the taxidermic specimens of English naturalist Charles Waterton, public visiting at scientific estates constituted a fairly common cultural practice.

Despite the middle-class emphasis on separate spheres, among these grand domestic spaces, private and public, family and society, intermingled in ways that escaped a clear separation. M. Jeanne Peterson (1989) poignantly demonstrated this in her detailing of the familial life of the scientific Paget clan, among whom, Peterson argued, gifted gentlewomen who engaged in scientific pursuits enjoyed social and familial prestige as opposed to "professional" notoriety—"professional" a term that pejoratively implied remunerative service unbecoming of elites. Even so, the genteel domesticity and respectability of private estates often permeated into public, academic spaces. Familial assemblages adapted domestic rituals like the afternoon teas enjoyed by laboratory staff, guests, and wives at Cambridge's Cavendish Laboratory, a tradition inaugurated during John William Strutt, third Baron Rayleigh's directorship, continued by his successor, J. J. Thomson, and replicated in other laboratories (Gould 1997, 144–5). Traveling scientific families in the field-based sciences similarly reproduced domestic elements at camps and field stations, ameliorating otherwise arduous, if not highly masculinized, settings (Pang 1996).

Such forums as the tea and meal breaks in laboratories and fieldwork camps sustained domestic traditions of intellectual sociability, drawing upon such models as the elite Parisian salons of the seventeenth century and the middle-class English conversaziones of the eighteenth and nineteenth centuries. Such intimate gatherings promoted discussions traversing an array of philosophical and political topics, yet they also purposefully included scientific investigations, notably in the area of psychical research amid the nineteenth-century séance craze. The notorious coterie, "the Souls," of Scottish science philosopher Arthur James Balfour illustrated well the eccentric dynamics of country-house soirees. Although supplanted by more semi-public and public venues, middle-class homes staged similar gatherings in designated conversazione rooms, a noted example being that of Scottish surgeon John Hunter. Arguably, the attenuation of these practices among research facilities and field sites created a domesticated form of "shop talk" (Lambert 1984; Terrall 1995; Alberti 2003; Secord 2007).

The reproduction of domestic traditions in public research spaces also adhered to gendered prescriptions. Shapin (1988, 393) argued how gentlemanly codes guarding privacy in domestic realms extended to the proceedings within public scientific meeting places, such as the rooms of Gresham College where the young Royal Society met. As Simon Schaffer (1998) showed, among a range of country houses in Victorian Britain, spaces allocated for gentlemen's experiments in the physical sciences were designated masculine domains, often dangerous, yet disciplined, and set in isolation from the rest of the household. The hierarchical roles of lord and servant, with their attendant codes of deference, translated directly to analogous roles associated with the management of laboratory technicians. In cases like Rayleigh's where manor lord became university professor, gentlemanly authority and deference extended to academic spaces. Christian mandates, particularly strong in the wake of the Evangelical Revival, infused domestic industry with such qualities as piety, asceticism, and paterfamilias duty (Shapin 1994, 126–92). According to Arthur Balfour, his home constituted "a Temple of Research," invoking a traditional association of the learned gentleman with Christian moral uprightness (Opitz 2006). Paul White's (2003) analysis of paleontologist Thomas Huxley's career development in science details the status of companionate marriage within constructions of the "man of science" identity in Victorian England.

Complementing this trajectory of scholarship, several scholars have analyzed the gendered socialization of scientists into their career identities and adoption of research practices. In her pioneering anthropological study, Sharon Traweek (1988, 145–52) argued that the organization of labor among particle physics research groups in Japan, such as the group at Kō-enerugī butsurigaku kenkyūjo (KEK) accelerator in Tsukuba, was consistent with the cultural model of the *ie* or household, including its tenet of *amae* or interdependence between familial generations. Building upon Traweek's approach, others have revealed the gendered interplay of family dynamics and scientists' mobility, as in Helena Pettersson's (2011) fine study of a group of transnational plant scientists based at a Swedish plant institute. Focusing on one family, Staffan

Bergwik (2014) showed how the presence of the Swedish chemist Svante Arrhenius's family, whether physically or representationally (through the showing of family photographs), during his travels in the early twentieth century replicated and reinforced a "gendered lifestyle" promoting his stability, credibility, and status while away from home.

Domesticity by Design

Newly built independent, academic, or governmental research institutions often included domestic spaces. Following the early nineteenth-century German prototype, living quarters sufficiently commodious to house directors' families appeared among the plans for chemical and physical laboratories (Nye 1996, 1–27). Numerous examples abound in the literature, notably the Physikalisch-Technische Reichsanstalt in Charlottenburg (Cahan 1989), the cluster of research institutes forming the Mediziner-Viertel in Vienna (Rentetzi 2005), and the Nobel Institute in Stockholm (Bergwik 2014). Architectural designs structured research and social interactions along gendered lines, and the provisions for male directors' wives to reside on-site allowed them to lend ready, complementary support to their husbands, whether as research assistants or social hostesses. Apart from this standard residential prototype, as Graeme Gooday (2008) highlighted, laboratory designs also often took their cues from specialized residential spaces like kitchens, with the University of Oxford's chemistry building, modeled after the kitchen of Glastonbury Abbey, serving as an exemplar (Forgan 1989). The domestic architectural imprint on research buildings could sometimes be wholesale. During the extended modern decline of the European landed aristocracy, architects repurposed vacant grand residences as dedicated research sites. Two noted examples from the early twentieth century are the installation of Britain's National Physical Laboratory within a royal palace, Bushy House, in London, and the John Innes Horticultural Institution, devoted to genetics research, within the benefactor's Manor House at Merton Park. These instances of modern research facilities housed within former residences add to further ones in which country manors hosted agricultural research, most notably the experiment station at Sir John Bennett Lawes's Rothamsted estate in Harpenden, England (Russell 1966).

The "new" nineteenth-century German style of laboratory design, incorporating both living and working spaces, was really not so new. As Paula Findlen (1994, 41) noted, gentlemen's possession of museums within their stately homes constituted a fairly common practice promoting upward social mobility in early modern Italy. Private collections could become public collections, as in the case of Ulisse Aldrovandi's *studio*, originally distributed among carefully gendered spaces within his family's Bologna palace, transformed into a public *galleria* at the government's Palazzo Publico (Findlen 1994, 17–31; Findlen 1999). Also in seventeenth-century Bologna, in human and comparative anatomy, household studios proliferated at a time when the university curriculum repressed public, practical demonstrations (Messbarger 2010, 63–6). Challenging the design of Tycho Brahe's Uraniborg complex, in which the subterranean location of alchemical experiments safeguarded their secrecy, the German chemist Andreas Libavius designed, though never built, a *domus chemiae* ("house of chemistry") that purported to embody an open, civic humanism, with family and alchemy living beneath the same roof, above ground (Hannaway 1986; Newman 1999). The construction of Paris's Muséum National d'Histoire Naturelle at the end of the eighteenth century implemented the juxtaposition of the director's private residence with indoor spaces for research and exhibits as well as outdoor gardens (Outram 1996). Alongside the teaching laboratory movements of the nineteenth century, as Graeme Gooday (2008, 789–91) noted, instances of home-based laboratories persisted across the disciplines.

Domestication through Popularization

Much of what is known about home-based research has been mediated through such familiar genres as scientists' life-and-letters and popular science books, and in light of this, the significance of the household for scientific practice must be interpreted with respect to authors' literary agendas, shaped as they are by social and religious interests. So, for example, although science doubtlessly occurred within, and around, artisans' homes, as Anne Secord (1994) argued, popular nineteenth-century English didactic texts, originating among the middle classes, prescribed the "homely virtue" of domestic natural history pursuits against the vices of sacrilegious and political (if not drunken) agitation among artisan botanists' Sunday meetings in public houses. A robust advice literature, fueled by Evangelical, temperance, and Sabbatical movements, popularized images of home-based science in service to moral ends, with titles circulating like James Cash's *Science in the Cottage* (1873). Pictorial scenes of familial gatherings around sites of invention and creativity abounded, a fine example being the frontispiece of select editions of Cecilia Brightwell's biographical *Heroes of the Laboratory and the Workshop* (1859) (Figure 18.1). Yet, as Secord stressed, the pervasiveness of these popularized representations served to domesticate scientific activity normally occurring in rather unseemly places, like pubs, *outside* of the virtuous confines of homes.

The household experience of post-Newtonian science was no more commonplace than within informal educational arenas, especially among sectors still awaiting the introduction of formalized school science instruction and yet benefiting from the rapid expansion of popular science publishing. A famous example of the budding "familiar format" genre is Jane Marcet's *Conversations on Chemistry* (1806), renowned for inspiring the young bookbinder, Michael Faraday, to engage in serious pursuit of the field (Fyfe 2004). Translations and expositions of Newtonian texts, especially those authored by women, greeted European reading markets in the eighteenth and nineteenth centuries (Myers 1989; Mullan 1993; Findlen 1995; Secord 1985). A parallel literature emerged for natural history, ranging from botany "for ladies" (Shteir 1996), to companions for amateur naturalists pursuing marine studies. These genres drew upon tropes of domesticity and emphasized the virtues of domestic-based learning, both inside and out-of-doors (Page and Smith 2011). As Sally Gregory Kohlstedt (1990) suggested, domestic parlors and home-schooling primers served as resources for scientific learning in private settings alongside the rise of public schools in the American republic. With the emergence of the Nature Study educational movement at the end of the nineteenth century, home gardens served a new didactic function intended to complement schools' hands-on science curriculum (Kohlstedt 2010). "Portable laboratories" and other scientific kits for youthful education and recreation also promoted science at home (Gee 1989). In efforts to "domesticate nature" in the

Figure 18.1 George Stephenson, drawn by John Absolon, appearing as the frontispiece to evangelical Christian writer Cecilia Lucy Brightwell's *Heroes of the Laboratory and the Workshop* (1859). Reproduced with permission of University of Florida, Smathers Library Special and Area Studies Collections, Baldwin Library of Historical Children's Literature.

Figure 18.2 Filippo Pelagio Palagi, *Newton scopre la teoria della rifrazione della luce* ("Newton discovers the theory of light refraction"), oil painting, 1827. By permission of Musei Civici d'Arte e Storia, Brescia.

new teaching laboratories, outdoor nature was converted into objects for controlled, indoor study through the use of microscopes, terraria, and aquaria (Gooday 1991). Given the emphasis on instilling familiarity with science among nineteenth-century middle-class households in Britain, Melanie Keene (2014) recently proposed "familiar science" as an apt category for guiding future historical research.

The familiarity of domestic spaces and objects has infused scientific discovery and its representation in certain canonical ways. Legendary, though certainly apocryphal, is Archimedes' "Eureka!" moment while stepping into a (public) bathtub, the water displaced by his body suggesting to him a method for determining the density (and authenticity) of Hiero's gold crown. Even so, the lure of bathwater, dishwater, and similar domestic sources of natural wonders strongly shaped the storytelling of scientific discovery as well as its artistic rendition. Schaffer (2004, 158–9) noted the role of soap bubbles in domesticating (and making commodities out of) scientific principles, exemplified by the early nineteenth-century painting, *Newton Discovers the Theory of Light Refraction*, by the Bolognese designer Filippo Pelagio Palagi, which portrays a carefree child creating bubbles at a sunlit window while Newton sits at his writing desk working at optics (Figure 18.2). Chemists noted how Pockels's observations of the behavior of surface films amid "greasy washing-up water" joined a longer tradition of kitchen-sink discoveries (Giles and Forrester 1971, 48). Such imagery showcased the mundaneness of science yet also calls to mind an earlier form of messy kitchen science, like anatomical dissection, summarized by Claude Bernard when likening the physiological laboratory to a "long and ghastly kitchen" (Bernard 1865, 28, as quoted in Latour 1992, 295; see also Bucchi 2013). Of course, other domestic sites, including bedrooms and boudoirs, also offered convenient places for experiments and demonstrations (Opitz 2006, 145; Keene 2014, 62–4). In the early modern period, the ubiquity of domestic experimental space inspired Sir Francis Bacon's utopian "Salomon's House," as well as Margaret Cavendish's plentiful domestic metaphors appearing throughout her philosophical writings (Shanahan 2002, 226–31).

Future Research

Although domestic spaces tend to be positioned within the historiography of science as antecedents to the new public and semi-public spaces that emerged during periods of professionalization—private residences gradually being replaced by dedicated research facilities in contexts like the academy—the dominant narrative arc has shifted from one of teleological progress to one of parallel developments: domestic sites continued to retain important, albeit changing, functions alongside those of newly emerging venues. A growing body of studies has also complicated the relationship between public and private spheres where science is practiced and performed, emphasizing a multidirectional transit of knowledge between sites and the (gendered) communities occupying them, with profound implications for the shaping of knowledge "beyond the academy" (von Oertzen, Rentetzi, and Watkins 2013). Studies are increasingly revealing the role of domesticity, constitutive of values, rituals, social forms, and architectural designs—shaped as they were by gender and class norms—on the making of scientific knowledge, within both private and public spheres. Domestic spaces—their character, layout, occupants, activities, roles, and relation to other sites—continue to beckon for further historical scrutiny. The field awaits new research on these spaces within geographies outside Europe and the United States, as well as the establishment of broader narratives on the significance of domesticity and domestic spaces for knowledge production within the history of science. New directions being forged by scholars include deeper analyses of scientific households, especially the interplay between family dynamics, gender, and scientists' careers—involving such elements as their identity formation, professional advancement, and geographical mobility.

References

Abir-Am, Pnina and Dorinda Outram (eds.) 1987. *Uneasy Careers and Intimate Lives: Women in Science, 1789–1979.* New Brunswick, NJ: Rutgers University Press.

Alberti, Samuel J.M.M. 2001. "Amateurs and professionals in one county: Biology and natural history in late Victorian Yorkshire." *Journal of the History of Biology,* 34: 115–47.

Alberti, Samuel. 2003. "Conversaziones and the experience of science in Victorian England." *Journal of Victorian Culture,* 8: 208–30.

Alter, Peter. 1987. *The Reluctant Patron: Science and the State in Britain, 1850–1920.* Translated by Angela Davies. Revised edition. Oxford: Berg.

Becker, Barbara J. 2011. *Unravelling Starlight: William and Margaret Huggins and the Rise of the New Astronomy.* Cambridge: Cambridge University Press.

Bergwik, Staffan. 2014. "An assemblage of science and home: The gendered lifestyle of Svante Arrhenius and early twentieth-century physical chemistry." *Isis*, 105: 265—91.

Bernard, Claude. 1865. *Introduction a l'étude de la médecine expérimentale*. Paris: Baillière.

Bray, Francesca (ed.) 2008. *Constructing Intimacy: Technology, Family and Gender in East Asia*. Special issue of *East Asian Science, Technology and Society*, 2, No. 2.

Brightwell, C.L. 1859. *Heroes of the Laboratory and the Workshop*. London: Routledge, Warnes & Routledge.

Browne, Janet. 2002. *Charles Darwin: The Power of Place*. New York: Alfred Knopf.

Bucchi, Massimiano. 2013. *Il pollo di Newton: La sienza in cucina*. Parma: U. Guanda.

Cahan, David. 1989. *An Institute for an Empire: The Physikalisch-Technische Reichsanstalt, 1871–1918*. Cambridge: Cambridge University Press.

Carroll, Victoria. 2004. "The natural history of visiting: Responses to Charles Waterton and Walton Hall." *Studies in History and Philosophy of Biological and Biomedical Sciences*, 35: 31–64.

Cash, James. 1873. *Where There's a Will, There's a Way! Or, Science in the Cottage: An Account of the Labours of Naturalists in Humble Life*. London: Robert Hardwicke.

Chapman, Allan. 1998. *The Victorian Amateur Astronomer: Independent Astronomical Research in Britain, 1820–1920*. New York: John Wiley & Sons.

Coen, Deborah R. 2007. *Vienna in the Age of Uncertainty: Science, Liberalism, and Private Life*. Chicago: University of Chicago Press.

Cooper, Alix. 2006. "Homes and households." In *The Cambridge History of Science: III. Early Modern Science*, edited by Katharine Park and Lorraine Daston, 224–37. Cambridge: Cambridge University Press.

Cooter, Roger and Stephen Pumphrey. 1991. "Separate spheres and public places: Reflections on the history of science popularization and science in public culture." *History of Science*, 32: 237–67.

Cowan, Ruth Schwartz. 1983. *More Work for Mother: The Ironies of Household Technology from the Open Hearth to the Microwave*. New York: Basic Books.

Davidoff, Leonore and Catherine Hall. 1987. *Family Fortunes: Men and Women of the English Middle Class, 1780–1850*. Chicago: University of Chicago Press.

de Chadarevian, Soraya. 1996. "Laboratory science versus country-house experiments: The controversy between Julius Sachs and Charles Darwin." *British Journal for the History of Science*, 29: 17–41.

Drouin, Jean-Marc and Bernadette Bensaude-Vincent. 1996. "Nature for the people." In *Cultures of Natural History*, edited by N. Jardine, J.A. Secord, and E.C. Spary, 408–25. Cambridge: Cambridge University Press.

Eamon, William. 1994. *Science and the Secrets of Nature: Books of Secrets in Medieval and Early Modern Culture*. Princeton, NJ: Princeton University Press.

Findlen, Paula. 1994. *Possessing Nature: Museums, Collecting, and Scientific Culture in Early Modern Italy*. Berkeley: University of California Press.

Findlen, Paula. 1995. "Translating the New Science: Women and the circulation of knowledge in Enlightenment Italy." *Configurations*, 3: 167–206.

Findlen, Paula. 1999. "Masculine prerogatives: Gender, space, and knowledge in the early modern Museum." In *The Architecture of Science*, edited by Peter Galison and Emily Thompson, 29–57. Cambridge, MA: MIT Press.

Forgan, Sophie. 1989. "The architecture of science and the idea of a university." *Studies in History and Philosophy of Science*, 20: 405–34.

Fyfe, Aileen. 2004. "Introduction." In [Jane Marcet], *Conversations on Chemistry, in which the Elements of that Science are Familiarly Explained and Illustrated by Experiments*, I: xxi–xxvii. Bristol: Thoemmes Continuum.

Gee, Brian. 1989. "Amusement chests and portable laboratories: Practical alternatives to the regular laboratory." In *The Development of the Laboratory: Essays on the Place of Experiment in Industrial Civilization*, edited by Frank A.J.L. James, 37–60. New York: American Institute of Physics.

Giles, C.H. and S.D. Forrester. 1971. "The origins of the surface film balance: Studies in the early history of surface chemistry, Part 3." *Chemistry and Industry* (9 January): 43–53.

Gooday, Graeme. 1991. "'Nature' in the laboratory: Domestication and discipline with the microscope in Victorian life science." *British Journal for the History of Science*, 24: 307–41.

Gooday, Graeme. 2008. "Placing or replacing the laboratory in the history of science?" *Isis*, 99: 783–95.

Gould, Paula. 1997. "Women and the culture of university physics in late nineteenth century Cambridge." *British Journal for the History of Science*, 30: 127–49.

Habermas, Jürgen. 1962. *Strukturwandel der Öffentlichkeit: Untersuchungen zu einer Kategorie der bürgerlichen Gesellschaft*. Neuwied am Rhein: Luchterhand.

Hannaway, Owen. 1986. "Laboratory design and the aim of science: Andreas Libavius and Tycho Brahe." *Isis*, 77: 585–610.

Harkness, Deborah E. 1997. "Managing an experimental household: The Dees of Mortlake and the practice of natural philosophy." *Isis*, 88: 247–62.

Hitt, Jack. 2012. *Bunch of Amateurs: A Search for the American Character*. New York: Crown.

Hunter, Lynette and Sarah Hutton (eds.) 1997. *Women, Science, and Medicine, 1500–1700: Mothers and Sisters of the Royal Society*. Stroud: Sutton.

Jones, Bessie Zaban and Lyle Gifford Boyd. 1971. *The Harvard College Observatory: The First Four Directorships, 1839–1919*. Cambridge, MA: Belknap Press.

Jones, Claire G. 2009. *Femininity, Mathematics and Science, 1880–1914*. Basingstoke: Palgrave Macmillan.

Keene, Melanie. 2014. "Familiar science in nineteenth-century Britain." *History of Science* 52: 53–71.

Kohlstedt, Sally Gregory. 1976. "The nineteenth-century amateur tradition: The case of the Boston Society of Natural History." In *Science and its Public: The Changing Relationship*, edited by Gerald Holton and William A. Blanpied, 173–90. Dordrecht: D. Reidel.

Kohlstedt, Sally Gregory. 1990. "Parlors, primers, and public schooling: Education for science in nineteenth-century America." *Isis*, 81: 425–45.

Kohlstedt, Sally Gregory. 2010. *Teaching Children Science: Hands-On Nature Study in North America, 1890–1930*. Chicago: University of Chicago Press.

Lambert, Angela. 1984. *Unquiet Souls: A Social History of the Illustrious, Irreverent, Intimate Group of British Aristocrats Known as "The Souls."* New York: Harper & Row.

Latour, Bruno. 1992. "The costly ghastly kitchen." In *The Laboratory Revolution in Medicine*, edited by Andrew Cunningham and Perry Williams, 295–303. Cambridge: Cambridge University Press.

Leavitt, Sarah A. 2002. *From Catharine Beecher to Martha Stewart: A Cultural History of Domestic Advice*. Chapel Hill: University of North Carolina Press.

Lightman, Bernard. 2007. *Victorian Popularizers of Science: Designing Nature for New Audiences*. Chicago: University of Chicago Press.

Lindsay, Debra. 1998. "Intimate inmates: Wives, households, and science in nineteenth-century America." *Isis*, 89: 631–52.

Livingstone, David N. 2003. *Putting Science in its Place: Geographies of Scientific Knowledge*. Chicago: University of Chicago Press.

Livingstone, David N., and Charles W.J. Withers (eds.) 2011. *Geographies of Nineteenth-Century Science*. Chicago: University of Chicago Press.

Lykknes, Annette, Donald L. Opitz, and Brigitte van Tiggelen (eds.) 2012. *For Better or for Worse? Collaborative Couples in the Sciences. Science Networks: Historical Studies.* Basel: Birkhäuser.

Messbarger, Rebecca. 2010. *The Lady Anatomist: The Life and Work of Anna Morandi Manzolini.* Chicago: University of Chicago Press.

Mullan, John. 1993. "Gendered knowledge, gendered minds: Women and Newtonianism, 1690–1760." In *A Question of Identity: Women, Science, and Literature*, edited by Marina Benjamin, 41–56. New Brunswick, NJ: Rutgers University Press.

Myers, Greg. 1989. "Science for women and children: The dialogue of popular science in the nineteenth century." In *Nature Transfigured: Science and Literature, 1700–1900*, edited by John Christie and Sally Shuttleworth, 171–200. Manchester: Manchester University Press.

Newman, William R. 1999. "Alchemical symbolism and concealment: The chemical house of Libavius." In *The Architecture of Science*, edited by Peter Galison and Emily Thompson, 59–77. Cambridge, MA: MIT Press.

Nye, Mary Jo. 1996. *Before Big Science: The Pursuit of Modern Chemistry and Physics, 1800–1940.* London: Prentice Hall.

Nye, Mary Jo. 1997. "Aristocratic culture and the pursuit of science: The de Broglies in modern France." *Isis*, 88: 397–421.

Nyhart, Lynn K. 1996. "Natural history and the 'new' biology." In *Cultures of Natural History*, edited by N. Jardine, J.A. Secord, and E.C. Spary, 426–43. Cambridge: Cambridge University Press.

Opitz, Donald L. 2006. "'This house is a temple of research': Country-house centres for late-Victorian Science." In *Repositioning Victorian Sciences: Shifting Centres in Nineteenth-Century Scientific Thinking*, edited by David Clifford, Elisabeth Wadge, Alex Warwick, and Martin Willis, 143–53. London: Anthem Press.

Opitz, Donald L. 2011. "Cultivating genetics in the country: Whittingehame Lodge, Cambridge." In *Geographies of Nineteenth-Century Science*, edited by David N. Livingstone and Charles W.J. Withers, 73–98. Chicago: University of Chicago Press.

Opitz, Donald L., Staffan Bergwik, and Brigitte van Tiggelen (eds.) 2015. *Domesticity in the Making of Modern Science.* Basingstoke: Palgrave Macmillan.

Outram, Dorinda. 1996. "New spaces in natural history." In *Cultures of Natural History*, edited by N. Jardine, J.A. Secord, and E.C. Spary, 249–65. Cambridge: Cambridge University Press.

Page, Judith W. and Elise L. Smith. 2011. *Women, Literature, and the Domesticated Landscape: England's Disciples of Flora, 1780–1870.* Cambridge: Cambridge University Press.

Pandora, Katherine. 2001. "Knowledge held in common: Tales of Luther Burbank and science in the American vernacular." *Isis*, 92: 484–516.

Pang, Alex Soojung-Kim. 1996. "Gender, culture, and astrophysical fieldwork: Elizabeth Campbell and the Lick Observatory–Crocker eclipse expeditions." *Osiris*, 11: 17–43.

Perrot, Michelle (ed.) 1990. *A History of Private Life: IV. From the Fires of Revolution to the Great War.* Cambridge, MA: Belknap Press.

Peterson, M. Jeanne. 1989. *Family, Love, and Work in the Lives of Victorian Gentlewomen.* Bloomington: Indiana University Press.

Pettersson, Helena. 2011. "Gender and transnational plant scientists: Negotiating academic mobility, career commitments and private life," *Gender*, 1: 99–116.

Pycior, Helena M., Nancy G. Slack, and Pnina Abir-Am (eds.) 1996. *Creative Couples in the Sciences.* New Brunswick, NJ: Rutgers University Press.

Rentetzi, Maria. 2005. "Designing (for) a new scientific discipline: The location and architecture of the Institut für Radiumforschung in early twentieth-century Vienna." *British Journal for the History of Science*, 38: 275–306.

Richmond, Marsha L. 2006. "The 'domestication' of heredity: The familial organization of geneticists at Cambridge, 1895–1910." *Journal of the History of Biology*, 39: 565–605.

Rossiter, Margaret. 1982. *Women Scientists in America: Struggles and Strategies to 1940*. Baltimore: Johns Hopkins University Press.

Rossiter, Margaret. 1993. "The ~~Matthew~~ Matilda effect in science." *Social Studies of Science*, 23: 325–41.

Russell, Sir E. John. 1966. *A History of Agricultural Science in Great Britain, 1620–1954*. London: George Allen and Unwin.

Schaffer, Simon. 1998. "Physics laboratories and the Victorian country house." In *Making Space for Science: Territorial Themes in the Shaping of Knowledge*, edited by Crosbie Smith and Jon Agar, 149–80. Basingstoke: Macmillan.

Schaffer, Simon. 2004. "A science whose business is bursting: Soap bubbles as commodities in classical physics." In *Things that Talk: Object Lessons from Art and Science*, edited by Lorraine Daston, 147–92. New York: Zone Books.

Schiebinger, Londa. 1989. *The Mind Has No Sex? Women in the Origins of Modern Science*. Cambridge, MA: Harvard University Press.

Secord, Anne. 1994. "Science in the pub: Artisan botanists in early nineteenth-century Lancashire." *History of Science*, 32: 269–315.

Secord, James A. 1985. "Newton in the nursery: Tom Telescope and the philosophy of tops and balls, 1761–1838." *History of Science*, 23: 127–51.

Secord, James A. 2007. "How scientific conversation became shop talk." In *Science in the Marketplace: Nineteenth-Century Sites and Experiences*, edited by Aileen Fyfe and Bernard Lightman, 23–59. Chicago: University of Chicago Press.

Serrano, Elena. 2012. "Science for women in the Spanish enlightenment, *1753–1808*." Ph.D. thesis. Barcelona: Universitat Autònoma de Barcelona.

Shanahan, John. 2002. "The indecorous virtuoso: Margaret Cavendish's experimental spaces." *Genre*, 35: 221–52.

Shapin, Steven, and Simon Schaffer. 1985. *Leviathan and the Air-Pump: Hobbes, Boyle, and the Experimental Life*. Princeton, NJ: Princeton University Press.

Shapin, Steven. 1988. "The house of experiment in seventeenth-century England." *Isis*, 79: 373–408.

Shapin, Steven. 1994. *A Social History of Truth: Civility and Science in Seventeenth-Century England*. Chicago: University of Chicago Press.

Shapin, Steven. 1995. "Here and everywhere: Sociology of scientific knowledge." *Annual Review of Sociology*, 21: 289–321.

Sheffield, Suzanne Le-May. 2006. "Gendered collaborations: Marrying art and science." In *Figuring it Out: Science, Gender, and Visual Culture*, edited by Ann B. Shteir and Bernard Lightman, 240–64. Lebanon, NH: Dartmouth College Press.

Shelley, Mary. 1992. *Frankenstein; or, the Modern Prometheus*. London: Dent.

Shteir, Ann B. 1996. *Cultivating Women, Cultivating Science: Flora's Daughters and Botany in England, 1760–1860*. Baltimore: The Johns Hopkins University Press.

Smith, Crosbie and Jon Agar (eds.) 1998. *Making Space for Science: Territorial Themes in the Shaping of Knowledge*. Basingstoke: Macmillan.

Stage, Sarah and Virginia B. Vincenti (eds.) 1997. *Rethinking Home Economics: Women and the History of a Profession*. Ithaca, NY: Cornell University Press.

Terrall, Mary. 1995. "Gendered spaces, gendered audiences: Inside and outside the Paris Academy of the Sciences." *Configurations*, 5: 207–32.

Terrall, Mary. 2014. *Catching Nature in the Act: Réaumur and the Practice of Natural History in the Eighteenth Century*. Chicago: University of Chicago Press.

Thoren, Victor E. 1990. *The Lord of Uraniborg: A Biography of Tycho Brahe*. Cambridge: Cambridge University Press.

Traweek, Sharon. 1988. *Beamtimes and Lifetimes: The World of High Energy Physics.* Cambridge, MA: Harvard University Press.

Von Oertzen, Christine, Maria Rentetzi, and Elizabeth S. Watkins (eds.) 2013. *Beyond the Academy: Histories of Gender and Knowledge.* Special issue of *Centaurus*, 55, No. 2.

White, Paul. 2003. *Thomas Huxley: Making the "Man of Science."* Cambridge: Cambridge University Press.

Winterburn, Emily Jane. 2011. "The Herschels: A scientific family in training." Ph.D. thesis. London: Imperial College.

CHAPTER NINETEEN

Commercial Science

PAUL LUCIER

Twenty-five years ago, industrial research was an important topic in the history of American science and technology. Several detailed case studies, survey articles, and extended reviews discussed the complex evolution of laboratories within the corporate organizational structure and, equally important, the complicated nature of the scientific research done within those laboratories. This impressive output brought to light the inadequacy of longstanding sociological theories about the practice of science and of the psychological assumptions about the personae of the scientist, as well as the problems in prevailing models of the relations between science and technology. Such significant results seemed to bode well for continued historical research on the relations of science, technology, and industry (Wise 1985a; Dennis 1987; Smith, Jr. 1990; Hounshell 1996). But it was not to be; the strong start was not sustained. Perhaps it was a sense that there was nothing left of theoretical interest to say about industrial research or about laboratories in general, which along with other institutional histories have looked rather dull and outdated since the cultural turn in history (Kohlstedt 1985; Gooday 2008; Kohler 2008). Or perhaps the sudden freeze on industrial research was the unfortunate collateral damage of a noticeable chill in the post-Cold War relations between historians of science and historians of technology (Alexander 2012). The divisiveness of the disciplines left studies of industrial research in an abandoned no-scholar's-land. A 1996 review of the scholarship by the Harvard Business School was subtitled, aptly enough, "the end of an era" (Rosenbloom and Spencer 1996).

It may be too early to declare the start of a new era or a rapprochement between the two scholarly communities, but there is hope for future research on the relations among science, technology, and industry under the new rubric "commercial science." A recent generation of scholarship is distinguishing itself by taking different approaches to some of the enduring themes embedded within the previous studies. Early modern historians, for example, have undertaken extensive re-evaluations of the circulation and accumulation of materials, the operations and productions of

A Companion to the History of Science, First Edition. Edited by Bernard Lightman.
© 2016 John Wiley & Sons Ltd. Published 2020 by John Wiley & Sons Ltd.

chemical laboratories, and the global exchanges of facts and knowledge (Smith 1994; Smith and Findlen 2002; Cook 2007; Raj 2007; Roberts, Schaffer, and Dear 2007; Klein and Spary 2010). Historians of the nineteenth century, likewise, have begun to re-examine the role of government in creating new departments and bureaux that required scientific expertise and in reforming patent offices and patent laws that fostered science-based inventions and innovations (Macleod 2007; Macleod 2012; Lucier 2008; Arapostathis and Gooday 2013). These scholarly trends are positive; nonetheless, the main driver for a resurgent interest in the relations between science and industry seems to be the recent and perhaps radical change in the funding of academic research and in the relations between universities and corporations. In the twenty-first century, university-based science, and the university itself, seem to be taking a commercial turn, and, consequently, scholars have become increasingly worried about the moral and ethical entanglements of money and research, interest and objectivity, and science and capitalism (Greenberg 2007; Radder 2010; Mirowski 2011). Among historians of science and technology, the critical revisions of the so-called linear model (to be discussed below) seem to be as much a search for new economic and business frameworks for situating the pursuit of knowledge alongside (or within?) the pursuit of profit as they are symptoms of a deeper despair over the supposedly lost purity of the scientific vocation and the pernicious commodification and privatization of science (Grandin, Wormbs, and Widmalm 2004). In short, the topic of commercial science has become hot.

This chapter will address commercial science by examining the places and practices of science within the political economy of the United States, arguably the most dynamic capitalist society of the late nineteenth century and certainly the largest by the early twentieth century. It will begin with a summary of that once-exciting work on early twentieth-century industrial research and then move on to a discussion of science in the age of industrial capitalism. The choice of this period is strategic, for it not only captures several important developments within capitalism, for instance, the rise of very large industrial corporations, but also important changes within science, such as the rise of research universities. In addition, attention to the age of industrial capitalism might serve as a chronological bridge between the somewhat more developed scholarship on early modern science and commerce and the burgeoning field of post-WWII studies of technoscience. Finally, a focus on commercial science of the nineteenth century might bring together, once again, historians of science and historians of technology.

Science in Twentieth-Century Corporate Laboratories

In late 1900, General Electric established the Research Laboratory in Schenectady, New York. It was, arguably, a new place for science, the world's first laboratory dedicated to research that was set up within a corporation. On the other hand, the novelty of the laboratory was little noticed among the general public, within GE, or by its new director, Willis Whitney, a 32-year-old physical chemist who had received his PhD at the University of Leipzig in 1896 under Wilhelm Ostwald. Whitney had been teaching at his undergraduate alma mater, MIT, and like many of the faculty there he had also been consulting, in his case, three days a week for GE. Still, it was an important personal and professional decision for Whitney to leave MIT to become a corporate employee (Birr 1957; Reich 1985; Wise 1985b; Kline 1992).

GE hired Whitney to work full-time on a pressing problem with its most profitable product—the light bulb. Thomas Edison's original patents for the carbon filament light bulb had expired, and GE feared its electric lamp business would become obsolete in the face of increasing competition from European firms that had introduced a metal filament light bulb. Whitney's research was directed toward finding a replacement filament, and he set about staffing his new laboratory with engineers and scientists, including two PhD chemists, Irving Langmuir and William Coolidge. It took ten years for Whitney's research team to invent a ductile tungsten filament and another three years to invent a gas-filled light bulb in which to burn it. Yet by the eve of World War I, the new light bulb had restored GE to a dominant market position, and secured some autonomy for Whitney and the Research Laboratory.

From the start, Whitney had been under constant pressure from the corporate office to prove the commercial value of science. At the same time, he had to convince his scientific staff of the intrinsic merit of the work. The dual demands took a heavy toll on Whitney's health—he suffered several breakdowns—but he continued as director until 1932, when Coolidge took over. According to Whitney's sympathetic biographer George Wise (1985b), Whitney became a deft manager of both research and business. The dual tasks represented a new role for the scientist in industry, and by 1915, Whitney had recruited 250 staff members for the Research Laboratory (Wise 1980).

GE was not the only large American corporation to hire university-trained scientists; in the years before World War I, DuPont (1903), Westinghouse (1904), AT&T (1909), and Kodak (1912) all established research laboratories (Jenkins 1976; Hounshell and Smith 1988; Kline and Lassman 2005). These pioneering laboratories, though, were not modeled on GE's, although their purposes were nearly identical—to thwart competition by improving an existing product or process. Each research laboratory was able to demonstrate the commercial value of its science by improving, respectively, dynamite, electric power, long-distance telephony, and photography. And similar to Whitney at GE, C. E. Kenneth Mees (PhD chemistry) at Kodak and Frank Jewett (PhD physics) at AT&T became exemplary directors of industrial research.[1]

The term "industrial research" itself was coined during World War I and came into common use in America during the 1920s. The Jazz Age was the "golden era" of industrial research. Between 1919 and 1936, US corporations established over 1100 industrial research laboratories. In 1921, these laboratories employed approximately 2700 scientists and engineers. By 1927, their numbers had increased to over 6300, and even in the depths of the Great Depression, 1933, research laboratories continued to grow to nearly 11,000 employees. On the eve of World War II, there were over 27,000 scientists and engineers at work in industry (Hounshell 1996, 36).

By that time, directors of industrial laboratories had achieved a degree of autonomy by showing corporate managers that investment in science could pay dividends. Having secured a permanent place within the corporate hierarchy, laboratory directors could then set their own agendas and develop their own identities, even to the degree that their research sometimes conflicted with the demands of the company. Still, industrial laboratories were not university-departments-in-exile. They were places to do research that was regarded by scientists as basic and by corporations as profitable. In 1932, Irving Langmuir's innovative research on surface chemistry (part of the original light bulb project) won him a Nobel Prize in chemistry. And in 1937, Clinton J.

Davisson of Bell Labs won a Nobel in physics for his work on electron diffraction.[2] Such international acclaim seemed to confirm—to contemporaries at least—that industrial research was on par with university research.

One of those who made such a comparison was the historian Richard Harrison Shryock. In his now-classic article from 1948, "American indifference to basic science during the nineteenth century," Shryock heralded Langmuir's Nobel as proof of the fundamental nature of industrial research.[3] In a sweeping condemnation of all that had been done before the rise of industrial research laboratories, Shryock declared that nineteenth-century American science had been blighted by commercial demands (Shryock 1948). In other words, nineteenth-century commercial science was theoretically inferior to twentieth-century industrial research. The emergence of world-class basic science in the first decades of the twentieth century was thus due in large part to the demands of corporate America. This interpretation has become the standard history: big business was now *doing* basic science, although it is worth noting that these American corporations were large and rich before they started doing research, not because of it (Edgerton 2012, 321). Nonetheless, industrial research represented the final step in the evolution of the fully integrated corporation, an evolution that had begun with the legal and political innovations of limited liability and joint stock, proceeded through trusts and mergers, and seemed to culminate in the early twentieth century with the incorporation of basic science. Scientific knowledge was the new source of raw material for inventions and innovations.

Such a seductive explanation for the rise of industrial research rests to large extent on what it meant to *do* research in industry.[4] In the post-World War II era, Shryock's fellow academics did not celebrate the corporate laboratory nearly as much as he did because they did not consider industrial research to be the same thing as basic science. They regarded industrial research as applied science, and a good deal of scholarly work in the 1950s, 1960s, and 1970s, went into distinguishing what was done in the university laboratory from what was done in the corporate laboratory. If the two activities were indeed different, as the academics' argument went, then those who did industrial research must be different from university scientists (Shapin 2008). Here was one starting point for the long scholarly debate about the normative structure of science.

The norms of science, so famously articulated by the sociologist Robert K. Merton in 1942, speak directly to the difficulty among some university-based scholars to understand the industry-based scientist. Merton described the proper behavior of a scientist by four characteristics: organized skepticism, universalism, disinterestedness, and, most importantly, communism, the shared recognition that science was a public good (Merton [1942] 1973). As another sociologist Steven Shapin (2008) has explained, the Mertonian norms reified a conflict, an unavoidable clash of values between the university-based scientist, whose disinterestedness and public goodness were demonstrated by open publication of results for non-pecuniary rewards from peers, and the industry-based scientist, whose interest in proprietary research (or worse secrecy), was substantiated in patents and a paid salary. According to this theory, the place where research was conducted—the corporation vs the university—determined not only the type of research—published vs patented, open vs proprietary—but also the type of researcher—independent-thinking scientists vs corporate-controlled employees (Weart 1976; Shapin 2008).

One pesky problem with these premises and their inferences was a lack of factual information about what scientists actually did in corporate laboratories as opposed to what academics thought they did. This absence of evidence motivated historians in the 1980s to examine more closely the topic of industrial research. It is somewhat ironic, then, that those scholars reached conclusions very like those of Shryock and of most Americans in the 1930s, namely, that scientists in corporate and university laboratories seemed to be studying the same phenomena using the same instruments and materials and thereby reaching the same results (Reich 1985; Wise 1985a). Thus, the different places where science was being done did not have a differential effect on the practices or theories of science. The laboratories were alike; the research and the researchers were much of a muchness.

That conclusion might not strike historians of the early modern period as very surprising. Scholars who have studied seventeenth- and eighteenth-century chemical laboratories in all their various forms and locations have found very strong correspondences in their material cultures, their methods of manipulation and making, and their ways of knowing. But such a conclusion might worry twenty-first-century commentators, especially those academics who issue dire warnings that the university has become indistinguishable from the corporation (Kleinman and Vallas 2001). What separates the seemingly acceptable promiscuity of early modern science and commerce from today's acute concerns over the commercialization of science is the nineteenth-century ideology of pure science.

Science in Nineteenth-Century Capitalist Enterprises

In its rise to cultural prominence in the latter decades of the nineteenth century, pure science was often paired with applied science, and the comparisons and contrasts between the two became subjects of lively and important debate, one that is only now being studied on its own terms (Bud 2012). For much of the post-World War II era, historians of science and historians of technology reworked the terms of the nineteenth-century debate in order to suit other twentieth-century political, economic, and disciplinary agendas (Staudenmaier 1985; Kline 1995). It is worthwhile to attend, at least briefly, to those agendas in order to explain why commercial science got written out of the story.[5]

In the United States, the history of technology was a discipline formed in the late 1950s in partial response to a mistaken identity, namely, applied science = technology. The origin of that equation has often been traced to science policymakers at the end of World War II, specifically Vannevar Bush and his *Science, The Endless Frontier* (1945). In its postwar formulation, applied science was not a contrast to pure science or even comparable to it as a type of knowledge; rather, applied science was a result or product of pure science, which, in the postwar era, had been recast as basic science. Technology meant commercial or military products and processes, and those inventions and innovations were supposedly dependent on basic science; they were science-based. That characterization was transformed into the so-called linear model, a sort of idealized conveyor belt that begins with basic scientific theories and discoveries, moves along with their application in the form of patentable inventions, then proceeds to the development of innovations (marketable and profitable products and processes), and culminates with entirely new industries. In the two decades immediately following World War II, the linear model informed both US science policy and corporate planning, hence the willingness of the US military and multinational corporations to sink large sums of money into research laboratories staffed with scientists. But beginning in the late 1960s and continuing to the present, the linear model and its underlying equation have come under increasing criticism from politicians, policymakers, and scholars, especially from the newly organized historians of technology (Stokes 1997; Grandin, Wormbs, and Widmalm 2004). They not only rejected the formula, but they began to invest a good deal of time and effort, as well as their own identity, into explorations and explications of the original nineteenth-century meanings of applied science.[6] In the process, they laid claim to an epistemological autonomy for applied science and vested such knowledge in engineers and inventors, who, they argued, were largely responsible for the tremendous technological developments of industrial America (Alexander 2012).

Historians of science in America meanwhile redefined the terms of the nineteenth-century pure and applied debate around a different identity, namely, pure science = professional science. The origin of that equation can also be traced to the 1960s, when scholars began to purpose models for the growth of the early American scientific community. Accordingly, American scientists began to professionalize when they started to self-identify themselves by their expertise, self-segregate themselves into selective organizations, and self-replicate themselves through education and credentialing. By the end of the nineteenth century, American scientists had become professionals through the establishment of legitimate social roles as salaried professors within universities, institutions which recognized and rewarded them for the originality and importance of their research, as judged by the scientists themselves. The professionalization of American science thus coincided with, if not caused, the rapid maturation of mathematical and laboratory-based physical sciences (as opposed to traditional natural history or observational field sciences), the building of graduate schools and PhD programs, and the creation of the modern American research university (Daniels 1967; Servos 1986; Kohler 1990).

What has been lost in these modern disciplinary moves around pure and applied is commercial science; it simply did not fit within the dominant narratives being constructed by historians of science or historians of technology. On the one hand, commercial science represents a very different kind of professionalization; and on the other, it looks a lot like applied science, yet its practitioners were men of science. From either vantage, pure or applied, commercial relations were at the center of fundamental questions about the identity and interests of nineteenth-century American science (Lucier 2009).

The ideology of pure science emerged in the United States in response to the crass and greedy extravagances of the Gilded Age (Daniels 1967; Lucier 2012). It was part of a larger political and cultural reform movement whose distinguished leaders espoused higher motivational goals for human inquisitiveness than the mere accumulation of wealth and position—or, at least, that was the argument. What made science pure and putatively noble, like fine arts and literature, was its *non-commercial* character. According to one of the most outspoken reformers, the physicist Henry Rowland, the practitioners of pure science, the newly christened *scientists*, made investigations into nature purely for the love of science (Rowland 1883). No longer would scientists need to "sell" themselves and their science on the basis of usefulness or practicality as they had done in the past. In his now oft-quoted address before the Physics Section of the American Association for the Advancement of Science in August 1883, "A plea for pure science," Rowland explained that the objective of scientists were ideas, that is, theories of nature, not theories of things, and the proper place for the pursuit (or preservation) of pure science was a well-endowed university, such as Johns Hopkins, where Rowland himself was a professor. The places, practices, and persons that did not meet Rowland's strict non-commercial definition were, by implication, impure, meaning corrupted by the immediacy of demands and patrons that were not ideally scientific.

If commercialization led to corruption, then there was much in need of reform, beginning with the public's conception of science. To Rowland and other purists, like the English gentleman of science Thomas Henry Huxley, the public needed to know that theories of nature had to be discovered before their application to any practical problems.[7] Pure science carried a temporal and substantive primacy, hence the rationale behind the argument that inventions, innovations, and late nineteenth-century industry itself were literally science-based (Forman 2007). According to Rowland, a good example of the application of physics was electric light, but Gilded Age Americans could not comprehend the value of pure physics; they could only understand the cash and convenience of light bulbs and other inventions. Rowland complained that Americans confounded the applications of science with the science itself, and, not surprisingly, they showered fame and fortune on inventors like Edison, and not on scientists like him.

For Rowland, Huxley, and other purists, the kind of work that went into inventing electric light was not the same thing as scientific research. But many of their contemporaries *did* think inventing and researching were equivalent, including the inventor of the telephone, Alexander Graham Bell. In theory and in practice, Bell believed that commercial demands had a *stimulating* effect on nineteenth-century American science (Shryock 1948; Bruce 1990). For Bell, research on telephones or electricity was no less genuine than the research in the physics of sound or light, regardless of the motivation of the researcher. In fact, the ability to meet the demands of *both* commerce and science marked the highest achievement, a belief Willis Whitney would certainly share. Bell was also a major financial backer of the journal *Science*, which had been started under Edison's auspices, but which he had relaunched in 1883. Bell and the editors of *Science* helped to craft and spread the new message of applied science, meaning the kind of genuine knowledge produced in the investigation of telephones, electric light, and other inventions, the meaning historians of technology later adopted.

As much as Rowland resented Edison's fame, neither inventors nor the American public were the major targets of his censure; it was the men of science who did commercial work. They were not behaving as Rowland's newly-defined non-commercial scientists should; but rather they were being professionals, selling their science to meet the demands of new industries. According to Rowland, their motivation was money, not science. In late nineteenth-century America, the most lucrative kinds of commercial work were consulting and testifying in courts of law as an expert witness, usually in cases involving patents.

In its most general outlines, commercial science encompassed a variety of practices requiring scientific expertise and experience for which men of science got paid. Basically, it was career-making, earning a livelihood by doing science. In Victorian Britain, career-making often characterized lower-status occupations such as selling specimens, instrument-making, or curating collections. Higher-status gentlemen of science did not resort to such commerce (Secord 2000; Endersby 2008). For them, the only respectable means to an income was scientific authorship, which meant some combination of writing, reviewing, editing, translating, and public lecturing.[8] Scholars often cite T. H. Huxley as the prime example of a young gentleman of science out to make his reputation and his livelihood from proper scientific authorship (Lightman 2007, 353–97). In the United States, scientific authorship was not as popular, and hence not as profitable, except for public lecturing, for which there was literally a system in operation by mid-century. American men of science booked at lyceums and theatres for 50–100 lectures a season (roughly four months from October through January) for which they could expect to be paid $50 to $150 per lecture. The income was substantial, but regular engagements could not be counted on (Lucier 2009).

A more bankable kind of commercial science was consulting, and consulting had a clear connection to economic development. In the United States, the practice had first developed in the late 1830s as a form of public patronage. In contrast to aristocratic or government patronage, men of science sought and began receiving commissions directly from the public, or more precisely, from a particular segment of the public—capitalists. These individuals (invariably men) had upstanding reputations, access to substantial financial and political resources, and plans for improvement. The projects on which capitalists might consult men of science were on the scale of mines and manufactories. For these kinds of private enterprises, consultants provided the expertise about where to dig for coal and other mineral resources or the equipment and procedures for making chemicals like paints, lubricants, and lamp oils. In return they received fees, and these consulting fees could be considerable, on the order of $500 to $1500 for a single commission, an income near that of a professor's annual salary.[9]

By mid-century, consulting had developed into a professional practice akin to medicine or law. The most sought-after men of science were geologists and chemists, a clear reflection of the importance of mining and chemical manufacturing to the growing American economy. In their practice of commercial science, consultants displayed neither impurity nor indifference to basic science. Consulting represented and reinforced the widely held assumption that science was useful and therefore should be used. These professional men of science embodied autonomy, authority, and honor, the virtues of a moral middle class. Consulting had become a central part of the practice and identity of American men of science. In short, it epitomized the business of mid-century American science (Lucier 2008).

By the second half of the nineteenth century, the practice itself as well as the businesses seeking scientific expertise began to change. The companies grew larger in terms of their organization, income, and most significantly investors. They were no longer family-run businesses or partnerships; they were joint-stock, limited liability corporations, whose operations, it seemed to many Americans, were synonymous with speculation and corruption. The consultants themselves were no longer serving in short-term, advisory roles; they were being hired and held on long-term retainers, often working exclusively for one major company. And in that work, men of science were often involved in patenting inventions, and increasingly those patents involved science.

By legal definition in both the US and Britain, patents could not be granted for scientific laws or natural objects. They could be granted for discoveries, but it was a matter of much controversy whether a particular claim to novelty and originality was a discovery in science or in art. In the late nineteenth century, patents were inscribed with broader and broader claims to chemical and physical principles, and these claims became the sources of extended and expensive legal battles. Both sides in these cases often turned to men of science as expert witnesses to testify as to the scientific nature of these principles. The witness-box quarreling made for good newspaper copy, but it left the public face of science bloodied and the authority of men of science suspect (Lucier 1996).

Patents were challenging subjects for nineteenth-century men of science, as well as for present-day scholars who study them. Most nineteenth-century men of science held themselves aloof from patents, but the reasons were complicated. For some, the aversion had to do with the time-consuming legal processes required to file patent claims and defend them against would-be rivals; for others, it was a reaction to the distastefulness of public controversies. For most, the detachment reflected ethical qualms; patents smacked of interest, and interest did not suit the disinterestedness displayed by most proper men of science, including the professional consultants. Men of science agreed that knowledge could not be proprietary and still be science (Macleod 2012). But not all men of science regarded patents as proprietary knowledge; after all, one of the purposes of a patent was to encourage patentees to make public a discovery that otherwise might have remained secret in return for government protection of the invention. Moreover, many men of science held the belief that patents represented useful knowledge. William Thomson, Lord Kelvin, was probably the most famous man of science to secure patents for discoveries derived from his scientific research on specific industrial problems, notably those related to the laying of undersea telegraph cables, from which he earned several thousand pounds sterling a year (Smith and Wise 1989). Likewise, Louis Pasteur derived substantial revenues beginning in 1857 from patents he held on chemical processes for manufacturing and preserving wine, vinegar, and beer (Geison 1995). And there were many others; some familiar to historians, such as the English chemist Edward Frankland, and some less so, like the Canadian geologist Abraham Gesner, who patented a lamp oil called Kerosene (Russell 1996; Lucier 2008).

In many ways, patents are key to understanding the relations between science and commerce in nineteenth-century America as well as the relations between the history of science and the history of technology in the twenty-first century. A good illustration of their significance can be found in the case of Edison and electric light. The electrical industry is the exemplar of science-based industry in the United States, or, at least, many historians of science have asserted so. Yet it is historians of technology who have most often lionized Edison, the most prolific inventor of all time in terms of the number of patents, more than 1000 in the US alone (Hughes 1989; Israel 1998). Edison was also the founder in 1876 of a new kind of laboratory at Menlo Park, New Jersey, where he and a select group set to work on inventions relating to telegraphy and incandescent lighting. By the time Edison built his larger laboratory at West Orange, New Jersey in 1887, he was justly celebrated for his light bulb and his integrated system for generating and distributing electricity. Many scholars credit the "Wizard of Menlo Park" with having invented a method of invention, and Edison himself referred to his Menlo Park laboratory as an "invention factory." His laboratories were the first to separate the process of invention from the process of manufacturing. Prior to Edison, invention was regarded as an unpredictable activity of a single creative individual; Edison treated invention as a systematic process, which could be organized by a dedicated team. At Menlo Park and West Orange, he gathered together machinists, mechanics, skilled craftsmen, and a few men of science—chemists, physicists, and mathematicians. In its diversity of experts and expertise, Edison's invention factory looks much like a twentieth-century corporate laboratory, and scholars have duly identified Edison with the origin of industrial research. But these same scholars do not regard Menlo Park or West Orange as places for doing science, in large part because they maintain the motivational distinctions between men of science and inventors.[10] If there is an apparent paradox to Edison's non-scientific laboratories, it may be explained by the confusion (both then and now) about the content and meaning of nineteenth-century patents. While the laboratories at GE, DuPont, Westinghouse, AT&T, Kodak, and all other large twentieth-century corporations could pursue scientific research *and* patents at the same time and by the same team members (in fact, patents were critical to corporate strategy—defending market share by preventing potential rivals from inventing and/or innovating competing products and processes), in the late nineteenth century the patenting of scientific research was still a very debatable and divisive endeavor. Still, a scientist of Rowland's stature knew there was plenty of research in those patents, as well as potentially lots of money, which is what nineteenth-century commercial science was basically all about.[11]

Conclusion

In the United States today, the vast majority of university-educated scientists—almost 75 percent—work in industry. Only 18 percent work in colleges and universities; the remaining 6 percent work in government. Such statistics have led the historian of technology David Hounshell (1996, 14) to assert that "the development of industrial R&D must surely rank as one of the [twentieth] century's defining characteristics." Yet the defining characteristics of American science have been anything but industrial. As the historian of technology David Edgerton (2012, 326) has emphasized, the history of science in the twentieth century is an account *of* (and often *by* and *for*) that small minority of university-based scientists, and despite its importance, "[o]ur knowledge of scientists pursuing standard careers in industry," Edgerton concluded, "is very limited."

Likewise, our knowledge of nineteenth-century careers is limited. Historians of science still focus on universities and the rise of science within those rarified settings. Yet most American men of science found some measure of employment off-campus, and even those who remained in college often consulted, a professional practice that added much more to science, invention, and industry than merely supplementing a meager teaching salary. Consulting engaged men of science in commercial concerns; it introduced them to problems in the state of the art, what we call cutting-edge technologies today, and often presented these men of science with ethical challenges about their disinterested motives and commitments to useful knowledge.

Edgerton (2012) urged scholars to follow the money when investigating twentieth-century science. The meaning of money is also a good way to understand nineteenth-century American science and especially the debate over pure and applied. Assumptions about money, motive, and meaning are what have led historians of science to create the pure = professional narrative and historians of technology to reject the applied = technology formulation. Closer attention to the relations between identity and income may bring historians from both camps to come together on commercial science. Commercial science found expression in the short-term consulting engagements of mid-nineteenth-century men of science, in the patenting of certain kinds of research by the end of the century, in the steady employment of scientists in twentieth-century industrial laboratories, and in the increasingly fluid careers of twenty-first-century academic entrepreneurs.

Endnotes

Funding for this research/article was provided in part by a fellowship from the National Endowment for the Humanities.

1   The emergence of these charismatic directors of industrial research was one of the key features of Philip Mirowski's "The captains of the erudition regime," one of three regimes of science organization, funding, and thought-styles in the twentieth century (Mirowski 2011, 98–105).
2   In 1925, AT&T formally incorporated Bell Labs. Clinton J. Davisson had published his research in *Physical Review* in 1927 (Kevles 1995, 188–9).
3   Indifference was the bugbear of so many historians of nineteenth-century American science, and no one worked harder to banish it than Nathan Reingold; see, for example, Reingold ([1972] 1991).
4   Michael Aaron Dennis emphasized the distinction between *doing* science in a twentieth-century industrial research laboratory and merely *using* science, as many nineteenth-century companies supposedly did when they purchased the professional services of a consultant or the patents of an inventor. Such a distinction does not hold up to historical investigation, much as the distinction between science and technology in the corporate laboratory does not hold up. Ironically, that latter point was made by Dennis himself (Dennis 1987, 481–2, 490). On doing science in nineteenth-century company contexts, see Lucier (2008) and Lucier (1995).
5   The best explanation for why (and how) the history of technology community treated technological knowledge as different from scientific knowledge remains Staudenmaier (1985).
6   It is important to note that in the original, circa 1880s, formulation of the linear or science-based model, Rowland and other purists did not specify who should pay for pure science—presumably the patronage of wealthy individuals—but, significantly, not state or federal government, for that would be corrupt (because political) science as much as commercially funded science (Mirowski 2011, 46–56).
7   For the English gentleman of science, T. H. Huxley, there was only one kind of science, hence no need for either modifier—pure *or* applied (Huxley 1881; Gooday 2012).
8   Of course, not all kinds of scientific writing were respectable (Secord 2000; Lightman 2007); and a few gentlemen of science did also consult (Morrell 2005).

9   Larger-scale improvements like railroads and canals required some form of govern-
     ment funding and often government-subsidized expertise, such as geological surveys
     (Larson 2001; Lucier 2008).

10   Thomas P. Hughes (1989) explained the independence of inventors such as Edison by
     their focus on the front edge of technology, whereas scientists worked on outmoded
     and inapplicable theory.

11   Rowland had worked with Edison on electric lighting, and he had watched as other
     physicists, namely George Barker of the University of Pennsylvania, had prepared,
     patented, and promoted inventions (Hounshell 1980, 614; Israel 1998, 463–72).

References

Alexander, Jennifer Karns. 2012. "Thinking again about science and technology." *Isis*, 103:
    518–26.

Arapostathis, Stathis, and Graeme Gooday. 2013. *Patently Contestable: Historical Trials of Elec-
    tricity, Identity, and Inventorship*. Cambridge, MA: MIT Press.

Birr, Kendall. 1957. *Pioneering in Industrial Research: The Story of the General Electric Research
    Laboratory*. Washington, DC: Public Affairs Press.

Bruce, Robert V. 1990. *Bell: Alexander Graham Bell and the Conquest of Solitude*. Ithaca, NY:
    Cornell University Press.

Bud, Robert. 2012. "Focus: Applied science." *Isis*, 103: 515–63.

Cook, Harold J. 2007. *Matters of Exchange: Commerce, Medicine, and Science in the Dutch
    Golden Age*. New Haven, CT: Yale University Press.

Daniels, George H. 1967. "The pure-science ideal and democratic culture." *Science*, 156: 1699–
    1705.

Dennis, Michael Aaron. 1987. "Accounting for research: New histories of corporate laboratories
    and the social history of American science." *Social Studies of Science*, 86: 245–67.

Edgerton, David. 2012. "Time, money, and history." *Isis*, 103: 316–27.

Endersby, James. 2008. *Imperial Nature: Joseph Hooker and the Practices of Victorian Science*.
    Chicago: University of Chicago Press.

Forman, Paul. 2007. "The primacy of science in modernity, of technology in postmodernity,
    and of ideology in the history of technology." *History and Technology*, 23: 1–152.

Geison, Gerald L. 1995. *The Private Science of Louis Pasteur*. Princeton, NJ: Princeton Univer-
    sity Press.

Gooday, Graeme. 2008. "Placing or replacing the laboratory in the history of science." *Isis*, 99:
    783–95.

Gooday, Graeme. 2012. "'Vague and artificial': The historically elusive distinction between pure
    and applied science." *Isis*, 103: 546–54.

Grandin, Karl, Nina Wormbs, and Sven Widmalm, eds. 2004. *The Science–Industry Nexus: His-
    tory, Policy, Implications*. Sagamore Beach, MA: Science History Publications.

Greenberg, Daniel S. 2007. *Science for Sale: The Perils, Rewards, and Delusions of Campus Cap-
    italism*. Chicago: University of Chicago Press.

Hounshell, David A. 1980. "Edison and the pure science ideal in nineteenth-century America."
    Science, 207: 612–17.

Hounshell, David A. 1996. "The evolution of industrial research in the United States." In
    Engines of Innovation: U.S. Industrial Research at the End of an Era, edited by Richard S.
    Rosenbloom, and William J. Spencer, 13–85. Boston: Harvard Business School.

Hounshell, David A., and John Kenly Smith, Jr. 1988. *Science and Corporate Strategy: Du Pont
    R&D, 1902–1980*. Cambridge: Cambridge University Press.

Hughes, Thomas P. 1989. *American Genesis: A Century of Invention and Technological Enthusiasm, 1870–1970.* New York: Viking Penguin.

Huxley, Thomas Henry. 1881. "Science and culture." In *Science and Culture and Other Essays,* by Thomas Henry Huxley, 1–23. London: Macmillan.

Israel, Paul. 1998. *Edison: A life of invention.* New York: John Wiley & Sons.

Jenkins, Reese. 1976. *Images and Enterprise: Technology and the American Photographic Industry, 1839–1925.* Baltimore: Johns Hopkins University Press.

Kevles, Daniel J. 1995. *The physicists: The history of a scientific community in modern America.* Cambridge, MA: Harvard University Press.

Klein, Ursula, and Emma Spary (eds.) 2010. *Materials and Expertise in Early Modern Europe: Between Market and Laboratory.* Chicago: University of Chicago Press.

Kleinman, Daniel Lee, and Steven P. Vallas. 2001. "Science, capitalism, and the rise of the 'Knowledge Worker': The changing structure of knowledge production in the United States." *Theory and Society,* 30: 451–92.

Kline, Ronald R. 1992. *Steinmetz: Engineer and Socialist.* Baltimore: Johns Hopkins University Press.

Kline, Ronald R. 1995. "Construing 'Technology' as 'Applied Science': The public rhetoric of scientists and engineers in the United States, 1880–1945." *Isis,* 86: 194–221.

Kline, Ronald R., and Thomas C. Lassman. 2005. "Competing research traditions in American Industry: Uncertain alliances between engineering and science at Westinghouse Electric, 1886–1935." *Enterprise & Society,* 4: 601–45.

Kohler, Robert E. 1990. "The Ph.D. machine: Building on the collegiate base." *Isis,* 81: 638–62.

Kohler, Robert E. 2008. "Lab history: Reflections." *Isis,* 99: 761–8.

Kohlstedt, Sally Gregory. 1985. "Institutional history." *Osiris,* 1: 17–36.

Larson, John Lauritz. 2001. *Internal Improvement: National Public Works and the Promise of Popular Government in the Early United States.* Chapel Hill: University of North Carolina Press.

Lightman, Bernard V. 2007. *Victorian Popularizers of Science: Designing Nature for New Audiences.* Chicago: University of Chicago Press.

Lucier, Paul. 1995. "Commercial interests and scientific disinterestedness: Consulting geologists in antebellum America." *Isis,* 86: 245–67.

Lucier, Paul. 1996. "Court and controversy: Patenting science in the nineteenth century." *British Journal for the History of Science,* 29: 139–54.

Lucier, Paul. 2008. *Scientists and Swindlers: Consulting on Coal and Oil in America, 1820–1890.* Baltimore: Johns Hopkins University Press.

Lucier, Paul. 2009. "The professional and the scientist in nineteenth-century America." *Isis,* 100: 699–732.

Lucier, Paul. 2012. "The origins of pure and applied science in Gilded Age America." *Isis,* 103: 527–36.

Macleod, Christine. 2007. *Heroes of Invention: Technology, Liberalism, and British Identity.* Cambridge: Cambridge University Press.

Macleod, Christine. 2012. "Reluctant entrepreneurs: Patents and state patronage in new technosciences, circa 1870–1930." *Isis,* 103: 328–39.

Merton, Robert K. [1942] 1973. "The normative structure of science." In *The Sociology of Science: Theoretical and Empirical Investigations,* edited by Robert K. Merton, 267–78. Chicago: University of Chicago Press.

Mirowski, Philip. 2011. *Science-Mart: Privatizing American Science.* Cambridge, MA: Harvard University Press.

Morrell, Jack. 2005. *John Phillips and the Business of Victorian Science.* Aldershot: Ashgate.

Radder, Hans (ed.) 2010. *The Commodification of Academic Research: Science and the Modern University*. Pittsburgh: University of Pittsburgh Press.

Raj, Kapil. 2007. *Relocating Modern Science: Circulation and the Construction of Knowledge in South Asia and Europe, 1650–1900*. Basingstoke: Palgrave Macmillan.

Reich, Leonard S. 1985. *The Making of American Industrial Research: Science and Business at GE and Bell, 1876–1926*. Cambridge: Cambridge University Press.

Reingold, Nathan. [1972] 1991. "American indifference to basic research: A reappraisal." In *Science American Style*, edited by Nathan Reingold, 54–75. New Brunswick, NJ: Rutgers University Press.

Roberts, Lissa, Simon Schaffer, and Peter Dear (eds.) 2007. *The Mindful Hand: Inquiry and Invention from the Late Renaissance to Early Industrialization*. Amsterdam: Koninklijke Nederlandse Akademie van Wetenschappen.

Rosenbloom, Richard S., and William J. Spencer (eds.) 1996. *Engines of Innovation: U.S. Industrial Research at the End of an Era*. Boston: Harvard Business School.

Rowland, Henry A. 1883. "A plea for pure science." *Science*, 2: 242–50.

Russell, Colin A. 1996. *Edward Frankland: Chemistry, Controversy, and Conspiracy in Victorian England*. Cambridge: Cambridge University Press.

Secord, James A. 2000. *Victorian Sensation: The Extraordinary Publication, Reception, and Secret Authorship of* Vestiges of the Natural History of Creation. Chicago: University of Chicago Press.

Servos, John W. 1986. "Mathematics and the physical sciences in America, 1880–1930." *Isis*, 77: 611–21.

Shapin, Steven. 2008. *The Scientific Life: A Moral History of a Late Modern Vocation*. Chicago: University of Chicago Press.

Shryock, Richard Harrison. 1948. "American Indifference to basic science during the nineteenth century." *Archives Internationales d'Histoire des Sciences*, 28: 3–18.

Smith, Crosbie, and M. Norton Wise. 1989. *Energy and Empire: A Biographical Study of Lord Kelvin*. Cambridge: Cambridge University Press.

Smith, John Kenly Jr. 1990. "The scientific tradition in American industrial research." *Technology and Culture*, 30: 121–31.

Smith, Pamela H. 1994. *The Business of Alchemy: Science and Culture in the Holy Roman Empire*. Princeton, NJ: Princeton University Press.

Smith, Pamela H., and Paula Findlen (eds.) 2002. *Merchants and Marvels: Commerce, Science, and Art in Early Modern Europe*. London: Routledge.

Staudenmaier, John M. 1985. *Technology's Storytellers: Reweaving the Human Fabric*. Cambridge, MA: MIT Press.

Stokes, Donald E. 1997. *Pasteur's Quadrant: Basic Science and Technological Innovation*. Washington, DC: Brookings Institution Press.

Weart, Spencer. 1976. "The rise of 'prostituted' physics." *Nature*, 262: 13–17.

Wise, George. 1980. "A new role for professional scientists in industry: Industrial research at General Electric, 1900–1916." *Technology and Culture*, 21: 408–29.

Wise, George. 1985a. "Science and technology." *Osiris*, 1: 229–46.

Wise, George. 1985b. *Willis R. Whitney, General Electric, and the Origins of U.S. Industrial Research*. New York: Columbia University Press.

CHAPTER TWENTY

The Field

ROBERT E. KOHLER AND JEREMY VETTER

"Field" as a spatial term and concept has a deep and varied history in warfare, agriculture, sport, and territorial rule. Its application to science is more recent, dating roughly from the time when the institutions, categories, and practices of science were taking their modern shape in the middle decades of the nineteenth century. Although the term's detailed history has yet to be worked out, Google Ngrams of usage in English-language books afford a rough silhouette of that history. The term "field science" first came into steady use around 1870, after a few initial bumps, and increased with marked ups and downs to an apogee around 1960 before declining. "Field method" and "field methods" followed a similar, if smoother, trajectory. (A related usage, "field naturalist," appeared in the 1830s with the first wave of amateur field clubs, in Britain.) Despite the occasional deployment of "field" as a category of analysis for earlier periods (e.g., Cooper 1998; Te Heesen 2005), the idea and the category became necessary only when the laboratory form of organization and practice escaped its limited domain in chemistry and was adopted by disciplines of all sorts, thereby becoming the premier place where high-status, modern science was carried out. That rise is reflected in the use of the term "laboratory science," which appeared in the 1880s and grew steadily to a plateau about 1930.

In this great spatial and epistemic reorganization, "field" became the "other": the not quite modern kind of science that was associated with amateur field activities and practices that were not properly quantitative, experimental, hypothetico-deductive, and analytically rigorous. Before the laboratory revolution of the 1840s to 1880s there was no call for such a categorical distinction, because laboratory modes of organization and practice were not yet the universal default. The pecking order of sciences did not yet depend crucially on place. Once the new world of scientific disciplines and careers had arrived, however, the distinction simply described the new reality: that the science that filled journals and made careers was the science done in places designed specifically for the purpose, especially labs, museums, and experimental gardens. And when ascendant social practices call forth new social categories, new categories will be

A Companion to the History of Science, First Edition. Edited by Bernard Lightman.
© 2016 John Wiley & Sons Ltd. Published 2020 by John Wiley & Sons Ltd.

created as well for practices now recast as outmoded or second best. "Field" was one such category.

This epistemic hierarchy of place long outlasted the historical situation that created it. The connotation of "field" as it emerged in a world of labs was less "not-lab" than it was "*pre*-lab," and pre-lab less in the sense of historical precedence than of methodological procedure. In the modern view, field practices of collecting and describing were "mere" preliminaries to proper science. The field was a useful, even essential, source of raw materials, but it was a place to which one resorted in expectation of moving expeditiously to places designed for science. Another mirror concept of field as "post-lab" developed in the later nineteenth century. Here the field was where the "pure," universal knowledge produced in custom-built places was "applied" in particular situations. Whether conceived as pre- or post-lab, "field" became epistemically secondary. And despite the mixed and varied realities of field science, a historically rooted hierarchy of place has persisted.

The Field as Category and Place

What, then, is "field" exactly, and how is the concept best used? "Field" as applied to science is a portmanteau category, encompassing all the diverse natural and social places where science is done—forests, prairies, mountains, deserts, caves, marshes, agricultural and pastoral landscapes; villages, cities, suburbs, encampments, roads, and paths; as well as the world's layered elements of water, earth, and air from tectonic or ocean depths to outer space. For scientists, "field" means, operationally, the particular place in which each works: it is an actor's category. As an all-encompassing term, in contrast, "field" is not an actor's but a historian's category: a frame for comparative analysis of place and practice across the field sciences. Scientists do not experience "field" in general, only some particular field, so find little meaning in a concept of the whole picture. Historians, in contrast, concern themselves with all fields, so find the generic rubric of use.

It is usual to think of "field" in contrast to "lab"—*Webster's Dictionary* defines this usage as "the sphere of practical operation outside a laboratory, office, or factory"— and historically that makes sense. However, the contrast misleadingly suggests a static and cleanly dichotomous view. Field and lab are not separate or incompatible worlds— far from it. Many, perhaps most, field sciences have a lab, office, or museum side as well (Kohler 2006; Brinkman 2010) and practitioners of lab sciences may routinely go afield. There is no clear-cut distinction between built and natural places. Social scientists treat and use cities as both laboratory and field (Gieryn 2006). Built environments can be field as well: for example, expeditionary ships and encampments (Rozwadowski 1996; Sorrenson 1996; Pang 2002; Adler 2014). Field labs may be opened to local conditions of sun, shade, and weather (De Bont 2015; Vetter 2012); and indoor labs can be "naturalized" to mimic particular natural conditions, if imperfectly (Kohler 2002b; Kingsland 2009). As Christopher Henke and Thomas Gieryn (2008, 364) note, "'placeless' places are not necessarily 'faceless' ones." Nor do research methodologies cleanly distinguish field and lab. Intensive observation is, to be sure, the prevalent method of field science, as experimental setup and manipulation are hallmarks of lab science. Yet exact measurement and experiments can be done in the field: they just take more ingenuity and improvising to arrange than in places designed for the purpose (Kohler 2002a).

The distinctive meaning of field in science thus lies not in any characteristics of place or of use as such, but rather in place and use combined. Places have meaning as they are imagined, selected, and used for scientific (or other) purposes; uses have meaning as they are deployed in one sort of place or another. "Field" is not an essential, but a relational category—as it is also in "fields" of battle, sport, or agriculture. Fieldwork is in some cases best treated as a kind of land use: a unit of place and human activity, as in environmental history.

One defining characteristic of field places is that they are multipurpose: because they are inhabited or used by diverse actors whose pursuits range from complementary to science to competitive with or even hostile to it. Field scientists interact with landowners, farmers, mineral and timber harvesters, hunters and fishers, travelers and recreationists, indigenous peoples, and reclusive sects, among others; and may be accepted, ignored, or ejected. And, through varied use, field places tend to become in time more multipurpose (Alagona 2012; Bocking 2012; Lachmund 2013). Knowing a place scientifically may also make it more attractive to those pursuing ends of recreation, resource extraction, or social "improvement," as scientists may be attracted to places already partially known through these other uses. Examples include parks, preserves, field stations, fishing grounds, and ethnic enclaves. Many of the distinctive features of science in the field follow from the human diversity of places both natural and social.

For example, access is a far more complicated matter in the field than in sole-purpose places, which are easily opened to approved users (scientists, their helpers) and closed to others. There are the sheer logistical problems of travel to and residing in distant or inhospitable places. Human communities may resist being subjects of study by strangers who come to reside among them, and animal communities are often elusive and sometimes dangerous up close. Scientists' own communal mores have impeded access. Because work in the field can be rough hand-labor, it was once considered demeaning for men of science and best left to practical occupations of collectors, taxidermists, or fishermen. Before going afield, zoologists and ethnographers among others had first to persuade the world that the field was a legitimate and proper place for their professional labors (Kuklick 1997, 2011; Kohler 2006).

All places have politics, and in science that is especially the case in the field, where the politics of competing uses is not an externality, as in labs, but an inextricable element of the objects and the methods of scientific study. This is self-evidently so in sciences of human communities but is also so in places of "second nature," in which humans are actors in natural phenomena (Schneider 2000; Buhs 2004; Henke and Gieryn 2008, 365–9). The ecology of any place of rich and contested pickings may include harvesters, foragers, residents, developers, recreationists, officials, and activists—along with scientists. Leave their politics out of the science of such places, and the science is incomplete.

A further consequence of diverse use is that, in the field, vernacular practices and meanings may enter and reshape those of expert science—a categorical breach that rarely if ever happens in places of single use. For example, field scientists have drawn on vernacular sources of epistemic authority: like the heroism of explorers that once gave authenticity to the scientific reports of glaciologists and lone anthropologists, whose unwitnessed science could only be taken on trust (Hevly 1996; Kuklick 1997; 2011). No one would risk life and limb to get facts and then not tell them truly: that was the logic of authenticity. Field biologists have, over time, shared the means and ends of science and outdoor recreation with amateur naturalists, bird watchers, and collectors of various things (Barrow 1998; Alberti 2001).

Such cases are hardly rare: practices of mining have been adapted to ends of stratigraphic geology (Rudwick 1976), of hunting and camping to taxonomy and ecology (Kohler 2006), of social services to social survey (Bulmer, Bales, and Sklar 1991). This mixing of defining activities has blurred the categorical boundaries between science and not-science, and made scientific identity more open-ended and mutable than it is in single-purpose places. Gatherers of data are mistaken for gatherers of other kinds (foragers, pleasure seekers, spies), and observers of human communities assume or are assigned roles that their subjects can make sense of.

A second defining characteristic of the field is that phenomena there are studied in the situations in which they normally play out. In the field the essential unity of activities and situations is self-evident, as it is obscured in single-purpose places. The design of lab, museum, office, or factory is precisely to isolate phenomena from the messy situations that impair experimental, cause-and-effect analysis of one variable at a time. The premise of science in the field, in contrast, is that phenomena are by nature situated and cannot be removed from the situations that give them pattern and meaning. Situated field study shifts the focus from our own tidy experiments to nature's (*and* our own) messy ones. Collectors as well, though they remove things from the field for study indoors, attach precise accounts of objects' situations—in labels and field notes—because without these the objects' meanings would be lost.

In the field, situated observers and collectors strive to be intimately involved in the places and phenomena they observe: inside their objects of study, so to speak. In behavioral sciences of the field, observers become "participant observers," and the ethnographic term is widely applicable. Participation entails being on the spot, observing and recording who is doing what with whom and in what situations. It may also mean active participation in the goings-on, though usually in a limited way. The ideal is for observers to know situations, actors, and activities as intimately as their subjects do. In the case of human communities, residents in turn may become participants in the science, assisting in gathering and interpreting field data, as in paleontology and anthropology (Knell 2000; Schumaker 2001) or, more recently, in agricultural field trials, where field hands collect and grade "data" (saleable produce), and farmers decide when experiments end (Henke 2000). Such transgressions of the rights of expertise are difficult in "placeless" places. Where access is controlled and contexts are eliminated—as in labs, offices, and factories—the idea of situated or participant science is unthinkable. That fundamental difference is what makes the field a real and useful category.

Doing field science is for the same reasons unusually "experiential": that is, it is experienced not as different and separate from other activities of everyday life but as another such activity. As field practices borrow from vernacular activities, so are they experienced as continuations of everyday life. As scientists reside in the places they study, so must they experience their work as residents do. Scenes of science are also scenes of life. Geologists' visits to the special sites where deep buried history has been laid bare may be experienced as quasi-religious pilgrimage (Rudwick 1996). Ethnographic surveys, specimen collecting, and ecological study may feel like recreation or exotic travel—science as a vacation (Kohler 2007; Anker 2007). Work in deep-ocean vessels, polar stations, expedition ships, observatories, and space stations may be experienced as adventure, extreme sport, or quasi-monastic retreat. The experiential quality of field science is one of its most distinctive features, and so far one of the least studied.

Although historians may envision the field as postcards, actual field experience is more typically cinematic. Fieldwork is a seasonal to-and-fro between field and laboratory or museum—a scientific transhumance. As situations and objects move, observers move with them; and the sciences of large scale or deep time—taxonomy, stratigraphic and tectonic geology, biogeography, geophysics and the sciences of ocean and atmosphere, social and ethnographic surveys—are of necessity sciences on the move. The field is thus a category of people and things in place *and* in motion.

When place first became a topic of serious interest to historians of science, in the 1980s, it was for the specific, epistemic, purpose of deconstructing the universal claims of scientific knowledge: by showing in vivid particularity how knowledge presented as placeless and disembodied—what everyone knows—was in fact created by someone, someplace, for some reason. Initially it was laboratories that were thus used to deconstruct disembodied science—not surprisingly, since historians then mainly studied lab science; and because exposing the placelessness of labs as a discursive fiction was the acid test of deconstruction (Henke and Geiryn 2008). Putting field science in its place would have been too easy and obvious. So, when historians took up the field sciences, in the 1990s, place was a central concern but less for its epistemological use (that case had been made) than for its concrete uses in cultural, imperial, and global history, and in studies of vernacular knowledge practices and the politics of uses and abuses of nature. This range of topics has since attracted scholars of diverse sorts to field science, including historical geographers, students of literature and the arts, and environmental historians. The epistemic monoculture of deconstruction has become the polyculture of field studies.

Many Fields

Historians' interest in the field as place and practice has taken them into a widening range of scientific disciplines. Pioneering studies of field and place were, not surprisingly, of sciences traditionally associated with fieldwork: natural history (Allen 2004), geology and paleontology (Rudwick 1985), anthropology (Stocking 1983), and ecology (Tobey 1981; Cittadino 1993). This initial range of interest is apparent in the topical volumes from the mid-1990s that marked the wider recognition of "the field" as a category (Jardine, Secord, and Spary 1996; Kuklick and Kohler 1996a). A follow-up survey (Vetter 2010) takes in a different and a wider range of disciplines, including archaeology, oceanography, climatology, horticulture, and human ecology. Ethology is another apt subject for study of field as place (Burkhardt 1999; Montgomery 2005; Rees 2009).

The diversity of field sciences may be lumped for comparative analysis into social-historical types: mapping, surveying, observing, collecting, and expeditioning are some that have been proposed (Edney 1997; Knell 2000; Kohler 2007; Endersby 2008). Expeditionary sciences are an especially rich group, because of their

Figure 20.1   Field party for the US Geological and Geographical Survey of the Territories (Hayden Survey) in camp at Red Buttes, Wyoming, 1870. Photograph by William Henry Jackson. From USGS Photographic Library, photo jwh00282, libraryphoto.cr.usgs.gov/index.html.

association with European and American exploration and expansion (Naylor and Ryan 2010; Dritsas 2011; Nielsen, Harbsmeier, and Ries 2012; Kennedy 2013). Based at first on wind, water, wood, and animal power, but increasingly fueled by fossil-carbon technologies—ships, railroads, trucks, planes—expeditioning created vast infrastructures and landscapes of hybrid places that, like older outposts or encampments, are equally built and field environments (Sorrenson 1996; Vetter 2004). It is in such border places that the mingled nature and culture of the field are best studied (Figure 20.1).

The geographical scope of "field" has likewise expanded: from the "natural laboratory" of protected areas (Rumore 2012) to the "living laboratory" of heavily humanized landscapes (Tilley 2011); and from the tropics (Christen 2002) to the poles (Bravo and Sörlin 2002; Benson and Rozwadowski 2007; Powell 2007; Bocking 2013; Farish 2013). This horizontal expansion of "field" has been matched by vertical extension, as ocean (Rozwadowski 2005; Doel, Levin, and Marker 2006) and atmosphere (Fleming, Jankovic, and Coen 2006) have become prime areas for scientists, and thus for history. *Verticality* is a key emerging concept for thinking about place in the field. Whereas a horizontal view takes students of place across airy landscapes of towns, forests, farms, and ranches, the vertical view upward or downward takes us away from human habitation into depths and heights in which no one lives (for long) yet which are vital to global economy and polity (Braun 2000; Bigg, Aubin, and Felsch 2009).

The investigation of vertically differentiated places—mountains, seas, space—has been conducted not only *by* human observers but sometimes *on* human subjects (including the field scientists themselves) at such places as the Chilean Andes, Mount Everest, Sealab II, and Skylab (Tracy 2012; Heggie 2013; Karafantis 2013). And places that are imagined but not directly experienced—paleo-environments, deep sea, and even distant planets—have been transformed by empirical study into places in the "field" (Rudwick 1992; Rozwadowski 2005; Lane 2011). Studies of "extreme" places extend and complicate the more traditional concepts of place derived from the "second natures" of ethnographers' villages, cities, and working rural landscapes. In a world of diverse "fields" the categorical divide between natural and social sciences fades, as do distinctions between history of science, historical geography, and environmental history.

Places and Practices

As the places of field science have diversified, so have its tools and practices. Although physical instruments have long been used by field scientists, the tendency in modern times has been to better adapt these to actual situations in the field. The reason for going afield, after all, is to study phenomena in the natural or social contexts in which they normally play out. In field biology, for example, early efforts to import laboratory instruments and experimental procedures had mixed success. Yet, by the mid-twentieth century an array of robust hybrid instruments and practices had been developed; for example, devices for remote sensing and tracking of animals on the move (Kohler 2002a; Kohler 2002b; Lewis 2004; Benson 2010).

Nonetheless, the particularity of place and time in the field makes it virtually impossible to exactly replicate measurements or observations. This circumstance has entailed a "fieldworkers' regress," in which interpretations can always be contested and attributed to uncontrolled variability of place. In noisy places it is easy to redefine signal as noise. For example, a controversy among primatologists over the meaning of infanticide seemed closed for a time, yet kept reopening (Rees 2009). Epistemic uncertainty of place may thus be used in skirmishing over scientific turf, even where facts are reasonably secure. Environmental and conservation science is another multiuse and multidisciplinary area conducive to open-ended contests over practices and meanings (Bocking 2012).

Accounting methods constitute another family of practices that depend less on physical instruments than on the paper tools of data management: inventory, counting, sorting, tabulating, and calculating (Te Heesen 2005). The story here is less one of adapting lab tools to field conditions than of intensifying accounting tools that had always been the stock-in-trade of taxonomic and geographical science. In a world of humanized, working landscapes, few places escape the web of human inventory and accounting (Höhler and Ziegler 2010), and practices of commerce and science mingle in ways that are both productive and less wholesome. Key concepts of ecology, such as carrying capacity and beneficial fire, have emerged from range and forest management, even as these worldly practices became more ecological (Young 1998; Way 2011). Cetologists' field practices came to depend on the whaling industry (Burnett

2012); while commercial whaling has been disguised as science. Accounting and eco-nomic practices of ecosystem ecology and cost–benefit econometrics are, together, core practices of stream and landscape restoration (Brock 2004; Lave 2012). A newer and growing family of field practices is one that might be called "surveillance" prac-tice. These range in scale from radiotelemetry of wildlife at local sites to surveillance of the entire globe, especially of the places where people do not live yet highly value—oceans, poles, lithosphere, atmosphere, space. Unlike older types of hands-on survey and inventory, surveillance practices are more typically hands-off and unseen, per-formed by sophisticated machines that are remotely and often covertly controlled, such as satellites, cameras, or sensors. They are further removed from immediate experience by computer analysis and mathematical modeling. Though remote, surveillance prac-tices are practices of place—global place—and they may be fundamentally changing the meaning of field and fieldwork (Edwards 2010; Launius, Fleming, and DeVorkin 2010; Howe 2014). In glaciology, for example, a concept of polar warming based on local field observations gave way to a concept based on computer modeling (Sörlin 2011). Yet, however striking the shift to mathematical modeling, climate science still requires local field sites that have "barely enough data to test models or theories," as in vast swathes of Antarctica (Howkins 2011, 190). Whether place is becoming less important in surveillance sciences, or just important in new ways, remains to be seen.

Practitioners

Finally, historians of field science have extended the range of their interest in practition-ers to include those who were once regarded as marginal to science or even outside it. Women scientists were the first such group to move from the wings to center stage, as known and unknown individuals were taken up and reasons for their relative invisibil-ity explored (Oreskes 1996). Cases range from field naturalists to primatologists and eugenics fieldworkers (Bonta 1991; Bix 1997; Strum and Fedigan 2000). Arguably, women were historically more central players in field than in lab sciences, because field environments were once less stringent in matters of social identity and participation (Kuklick and Kohler 1996b, 10–13). Where, when, and for whom this generalization holds true needs further study.

Historians have also examined the role played by lay practitioners, whose forms of knowledge derive from personal life experience in particular places. These ways of knowing are variously termed vernacular, local, residential, folk, or indigenous, depending on the situation; however, the best catch-all term may be "experiential," since all derive from the experience of daily living (Vetter 2011, 131–4). Visiting tax-onomists will know from work in collections how to classify plants or animals, yet depend on residents with experiential knowledge of the locale to find them. Visiting ethnographers and sociologists likewise depend on informants with a living knowledge of local societies. Because field scientists deal firsthand with phenomena in context, they can put to their own use the varied experiential knowledge of hunters, farm-ers, ranchers, fishermen, mariners, prospectors, naturalists, collectors of local facts and artifacts, and others who live where visiting scientists work (Burns 2008; Reidy 2008; Vetter 2008; Keiner 2009; Frehner 2011). Learning from individuals with experiential knowledge has been especially critical in cross-cultural and colonial encounters (Green Musselman 2003; Fan 2004).

In the study of ephemeral phenomena such as earthquakes or extreme weather events, resident observers have the advantage over experts, who cannot be everywhere at once and cannot make careers from rare occurrences (Coen 2013; Valencius 2013). Politics has in some cases enabled local residents to achieve status as practitioners of "citizen science" (Fan 2012), as scientists have used fieldwork to help create national, regional, or "creole" political identities (McCook 2002; Naylor 2010; Shen 2014). Field scientists also may attain through intensive local work a kind of experiential knowledge that, combined with the "cosmopolitan" knowledge of books and class-rooms, forms what might be termed "residential science" (Kohler 2011). Varieties of expert vernacular science constitute a developing area in history of science, and the field sciences afford exceptional opportunities for their study.

Conclusion

The expanding scope and popularity of field science constitute one of the most notable developments in contemporary history of science. Always a substantial and respected, yet somewhat ghettoized subject, field sciences are now center stage in the study of general issues such as lay practitioners, situated and big-data practices, and place. Why this trend occurred when and how it did we cannot say, but James Secord has pointed to a likely explanation when he argued that *materiality* is what underlies all modern sciences and their histories. "The key," Secord writes, "is our new understanding of scientific knowledge as practice. All evidence from the past is in the form of material things" (Secord 2004, 665). And, one might add, it is material things *in place*. For what is more material than place, and where is place more vital to science than in the field? Where are tools, practices, and identities more powerfully shaped by material situations? It is perhaps little wonder that field sciences have moved into history's limelight.

And if, in a world of anonymous global information, place becomes the one attribute that reliably makes science credible, then field science is likely to remain in the limelight as a model of authentic, situated practice (Henke and Gieryn 2008, 369). Once the "view from nowhere" and then the "view from everywhere," science may come to be the view from somewhere. Many labs, factories, offices, and museums are "truth spots." *All* the field is somewhere.

References

Adler, Antony. 2014. "The ship as laboratory: Making space for field science at sea." *Journal of the History of Biology*, 47: 333–62.

Alagona, Peter S. 2012. "A sanctuary for science: The Hastings Natural History Reservation and the origins of the University of California's Natural Reserve System." *Journal of the History of Biology*, 45: 651–80.

Alberti, Samuel J. M. M. 2001. "Amateurs and professionals in one county: Biology and natural history in late Victorian Yorkshire." *Journal of the History of Biology*, 34: 115–47.

Allen, David E. 2004. *The Naturalist in Britain: A Social History*. London: Penguin, 1976; reissued Princeton: Princeton University Press.

Anker, Peder. 2007. "Science as a vacation: A history of ecology in Norway." *History of Science*, 45: 455–79.

Barrow, Mark V. 1998. *A Passion for Birds: American Ornithology after Audubon.* Princeton, NJ: Princeton University Press.

Benson, Etienne. 2010. *Wired Wilderness: Technologies of Tracking and the Making of Modern Wildlife.* Baltimore: Johns Hopkins University Press.

Benson, Keith R. and Helen M. Rozwadowski (eds.) 2007. *Extremes: Oceanography's Adventures at the Poles.* Sagamore Beach, MA: Science History Publications.

Bigg, Charlotte, David Aubin, and Philipp Felsch (eds.) 2009. "The laboratory of nature—science in the mountains." *Science in Context*, 22: 311–531.

Bix, Amy Sue. 1997. "Experiences and voices of eugenics field-workers: 'Women's work' in biology." *Social Studies of Science*, 27: 625–68.

Bocking, Stephen. 2012. "Science, salmon, and sea lice: Constructing practice and place in an environmental controversy." *Journal of the History of Biology*, 45: 681–716.

Bocking, Stephen. 2013. "Situated yet mobile: Examining the environmental history of Arctic ecological science." In *New Natures: Joining Environmental History with Science and Technology Studies*, edited by Dolly Jorgensen, Finn Arne Jorgensen, and Sara B. Pritchard, 164–78. Pittsburgh, PA: University of Pittsburgh Press.

Bonta, Marcia Myers. 1991. *Women in the Field: America's Pioneering Women Naturalists.* College Station, TX: Texas A&M University Press.

Braun, Bruce. 2000. "Producing vertical territory: Geology and governmentality in late Victorian Canada." *Ecumene*, 7: 7–46.

Bravo, Michael and Sverker Sörlin (eds.) 2002. *Narrating the Arctic: A Cultural History of Nordic Scientific Practices.* Canton, MA: Science History Publications.

Brinkman, Paul D. 2010. *The Second Jurassic Dinosaur Rush: Museums and Paleontology in America at the Turn of the Twentieth Century.* Chicago: University of Chicago Press.

Brock, Emily. 2004. "The challenge of reforestation: Ecological experiments in the Douglas fir forest, 1920–1940." *Environmental History*, 9: 57–79.

Buhs, Joshua Blu. 2004. *The Fire Ant Wars: Nature, Science, and Public Policy in Twentieth-Century America.* Chicago: University of Chicago Press.

Bulmer, Martin, Kevin Bales, and Kathryn Kish Sklar (eds.) 1991. *The Social Survey in Historical Perspective*, 1880–1940. Cambridge: Cambridge University Press.

Burkhardt, Richard W. 1999. "Ethology, natural history, the life sciences, and the problem of place." *Journal of the History of Biology*, 32: 489–508.

Burnett, D. Graham. 2012. *The Sounding of the Whale: Science and Cetaceans in the Twentieth Century.* Chicago: University of Chicago Press.

Burns, J. Conor. 2008. "Networking Ohio Valley archaeology in the 1880s." *Histories of Anthropology Annual*, 4: 1–33.

Christen, Catherine A. 2002. "At home in the field: Smithsonian tropical science field stations in the U.S. Panama Canal Zone and the Republic of Panama." *The Americas*, 58: 537–75.

Cittadino, Eugene. 1993. "A 'marvelous cosmopolitan preserve': The Dunes, Chicago, and the dynamic ecology of Henry Cowles." *Perspectives on Science*, 1: 520–59.

Coen, Deborah R. 2013. *The Earthquake Observers: Disaster Science from Lisbon to Richter.* Chicago: University of Chicago Press.

Cooper, Alix. 1998. "From the Alps to Egypt (and back again): Dolomieu, scientific voyaging, and the construction of the field in eighteenth-century natural history." In *Making Space for Science: Territorial Themes in the Shaping of Knowledge*, edited by Crosbie Smith and Jon Agar, 39–63. London: Macmillan.

De Bont, Raf. 2015. *Stations in the Field: A History of Place-Based Animal Research, 1870–1930.* Chicago: University of Chicago Press.

Doel, Ronald E., Tanya J. Levin, and Mason K. Marker. 2006. "Extending modern cartography to the ocean depths: Military patronage, Cold War priorities, and the Heezen–Tharp mapping project, 1952–1959." *Journal of Historical Geography*, 32: 605–26.

Dritsas, Lawrence. 2011. "Expeditionary science: Conflicts of method in mid-nineteenth-century geographical discovery." In *Geographies of Nineteenth-Century Science*, edited by David N. Livingstone and Charles W. J. Withers, 255–77. Chicago: University of Chicago Press.

Edney, Matthew H. 1997. *Mapping an Empire: The Geographical Construction of British India, 1765–1843*. Chicago: University of Chicago Press.

Edwards, Paul N. 2010. *A Vast Machine: Computer Models, Climate Data, and the Politics of Global Warming*. Cambridge, MA: MIT Press.

Endersby, Jim. 2008. *Imperial Nature: Joseph Dalton Hooker and the Practices of Victorian Science*. Chicago: University of Chicago Press.

Fan, Fa-ti. 2004. *British Naturalists in Qing China: Science, Empire, and Cultural Encounter*. Cambridge, MA: Harvard University Press.

Fan, Fa-ti. 2012. "'Collective monitoring, collective defense': Science, earthquakes, and politics in communist China." *Science in Context*, 25: 127–54.

Farish, Matthew. 2013. "The lab and the land: Overcoming the Arctic in Cold War Alaska." *Isis*, 104: 1–29.

Fleming, James Rodger, Vladimir Jankovic, and Deborah R. Coen (eds.) 2006. *Intimate Universality: Local and Global Themes in the History of Weather and Climate*. Sagamore Beach, MA: Science History Publications.

Frehner, Brian. 2011. *Finding Oil: The Nature of Petroleum Geology, 1859–1920*. Lincoln, NE: University of Nebraska Press.

Gieryn, Thomas F. 2006. "City as truth-spot: Laboratories and field-sites in urban studies." *Social Studies of Science*, 36: 5–38.

Green Musselman, Elizabeth. 2003. "Plant knowledge at the Cape: A study in African and European collaboration." *International Journal of African Historical Studies*, 36: 367–92.

Heggie, Vanessa. 2013. "Experimental physiology, Everest and oxygen: From the ghastly kitchens to the gasping lung." *British Journal for the History of Science*, 46: 123–47.

Henke, Christopher R. 2000. "Making a place for science: The field trial." *Social Studies of Science*, 30: 483–511.

Henke, Christopher R. and Thomas F. Gieryn. 2008. "Sites of scientific practice: The enduring importance of place." In *The Handbook of Science and Technology Studies*, edited by Edward J. Hackett, Olga Amsterdamska, Michael Lynch, and Judy Wajcman, 353–76. Cambridge, MA: MIT Press.

Hevly, Bruce. 1996. "The heroic science of glacier motion." *Osiris*, 11: 66–86.

Höhler, Sabine and Rafael Ziegler (eds.) 2010. "Nature's accountability: Stocks and stories." *Science as Culture*, 19: 417–584.

Howe, Joshua P. 2014. *Behind the Curve: Science and the Politics of Global Warming*. Seattle: University of Washington Press.

Howkins, Adrian. 2011. "Melting empires? Climate change and politics in Antarctica since the International Geophysical Year." *Osiris*, 26: 180–97.

Jardine, Nicolas, James A. Secord, and Emma C. Spary (eds.) 1996. *Cultures of Natural History*. Cambridge: Cambridge University Press.

Karafantis, Layne. 2013. "Sealab II and Skylab: Psychological fieldwork in extreme spaces." *Historical Studies in the Natural Sciences*, 43: 551–88.

Keiner, Christine. 2009. *The Oyster Question: Scientists, Watermen, and the Maryland Chesapeake Bay since 1880*. Athens, GA: University of Georgia Press.

Kennedy, Dane. 2013. *The Last Blank Spaces: Exploring Africa and Australia*. Cambridge, MA: Harvard University Press.

Kingsland, Sharon E. 2009. "Frits Went's Atomic Age greenhouse: The changing landscape on the lab–field border." *Journal of the History of Biology*, 42: 289–324.

Knell, Simon J. 2000. *The Culture of English Geology, 1815–1851: A Science Revealed through Its Collecting*. Aldershot: Ashgate.

Kohler, Robert E. 2002a. "Labscapes: Naturalizing the lab." *History of Science*, 40: 473–501.

Kohler, Robert E. 2002b. *Landscapes and Labscapes: Exploring the Lab–Field Border in Biology*. Chicago: University of Chicago Press.

Kohler, Robert E. 2006. *All Creatures: Naturalists, Collectors, and Biodiversity, 1850–1950*. Princeton, NJ: Princeton University Press.

Kohler, Robert E. 2007. "Finders, keepers: Collecting sciences and collecting practice." *History of Science*, 45: 428–54.

Kohler, Robert E. 2011. "Paul Errington, Aldo Leopold, and wildlife ecology: Residential science." *Historical Studies in the Natural Sciences*, 41: 216–54.

Kuklick, Henrika. 1997. "After Ishmael: The fieldwork tradition and its future." In *Anthropological Locations: Boundaries and Grounds of a Field Science*, edited by Akhil Gupta and James Ferguson, 47–65. Berkeley, CA: University of California Press.

Kuklick, Henrika. 2011. "Personal equations: Reflections on the history of fieldwork, with special reference to sociocultural anthropology." *Isis*, 102: 1–33.

Kuklick, Henrika and Robert E. Kohler (eds.) 1996a. *Science in the Field*. Vol. 11 (2nd ser.), *Osiris*. Chicago: University of Chicago Press.

Kuklick, Henrika and Robert E. Kohler. 1996b. "Introduction." In *Science in the Field*, edited by Henrika Kuklick and Robert E. Kohler, 1–14. Chicago: University of Chicago Press.

Lachmund, Jens. 2013. *Greening Berlin: The Co-Production of Science, Politics, and Urban Nature*. Cambridge, MA: MIT Press.

Lane, K. Maria D. 2011. *Geographies of Mars: Seeing and Knowing the Red Planet*. Chicago: University of Chicago Press.

Lave, Rebecca. 2012. *Fields and Streams: Stream Restoration, Neoliberalism, and the Future of Environmental Science*. Athens, GA: University of Georgia Press.

Launius, Roger D., James Rodger Fleming, and David H. DeVorkin (eds.) 2010. *Globalizing Polar Science: Reconsidering the International Polar and Geophysical Years*. Basingstoke: Palgrave Macmillan.

Lewis, Michael L. 2004. *Inventing Global Ecology: Tracking the Biodiversity Ideal in India, 1947–1997*. Athens, OH: Ohio University Press.

McCook, Stuart. 2002. *States of Nature: Science, Agriculture, and Environment in the Spanish Caribbean, 1760–1940*. Austin: University of Texas Press.

Montgomery, Georgina. 2005. "Place, practice and primatology: Clarence Ray Carpenter, primate communication and the development of field methodology, 1931–1945." *Journal of the History of Biology*, 38: 495–533.

Naylor, Simon. 2010. *Regionalizing Science: Placing Knowledges in Victorian England*. London: Pickering & Chatto.

Naylor, Simon and James R. Ryan (eds.) 2010. *New Spaces of Exploration: Geographies of Discovery in the Twentieth Century*. London: I.B.Tauris.

Nielsen, Kristian H., Michael Harbsmeier, and Christopher J. Ries (eds.) 2012. *Scientists and Scholars in the Field: Studies in the History of Fieldwork and Expeditions*. Aarhus: Aarhus University Press.

Oreskes, Naomi. 1996. "Objectivity or heroism? On the invisibility of women in science." *Osiris*, 11: 87–113.

Pang, Alex Soojung-Kim. 2002. *Empire and the Sun: Victorian Solar Eclipse Expeditions*. Stanford: Stanford University Press.

Powell, Richard C. 2007. "'The rigours of an Arctic experiment': The precarious authority of field practices in the Canadian High Arctic, 1958–1970." *Environment and Planning A*, 39: 1794–811.

Rees, Amanda. 2009. *The Infanticide Controversy: Primatology and the Art of Field Science.* Chicago: University of Chicago Press.

Reidy, Michael S. 2008. *Tides of History: Ocean Science and Her Majesty's Navy.* Chicago: University of Chicago Press.

Rozwadowski, Helen M. 1996. "Small world: Forging a scientific maritime culture for oceanography." *Isis,* 87: 409–29.

Rozwadowski, Helen M. 2005. *Fathoming the Ocean: The Discovery and Exploration of the Deep Sea.* Cambridge, MA: Harvard University Press.

Rudwick, Martin S. J. 1976. "The emergence of a visual language for geological science." *History of Science,* 14: 149–95.

Rudwick, Martin J. S. 1985. *The Great Devonian Controversy: The Shaping of Scientific Knowledge among Gentlemanly Specialists.* Chicago: University of Chicago Press.

Rudwick, Martin S. J. 1992. *Scenes from Deep Time: Early Pictorial Representations of the Prehistoric World.* Chicago: University of Chicago Press.

Rudwick, Martin. 1996. "Geological travel and theoretical innovation: The role of 'liminal' experience." *Social Studies of Science,* 26: 143–59.

Rumore, Gina. 2012. "Preservation for science: The Ecological Society of America and the campaignx for Glacier Bay National Monument." *Journal of the History of Biology,* 45: 613–50.

Schneider, Daniel W. 2000. "Local knowledge, environmental politics, and the founding of ecology in the United States: Stephen Forbes and 'The lake as microcosm' (1887)." *Isis,* 91: 681–705.

Schumaker, Lyn. 2001. *Africanizing Anthropology: Fieldwork, Networks, and the Making of Cultural Knowledge in Central Africa.* Durham, NC: Duke University Press.

Secord, James A. 2004. "Knowledge in transit." *Isis,* 95: 654–72.

Shen, Grace Yen. 2014. *Unearthing the Nation: Modern Geology and Nationalism in Republican China.* Chicago: University of Chicago Press.

Sörlin, Sverker. 2011. "The anxieties of a science diplomat: Field coproduction of climate knowledge and the rise and fall of Hans Ahlmann's 'polar warming.'" *Osiris,* 26: 66–88.

Sorrenson, Richard. 1996. "The ship as a scientific instrument in the eighteenth century." *Osiris,* 11: 221–36.

Stocking, George W. (ed.) 1983. *Observers Observed: Essays on Ethnographic Fieldwork.* Madison: University of Wisconsin Press.

Strum, Shirley C. and Linda Marie Fedigan (eds.) 2000. *Primate Encounters: Models of Science, Gender, and Society.* Chicago: University of Chicago Press.

Te Heesen, Anke. 2005. "Accounting for the natural world: Double-entry bookkeeping in the field." In *Colonial Botany: Science, Commerce, and Politics in the Early Modern World,* edited by Londa Schiebinger and Claudia Swan, 237–51. Philadelphia: University of Pennsylvania Press.

Tilley, Helen. 2011. *Africa as a Living Laboratory: Empire, Development, and the Problem of Scientific Knowledge, 1870–1950.* Chicago: University of Chicago Press.

Tobey, Ronald C. 1981. *Saving the Prairies: The Life Cycle of the Founding School of American Plant Ecology, 1895–1955.* Berkeley: University of California Press.

Tracy, Sarah W. 2012. "The physiology of extremes: Ancel Keys and the International High Altitude Expedition of 1935." *Bulletin of the History of Medicine,* 86: 627–60.

Valencius, Conevery Bolton. 2013. *The Lost History of the New Madrid Earthquakes.* Chicago: University of Chicago Press.

Vetter, Jeremy. 2004. "Science along the railroad: Expanding fieldwork in the US Central West." *Annals of Science,* 61: 187–211.

Vetter, Jeremy. 2008. "Cowboys, scientists, and fossils: The field site and local collaboration in the American West." *Isis,* 99: 273–303.

Vetter, Jeremy (ed.) 2010. *Knowing Global Environments: New Historical Perspectives on the Field Sciences*. New Brunswick: Rutgers University Press.

Vetter, Jeremy. 2011. "Lay participation in the history of scientific observation." *Science in Context*, 24: 127–41.

Vetter, Jeremy. 2012. "Labs in the field? Rocky Mountain biological stations in the early twentieth century." *Journal of the History of Biology*, 45: 587–611.

Way, Albert G. 2011. *Conserving Southern Longleaf: Herbert Stoddard and the Rise of Ecological Land Management*. Athens, GA: University of Georgia Press.

Young, Christian C. 1998. "Defining the range: The development of carrying capacity in management practice." *Journal of the History of Biology*, 31: 61–83.

CHAPTER TWENTY-ONE

The Laboratory

CATHERINE M. JACKSON

The laboratory is the iconic space of modern science. Such is the power of this status that the historical role of the laboratory in the practice of science and the production of scientific knowledge is easily taken for granted, thereby obscuring the laboratory's changing relationship to our understanding of, and control over, the natural world. The laboratory as we know it today is a product of modernity, and especially of the rise of science as a professional and highly institutionalized activity in the late-nineteenth and early-twentieth centuries. Our view of the laboratory's significance, one might argue, has tracked our historical and philosophical understanding of the very nature of science. With this in mind, this chapter pursues two interconnected goals. First, it presents a short history of the laboratory, beginning with a brief descriptive overview of changes in the laboratory up to about 1800, before focusing on the development of the laboratory during the nineteenth and twentieth centuries. And second, it analyzes how historical approaches to the laboratory have changed over recent decades, showing how these approaches reflect different views concerning the nature of science and varying conceptions of what it means to do science, or to be a scientist.

A Brief History of the Laboratory

It is difficult to be definitive about when something called a laboratory became inevitably and exclusively scientific. According to Steven Shapin (1988, 277) the word laboratory (or *elaboratory*) was not in common English usage before the mid-seventeenth century. The association between the laboratory and science, moreover, remains distinctly Anglophone: in Italy, for example, the word *laboratorio* continues to refer generically to workshops in a wide range of artisanal and artistic as well as scientific settings. Principally associated with the practice of alchemy, early laboratories were places for the preparation of medicines, the production of reagents such as acids, and, above all, the pursuit of the elusive philosopher's stone (the substance believed capable of transmuting base metals into gold and silver). Staffed by apothecaries, alchemists, and a range of laboratory servants responsible for everything from keeping the fires

A Companion to the History of Science, First Edition. Edited by Bernard Lightman.
© 2016 John Wiley & Sons Ltd. Published 2020 by John Wiley & Sons Ltd.

burning and doing the washing up to guarding the key to the door, laboratories like this were the private, often secret domain of princes and noblemen, and occasionally their wives. Despite its connection to philosophy and theology—made manifest in books on theosophical alchemy and other manuscripts—the laboratory before about 1650 was a place of essentially practical knowledge based on manual labor. As a result, laboratory work was certainly "not a suitable activity for the independent and leisured free man" and its products, though undeniably useful, did not include certain or scientific knowledge (Smith 2006, 293).

By the end of the seventeenth century some laboratories—mainly connected to academic institutions such as universities, botanic gardens, and academies (Klein 2008, 770)—had become the location of a new kind of experimental science, home to a "new active mode of doing philosophy" (Smith 2006, 305). But not everything done in these laboratories met this definition of science as a philosophical enterprise, nor were all laboratories spaces within which experiments carried out by appropriate persons gave access to legitimate truths about nature: many more were the workplaces of apothecaries, textile dyers, and mining engineers. Laboratories continued to proliferate throughout the eighteenth century in both academic and industrial settings. But, whatever their institutional location, these laboratories had some important common features. They were almost exclusively the site of chemical operations such as distillation, combustion, smelting, dissolution, and precipitation. And they were far more likely to be concerned with commercial production and its regulation than with anything we might characterize as pure scientific research (Klein 2008, 770).

The relationship between the development of the laboratory and the rise of experimental natural philosophy is, therefore, not a simple one. And, although the venue of scientific and particularly experimental activity has long been identified as important in the history of science (Shapin 1988, 373–4), we remain far from having a coherent view—let alone a detailed historical understanding—of how histories of the laboratory and of experiment intersect, where they diverge, and what should be their proper place within broader history of science (Gooday 2008; Klein 2008; Kohler 2008). This short chapter cannot hope to answer questions of such magnitude. But, by drawing attention to them at this stage, I hope to alert the reader to current historical and historiographical concerns that will form the basis of a more detailed discussion of the field of laboratory studies in the second section, leading to some proposals for potentially fruitful future directions of enquiry that are set out in the conclusion.

The uniquely *chemical* associations of the laboratory before 1800 provide a salient reminder of the important distinction between a history of the laboratory and a history of experiment. Many of the classic historical studies of experiment describe events that did not take place in the laboratory at all, but rather in learned societies, salons, private houses, and anatomical theaters[1]—not to mention the wider terrain that is home to what are now called the field sciences (see Chapter 20 and other chapters in Part II). So, while experiment implied a certain control over nature that we have come to associate mainly (though certainly not exclusively) with the laboratory (Latour 1983), the laboratory around 1800 was merely one among many potential locations of scientific activity (Crosland 2005, 233). Nor should we forget that the modern laboratory remains home to much that is not experimental: teaching, testing, assaying, and analysis are all important laboratory activities (Gooday 2008, 784–5). These observations, in turn, raise two important questions. If the laboratory was not primarily a place of

Figure 21.1 Frontispiece from William Lewis (1714–1781). 1763. *Commercium Philosophico-Technicum; or the Philosophical Commerce of Arts: Designed as an attempt to improve arts, trades, and manufactures.* London: Baldwin.

experiment but rather somewhere necessary for doing "chemical work" (Klein 2008, 771), what was it about the practice of chemistry in particular that relied upon the facilities of a laboratory? And what subsequently led practitioners of other sciences to move into the laboratory, ultimately so weakening the association between the laboratory and chemistry that studies of nineteenth and twentieth-century laboratories have, until very recently, tended to focus on physics and the life sciences?[2]

An inspection of the chemical laboratory around 1800 is an instructive place to begin answering these questions. Fortunately, several excellent historical studies provide detailed descriptions—together with numerous potent images—of chemical laboratories of this period and of the work carried out there (Holmes 1989; Beretta 2004; Klein and Lefèvre 2007; Klein and Spary 2010). Fireplaces and furnaces are prominent in all these laboratories, reflecting the dominance in this period of heat and fire as means of effecting chemical change. We should also note some important absences. Eighteenth-century laboratories such as that illustrated in William Lewis's (1763) *Philosophical Commerce of Arts* contained apparatus and instruments for performing a range of manipulations and measurements (Figure 21.1). But they were not equipped with the glass reagent bottles and wide range of glass apparatus that later became central to the material culture of the chemical laboratory (Jackson 2015a; Jackson 2015b). Nor was large-scale pedagogical activity among the major purposes of laboratories like this. Although apprentices certainly trained in the laboratories of apothecaries, perfumers, textile manufacturers, gunsmiths, and assayers, they did so in relatively small numbers. The chemical laboratory around 1800 had long since ceased to be a secret, or even a necessarily private, space (Hannaway 1986, 599) but it had much more in common with alchemical kitchens than with the institutional laboratories which became essential tools of disciplinary training and research in late nineteenth-century science (Jackson 2011).

Doing chemistry—especially doing chemical experiments—was a smelly, dangerous business, best kept below stairs, well away from polite society. Here is a plausible explanation for why eighteenth-century chemists—whether they were doing experiments or producing commercial products—worked in laboratories, while most natural philosophers did not (Klein 2008). In the famous Society of Arcueil, for example, the great Berthollet performed chemical experiments in a laboratory located at the bottom of the garden and not in the drawing room, library, or even the kitchen of his palatial country house.[3] The laboratory was not, in the first instance, a place of experiment. It was the particular space in which the technical hazards of chemical practice, whatever its purpose and setting, could be made to conform to social norms.

The combined demands of chemical practice and pedagogy acquired a new urgency during the first half of the nineteenth century. First, changes in chemistry's disciplinary status and its move from the medical to the philosophical faculty of the reformed German research university necessarily brought chemistry within the polite, scholarly realm (Meinel 1983; Meinel 1988; Meinel 2000). And second, developments in chemists' investigative approach to organic nature—beginning, famously, with Justus Liebig's introduction of large-scale laboratory training in Giessen (Morrell 1972)— exposed a rapidly increasing number of students to the potentially incendiary combination of naked flame and organic substance. In exactly the period when new theories of chemical composition and constitution were providing ever-increasing legitimacy to chemistry as a philosophical endeavor, the practice of chemistry had never been more dangerous (Jackson 2011, 56–7).

Several studies have demonstrated the role of industrialization and competition in driving the nineteenth-century construction of institutional chemical laboratories in various German states (Borscheid 1976; Tuchman 1993). During the middle decades of the century it became increasingly common for academic chemical laboratories to be housed within grand, purpose-built institutes. But, as I have argued elsewhere, the interior design and fittings of these laboratories also changed rapidly in this period in response to the introduction of new investigative methods and novel glass apparatus, as well as further developments in chemical pedagogy (Jackson 2011). Training in Liebig's famous Giessen laboratory focused almost exclusively on the technique of combustion analysis and the interpretation of the results this produced in terms of the composition of organic substances. Towards the end of the century, by contrast, student chemists learnt a wide range of techniques, including how to perform the dozens of reactions and manipulations upon which the practice of organic synthesis relied. And while many aspects of the Giessen laboratory—for example, its central tables, coke-fired furnaces, and limited ventilation—resembled earlier laboratories, late nineteenth-century chemical laboratories increasingly incorporated what remain ubiquitous modern features.

Some of these related to the storage and use of new laboratory apparatus, especially chemical glassware components that, when skillfully combined, allowed chemists to control the reaction, purification, and characterization of organic substances (Jackson 2015a; Jackson 2015b). But many others were concerned with maintaining a safe working environment, particularly for inexperienced chemists in training.

Ventilation was a major concern, leading to the introduction of effective fume hoods in August Hofmann's Berlin laboratory in the 1860s (Jackson 2011, 58). So, too, were fire and explosion. The directors of late nineteenth-century institutional chemical institutes adopted a range of strategies in managing these day-to-day hazards of practical organic chemistry. They certainly encouraged safe practices through teaching and the hierarchy of the laboratory as a social institution (Meinel 2000, 298–9). But they also addressed this central problem in material ways, at every level from the provision of individual items of apparatus through to the design, construction, and fitting of the laboratory building itself, as well as its location and garden setting (Jackson 2011, 57–8). In the institute built around 1900 for Emil Fischer, Hofmann's successor as professor of chemistry at the University of Berlin, for example, fire extinguishers were available at the door of each laboratory. The institute building, meanwhile, incorporated special rooms with reinforced walls containing protective steel enclosures called ballistic cabinets: dangerous procedures in which flammable substances were subjected to extremes of temperature and pressure inside glass apparatus were controlled by specific physical—and not merely social—means (Jackson 2011, 59–60).

By about 1870 chemical laboratories were an important feature of many German universities. Second only in cost to anatomical institutes, the institutional chemical laboratory was an essential tool of professional chemical training and cutting-edge research, mainly in the field of organic chemistry (Jackson 2011, 60–2). The early decades of the nineteenth century had seen a spectacular expansion in popular interest in chemistry, encouraged by the publication of manuals and treatises and the availability of "portable laboratories" including adequate apparatus for basic practical inorganic chemistry (Gee 1989). But the basics of practical organic chemistry could not safely be learnt at home, an important factor driving the institutionalization of chemistry during the second half of the nineteenth century.

These changes also provided the model for subsequent and much more widely studied aspects of laboratory history. David Cahan's work on the "institutional revolution" of the 1870s described the rapid institutionalization of physics, and especially the physics of precision measurement, in a newly unified Germany (Cahan 1985; 1989). In architectural terms, the physikalisch-technische Reichsanstalt (PTR) resembled existing institutional chemical laboratories. But, like many other state-funded laboratories across Europe from the late nineteenth century onwards, the PTR's main purpose was to create, maintain, and distribute the standards of measurement upon which late nineteenth-century science and technology relied (Schaffer 1992). One important consequence of this focus on metrology—first noted in Graeme Gooday's study of the "laboratory revolution" in British physics—was that, while training was a central function of institutional physics laboratories of the 1870s and 1880s, these laboratories were not, in fact, primarily places of experimental research (Gooday 1990).

Studies of the institutionalization of medicine and the life sciences, especially physiology, have revealed a pattern of late nineteenth-century laboratory building in academic settings broadly similar to that described here for physics (Cunningham and Williams 1992, and especially Lenoir 1992 in that volume). By the end of the century institutional laboratories had become defining features of all the major scientific disciplines. The institutional laboratory around 1900 owed its origins to the particular difficulties associated with practical chemistry, and it had been developed into an essential tool of training and research. It also played an equally important role in establishing the legitimacy, reliability, and productivity of practical science in all its forms, functioning as an essential badge of intellectual—and especially academic—credibility (Jackson 2011, 61–2).

The role and status of laboratories in the life and human sciences is complicated by their dependence on living subjects—both plant and animal—and, especially in the case of humans, by the requirement for the subjects' participation, whether willing or coerced (Kirk 2010; Lundgren 2013, and see also Chapter 9 of this volume). Robert Kohler (2002) has shown that biological sciences push at the boundary between field and laboratory. Human sciences similarly connect clinic and laboratory, thereby bringing laboratory science into contact with broader society in ways that highlight its moral and ethical dimensions.

This chapter has deliberately privileged the development of academic laboratories in the period after 1800. As a result, it has paid more attention to modern laboratories as sites of training and research than to their equally important but much less well-understood functions as places where industrial processes and military technologies were developed, product quality tested, forensic investigations pursued, and government policy and legislation supported and enforced. This decision does not reflect any lack of commitment to the importance of industrial and government laboratories, whether engaged in activities of primarily social, political, economic, or military significance. Rather, it is driven by a much lesser degree of clarity concerning the grand narrative of these developments. Despite a rapidly growing literature dealing with late nineteenth- and twentieth-century laboratories within or funded by industry, the military, and government (e.g., Meinel 2000; Reinhardt and Travis 2000; Burney 2002; Rooij 2011; Slayton 2012), there remains no clear framework for those developments—an important future direction we shall revisit in the next section.

Up to this point, this chapter has focused on laboratories located in Western Europe and North America. That geographical focus is, to a considerable extent, a consequence of the topic and it is certainly reflected in the existing secondary literature. It has been a major argument of this chapter that, while the early laboratory was a chemical workplace, the modern laboratory is a product of the professionalization and institutionalization of science in nineteenth-century Europe. Germany took a clear lead in these changes, followed first by other European countries and later by the United States. There are numerous excellent studies of the translation of laboratory training on the German model to Britain, France, and the United States (Rezneck 1970; Rossiter 1975; Rocke 2003) including a few that have focused on women's access to laboratory training and work (e.g., Richmond 1997; Micault 2013).

In recent years, valuable attention has been devoted to the spread of institutional laboratories beyond the West. In addition to studies focused mainly on the development of laboratory science in imperial settings in the nineteenth and twentieth centuries (Gooday and Low 1998; Arnold 2000, esp. chapter 5; Günergun 2009; Chakrabarti 2012; Clarke 2013; Günergun and Etker 2013), there is also a growing body of work dealing with postcolonial laboratories (e.g., Baytop 1997; Kikuchi 2013; Phalkey 2013). These histories increasingly give due credit to non-Western figures as active participants in the creation of their own future and they show that the laboratory became an established feature of science outside the West by the late nineteenth century. There is clearly much important work to be done in this area but non-Western laboratories are now firmly on the historical agenda.

This account of the origins of the modern laboratory has pointed to the historical contingency of what now appear to be inevitable connections between the laboratory and major areas of scientific activity. As a result, it has drawn attention to some important opportunities presented by the existing literature. By outlining how the chemical laboratory was transformed during the nineteenth century from a place of specifically chemical labor into a site of academic training and experimental research, and how other scientific disciplines adopted laboratories like this by the end of the century, for example, it indicates that—even in the most-studied case of academic laboratories— our understanding of the processes by which modern science became a laboratory activity remains imperfect. And it has delineated some important areas—most notably concerning non-Western laboratories, non-academic laboratories and twentieth- and twenty-first-century laboratories—where, despite many outstanding recent studies, we simply do not know enough to be able to construct a coherent macro-level account. Future work in all these areas will make an important contribution to our understanding of the nature of the modern scientific enterprise. Such work, moreover, will be greatly facilitated by knowing something of the origins of the field of lab studies and the range of approaches scholars have used to investigate the laboratory as a social, historical, scientific, and cultural entity.

Lab Studies as a Field: Past, Present, and Future

The previous section noted in passing some of the historiographical problems inherent in a methodological focus on the laboratory as a location of scientific, and especially experimental, activity. By choosing the laboratory (or any other site of science) as an analytical category, the historian exercises a series of commitments concerning the nature of science, its proper venues and practitioners, its performance and product. But because these commitments are frequently implicit there is a tendency for them to remain invisible. Articulating and understanding the consequences of such commitments is important if we are to realize the possibilities offered by lab studies and how these might best contribute to the history of science as a whole. The goal of this section is to sketch the driving forces behind the original field of lab studies and its more recent offshoot, the trend towards writing histories of science through its many and various locations.[4]

Unlike the laboratory, it is easy to say when the field of lab studies was born. Offspring of the late 1970s turn to practice, during which historical and philosophical enquiry were supplemented by methods drawn from anthropology and sociology, lab studies reached a first maturity in the 1990s. Drawing upon the work of Ludwig Fleck and Thomas Kuhn, and gaining momentum from the new sociology of scientific knowledge at the heart of the Edinburgh Strong Programme, some of the first lab studies involved a newly empirical, ethnographic approach to science and scientists.[5] Dissatisfied with existing macro-social studies of science, Bruno Latour and Steve Woolgar (1979, 15) famously sought "to provide a reflexive understanding of the detailed activities of working scientists" through an account "based on the experiences of close daily contact with laboratory scientists over a two year period."

Latour and Woolgar's *Laboratory Life* has since become a classic in the field. But, because it argued that what we call scientific facts are the negotiated outcome of interactions between scientists, it was construed by some—distressingly wrongly, according to Latour (1999)—as an attack on science. This perception was reinforced by later sociological studies of present-day laboratory practice—notably Harry Collins' (1985) *Changing Order*, which exposed the difficulty of relying on experimental replication as a method of establishing new facts. The field of lab studies, as it emerged around 1980, was primarily sociological and philosophical rather than historical. Its methods were empirical and its goals were significantly epistemological, and it was by no means hostile to science and scientists. What is it that scientists know? How is it that they know those things? Why is scientific knowledge stable enough to be useful yet capable of change? Questions like these motivated studies of present-day laboratories in the 1980s and early 1990s.

The same questions, moreover, could be translated from the present into the past— a move that opened the way for studies of historical laboratories including, to give an early example of outstanding importance, Owen Hannaway's (1986) comparison of Andreas Libavius' Chemical House with Tycho Brahe's Uraniborg laboratory. But, as Steven Shapin (1988, 276–8) soon pointed out, historical studies of laboratories and other sites of experimentation involved considerable sensitivity to language and custom as well as to social structure and scientific practice. Shapin's much-cited essay "The House of Experiment in seventeenth-century England" referred to a large variety of locations, almost none of which were called laboratories.

Numerous scholars from differing disciplines participated in the development during the late 1980s and early 1990s of what Kohler (2008, 762 and n. 3) has called "diversely fruitful theoretical frameworks for understanding laboratories." Many of these frameworks took due account of the historical complexity of the laboratory, and they included those studies of the laboratory as primary site of the institutionalization of various scientific disciplines in a range of national settings discussed in the previous section. When Karin Knorr Cetina identified "laboratory studies" with "the cultural approach to the study of science" in 1995, she envisaged a broad and productive future for the new field as a component of both history of science and the new discipline of science studies. Robert Kohler—himself a noted historian of the laboratory—has since regretted lab studies' failure to fulfill this early promise, using his 2008 review essay to urge for re-invigoration of the field (762). I agree with Kohler about both the present state of the field and the desirability of change. My analysis of how we have come to this point is somewhat different, however, and this leads me in the final section to make some alternative proposals for future lab studies.

The Laboratory at the Intersection: Practice, Pedagogy, and Material Culture

In 2008, Robert Kohler (763–4) suggested two possible reasons for what he saw as the decline in lab studies since the mid-1990s. The first was that scholars could see nothing further of theoretical interest to be derived from such studies. The second related the decline in lab studies to the general unpopularity of institutional history. What we lack, Kohler claimed, are studies of the systematic integration of the laboratory with its wider social setting. As a result, Kohler (2008, 761) argued that future lab studies should produce a "systematic, macrosocial history of the lab." There is certainly valuable work to be done in this direction. But I propose some additional, complementary possibilities for future lab studies, built on a rather different conception of the laboratory.

For Kohler, laboratories are primarily "cultural spaces" whose "conventions ... embody those of other important social institutions" and he is certainly right that it is insufficient to focus solely on "the science done in labs" (2008, 763–4). But, as I have shown elsewhere, laboratories are also essential material and pedagogical resources in the practice of modern science, irreducible to mere locations or venues of scientific activity (Jackson 2011). Studies of the laboratory therefore offer the possibility for a different kind of integration from that proposed by Kohler, for micro-histories that consider the totality of the laboratory, integrating their institutional essence with what goes on within their walls. Setting out the bones of this approach and indicating how it can be used to produce consciously historically, fully situated accounts of the production of scientific knowledge is the goal of this final section.

This focus on the production of scientific knowledge is deliberate and it runs counter to the reasons offered by Kohler for the decline of lab studies as a field. If we focus our attention for a moment on the questions at the heart of early lab studies (listed above), there seems to me to be plenty of interest, both theoretical and historical, to pursue. And if laboratory history as institutional history has ceased to attract practitioners this is perhaps because, somewhere in the transition from studies driven by primarily epistemological concerns to those conceptualized within cultural or social history, scholars have lost sight of the essential, and essentially historical, connection between the practice of science and its venue. Current studies address one or the other but they almost never integrate the two.[6] Indeed, only in this integrated form might this field properly be called lab studies.

An example from the history of chemistry will help me to explain why the approach I am proposing here is far from being a return to mere institutional history or narrow technical internalism, why the connection between practice and place matters, and what we might hope to learn from studies that achieve this integration. We have been repeatedly led to believe that the structural theory of organic molecules was both essential and sufficient in enabling the synthesis of organic compounds from the mid-nineteenth century onwards—this ability providing the foundation for the rise of vast chemical industries devoted to the production of chemical dyes, pharmaceuticals, agrochemicals, and a range of new synthetic materials. My recent studies of chemical practice in the laboratories of chemists, including August Hofmann, have shown otherwise, revealing the limitations of such theoretical knowledge and the absolute dependence of synthetic capability on practical expertise that was developed, learnt, taught, and deployed in a very particular place: the institutional chemical laboratory (Jackson 2014a; Jackson 2014b).

The laboratory as practical, material, and pedagogical resource, as well as social space, is the central core of my historical account of what Hofmann knew, how he knew it, and how this knowledge could provide the basis of a science that was stable enough to be useful in industrial settings and yet capable of rapid further development. This approach foregrounds in historical context exactly the questions driving early lab studies and it certainly depends on suitably adapted forms of the methods developed by early scholars in the field. But it also draws upon more recent approaches originating in studies of scientific pedagogy (Warwick 2003; Kaiser 2005; Warwick and Kaiser 2005) and the materiality of scientific—and especially chemical—practice (Klein and Lefèvre 2007; Klein and Spary 2010), including a focus on instrumentation (Holmes and Levere 2000; Morris 2002).

Marrying tools for investigating the laboratory with the insights provided by the ongoing study of science as practice produces important historical results. Some of these relate to important questions about the nature of scientific knowledge and, specifically, the relationship between practice and theory—as reflected in a new understanding of the role of practical knowledge of organic synthesis in generating and securing developing structural theory (Jackson 2014a; Jackson 2014b). But they also reveal fresh perspectives on the acquisition and dissemination of scientific knowledge and research expertise, as well as the undeniably collective nature of much scientific research in the modern period. Many excellent studies of scientific practice have been produced by the study of individual lab notebooks (Holmes, Renn, and Rheinberger 2003). Such notebooks, where they survive, are certainly important sources for the lab historian. In an age where historians are increasingly pressed to show the relevance of their work to current concerns, including the best means of fostering creativity and encouraging innovation, however, there is surely much to recommend an approach that makes it possible to incorporate both individual and collective contributions to the development and prosecution of science as a social, material, and technical activity.

Its focus on science may cause this approach to appear incompatible with—perhaps even hostile to—Kohler's broader social concerns. It is not. On the contrary, it is only by including the kind of approach I have described here alongside other, more purely social historical approaches that we shall come to understand the laboratory's changing position in the world—producing exactly that "systematic, macrosocial history of the lab" Kohler urged a new generation of historians to pursue. The laboratory—whether for teaching, testing, or research—is the material nexus of scientific practice, a manifestation of all that makes science something to be trusted and suspected, looked up to and laughed at. In seeking to understand the place of science in society, we could do a lot worse than begin by understanding science in its own social space.

Endnotes

1 Gooding 1985; Shapin and Schaffer 1985; Schaffer 1998; and see also Cunningham 2010 on anatomy as an experimental discipline.

2 James (1989) is exemplary of the relative neglect of chemical laboratories. This collection of 13 essays about the development of laboratory science in the nineteenth century contains only three essays about chemical laboratories, of which only two are concerned with the laboratory as a physical space.

3 Crosland (1967, 283) refers to Berthollet's laboratory but gives no indication that at least some of Berthollet's experiments were performed far away from the main house.

4 See, for example, the recent project "Sites of Chemistry" sponsored by the Society for the History of Alchemy and Chemistry.

5 Kuhn's (1977) *Essential Tension* is perhaps more significant than his famous (1996[1962]) *Structure of Scientific Revolutions* in this context. These developments were also fueled by the publication in 1979 of a new English edition of Ludwik Fleck's

(1935) *Genesis and Development of a Scientific Fact.* Golinski (2005[1998]) provides an excellent introduction to constructivism and the Edinburgh Strong Programme.

6   Compare, for example, Rooij's (2011) proposed laboratory taxonomy with the studies of practice presented in Holmes, Renn, and Rheinberger 2003. This separation reflects the failure of more recent scholarship to realize the possibilities Knorr Cetina (1992, 115) believed laboratory studies offered for considering "experimental activity within the wider context of equipment and symbolic practices." It also leaves the path open for a new integration of histories of scientific practice within lab studies.

References

Arnold, David. 2000. *Science, Technology and Medicine in Colonial India.* Cambridge: Cambridge University Press.

Baytop, Turhan. 1997. *Laboratuvar'dan Fabrika'ya, Türkiye'de İlaç Sanayii (1833–1954).* Istanbul: Bayer Türk Sağlık Ürünleri Bölümü. (In Turkish, title translates as: From Laboratory to Factory, Pharmaceutical Industry in Turkey).

Beretta, Marco. 2004. "Sur les Traces du Laboratoire de Lavoisier." *La Revue du Musée des Arts et Métiers,* 41: 14–23.

Borscheid, Peter. 1976. *Naturwissenschaft, Staat und Industrie in Baden, 1848–1914.* Stuttgart: Klett.

Burney, Ian. A. 2002. "Testing testimony: Toxicology and the law of evidence in early nineteenth-century England." *Studies in the History and Philosophy of Science,* 33: 289–314. DOI: 10.1016/S0039-3681(02)00002-X.

Cahan, David. 1985. "The institutional revolution in German physics, 1865–1914." *Historical Studies in the Physical Sciences,* 15: 1–65. DOI: 10.2307/27757549.

Cahan, David. 1989. *An Institute for an Empire: The Physikalische-Technische Reichsanstalt, 1871–1918.* Cambridge: Cambridge University Press.

Chakrabarti, Pratik. 2012. *Bacteriology in British India: Laboratory Medicine in the Tropics.* Rochester, NY: Rochester University Press.

Clarke, Sabine. 2013. "The Research Council system and the politics of medical and agricultural research for the British colonial empire, 1940–1952." *Medical History,* 57: 338–58. DOI: 10.1017/mdh.2013.17.

Collins, Harry. 1985. *Changing Order: Replication and Induction in Scientific Practice.* Chicago: University of Chicago Press.

Crosland, Maurice. 1967. *The Society of Arcueil: A View of French Science at the Time of Napoleon.* Cambridge, MA: Harvard University Press.

Crosland, Maurice. 2005. "Early laboratories c.1600–c.1800 and the location of experimental science." *Annals of Science,* 62: 233–53.

Cunningham, Andrew. 2010. *The Anatomist Anatomis'd: An Experimental Discipline in Enlightenment Europe.* Farnham: Ashgate.

Cunningham, Andrew, and Perry Williams (eds.) 1992. *The Laboratory Revolution in Medicine.* Cambridge: Cambridge University Press.

Fleck, Ludwik. 1979. *The Genesis and Development of a Scientific Fact.* Edited by T. J. Trenn and R. K. Merton, foreword by Thomas Kuhn. Chicago: University of Chicago Press.

Gee, Brian. 1989. "Amusement chests and portable laboratories: Practical alternatives to the regular laboratory." In *The Development of the Laboratory: Essays on the Place of Experiment in Industrial Civilisation,* edited by Frank A. J. L. James 1989, 37–59. Basingstoke: Macmillan.

Golinski, Jan. 2005 (1998). *Making Natural Knowledge: Constructivism and the History of Science.* Chicago: University of Chicago Press.

Gooday, Graeme. 1990. "Precision measurement and the genesis of physics teaching laboratories in Victorian Britain." *British Journal of the History of Science*, 23: 25–51. DOI: 10.1017/S0007087400044447.

Gooday, Graeme. 2008. "Placing or replacing the laboratory in the history of science." *Isis*, 99: 793–5. DOI: 10.1086/595772.

Gooday, Graeme J. N., and Morris F. Low. 1998. "Technology transfer and cultural exchange: Western scientists and engineers encounter late Tokugawa and Meiji Japan." *Osiris*, 13: 99–128. DOI: 10.1086/649282.

Gooding, David. 1985. "'In nature's school': Faraday as an experimentalist." In *Faraday Rediscovered: Essays on the Life and Work of Michael Faraday, 1791–1867*, edited by David Gooding and Frank A. J. L. James, 105–32. London: Macmillan.

Günergun, Feza. 2009. "Chemical laboratories in nineteenth-century İstanbul: A case-study on the laboratory of the Hamidiye Etfal children's hospital." In *Spaces and Collections in the History of Science*, edited by Marta C. Laurenco and Ana Carneiro, 91–101. Lisbon: Museum of Science of the University of Lisbon.

Günergun, Feza and Şeref Etker. 2013. "From Quinaquina to 'Quinine Law': A bitter chapter in the Westernisation of Turkish medicine." *Osmanlı Bilimi Araştırmaları*, 14: 41–68.

Hannaway, Owen. 1986. "Laboratory design and the aim of science: Andreas Libavius versus Tycho Brahe." *Isis*, 77: 585–610. DOI: 10.1086/354267.

Holmes, Frederic L. and Trevor H. Levere (eds.) 2000. *Instruments and Experimentation in the History of Chemistry*. Cambridge, MA: MIT Press.

Holmes, Frederic L., Jürgen Renn, and Hans-Jörg Rheinberger (eds.) 2003. *Reworking the Bench: Research Notebooks in the History of Science*. Dordrecht: Kluwer.

Holmes, Frederic L. 1989. *Eighteenth-Century Chemistry as an Investigative Enterprise*. Berkeley: University of California Press.

Jackson, Catherine M. 2011. "Chemistry as the defining science: Discipline and training in nineteenth-century chemical laboratories." *Endeavour*, 35: 55–62. DOI: 10.1016/j.endeavour.2011.05.003.

Jackson, Catherine M. 2014a. "The curious case of coniine: Constructive synthesis and aromatic structure theory." In *Objects of Chemical Inquiry*, edited by Ursula Klein and Carsten Reinhardt, 61–101. Sagamore Beach, MA: Science History Publications.

Jackson, Catherine M. 2014b. "Synthetical experiments and alkaloid analogues: Liebig, Hofmann, and the origins of organic synthesis." *Historical Studies in the Natural Sciences*, 44: 319–63. DOI: 10.1525/hsns.2014.44.4.319.

Jackson, Catherine M. 2015a. "The 'wonderful properties of glass': Liebig's Kaliapparat and the practice of chemistry in glass." *Isis*, 106, No. 1: 43–69.

Jackson, Catherine M. 2015b. "Chemical identity crisis: Glass and glassblowing in the identification of organic compounds." *Annals of Science*, 72, No. 2: 187–205.

James, Frank A. J. L. (ed.) 1989. *The Development of the Laboratory: Essays on the Place of Experiment in Industrial Civilisation*. Basingstoke: Macmillan.

Kaiser, David (ed.) 2005. *Pedagogy and the Practice of Science: Historical and Contemporary Perspectives*. Cambridge, MA: MIT Press.

Kikuchi, Yoshiyuki. 2013. *Anglo-American Connections in Japanese Chemistry: The Lab as Contact Zone*. New York: Palgrave Macmillan.

Kirk, Robert G. W. 2010. "A brave new animal for a brave new world: The British Laboratory Animals Bureau and the Constitution of International Standards of Laboratory Animal Production and Use, circa 1947–1968." *Isis*, 101: 62–94. DOI: 10.1086/652689.

Klein, Ursula and Wolfgang Lefèvre. 2007. *Materials in Eighteenth-Century Science: A Historical Ontology*. Cambridge, MA: MIT Press.

Klein, Ursula and Emma Spary (eds.) 2010. *Materials and Expertise in Early-Modern Europe: Between Market and Laboratory*. Chicago: University of Chicago Press.

Klein, Ursula. 2008. "The laboratory challenge: Some revisions of the standard view of early modern experimentation." *Isis*, 99: 769–82. DOI: 10.1086/595771.

Knorr Cetina, Karin. 1992. "The couch, the cathedral, and the laboratory: On the relations between experiment and laboratory in science." In *Science as Practice and Culture*, edited by Andrew Pickering, 113–38. Chicago: University of Chicago Press.

Knorr Cetina, Karin. 1995. "Laboratory studies: The cultural approach to the study of science." In *Handbook of Science and Technology Studies*, edited by Sheila Jasanoff, Gerald E. Markle, James C. Peterson, and Trevor Pinch, 140–66. Beverly Hills, CA: Sage.

Kohler, Robert E. 2002. *Landscapes and Labscapes: The Field-Lab Border in Biology*. Chicago: University of Chicago Press.

Kohler, Robert E. 2008. "Lab history: Reflections." *Isis*, 99: 761–8. DOI: 10.1086/595769.

Kuhn, Thomas S. 1977. *The Essential Tension*. Chicago: University of Chicago Press.

Kuhn, Thomas S. 1996 [1962]. *Structure of Scientific Revolutions*. Chicago: University of Chicago Press.

Latour, Bruno. 1983. "Give me a laboratory and I will raise the world." In *Science Observed: Perspectives on Social Studies of Science*, edited by Karin Knorr Cetina and Michael Mulkay, 141–70. London: Sage.

Latour, Bruno. 1999. *Pandora's Hope: Essays on the Reality of Science Studies*. Boston, MA: Harvard University Press.

Latour, Bruno and Steven Woolgar. 1979. *Laboratory Life: The Social Construction of Scientific Facts*. London: Sage.

Lenoir, Timothy. 1992. "Laboratories, medicine and public life in Germany, 1830–1849: Ideological roots of the institutional revolution." In *The Laboratory Revolution in Medicine*, edited by Andrew Cunningham and Perry Williams, 14–71. Cambridge: Cambridge University Press.

Lundgren, Frans. 2013. "The politics of participation: Francis Galton's Anthropometric Laboratory and the making of civic selves." *British Journal for the History of Science*, 46: 445–66. DOI: 10.1017/S0007087411000859

Meinel, Christoph. 1983. "Theory or practice? The eighteenth-century debate on the scientific status of chemistry." *Ambix*, 30: 121–32.

Meinel, Christoph. 1988. "Artibus acadmicis inserenda: Chemistry's place in eighteen and early nineteenth century universities." *History of Universities*, 7: 89–115.

Meinel, Christoph. 2000. "Chemische Laboratorien: Funktion und Disposition." *Berichte zur Wissenschaftsgeschichte*, 23: 287–302. DOI: 10.1002/bewi.20000230306.

Micault, Natalie Pigeard. 2013. "The Curie's lab and its women (1906–1934)." *Annals of Science*, 70: 71–100. DOI: 10.1080/00033790.2011.644194.

Morrell, Jack B. 1972. "The chemist breeders: The research schools of Liebig and Thomas Thomson." *Ambix*, 19: 1–46.

Morris Peter J. T. (ed.) 2002. *From Classical to Modern Chemistry: The Instrumental Revolution*. Cambridge: Royal Society of Chemistry.

Phalkey, Jahnavi. 2013. *Atomic State: Big Science in Twentieth Century India*. New Delhi: Permanent Black Press.

Reinhardt, Carsten, and Anthony S. Travis. 2000. *Heinrich Caro and the Creation of Modern Chemical Industry*. Dordrecht: Kluwer.

Rezneck, Samuel. 1970. "The European education of an American chemist and its influence in 19th-century America: Eben Norton Horsford." *Technology and Culture*, 11: 366–88. DOI: 10.2307/3102198.

Richmond, Marsha L. 1997. "'A lab of one's own': The Balfour Biological Laboratory for Women at Cambridge University, 1884–1914." *Isis*, 88: 422–55. DOI: 10.1086/383769.

Rocke, Alan J. 2003. "Origins and spread of the 'Giessen Model' in university science." *Ambix*, 50: 90–115.

Rooij, Arjan van. 2011. "Knowledge, money and data: an integrated account of the evolution of eight types of laboratory." *British Journal for the History of Science*, 44: 427–48. DOI: 10.1017/S0007087410001330.

Rossiter, Margaret. 1975. *The Emergence of Agricultural Science: Justus Liebig and the Americans*, 1840–1880. New Haven, CT: Yale University Press.

Schaffer, Simon. 1992. "Late Victorian metrology and its instrumentation: A manufactory of Ohms." In *Invisible Connections: Instruments, Institutions, and Science*, edited by Robert Bud and Susan E. Cozzens, 23–56. Bellingham, WA: SPIE Optical Engineering Press.

Schaffer, Simon. 1998. "Physics laboratories and the Victorian country house." In *Making Space for Science: Territorial Themes in the Shaping of Knowledge*, edited by Crosbie Smith and Jon Agar, 149–80. Basingstoke: Macmillan.

Shapin, Steven. 1988. "The house of experiment in seventeenth-century England." *Isis*, 79: 273–304. DOI: 10.1086/354773.

Shapin, Steven, and Simon Schaffer. 1985. *Leviathan and the Air-Pump: Hobbes, Boyle and the Experimental Life*. Princeton, NJ: Princeton University Press.

Slayton, Rebecca. 2012. "From a 'dead albatross' to Lincoln labs: Applied research and the making of a 'normal' Cold War university." *Historical Studies in the Natural Sciences*, 42: 255–82. DOI: 10.1525/hsns.2012.42.4.255.

Smith, Pamela H. 2006. "Laboratories." In *The Cambridge History of Science: Volume 3, Early Modern Science*, edited by Roy Porter, Katherine Park, and Lorraine Daston, 290–305. Cambridge: Cambridge University Press.

Tuchman, Arleen Marcia. 1993. *Science, Medicine, and the State in Germany: The Case of Baden*, 1815–1871. Oxford: Oxford University Press.

Warwick, Andrew C. 2003. *Masters of Theory: Cambridge and the Rise of Mathematical Physics*. Chicago: University of Chicago Press.

Warwick, Andrew C., and David Kaiser. 2005. "Kuhn, Foucault, and the power of pedagogy." In *Pedagogy and the Practice of Science: Historical and Contemporary Perspectives*, edited by David Kaiser, 393–409. Cambridge, MA: MIT Press.

CHAPTER TWENTY-TWO

Modern School and University

HEIKE JÖNS

Modern schools and universities are key institutions for the production, transmission, discussion, and critique of scientific knowledge and education. Based on a relational concept of modernity (Taylor 1999), this chapter defines "modern" schools and universities as post-medieval institutions that were subsequently modernized at various occasions through new intellectual movements and related institutional reform (Burke 2000). By so doing, the chapter aims to understand how these institutions evolved in mutual interaction with the rise of modern science from the early modern period, beginning around 1450, to the origins of mass higher education in the 1960s.

In the five centuries under consideration, four key processes will be examined from an historical geographical perspective that links intellectual movements and individual achievements to institutional change and combines a focus on Europe and the United States with a global outlook on interactions between European and local educational traditions elsewhere. First, modern schools and universities were affected by *cycles of expansion and contraction* that linked to periods of economic, cultural, and political prosperity and crisis and displayed complex geographies at regional, national, and global scales (Meusburger 1998). Second, the formation of personal and institutional networks through shared learning experiences, academic mobility and communication, and government interventions was instrumental for developing a *core set of common practices* that evolved into relatively standardized national systems of primary, secondary, and tertiary education across the world (Postlethwaite 1995).

Third, an ongoing *professionalization of learning, teaching, and research* through a series of reforms saw the emergence of an increasingly differentiated landscape of educational institutions (Jarausch 1983; Porter 1996; Neave 2011). Fourth, their post-medieval history was shaped by a *complex transition from a humanistic to a scientific paradigm* because of the increasingly vital role that scientific innovations have played for the socioeconomic prosperity and geopolitical superiority of cities, nation states, and empires (Taylor 1999). The significant role of the sciences in shaping modern schools and universities is stressed by the chapter's argument that the

A Companion to the History of Science, First Edition. Edited by Bernard Lightman.
© 2016 John Wiley & Sons Ltd. Published 2020 by John Wiley & Sons Ltd.

long-term paradigmatic shift from an emphasis on humanistic to scientific education and research revolutionized these institutions in the nineteenth and twentieth centuries and explains why recent globalization processes in education and research are governed by discourses about the natural and technical sciences and their evaluation cultures.

Early Modern Universities and Schools

By 1451, at the end of the medieval long economy cycle, about 50 universities were operating across Europe, the world region that hosted the first *universitates magistrorum et scholarium*, or communities of teachers and students. These were modeled on the degree-awarding universities of Bologna and Paris since the twelfth century and served as centers for the translation, preservation, transmission, and debate of existing knowledge, including Greek and Arabic science (Verger 1992; Burke 2000; Chapter 13 of this volume). The beginning of the Renaissance (c.1450–1600) saw a gradual emancipation of education from churchly–religious interests because the "humanist movement" revived classical Greek and Roman traditions in learning, architecture, and art, thus considering humankind rather than God as the measure of all things (de Ridder-Symoens 2004). This led to a proliferation of scientific innovations, inquiries, and imaginations, as exemplified by the invention of the printing press (c.1450), early modern discoveries and expeditions (e.g., Columbus's journey in 1492), and the Copernican Revolution (1514) that introduced the heliocentric worldview.

The Renaissance emerged in economically prospering northern Italian city states such as Florence, Venice, and Genoa and transformed the University of Padua in the hinterland of Venice into Europe's most important center for the production of scientific knowledge in the sixteenth century (Taylor, Hoyler, and Evans 2008). The Renaissance polymaths Leonardo da Vinci (1452–1519) and Michelangelo (1475–1564), who made innovative contributions across the arts and sciences, also worked in Italian cities such as Florence, Milan, Venice, Bologna, and Rome, but they never attended or taught at university. Drawing on writings by the economist Thorstein Veblen and the sociologist Pierre Bourdieu, the historian Peter Burke (2000, 49–51) has argued that in the early modern period new ideas in the humanities and sciences were often produced by creative individuals outside of established universities because these were mostly opposed to the new intellectual movements. This either led to the foundation of new institutions, such as early modern academies and scientific societies, or meant that innovative ideas were promoted by members of new universities.

Martin Luther (1483–1546), the Augustinian monk, Catholic priest, and German professor of theology, provides a prime example because he launched the Protestant Reformation (1517) "when his university was only fifteen years old" (Burke 2000, 37). For Perkin (1984, 27), Luther's nailing of his 95 theses to the church door at the University of Wittenberg in 1517 marked no less than "the triumph of the university over the medieval Church and with it the demise of the medieval world order". Intellectual progress in established universities remained slow, as well as geographically and institutionally fragmented, and by the time humanist ideas had entered most university curricula, the Scientific Revolution, "a diverse array of cultural practices aimed at understanding, explaining, and controlling the natural world," was already under way (Shapin 1996, 3).

In some places, certain scientific views continued to be in conflict with the interests of some religious authorities well into the seventeenth century, as exemplified by Galileo Galilei's (1564–1642) famous inquisition trial by the Roman Catholic Church in 1633 that required him to "abjure professing the *physical truth* of Copernicanism" (Shapin 1996, 137). Like Galileo, who taught at the Universities of Pisa and Padua for 21 years (1589–1610), several protagonists of the seventeenth-century Scientific Revolution were at least for some time based in universities when developing "natural philosophy," the predecessor to the natural sciences. Isaac Newton (1642–1727), for example, became the second Lucasian Professor of Mathematics at the University of Cambridge, a position he held for 33 years (1669–1702) and that allowed him to write his seminal book *Philosophiae Naturalis Principia Mathematica* (1687), which laid the foundations for classical mechanics (Porter 1996; Shapin 1996).

Partly encouraged by considerable opposition to natural philosophy in some academic circles, supporters of this new, "self-conscious process of intellectual innovation" (Burke 2000, 38–9), which rejected both classical and medieval traditions, founded new institutions for the study of nature, such as the Accademia dei Lincei in Rome (1603), the Royal Society in London (1660), and the Académie Royale des Sciences in Paris (1666). These new institutions offered a possibility to practice new forms of knowledge production in new environments and to build their own epistemological communities (Burke 2000). Some universities such as Leiden (founded in 1575) also established botanical gardens, observatories, and laboratories in the late sixteenth and seventeenth centuries, but many of them did not keep up with the rapid developments in natural philosophy until the nineteenth century and therefore remained centers of learning and teaching, while scientific research began to flourish in other settings (Porter 1996; Burke 2000; Chapters 14, 15, 16, and 17 of this volume).

The new centers of scientific knowledge production that developed in northwestern Europe during the seventeenth century created a polycentric network of European universities and academies, reaching from Rome to Edinburgh and from Oxford to Wittenberg. London hosted the most renowned scientists in the seventeenth century, followed by Leiden, the intellectual powerhouse of Dutch hegemony (c.1609–1672), as well as Padua and Paris (Taylor, Hoyler, and Evans 2008). Connected by academic mobility and communication, early modern scientists and scholars formed a European community, but this "Republic of Letters," or "Commonwealth of Learning," remained socially and geographically highly selective, concentrating on a few outstanding individuals and major knowledge hubs (Chapter 25 of this volume).

Religious reform movements were instrumental for humanist and scientific ideas reaching a wider segment of the early modern population through improved education. The Reformation's impact on a broader revival of learning was most likely inspired by Luther's own experience that "[t]he common people, especially in the villages, are utterly ignorant of the Christian doctrine" (cited in Painter 1896, 137). In 1524, Luther wrote a letter to all German cities, encouraging the establishment of schools and the implementation of compulsory school attendance as "a means to more effective service in church and state" (Painter 1896, 143). Protestantism promoted popular education because it required the study of the bible and therefore created a demand for teachers and schools to teach basic reading skills. Consequently,

Protestant regions were supplied with schools for primary and secondary education of both boys and girls, who were either instructed together or separately, and with new universities (Painter 1896). Catholic countries followed suit and established their own schools, colleges, and universities, especially in Spain, France, Central and Eastern Europe (Frijhoff 1996). These were often run by the Jesuits, a Christian congregation founded in 1534 that, too, promoted education in reaction to poor knowledge among the clergy but also aimed to combat the Reformation (Painter 1896).

Religious struggles following the Reformation and Counterreformation shaped university life profoundly and in very different ways, as exemplified by the University of Heidelberg (Meusburger and Schuch 2012). Founded in 1386, the university experienced its first golden age as a center of humanism—and later Calvinism—from the introduction of the Reformation in 1556 until 1622. The university hosted Martin Luther as a guest lecturer as early as 1518 and exercised much influence across Europe, attracting exiled protestant professors from France, the Netherlands, and Italy and recruiting about two fifths of foreigners among all registered students in the 1600s (de Ridder-Symoens 1996; Wolgast 1996). This golden age ended during the Thirty Years' War (1618–1648), when Heidelberg became a prime target of alternate attacks by the Catholic League and Protestant forces, losing its famous *Bibliotheca Palatina* as loot to the Vatican in 1623. After a short period of recovery under the Calvinists, the troops of Louis XIV (1638–1715) destroyed Heidelberg almost entirely in 1693 during the Palatine Succession Wars (1688–1697). The town was subsequently converted to Catholicism so that reconstruction was shaped by returning Catholic orders and especially the Jesuits, who provided the new guard of university professors until the order's dissolution in 1773 (Wolgast 1996).

Increasing government intervention, religious intolerance, traditionalism, hereditary professorships, and a serious financial crisis meant that the University of Heidelberg, like several other European universities, experienced a period of intellectual stagnation and low reputation in the eighteenth century, from which it only recovered after Napoleon's reorganization of German lands had incorporated the area into the state of Baden in 1802 (Wolgast 1996). Other universities showed less longevity because 18 of the 34 German universities that existed in 1789 had been closed by 1815 (Rüegg 2004). As a consequence of the French Revolution and Napoleon's conquests, the overall number of European universities shrunk in the same period from 143 to 83. This included the abolishment of all 24 French universities that were subsequently replaced by specialized professional schools (Rüegg 2004).

From the sixteenth to the eighteenth centuries, English universities also experienced alternating periods of recovery (1540s–1650s) and decline (1670s–1800s). Henry VIII's (1491–1547) anticlerical wrath that led to the dissolution of monasteries in England, Wales, and Ireland impacted on the universities by banning the study of canon law in 1535, which equaled the abolishment of the largest postgraduate faculty and meant that the depleted universities had to reinvent themselves. They did this by educating a new lay clientele in addition to the secular clergy, the majority of which were "sons of smaller gentry, merchants, and yeomen" (Perkin 1984, 29). As a result, student numbers rose and reached their peak with an annual intake of 400 to 500 students at the time of the English Civil War (1642–1651), to which rebellion in the universities had contributed (Perkin 1984). After the restoration of the monarchy (1660), the universities saw another period of declining student interest that was linked to a reputation for excessive drinking habits, debauchery, and political rebellion. This also changed the social composition of students because the wealthiest preferred private tutors and the Grand Tour of the continent and the poorest were not able to afford the rising costs of college residence due to the "competitive extravagance of gentlemen undergraduates" (Perkin 1984, 30). In this period of crisis, the English Enlightenment largely bypassed eighteenth century Oxford and Cambridge, whereas the four Scottish universities, supported by local leaders for patriotic reasons, developed modern professional teaching and liberal education that produced some of the most outstanding Enlightenment thinkers (Anderson 2004).

The Enlightenment and Compulsory School Education

The Scientific Revolution provided the grounds for the Age of Enlightenment that began in the late seventeenth century and can be concisely characterized by the philosopher Immanuel Kant's (1724–1804) call for "man's emergence from his self-incurred immaturity" through the use of public reason and rationality (Kant 1784, 1). Enlightenment ideas were discussed in universities, academies, scientific societies, courts, and less formal settings such as the salon and the coffeehouse, which encouraged public discussion of lectures and newspapers and thus gave rise to the public sphere (Burke 2000). An increasing awareness that "searches for knowledge [need] to be systematic, professional, useful and co-operative" promoted the very idea of research and resulted in the foundation of research institutes (Burke 2000, 46). Academies became the center for knowledge-gathering expeditions, prizes, and scholarly networks based on the exchange of visits, letters, ideas, and publications. University curricula were transformed by emphasizing the medieval *quadrivium* (arithmetic, geometry, astronomy, music) over the *trivium* (grammar, logic, rhetoric) and by introducing new subjects of study. Eventually, natural philosophy parted from the *quadrivium* to "split into virtually independent subjects such as physics, natural history, botany and chemistry" (Burke 2000, 100).

For many European universities, such as Oxford and Heidelberg, the eighteenth century was a time of crisis, for others, such as Leiden and Göttingen, a period of prosperity. On the advice of the German mathematician and philosopher Gottfried Wilhelm von Leibniz (1646–1716), the governments of Prussia and Russia transformed Berlin and St. Petersburg into knowledge centers by founding The Royal Prussian Academy of Sciences in Berlin (1700), of which Leibniz was appointed first president, the St. Petersburg Academy of Sciences (1724), and—as Russia's first university—the University of St. Petersburg (1724; Burke 2000). Based on the analysis of career movements by eminent scientists, Taylor, Hoyler, and Evans (2008) found a proliferation of regional, often language-based scientific networks in the eighteenth century that centered on Paris, Berlin–Göttingen, London–Leiden–Edinburgh, and several northern Italian cities, thus illustrating a transition from Latin as the *lingua franca* in university teaching and learning to national and regional languages.

With the European colonization of the new world, the concept of the European university was transferred to Central, South, and North America from the sixteenth century onwards. The first university in the Americas was chartered in 1538 and had developed out of theological teaching in the Dominican convent of Santo Domingo on the island of Hispaniola (Roberts, Rodriguez Cruz, and Herbst 1996).

Subsequently, the Spanish crown established universities in Lima (1571) and Mexico City (1595), with all three institutions having received the privileges of the University of Salamanca in the mother country. More than two dozen universities were set up and closely controlled by the Spanish colonial government in the Americas until the early nineteenth century. In Brazil, the Portuguese rulers did not establish any universities in colonial times so that, before the first Brazilian university was established through a merger of existing institutions in Rio de Janeiro in 1922, students had to complete their higher education in the mother country or elsewhere (Roberts, Rodriguez Cruz, and Herbst 1996).

North of the Rio Grande, Anglophone Americans drew on different English and Scottish universities when founding their first college, named Harvard after its main benefactor, in 1636. Harvard received its charter in 1650 and was followed by eight other colonial colleges of which the latter seven formed the nucleus of the now famous Ivy League universities along the east coast: College of William and Mary in Virginia (chartered as a college in 1693), Yale University (1745), Princeton University (1746), Columbia University (1754), University of Pennsylvania (1755), Brown University (1764), Rutgers University (1766), and Dartmouth College (1769; Roberts, Rodriguez Cruz, and Herbst 1996). Initially struggling because of poverty and a shortage of qualified teachers, these American colleges were characterized by diversity and tolerance and "probably contributed more to the general educational level of the societies of which they were a part than did their Spanish equivalents" (Roberts, Rodriguez Cruz, and Herbst 1996, 281).

The institutional structure of European schooling had not much changed from the Middle Ages until the 1700s, whereas the curriculum was increasingly shaped by humanist ideas and divided into Catholic, Protestant, and Calvinist influences. Elementary schooling was organized by parishes, cities, or private individuals, whereas secondary schools in the form of grammar schools, high schools, gymnasia, and Latin schools—the main gateways to university—could also be organized by religious orders and chapters (de Ridder-Symoens 2004). Secondary schools were structured into six to eight classes and the age for entering higher education was raised to 17–18 years; private tuition could be found in the homes of the nobility; girls were still finishing their education at convent schools; the guilds continued to train merchants, financiers, skilled artisans, and artists; and new institutions included Sunday schools for poor children, private day and boarding schools for the elites, court schools that competed with the universities, as well as vocational and specialized professional schools that flourished in the eighteenth century. The provision of schools was characterized by a quite distinct urban–rural divide that was especially large in Catholic countries, where the church remained in charge of schooling, while Protestant education became organized by the state. By 1700, about 80 per cent of the Protestant population was literate but only 45 per cent in Catholic countries (de Ridder-Symoens 2004).

The seventeenth and eighteenth centuries produced a number of theoretical treatises about the nature and value of education. In England, John Locke (1632–1704) developed a theory that assumed a blank state of mind at birth *(tabula rasa)* and the subsequent acquisition of all knowledge through experience, which stressed the need to shape the minds of young children early (Moseley 2007). Locke's ideas were highly influential for Enlightenment thinkers such as Immanuel Kant and contributed to the ideological context that paved the way for compulsory school education provided by the state. In 1717, Frederick William I of Prussia published an edict that has become known as the first outline of a national system of compulsory school education in Europe because it instructed parents to send their children to existing schools (Van Horn Melton 1988). Inspired by the Pietist movement that made universal literacy a prime goal of education reform, Frederick William I provided a second edict on Prussian elementary schools in 1736 and set up a royal endowment that marked an historic step for both compulsory attendance and state-funded schools because it created 884 rural schools in East Prussia from 1736 to 1742 (Van Horn Melton 1988).

Inspired by the Prussian model, the "most ambitious school reform of the [eighteenth] century" (Van Horn Melton 1988, 199) was launched in 1774 by the government of Maria Teresa (1717–1780), empress of Austria and Hungary from 1740 to 1780, through the introduction of compulsory education from the age of six to twelve for all children living within the territories of the Habsburg Empire. While both reform projects set important precedents and were successful in the long term, they took time to be fully implemented and generated large regional disparities of literacy. At the *fin de siècle*, Prussia displayed the most favorable ratio of illiterates among the total population compared to other European countries and the United States, but in the mid-nineteenth century, regional disparities ranged from 70 percent school attendance in Posen to 95 percent in Saxony, whereas school attendance in Austria–Hungary varied between around 15 percent in Galicia and the Bukovina and almost 100 percent in Vorarlberg, Tyrol, Salzburg, Bohemia, and Moravia (Meusburger 1998).

Elsewhere in Europe, compulsory education became an important vehicle for nationalism in the second half of the nineteenth century (Brockliss and Sheldon 2012). In France, the state had intended to construct compulsory elementary schools after the Revolution of 1789 because literacy was seen as vital for spreading liberal ideas, but this reform was only fully implemented after 1830 (Painter 1896). In England, boarding and other privately funded preparatory schools delayed initiatives for the provision of state-funded mass education until the 1830s, with compulsory school education eventually being implemented by the government in 1870 (Painter 1896). Germany and France pioneered the sharp division of secondary and higher education through school-leaving examinations, the *Abitur* at the *Gymnasium* (1788) and the baccalaureate at the *lycée* (1809), which marked bourgeois status and served as portals to higher learning (Anderson 2004).

Colonial governments and Christian missionaries brought aspects of European schooling to the Americas and Australia, to Asian countries, such as India and China, and to Africa, thus explaining, together with international standardizations of national statistics in the nineteenth century, the emergence of diverse but relatively comparable national systems of primary, secondary, and tertiary education in the nineteenth and twentieth centuries (Meusburger 1998; Brockliss and Sheldon 2012). These education systems were initially quite selective and became only gradually more inclusive, as exemplified by the admission of native peoples to colleges and universities in the Americas, which remained the exception until the twentieth century (Roberts, Rodriguez Cruz, and Herbst 1996).

The Research University Revolution

Prussia also played an important role in the reform of the university. After the country's decisive defeat against the troops of Napoleon I (1769–1821) in the battle of Jena in

1806, Napoleon closed the University of Halle. The institution's leaders appealed to Prussia's King Frederick William III (1770–1840) to restore the institution in Berlin, to which the King famously replied "That's right! That's fine! The state must replace by intellectual powers what it has lost in material ones" (cited in Perkin 1984, 33–4). Subsequently, the comprehensive reform of the Prussian education system was spearheaded by Wilhelm von Humboldt (1767–1835), the elder brother of the renowned naturalist, geographer, and scientific traveler Alexander von Humboldt (1769–1859), who served as director of the Department for Ecclesiastical Affairs and Education in the Ministry of the Interior from March 1809 to June 1810 (Rüegg 2004).

For the conceptualization of the University of Berlin, which opened in 1810, Wilhelm von Humboldt drew on the examples of the Universities of Göttingen and Halle, two leading centers of scientific knowledge in the eighteenth century, and on the liberal ideas of his advisor Friedrich Schleiermacher (1768–1834), who had been professor of theology in Halle from 1804 until the institution's closure in 1806 and gained this position again at the new University of Berlin (Charle 2004; Paletschek 2010). The Humboldtian ideal of *Wissenschaft*, expressed in an unpublished programmatic script of ten pages (1809/10), was based on humanist principles of education and promoted freedom *(Freiheit)* and solitude *(Einsamkeit)* in learning, teaching, and research (*Lehrfreiheit und Lernfreiheit*); the unity of research and teaching (*Einheit von Forschung und Lehre*); personal responsibility (*Selbstverantwortung*); academic self-government (*akademische Selbstverwaltung*); and pure learning and research in science and scholarship (*reine Wissenschaft*) (Perkin 1984; Anderson 2004).

Professors became specialized in one discipline, could act fairly autonomously as chairs (*Ordinarien*), and were supported in their research and teaching by both post-doctoral researchers (*Privatdozenten*) and research students who were based in the professors' research institutes and university laboratories. Since promotion criteria were based on achievements in advancing academic knowledge through research, many new professorships were established that transformed the original four faculties through an array of new specialized disciplines (Perkin 1984). The emphasis on pure learning strengthened the philosophical faculty that united humanist and scientific disciplines at the time vis-à-vis the professional faculties of law, theology, and medicine (Anderson 2004). Inspired by ancient Greek role models, pure learning was thus not confined to scientific research, as often assumed when the German term *Wissenschaft* that addresses all forms of academic knowledge production is limitedly translated as "science" instead of "science and scholarship" (Anderson 2004, 57), but it meant that practical knowledge of the applied sciences, such as engineering, had to be taught outside of the neo-humanist university in newly founded *Technischen Hochschulen* (technical colleges) until the late nineteenth century (Perkin 1984).

Conventionally, the Humboldtian blueprint for the research university has been stylized as the role model for the emergence of the successful German research university in the nineteenth century (Charle 2004; Rüegg 2004). Recent scholarship, however, has argued that Wilhelm von Humboldt's reform was part of a complex process of university reforms in several German states that often had already begun, as in the case of the University of Freiburg, in the final decades of the eighteenth century (Paletschek 2001; 2010; Anderson 2004). According to the historian Sylvia Paletschek (2010, 221), the Humboldtian model of the research university is a scholarly construction that originated at the *fin de siècle* when von Humboldt's, Schleiermacher's, and

Fichte's ideas were constructed as canonical to trace the then world-leading status of German universities straight back to the Prussian archetype. The formation of the German research university thus needs to be regarded as a long-term process that was only completed by the 1880s (Paletschek 2010).

Up until World War I, the German system of research universities, led by Berlin, Göttingen, Leipzig, and Heidelberg, represented the leading scientific research network in Europe because it "allowed scientific research to be a professional, bureaucratically regulated activity" (Rüegg 2004, 17; Taylor, Hoyler, Evans 2008). Despite the focus on intellectual freedom and autonomy, the state played an important role for the German research university because it provided enormous financial resources, employed professors as civil servants, and absorbed about two thirds of the graduates as public sector workers (Perkin 1984). Accordingly, other European countries also founded ministries to implement some degree of national control in higher education, ranging from the allocation of public funds via the governance of university access, curricula, and exams to the appointment of academics (Rüegg 2004). This nationalization of higher education in the nineteenth century contributed to the formation of national identities, for example, in Greece, Bulgaria, and Romania, where new universities in the capital cities served as an important symbol of national independence (Charle 2004).

In a complex process of university expansion and reform that was aided by international mobility of students, academics, and government experts, most European countries began to transform their universities along German lines from the 1830s onwards and had adopted basic principles of the research university by the 1900s (Anderson 2004; Charle 2004; Rüegg 2004). Romania was one of the few countries orientated more towards the French model of specialized schools and faculties because many students and professors were educated in Paris, but the French faculties were themselves regrouped to German-style universities in 1896 (Charle 2004). The Italian university system adopted centralized control from France, research orientation from Germany, and an emphasis on the professional faculties of law and medicine from Central European role models, whereas the Turkish government introduced a European university system in the 1930s by appointing mainly German professors to the reorganized University of Istanbul (Charle 2004).

As the first institutions in the world that provided doctoral research training, German universities also became a role model for the reform of existing institutions and the foundation of new universities outside Europe, especially in the United States and Japan (Clark 2006). Many Americans who studied or received their doctorates at German universities in the second half of the nineteenth century were subsequently instrumental for reforming US universities according to different German prototypes. In 1876, John Hopkins University in Baltimore was newly founded as the first US research university with a graduate school, an American innovation based on the unity of research and teaching, as practiced at the time in the University of Heidelberg and elsewhere in Germany (Shils and Roberts 2004). The new graduate programs proliferated at US universities in the 1890s and soon met the quality standards of their German counterparts (Clark 2006). This reduced the need for academic mobility to Europe before World War I and accelerated the rise of US research universities to both global knowledge centers and new exemplars for universities across the world.

Many places in Africa and Asia-Pacific can look back on much longer traditions of higher learning than Europe, but degree-awarding universities in the genealogy of Bologna and Paris—and later research universities—were mostly established through the import of European models in the nineteenth and twentieth centuries. Exceptions include Al Azhar University in Egypt, an Islamic theological university founded in 970 and sometimes designated the oldest worldwide, and the Catholic University of Santo Tomás, which was founded by the archbishop of Manila during the Spanish colonial period in the Philippines in 1611 (Shils and Roberts 2004). In India, the British colonizers decided in 1835 to promote modern, Western education through the foundation of educational institutions and, in 1857, also through the first universities set up in Calcutta, Bombay, and Madras (Seth 2007). Supported by both colonizers and nationalists, modern Western knowledge was to some extent shaped by local knowledge traditions, but it quickly became normalized and, despite its European origins, considered as global and "universal rather than parochial" (Seth 2007, 183).

Along these lines, the first Japanese institutions of higher education were founded as schools for Western studies in the second half of the nineteenth century. The University of Tokyo was chartered in 1877 and initially excluded Japanese and Chinese classics (Shils and Roberts 2004). In an attempt to modernize China, the Guangxu Emperor of the Qing Dynasty approved three institutions of modern higher education in the 1890s that were—in the case of the first two—co-founded by American scholars and grew to become Tianjin, Jiaotong, and Peking University. New universities chartered in Australia, Pakistan, Lebanon, New Zealand, and South Africa since the mid-nineteenth century and in many other Asian and almost all African countries during the twentieth century—another exception being the University of Sierra Leone (1876)—contributed to the unquestioned universalization and globalization of Western epistemology, and especially of new scientific and technological knowledge (Shils and Roberts 2004; Seth 2007).

University Expansion and Technoscientific Innovation

Industrialization in the nineteenth century created a growing demand for skilled professional labor, which resulted in an increasing number of students being attracted to the natural and medical sciences. In the decades 1860 to 1940, emerging national systems of higher education became the engines of national and regional economic growth and henceforth expanded in regard to the number of institutions, students, and academics; they diversified through new subjects and institutions, especially in technical and commercial education and teacher training; they became more inclusive of a wider social range of students and academics; and they adjusted university study to the needs of modern careers in industrial societies, which went hand-in-hand with the proliferation of new professorships, laboratories, journals, conferences, research awards, and supporting institutions (Jarausch 1983; Anderson 2004). From 1860 to 1930, participation rates in higher education rose in Germany from around 0.5 percent to 2.6 percent and in Britain from around 0.3 percent to 1.9 percent, while the United States' larger and more diverse system of universities and colleges already enrolled 3 percent of 18- to 21-year-olds in 1860 and reached 15 percent of that age cohort in 1930, thus paving the way for "a large, diversified, middle-class and professional system of higher learning" (Jarausch 1983, 10).

The rise of student numbers beyond population growth was linked to more inclusive admission policies, especially for women. For centuries, women had been educated to fulfill their roles as wives and mothers but this changed during the nineteenth century. Apart from individual female pioneers, equality in education was most advanced in the United States. The women colleges Salem (founded in 1772) and Wesleyan (chartered in 1836) and the co-educational Oberlin College (founded in 1833; accepting women from 1837) were the first institutions of higher learning to accept women as students (Solomon 1985). The admission of women to universities proliferated after the American Civil War (1861–1865) so that by 1900 about two thirds of US colleges and universities were co-educated (Painter 1896).

In Europe, female higher education was delayed by a lack of secondary schooling for women. Calls for the formal admission of women to university multiplied in the 1860s because teaching became a recognized profession for both men and women and female visiting students from the United States had made "the idea of women in university classes familiar" (Anderson 2004, 257). Formal admission of female students began in France and Switzerland in the 1860s and was subsequently taken up across Europe before arriving in Germany only in the 1900s (Anderson 2004). The most progressive European country for female students became Britain, where the first women's college was founded with Girton at Cambridge in 1869 and women students were permitted after 1870. Female students were particularly welcome at the newly founded civic universities, where they represented a higher share and were awarded degrees for their examinations much earlier (in London since 1877) than in Oxford and Cambridge (1920 and 1948 respectively) (Anderson 2004). In 1908, Marie Curie (1867–1934) was appointed the first female university professor in France, after she had taken over the university laboratory of her deceased husband Pierre Curie (1859–1906). She also became the first female Nobel Prize Winner, who shared the prize in physics with her husband and Henri Becquerel (1852–1908) in 1903 and received a second prize in Chemistry on her own in 1911 (Anderson 2004).

In Britain, the world's hegemonic power from around 1815 to 1873, the university system expanded considerably during the nineteenth century. The University of London (1828/1836) was founded as the first British university since the sixteenth century, followed by Durham (1834) and the three Irish Colleges of Belfast, Cork, and Galway (founded in 1845) (Anderson 2006). In the 1870s and 1880s, seven "redbrick" universities were set up as university colleges with philanthropic support in the new economic powerhouses of the Victorian era, such as Manchester, Birmingham, and Liverpool, to serve the predominantly local clientele of an urban industrial society by emphasizing the natural and technical sciences and local needs (Anderson 2006). Oxford and Cambridge retained their leading position in England after confessional restrictions were removed in 1870, but many dons remained skeptical towards the innovation of the research university so that it required three Royal Commissions and the example of thriving US universities to transform these ancient centers of learning into modern research universities (Heffernan and Jöns 2013).

The historian Laurence Brockliss has argued that Anglo-American universities were more successful with putting the modern research university into practice than German universities because the former kept corporate autonomy and non-professional education as their primary mission (Rüegg 2004, 12). Recent scholarship adds to this the important role of academic travel through the implementation of research sabbaticals, a concept pioneered by Harvard University in 1880 and taken over by Cambridge—presumably as the first European university—in 1926 (Heffernan and Jöns 2013). By enabling its academics to focus on study leave at regular intervals, the University of Cambridge encouraged overseas journeys across all disciplines that increasingly targeted state-of-the-art laboratories, libraries, and academic expertise in the United States. This made travel the key research technique, created close ties between British and American universities, and thus contributed to the emergence of an Anglo-American hegemony in science and higher education (Heffernan and Jöns 2013).

Before World War I, British universities were firmly embedded within multilateral imperial networks spanning British and English-language universities in the settler colonies of Canada, Australia, New Zealand, and South Africa, thereby constituting a geographically expansive, closely knit but highly selective, exclusionary, and unequal "British academic world" (Pietsch 2013). These imperial networks were also responsible for the influence and often mutual enrichment of English and Scottish education in North America, India, Australia, and South Africa, which was reinforced through the awarding of external degrees by the University of London. French education left traces in French North and West Africa, Syria, and Indo-China (Shils and Roberts 2004), but all these imperial ties began to erode with advancing decolonization, nationalization, and Americanization after World War II.

The two World Wars interrupted research and teaching in most European universities and enrolled academics across all disciplines in war service, where they often provided vital expertise. World War I (1914–1918) paved the way for more nationalistic science policies because it harmed the international academic community by "leaving behind resentment, malicious prejudices and apparently irreconcilable differences" (Hammerstein 2004, 642). World War II (1939–1945) had even more devastating effects, especially in Germany, where in addition to the physical destructions, the Nazi regime had expelled or killed one third of their academics, while another third had to be removed during denazification (Hammerstein 2004). Both World Wars fuelled a need for scientific innovation and thus paved the way for the emergence of the technosciences as the main competitive arena in which the Americanization and Sovietization of research and education unfolded during the Cold War between a capitalist West and a communist East from 1945 to 1989.

Most famously, the Sputnik shock of 1957, when the Soviet Union successfully launched the first space mission, channeled new dimensions of public investment into the technosciences and gave priority to the US' first mission to the moon. The predominance of the sciences in governing university policies thus emerged during World War II and the subsequent Cold War, when American hegemony (c.1945–1971) developed a sophisticated technoscientific complex with the US research university at its core (Etzkowitz 2008). Attracted by superior educational and research infrastructure, thousands of students and academics left their European homeland in a new postwar "brain drain" to seek education and employment in North America and elsewhere, while US institutions such as the Carnegie and Rockefeller Foundations created US-centered knowledge networks in both developed and developing countries to consolidate state power during the Cold War, thereby promoting English as the new *lingua franca* of scientific research (Parmar 2002).

In the COMECON countries of Central and Eastern Europe, university systems were reorganized in the immediate postwar decade according to the Soviet model:

higher education became subordinated to centralized state planning and the needs of the economy; curricula were controlled by central administration; and universities played a minority role in a much wider sector of tertiary education focusing on the training of researchers, academics, and secondary school teachers (Neave 2011). Within both the Western and Eastern blocs, international mobility and collaboration of students, researchers, and academics flourished, but exchanges remained highly restricted and controlled, intensifying only briefly in the wake of reform periods such as the Prague Spring in 1968 (Jöns 2009).

The period 1946 to 1970 saw major waves of university expansion in most developed and developing national economies that created 45 percent of all universities existing in 1970 (Figure 22.1F) and initiated the transition from elite to mass higher education. This worldwide expansion had begun and was particularly strong in South America as well as in East, Southeast, and South Asia, where countries such as Brazil, Colombia, the Philippines, Korea, India, and especially Japan in the late 1940s; Argentina, Brazil, India, China, and especially Indonesia in the 1950s; and Peru, Brazil, Argentina, India, and Indonesia in the 1960s used new universities for cultural and economic development, nation state building, and—in former colonies— also as a symbol of national independence (Figure 22.1A–E).

In the United States and Western European countries, expansion was driven by debates about the future demand for university graduates in the growing professions, particularly in science and technology, and the need to increase the participation of geographically and economically disadvantaged strata of the population (Anderson 2006; Neave 2011). The existing centers of academic knowledge production in northwestern Europe and the northeastern United States were most affected by university expansion in the 1960s and thus belatedly (Figure 22.1D and E). In Europe, participation rates rose from 4 to 5 percent of the age cohort in the 1950s to 15 to 25 percent in the 1970s, while the US system had already achieved mass higher education with a participation rate of more than 30 percent in 1960 but expanded further to reach 50 percent by 1970 (Trow 1974). Creating a demand for more inclusive academic practices, the advent of mass higher education also fueled the widespread student revolts of the late 1960s.

Five centuries passed between the humanist Agricola (1443–1485) advising that "Natural philosophy is not as necessary as moral philosophy; it is scarcely more than a means of culture" (cited in Painter 1896, 127) and US president John F. Kennedy (1917–1963) inviting his audience at the National Academy of Sciences on October 22, 1963 "to encourage other gifted young men and women to move into these high fields [of science] which require so much from them and which have so much to give to all of our people" (Kennedy 1965, 319). From the renewal of universities following the Reformation to the rise of US universities as leading global knowledge centers in the twentieth century, the humanities were either more important than the sciences or of similar significance for thriving universities. After a long period of gradually altering priorities, this has changed in the twenty-first century because an ongoing global shift of economic power towards Asia-Pacific has encouraged the Chinese government to emulate the still world-leading US research universities by prioritizing university funding, mobility programs, and international collaboration in the natural and technical sciences (Jöns and Hoyler 2013). The most recent emergence of new global academic knowledge centers is, therefore, for the first time in history, led by the sciences.

A Universities founded 1946-1950, N = 150

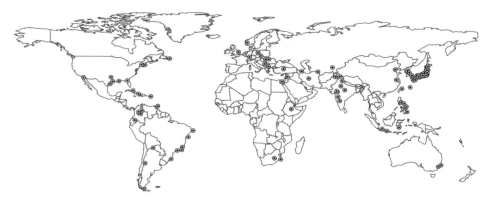

B Universities founded 1951-1955, N = 80

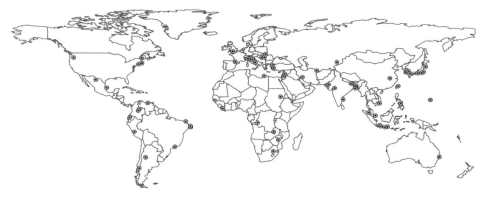

C Universities founded 1956-1960, N = 148

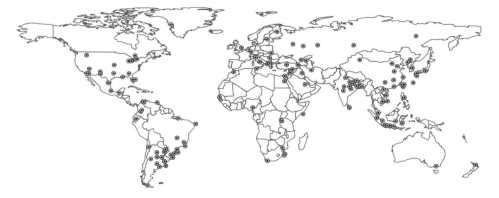

Figure 22.1 Global geographies of university expansion, 1945–1970. Author's map design based on information published in *The World of Learning 1970–1971*, 21st edition, 1971. London: Europa Publications.

D Universities founded 1961-1965, N = 204

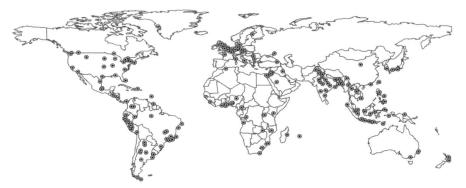

E Universities founded 1966-1970, N = 103

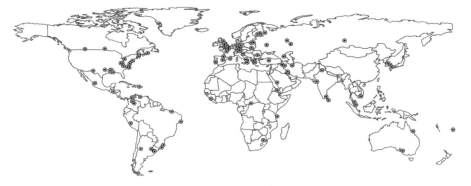

F Origin of all universities exisiting in 1970-71, N = 1512

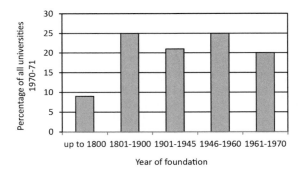

Figure 22.1　(Continued)

Endnote

1   I gratefully acknowledge very helpful remarks by Bernard Lightman, Robert Anderson, Chay Brooks, Raja Herrmann, and Peter Meusburger on an earlier version of this chapter. Emily Rodbourne kindly supported the creation of the database on all universities existing in 1970. The views expressed in this article and any errors are solely my responsibility.

References

Anderson, Robert D. 2004. *European Universities from the Enlightenment to 1914.* Oxford: Oxford University Press.

Anderson, Robert. 2006. *British Universities Past and Present.* London: Continuum.

Brockliss, Laurence and Nicola Sheldon (eds.) 2012. *Mass Education and the Limits of State Building, c.1870–1930.* Basingstoke: Palgrave Macmillan.

Burke, Peter. 2000. *A Social History of Knowledge: From Gutenberg to Diderot.* Cambridge: Polity Press.

Charle, Christophe. 2004. "Patterns." In *A History of the University in Europe: Volume III: Universities in the Nineteenth and Early Twentieth Centuries (1800–1945)*, edited by Walter Rüegg, 33–80. Cambridge: Cambridge University Press.

Clark, William. 2006. *Academic Charisma and the Origins of the Research University.* Chicago: The University of Chicago Press.

de Ridder-Symoens, Hilde. 1996. "Mobility." In *A History of the University in Europe: Volume II: Universities in Early Modern Europe (1500–1800)*, edited by Hilde de Ridder-Symoens, 416–48. Cambridge: Cambridge University Press.

de Ridder-Symoens, Hilde. 2004. "The changing face of centres of learning, 1400–1700." In *Schooling and Society: The Ordering and Reordering of Knowledge in the Western Middle Ages*, edited by Alasdair A. MacDonald and Michael W. Twomey, 115–38. Leuven: Peeters.

Etzkowitz, Henry. 2008. *The Triple Helix: University–Industry–Government Innovation in Action.* New York: Routledge.

Frijhoff, Willem. 1996. "Patterns." In *A History of the University in Europe: Volume II: Universities in Early Modern Europe* (1500–1800), edited by Hilde de Ridder-Symoens, 43–110. Cambridge: Cambridge University Press.

Hammerstein, Notker. 2004. "Universities and war in the twentieth century." In *A History of the University in Europe: Volume III: Universities in the Nineteenth and Early Twentieth Centuries (1800–1945)*, edited by Walter Rüegg, 637–72. Cambridge: Cambridge University Press.

Heffernan, Michael and Heike Jöns. 2013. "Research travel and disciplinary identities in the University of Cambridge, 1885–1955." *British Journal for the History of Science*, 46: 255–86.

Jarausch, Konrad H. 1983. "Higher education and social change: some comparative perspectives." In *The Transformation of Higher Learning 1860–1930*, edited by Konrad H. Jarausch, 9–36. Chicago: The University of Chicago Press.

Jöns, Heike. 2009. "'Brain circulation' and transnational knowledge networks: studying long-term effects of academic mobility to Germany, 1954–2000." *Global Networks*, 9: 315–38.

Jöns, Heike and Michael Hoyler. 2013. "Global geographies of higher education: the perspective of world university rankings." *Geoforum*, 46: 45–59.

Kant, Immanuel. 1784. "An answer to the question: 'What is Enlightenment?'" In *An Answer to the Question: "What is Enlightenment?"* translated by H. B. Nisbet, 1–11. Published in 2009. London: Penguin Books.

Kennedy, John F. 1965. "A century of scientific conquest." In *The Scientific Endeavor*, edited by National Academy of Sciences, 312–19. New York: Rockefeller Institute Press.

Meusburger, Peter. 1998. *Bildungsgeographie: Wissen und Ausbildung in der räumlichen Dimension*. Heidelberg: Spektrum Akademischer Verlag.

Meusburger, Peter and Thomas Schuch (eds.) 2012. *Wissenschaftsatlas of Heidelberg University*. Knittlingen: Verlag Bibliotheca Palatina.

Moseley. Alexander. 2007. *John Locke*. London: Bloomsbury.

Neave, Guy. 2011. "Patterns." In *A History of the University in Europe: Volume IV: Universities Since 1945*, edited by Walter Rüegg, 31–69. Cambridge: Cambridge University Press.

Painter, Franklin V. N. 1896. *A History of Education*. New York: D. Appleton.

Paletschek, Sylvia, 2001. "Verbreitete sich ein 'Humboldtsches Modell' an den deutschen Universitäten im 19. Jahrhundert?" In *Humboldt international. Der Export des deutschen Universitätsmodells im 19. und 20. Jahrhundert*, edited by Rainer Christoph Schwinges, 75–104, Basel: Schwabe.

Paletschek, Sylvia. 2010. "Eine deutsche Universität oder Provinz versus Metropole? Berlin, Tübingen und Freiburg vor 1914." In *Die Berliner Universität im Kontext der deutschen Universitätslandschaft*, edited by Rüdiger vom Bruch, 213–42. Munich: Oldenbourg.

Parmar, Inderjeet. 2002. "American foundations and the development of international knowledge networks." *Global Networks*, 2: 13–30.

Perkin, Harold. 1984. "The historical perspective." In *Perspectives on Higher Education: Eight Disciplinary and Comparative Views*, edited by Burton R. Clark, 17–55. Berkeley: University of California Press.

Pietsch, Tamson. 2013. *Empire of Scholars: Universities, Networks and the British Academic World 1850–1939*. Manchester: Manchester University Press.

Porter, Roy. 1996. "The Scientific Revolution and universities." In *A History of the University in Europe: Volume II: Universities in Early Modern Europe (1500–1800)*, edited by Hilde de Ridder-Symoens, 531–62. Cambridge: Cambridge University Press.

Postlethwaite, T. Neville. 1995. *International Encyclopedia of National Systems of Education*. Second edition. Oxford: Pergamon.

Roberts, John, Agueda Maria Rodriguez Cruz, and Jürgen Herbst. 1996. "Exporting models." In *A History of the University in Europe: Volume II: Universities in Early Modern Europe (1500–1800)*, edited by Hilde de Ridder-Symoens, 256–82. Cambridge: Cambridge University Press.

Rüegg, Walter. 2004. "Themes." In *A History of the University in Europe: Volume III: Universities in the Nineteenth and Early Twentieth Centuries (1800–1945)*, edited by Walter Rüegg, 3–31. Cambridge: Cambridge University Press.

Seth, Sanjay. 2007. *Subject Lessons: The Western Education of Colonial India*. London: Duke University Press.

Shapin, Steven. 1996. *The Scientific Revolution*. Chicago: The University of Chicago Press.

Shils, Edward and John Roberts. 2004. "The diffusion of European models outside Europe." In *A History of the University in Europe: Volume III: Universities in the Nineteenth and Early Twentieth Centuries (1800–1945)*, edited by Walter Rüegg, 163–230. Cambridge: Cambridge University Press.

Solomon, Barbara M. 1985. *In the Company of Educated Women: A History of Women and Higher Education in America*. New Haven, CT: Yale University Press.

Taylor, Peter J. 1999. *Modernities: A Geohistorical Interpretation*. Cambridge: Polity Press.

Taylor, Peter J., Michael Hoyler, and David M. Evans. 2008. "A geohistorical study of 'the rise of modern science': Mapping scientific practice through urban networks, 1500–1900." *Minerva*, 46: 391–410.

Trow, Martin. 1974. "Problems in the transition from elite to mass higher education." In *Twentieth Century Higher Education: Elite to Mass to Universal*, edited by Michael Burrage, 88–142. Published in 2010. Baltimore: Johns Hopkins University Press.

Van Horn Melton, James. 1988. *Absolutism and the Eighteenth-Century Origins of Compulsory Schooling in Prussia and Austria*. Cambridge: Cambridge University Press.

Verger, Jacques. 1992. "Patterns." In *A History of the University in Europe: Volume I: Universities in the Middle Ages*, edited by Hilde de Ridder-Symoens, 35–74. Cambridge: Cambridge University Press.

Wolgast, Eike. 1996. "Die Universität Heidelberg: Historische Entwicklung." In *Heidelberg: Geschichte und Gestalt*, edited by Elmar Mittler, 284–320. Heidelberg: Winter.

PART III

Communication

CHAPTER TWENTY-THREE

Manuscripts

JOYCE VAN LEEUWEN

Before the invention of the printing press, the communication of knowledge relied on handwritten sources. The medieval period is of supreme significance for the transmission processes of texts from the ancient world, and it is due to the activities of the scribes from this period that we can still read these texts today. This chapter offers an introduction to paleography and manuscript studies by explaining the different steps in the making of manuscripts as well as in working with these sources. Moreover, it will specifically focus on the transmission of scientific texts, and describe how textual and visual arguments are structured in manuscripts. Whereas the copying practices of literary texts were primarily concerned with preserving these texts for later generations, we notice that scribes of scientific texts actively engaged with their sources and adapted them to the specific needs and interests of their time. In the past, scholars often distrusted these texts as they did not faithfully render the intentions of their original authors. Recent scholarship, however, takes a somewhat different approach and values scientific texts as being part of a creative tradition. The images and diagrams that are contained in medieval manuscripts are especially studied in this context. They convey important information about the frameworks in which scientific texts were transmitted to us and shed light on the processes of appropriation and transformation of scientific knowledge. This chapter will focus on the transmission processes of scientific texts from the ancient world and the interpretation of these sources in Greek, Arabic and Latin traditions.

Transmission of Texts

Of all the different writing surfaces that have been used throughout history, three are predominant for ancient and medieval books: papyrus, parchment, and paper. Papyrus was manufactured from the papyrus plant that grew almost exclusively in the Nile Delta in Egypt. After the preparation process, in which sheets were formed by placing two layers of thin strips cut from the pith of the plant crosswise on top of each other, the fibers would still be visible in the final product. The side of the papyrus roll where

A Companion to the History of Science, First Edition. Edited by Bernard Lightman.
© 2016 John Wiley & Sons Ltd. Published 2020 by John Wiley & Sons Ltd.

the fibers were oriented horizontally formed the inner side of the roll designated for writing. Papyrus was the most widely used writing support in the ancient world. However, it gradually lost its leading position to parchment, which would become the most important medium for the medieval period.

Parchment is animal skin that has gone through a special process of cleaning and smoothing to make it suitable as a writing support.[1] Even though the use of animal skin as a writing surface has a much longer history, the term "parchment" (*pergamēnum*) derives from its use and perfection in the second century BCE at the court of the Attalids in Pergamum. The quality of parchment depended both on the accuracy of the preparation process and on the type of skin that was used. Here young animals, especially calves, were preferred to other animals such as cows, goats, and sheep. Because of the cost intensity of making a book out of parchment, often all parts of the skin were used including those that were defective. In some manuscripts of lesser quality, for example, bone holes are still visible.

The expense, together with the shortage of writing material that occurred from time to time, led to the development of a procedure in which parchment was re-used by scraping off the original text and putting a new text in its place. Parchment that received such a treatment is called "palimpsest," from the Greek term *palimpsēstos*, "scraped again". The decision as to which texts would be erased and written over by new ones could depend on several different factors. Of especial significance in this context are the personal taste and interests of a scribe or commissioner of a manuscript. A text that did not have immediate relevance to its owners ran a higher risk of being sacrificed in favor of another text. This applies to works that might be obscure or had become obsolete, such as technical or legal texts. An exciting example of the discovery of such a recycled manuscript is the Archimedes Palimpsest. Several works by the Greek mathematician Archimedes have been found below the text of a thirteenth-century prayer book. The palimpsest is of extraordinary importance to the history of science, as it contains two unique texts by Archimedes, the *Method* and *Stomachion*, as well as the only Greek version of *On Floating Bodies*. Advances in imaging technology and multispectral imaging have made it possible to retrieve the lower text and to disclose these mathematical works to the scientific community. The edition of the Archimedes Palimpsest by Netz, Noel, Wilson, and Tchernetska has recently been published (2011).

Paper, the third widespread writing support, was invented in China, supposedly at the beginning of the second century. It was exported to the Islamic world in the middle of the eighth century as the Arabs learned the technique of paper-making from Chinese prisoners of war. From there it spread to Europe, where it gained increasing importance from the twelfth century onwards. The early introduction of paper in the Islamic world explains why there are relatively few Arabic manuscripts on parchment when compared to codices from the Greco-Roman world. Although there was a rapid decline in the use of parchment after the discovery of paper-making, parchment continued to be used for manuscripts of especial value. Arabic paper differed from paper developed in Western Europe in that it had long fibers and generally lacked a watermark—that is, a specific pattern in the paper by which it could be identified. The presence of watermarks in occidental paper provides paleographers with an important tool in assessing the date and place of origin of manuscripts. Since each papermaker used unique watermarks that were regularly replaced, we are often able to identify and date paper within a time-frame of about thirty years. Briquet's survey of water-marks remains an important reference for scholars working with manuscripts to this day (1907).

In the ancient world, books were made in two different forms, rolls and codices. The roll is the oldest book format and consists of several sheets that were glued together. Usually only the inner side of the roll would be inscribed, and writing and reading involved a constant unrolling and re-rolling. The codex, on the other hand, resembles our modern book and is made of a number of folded sheets. A major event in the transmission of texts was the transition from roll to codex that took place between the second and fourth centuries. Ancient literature as a whole was transferred from one form to the other; this process involved considerable losses, as many unpopular texts were not copied into codex form. The codex had many advantages over the roll: in a codex both sides of the writing surface could be used, which means that less material was wasted; a codex was much easier to handle than a roll, especially if one needed to consult a specific passage; using a codex involved less wear and tear of the material; finally, a codex had a larger capacity—it could contain the contents of several rolls. In the Islamic context the roll was of no great importance as the transition from roll to codex was largely completed by the end of the fourth century, well before Islam appeared.

In the same period that the codex largely replaced the roll, we also see the development of uncial script in the Greco-Roman world. Uncial is a majuscule script (written entirely in upper-case letters) that flourished between the fourth and eighth centuries. It was the common script for literary texts, although it was especially favored in biblical manuscripts. The subsequent development of minuscule script (written in lower-case letters) brought some clear advantages for writing and reading manuscripts: minuscule could be written more quickly than majuscule; it saved space, which becomes especially relevant in lean times when people were looking for more economic uses of expensive parchment; and, it was easier to read, as accents and punctuation marks were used. In Greek writing, minuscule script was used from the ninth century onwards. An early example of minuscule from this period is a manuscript in the Bodleian Library in Oxford, MS D'Orville 301, containing Euclid's geometric treatise the *Elements* (Figure 23.1). It was written by Stephanos in the year 888 and is therefore one of the first precisely dated Greek manuscripts. In the Latin West we notice a similar development in the emergence of the Carolingian minuscule at the end of the eighth century during the reign of Charlemagne. It spread rapidly in the century thereafter and remained the dominant script until the twelfth century. Minuscule did not, however, replace the use of earlier scripts on every occasion. Majuscule script for instance was still used for expensive illuminated copies for the imperial library or in liturgical books. The change to minuscule script is another decisive moment in the transmission of ancient texts. As soon as texts were copied into minuscule, the majuscule exemplars were no longer needed and were often discarded. This also implies that those authors and texts that were not popular enough to make the step to minuscule are now completely lost to us.

Turning to Arabic script, the style used in early Islamic times is usually known as Kufic. The term applies to a variety of old scripts and was mainly used for copying the Qurʾān. Since there is no distinction between upper- and lower-case letters in Arabic, there was no equivalent of the transition from majuscule to minuscule

Figure 23.1   MS D'Orville 301, f.51r. Greek manuscript containing Euclid's *Elements*, dated 888. By permission of The Bodleian Libraries, University of Oxford.

script in the Islamic world. However, we may recognize a similar phenomenon in the appearance of proportional scripts between the tenth and thirteenth centuries. Known as the "Six Pens," there are six proportional scripts grouped in pairs of distinct sizes.

In Greek, Latin, and Arabic writing there have been many different styles and varieties, some of which are characteristic of a time period or region. Certain trends are well described and allow a precise dating of a manuscript on the basis of a specific script. Paleographers should, however, always be cautious when dating a manuscript, since scribes often imitated older styles. Further codicological properties, such as a colophon or subscription, may convey important additional information in assessing the date of a manuscript.

In general, there are differences in writing between professional copyists and scholars–amateurs. The former were commissioned to make a manuscript copy and wrote in a fine and legible hand, whereas the latter were transcribing books for their personal use. Such working copies are often characterized by cursive handwriting and a great number of corrections and annotations. The final appearance of a manuscript thus depends, amongst other factors, on its writer and audience. The illustrations that are contained in many scientific manuscripts also point to these different contexts. Whereas in some manuscripts we find sketchy diagrams that offer the writer's personal understanding of the text, others are lavishly illustrated codices especially made for, for example, an imperial library. A well-known example of a luxury manuscript is the Vienna Dioscurides, a sixth-century copy of *De Materia Medica* by Dioscurides in Greek. This codex, made for Princess Julia Anicia, is famous for its illustrations of plants and animals. The book was widely read in antiquity, as is evident from the many translations in, for instance, Latin, Syriac, and Arabic.

With the invention of printing in Europe around 1450 it was no longer necessary to make handwritten copies of manuscripts. Nevertheless, for a period, printed texts and handwritten manuscripts were both still being produced. Due to the relatively late introduction of printing technology in the Islamic world, where it was employed on a larger scale only some centuries after its discovery in Europe, we see that there is a much greater number of Islamic manuscripts.

The development of printing technology led to the publication of many editions of ancient texts. These editions were for the most part, based on a single or few manuscripts at hand and so do not meet modern standards of editing ancient texts that take into account all manuscript copies of a particular text.

Textual Criticism

We have already seen several factors that impacted the survivability of manuscripts. The transition from roll to codex and the replacement of majuscule script by minuscule in the Greco-Roman world were two key events which resulted in the loss of many ancient texts. In addition, many books have been destroyed over time, either accidentally, for example through fire or damage by mold, or on purpose. This includes actions like those by the Catholic Church implementing the banning of books listed on the Index of Prohibited Books, but also examples of destruction on a smaller scale. The Archimedes Palimpsest exemplifies the fate of many manuscripts in which the original text was erased and a new text put in its place. Such recycled manuscripts, however, are not evidence of a desire to destroy specific books. In most cases, texts selected for destruction had become obsolete or were of minor interest to the copyist or commissioner of a manuscript. The loss rates of ancient literature may look somewhat different for scientific texts when compared to literary works. Whereas Homer's epics address large audiences at all times, the situation is less favorable in, for example, the case of an obscure astronomical text. Technical treatises generally ran a higher risk of not being copied at some time because their contents were regularly updated and older versions had lost their relevance.

All manuscripts containing the works of ancient authors are later copies, far removed in time from the originals. We possess almost no manuscripts older than the ninth century. The number of manuscript copies that have survived vary between authors, in part due to accidents of transmission, but also as a consequence of the interest in an author at a certain point in time. The fact that Euclid's *Elements* is transmitted to us in over a hundred Greek manuscripts is certainly related to the ancient mathematician's all-time popularity. The challenge posed to the modern editor is to collect the extant manuscript material of a certain work and try to reconstruct the original text.

Textual criticism is the field of scholarship concerned with the recovery of a text by means of examining all textual evidence. To begin with, the editor takes into account the external evidence that can be gained from an autopsy of a manuscript. What does the script reveal about the period in which a manuscript was written? In fortunate circumstances, we might even be able to identify a copyist based on the handwriting. In cases when the paper has watermarks, what are the likely origins of a manuscript and to what period can it be dated? A close analysis of the outer appearance of a manuscript already provides important information that needs to be considered before turning to the internal evidence.

After collecting the manuscripts that preserve a certain text, the editor will try to assess the relationships between the manuscripts. Based on a comparison of each manuscript with a transcription or an earlier edition of the text, a list of errors can be made. Generally we see that a copy contains the same errors as its exemplar—although we may not exclude the possibility that some copyists were responsible for corrections—as well as a number of additional errors. The agreement in errors enables the grouping of manuscripts, after which step specific affinities can be detected. Finally, the results of this examination are presented in a family tree, a so-called *stemma codicum*, which represents the affiliations between the manuscripts. There are many factors that may complicate the construction of such a stemma. The fact that not all manuscript copies are extant sometimes makes it difficult to determine precise relationships. We do not know, of course, how many manuscripts are lost and which positions in the stemma they would have occupied. Contamination is a further practice in the transmission of texts. It occurs when readings originating from different sources are combined, for instance when a manuscript is copied from two exemplars at the same time, or when a manuscript is emended at a later point in time by using a manuscript from another branch of the tradition. Sometimes an editor cannot trace at which stage in the tradition contamination took place, and such contaminated manuscripts cannot be allocated a position in the stemma.

The aim of textual criticism is to discover as far as possible the text of the archetype, that is the earliest reconstructable form, which is, or most closely approximates to, the original text. The constructed stemma will reveal which manuscripts are most important for a reconstruction of the text. Those manuscripts that are closer in time to the archetype usually contain fewer errors and corruptions, and are thus more likely to represent the author's words. The editor decides which manuscripts will be eliminated and which will be considered in a critical edition; at least one manuscript from each branch of the textual tradition should be adopted.

The next step in restoring the archetype is to decide for each case in which the manuscripts contain divergent readings which variant represents the authentic text.

Here we can rely on some general principles, although the actual circumstances may be much more complicated (West 1973 offers a very good introduction to textual criticism of Greek and Latin texts). When the majority of the manuscripts deriving from the same lost source agree on a particular reading, this reading was very likely also contained in the exemplar. However, if there are, for example, only two copies of the archetype, we are not always able to reconstruct the authentic text, namely when these two copies disagree in a particular reading and neither of them can be considered to be a plain error. Another principle in textual criticism is known as *lectio difficilior*. Generally, the difficult reading is more likely to represent the authentic text. This is because scribes had the tendency to simplify an unfamiliar word rather than complicate it.

What does a critical edition of a text finally look like? On the upper part of the page we find the text as reconstructed by the editor; a critical apparatus is printed below. A critical apparatus contains information on variant readings in other manuscripts and conjectures of corrupt or contaminated passages that have been proposed by the editor or earlier scholars. Manuscripts are assigned an abbreviation, a so-called *siglum*, and they are referred to by this abbreviation in order to save space in the critical apparatus. Since a reconstruction of the text depends in part on editorial choices, which differ from one person to the next, the apparatus offers important source material for those who are working with the text. Many scholars work with editions only and do not themselves check the manuscript material, so the critical apparatus is an essential tool to find out whether there are any corruptions or insecurities in the text. Often there is more than one possible reading and in each case scholars can decide for themselves which reading they think should be preferred.

In recent years textual criticism has opened itself up to a wider audience. Many manuscript collections have now been digitized, facilitating access by scholars. High-quality images reduce the need to visit libraries and archives, although in-situ analysis of a manuscript gives the best information on external features such as manuscript material, types of ink, or different hands. Along with digitization of manuscripts, a number of initiatives are under way to prepare digital editions of ancient texts.

Visual Images

A conspicuous feature of scientific texts is the presence of diagrams and images. Visualizations in manuscripts may be integrated in the main body of the text, where they occasionally occupy a full page. Or diagrams may be inserted in the margins of a manuscript, sometimes in addition to diagrams placed in the openings of a text. Such variant diagrams are on a par with marginal annotations, reflecting a commentator's reading of the text. We find different levels of expertise in the execution of the diagrams. Some are clumsy freehand drawings by scribes who were not very talented draughtsmen, whereas others are carefully elaborated with the help of instruments like compass and ruler. The copyist of the text and the illustrator were not always one and the same person: particularly in the later medieval period there was a division of labor between the two. After having made a copy of the text, the scribe would hand over the manuscript to an illustrator who was responsible for the visual aspects. The windows that are present in some manuscripts attest this division of labor between copyist and illustrator: these are blanks in the text provided for illustrations that were never executed.

A majority of scientific manuscripts contain diagrams. These diagrams are often essential to the understanding of the argument of the text. Therefore, it is remarkable to see that modern critical editions focus almost exclusively on the text of manuscripts and neglect visual components. Editors generally include their own diagrams, remote from the ancient tradition and interpreting the text in modern terms. The critical apparatus typically contains much information on different textual variants, but notes next to nothing on the diagrams. For the users of modern editions of ancient scientific texts, it often remains unclear whether a certain diagram is also contained in the manuscript tradition and what it looks like. Even though Neugebauer (1975) called for the critical assessment of diagrams, the practice has not been adopted by the scholarly world until recently. We are now witnessing a slow increase in the number of critical editions of ancient scientific texts that take into account the diagrams that are contained in the manuscripts. This interest was initiated by Netz's research on the relation between the text and diagrams in Greek mathematical works (1999).

In scientific texts, there is often a close connection between text and diagrams, and many authors explicitly refer to visualizations that illustrate certain principles or geometrical objects. Moreover, a common practice in the mathematical tradition is the use of letter labels in the text to refer to accompanying diagrams. These letter labels are only relevant in relation to a diagram and we are often unable to understand a text without the corresponding diagram. This implies that many of the earliest scientific texts were illustrated. However, it does not imply that we can be certain in all cases of the actual form of the original diagrams. Whereas diagrams in mathematical manuscripts may be traced back to the author's lifetime, the situation is different for those scientific texts in which we find a looser connection between text and image. In medical texts, for example, an image may be helpful to illustrate a certain procedure, but it is not necessarily required in order to understand its textual description. It is in those works in which diagrams and images are not specifically defined by the text that we find the greatest variety. Images could be transferred from one tradition to the next, when illustrators borrowed images that were originally included in a different genre of texts. It also happened that scribes and illustrators themselves devised new images to a text, that they added variant diagrams, or adapted the ones they found in their source to fit a specific context. The next section will discuss some of the visual strategies that underlie the creative treatment of scientific texts.

Despite many possible complications in the transmission of diagrams, for many texts we are still able to recover the diagram in the archetype. Picture criticism largely proceeds along similar lines as textual criticism (see Weitzmann 1970 for a first introduction to picture criticism). Errors are again the guiding principle in grouping manuscripts. Those manuscripts that share the same set of diagrammatic errors in all probability derive from the same exemplar. Since each new copy will contain some additional errors, we see that in general the further removed in time a diagram is from the archetype, the more flawed it will be, and the diagram that corresponds best to the text is most likely to resemble the archetype. The results from a critical examination of manuscript diagrams may again be presented in a stemma. This stemma, however, may well differ from the one established for the text as there are numerous scenarios which could disturb a linear relationship between the transmission of a text and its visualizations. Diagrams and images were not always copied from the same source as the text, or they may have been added by different illustrators independently of each other, which means that we may distinguish more than one diagram tradition. Because of the fact that our earliest manuscript evidence is often centuries removed in time from the author himself, it is often impossible to discover what the original diagrams and images exactly looked like. In working with diagrams one should also remain attentive to the possibility that diagrams have undergone deliberate alteration in the course of transmission. Later scribes and commentators might add variant diagrams or enrich existing illustrations with decorative elements. The latter occurs to a greater extent in those cases in which diagrams are less necessitated by the text, than in, for example, mathematical texts where reasoning depends on diagrams that are precisely defined by the text.

Recent research has focused particularly on the critical assessment of diagrams in ancient mathematical texts (Netz 2004; Saito 2006). Ancient diagrams differ systematically from what we find in mathematical texts today. Their most remarkable feature is that they are more schematic than their modern counterparts. This means that they do not give a metrically correct representation of the geometrical objects. When the text for example defines a right angle, we should not expect to find an angle of exactly 90 degrees in the corresponding diagram. This property is irrelevant in ancient diagrams and the angle could equally be depicted by a right, an acute or an obtuse angle. Rather than focusing on these metrical qualities, ancient diagrams convey topological qualities only. Look again at Figure 23.1 to find an example of this fundamental principle in ancient mathematics. The diagram refers to proposition III.13 of Euclid's *Elements*. This proposition demonstrates that a circle cannot touch another circle at more than one point, whether it touches it internally or externally. The illustrator of this manuscript completely ignored metrical properties in the diagram—the internal and external figures are clearly not circles. The intention of the diagram is to represent its topological details: in both cases we assume that there are two points where the figures touch each other. This principle applies to ancient mathematical diagrams in general: they reliably depict topological properties, whereas metrical qualities are less relevant as these may alter depending on the accuracy with which a diagram was executed.[2] The fact that the practices of ancient diagrams are systematically different from later diagrammatical notions is an important basis for comparing various visual strategies in scientific texts and studying the different functions of diagrams and images over time.[3]

Tradition and Innovation

The previous sections focused on the transmission processes of ancient texts and the means by which textual criticism attempts to restore the original text. For the sake of simplicity we assume that medieval scribes tended to copy texts as faithfully as possible in order to preserve their contents for later generations. Even though the conservation of texts is the main objective in many copying practices, there is another aspect to it, which is especially relevant for scientific treatises. A characteristic of the transmission of scientific texts from antiquity is the creative treatment of these texts at different times. Scientific works were often not copied as a whole, as literature was, but were handed down in excerpts, compendia, and anthologies. Scribes and commissioners of manuscripts would select those passages that were of most interest to them and adapt these to the specific aims and contexts of their projects. Furthermore, translations of scientific texts into different languages involve many forms of alteration, intended and unintended. The transmission of scientific texts is thus affected by processes of transformation of their contents. Examples of such transformations are the Arabic revisions of Greek mathematical works mentioned below (Sidoli 2007), or the pedagogical framework in which Roman astronomical texts were transmitted in the Carolingian Renaissance (Eastwood 2007). Stepping aside from the strictly philological approach that aims to reconstruct the author's original words and considers later emendations and alterations to be the result of corruptions in the tradition, recent scholarship increasingly emphasizes scientific texts as being part of a creative tradition and analyzes the historical contexts in which the transformations took place.

Transformations did not only occur at the level of the text, but also in the relations between word and image. For example, images were added to a text that was originally not illustrated, or diagrams were altered on the basis of increased knowledge. Various approaches that are employed in Ottoman maps effectively illustrate strands of conservation and innovation. Some of these maps were regularly updated in the process of copying, and new maps were added, whereas others preserve their foreign origins as translations of Latin atlases of the seventeenth century (Brentjes 2009). The Arabic tradition of Dioscurides's pharmacological treatise *De Materia Medica* provides further evidence of textual and visual adaptations. Different traditions were developed in Islam to take account of new species and genera unknown to the Greeks (Rogers 2007). In the course of the translation process, the names of plants were adapted to Arabic terminology. Along with these ongoing textual revisions, the illustrations were also transferred to the Islamic sphere. In a study of the author portrait in thirteenth-century Arabic manuscripts, Hoffman has shown that illustrators deliberately combined a range of visual traditions (1993). A 1229 Arabic manuscript of *De Materia Medica*, now in the Topkapi Palace Library (Ahmet III, 2127), contains a portrait of Dioscurides on a double frontispiece. The figure of Dioscurides on the right side of the frontispiece is depicted in Greek Byzantine style and can be traced back to a Byzantine model. It makes a clear contrast with the Islamic style for the student figures on the left side of the frontispiece, as Hoffman points out. The deliberate fusion of visual models by the illustrator of the Topkapi Dioscurides highlights a preeminent aspect of the transmission process from Greek to Arabic: the Greek origins of the text as well as its adaptation to an Islamic context are both visible.

A further example taken from ancient medicine shows that images often offer a reflection of the contexts in which they were made. The author of a Hellenistic commentary on the Hippocratic treatise *On the Articulations* prescribes images in order to exemplify the procedures described in the text (Nickel 2005). Certain aspects of the images we possess, however, cannot be traced back to Hellenistic times. The commentary has been handed down to us in a single manuscript from the Byzantine period, which is evidenced by the images contained in it, as Nickel explains. The architectural details in the background of some of the images are later additions that resemble Byzantine taste. Other occurrences in which the image does not correspond with the description in the text reveal that the visualizations in their present form are not likely to be authentic. Despite the fact that these images do not have their origins in antiquity, they are still significant as a later commentary on the text and are instructive in identifying characteristic aspects of Byzantine art and culture.

The significance of analyzing various diagram traditions for obtaining historical insights can be exemplified by the mathematical sciences. In a study of a number of manuscript diagrams in Greek, Arabic, and Latin sources of Aristarchus's *On the Sizes and Distances of the Sun and Moon*, Sidoli (2007) has pointed to the different contexts in which this text was interpreted by medieval and early modern scholars. Proposition 13 of *On Sizes* serves as a case study that clarifies the deliberate changes that were made in the text and diagrams. The oldest extant Greek manuscript of the treatise dates from the ninth century, which is approximately at the same time as an Arabic edition was composed by Thābit ibn Qurra. His edition followed a common pattern of ninth-century Arabic practice, in which Greek mathematical texts were preserved in innovative ways. Thābit made a revision of *On Sizes*; he offered both a mathematically corrected version of the text and drew new diagrams for his revision. This Arabic adaptation of the treatise in which the Greek text and diagrams were freely altered differs from its treatment in early modern times. The sixteenth-century Latin translation of Aristarchus's work by Frederico Commandino, for example, still displays a close connection to the Greek text, but the diagrams were all redrawn and reflect Commandino's mathematical understanding of it. Sidoli has shown that the wide variety of diagrams made by competent scholars shed light on the transmission processes of *On Sizes* and the ways in which this text was appropriated in ninth-century Baghdad and sixteenth-century Italy.

The transmission of Roman astronomy provides a rich source for illustrating the creative treatment of scientific texts. In the Carolingian Renaissance we notice that certain astronomical works contain diagrams that diverge from the descriptions in the text. Eastwood (2007) has convincingly shown that these diagrammatic changes should not be understood as corruptions, but are due to intentional revisions and re-interpretations of the text by Carolingian scholars. Their main goal in altering the diagrams was to further the reader's understanding of the texts. Such educational intentions are visible in the transformations of Pliny's *Natural History* in the ninth century. The many clarifications and simplifications that were added to Pliny's text are evidence of the pedagogical framework in which the text was received. As Eastwood argues, this is further reinforced by the presence of diagrams. Whereas Pliny did not make any explicit visual arguments in his text, diagrams were added by the Carolingian scholars working on the text. These diagrams had an important function as a learning device in the ninth-century teaching of astronomy. A later fifteenth-century manuscript of Pliny's text shows a completely different context of transmission. At this time the Renaissance scholar Giovanni Pico della Mirandola commissioned a luxurious copy of the *Natural History* with illustrations, in the devising of which he himself played an important role (McHam 2006). Some of the illustrations agree with the standard Plinian repertory of images at this time, but others depict less familiar scenes. The latter turn out to be Mirandola's personal choice and reflect his own interests and erudition. Unlike the pedagogical role the Plinian diagrams played in the instruction of science in the Carolingian Renaissance, we now find highly sophisticated diagrams that were specifically designed for a learned audience.

These examples illustrate some of the transmission frameworks of ancient scientific texts. By examining scientific treatises in their respective historical contexts, current research makes an important contribution to our understanding of the appreciation of these texts and the creative engagement with them at different times. Moreover, transformations in both text and images that accompany a text are crucial in assessing the conception and development of scientific knowledge over time.

Endnotes

1 In this chapter my descriptions are necessarily simplified; see Clemens and Graham (2007) for an excellent and thorough introduction into all aspects of the field of manuscript studies. For specific information on Greek and Latin paleography, I refer to Thompson (1912) and on Arabic manuscripts see Déroche (2006) and Gacek (2009).

2 See De Young (2005) on diagrams in the Arabic Euclidean tradition, which seem to mirror this practice of Greek manuscripts.

3 See, e.g., van Leeuwen (2014) for an analysis of mechanical diagrams in the Byzantine manuscript tradition and their counterparts in printed editions from the early modern period.

References

Brentjes, Sonja. 2009. "The interplay of science, art and literature in Islamic societies before 1700." In *Science, Literature and Aesthetics. History of Science, Philosophy and Culture in Indian Civilization*, edited by Amiya Dev, 453–84. New Delhi: Centre for Studies in Civilizations.

Briquet, Charles-Moïse. 1907. *Les filigranes: Dictionnaire historique des marques du papier dès leur apparition vers 1282 jusqu'en 1600*. 4 vols. Leipzig: Hiersemann.

Clemens, Raymond and Timothy Graham. 2007. *Introduction to Manuscript Studies*. Ithaca, NY: Cornell University Press.

Déroche, François. 2006. *Islamic Codicology: an Introduction to the Study of Manuscripts in Arabic Script*. London: Al-Furqān Islamic Heritage Foundation.

De Young, Gregg. 2005. "Diagrams in the Arabic Euclidean tradition: a preliminary assessment." *Historia Mathematica*, 32: 129–79.

Eastwood, Bruce S. 2007. *Ordering the Heavens: Roman Astronomy and Cosmology in the Carolingian Renaissance*. Leiden: Brill.

Gacek, Adam. 2009. *Arabic Manuscripts: A Vademecum for Readers*. Leiden: Brill.

Hoffman, Eva R. 1993. "The author portrait in thirteenth-century Arabic manuscripts: A new Islamic context for a late-antique tradition." *Muqarnas*, 10: 6–20.

McHam, Sarah Blake. 2006. "Erudition on display: The 'scientific' illustrations in Pico della Mirandola's manuscript of Pliny the Elder's *Natural History*." In *Visualizing Medieval Medicine and Natural History, 1200–1550*, edited by Jean A. Givens, Karen M. Reeds, and Alain Touwaide, 83-114. Aldershot: Ashgate.

Netz, Reviel. 1999. *The Shaping of Deduction in Greek Mathematics: a Study in Cognitive History*. Cambridge: Cambridge University Press.

Netz, Reviel. 2004. *The Works of Archimedes: Translated into English, together with Eutocius' Commentaries, with Commentary, and Critical Edition of the Diagrams*. Cambridge: Cambridge University Press.

Netz, Reviel, William Noel, Nigel Wilson, and Natalie Tchernetska, eds. 2011. *The Archimedes Palimpsest*. 2 vols. Cambridge: Cambridge University Press.

Neugebauer, Otto. 1975. *A History of Ancient Mathematical Astronomy*. 3 vols. Berlin: Springer.

Nickel, Diethard. 2005. "Text und Bild im antiken medizinischen Schrifttum." *Akademie-Journal*, 1: 16–20.

Rogers, Michael J. 2007. "Text and illustrations. Dioscorides and the illustrated herbal in the Arab tradition." In *Arab Painting: Text and Image in Illustrated Arabic Manuscripts*, edited by Anna Contadini, 41-7. Leiden: Brill.

Saito, Ken. 2006. "A preliminary study in the critical assessment of diagrams in Greek mathematical works." *Sciamus*, 7: 81–144.

Sidoli, Nathan. 2007. "What we can learn from a diagram: The Case of Aristarchus's *On The Sizes and Distances of the Sun and Moon*." *Annals of Science*, 64, No. 4: 525–47.

Thompson, Edward Maunde. 1912. *An Introduction to Greek and Latin Palaeography*. Oxford: Clarendon Press.

Van Leeuwen, Joyce. 2014. "Thinking and learning from diagrams in the Aristotelian *Mechanics*." *Nuncius*, 29: 53–87.

Weitzmann, Kurt. 1970. *Illustrations in Roll and Codex: A Study of the Origin and Method of Text Illustration*. Princeton, NJ: Princeton University Press.

West, Martin L. 1973. *Textual Criticism and Editorial Technique: applicable to Greek and Latin texts*. Stuttgart: Teubner.

CHAPTER TWENTY-FOUR

The Printing Press

NICK WILDING

De-centering Gutenberg from the Print Revolution

There is no dearth of reliable, erudite, and thorough surveys on the relationship between print and science (Thornton and Tully 1975; Frasca-Spada and Jardine 2000; Thornton, Tully, and Hunter 2000). These generally document and comment upon the contemporaneous adoption of print and transformation of science in early modern Western Europe, positing a range of weak to strong claims for their causal relationship. Rather than attempt to replicate or replace such work, this chapter will attempt to highlight the areas of this relationship that still lack adequate treatment, and guide students towards what might now be the most promising avenues of future research. In particular, I argue that inadequate attention has been paid to the changing technologies of print from the fourteenth to twenty-first centuries; that the roles of facsimiles in paper, microfilm, and digital media on the history of science deserves more thought, and that print, rather than merely offering a site of representation for scientific knowledge, in fact provides one of its enabling conditions. Furthermore, I propose that some forms of both scientific practice and practitioners exist solely in print, and that these cases might provide useful material for us to think about our methodologies.

Points of origin are always a good place to start when looking at historical narratives. The presumed effect of the typographical revolution of fifteenth-century Europe is largely dependent on how we characterize that episode, but this is also a circular process: by teleologically projecting certain supposedly essential elements of typography onto the history of printing, Gutenberg is used to underwrite a revolutionary model. One of the most pressing historiographical questions in this field is still considered to be the contribution and legacy of Elizabeth Eisenstein's *Printing Press as an Agent of Change* (1979). Generally, historians have either embraced, questioned, or modified models of fixity, replication, and multiplication they trace to this work. One response to the work, mirrored across other disciplines' intersections with the history of the book, has been to move away from claims of an intrinsic effect of the print revolution

A Companion to the History of Science, First Edition. Edited by Bernard Lightman.
© 2016 John Wiley & Sons Ltd. Published 2020 by John Wiley & Sons Ltd.

to an analysis of the processes of making meaning by charting the historical evidence left of actual reading. We will return to this trend, but first it might be worth considering again whether Gutenberg's revolution ever took place. I am not interested here in offering an account that tempers the effect of print with arguments of cultural continuity from scribal culture: this much has been well documented. The question requiring more urgent attention is the specificity of Gutenberg's invention.

First, it is worth pointing out that science, with whatever historiographical caveats we may throw at that term, was in print well before Gutenberg. Egyptian amulets from the fourteenth century were printed on paper with woodblocks. A few survive with magic number squares, evidence of some degree of sophisticated mathematical culture (Schaefer 2006; Muehlhaeusler 2010). Given the "delayed" impact of print in the Ottoman Empire from the fifteenth to the eighteenth centuries, it has been assumed that an essential enmity reigned between Islam and print, based on the cultural importance of calligraphy. The printed amulets have only recently received any attention, and, as well as being unexpected and interesting, they may well yet still provide the missing link of technology transfer from Chinese or Korean to European print. Whether or not Gutenberg's was an independent invention, it is clear that any account of the relationship between print and science should no longer start with him.

The precise nature of Gutenberg's contribution to the development of the common press and replicable type has also undergone modification in recent generations, though the fraught debates have been restricted to a relatively small world of incunabalists (specialists in print up to 1500), and made little impact on the models employed by historians of science. Almost every technological element in the printshop has been re-examined, and it now seems likely that several once considered essential were not present in the first European printed books. Metal type for Gutenberg's 42-line Bible, for example, was probably not made from a matrix whose impression was created using a punch; more likely it was cast from sand molds with the type impressions created individually from a variety of elements (Agüera y Arcas 2003). Individual pieces of type were not, in this early stage of moveable metal type printing, identical with each other; the punch/matrix solution only came later.

This is not just a question of extending the timeline of invention: perhaps more surprising is the brilliantly argued hypothesis that the printer(s) of the Catholicon, a long Latin dictionary published in 1460, c.1469 and 1472, produced stereotyped two-line slugs from the standing type of the first edition for those that came later (Needham 1982). The method of this specific process remains unknown, and in general, stereotyping was not rediscovered until the eighteenth century.

The disjointed and non-linear development of early printing is visible, too, in the use of xylography (woodblock printing) alongside moveable type. The figure of Regiomontanus (1436–1476), longtime hero of historians of print and science because of his technical interventions and strong presence in the printshop, is central here (Lowood and Rider 1994). But the case of Regiomontanus offers evidence not merely of the triumph of print and the virtuosity of the scientific author–printer: his own productions included examples of mixed print and manuscript charts, with numerals added to printed tables by hand. Perhaps of more interest to narratives of print revolution are the wooden blockbook reprints of typographically printed calendars from 1474–1475, published by Hans Sporer just after Regiomontanus's death in 1476 (Kremer 2013). These were on the market alongside new typographical reprints, yet we should think of woodcut printing not as Neanderthals outsmarted by *Homo sapiens*, but as interbreeding hominids.

The first decades of what we usually think of as an autochthonic and autogenous miracle were in reality marked by bricolage, hybridity, and experimentation. Even successful experiments did not lead to common practice, though, and one challenge for historians of the book is to attempt to reconstruct possible modes of production from the evidence supplied by imprinted paper alone. Many questions remain open, and historical reconstruction is probably the best way to solve them satisfactorily. The techniques used by the earliest printers of geometrical books, for example, are still not fully understood. In certain cases, most notably that of astronomy, typographic innovation occurred with astonishing rapidity and variety. Michael H. Shank's work (2012) on Regiomontanus proposes that he not only printed his own works, but also pioneered new techniques establishing what would swiftly become a standardized astronomical visual vocabulary. Shank shows that Regiomontanus was at the avant-garde of typographical invention, casting metal plates for his diagrams from molds of engraved copper plates. Here, xylography would ultimately, or at least temporarily, triumph, with woodcuts rather than metal plates becoming the usual tool for scientific illustration until the gradual introduction of other forms of graphic reproduction from the sixteenth through twenty-first centuries, but the lasting effect of incorporating scientific diagrams into the text remains with us still.

Similar typographical experimentation is visible in one of Regiomontanus's publishers, the celebrated Erhard Ratdolt (1442–1528), who not only inherited Regiomontanus's technique of casting images and perhaps also ornamental initials in metal, but also instituted the practice of the title page and ornamental border. Book historians have, by and large, tended to embrace a teleological model of the technological transformations of print technique and machinery, projecting backwards models that would later prove to be successful. Ratdolt's innovations are a useful reminder here: his editio princeps of Euclid's *Elements* (Venice 1482) included, in some dedicatory copies, a page printed in gold leaf, using a technique only reconstructed a few decades ago (Carter, Hellinga, and Parker 1983). Renzo Baldasso (2013) suggests that Ratdolt also adapted Regiomontanus's technique for creating solid line tabulation, using metal strips set into plaster or lead for the production of geometrical lines in his Euclid. Such technological dead-ends and temporary oblivions do not strike historians of technology as unusual, but bibliographers and historians of print have largely relied on a revolutionary model of print that is only now being modified, and historians of technology have failed to address the question of print.

The print revolution, if we still want to use that term, did not leap fully armed from a German businessman's brain. Replicable moveable metal type was only one choice, and not the first, among printers; its general success requires historical explanation, not more just-so stories. The parallel, or perhaps intertwining developments of Korean, Chinese, and Egyptian print technologies, which follow the occidentalization of paper itself, deserves closer scrutiny, for Gutenberg still plays a central role in various narratives of European technological supremacy (Carter 1925).

Perhaps the fluidity of modes of production does not much matter. We are still left with the quantitative effect of print on culture. In the case of science, though, this still has to be proved. Take the first scientific text to be printed, Lucretius's *De Rerum Natura*. The account of Poggio Bracciolini's rediscovery of the sole

surviving manuscript in a monastery still serves as an emblem for the light of humanism chasing out the dark ages. There are problems with this story, though: not only
were there medieval readers of Lucretius, and several manuscripts in existence at the
time of Poggio's find, but humanism and scribal culture, rather than the hand press,
secured the survival of the text. Several dozen copies were made and circulated, as
their marginalia attests (Brown 2010; Palmer 2012). Print certainly produced many
editions of this work, but also its censorship. It is hard to see how quantitative arguments necessarily become qualitative under their own steam. The expansion of both
private and institutional libraries, the rise in literacy, and the increase in textual production are all long processes which may be traced earlier than Gutenberg.

From Book to Text and Back Again

The effects of this rewriting of the "print revolution" on the history of science remain
to be seen; despite strong and sustained criticism of Eisenstein's characterization of the
change wrought by the invention of print as ushering in a new epistemology of fixity
and replicability, with immediate consequences for the sciences, historians of science
have been slow to follow literary scholars and bibliographers in adopting new models
to interrogate the printed book. This is puzzling, as one of the most productive and
exciting developments within the discipline over at least the last half-century has been
the turn away from a model of science conceived of as a disembodied history of ideas
and towards more material concerns. We routinely study procedures and protocols,
experiments and exercises, techniques and technologies. Paradoxically, though, this
movement seems to have produced an even deeper idealization of books as mere tex-
tuality, somehow divorced from the media that necessarily carries the text. Even when
texts do appear as actors alongside other material objects, they become (in a way many
object-actors do not) unchanging and dehistoricized. Yet were we to look for the single scientific object which survived in greatest numbers from the early modern, or
even modern, periods, we would be hard pressed to find a rival to the book. How did
the scientific book become reduced to a text, and how might it re-enter the world of
scientific objects?

Part of the answer is philosophical, part technological, though their connection
remains unclear. The twentieth-century shift away from a disembodied intellectual
history towards a world of things left books in a precarious situation. One of the most
potent models for reconsidering the printed book within the history of science has
been Bruno Latour's (1987) notion of scientific inscription, producing "immutable
mobiles": objects carrying texts and images that can travel through the cycle of scientific production. While Latour is interested in the process of scientific production,
his depiction of print as a final stage of inscription seems to rely on an Eisensteinian
idea of fixity, itself derived from Marshall McLuhan. It is the very immutability of the
printed object that grants it agency and power in Latour's scheme. There are several
responses available to such a model, all of which have been immensely productive.
The most obvious, but perhaps the hardest to undertake, is a reception study. The
results can be spectacular, as in Owen Gingerich's (2002 and 2004) census of the first
two editions of Copernicus's *De Revolutionibus*, which established clear lines of transmission through marginalia between intellectual communities. This is part of a larger
shift within the world of bibliography away from purely technical description towards

what has been termed a "sociology of texts" (McKenzie 1999). Key to this enterprise is the precept, first formulated in manuscript studies, that textual relations are social relations. The study of marginalia, as a seminal essay by Jardine and Grafton (1990) showed, not only offers historians a way of watching readers in action, it supersedes the sterile model of textual autopoesis upon which much intellectual history, especially the Cambridge School of political theory, relied.

Another response to the "immutable mobile" is that of Adrian Johns (1998), who elegantly showed how the very terms we assign to print's agency, such as credit and fixity, are themselves the product of complex social negotiations with a recoverable history. We might also pay more attention to the conditions of production and dissemination of individual editions and issues: the printed book usually contains several competing layers of inscription, as editors, compositors, printers, patrons, and authors vie for textual control. The result is often a high number of variants within an edition, and only painstaking bibliographical work can uncover and reconstruct such disputes from the marks left on pages. We might also reconsider the ethnocentricity of Latour's term "inscription," which invariably, and tautologically, refers to European practices and notions of writing, even while attempting to avoid such biases as a general principal (Schaffer 2007). Print's crucial role in affirming European technical superiority has its origins in the story it tells of itself as self-generated and unique. Even in the various global turns of recent history of science, celebrating hybridity and interconnectivity, we tend to rely on a Eurocentric model of book production that ignores even the swift globalization of the press. There was a European-style press in Mexico in 1539; by 1541 an account of the Guatemalan earthquake had been published there, and natural philosophy soon followed. Rare are the examples that break this mold, though they are becoming increasingly accessible (Berry 2006; Raj 2007; Schaffer, Roberts, Raj, and Delbourgo 2009; Schäfer 2011; Suarez and Woudhuysen 2013).

Locating the Scientific Book

Perhaps we should not fret too much about origins and essences, and move instead to the realm of practice. But how do we define historically the subject of this chapter, the scientific book? It is no longer necessary to rehearse the standard disclaimer of the anachronism of pre-nineteenth-century science: it is clear that a modern definition of science will not do. Generally, we offer what we call generous or broad definitions of science, and hope the problem is swept away by the flood of titles considered. An alternative is to accept an historical elite's definition, and see what books were used within university courses on natural philosophy. But might not books themselves, or rather, an historical reconstruction of the ways they were defined, not tend to reconfigure the question? Was there, say, such a thing as a scientific book in the early modern period? If so, how did readers identify and locate them? What were the paths that led to and defined this object?

One good place to start is with catalogues of book fairs. For several centuries, these catalogues provided the initial announcement of the titles of new books; they were collected, reprinted and annotated, used as reference works or shopping lists (Serrai 2000–1999; Maclean 2009; Maclean 2012; Blair 2010; Nuovo 2013). Not only do these catalogues give an unfiltered glimpse of what new or recycled stock was considered profitable for export or sale, they also provide a rough taxonomy, adopted from other bibliographic sources, placing books we now deem "scientific" alongside their contemporary traveling companions.

Take the case, for example, of Galileo's *Sidereus Nuncius*, a handy case study since a thorough census has recently been compiled (Needham 2011). It was first advertised in the Easter 1610 Fair catalogue for Frankfurt, alongside Guidobaldo Dal Monte's *Problematum Astronomicorum libri septem* (Venice 1609). So far so good, and the name of the section in which these books are located, "Philosophical Books and of Other Arts" (Libri Philosophici & aliarum artium) is encouraging, if vague. Other books within this section include the second edition of Magini's *Ephemeridum coelestium motuum continuatio* (Frankfurt 1610), a reprint of Aldrovandi's *Ornithologiae hoc est, de auibus historiae libri 12* (Frankfurt 1610), first printed in 1599, and a reprint of Brahe's *Astronomiae instauratae progymnasamata* (Frankfurt 1610). Also within the group are some books that would not now be considered science, but which we now admit to our historically generous fold, such as the third edition of Caspar Bartholin's work on astrological medicine, the *Exercitatio de stellarum* (Wittenberg, 1609), or works on alchemy, such as the *Artis auriferae, quam chemiam vocant, volumina duo* (Basel 1610). Thus far, the fair catalogue's category seems pretty stable and useful for our framing of the scientific book. Yet the category also contains general works of philosophy, ethics, logic and dialectic, as well as grammar, rhetoric, pedagogy, lexicography, chronology, epistolography, and even Aesop's *Fables*, Erasmus's *Colloquia*, and Machiavelli's *Art of War*. No subdivisions are made, so we cannot distinguish, using this evidence, between a (natural) philosophical work and these other arts. All we know is that these books were not considered, by the compilers of the fair catalogue at any rate, or the publishers submitting their information, to be Latin theological (Protestant, Catholic or Calvinist), legal, medical, historical, political, geographical, poetical, or musical. They could clearly not be entered into the other sections, German-language or vernacular books.

How else, then, might we approach and contextualize what we regard as an undoubtedly scientific book? Within a collection, we might be able to trace, using printed, manuscript or card catalogues, how books were arranged. The catalogue of the printed works in the library of Francesco Barberini, for example, was arranged in alphabetical order by author, but listed shelf marks for each volume (Holstenius 1681). When the collection entered the Vatican Library in 1902, new shelfmarks were assigned. Sticking with the case of the *Sidereus Nuncius* (Barberini shelfmark: LII.C.31, Vatican shelfmark, Stamp.Barb.N.XII.8, newly discovered after my detection of a cataloguing error, and the sole surviving example of this book in the Vatican), we can see that it originally sat alongside Galileo's *Il Saggiatore*, his *Discorso intorno alle cose che stanno in sù l'acqua, ò in quella si muovano*, the Letter to the Grand Duchess Cristina, and Kepler's *Dissertatio cum Nuncio Sidereo*. Whether this represented a conceptual or purchasing order is unclear; other authors' works are not generally kept together, and those of Galileo are broken into two groups, with the *Dialogo* and *Discorsi* set apart.

In the Bodleian Library, Thomas Hyde's catalogue (1674) erroneously distinguished between a quarto (i.e. Venetian) and an octavo (i.e. Frankfurt) copy. In fact, both copies were of the second, Frankfurt edition. One was located alongside the second edition of John Dee's *Propædeumata aphoristica* (1568), Joannes Meursius's *Orchestra* (1618), Thomas Lydiat's *Solis et lunæ periodus* (1620) and a 1648 work on the Peace of Westphalia, the *Instrumentum pacis*. The Bodleian's other copy was between Otho Casmannus's *Cosmopœia et ouranographia Christiana* (1611) and Kepler's *Dissertatio cum Nuncio Sidereo* (1611). These works had been shelved together since at least 1620, when they were all given the same shelf mark in an earlier catalogue (James 1620). Clearly these examples show a certain disorder to library organization, but already in the sixteenth century, some libraries had begun to organize their books using scientific categories. Wittenberg University Library in 1536 had a section containing "Mathematics, Cosmography and Geography translated from Greek;" by 1583, Cambridge University Library used the categories "Astronomy, Cosmography, Geometry and Arithmetic" (Suarez and Woudhuysen 2010, 1136).

These examples provide a sense of the radically different ways in which what we now consider to be an iconically "scientific" book was classified. The companionship of Kepler makes sense to us, but others do not. A similar sense of historical bewilderment or curiosity might be elicited by examining other titles bound with the *Sidereus*. Of the 83 copies examined in Needham's census (2011) of the *Sidereus*, over a quarter were either still bound with other works or showed clear evidence of having previously been in *sammelbände*. The contexts in which early owners placed their copies of the *Sidereus* were wide ranging, with over 35 other editions identified. Of these, nine titles occurred more than once, the most frequently recurring being Francesco Sizzi's highly critical *ΔIANOIA astronomica, optica, physica* (1611), present in no less than nine copies, over 10 percent of the total. Kepler's more supportive *Dissertatio* and *Narratio*, each published in two editions, are present in seven and four copies, La Galla's *De phoenomenis in orbe lunae ... disputatio* (1612) is in four, as is De Dominis's *De radiis visus et lucis...tractatus* (1611). These two last titles were published by the same printer as the *Sidereus Nuncius*, who succeeded in launching a successful career from the controversy surrounding the book. Some copies contained multiple works, setting up complex dialogues. The Milan Brera copy, for example, is bound with Sizzi's *ΔIANOIA*, Galileo's *Discorso ... intorno alle cose che stanno in sù l'acqua*, Kepler's *Dissertatio* (Prague, 1610) and the same author's *Dioptrice* (1611); the Bodleian Library's copy is now bound with Galileo's *Il Saggiatore* (1623), La Galla, Sizzi, Wedderburn's *Quatuor problematum ... confutatio* (1610) and Mario Guiducci's *Lettera [...]al Tarquinio Galluzzi* (1620); Corpus Christi's copy puts it with Leonard Digges' *A prognostication everlasting* (1556), John Field's *Ephemerides for 1558–1560* (1558), Peter Apian's *Cosmographia* (1564), Sizzi, and La Galla.

It is worth looking at a few other *sammelbände* to see in what other contexts the *Sidereus Nuncius* was placed by contemporary readers. A copy in the Palatine Library of Parma, bound after 1700, situates it alongside mathematical, geometrical, astronomical, theological, catoptrical, and military works.[1] One in Augsburg places it alongside martyrologies, mineralogies, numismatics, and Roman grain distribution (Needham 2011, 217, n.3). In some cases, the textual relations evinced by *sammelbände* demonstrate little more than the need to bind similar formats and dimensions together. But frequently, debates and dialogues are consciously constructed by collectors and readers, and such instances can provide crucial evidence for the original zones of reception of scientific works.

If the evidence of fair catalogues, library catalogues, and binding offer different answers to our search for the best historical context for a scientific book, perhaps we should withdraw from what we might term "reception" studies, and look more closely at the book's site of production. Here, again, the case of the *Sidereus* is revealing. Tommaso Baglioni put his name to a couple of dozen books in the decade around the publication of the *Sidereus*. These included, in 1607, Galileo's *Difesa*, printed on Niccolò Polo's presses. We might assume, then, that the scientific book is a book printed by a scientific printer or published by a scientific publisher. Baglioni is certainly a contender, putting his name to works of navigation, agriculture, and medicine. But there are two problems: first is the general point that few printers or publishers specialized to a degree that may be considered useful for modern historians. Certainly, some presses had reputations for some kinds of work, but many were non-specialists, even if there were a discreet field in which to specialize. Baglioni also put his name on political, legal, and religious books in these years. Second, Baglioni did not in fact print the *Sidereus*: his name is a front for his boss, the radical and excommunicated printer, Roberto Meietti. Most of his "publications" in these years are in fact re-issues of old Meietti stock. Meietti, using Niccolò Polo's presses, printed mainly controversial political material, especially pamphlets, using a variety of aliases to cover his tracks. Using bibliographical evidence such as the recurrence of ornamental woodcuts, we can begin to reconstruct something of the typographical culture in which the *Sidereus* was produced. Far from representing the output of a scientific publisher, its closest relatives were actually works of mild pornography by a condemned forger, and volatile anti-Jesuit tracts (Wilding 2014).

If the scientific book cannot be identified or isolated by focusing on its means of dissemination, might there, though, be something specific to it at the typographic level? The visual culture of science was both enabled and thwarted in its paper publication. Much energy has been spent, in fact, in restoring the practices of observation and experiment that are presumed to pre-exist their bookish representation. We might however, have been too quick to subscribe to an epistemological separation based on a naive ontology of observation and representation. There is ample evidence to suggest that the culture of the manuscript and printed book actively informs visual epistemologies. From this perspective, print does not so much represent the world, as make it possible. In Galileo's observation notes on the satellites of Jupiter that became part of the *Sidereus*, for example, Galileo observes through the medium of woodcuts, suggesting: "have them cut in wood all of them in one piece, the stars white and the rest black then saw it into pieces."[2] Such a comment is striking in that it both suggests the deep commitment to, and knowledge of, print culture by natural philosophy, and, in this specific case, demonstrates the process of social and technical negotiation necessary to producing scientific books. Galileo's suggested mode of graphically depicting the process of establishing the satellites' periods did not, in fact, make it to the printed book: there, instead, moveable type was used, necessitating the production of many new asterisks, and setting considerable compositorial challenges. Galileo's white-on-black woodblock solution seems a good one, and it is not clear why, or by whom, it was rejected. The technical challenges posed by scientific publication have been studied in isolation, but we still lack a synthetic or even comparative analysis. Was the printing of geometry, for example, harder, more expensive, or more prone to error than printing music or martyrologies? To what extent were conventions shared or re-invented between print communities? In what circumstances did typographical innovation occur?

Facsimiles and Digitization

The technological aspect of the dematerialization of books lies, perhaps, in the shift of modes of reproduction of our sources. With canon-forming surveys such as Bern Dibner's *Heralds of Science* (1955) and Carter and Muir's *Printing and the Mind of Man* (1967) as foundations, the first large-scale transmutation of scientific books from print to another format was the ambitious Landmarks of Science series from Readex, which made available nearly 9,000 titles.[3] While such a publishing venture might have succeeded in liberating the history of science from institutions wealthy enough to possess large special collections, the impact of Landmarks is still hard to ascertain, and remains unstudied: series such as these are difficult to catalogue and use, and are probably already obsolete. Digitization of Landmarks would presumably be quick and cheap, but many institutions have embraced digitization of their own collections as both an outreach and conservation opportunity. Many such collections are now freely available, and have become essential tools for historians of science. With the introduction of color and higher resolutions of images, it is usually possible to have some sense of the nature of an original book. Some publishers and institutions are still seeking opportunities to monetarize their digitization, the most promising of which seems to be ProQuest's Early European Books, featuring a total of more than 20,000 books from the Royal Library, Copenhagen, the National Library, Florence, the National Library of the Netherlands, and the Wellcome Library, London. Whereas it is sometimes impossible to know what copy is being viewed with some digitization projects, Early European Books offers copy-specific cataloguing, and has selected collections containing large numbers of books with interesting scientific provenances. It remains to be seen whether this will last longer than Landmarks and attain the essential status of similar series, such as Early English Books Online, Eighteenth Century Collections Online and Nineteenth Century Collections Online. We should remember that in all these forms of reproduction, as in print itself, the object under scrutiny is usually a single copy, and not necessarily the best, or most interesting, copy in an edition (McKitterick 2013).

Given the amount of energy and thought going into the construction of very many digitization projects by historians of science, both of manuscript and printed material, there is surprisingly little reflection on the impact of digital humanities on the profession. The way in which we conceive of and use books has surely changed, not just as an issue of online access, but more profoundly by accepting the logic of surrogacy. High-resolution facsimiles are certainly a pleasure to use, especially when considered as surrogates for microform, but we lack a critical engagement with a wider discourse on the kinds of bibliographical evidence that are lost in media translation. Such a criticism may sound reactionary, but it is not: the digitization of books is part of their ongoing history. A book's production never ceases. Digital surrogates are not inferior simulacra of the authentic object: they supplement and alter that object. Much of the evidence that is lost in translation between media actually concerns what we would now regard as the inauthenticity of the original, such as copies' sophistication, restoration, washing, and *remboîtage* (rebinding to deceive).

It might be far-fetched to imagine that higher levels of resolution and full color reproduction might resubstantiate the object for historians, but it may help in the move away from thinking about books as pure texts. One discernible trend in the historiography of the history of science includes a move towards considering books as things, part of the material world of objects among which humans live, love, think, and make sense of themselves (Ago 2013). The use, rather than mere production, of books thus becomes paramount, but this does not mean that we are condemned to endless reception studies: one very promising field that is now opening is the study of scientists' own reading cultures, which, despite the best rhetoric of the New Science to convince us that books had been rejected in favor of nature itself, has generally comprised a large part of scientific practice.

The World of Print

Where, though, does the rematerialization of the book leave it as an object for the historian of science? Might we reformulate some general remarks, despite and because of the criticisms of Eisenstein's thesis, of the epistemological spaces peculiar to the printed page? The continuity of certain forms of representation within typographical culture, undeniably borrowed from scribal culture, but nevertheless with their own histories, might provide examples of just such an opportunity. The most obvious place to look is in printed images, and the evidence seems strong for an argument that there exists a feedback loop between printed representations and conceptual frameworks. Isabelle Pantin (2001), for example, has charted the changes and continuities across several genres of astrological figures, maps, and diagrams in the sixteenth century. Woodcuts and engravings, she found, shifted from iconographically laden representations of constellations to star maps accompanied by verbal descriptions of their etiological myths. The generally woodcut images, freed from narrative clutter, could now introduce elements such as dimensions, both between and of stars. The cosmos of the *Sidereus Nuncius*, then, was made possible not just through the use of the astronomical telescope or a commitment to Copernicanism, but through the use of woodblock star charts. One might even posit the logical limit of the instrumental role of print in the case of the illustrations of Galileo's *Istoria e dimostrazioni intorno alle macchie solari e loro incidenti* (Rome, 1613), where both the conceptual vocabulary ("macchia", meaning a "stain," "blot," or "smudge") and graphic representation, using detailed engravings, posit a strong congruency between the act of depiction and the construction of a fact. Misprints become retrospective solar activity.

There are even realms of scientific practice that exist only in books: perpetual motion machines, for example, always and only ever work on the page (Schaffer 1995). Museums are another space where textual and graphic descriptions perform more work than mere representation, securing heterotopic epistemologies (Findlen 1994). We might think of such cases as peculiarly ironic or anomalous, but it is worth considering, too, a broader category in its relationship to the making of natural philosophical and scientific knowledge: anonyms and pseudonyms, or actor identities which exist only in print. One central thrust of social-constructivist thought has been to trace debates over actors' capabilities to produce reliable knowledge through analyses of their accumulation and deployment of credibility. What is lacking in such an account is an historical inquiry into the underlying assumptions of our accounts of the construction of social identity. The prerequisite for any socioepistemology is a strong sense of what constitutes the social, yet here we generally argue that actors pre-exist their representations. Rhetorics may, in this model, construct particular identities, but those identities are always refractions of a real, if unknowable, historical agent. How, then, do we deal with agents who did not historically exist at this level of being, such as pseudonyms? The problem may seem at first sight marginal, but in fact permeates the historical record (Secord 2000; Wilding 2014). If authors do not pre-exist their words, but are brought into being by them, why do we impose an ahistorical model of subjecthood on them as the starting point for our constructivist studies? Here, per-haps, we see most clearly how the history of the book and the skills of bibliography, with its attention to apparent material self-contradictions and discrepancies, might be useful to the history of science. The social is populated by bibliographical ghosts and avatars as well as "real" authors, and we need to understand historically how these fig-ures have participated in the making of knowledge. That they exist only in the realm of print should not force us into a stance of dull realism, for the pluralities and poten-tialities were alive and meaningful to their equally fictitious contemporaries, and this is the reality of the past.

As James Secord (2000) and Adrian Johns (1998) have argued, it may well be the nineteenth-century shift from hand- to steam-presses, rather than Gutenberg's contributions, that constitute the more decisive print-revolution. Secord's "industrial revolution in communication" (2000, 24) was part of a wider movement to reimag-ine the technologies of communication, the politics of knowledge production, and the relationship between the scientist–author and a scientific audience. It is not by chance that so many of the instruments and devices described in John Tresch's *The Romantic Machine* (2012) strove to bring new media and forms of representation into exis-tence. Indeed, we would do well to follow Tresch's lead in approaching the history of scientific print less as a single technological innovation than as an ongoing and beauti-ful experiment that simultaneously and symbiotically transforms both knowledge and the knower.

Endnotes

1   The works (not recorded in Needham's census) are: 1. Siri, Vittorio, *Propositiones mathematicae* (Parma, 1634); 2. Scoto, Paolo, *Problemata et theoremata geomet-rica, e mechanica publicè demonstranda…* (Bologna, 1633); 3. Montanari, Gemini-ano, *Ephemeris lansbergiana ad longitudinem almae studiorum matris Bononiae ad annum 1666… Addita in fine Ephemeride motus solis eiusdem anni ex tabulis excel-lentiss. d. Io. Dominici Cassini … vna cum eiusdem d. Cassini epistola responsiua ad authorem* (Bologna, no date); 4. *Sidereus Nuncius*; 5. Rondelli, Geminiano, *Urania custode del tempo.* (Bologna, 1700); 6. Rosati, Francesco Maria, *Chrysopyrrhina helio-physis* (Parma); 7. Fabri, Honoré [pseud.], *Opusculum geometricum de linea sinuum et cycloide auctore Antimo Farbio* (Rome, 1659); 8. Viperano, Giovanni Antonio, *De diuina prouidentia libri tres.* (Rome, 1588); 9. Lascaris, Gaspare, *Vsus speculi plani* (Rome, 1644); 10. Manacci, Marcello, *Compendio d'instruttioni per gli bombardieri* (Parma, 1640).

2   'faransi i[n]tagliar i[n] legno tutte in u[n] pezzo, et le stelle bia[n]che il resto nero poi si seghera[n]no i pezzi].' BNCF, Mss. Gal. 48, f.30v. First noted by Pantin 1992, xxviii.

3   I, 1967–1975, microfiche; II – Monographs, 1976–2013, microfiche; II – Journals, 1976–2013, microfilm.

References

Ago, Renato. 2013. *Gusto for Things: A History of Objects in Seventeenth-Century Rome.* Chicago: University of Chicago Press.

Agüera y Arcas, Blaise. 2003. "Temporary matrices and elemental punches in Gutenberg's DK type." In *Incunabula and their Readers: Printing, Selling and Using Books in the Fifteenth Century*, edited by Kristian Jensen, 1–12. London: The British Library.

Baldasso, Renzo. 2013. "Printing for the Doge: On the first quire of the first edition of the Liber Elementorum Euclidis." *Bibliofilía*, 115: 525–52.

Berry, Mary Elizabeth. 2006. *Japan in Print: Information and nation in the Early Modern Period.* Berkeley: University of California Press.

Blair, Ann. 2010. *Too Much to Know: Managing Scholarly Information Before the Modern Age.* New Haven, CT: Yale University Press.

Brown, Alison. 2010. *The return of Lucretius to Renaissance Florence.* Cambridge, MA: Harvard University Press.

Carter, Thomas Francis. 1925. *The Invention of Printing in China and Its Spread Westward.* New York: Columbia University Press.

Carter, John, and Percy Muir. 1967. *Printing and the Mind of Man: A Descriptive Catalogue Illustrating the Impact of Print on the Evolution of Western Civilization during Five Centuries.* London: Cassell.

Carter, Victor, Lotte Hellinga, and Tony Parker. 1983. "Printing with gold in the fifteenth century." *British Library Journal*, 9: 1–13.

Dibner, Bern. 1955. *Heralds of Science; As Represented by Two Hundred Epochal Books and Pamphlets Selected from the Burndy Library.* Norwalk, CT: Burndy Library.

Eisenstein, E. L. 1979. *The Printing Press as an Agent of Change: Communications and Cultural Transformations in Early-modern Europe.* Cambridge: Cambridge University Press.

Findlen, Paula. 1994. *Possessing Nature: Museums, Collecting, and Scientific Culture in Early Modern Italy.* Berkeley: University of California Press.

Frasca-Spada, Marina, and Nick Jardine (eds.) 2000. *Books and the Sciences in History.* Cambridge: Cambridge University Press.

Gingerich, Owen. 2002. *An Annotated Census of Copernicus' De Revolutionibus: (Nuremberg, 1543 and Basel, 1566).* Leiden: Brill.

Gingerich, Owen. 2004. *The Book Nobody Read: Chasing the Revolutions of Nicolaus Copernicus.* New York: Walker & Co.

Holstenius, Lucas. 1681. *Index bibliothecae qua Franciscus Barberinus.* Rome.

Hyde, Thomas. 1674. *Catalogus impressorum librorum Bibliothecæ Bodleianæ in Academia Oxoniensi.* Oxford.

James, Thomas. 1620. *Catalogus universalis librorum in Bibliotheca Bodleiana.* Oxford.

Jardine, Lisa and Grafton, Anthony. 1990. "'Studied for Action': How Gabriel Harvey read his Livy." *Past & Present*, 129: 30–78.

Johns, Adrian. 1998. *The Nature of the Book: Print and Knowledge in the Making.* Chicago: University of Chicago Press.

Kremer, Richard L. 2013. "Hans Sporer's xylographic practices: A census of Regiomontanus's blockbook calendars." *Bibliotheck und Wissenschaft*, 46: 161–87.

Latour, Bruno. 1987. *Science in Action: How to Follow Scientists and Engineers Through Society.* Cambridge, MA: Harvard University Press.

Lowood, Henry, and Robin Rider. 1994. "Literary technology and typographic culture: The instrument of print in early modern culture." *Perspectives on Science*, 2: 1–37.

Maclean, Ian. 2009. *Learning and the Market Place: Essays in The History of the Early Modern Book*. Leiden: Brill.

Maclean, Ian. 2012. *Scholarship, Commerce, Religion: The Learned Book in the Age of Confessions, 1560–1630*. Cambridge, MA: Harvard University Press.

McKenzie, D. F. 1999. *Bibliography and the Sociology of Texts*. Cambridge: Cambridge University Press.

McKitterick, David. 2013. *Old Books, New Technologies: the Representation, Conservation and Representation of Books since 1700*. Cambridge: Cambridge University Press.

Muehlhaeusler, Mark. 2010 "Math and magic: a block-printed wafq amulet from the Beinecke Library at Yale." *Journal of the American Oriental Society*, 130: 607–18.

Needham, Paul. 1982 "Johann Gutenberg and the Catholicon Press." *The papers of the Bibliographical Society of America*, 76, 395–456.

Needham, Paul. 2011. *Galileo Makes a Book: The First Edition of* Sidereus Nuncius. Berlin: Akademie Verlag.

Nuovo, Angela. 2013. *The Book Trade in the Italian Renaissance*. Leiden: Brill.

Palmer, Ada. 2012 "Reading Lucretius in the Renaissance." *Journal of the History of Ideas*, 73: 395–416.

Pantin, Isabelle (ed. and trans.) 1992. *Galileo Galilei. Sidereus Nuncius, Le Messager Celeste*. Paris: Les Belles Lettres.

Pantin, Isabelle. 2001. "L'illustration des livres d'astronomie à la Renaissance: L'évolution d'une discipline à travers ses images." In *Immagini per conoscere : dal Rinascimento alla rivoluzione scientifica: atti della giornata di studio, Firenze, Palazzo Strozzi, 29 ottobre 1999*, edited by Claudio Pogliano and Fabrizio Meroi, 3–41. Firenze: Olschki.

Raj, Kapil. 2007. *Relocating Modern Science: Circulation and the Construction of Knowledge in South Asia and Europe, 1650–1900*. Basingstoke: Palgrave Macmillan.

Schaefer, Karl R. 2006. *Enigmatic Charms: Medieval Arabic Block Printed Amulets in American and European Libraries and Museums*. Leiden: Brill.

Schäfer, Dagmar. 2011. *The Crafting of the 10,000 Things: Knowledge and technology in Seventeenth-Century China*. Chicago: University of Chicago Press.

Schaffer, Simon. 1995. "The show that never ends: Perpetual motion in the early eighteenth century." *The British Journal for the History of Science*, 28: 157–89.

Schaffer, Simon. 2007. "'On Seeing Me Write': Inscription devices in the South Seas." *Representations*, 97: 90–122.

Schaffer, Simon, Lissa Roberts, Kapil Raj, and James Delbourgo (eds.) 2009. *The Brokered World: Go-Betweens and Global Intelligence, 1770–1820*. Uppsala Studies in History of Science 35. Sagamore Beach MA: Watson Publishing International.

Secord, James A. 2000. *Victorian Sensation. The Extraordinary Publication, Reception, and Secret Authorship of* Vestiges of the Natural History of Creation. Chicago: University of Chicago Press.

Serrai, Alfredo. 1988–1999. *Storia della bibliografia*. 11 vols. Rome: Bulzoni.

Shank, Michael H. 2012. "The geometrical diagrams in Regiomontanus's edition of his own *Disputationes* (c.1475): Background, production, and diffusion." *Journal of the History of Astronomy*, 43: 27–55.

Suarez, Michael and H. R. Woudhuysen (eds.) 2010. *The Oxford Companion to the Book*, 2 vols. Oxford: Oxford University Press.

Suarez, Michael and H. R. Woudhuysen (eds.) 2013. *The Book: A Global History*. Oxford: Oxford University Press. (Concise edition of Suarez and Woudhuysen 2010)

Thornton, J. L. and R. I. J. Tully. 1975. *Scientific Books, Libraries and Collectors: A Study of Bibliography and the Book Trade in Relation to Science* (3rd revised edition, reprinted with minor corrections). London: Library Association.

Thornton, J. L., R. I. J. Tully, and A. Hunter. 2000. *Thornton and Tully's Scientific Books, Libraries, and Collectors: A Study of Bibliography and the Book Trade in Relation to the History of Science* (4th edition, considerably revised and rewritten.). Aldershot: Ashgate.

Tresch, John. 2012. *The Romantic Machine: Utopian Science and Technology after Napoleon.* Chicago: University of Chicago Press.

Wilding, Nick. 2014. *Galileo's Idol: Gianfrancesco Sagredo and the Politics of Knowledge.* Chicago: University of Chicago Press.

CHAPTER TWENTY-FIVE

Correspondence Networks

BRIAN OGILVIE

The role of correspondence in scientific exchange is obvious to any historian who has worked in an archive. However, an overview of correspondence and correspondence networks in the history of science cannot limit itself to science or scientists, particularly before the nineteenth-century professionalization of (much) science. Natural philosophers, naturalists, chemists, mathematicians, and other students of nature were part of a broader world of intellectual and cultural exchange, a world sometimes designated "the Republic of Letters" (Grafton 2009).

It is useful to distinguish correspondence and correspondence *networks*. Scientists' letters have long been used as historical sources. But correspondence networks, connecting many people into more or less tightly knit webs of exchange, have recently become distinct objects of historical inquiry in themselves. David Kronick notes an earlier, loose use of "network" as a synonym for mere correspondence, but prefers to save the term for more formally organized communications systems, with "gatekeepers" regulating access (Kronick 2001, 32). The study of networks offers the possibility of identifying "invisible colleges": groups of like-minded thinkers who, although not formally organized like early modern scientific academies, were in regular contact (Kronick 2001, 39–41).

Even simple network analyses can be illuminating. In a suggestive study using limited data, Peter Taylor, Michael Hoyler, and David Evans (2008) examine European scientists' career paths. Adopting a dynamic perspective, the authors trace *shifts* in the places where their subjects were active, with movement from city to city as the basic unit of analysis. For the sixteenth century they identify a small network with Padua as its main hub, with a London–Oxford–Cambridge network as a subsidiary part and a separate two-node Wittenberg–Jena network. The seventeenth century had a more complicated network, with hubs in Padua, Paris, Leiden, London, and Jena connecting to various provincial networks, some of them in contact with more than one scientific metropole. The eighteenth century saw the decline of Padua and Jena and the rise of Berlin and Göttingen, with national boundaries being far more pronounced. And in

A Companion to the History of Science, First Edition. Edited by Bernard Lightman.
© 2016 John Wiley & Sons Ltd. Published 2020 by John Wiley & Sons Ltd.

the nineteenth century, Germany dominated the European network, its major hub in Berlin, with a separate British network comprising London, Oxford, Cambridge, and Edinburgh.

This analysis is far from definitive. The authors limit their analysis to the thousand "top scientists" in Robert Gascoigne's *Chronology of the History of Science* (1987), divide their data set arbitrarily at century boundaries, and consider only career moves, not correspondence or travel. Nonetheless, it offers a starting point for considering science not only in terms of "the space of places" but also in terms of "the space of flows," to adopt Manuel Castells's terminology (Castells 2010, 440–59). Focusing on the present, Castells considers the base layer of the space of flows to be an electronic infrastructure. The next layer comprises the nodes and hubs of the network of exchanges. But before the twentieth century, the base layer of the "space of scientific flows" was in fact an infrastructure of post roads, post offices, and post riders (and from the seventeenth century on, post coaches) transporting letters and specimens, as well as people, from node to node.

Indeed, early modern Europe saw what Wolfgang Behringer has described as a "communications revolution" (Behringer 2003, 2006; cf. the nineteenth-century "industrial revolution in communication" discussed by Secord 2000, 24–34). In the first half of the fifteenth century, the Visconti dukes of Milan established a regular network of postal couriers in their domain. In the sixteenth century, this system was extended throughout Western and Central Europe by Habsburg rulers, who were anxious to connect their far-flung possessions. Unlike the ancient Roman *cursus publicus*, generally reserved for official business, early modern postal services were open to anyone who could pay. War and political unrest could disrupt the Imperial post, but over the course of the seventeenth and eighteenth centuries, the system became more extensive, more reliable and faster. It was soon joined by French, English, and other services that promised rapid, reliable delivery of letters and small packages to those who could afford carriage (the public character of the Royal Mail was affirmed by Charles I in 1635). By 1800, Western and Central Europeans, including residents of the British Isles, could rely on the national posts to carry their letters within each country, while postal treaties between states ensured an international flow of mail.

Correspondence between Europe and the rest of the world was more expensive and less certain. Colonial powers had courier networks for official correspondence, some of which expanded into full-fledged postal systems. As commerce grew, the post followed in its wake. Newly independent nations in the Americas established national postal services and, increasingly, signed postal treaties with one another and European powers. By the second half of the nineteenth century, business interests were urging international cooperation to simplify the use of long-distance communications, leading to two international communications unions: the International Telegraph Union (1865) and the General Postal Union (1874, later Universal Postal Union). The latter standardized rates for international postage and provided mechanisms for settling disputes between the postal services of member nations.

In sum, as with much of the rest of the history of science, the development and expansion of scientific correspondence networks has been intimately bound up with the history of the sovereign state, of colonial expansion and exploitation, and of international conflict and cooperation. Correspondence has created and sustained national and international communities of scientists, transmitting ideas and material while sustaining affective bonds between collaborators in the scientific enterprise. Situated between the immediacy of conversation and the finality of publication, it has served an important, if changing, role in the creation of knowledge. At the same time, it has provided a mechanism by which ideas from the periphery of science have reached the center while sometimes effacing the identities and interests of their contributors.

The Republic of Letters, Sixteenth–Eighteenth Centuries

The early modern communications revolution provided the space of flows for the Republic of Letters. This republic was an imagined community of writers that, at least in its ideals, transcended political and confessional boundaries (Bots and Waquet 1997). Epistolary exchange was its lifeblood. The term "*respublica literaria*" (literary republic) was first used in 1417 but came into fashion toward the end of the fifteenth century (Waquet 1989, 475–6). From the late fifteenth through the early seventeenth century, the Republic of Letters was based in a regulative fiction of free exchange of ideas between sincere friends, inculcated in Latin schools and in correspondence manuals (Dibon 1978; Ogilvie 2011). Of course, it was rife with personal animosities as well as religious, political, and intellectual conflict, but its members professed the value of its irenic norms even as they violated them in practice (Landtsheer and Nellen 2011).

In the seventeenth century, this Latin-oriented *respublica literaria*, itself never wholly Latinophone, was succeeded by a vernacular *république des lettres* or *Gelehrten-republik* (Goodman 1994; Goldgar 1995). A key factor distinguishing it from the earlier republic was the learned journal or review (see Chapter 28 for a fuller discussion). The *Journal des sçavans* (1665–) and the *Philosophical Transactions* (1665–) were soon joined by a host of others. Periodical publications helped sharpen the distinction between creators and consumers of learning. They also placed a new emphasis on critique as a value of the republic (Daston 1991).

However, we should not draw too sharp a line between letters and journals. Even in the age of the periodical press, letters could be circulated widely. Melchior Grimm's literary correspondence was a kind of circular manuscript journal that informed the courts of central Europe about intellectual and cultural developments in Paris, the philosophical capital of eighteenth-century Europe (Grimm and Meister 1877; Scherer 1968; Van Damme 2005). "Publishing" only in manuscript allowed Grimm to restrict access to his reports, thereby increasing their value to his aristocratic subscribers.

Nor should we think of letters as comprising a space of flows distinct from those of printed works. Rather, letters, books, and (after 1665) periodicals were mutually interdependent. Letters sometimes moved into print themselves, by design or by chance. However, the letter, in its immediacy and its intimacy, was nimble in ways that print, with its technological and capital demands, could not match.

Within the early modern Republic of Letters, many distinct scientific correspondence networks have been identified and studied (see Harris 2006). Most studies have focused on the correspondence of individual scholars or facilitators (Berkvens-Stevelinck, Bots, and Häseler 2005). Scholars have tended to describe networks in terms of the correspondence of a single individual. In an article on Jan Amos Comenius, Vladimír Urbánek refers to the "correspondence networks" of Marin Mersenne,

Samuel Hartlib, Louis de Geer, Mikuláš Drabík, and Comenius himself (Urbánek 2014). However, it is clear from Urbánek's account that many names recurred frequently in all these men's correspondence, and that rather than comprising distinct networks, they were important nodes in a much broader, loosely integrated network. In his study of the antiquary and naturalist Esprit Calvet (1728–1810) of Avignon, L. W. B. Brockliss notes that "the Republic of Letters in the second half of the eighteenth century was formed by an indefinite number of mini-Republics." Calvet and Jean-François Séguier, who corresponded with one another, each shared over 10 percent of their approximately 300 correspondents. But Calvet shared a significant number of correspondents with other scholars with whom he had no direct connection (Brockliss 2002, 390–91). The same observation can be applied more generally to correspondence networks from the sixteenth century to the present.

However, research into the strength and interconnectedness of these networks is still in its early stages (see O'Neill 2015). What is clear is that many, if not most, participants in the early modern Republic of Letters were connected to one another by only a few degrees of separation (Ultee 1987, 104). Many individuals served as significant nodes. In the late sixteenth and the early seventeenth century, correspondence between Carolus Clusius and Italian horticulturists and naturalists often passed through the hands of Gian Vincenzo Pinelli in Verona (Egmond 2013). In the early seventeenth century, Marin Mersenne coordinated a vast correspondence that brought together many innovators in natural philosophy and mathematics (Bots 2005; Grosslight 2013). Henry Oldenburg, secretary of the Royal Society and editor of the *Philosophical Transactions*, played an analogous role later in the century (Iliffe 2009), as did Benjamin Franklin in the eighteenth century (Winterer 2012, 609). These were only a few of the "great cultural intermediaries of the Republic of Letters" (Berkvens-Stevelinck et al. 2005).

The "great intermediaries" corresponded regularly with hundreds of their contemporaries. Other correspondence circles could be much narrower. Calvet may have had over 300 unique correspondents, but he received ten or more letters from only 48. His personal network, like many others, was "protean and unstable" (Brockliss 2002, 79). Nonetheless, exchanges between some individuals could go on for decades. This was due not only to genuine friendship or interest, but also because the norms of the Republic of Letters insisted on reciprocity. A letter, even one from an unknown correspondent, placed one under an obligation to be ignored at peril, though a tepid response could, in turn, put an end to an unwanted exchange (Brockliss 2002, 94; Ogilvie 2011).

Since full participation in universities, academies, and other formal institutions for science was often restricted by gender, social standing, location, and language, correspondence networks provided informal ways for women, artisans, collectors, and provincials to contribute to scientific knowledge making. Noblewomen in Habsburg Vienna and the Low Countries exchanged exotic plants and knowledge about them with medically trained naturalists, as did male garden enthusiasts (Egmond 2010, 45–71; Ogilvie 2006, 67–9). Other women engaged in sustained correspondence with men on natural philosophy and mathematics (Schiebinger 1989, 41–7, 88; Del Centina 2012). As the eighteenth-century mathematician and natural philosopher Émilie du Châtelet discovered, many men were unwilling to consider a woman to be an equal correspondent, or even a colleague at all (Terrall 1995). Nonetheless, correspondence networks brought many men and women informally into the scientific enterprise, a process that would continue into the nineteenth and twentieth centuries (Harvey 2009).

Such informal, or informally institutionalized, circles of exchange were sometimes complemented by more formal attempts to establish regular correspondence networks. Gathering information from afar was one of the goals of the Accademia dei Lincei (Academy of the Lynx-eyed), founded in Rome in 1603 by Federico Cesi (1585–1630): the Academy's statutes explicitly called for maintaining correspondence with other learned men (Ubrizsy Savoia 2011, 197). The Montmor Academy, established in 1657, included in its formal regulations this stipulation: "The Assembly requests that those who have the opportunity to do so should engage in correspondence with scholars in France and foreign countries, in order to learn from them what projects they have completed in the arts and sciences, and those which they have discovered or intend to publish" (as quoted in Kronick 2001, 37; the French text is in Bigourdin 1917).

The hundreds of thousands of surviving letters from the early modern Republic of Letters should not mislead us as to its total size. In an age before the professionalization and specialization of the academy, and when university education was limited to an almost exclusively male elite, the republic was small. Early modern guides to its luminaries, such as Jean-Pierre Nicéron's *Mémoires pour servir à l'histoire des hommes illustres de la République des lettres* (40 vols., 1727–45), list several hundred; Christian Gottlieb Jöcher's *Allgemeines Gelehrten-Lexikon* (begun in the 1710s and expanded in the 1750s) lists several thousand, but his list includes classical antiquity and the Middle Ages (Ultee 1987). At any given moment, the active participants in the scientific provinces of the Republic of Letters would have fit in a large auditorium or a small stadium. And as Lindsay O'Neill has shown in a rich recent study that develops many of the themes on which this chapter touches, personal connections were still its main binding agent (O'Neill 2015).

Exchanging Information, Opinions, and Objects

The information transmitted in epistolary exchanges took on varied forms. Scholars in small towns requested copies of books from their correspondents in more cosmopolitan cities. If a book was unavailable, they might request hand-copied excerpts. Botanists exchanged lists of plants or seeds that they had in their gardens and could transmit to their correspondents; they also exchanged the objects themselves, as did those interested in metals, figured stones, and other natural objects. In these instances, correspondence networks served to transmit knowledge and material objects (Ogilvie 2006, 80–82; Cooper 2007, 109–15; Harris and Anstey 2009).

Correspondence was also a place where knowledge was elaborated. Naturalists described things they observed, sometimes including sketches or elaborate drawings with their written account. The apothecary Giovanni Pona of Verona sent a lengthy description of his ascent of Mount Baldo to his correspondent Carolus Clusius, a description that Clusius later revised and published in his *History of Less Common Plants* (Ogilvie 2006, 279 n.8). Natural philosophers, mathematicians, astronomers, and chemists discussed theoretical matters or requested that their correspondents conduct observations or experiments on their behalf.

Conrad Gessner's medical correspondence frequently contained observations on remedies that he had applied to patients, and on the curative properties of herbs and other simple medicines. Gessner jokingly chided correspondents who neglected to include such "spices" (*condimenta*) in their letters, rhetorically underscoring his own attention to them. But beyond individual observations, Gessner used his correspondence network to coordinate empirical data gathering on plague and other contagions, and he encouraged his correspondents to compare their observations with one another. This correspondence, organized under headings in Gessner's notes, provided him with material that could be elaborated systematically (Delisle 2013).

As Dirk van Miert has observed, from the last quarter of the seventeenth century, learned periodicals and scientific academies took on some functions that had earlier been served exclusively by correspondence (van Miert 2013, 7). They circulated observational reports, offered synopses and critiques of newly published books, and allowed for a more public exchanges on questions in natural philosophy (bearing in mind, however, that some earlier correspondence had circulated widely, e.g. in Mersenne's circle). But private correspondence continued to serve those purposes. Some writers were less circumspect in letters than in published works (Kronick 2001, 30). But that depended on circumstances; the ideal of sincere friendship continued to structure epistolary exchanges, while printed polemics were common (Goldgar 1995).

Moreover, the practice of the new scientific academies of appointing foreign or corresponding members encouraged an epistolary flow between their metropolitan centers and the larger world. The Académie Royale des Sciences had not originally included corresponding members, but a few were appointed in the late seventeenth century. The 1699 reorganization of the Académie included a formal category of corresponding members, by far the largest number of individuals associated with it, each of whom was assigned to a resident academician who managed their correspondence (McClellan 1981). Correspondents did not necessarily communicate frequently with the Académie: later regulations stipulated that correspondents who did not communicate at least once every three years would be struck from the membership, a period later shortened to one year. The academician René-Antoine Ferchault de Réaumur maintained his own private correspondence network, which provided him with useful information on investigations ranging from silk production to the investigations of Abraham Trembley and Charles Bonnet on freshwater polyps (Terrall 2014).

Scientific Correspondence in the Nineteenth Century

Two nineteenth-century developments dramatically changed the nature of postal service and, thereby, the ease with which scientific correspondence circulated. The first was the reorganization of postal fees and accounting proposed by the Englishman Rowland Hill in the 1830s and adopted in much of the developed world over the next two decades. The second was the creation of a General (later Universal) Postal Union whose signatory states agreed that their national postal services would deliver mail that had been dispatched in another country. Along with faster transport technologies—steam locomotives, clipper ships, and steamships—both had dramatic effects on scientific correspondence.

The 1840s and 1850s saw a dramatic decline in the cost and inconvenience of sending a letter. Like other postal services, the Royal Mail had charged a letter's recipient according to the number of sheets it contained and the distance it traveled. Rowland determined that this system exaggerated the costs of transport, required cumbersome accounting, and delayed delivery if the postman could not find the recipient to collect the fee. His alternative, in which domestic letters were charged a fee by weight only, paid by the sender with a stamp to signify payment, was enacted in 1840. Because letters up to one-half ounce cost only a penny, a quarter or less of the previous rate for letters sent beyond London (and half the rate for letters within the metropolis), it was popularly called the Penny Post (Codding 1964, 7–11).

Prepaid postage spread quickly beyond Great Britain, becoming general in the second half of the century. It allowed scientists in much of the world to write to their compatriots without needing to worry about placing an undue burden on them— the sender would pay for his or her own prolixity. However, provincial naturalists who sent specimens to centrally located experts such as Charles Darwin and Joseph Hooker may have ended up bearing more of the substantial costs of building large research collections.

As correspondence became easier, the flow of ideas and material in the scientific world increased in volume and significance. And as Janet Browne has pointed out, new professional institutions of science often blurred the bounds between official and private exchange. William and Joseph Hooker at Kew Gardens, Spencer Baird at the Smithsonian Institution, Georges Cuvier at the Muséum d'histoire naturelle, and their counterparts at other institutions used correspondence to obtain scientific material and to arrange support for their research endeavors (Browne 2014). Organizers of nineteenth-century biological surveys took pains to ensure access to postal service and transportation for their field expeditions (Kohler 2006, 155ff). And by writing to a well-known scientist, a new practitioner might attempt to seek a name for himself—as Alfred Russel Wallace had done in his fateful 1858 letter to Charles Darwin (Browne 2014). Wallace's letter came not from within England but from Southeast Asia. It made its way to Down House in Kent via a colonial communications network.

Europe and the Wider World

While the early modern communications revolution was transforming the space of flows within Western and Central Europe, European states were building administrative and commercial networks linking them with parts of Africa, Asia, and the Americas. Some of these networks were highly developed. Spain founded an extensive clearing-house of goods and information in Seville, the Casa de la Contratación, which served as a central point of communication between officials in Madrid and Spain's American and Pacific colonies (Barrera-Osorio 2006; Portuondo 2009). The central node allowed metropolitan authorities to gather, but also to control, information about the geography, natural history, and commercial products of colonial territories.

The English colonial empire lacked such a central communications center, but the (theoretically) obligatory passage of ships from the Atlantic colonies via customs inspection at the Legal Quays and, later, sufferance wharves, as well as the central role of London as an entrepôt, largely compensated (Ford 1959). The amalgamation of Dutch commercial-colonial expansion in the early seventeenth century in two monopolies—the immensely successful Dutch East India Company and its poor relation, the West Indian Company—led to *de facto* centralization of information coming from southeast Asia, the Cape, and Surinam to Amsterdam and a few other Dutch port cities (Cook 2007, 175–202). As with Spain, this control allowed company officials to effectively censor sensitive and commercially valuable material, like Georg Eberhard Rumphius's herbal, which remained unpublished for half a century (Rumphius 1999; Cook 2007, 331–32).

Missionary activity led Jesuits and other religious orders to create their own networks extending far beyond Europe. Jesuit missions in China, Japan, Canada, and elsewhere sent regular reports to the Order's headquarters in Rome, many of which were published. Other letters, remaining in manuscript, accompanied natural history specimens or contained observations on natural phenomena (see Harris 1996; Hsia 2009, 14–18). Protestant missionary activity was often less coordinated, but there were significant exceptions: for example, Lutheran Pietist missions across the globe gathered natural and cultural objects and returned them to the movement's headquarters in Halle, Saxony (Wilson 2000; Whitmer 2013).

Early modern colonial networks did not constitute a single space of flows; they were largely distinct. As Caroline Winterer has astutely observed, the historiographical concept of "the Atlantic world" obscures the difficulties that those in the western hemisphere had in engaging in intellectual exchanges across political and cultural borders (Winterer 2012). In the case of British Americans, the Republic of Letters was centered on exchanges with England, and above all, London. In the seventeenth century, John Winthrop, Jr., exchanged some 5000 letters with English and some continental scholars within what Walter Woodward has dubbed a "republic of alchemy" (Woodward 2010; Winterer 2012, 609). By the late eighteenth century, London appeared in some ways to be "the capital of America," a key node connecting British colonies with each other as well as the metropole (Flavell 2010).

British Americans could participate in these networks because the Atlantic colonies were linked to the metropole by a regular postal service. Where the posts did not exist, colonial residents might be only weakly connected to European colleagues. Regular postal service between Saint-Domingue and the French metropole began only in 1763 (McClellan 2010, 81); before that, scientific correspondence between the island and Paris depended on private carriers.

Early modern scientific correspondence between Europe and its colonies was mostly, then, a matter of exchange between each European state and its colonial possessions. Religious missions were the main exception. From the late eighteenth century, European industrialization altered this landscape significantly. European factories required raw materials, often sourced in colonial possessions; their products, in turn, were shipped out to colonial markets. These bidirectional flows were secured by significant military investments by colonial powers, accompanied by commercial interests. As a consequence, the landscape of scientific correspondence underwent significant transformations beginning in the late eighteenth century.

Charles Darwin's correspondence during the voyage of H.M.S. *Beagle* illustrates the opportunities and limitations of colonial correspondence. Darwin carried on an extensive correspondence during the five years of the voyage (Burkhardt and Smith et al. 1985–2015 continuing; Darwin 2008). But the rhythm of his exchange was dictated by the availability of Royal Navy dispatch ships and merchant packets to carry his letters. If exchanges were relatively frequent during the years that the *Beagle* was charting the east coast of South America, they became slower and more irregular when she entered the Pacific. It took roughly five months for correspondence to travel between England and the Pacific coast of South America (Darwin 2008, 337, 365). Moreover, postage was expensive, and during the years of the voyage, the recipient paid for it. Darwin's family could bear the cost, but not all travelers were so fortunate.

Such difficulties plagued colonial correspondents much longer than their metropolitan counterparts. Improvements came piecemeal, connected to business and administrative interests Only in 1898 did the British extend the Penny Post to their colonial empire; in consequence, the volume of imperial mail more than doubled in under a decade (Jeffery 2006, 51–2). By that time, the Universal Postal Union had regularized international communications across the globe.

The Universal Postal Union and Beyond

Business interests led to the development of international treaties coordinating telegraphic and postal communications across borders: the International Telegraph Union (1865) and the General Postal Union (1874), later renamed the Universal Postal Union (Pollard 1997). The latter replaced separate postal treaties between individual states. It specified a uniform international postage rate between member states and fixed transit fees for letters and parcels that passed through third countries. In 1897, the UPU added natural history specimens to the list of permitted commercial samples (Codding 1964, 39). Though war occasionally interrupted the mails, the UPU streamlined and accelerated the exchange of scientific information.

The twentieth century saw further significant transformations in scientific correspondence and communication. On the one hand, colonial and postcolonial rail networks, steam liners, and airplanes contributed to a denser, faster postal system, which functioned increasingly smoothly despite the interruptions of war. On the other hand, electric communications began to take on some of the functions of postal correspondence. The telegraph could not supplant detailed letters. But the telephone began to make inroads as it became cheaper and more widespread. Such ephemeral communications technologies left fewer material traces for historians than correspondence. Their prevalence has been compensated to some degree by the bureaucratic nature of "big science," especially since the wartime projects of the 1940s; such projects, involving hundreds or thousands of participants, generated their own internal space of flows in the form of working papers, memos, and manuals (Galison 1987). Nonetheless, historians of recent science face a different landscape of correspondence than those who work on earlier periods.

Electronic mail has further complicated this landscape. Email can be less ephemeral than telegrams and phone calls, but it is abysmally preserved, except for government agencies and businesses that maintain email archives due to legal obligation or financial interest. As formats become obsolete and storage media decay, even existing email archives risk disappearing. Historians of contemporary science, and those who wish to assist future historians, should encourage scientists, their employers, and funding agencies to develop guidelines for the preservation of electronic materials, including correspondence (see Thomas and Martin 2006).

Corpora of Correspondence, Archival and Published

Since the sixteenth century, amateurs and scholars have collected and published scientific letters (Maclean 2008). The nineteenth-century archival turn in historiography and the Victorian vogue for epistolary biographies increased the volume of letters that were published, often with a heavy editorial hand. As history of science became a distinct discipline, historians and archivists began to pay more attention to comprehensiveness and editorial precision. Some publication projects retained the focus on the letters of famous individuals, such as the project to publish all the letters by the early microscopist Antoni van Leeuwenhoek, accompanied by only a handful addressed to him (Leeuwenhoek 1939). But others focused on correspondence as exchange, such as the edition of Henry Oldenburg's correspondence by A. Rupert Hall and Marie Boas Hall (Oldenburg 1965). The Darwin correspondence project is pursuing the same approach in even more meticulous detail, creating a portrait of Darwin not only as an innovative thinker but also as a significant node in an extended Victorian scientific network (Burkhardt and Smith et al. 1985–2015 continuing).

The Darwin project is now also publishing online annotated transcriptions of the correspondence. Finding aids allow exploration of the corpus by individual name, chronology, and subject keywords, while the full text of many letters can be searched. Similar projects have made, or are making, the correspondence of other important figures online: Benjamin Franklin, Albrecht von Haller, Hans Sloane, Joseph Banks, and many others.

Such full-text, annotated projects require extensive editorial resources and financial support. Other projects, such as the Clusius letter project at Leiden University, more modest in some regards yet more ambitious in others, aim to make high-resolution, professionally scanned images of correspondence available in electronic archives. Catalogued by author, addressee, place, and date, such databases allow researchers to consult material without traveling to archives; on the other hand, because the material is neither transcribed nor annotated, it is less accessible than a digital edition. At the same time, though, digital archives offer material for future editions, including the possibility of crowdsourcing some editorial contributions.

Digital Humanities and the Future of Research on Correspondence Networks

As online corpora suggest, the advent of humanities computing (now known as digital humanities) has brought new capacities for representing and analyzing scientific correspondence networks. These capacities are being employed in several collaborative research projects on scientific and learned correspondence, primarily in the seventeenth and eighteenth centuries, including Mapping the Republic of Letters (Stanford University), Cultures of Knowledge (Oxford University), and Circulation of Knowledge and Learned Practices in the 17th-century Dutch Republic (Leiden University).

These projects aim to transcend the limits of individual corpora of correspondence. As the Cultures of Knowledge website notes,

> While recent years have witnessed an efflorescence of online catalogues and digital editions of letters, these have tended to exist in the form of discrete silos around carefully defined collections and correspondents, making the large-scale exploration and analysis

so essential to a full appreciation of early modern epistolarity extremely difficult (Cultures of Knowledge n.d.).

Hence these projects are engaged in multistage processes. They must create uniform, machine-readable catalogues of correspondence, with appropriate metadata for analyzing intellectual exchange. And they must devise digital tools to query their databases and to present the results visually.

A few caveats are in order. Such projects depend on the interests and generosity of funding organizations, and thus reproduce contemporary inequalities in resources and interest, geographically and chronologically. Moreover, scientific correspondence has not survived in a random fashion. Already during their lifetimes, the papers and letters of well-known natural philosophers, naturalists, chemists, physicists, geologists, biologists, men and women of letters—in short, of famous republicans of letters— were being preserved while those of their lesser-known contemporaries disappeared. Women, provincials, non-Europeans, and artisans are less well represented than men from the upper echelons of European society, particularly those who resided, at least part of the time, in capital cities and participated in scientific academies and societies.

Despite these limits, digital correspondence projects will allow us to establish a clearer cartography of the space of flows of early modern science than our present-day sketch maps. On one level, they will allow scholars to dig more deeply into how correspondence helped individual scientists or small groups to better coordinate observations, exchange information, and debate interpretations. On another, they will help refine the flyover view of this space, showing how the geography and density of epistolary exchanges has shifted. In so doing, they will continue to illuminate the central role that correspondence has played in the practice of science from the early modern era to the present. From the Republic of Letters to the email inbox, it is difficult to overstate how epistolary exchange has created and sustained communities of inquiry, providing opportunities to obtain material, debate ideas, and disseminate discoveries. For half a millennium, correspondence has been one of the primary means by which local scientific knowledge has been made global.

References

Barrera-Osorio, Antonio. 2006. *Experiencing Nature: The Spanish American Empire and the Early Scientific Revolution*. Austin: University of Texas Press.

Behringer, Wolfgang. 2003. *Im Zeichen des Merkur: Reichspost und Kommunikationsrevolution in der Frühen Neuzeit*. Göttingen: Vandenhoeck & Ruprecht.

Behringer, Wolfgang. 2006. "Communications revolutions: A historiographical concept." *German History*, 24, No. 3: 333–74.

Berkvens-Stevelinck, Christiane, Hans Bots, and Jens Häseler (eds.) 2005. *Les grands intermédiaires culturels de la République des Lettres: Études de réseaux de correspondances du XVIe au XVIIIe siècles*. Paris: Honoré Champion.

Bigourdin, Guillaume. 1917. "Les premières sociétés scientifiques de Paris au XVIIe siècle: Les réunions du P. Mersenne et l'Académie de Montmor." *Comptes rendus hebdomadaires des séances de l'Académie des sciences*, 164: 129–34.

Bots, Hans. 2005. "Marin Mersenne, 'Secrétaire Général' de la République des Lettres." In *Les grands intermédiaires culturels de la République des Lettres: Études de réseaux de correspondances du XVIe au XVIIIe siècles*, edited by Christiane Berkvens-Stevelinck, Hans Bots, and Jens Häseler, 165–81. Paris: Honoré Champion.

Bots, Hans and Françoise Waquet. 1997. *La République des Lettres*. Paris: Belin-De Boeck.

Brockliss, L. W. B. 2002. *Calvet's Web: Enlightenment and the Republic of Letters in Eighteenth-Century France*. Oxford: Oxford University Press.

Browne, Janet. 2014. "Corresponding naturalists." In *The Age of Scientific Naturalism: Tyndall and His Contemporaries*, edited by Bernard Lightman and Michael S. Reidy, 157–69. London: Pickering & Chatto.

Burkhardt, F. B. and Smith, S., et al. (eds.) 1985–2015 continuing. *Correspondence of Charles Darwin*. 17 vols. Cambridge: Cambridge University Press.

Castells, Manuel. 2010. *The Rise of the Network Society*. 2nd ed. Chichester: Wiley-Blackwell.

Codding, George A. 1964. *The Universal Postal Union: Coordinator of the International Mails*. New York: New York University Press.

Cook, Harold J. 2007. *Matters of Exchange: Commerce, Medicine, and Science in the Dutch Golden Age*. New Haven, CT: Yale University Press.

Cooper, Alix. 2007. *Inventing the Indigenous: Local Knowledge and Natural History in Early Modern Europe*. Cambridge: Cambridge University Press.

Cultures of Knowledge. n.d. http://www.culturesofknowledge.org/?page_id=28 Accessed September 25, 2015.

Darwin, Charles. 2008. *The Beagle Letters*. Edited by Frederick Burkhardt. Cambridge: Cambridge University Press.

Daston, Lorraine. 1991. "The ideal and reality of the Republic of Letters in the Enlightenment." *Science in Context*, 4: 367–86.

Del Centina, Andrea. 2012. "The correspondence between Sophie Germain and Carl Friedrich Gauss." *Archive for History of Exact Sciences*, 66, No. 6: 585–700.

Delisle, Candice. 2013. "'The Spices of Our Art': Medical observation in Conrad Gessner's Letters." In *Communicating Observations in Early Modern Letters (1500–1675): Epistolography and Epistemology in the Age of the Scientific Revolution*, edited by Dirk van Miert, 27–42. London: Warburg Institute.

Dibon, Paul. 1978. "Communication in the *Respublica Literaria* of the seventeenth century." *Res publica litterarum*, 1: 43–55.

Egmond, Florike. 2010. *The World of Carolus Clusius: Natural History in the Making, 1550–1610*. London: Pickering & Chatto.

Egmond, Florike. 2013. "Observing nature: The correspondence network of Carolus Clusius (1526–1609)." In *Communicating Observations in Early Modern Letters (1500–1675): Epistolography and Epistemology in the Age of the Scientific Revolution*, edited by Dirk van Miert, 43–72. London: Warburg Institute.

Flavell, Julie. 2010. *When London Was Capital of America*. New Haven, CT: Yale University Press.

Ford, Leslie. 1959. "The development of the Port of London." *Journal of the Royal Society of Arts*, 107, No. 5040: 821–35.

Galison, Peter. 1987. *How Experiments End*. Chicago: University of Chicago Press.

Gascoigne, Robert Mortimer. 1987. *A Chronology of the History of Science, 1450–1900*. New York: Garland.

Goldgar, Anne. 1995. *Impolite Learning: Conduct and Community in the Republic of Letters, 1680–1750*. New Haven, CT: Yale University Press.

Goodman, Dena. 1994. *The Republic of Letters: A Cultural History of the French Enlightenment*. Ithaca, NY: Cornell University Press.

Grafton, Anthony. 2009. "A sketch map of a lost continent: The Republic of Letters." In Anthony Grafton, *Worlds Made By Words: Scholarship and Community in the Modern West*, 9–34. Cambridge, MA: Harvard University Press.

Grimm, Friedrich Melchior and Jacques-Henri Meister. 1877–1882. *Correspondance littéraire, philosophique et critique.* Edited by Maurice Tourneaux. 16 vols. Paris: Garnier Frères.

Grosslight, Justin. 2013. "Small skills, big networks: Marin Mersenne as mathematical intelligencer." *History of Science*, 51: 337–74.

Harris, Stephen A., and Peter R. Anstey. 2009. "John Locke's seed lists: A case study in botanical exchange." *Studies in History and Philosophy of Biological and Biomedical Sciences*, 40: 256–64.

Harris, Steven J. 1996. "Confession-building, long-distance networks, and the organization of Jesuit Science." *Early Science and Medicine*, 1, No. 3: 287–318.

Harris, Steven J. 2006. "Networks of travel, correspondence, and exchange." In *The Cambridge History of Science, Vol. 3, Early Modern Science*, edited by Katharine Park and Lorraine Daston, 341–62. Cambridge and New York: Cambridge University Press.

Harvey, Joy. 2009. "Darwin's 'Angels': The women correspondents of Charles Darwin." *Intellectual History Review*, 19, No. 2: 197–210.

Hsia, Florence C. 2009. *Sojourners in a Strange Land: Jesuits and Their Scientific Missions in Late Imperial China.* Chicago: University of Chicago Press.

Iliffe, Rob. 2009. "Making correspondents network: Henry Oldenburg, philosophical commerce, and Italian science, 1660–72." In *The Accademia del Cimento and Its European Context,* edited by Marco Beretta, Antonio Clericuzio, and Lawrence M. Principe, 211–28. Sagamore Beach, MA: Science History Publications.

Jeffery, Keith. 2006. "Crown, communication and the colonial post: Stamps, the monarchy and the British Empire." *Journal of Imperial and Commonwealth History*, 34, No. 1: 45–70.

Kohler, Robert E. 2006. *All Creatures: Naturalists, Collectors, and Biodiversity, 1850–1950.* Princeton, NJ: Princeton University Press.

Kronick, David. 2001. "The commerce of letters: Networks and 'Invisible Colleges' in seventeenth- and eighteenth-century Europe." *Library Quarterly*, 71, No. 1: 28–43.

Landtsheer, Jeanine de, and Henk Nellen (eds.) 2011. *Between Scylla and Charybdis: Learned Letter Writers Navigating the Reefs of Religious and Political Controversy in Early Modern Europe.* Leiden: Brill.

Leeuwenhoek, Antoni van. 1939-1999. *Alle de Brieven.* 15 vols. Amsterdam: Swets & Zeitlinger.

Maclean, Ian. 2008. "The medical Republic of Letters before the Thirty Years War." *Intellectual History Review*, 18, No. 1: 15–30.

McClellan, James E., III. 1981. "The Académie Royale des Sciences, 1699–1793: A statistical portrait." *Isis*, 72, No. 4: 541–67.

McClellan, James E., III. 2010. *Colonialism and Science: Saint Domingue in the Old Regime; with a New Foreword by Vertus Saint-Louis.* Chicago: University of Chicago Press.

O'Neill, Lindsay. 2015. *The Opened Letter: Networking in the Early Modern British World.* Philadelphia: University of Pennsylvania Press.

Ogilvie, Brian W. 2006. *The Science of Describing: Natural History in Renaissance Europe.* Chicago: University of Chicago Press.

Ogilvie, Brian W. 2011 [2012]. "How to write a letter: Humanist correspondence manuals and the late Renaissance community of naturalists." *Jahrbuch für Europäische Wissenschaftskultur/Yearbook for European Culture of Science*, 6: 13–38.

Oldenburg, Henry, et al. 1965–1986. *The Correspondence of Henry Oldenburg.* Edited by A. Rupert Hall and Marie Boas Hall. 13 vols. Madison, WI: University of Wisconsin Press [imprint varies].

Pollard, Sidney. 1997. "The integration of European business in the 'long' nineteenth century." *VSWG: Vierteljahrschrift für Sozial- und Wirtschaftsgeschichte*, 84, No. 2: 156–70.

Portuondo, María M. 2009. *Secret Science: Spanish Cosmography and the New World.* Chicago: University of Chicago Press.

Rumphius, Georg Everhard. 1999. *The Ambonese Curiosity Cabinet*. Edited by E. M. Beekman. New Haven, CT: Yale University Press.

Scherer, Edmond Henri Adolphe. 1968. *Melchior Grimm: L'Homme de lettres, le factotum, le diplomate: Avec un appendice sur la correspondance secrète de Métra*. Genève: Slatkine Reprints.

Schiebinger, Londa. 1989. *The Mind Has No Sex? Women in the Origins of Modern Science*. Cambridge, MA: Harvard University Press.

Secord, James A. 2000. *Victorian Sensation: The Extraordinary Publication, Reception, and Secret Authorship of Vestiges of the Natural History of Creation*. Chicago: University of Chicago Press.

Taylor, Peter J., Michael Hoyler, and David M. Evans. 2008. "A geohistorical study of 'The Rise of Modern Science': Mapping scientific practice through urban networks, 1500–1900." *Minerva*, 46: 391–410.

Terrall, Mary. 1995. "Émilie du Châtelet and the gendering of science." *History of Science*, 33, No. 3: 283–310.

Terrall, Mary. 2014. *Catching Nature in the Act: Natural History in the Eighteenth Century*. Chicago: University of Chicago Press.

Thomas, Susan and Janette Martin. 2006. "Using the papers of contemporary British politicians as a testbed for the preservation of digital personal archives." *Journal of the Society of Archivists*, 27, No. 1: 29–56.

Ubrizsy Savoia, Andrea. 2011. "Federico Cesi (1585–1630) and the Correspondence network of his Accademia dei Lincei." *Studium*, 4, No. 4: 195–209.

Ultee, Maarten. 1987. "The Republic of Letters: Learned correspondence, 1680–1720." *Seventeenth Century*, 2, No. 1: 95–112.

Urbánek, Vladimír. 2014. "Comenius, the Unity of Brethren, and correspondence networks." *Journal of Moravian History*, 14, No. 1: 30–50.

Van Damme, Stéphane. 2005. *Paris, capitale philosophique: De la Fronde à la Révolution*. Paris: Odile Jacob.

Van Miert, Dirk. 2013. "Introduction." In *Communicating Observations in Early Modern Letters (1500-1675): Epistolography and Epistemology in the Age of the Scientific Revolution*, edited by Dirk van Miert, 1–7. London: Warburg Institute.

Waquet, Françoise. 1989. "Qu'est-ce que la République des lettres? Essai de sémantique historique." *Bibliothèque de l'Ecole des Chartes*, 147: 473–502.

Whitmer, Kelly Joan. 2013. "What's in a name? Place, peoples and plants in the Danish-Halle Mission, c. 1710–1740." *Annals of Science*, 70, No. 3: 337–56.

Wilson, Renate. 2000. *Pious Traders in Medicine: A German Pharmaceutical Network in Eighteenth-Century North America*. University Park, PA: University of Pennsylvania Press.

Winterer, Caroline. 2012. "Where is America in the Republic of Letters?" *Modern Intellectual History*, 9, No. 3: 597–623.

Woodward, Walter. 2010. *Prospero's America: John Winthrop, Jr., Alchemy, and the Creation of New England Culture, 1606–1676*. Chapel Hill, NC: University of North Carolina Press.

CHAPTER TWENTY-SIX

Translations

MARWA ELSHAKRY AND CARLA NAPPI

This chapter looks into some of the ways in which both meditations on, and acts of, translation have shaped the historiography and the history of science, medicine, and technology from antique to present times. It considers how translation has affected historiographic traditions in general, and the historiography of science in particular. We follow two turns to translation that tied universal history together with the history of science. The first section of the chapter examines the "first turn" by looking at the historiography of the Arabic sciences in the nineteenth and twentieth centuries, paying special attention to modern European and late Ottoman histories of science. It focuses on key texts written in Arabic, French, and English that prefigured or discussed the idea of a "translation movement" through the Arabic sciences, and traces how translation became important for understanding both the history of science and universal histories at that time. The second section looks at a "second turn" to translation studies and how it has been used by scholars in the last few decades to help us rethink canonical chronologies and geographies in the history of science. Drawing from examples in the historiography of East Asian sciences it considers how and why translation became particularly important for scholars who work on places and periods that fall beyond the usual borders of the history of science as well as for those searching for new methodological and analytical approaches to texts and practices within the history of science generally. It looks at a variety of antique to early modern encounters with translation in East Asia and outlines some of the practical ways in which translation has shaped our ideas about the historical transmission, preservation, and even innovation of ideas and practices of science in that context.

The First Turn: Modern Histories of "Arabic Science" in the Nineteenth and Twentieth Centuries

The nineteenth century gave rise to a plethora of new vernacular histories across the world. At the same time, it also witnessed a popular return to universal history. In

A Companion to the History of Science, First Edition. Edited by Bernard Lightman.
© 2016 John Wiley & Sons Ltd. Published 2020 by John Wiley & Sons Ltd.

both cases, translation played a crucial role. This section begins by examining a variety of nineteenth- and twentieth-century histories of the Arabic sciences to see how these different, but still curiously connected, historiographical impulses took shape.

By the mid-nineteenth century, new histories of the Arabs and of Islam were being published in the works of orientalists, Arabists, and others (e.g. Irving 1850; Muir 1861). Louis-Pierre-Eugène Sédillot's *Histoire des arabes* (1854) was among the first to focus so concretely on the history of an Arabic canon of science and philosophy. He had a longstanding interest in Arabic and Persian astronomical works in particular (e.g. Sédillot 1834–35; Sédillot 1841; Sédillot 1847), and he structured his history around an account of a grand historical narrative of rise and fall, or the "grandeur et décadence des arabes en Orient" (Sédillot 1854, 164–232). Much of this concerned dynastic succession: from the 'Abbasids and the Seljuks to the different dynasties in Spain and the Maghreb. In another key section, he offered a "tableau" account of "Arab civilization," focusing on arts and sciences and beginning with the Baghdad school, emphasizing their various "inventions" in particular (Sédillot 1854, 332–441). The categories he outlines are themselves revealing. They included sciences that would have been recognized as "sciences" by their contemporary readers but excluded others. Hence, astronomy, but not astrology, is covered, just as logic is favored over jurisprudence (Leaman 1980; Gutas 2002).

The idea that "Arab civilization" played a critical part in the universal progression of knowledge was a key motif. A prize-essay for the Bombay Education Society's press, "The Reciprocal Influence of European and Muhammadan Civilization," published in 1871, makes the point even more explicitly: "[T]he epoch which goes in Europe by the name of the Dark Ages, and which was really an epoch of ignorance and servitude, embraces the most brilliant period of the history of the Arabs." It was not until the twelfth century that "many Arabic books were translated into Latin, which facilitated the progress of science." As he put it, "When two or more nations come into long and close contact with each other, it is a natural consequence that they will, to a certain extent, influence each other in many things; the stronger and more cultivated will not only bestow its civilization and science, but will from its language engraft many words, and even whole locutions, on the weaker nation" (Rehatsek 1871, 64–9). This question of "reciprocal influence" would, in later universal histories, be presented as the story of translation itself.

The book that had the greatest impact on Arabic histories of the nineteenth century and after, however, was Gustave Le Bon's 1871 *La Civilisation des Arabes.* Le Bon was a French social scientist and amateur physicist. He was the author of several well-known and internationally circulated works, covering such diverse subjects as psychology, physics, socialism, and racial science. His popular study of crowd psychology, *La Psychologie des Foules* (1895), was published in Arabic in 1909 (Mitchell 1988), but many of his ideas made their way into a variety of late nineteenth-century Arabic works. It was his history of *Arab Civilization,* part of his "civilization" series, which proved the most popular in summary or digest form.

The term "civilization" was then fast becoming a marker for historical studies worldwide. For Le Bon, his understanding of civilization also reflected his conception of "race," another category of human difference then similarly gaining new institutional and intellectual form. Reflecting these interests, Le Bon begins his analysis of Arab civilization with a discussion of "milieu," a term he also connects to "race."

For him, as for many others of the time, the idea of race itself included climactic, geographic, physiological, and linguistic considerations, and even psychological and ethical, or moral, implications. This included the various "psychological factors in the classification of race," such as the virtues (and vices) of any kind of racial or collectivist mentality, or solidarity itself.

Curiously, in his discussion of "Arab civilization," he comes close to the ideas of Ibn Khaldun, the fourteenth-century historiographer and historian, and his history of the Arabs in particular, written centuries earlier but only translated into French in the early nineteenth century. A much cited text, Le Bon was clearly familiar with the *Muqaddima*. In a sense, Ibn Khaldun had already made an analogous argument to the one Le Bon made much later while thinking about the historical development of an Arab "civitas," set against the various Berber tribes they settled (or failed to): the *barbarah* were in fact morally superior to their Arab civilizers but they lost their *ta'assub*, or what we might term a kind of group "*thymos*" ($\theta\upsilon\mu\acute{o}\varsigma$) or consciousness, in the formation of new urban, political collectivities, and in the process of developing the civil sciences, arts and crafts, and trade in particular.

When Le Bon lamented "humanity is about to enter an Iron Age, where anything weak must inevitably perish," he was arguing something similar. According to him, when the Arabs had long ago conquered the East they did not harm their subjects as they shared a common racial tie (or a kind of "thymic" collective mentality). But as "anyone who has penetrated the East knows," he wrote, the current "commercial deceptions," betrayed "the low civilized veneer" of this new conquest (Le Bon 1871, 565–6). Writing during the commercial, and just before the colonial, expansion of European empires into Ottoman lands, this formulation no doubt would have explained his popularity among Arabic readers of the late nineteenth century, and after too.

Much of Ibn Khaldun's introduction was concerned with the development of the Arabic sciences and crafts in particular. We might find similar resonances in Le Bon's *History of Arab Civilization*. Yet he also clearly drew from more recent and local predecessors. Like Sédillot before him, Le Bon was interested in questions of dynastic succession, the conquest and fall of empires, and he also concentrated mostly on "the rise and fall of Arab civilization" while pursuing the question of "origin of their knowledge and educational methods" and then their later decline. He even cites Sédillot when crediting the Baghdad school with the invention of an "experimental method." Le Bon also covered a broad range of subjects that were similarly organized around contemporary disciplines: mathematics, astronomy, geography, the natural and physical sciences, philosophy, the visual and industrial arts, architecture, and commerce. Yet, he does not give much attention to any of the other classical sciences dealt with in earlier Arabic works: the prophetic traditions, alchemy, dream interpretation, or the reading of talismans for instances—all subjects Ibn Khaldun discussed at length. In a sense, the transmission and translation of words, ideas, and texts on histories of the Arabic sciences worked in multiple directions here, and they crisscrossed disparate times and places as much as texts and contexts, creating various forms of heterological and homological address along the way (Sakai 1997). Translation figured somewhat more directly in Le Bon's work too: the translation of Greek science and philosophy was a theme he referenced often. In fact, it was precisely this subject that gained him numerous citations in Arabic histories of the Arabs in the late nineteenth and early twentieth century.

Jirji Zaydan (1861–1914) was among the first authors to compose modern Arabic histories of the Arabs, and for him, translation was itself the core theme behind this story, if not behind the general, humanist and universal transmission of knowledge (and therefore of power) across "civilizations" and over time. Zaydan had a long-standing interest in questions of language, and the future of Arabic prose (Zaydan 1886), and he seized upon the idea of the value of translation for the history—or historical evolution—of languages generally: in essence, he saw the progress and evolution of Arabic in particular as fundamentally inseparable from a series of translation movements. Viewing the Arabs (and Arabic) as descending from Hammurabian Babylonia (and their linguistic and legal codes), he argued that they both continued the "civilizational contributions" of ancient Eastern empires, and through the linguistic and intellectual innovations that they added to those, marked a new step in the world history of civilizations. Of special interest to him was the idea that many of the "civilizational" contributions of ancient Mesopotamia and the Fertile Crescent were in fact preserved through this linguistic heritage (Zaydan 1893). His views on the rise of classical or Qurʾānic Arabic followed similar lines, and like other Arabic language reformers of the time, he was fond of pointing out the various non-Arabic linguistic borrowings it contained. He also thought that while the Qurʾān had standardized Arabic and given it greater specificity and scope, its more dogmatic standardization of the language had also ultimately produced linguistic stasis and contributed to civilizational stagnation in the long run.

Zaydan both argued and advocated for the importance of linguistic borrowings and constant translation for understanding the history of Arabic, of the Arabs, and of Islam. The context behind these ideas—and his own works in translation—was also critical here. The struggle with British and other European diplomats, creditors, and colonial officials was reaching its climax in the years he was writing. For Zaydan, tracing the long gestation of Arabic before and beyond the Qurʾān was no doubt tied to his desire to create a new Arab historical consciousness that was bound up with the potential rise of a new kind of Arab nationalism. Yet, as with many other intellectuals in Egypt at this time, this did not preclude his participation in new forms of colonial rule there, and he had earlier offered his translation services to British officials in the early years of the occupation: with his friend Jabr Dumit, who also wrote on the philosophy of language and on the history of Arabic, he joined the British army as a translator during the Wolseley expedition to the Sudan in 1885. Then and afterward, Zaydan remained a translator, and translation was key to his career and thought. His journal of history, science, and literature, *al-Hilal* (founded in 1892), for instance, featured numerous articles on the history and philosophy of language that were partial, summary, or full translations from a variety of studies published worldwide around the time.

Zaydan drew from the works of orientalists in particular. His interest in orientalism had also developed early: in Beirut in the 1880s, his involvement in various literary and scientific societies (al-Majmaʿ al-ʿilmi al-Sharqi and Shams al-Birr), as well in Freemasonry, led him to an interest in the deep antiquity of the Orient, and it was during his Beirut years that he took up the study of both Hebrew and Syriac. Shortly after Carl Brockelmann published his *Geschichte der Arabischen Literatur* (1898–1902), he decided to teach himself German. He also corresponded regularly with a number of noted orientalists. From his reading notes (now held at the archives in the American University of Beirut), we see how over many years, in a neat hand, annotated in

German and French as well as Arabic, Zaydan tracked the latest scholarship, plotted language change over time, and began to outline the structure of his later writings. In this way, Zaydan borrowed from the bibliographic resources of a rapidly growing global network of Orientalist and Arabist scholars. In the list of sources he offers in his essays on *Tarikh al-'Arab qabl al-Islam* (*The History of the Arabs Before Islam*) it is striking how relatively few canonical Arabic sources he utilized, in fact, and how many of the Arabic authors he cites were the same ones then being extensively discussed by contemporary European and American scholars—al-Mas'udi, al-Suyuti, and Ibn Khaldun (Zaydan 1982, vol. 10, 36–9).

Like the many histories of the Arabs, of Arabic, and of Islam he followed in these works, Zaydan also emphasized the role of translation in the formation of an Arabic canon. In particular, he stressed the importation of Greek thought. But his conception of the Arabic "sciences" was still somewhat broader than those of the orientalists he followed, often enfolding discussions of literary, philosophical, and metaphysical works under the rubric. Yet his writings on the "Arabic sciences" were also typically organized around biographies of then widely cited figures, and he often classed them as falling under either the "Islamic" or the "imported" sciences to once again emphasize the role of translation.

As Arabic scholars adapted, appropriated, or rejected new histories of the Arabic sciences, the history of science itself was also drawing from this same vein. Though we do not often think of it this way, the early disciplinary history of science and orientalism had much in common. Historians of science writing in the first few decades of the twentieth century had more than a few affinities with orientalists of the era. Orientalism had created, in fact, a new kind of international network of scholarship and correspondence in which many historians of science around the world took part. For example, consider how we might compare the exchange of ideas and letters between Jirji Zaydan and the Russian orientalist Ignaty Krachkovsky, on the one hand, and between Krachkovsky and George Sarton, the noted Belgian scholar and the first man to hold a chair in history of science in the United States, on the other. In a sense, both offer examples of various convergences, and divergences, over the exact texts, borders, and references to a rise and fall of an Arabic or Islamic canon of arts and sciences.

Sarton was a key figure in the early, disciplinary history of the history of science, and he was as much an orientalist as a historian of science, as he himself admitted later in life. Founder of the journals *Isis* and *Osiris*, his choice of titles speaks to these affinities. He launched *Isis* shortly before the First World War, and then and later, held a deep commitment to an ecumenical history of science as the vehicle for a "new humanism." Like many historians of science who formed part of an emerging international network of historians of science, Sarton's vision of humanism and of the history of science formed part of his commitment to both universalism and internationalism. He was among the first to articulate a program for this all-encompassing vision. His narrative of the world history of science reflected these concerns, and more than anyone else in the first half of the twentieth century, he helped to popularize a timeline that stretched from ancient Mesopotamia to modern Europe. He connected ancient and medieval with modern histories and saw their development as taking place between a series of translation movements and cross-cultural intellectual or material contacts between the "East," the original home and seat of ancient knowledge and civilization, and the "West," the apex of this narrative, if not of the collective history of humanity itself. Sarton's views on the relation between Islam and Europe, moreover, also refracted the views of other internationalists of the time, particularly the fellow Belgian historian, Henri Pirenne. For Pirenne, whose views have been termed the "Pirenne thesis," the rise of Islam encapsulated and made Europe after Charlemagne (Pirenne 1935). For Sarton, it was the "translation movement" under the 'Abbasids that did the converse: it helped to revive Europe, and became the very source of its own Renaissance, serving as a prime mover for the universal history of humanity along the way. We might say that the rise and fall of Muslim civilization itself was therefore depicted as both encapsulating and then liberating Europe, allowing it to play a role once again on the world stage of history.

Sarton was also an Arabist: he corresponded regularly with Arabists and with Arabic-writing scholars from the region, often in Arabic. The network of these affiliations has not been studied; yet his work is unimaginable without them. They included book dealers, teachers, and translators as well as 'ulama writing in Arabic around the world. He drew upon both their work and the work of orientalists to highlight similar themes too. He helped to popularize, if not institutionalize, the contributions of the Arabic sciences to the new universal history of science. He saw the Arabic sciences, particularly through translation and as developed under the 'Abbasids, as providing a crucial link between the evolution of science from ancient Mesopotamia to the modern West and as key to the story of preserving the spirit of Greek rationalism to which they were heir. In this way, the old question of the relation of the orient to the occident was rearticulated, and it structured his periodization too (Sarton 1931).

Sarton also promoted a more general and all-encompassing view of the progress of science over time. As he put it, "the progress of science is naturally an acceler-ated one (hence if we look backwards the acceleration is negative)." This progressive development from ancient to medieval science worked by passing on oral or written or "manual" "traditions," the last one he described as "an underground river which remains hidden for long stretches," adding, "yet we can be reasonably certain that the river emerging from the earth at point B is the same as disappeared at another point A many miles distant" (Sarton 1952, 26–7). In his 1952 *Guide to the History of Science* he provides two diagrams for this story of visible and invisible influences:

> We might attempt a graphical representation of these views. The tradition of each single idea or fact might be symbolized by a line, more or less regular, with ups and downs. Some of these lines are interrupted because the tradition has ceased for a time to be visible. Sometimes the lines cross and their intersections may be indifferent or they may correspond to a knot or a new discovery.

This is followed by a brief sketch (Figure 26.1). In a following image (Figure 26.2), he sketches out a more visible "transmission" of ideas, in which a "Greek" linguistic, oral, and written tradition run in parallel with an "Arabic" one in the center:

> Should we wish to represent the whole tradition, not only the development of single ideas or inventions, but the scientific pattern in its totality, the graph would be very dif-ferent, something like this (Fig 2). The roots of western science, the graph reminds us, are Egyptian, Mesopotamian, and to a much smaller amount, Iranian and Hindu. The central line represents the Arabic transmission which was for a time, say, from the ninth

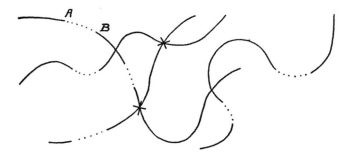

Figure 26.1   Graphic representation of single discoveries, ideas, or inventions (Sarton 1952, 26).

to the eleventh century, the outstanding stream, and remained so until the fourteenth century, one of the largest streams of medieval thought (Sarton 1952, 26–7).

He added: "the Arabic tradition was a continuation and revivification not only of Greek science but also of Iranian and Hindu ideas," though this transmission—or translation of civilizations—was now only "imperfectly known." The Graeco-Arabic vein thus provided the critical medieval bridge between the Ancients and the Moderns and between (the Near and Far) East and the West. "That network, the Oriental-Greek-Arabic, is *our* network." He also advocated for a greater disciplinary infusion between scholars of the Arabic sciences and historians of science: "The neglect of Arabic science and the corresponding misunderstanding of our own medieval traditions was partly due to the fact that Arabic studies were considered a part of Oriental studies. The Arabists we left alone." And whereas "much in the field of orientalism is definitely exotic as far as we are concerned" he reassured his readers that both "the religious Hebrew traditions and the scientific Arabic ones are not exotic." Indeed, "they are an integral part of our network today." He even claimed them as "part and parcel of our spiritual existence." Writing in the aftermath of World War II, he wrote: "Arabic culture is of a singular interest to the student of human traditions in general, to those whose greatest task it seems to them is the rebuilding of human integrity in

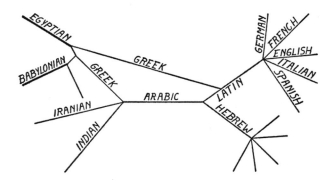

Figure 26.2   Diagram of the scientific pattern in its totality (Sarton 1952, 27).

the face of national and international disasters, because it was, and to some extent still is, a bridge, the main bridge between East and West" (Sarton 1952, 26–7).

Having looked, all too briefly and somewhat schematically, at a variety of examples of nineteenth- and twentieth-century histories of science that focused on the mutually formative role of Arabic, its arts and sciences, and of Arab civilization more generally, we get some sense of how central the idea of the translation of ideas and practices across civilizations was for the foundational concepts and chronologies of the discipline itself. In a sense, both the process and the metaphor of translation offered novel conceptual as much as political grounds for these early histories of the Arabic sciences, marking the movement of ideas with the progress of humanity across linguistic, civilizational, and epochal boundaries. This section has traced a few of these crisscrossed, sometimes interconnected, histories by focusing on both translation practices and historical studies on translation in nineteenth- and twentieth-century discussions of the Arabic sciences. Next, we turn to more recent histories, while following a variety of examples of late antique and early modern sciences in East Asia.

The Second Turn: Antique to Early Modern Encounters with Translation in East Asia and Beyond

Thinking about, and with, translation has helped historians of science to re-imagine their discipline in recent decades, expanding its scope to be more global while simultaneously using that expanded scope to critically examine what "global" might mean as a historiographical concept. Historians of science, medicine, and technology in East Asia have taken a leading role in this re-imagining, as much of the historiography of the field in the past several decades has dealt extensively with issues of translation. In moving to this second section of the chapter, we take advantage of this robust historiography to explore how a history of, and with, translation can inform some of our broader methodological concerns within the history of science. Following this recent scholarship, we consider how the vast range of texts and practices covered both schematically and thematically attest to the many ways in which we might both conceptualize and trace instances of translation. Taking this more expansive view of "translation" itself, we suggest here, can help us reframe and redefine the very notion of translation by broadening its scope from the understanding of translation that animated the nineteenth- and twentieth-century scholarship described above—which by and large focused on text-based language-to-language translation—to include instances of the transformation of both knowledge and practice across non-linguistic, or even extra-linguistic, frames, including forms of visual representation and imaging techniques that carry meaning beyond words. This section will now turn briefly to instances of both these approaches, concentrating on the latter in particular, before then moving on to consider the ways in which translation figured in the historiography of science in East Asia.

Textual translations in East Asian contexts include numerous instances of both inter- and intra-lingual translations. Considering these works invites us to consider examples of translation that render texts in new forms, and that highlight the problem of what many translators refer to as textual fidelity. Many scientific, medical, and technological texts in translation could and did take a number of forms that did not necessarily appear to be direct, faithful reproductions of a text from one language into another. Early authors used commentary practices as a means of translating texts into new contexts, giving new meanings to old words and coining new terms. These practices can help challenge—or at least historicize –the very notion of "fidelity" to an original work, and also allow us to reconsider the very meanings we ascribe to any "original" text. There were also many examples of textual transmissions that we might class as forms of conceptual translations. Many early medical texts in China offered deliberations on human bodies and their relationships by incorporating or providing readings and commentaries of classical cosmological texts. We might understand these texts as translations. The early medical treatise *Huangdi neijing* (*Inner Canon of the Yellow Emperor*), for example, can be understood not just as a canonical medical work (Unschuld 2003; Unschuld, Tessenow, and Zheng 2011), but as a translation of broader cosmological principles of the systematic correspondence of bodies into the context of theoretical writings on the human body. Later scholars similarly translated classic Chinese cosmological, medical, botanical, and other technical texts into local contexts for Korean and Japanese readerships, and these often contained translations of both the ingredients and the bodies of the Chinese originals in equivalents or examples that were more readily available or identifiable to their local or intended audience (Suh 2008, 2013). In the Chinese context, the use of commentaries was a longstanding form of translation—and vice versa—and they can be found in examples of science translations well into the twentieth century. Scholars who rendered the work of European writers like Darwin, Huxley, and others into Chinese in the late nineteenth century, for instance, often adopted a form of translation that simultaneously offered an analysis and a judgment of these scholars' ideas while referring to concepts, objects and events, that went well beyond the concerns or contexts of the original authors whom they translated for their Chinese readers (Schwartz 1964; Pusey 1983; Jones 2011; Hill 2013).

Authors might also produce a kind of translation by moving a text into a new medium. We might consider forms of technical or diagrammatic illustrations in East Asia as examples of translations that carried meanings across both texts and images. In early Chinese exegeses of the *Yijing* (*Classic of Changes*), for instance, a relatively opaque set of symbols was repeatedly translated for contemporary readers, as these textual renditions of these re-imaged trigrams served as the basis for their continued, practical use. This can be understood as part of a larger context of translating divinatory images—including cracks in oracle bones and arrangements of plant materials—into textual or discursive commentaries. Looking at the use of images as tools of the translator has also significantly changed the ways in which we understand the role of the sciences and early modernity in Tibet and the Buddhist world (Gyatso 2015). In later works, other imaging practices of all sorts were used when producing scientific, technical, and medical knowledge. Recent authors have shown the deep imbrication of the histories of science and the history of art, for example, when charting the adoption and translation of linear perspective into an eighteenth-century Chinese context (Kleutghen 2015), when considering the ways in which the botanical sciences were translated through a variety of novel visual imaging technologies in nineteenth-century Japan (Fukuoka 2012), or, to invoke an example from beyond East Asia, when understanding early modern science as it was shaped by experience in the Americas (Safier 2008; Bleichmar 2012). Focusing on images to trace the translation of scientific and medical discourses across European-language and Chinese texts can also, as some have shown, illuminate how the very ideas of a "China" or "human body" have been translated across linguistic and cultural contexts (Heinrich 2008).

We might also consider the role of translations in terms of the forms of tacit or embodied knowledge that shaped practices of science in China and elsewhere in the region. Inspired by new works on the translation of embodied techniques in other examples of early modern science (Smith 2006; Long 2011), recent histories of East Asian science, medicine, and technology have similarly begun to pay attention to this phenomenon, including studies of the translation of Indian medical and healing practices in medieval China (Salguero 2014) or of contemporary healing in the Chinese-speaking world (Zhan 2009) or in Tibet and Nepal (Craig 2012). Considering these few examples all too briefly here, we might then see how science translations operated at a variety of practical levels.

Yet another approach to situating scientific translation in time is to look at the ways that translation has created its own temporalities and modes of time-travel for its practitioners. The times and temporalities of the history of science can often look quite different depending on the geographic region under discussion. The periodization of historical time can be especially problematic for historians of science who aim to take a comparative, comprehensive, and polyvocal approach to the subject, exemplified in the recent turn to global and transnational studies. This has brought a concomitant turn toward a focus on translation not just as an important feature of the history of science, but as a field of practices that are in many ways constitutive of that history. A focus on translation can also show us how traditional modes of shaping historical time have been forged or broken. This could take a number of forms: some authors reanimated long-neglected texts by translating them into a new language for a new audience, and some created conversations across centuries by rendering classic works into new languages and commentary traditions. In each case, translation enabled the creation of new dialogues and relationships across time. We now have ample accounts of this phenomenon as embodied in medieval or early modern translations of scientific (and other related) texts across Latin, Syriac, Arabic, Greek, and European linguistic contexts (Gutas 1998; Mavroudi 2002; Morrison 2007; Morrison 2014; Saliba 2007; Burnett 2009; Martin 2014; Smith 2015). In China, as well, the work of translators across classical and early modern texts and languages created conversations that crisscrossed time and space (Wardy 2000). In some ways, we might also consider how translation itself was called in to construct novel experiences or categories of time, and of historical time in particular: many texts in translation make genealogical claims of transmission that are themselves often ways of constructing new temporalities, whether in terms of chronologies or of periodizations, for the particular history of science that they serve.

Focusing on translation might also help us re-think the role of space in the history of science, whether in terms of political geographies or of linguistic and social topographies. This might help break down not only classical periodizations—ancient, medieval or early modern, and modern—but also older, civilizational categories like "the East" and "the West," alongside traditionally conceived area studies geographies—the orient or East Asia, for instance (Hart 2000). Re-evaluating the activities of translators who seemed comfortable moving across Greek, Syriac, Arabic, German, French and English, or Latin and Chinese, might offer us a way to map the spaces of the history of science in more expansive terms.

A renewed focus on translation has also introduced new discussions around the idea of something like a global "early modernity," bringing a number of area studies fields into dialogue with the history of science, medicine, and technology. Work in this vein often problematizes categories of both space and time. While some authors wonder whether the terminology of "early modernity" can be usefully applied beyond European contexts, others have moved on from those concerns to consider how the increasing mobility and exchange among groups of people and intellectual traditions transformed the sciences between roughly 1500 and 1800 (Raj 2006; Raj 2013). This period is often noted as one in which translations of various kinds took place with renewed gusto—from textual renderings into and from Latin and/or European vernaculars, to new modes of periodizing antiquity and modernity, to transmissions of tacit knowledge or embodied practice, as briefly outlined above.

Many early modern translators created and sustained novel relations between global networks of texts, ideas, goods, and people that were said to have shaped a global early modernity. Many scholars have now pointed out how the work of translators was crucial for establishing and sustaining early modern commercial relations (Cook 2007; Cook and Dupre 2012). Similarly, scholars of Tokugawa Japan have looked carefully at the ways that commercial networks brought Japanese scholars into contact with Dutch merchants—creating a robust literature in translation to and from Japanese and European-language works on animals, plants, and people—all while also rendering Chinese texts into local Japanese contexts (Vande Walle and Kazaya 2001; Fukuoka 2012). In Ming China (1368–1644), commerce traveled through networks where money was not the only form of power being negotiated. Scholars have studied the forms of translation undertaken by Jesuits at the Ming court, a context wherein the rendering of science, medicine, and technology from European languages into Chinese was often a means of translating broader religious conceptions or cultural values alongside discussions of astronomy and medicine (Elman 2005; Hsia 2009; Hart 2013).

Imperial patronage relationships also sustained, and were in turn sustained by, early modern translation practices devoted to science, medicine, and technology. Translation of texts and ideas about the nature and transformations of bodies—be they planetary, human, or otherwise—was an integral activity of scholars at the Qing (1644–1911) court. During the high Qing period at the court of the Kangxi Emperor (r. 1661–1722), Jesuits rendered anatomical, astronomical, and other forms of knowledge from European contexts into the Manchu language, creating a corpus of translated material that was inaccessible to readers for whom this language of the Qing rulers was not legible (Jami 2012). Kangxi was personally involved in many of these efforts to varying degrees, whether by commissioning translations of texts on various topics whose significance struck him for one reason or another, assigning language tutors to the Jesuit translators to facilitate their work on Manchu, or by personally proofreading manuscript copies of the resulting Jesuit products. Beyond the Qing empire, translation was also central in mediating scientific exchange in other early modern court contexts across the Mediterranean and Indian Ocean worlds (Subrahmanyam 2012).

As we move into the nineteenth century and beyond, the translation of, and with, scientific texts expanded the number and kinds of connections forged by translators. In the late Qing context, both individual scholars and scholarly collectives rendered ideas from European languages into Chinese, as Japanese texts became crucial for enabling the introduction and movement of ideas into a Chinese context (Schwartz 1964; Liu 1995; Wright 2000; Lackner and Vittinghoff 2004; Lackner, Amelung, and Kurtz 2001; Wu 2015). The translation of scientific ideas from English and European languages into Chinese was enabled by translating a range of kinds of texts, including magazines and works of fiction as well as treatises on geology, biology, and social sciences. It can be argued that translation helped make the modern sciences—catalyzed the notion of modernity itself—from the late nineteenth through the mid-twentieth centuries in East Asia as elsewhere.

Concluding Remarks

This chapter has looked closely at two moments in the historiography of science in which a concern with translation transformed the field in powerful and subtle ways. In the nineteenth and early twentieth centuries, translation was an important component of scholarly attempts to understand science, universal history, and the relationships between them. We explored this phenomenon using a case study in which this move was particularly important, looking carefully at modern histories of the "Arabic sciences" in which translation became an important methodological and historiographical tool, especially for those who wished to construct universal histories around the progress of science. Late in the twentieth century and into the twenty-first, a concern with translation also shaped the work of historians of science, some of whom explicitly articulated a more expansive view of translation in history, while others looked at phenomena that help us reimagine what translation might look like and where we might find evidence of it in our sources even when not explicitly articulating their work in those terms. This moment took on particular salience in the work of historians of East Asian science, technology, and medicine, where translation was a tool to disaggregate the notions of science and progress to construct a more poly-vocal history of science that appreciated the emergence of different ways of knowing across diverse geographical and temporal localities. As we move on from here, a rich account of the ways that translation has shaped the history and historiography of science remains to be written. Taking an expansive view of "translation" as a lens through which to look anew at the history of science will require us to attend to a broad range of sources and practices that are scattered across different area studies fields, disciplinary categories, languages and literatures, and scholarly canons. In taking up this challenge, we have an opportunity to transform not only how we practice the history of science as a discipline, but also how we conceive of the nature and practice of translation itself.

References

American University in Beirut, Jaffet Library Archives, Jirji Zaydan Collection AA:6.2.26.1; notebooks, see especially the file labeled "small notebooks."

Bleichmar, Daniela. 2012. *Visible Empire: Botanical Expeditions and Visual Culture in the Hispanic Enlightenment*. Chicago: University of Chicago Press.

Borges, Jorge Luis. 1964. *Labyrinths: Selected Stories and Other Writings*. New York: New Directions.

Burnett, Charles. 2009. *Arabic into Latin in the Middle Ages: The Translators and Their Intellectual and Social Context.* Farnham: Variorum.

Cook, Harold J. 2007. *Matters of Exchange: Commerce, Medicine, and Science in the Dutch Golden Age.* New Haven, CT: Yale University Press.

Cook, Harold J., and Sven Dupre (eds.) 2012. *Translating Knowledge in the Early Modern Low Countries.* Wien: Lit Verlag.

Craig, Sienna R. 2012. *Healing Elements: Efficacy and the Social Ecologies of Tibetan Medicine.* Berkeley: University of California Press.

Elman, Benjamin A. 2005. *On Their Own Terms: Science in China, 1550–1900.* Cambridge, MA: Harvard University Press.

Fu, Liangyu. 2013. "Indigenizing visualized knowledge: Translating Western scientific illustrations in China, 1870–1910." *Translation Studies*, 6, No. 1: 78–102.

Fukuoka, Maki. 2012. *The Premise of Fidelity: Science, Visuality, and Representing the Real in Nineteenth-Century Japan.* Stanford, CA: Stanford University Press.

Gordin, Michael. 2012. "Translating textbooks: Russian, German and the language of translation of chemistry." *Isis*, 103, No. 1: 88–98.

Gordin, Michael. 2015. *Scientific Babel: How Science Was Done Before and After Global English.* Chicago: University of Chicago Press.

Green, Nile. 2003. "The religious and cultural roles of dreams and visions in Islam." *Journal of the Royal Asiatic Society of Great Britain & Ireland*, 13: 287–313.

Gutas, Dmitri. 1998. *Greek Thought, Arabic Culture: The Graeco-Arabic Translation Movement in Baghdad and Early Abbasid Society (2nd-4th/8th-10th Centuries).* London: Routledge.

Gutas, Dimitri. 2002. "The study of Arabic philosophy in the twentieth century: An essay on the historiography of Arabic philosophy." *British Journal of Middle Eastern Studies*, 29, No. 1: 5–25.

Gyatso, Janet. 2015. *Being Human in a Buddhist World: An Intellectual History of Medicine in Early Modern Tibet.* New York: Columbia University Press.

Hart, Roger. 2000. "Translating the untranslatable: From copula to incommensurable worlds." In *Tokens of Exchange: The Problem of Translation in Global Circulations*, edited by Lydia Liu, 45–73. Durham, NC: Duke University Press.

Hart, Roger. 2013. *Imagined Civilizations: China, The West, And Their First Encounter.* Baltimore: Johns Hopkins University Press.

Heinrich, Larissa. 2008. *The Afterlife of Images: Translating the Pathological Body between China and the West.* Durham, NC: Duke University Press.

Henderson, Felicity. 2013. "Faithful interpreters? Translation theory and practice at the early Royal Society." *Notes and Records: The Royal Society Journal of the History of Science*, 67, No. 2: 101–22.

Hill, Michael Gibbs. 2013. *Lin Shu, Inc.: Translation and the Making of Modern Chinese Culture.* Oxford: Oxford University Press.

Hsia, Florence. 2009. *Sojourners in a Strange Land: Jesuits and Their Scientific Missions in Late Imperial China.* Chicago: University of Chicago Press.

Irving, Washington. 1850. *Lives of Mahomet and his Successors.* Paris: Baudry's European Library and A. W. Galignani. Also 1868. *Mahomet and his Successors*, Two vols. New York: Putnam.

Jami, Catherine. 2012. *The Emperor's New Mathematics: Western Learning and Imperial Authority During the Kangxi Reign (1662–1722).* Oxford: Oxford University Press.

Jones, Andrew F. 2011. *Developmental Fairy Tales: Evolutionary Thinking and Modern Chinese Culture.* Cambridge, MA: Harvard University Press.

Kleutghen, Kristina. 2015. *Imperial Illusions: Crossing Pictorial Boundaries in the Qing Palaces.* Seattle, WA: University of Washington Press.

Lackner, Michael, and Natascha Vittinghoff. 2004. *Mapping Meanings: The Field of New Learning in Late Qing China.* Leiden: Brill.

Lackner, Michael, Iwo Amelung, and Joachim Kurtz (eds.) 2001. *New Terms for New Ideas: Western Knowledge and Lexical Change in Late Imperial China*. Leiden: Brill.

Leemans, Pieter De, and An Smets (eds.) 2008. *Science Translated: Latin and Vernacular Translations of Scientific Treatises in Medieval Europe*. Leuven: Leuven University Press.

Le Bon, Gustave. 1884. *La Civilisation des Arabes*. Paris: Firmin-Didot et cie.

Liu, Lydia. 1995. *Translingual Practice: Literature, National Culture, and Translated Modernity–China, 1900–1937*. Stanford, CA: Stanford University Press.

Long, Pamela O. 2011. *Artisan/Practitioners and the Rise of the New Sciences, 1400–1600*. Corvallis, OR: Oregon State University Press.

Martin, Craig. 2014. *Subverting Aristotle: Religion, History, and Philosophy in Early Modern Science*. Baltimore: Johns Hopkins University Press.

Mavroudi, Maria V. 2002. *A Byzantine Book on Dream Interpretation: The Oneirocriticon of Achmet and Its Arabic Sources*. Leiden: Brill.

Mitchell, Timothy. 1988. *Colonising Egypt*. Berkeley: University of California Press.

Montgomery, Scott. 2000. *Science in Translation: Movements of Knowledge through Cultures and Time*. Chicago: University of Chicago Press.

Morrison, Robert G. 2007. *Islam and Science: The Intellectual Career of Niẓām al-Dīn al-Nīsābūrī*. London: Routledge.

Morrison, Robert G. 2014. "A scholarly intermediary between the Ottoman Empire and Renaissance Europe." *Isis*, 105, No. 1: 32–57.

Muir, William. 1861. *Life of Mahomet*. London: Smith, Elder and Co.

Olohan, Maeve. 2014. "History of science and history of translation: Disciplinary commensurability?" *The Translator*, 20, No. 1: 9–25.

Paniotis, Manolis, and Kostas Gavroglu. 2012. "The sciences in Europe: Transmitting centers and appropriating peripheries." In *The Globalization of Knowledge in History*, edited by Jürgen Renn, 321–43. Berlin: Max Planck Research Library for the History and Development of Knowledge, Studies 1.

Pirenne, Henri. 1937. *Mahomet et Charlemagne*. Paris: F. Alcan and Bruxelles: Nouvelle société d'éditions.

Pusey, James Reeve. 1983. *China and Charles Darwin*. Cambridge, MA: Harvard University Press.

Raj, Kapil. 2006. *Relocating Modern Science: Circulation and the Constitution of Knowledge in South Asia and Europe, 1650–1900*. Basingstoke: Palgrave Macmillan.

Raj, Kapil. 2013. "Beyond postcolonialism … and postpositivism: Circulation and the Global history of science." *Isis*, 104, No. 2: 337–47.

Rehatsek, Edward. 1877. *Prize Essay on the Reciprocal Influence of European and Muhammadan Civilization: During the Period of the Khalifs and at the Present Time*. Bombay: Education Society's Press.

Rupke, Nicolaas. 2000. "Translation studies in the history of science: The examples of *Vestiges*." *British Journal for the History of Science*, 33, No. 2: 209–22.

Safier, Neil. 2008. *Measuring the New World: Enlightenment Science and South America*. Chicago: University of Chicago Press.

Sakai, Naoki. 1997. *Translation and Subjectivity*. Minneapolis: University of Minnesota Press.

Salguero, Pierce. 2014. *Translating Buddhist Medicine in Medieval China*. Philadelphia, PA: University of Pennsylvania Press.

Saliba, George. 2007. *Islamic Science and the Making of the European Renaissance*. Cambridge, MA: MIT Press.

Sarton, George. 1931. *The History of Science and the New Humanism*. New York: H. Holt.

Sarton, George. 1952. *A Guide to the History of Science: A First Guide for the Study of the History of Science, with Introductory Essays on Science and Tradition*. Waltham, MA: Chronica Botanica.

Schwartz, Benjamin I. 1964. *In Search of Wealth and Power: Yen Fu and the West*. Cambridge, MA: Belknap Press.

Sédillot, J.-J. 1834-1835. *Traité des instruments astronomiques des Arabes composé au treizième siècle par Aboul Hhassan Ali, de Maroc, intitulé Collection des commencements et des fins, traduit de l'arabe sur le manuscrit 1147 de la Bibliothèque royale par J.-J. Sédillot, et publié par L.-Am. Sédillot*, 2 volumes. Paris: Imprimerie Royale.

Sédillot. 1841. *Mémoire sur les instruments astronomiques des Arabes*. Paris: Imprimerie Royale.

Sédillot. 1847. *Prolégomènes des tables astronomiques d'Oloug-Beg, publiés avec notes et variantes et précédés d'une introduction*. Paris: Didot frères.

Smith, A. Mark. 2015. *From Sight to Light: The Passage from Ancient to Modern Optics*. Chicago: University of Chicago Press.

Smith, Pamela. 2006. *The Body of the Artisan: Art and Experience in the Scientific Revolution*. Chicago: University of Chicago Press.

Subrahmanyam, Sanjay. 2012. *Courtly Encounters: Translating Courtliness and Violence in Early Modern Eurasia*. Cambridge, MA: Harvard University Press.

Suh, Soyoung. 2008. "Herbs of our own kingdom: Layers of the 'local' in the materia medica of Chosŏn Korea." *Asian Medicine: Tradition and Modernity*, 4, No. 2: 395–422.

Suh, Soyoung. 2013. "A Chosŏn Korea medical synthesis: Hŏ Chun's *Precious Mirror of Eastern Medicine*." In *Chinese Medicine and Healing: An Illustrated History*, edited by T. J. Hinrichs and Linda L. Barnes, 137–9. Cambridge, MA: Belknap Press.

Unschuld, Paul U. 2003. *Huang Di Nei Jing Su Wen: Nature, Knowledge, Imagery in an Ancient Chinese Medical Text*. Berkeley: University of California Press.

Unschuld, Paul U., Hermann Tessenow, and Jinsheng Zheng. 2011. *Huang Di Nei Jing Su Wen: An Annotated Translation of Huang Di's Inner Classic– Basic Questions*. Berkeley: University of California Press.

Vande Walle, Willy, and Kazuhiko Kazaya (eds.) 2001. *Dodonaeus in Japan: Translation and the Scientific Mind in the Tokugawa Period*. Leuven: Leuven University Press.

Wardy, Robert. 2000. *Aristotle in China: Language, Categories and Translation*. Cambridge: Cambridge University Press.

Wright, David. 2000. *Translating Science: The Transmission of Western Chemistry into Late Imperial China, 1840–1900*. Leiden: Brill.

Wu, Shellen Xiao. 2015. *Empires of Coal: Fueling China's Entry into the Modern World Order, 1860–1920*. Stanford: Stanford University Press.

Yucesoy, Hayrettin. 2009. "Translation as self-consciousness: Ancient sciences, antediluvian wisdom, and the 'Abbasid translation movement." *Journal of World History*, 20: 523–57.

Zaydan, Jirji. 1886. *al-Falsafa al-lughawiya wa al-alfaẓ al-'Arabiya*. Beirut: n.p.

Zaydan, Jirji. 1904. *al-Lugha al-'Arabiya*. Cairo: Dar al-Hilal.

Zaydan, Jirji. 1982. *Tarikh al-'Arab qabl al-Islam*. In *Mu'allifat Jirji Zaydan*, 21 volumes. Beirut: Dar al-Jil, vol. 10.

Zhan, Mei. 2009. *Other-Worldly: Making Chinese Medicine Through Transnational Frames*. Durham, NC: Duke University Press.

CHAPTER TWENTY-SEVEN

Journals and Periodicals

AILEEN FYFE[1]

During the interwar years, the geneticist J. B. S. Haldane wrote over 80 original research articles for publication in journals varying from *Biometrika* and the *Journal of Genetics*, to *Nature* and the *Proceedings of the Royal Institution*. He also wrote popular books on evolution, genetics, and science and society; and articles for such newspapers as the *Daily Herald*, the London *Evening Standard* and the *News Chronicle*. Moreover, his research and views on science were reported, excerpted and glossed by others, especially in magazines and newspapers. Haldane's scholarly reputation depended upon his articles in research journals, but he engaged with a much wider range of publications, both at first- and second-hand (Bowler 2009, Ch. 11). This should remind us that, though we tend to associate professionalization with the increased importance of scientific journals, science continues to be published in many places.

The late seventeenth century is usually taken as the origin of scholarly journals, including those devoted to the sciences. However, an emphasis on origin myths has obscured the fact that the meaning and significance, as well as the format and style, of scientific authorship and reportage have changed considerably over the last four centuries. Most of the features we associate with the modern scientific journal, including originality of research, self-authorship, refereeing procedures and standardized rhetoric and structure, were nineteenth-century developments, while the emergence of English as the international language of science, professionalism among authors and editors, and profitability, are largely twentieth-century phenomena. Scientific research journals have never operated in a vacuum, and the most successful ones have always required connections to, and an understanding of, both science and commerce.

This chapter will open with a chronological survey of the sorts of magazines, newspapers and research journals that carried scientific knowledge. It will then take that survey as the foundation for some reflections on changing trends in scientific authorship; editing and reviewing; and distribution and readership. It will become clear that we currently know far more about certain elements of this story than about others.

A Companion to the History of Science, First Edition. Edited by Bernard Lightman.
© 2016 John Wiley & Sons Ltd. Published 2020 by John Wiley & Sons Ltd.

The Seventeenth Century

The association between early scientific periodicals and the new scientific societies and academies of the late seventeenth century, has long been recognized. It is only more recently that these publications have been situated in the wider book trade, drawing attention to the novelty of the periodical format. The practice of regularly printing news and information (or gossip) in "corantos" or "news-sheets" had only developed during the wars of religion earlier in the century, and "the periodical" as a genre barely existed (Pettegree 2014).

In the early modern learned world, substantial treatises and books were the respectable formats for appearing in print, assuming, of course, that a scholar had any such desire. Manuscript correspondence networks (see Chapter 25) were central to the communication of scholarship, and carried no taint of the commercial book trade. The same was true of the oral settings in which letters and specimens were discussed. It was far from obvious that scholars in the late seventeenth century would wish to edit or contribute to a learned journal, nor that booksellers or printers would wish to take on such a product (Johns 2000).

Nonetheless, a number of learned journals did appear in the late seventeenth century, of which the best-known are *Journal des sçavans* (Paris, 1665), *Philosophical Transactions* (London, 1665), *Giornale de literati d'Italia* (Florence, 1668), *Miscellanea curiosa medico-physica* (Schweinfurt, 1670), and *Acta medica et philosophica* (Copenhagen, 1673). Of these, only *Philosophical Transactions* focused specifically on natural knowledge: the German and Danish journals were heavily medical, while those of France and Italy covered all humanistic scholarship. Thirty-five journals focused on the sciences commenced between 1665 and 1699, but most did not last and have been little studied (Kronick 1976).

It is impossible to claim that the modern scientific journal sprang into existence in 1665. Firstly, the early issues of *Philosophical Transactions* were vastly different from the scientific journals of the twentieth century. Henry Oldenburg was not merely editor, but also author, translator or excerpter of almost all the content. Oldenburg saw it as a news-sheet for sharing the latest scholarly news across Europe, and included extracts from correspondence, book reviews, and the reports of other goings on in the republic of letters (Johns 1998). Secondly, the survival of *Philosophical Transactions* should not blind us to the fact that it was not the only model for a learned periodical. In England alone, Robert Hooke's *Philosophical Collections*, Joseph Moxon's *Mechanical Exercises*, John Houghton's *Collection of Letters for the Improvement of Husbandry and Trade*, the *Weekly Memorials for the Ingenious* and the *Athenian Mercury* offered alternative visions of both scholarship and the scholarly periodical (Johns 2000; Broman 2013). In later centuries, distinct genres of periodicals would emerge (such as review journals, abstract journals, society proceedings, substantive research journals, and popular science magazines), but in these early days, the conductors and publishers of periodicals routinely tried to combine several (or all) of these functions.

The Eighteenth Century

By 1790, at least a thousand scientific and technical journals had been established. Around a quarter of these were the transactions of the expanding number of Enlightenment learned academies and societies (see Chapter 16), but the majority were what we would term commercial journals, set up by printers, booksellers or editors with the hope of turning a profit from the learned and/or public culture of science. It is clear from their short life-spans that most failed in this regard; those supported by a learned society had a far better survival rate, with 46 percent lasting over a decade (Kronick 1976; Fyfe 2015). Nonetheless, the *Observations sur la physique, sur l'histoire naturelle et sur les Arts,* (Paris, 1773, known as Rozier's journal) showed that a commercial journal could be successful, with the right skills (McClellan III 1979).

Prior to the 1730s, there was little disciplinary specialization, though journals focusing on medical and agricultural topics were being established by mid-century. The general absence of specialization means that efforts to identify "scientific" periodicals have by definition ignored the appearance of natural knowledge in general periodicals and newspapers. For instance, the *Ladies' Diary* was notable for its mathematical puzzles (Costa 2002) and newspapers featured natural knowledge both as news and advertisements (Wigelsworth 2010). Along with spectacular demonstrations, coffee houses and salons, print was key to the public culture of science in the Enlightenment (Golinski 1992; Stewart 1992; Bensaude-Vincent and Blondel 2008). However, in contrast to the nineteenth century, the role of science in eighteenth-century print culture has yet to be extensively studied (Broman 1998; Broman 2000).

Over three-quarters of Kronick's list of scientific journals were issued in the German-speaking lands, apparently due to the political fragmentation and cultural rivalry of the various German lands. The total figure hides the reality that over half of these lasted less than three years, and there were rarely more than a handful in operation at any one time. Fewer journals were launched in Britain and France in this period, but they tended to last longer.

Some of the newly-established learned societies were content to have their activities reported in the local newspaper, but most were ambitious to issue serial publications as proof of their existence and worth. The American Philosophical Society began its *Transactions* in 1771, as did the Royal Society of Edinburgh in 1786. The Société Royale de Medicine (Paris) issued short-lived *Histoires et Mémoires* from 1776 and the Manchester Literary and Philosophical Society issued *Memoirs* from 1785. Such "transactions" or "memoirs" differed from the commercial journals in their emphasis on providing an authoritative record of the activities of the parent society, rather than communicating the latest news and observations from across the republic of letters. By the mid-eighteenth century, *Philosophical Transactions* had also narrowed its focus to the papers presented at Royal Society meetings (Moxham 2015). After almost 80 years as the private enterprise of successive clerks or secretaries of the Society, it had been formally taken over by the Society in 1752.

Other periodicals emerged which were dedicated specifically to reviewing, such as the *Critical Review* (London, 1726) and the *Gottingische Zeitung von gelehrten sachen* (1739). The first to deal solely with the sciences was probably the *Commentarii de rebus in scientia naturali* (Leipzig, 1752), and more would be created—especially for medicine—in subsequent decades. Such journals promised readers a way of keeping up to date with the constantly expanding number of publications.

The Nineteenth Century

The nineteenth century was an age of increased literacy and education in Europe and North America, which, coupled with significant technological improvements in the production of printed matter (see Fyfe 2012, and Chapter 24), meant that scientific knowledge, news and controversy now appeared in mass-market publications as well as in middlebrow magazines and scholarly research journals (Secord 2009; Topham 2009). The *Science in the Nineteenth-Century Periodical* project demonstrated that scientific content could be found in a wide variety of British periodicals (Cantor and Shuttleworth 2004; Cantor et al. 2004; Henson, Cantor, and Shuttleworth 2004). By the end of the century, magazines of popular science were published all over Europe, often linked with narratives of modernization and nationalism, as in Spain, Italy, and Greece (Papanelopoulou, Nieto-Galan, and Perdiguero 2009). Elsewhere in the world, efforts to introduce Western scientific concepts and practices often involved the publication of monthly magazines (as well as the establishment of reading rooms and science classes). In China, the *Shanghae Serial* (1857–58) and *Chinese Scientific and Industrial Magazine* (Shanghai, 1876–92) were produced by editors and translators associated with (British) Protestant missionary organizations (Elman 2005); but by the early twentieth century, it was American-based Republican Chinese students who advocated scientific and industrial development in *Ke Xue* (*Science*; Shanghai, 1915) (Wang 2002).

Yet, concomitantly with this broadening of the audience for science, scholarly research periodicals were becoming increasingly specialized, technical in their language, and limited in their intended readership. Recent scholarly attention to "popular science" has given us an understanding of science in general print culture that is far richer than that of the communication of research results (Meadows 1980a). It is also the case that we know more about Britain than about other countries (though see Kohlstedt 1980; Sheets-Pyenson 1985; Daum 1998; Fox 2012, Ch. 5–6).

In 1858, a team at the Royal Society began work on its *Catalogue of Scientific Papers*—which eventually claimed to include every scientific paper published between 1800 and 1900—by indexing 1400 periodicals (*Catalogue of Scientific Papers [first series]* 1867–72, viii). The task was so massive that although the author listing was completed up to 1900, only a part of the subject index was completed in the early twentieth century. Other ways of keeping up to date with the burgeoning scientific literature included a new wave of abstract journals, typically aimed at particular professional or disciplinary communities: *Pharmaceutisches Centralblatt* (1830), *Zoological Record* (1855), and *Physics Abstracts* (1895) (Manzer 1977). The effort involved in compiling these works, and their limited market, meant that many were ultimately taken over by learned societies. Yet learned societies were struggling to cope: continuing the *Catalogue of Scientific Papers* was beyond the resources of the Royal Society, so plans were made for an international, multi-society collaboration (which then lapsed during the First World War) (Csiszar 2010).

Society memoirs and transactions continued to be issued throughout the nineteenth century, and their titles show both increasing disciplinary specialization and international range. New learned societies routinely saw it as their duty to issue a journal, hence *Transactions of the Geological Society* (London, 1811), the *Transactions of the Seismological Society of Japan* (Yokohama, 1880), the *Anales de la Sociedad Científica Argentina* (Buenos Aires, 1876), and the *Actas de la Sociedad Científica de Chile* (Santiago, 1892).

Commercial journals also proliferated, accounting for some 60 percent of British titles. Rozier's journal had inspired a spate of imitators in the Napoleonic war years, usually issued monthly and in a smaller, cheaper format than the society transactions; the best-known include *Annales de Chimie* (Paris, 1789), *Philosophical Magazine* (London, 1798), *Annalen der Physik* (Leipzig, 1799), and the *American Journal of Science and Arts* (New Haven, 1818). By the end of the century, ambitious publishers were trying the weekly format; *Nature* (London, 1869) and *Science* (New York, 1880) have proven successful in the long run but both (especially the latter) struggled to find commercial success (Baldwin 2015; Vandome 2013).

While the transactions of the learned societies represented the reputable, authoritative and often ponderous faces of their scholarly fields, the commercial journals filled a different niche, publishing a variety of reviews, news, letters and meteorological reports. They were not necessarily the place where new research was first announced; but they aimed to offer observations that would be new to their readers, often because they had initially been announced in a distant city. Contributions could be submitted directly by scholars (e.g. letters to the editor), but much was excerpted, translated or reported by the editor or his assistants. Successful journals depended upon an editor with the right combination of commercial know-how, editorial skills and scholarly connections, as the experiences of William Nicholson and David Brewster illustrate (Brock 1984; Watts 2014). Given the marginal profitability of most of these commercial journals, it is possible that some publishers regarded them as loss-leaders: good for the firm's reputation, and a way of bringing talented authors to the firm.

The commercial journals demonstrated the potential appeal of a faster publication schedule, and some learned societies responded. The Royal Astronomical Society started issuing *Monthly Notices* in 1827, and various societies launched *Proceedings* (in addition to *Transactions*), including the Royal Society (1832). The French academy went furthest, launching *Comptes Rendus Hebdomadaires des Séances de l'Académie des Sciences* under the control of its secretary, François Arago, in 1835. Its rapid publication of news and innovation undercut commercial competitors.

Britain, France and Germany dominated scientific journal publishing in the nineteenth century, largely due to their lead in scientific research. In contrast to the late seventeenth and late twentieth centuries, publication was largely a national, not international, practice. Of the British commercial firms involved, Taylor & Francis was particularly noteworthy for acquiring a list of journals, including *Philosophical Magazine* and *Annals of Natural History* (Sheets-Pyenson 1981; Brock and Meadows 1998). The origins of the later American dominance of the scientific journal market might be traced to Silliman's *American Journal of Science*, *Scientific American* (1845) and *Science* (1880). There was science publishing elsewhere in the world, both of research and popular articles, most visibly in the British dominions and the newly independent nations of Latin America.

The Twentieth Century

The broadcast communications technologies of the twentieth century—radio, television, cinema—had significant consequences for the provision of scientific information for education, the reporting of scientific news, and the inclusion of scientific themes in entertainment (see chapter 30). The professionalization of science journalism can be seen in the creation of the "Science Service" news syndication service (1924), the National Association of Science Writers (New York, 1934); and the British

Association of Science Writers (1947). These journalists developed links with the scientific research community and sought respect through their accurate reporting. Nonetheless, they were few in number, and even at the end of the century, few newspapers had a dedicated science correspondent (Nelkin 1995; Bowler 2009). Although print-magazines such as *Scientific American* (1845), *National Geographic* (1888) and *New Scientist* (1956) continued to attract readers, the mass audiences watched such landmark television series as David Attenborough's *Life on Earth* (BBC, 1979) or Carl Sagan's *Cosmos* (PBS, 1980).

The new communications technologies had less effect on the communication of original scientific research, which was usually targeted at specialists, not broadcast. Researchers continued to communicate through personal correspondence, oral presentation at meetings and conferences, and publication in printed journals. There continued to be a mixed economy of journals sponsored by learned societies and those owned by commercial firms; but the two became less distinct in form and content, as society journals, especially in the USA, moved away from being entirely "transactions" of the sponsoring society; and as journals increasingly converged on standards of format, style, and structure (Atkinson 1998). This process was assisted by the creation of style guides, and of associations of journal editors to share best editorial practice. The Royal Society issued guidelines for its authors in 1936, while the (American) Conference of Biology Editors (1957, subsequently the Council of Science Editors) has issued style manuals and guides for editors since the 1960s. Nonetheless, *Nature* and *Science* retained an emphasis on news, along with their weekly periodicity, and continued to welcome short submissions and "letters", as well as commissioning reviews and reports. In contrast to earlier periods, in the twentieth century, even short announcements of research results were typically authored by their discoverers and contributed significantly to that researcher's esteem (Baldwin 2015).

In 1934, the *World List of Scientific Periodicals* listed over 36,000 titles, in contrast to the 1,400 indexed by the Royal Society in the early 1860s. English was the language of almost 14,000 of these titles, with another 11,000 titles being issued in either German or French (Sherrington 1934). Although these three languages dominated scientific research publications in the early twentieth century, the existence of journals in a further fifteen languages indicates the increasingly global extent of scientific publishing. The creation of new national academies around the world led to the launch of more journals, including *Proceedings of the Imperial Academy* (Tokyo, 1912) and *Science China* (Peking, 1950). Increasing disciplinary specialization had the same effect: by the early twentieth century, there were separate Turkish publications dealing with pharmacy, mathematics and engineering (Gunergun 2003; Gunergun 2007). The apparent flood of research publications meant that the voluntary efforts of learned societies to create bibliographical tools for scientists struggled more and more. From the 1960s, companies using mechanical and, later, electronic, systems learned how to make the provision of indexes, abstracts, and citation metrics profitable. The pioneer was Eugene Garfield's Institute of Scientific Information (Philadelphia, 1960), which published the first Science Citation Index in 1963 (Cronin and Atkins 2000).

In the postwar period, the USA was home to the largest number of scientific research journals, ahead of Britain, West Germany and the USSR. The authors and readers of US journals were predominantly American, but the post-war years saw the development of international English-language journals based in the Netherlands and in Britain. These journals, including Elsevier's *Biochimica et Biophysica Acta* (1947) and *Clinica Chimica Acta* (1956), were often attempts to carve out niches in new subfields which had yet to be fully recognized by the learned societies. Elsevier recruited international editorial boards, and relied heavily upon contributors and readers outside the Netherlands. A 1977 survey of chemistry journals found that 93 percent of authors in Dutch journals were foreign-domiciled; compared to 66 percent for British journals, 38 percent for US journals and just 3 percent for West German journals. Sales income for Dutch and British journals was similarly dependent on the international market. Around two-thirds of British and Dutch science journals in the 1970s were commercial operations, whereas only around a half of American journals were (Meadows 1980b; van Leeuwen 1980).

Circulations rose in the postwar decades, and, combined with rising prices and an increasing reliance on institutional subscribers to pay those prices, scientific journal publishing was finally profitable for both learned societies and commercial firms. The new *Biochimica et Biophysica Acta* returned a profit within eight years, and it seems to be in the late 1940s that the Royal Society's publishing finally started to show annual surpluses.

Scientific Authorship

In the absence of well-paid professional opportunities, authorship has long been a potentially valuable income stream for scholars (Fyfe 2005). The development of authorship as a profession in the late eighteenth and nineteenth centuries was dependent on periodicals, because editors and proprietors were willing to pay for a constant stream of fresh copy. Many nineteenth-century men of science wrote for magazines and reviews, for money as well as public fame (consider David Brewster and T. H. Huxley). Although it has often been assumed that the professionalization of science in the second half of the nineteenth century focused scientists' authorial activity upon research publications, many were still writing textbooks, popular books, and magazine articles well into the 1930s (Bowler 2009). It can only be postwar, if at all, that scientists have dedicated such a high proportion of their authorial energies to the writing of research articles.

Until the early nineteenth century, it was perfectly possible to acquire a high reputation as a natural philosopher or natural historian without publishing. If a scholar did embark on authorship, it was more likely to be a treatise than an article, and those who did write a paper were usually doing so for presentation to a learned society meeting, with publication being a separate and later possibility. Discoveries, and, consequently, attributions of priority, did certainly appear in print, but an editor or journalist often authored such reports.

In 1830, Charles Babbage's *Reflections on the Decline of Science in England* included a critique of the Royal Society and its membership policy. Babbage argued that the fact that only 109 of the 714 Fellows had ever published in *Philosophical Transactions* meant that most Fellows were not worthy of the honor of election (Babbage 1830). Babbage's equation of reputation with the authorship of journal articles was novel in 1830, but had become normal by the end of the nineteenth century (see Chapter 12). The procedures for election to learned societies now asked about publications, as did university appointments. In the twentieth century, co-authorship became more common, reflecting the development of large research teams, especially postwar.

The evolution of the style, structure and language of the scientific article has been well-studied (Bazerman 1988; Atkinson 1998; Gross, Harmon, and Reidy 2002). The highly personal narratives typical of the seventeenth century became increasingly impersonal, until the author's voice had disappeared from scholarly research prose by the mid-twentieth century. During the nineteenth century, scientific prose became more efficient, with shorter sentences and fewer clauses; and in the twentieth century, the use of noun–adjective complexes and their abbreviations (jargon) enabled even greater concision at the expense of comprehensibility to non-specialists. The now-common format of "Introduction, Methods, Results, Discussion" was emerging in the 1920s, and Atkinson reported it as utterly standard by 1975.

Editors and Reviewing Processes

Editors who owned their journals, like Oldenburg, had extensive powers to shape them, determining the structure and format, selecting (or writing) the copy, and excerpting, paraphrasing and translating as they pleased. When, as was increasingly common by the nineteenth century, an editor was paid to manage a journal owned by a bookseller or publisher, they were given significant control over the contents (but much less over the commercial aspects of the journal). Although the terms were not yet used, it is possible to see the emerging distinction between an "academic" editor and a "managing" or "publishing" editor by the mid-nineteenth century, though the balance of power between them could vary significantly.

Journals with a commitment to the rapid publication of news depended on highly efficient and decisive editorial processes, which a sole-editor could do better than a committee. Nonetheless, many editors recognized that they needed assistance to evaluate contributions in certain areas. As early as 1701, the *Journal des sçavans* had created an editorial team as a means to reconcile the broad scope of the journal with the limits of individual knowledge (Kronick 1978). Both *Philosophical Magazine* and *Annals of Natural History* had teams of co-editors by the 1850s, each editor taking responsibility in his own field. Lockyer achieved the same effect for *Nature* by seeking opinions informally from men of science in his personal network. Such methods of evaluation appear to have lasted well into the twentieth century at commercial journals.

At the learned societies in the eighteenth and nineteenth centuries, editorial decisions were usually made by a committee, known variously as the "Comité de Librairie" (from 1700) at the Académie Royale des Sciences, and as the "Committee of Papers" (from 1752) at the Royal Society. Such committees enabled the societies' journals to claim the collective authority of the society, and also provided breadth of scholarly expertise. In contrast to teams of independently acting co-editors, these editorial committees took collective decisions. The evaluation activities of these early editorial committees, coupled with the requirement for papers to be read at a meeting of the society before being considered for publication, are clearly precursors to modern reviewing processes; but to label them as "peer review" (Zuckerman and Merton 1971) would be to ignore significant differences.

For instance, the Royal Society's Committee of Papers had the power to seek additional expertise, but this was done by inviting another Fellow to join the meeting, where he could then vote on all papers under consideration. In contrast, the Comité de Librairie in the 1760s and 1770s did refer papers to reviewers; but these reports seem to have been written jointly, and extracts from them were sometimes published with the paper (McClellan III 2003). It was in 1832 that the Royal Society began to seek written reports from (usually) two separate referees, which were used privately to inform the decision of the Committee of Papers. While this looks more like peer review, it should be noted that even in the late nineteenth century, many papers (including most for the *Proceedings*) were not refereed; authors did not usually see these reports (Baldwin 2014b); and the papers were certainly not "blinded" before being sent to referees (Despaux 2011).

While the learned societies were developing processes for collective responsibility, fairness and expert judgment, editors of commercial journals seem to have felt no need to follow suit. The widespread application of peer review to learned journals of all types—and the characterization of it as a guarantee of quality control in the production of knowledge—is of much more recent origin, probably postwar. The influence went both ways, however, because during the mid-twentieth century, learned societies replaced their editorial committees with editors, presumably seeking a more streamlined editorial process. The creation of "associate editors" and "editorial boards" or "advisory boards" retained the impression of a scholarly community behind the journal, and provided a captive pool of referees, but were distinctly different from the collective decision-making committees of the eighteenth century.

Distribution and Readership

We know far too little about the distribution, circulation and readership of scientific journals, other than what arises in studies focused on specific individuals or organizations (e.g. Brockliss 2002). An alternative approach to readership is to use citation analysis, though this naturally focuses only on that subset of readers who themselves are scholarly authors (Allen, Jian, and Lancaster 1994; Qin 1994).

Scientific journals, and especially those issued by learned societies, have multiple modes of distribution. Firstly, like other print products, they travelled through national and international bookselling channels. Thus, in the seventeenth and eighteenth centuries, having a good agent in Leipzig was important to international success. Information on print runs is scarce, though figures of 500, 750 or 1000 are most-often mentioned for pre-1900.

Secondly, journal issues and, especially, separate copies of individual articles (preprints, offprints, reprints) were part of international correspondence networks, enclosed in a letter to a friend or colleague. A hundred copies was the standard number of offprints at the Royal Society for most of the nineteenth century, as it still was for at least some journals in the mid-twentieth century. If so many separate copies were circulating via scientific correspondence networks, it does raise the question of who the actual journal issues were aimed at. Further study is much-needed.

Thirdly, learned societies usually offered a significant fraction of the print run of their journals to their membership (it was two-fifths at the Royal Society in the mid-eighteenth century, rising to two-thirds in the early nineteenth century). At some societies, the cost of this was built into the membership fee; at others, the members bought their copies, perhaps at a reduced rate. Substantial numbers of learned society journals never went through the regular book trade, and there was a secure (captive) market (Fyfe 2015). This model is still used by many learned society journals today.

Fourthly, learned societies formed exchange networks with each other, on a global scale, as is illustrated by their annual reports. For instance, in 1877, the secretary of the Academy of Natural Sciences of Philadelphia reported that the latest volume of its *Proceedings* had been received at the Royal University Library, Strasburg; the Royal Bavarian Academy, Munich; the Literary and Philosophical Society of Liverpool; the Royal University of Würzburg; the Royal Society of Sciences of Upsalla; the Imperial Botanical Garden, St. Petersburg; the Royal Society of New South Wales and the South African Museum, Cape Town, among many others. The list of societies which sent publications to Philadelphia was equally global, and varied (*Proceedings of the Academy of Natural Sciences of Philadelphia* 1877, 330–35). These exchanges ensured that journals were available for reference in learned societies all over the world. However, it was not necessarily an effective means of building up a library, as the American Philosophical Society had discovered in the late eighteenth century. The Royal Society only exchanged one-for-one, so the APS's less frequent publishing schedule meant that its holdings of *Philosophical Transactions* were utterly incomplete (Sioussat 1949).

Conclusions

Our knowledge of the variety of ways in which the sciences have appeared in periodical form—whether research journal or monthly magazine—has improved dramatically over the last few decades. Much of this work has, however, been based upon the published record, so that we know far more about the contents of periodicals than about the authorial, editorial, and commercial processes that put them there. We also know more about scholarly authors and readers, and about certain intellectual topics, than about others. This is in large part due to the sources available to us; yet there must be more creative ways to address some of our unanswered questions, perhaps utilizing digital humanities. We have a reasonable grasp of how the meaning and rewards for scientific authorship have changed over time; but it would be interesting to see more studies taking full cognizance of the wide range of authorial activities that scholars have undertaken, including their journalistic work, their letter-writing, their popular books, and their writings on completely other topics. The bigger challenge will be to learn about the involvement of people other than scholars, be they ordinary readers and library users, or the editors, journalists, and booksellers who ran the commercial side of journal and periodical publishing, but who left few records behind. With regard to the development of editorial and commercial practices of scientific journals (and indeed, of scholarly journals at large), we may know the start- and end-points of the narrative, and the names of some of the key actors, but the journey, and the paths not taken, remain surprisingly obscure given the epistemological importance of communication to the making of scientific knowledge. Moreover, the current debates over the future of scientific journals, their pricing models, and their editorial review processes, mean that these are highly topical issues that we as historians ought to be able to contribute to.

Endnote

1 Discussion of the *Philosophical Transactions* in this chapter draws upon the as-yet unpublished work of the 'Publishing the Philosophical Transactions' project (AHRC grant AH/K001841/1), directed by Aileen Fyfe, with the assistance of Julie McDougall-Waters and Noah Moxham, and based at the University of St Andrews and the Royal Society.

References

Allen, Bryce, Qin Jian, and F. W. Lancaster. 1994. "Persuasive communities: A longitudinal analysis of references in the *Philosophical Transactions of the Royal Society, 1665–1990.*" *Social Studies of Science*, 24, No. 2: 279–310.

Atkinson, Dwight. 1998. *Scientific Discourse in Sociohistorical Context: The* Philosophical Transactions of the Royal Society of London, *1675–1975*. London: Routledge.

Babbage, Charles. 1830. *Reflections on the Decline of Science in England*. London: Fellowes.

Baldwin, Melinda. 2014. "Tyndall and Stokes: Correspondence, referee reports and the physical sciences in Victorian Britain." In *The Age of Scientific Naturalism: Tyndall and his Contemporaries*, edited by Bernard Lightman and Michael S. Reidy, 171–86. London: Pickering & Chatto.

Baldwin, Melinda. 2015. *Making "Nature": The History of a Scientific Journal*. Chicago: University of Chicago Press.

Bazerman, Charles. 1988. *Shaping Written Knowledge: The Genre and Activity of the Experimental Article in Science*. Madison: University of Wisconsin Press.

Bensaude-Vincent, Bernadette and Christine Blondel. 2008. *Science and Spectacle in the European Enlightenment*. Aldershot: Ashgate.

Bowler, Peter J. 2009. *Science for All: The Popularization of Science in Early Twentieth-century Britain*. Chicago: University of Chicago Press.

Brock, William H. 1984. "Brewster as scientific journalist." In *'Martyr of Science': Sir David Brewster, 1781–1863; Proceedings of a Bicentennial Symposium*, edited by Alison Morison-Low and John R. R. Christie, 37–44. Edinburgh: Royal Scottish Museum.

Brock, William H. and A. J. Meadows. 1998. *The Lamp of Learning: Taylor & Francis and the Development of Science Publishing*. London: Taylor & Francis.

Brockliss, Laurence. 2002. *Calvet's Web: Enlightenment and the Republic of Letters in Eighteenth-Century France*. Oxford: Oxford University Press.

Broman, Thomas. 1998. "The Habermasian public sphere and 'Science *in* the Enlightenment'." *History of Science*, 36: 123–49.

Broman, Thomas. 2000. "Periodical literature." In *Books and the Sciences in History*, edited by Marina Frasca-Spada and Nicholas Jardine, 225–38. Cambridge: Cambridge University Press.

Broman, Thomas. 2013. "Criticism and the circulation of news: The scholarly press in the late seventeenth century." *History of Science*, 51: 125–50.

Cantor, Geoffrey, Gowan Dawson, Graeme Gooday, Richard Noakes, Sally Shuttleworth, and Jonathan R. Topham. 2004. *Science in the Nineteenth-Century Periodical*. Cambridge: Cambridge University Press.

Cantor, Geoffrey and Sally Shuttleworth. 2004. *Science Serialized: Representations of the Sciences in Nineteenth-century Periodicals*. Cambridge, MA: MIT Press.

Catalogue of Scientific Papers [first series]. 1867–72. Edited by Henry White. 6 vols. London: Royal Society.

Costa, Shelley. 2002. "The 'Ladies' Diary': Gender, mathematics, and civil society in early-eighteenth-century England." *Osiris*, 17: 49–73.

Cronin, Blaise, and Helen Barsky Atkins. 2000. *The Web of Knowledge: A Festschrift in honor of Eugene Garfield*, ASIS&T Monograph Series. Medford, NJ: Information Today.

Csiszar, Alex. 2010. "Seriality and the search for order: Scientific print and its problems during the late nineteenth century." *History of Science*, 48: 399–434.

Daum, Andreas W. 1998. *Wissenschaftspopularisierung im 19. Jahrhundert: Bürgerliche Kultur, naturwissenschaftliche Bildung und die deutsche Öffentlichkeit, 1848–1914 [Popularizing Science in the Nineteenth Century: Civil Culture, Scientific Education, and the Public Sphere in Germany, 1848–1914]*. Munich: Oldenbourg Wissenschaftsverlag.

Despaux, Sloan Evans. 2011. "Fit to print? Referee reports on mathematics for the nineteenth-century journals of the Royal Society of London." *Notes & Records of the Royal Society*, 65: 233–52.

Elman, Benjamin. *On Their Own Terms: Science in China, 1500–1900*. Cambridge, MA: Harvard University Press, 2005.

Fox, Robert. 2012. *The Savant and the State: Science and Cultural Politics in Nineteenth-Century France*. Baltimore: John Hopkins University Press.

Fyfe, Aileen. 2005. "Conscientious workmen or booksellers' hacks? The professional identities of science writers in the mid-nineteenth century." *Isis*, 96: 192–223.

Fyfe, Aileen. 2012. *Steam-Powered Knowledge: William Chambers and the business of publishing, 1820–1860*. Chicago: University of Chicago Press.

Fyfe, Aileen. 2015. "Journals, learned societies and money: *Philosophical Transactions*, ca. 1750–1900." *Notes & Records*, 69: 277–299.

Golinski, Jan. 1992. *Science as Public Culture: Chemistry and Enlightenment in Britain, 1760–1820*. Cambridge: Cambridge University Press.

Gross, Alan G., Joseph E. Harmon, and Michael S. Reidy. 2002. *Communicating Science: The Scientific Article from the Seventeenth Century to the Present*. New York: Oxford University Press.

Gunergun, Feza. 2003. "Le premier journal pharmaceutique publié en langue Turque: *Eczacı* (1911–1914)." Paper read at 36th International Congress of History of Pharmacy, September 24–27, 2003, at Sinai/Romania.

Gunergun, Feza. 2007. "Matematiksel bilimlerde ilk Türkçe dergi: *Mebahis-i İlmiye*, 1867–69 ("An early Turkish journal on mathematical sciences: *Mebahis-i İlmiye*, 1867–69")." *Osmanlı Bilimi Araştırmaları / Studies in Ottoman Science*, 8, No. 2: 1–42.

Henson, Louise, Geoffrey Cantor, and Sally Shuttleworth (eds.) 2004. *Culture and Science in the Nineteenth-Century Media*. Aldershot: Ashgate.

Johns, Adrian. 1998. *The Nature of the Book: Print and Knowledge in the Making*. Chicago: University of Chicago Press.

Johns, Adrian. 2000. "Miscellaneous methods: Authors, societies and journals in early modern England." *British Journal for the History of Science*, 33, No. 2: 159–86.

Kohlstedt, Sally Gregory. 1980. "Science: The struggle for survival, 1880 to 1994." *Science*, 209, No. 4452: 33–42.

Kronick, David A. 1976. *A History of Scientific & Technical Periodicals: The Origins and Development of the Scientific and Technical Press, 1665–1790*. Metuchen, NJ: Scarecrow Press.

Kronick, David A. 1978. "Authorship and authority in the scientific periodicals of the seventeenth and eighteenth centuries." *Library Quarterly*, 48, No. 3: 225–75.

Manzer, Bruce M. 1977. *The Abstract Journal 1792–1920: Origin, Development and Diffusion*. Metuchen, NJ: Scarecrow Press.

McClellan III, James. 1979. "The scientific press in transition: Rozier's Journal and the scientific societies in the 1770s." *Annals of Science*, 36, No. 5: 425–49.

McClellan III, James. 2003. "Specialist control: The Publications Committee of the Académie Royale des Sciences (Paris), 1700–1793." *Transactions of the American Philosophical Society*, 93, No. 3: 1–134.

Meadows, Arthur Jack (ed.) 1980a. *The Development of Science Publishing in Europe*. Amsterdam: Elsevier.

Meadows, Arthur Jack. 1980b. "European science publishing and the United States." In *The Development of Science Publishing in Europe*, edited by A. J. Meadows, 237–50. Amsterdam: Elsevier.

Moxham, Noah. 2015. "Fit for print: developing an institutional model for scientific periodical publishing in England, 1665-ca. 1714." *Notes & Records*, 69: 241–260.

Nelkin, Dorothy. 1995. *Selling Science: How the Press Covers Science and Technology*. Revised edition. New York: W. H. Freeman.

Papanelopoulou, F., A. Nieto-Galan, and E. Perdiguero. 2009. *Popularizing Science and Technology in the European Periphery, 1800–2000*. Aldershot: Ashgate.

Pettegree, Andrew. 2014. *The Invention of News: How the World Came to Know About Itself*. New Haven, CT: Yale University Press.

Proceedings of the Academy of Natural Sciences of Philadelphia. 1877. Philadelphia: Academy of Natural Sciences.

Qin, Jian. 1994. "An investigation of research collaborations in the sciences through the *Philosophical Transactions*, 1901–1991." *Scientometrics*, 29, No. 2: 219–38.

Secord, James A. 2009. "Publishing science, technology and mathematics." In *The History of the Book in Britain, vol. 6: 1830–1914*, edited by David McKitterick, 443–74. Cambridge: Cambridge University Press.

Sheets-Pyenson, Susan. 1981. "From the North to Red Lion Court: The creation and early years of the *Annals of Natural History*." *Archives of Natural History*, 10: 221–49.

Sheets-Pyenson, Susan. 1985. "Popular science periodicals in Paris and London: The emergence of a low scientific culture, 1820–75." *Annals of Science*, 42: 549–72.

Sherrington, Charles. 1934. "Language distribution of scientific periodicals." *Nature*, 134 (20 October): 625.

Sioussat, George L. 1949. "The 'Philosophical Transactions' of the Royal Society in the libraries of William Byrd of Westover, Benjamin Franklin, and the American Philosophical Society." *Proceedings of the American Philosophical Society*, 93, No. 2: 99–113.

Stewart, Larry. 1992. *The Rise of Public Science: Rhetoric, Technology and Natural Philosophy in Newtonian Britain, 1660–1750*. Cambridge: Cambridge University Press.

Topham, Jonathan R. 2009. "Scientific books, 1800–1830." In *The Cambridge History of the Book in Britain, Vol. 5, 1695–1830*, edited by Michael Turner and Michael Suarez, 827–33. Cambridge: Cambridge University Press.

Vandome, R. 2013. "The Advancement of Science: James McKeen Cattell and the networks of prestige and authority, 1894–1915." *American Periodicals*, 23, No. 2: 172–87. doi: 10.1353/amp.2013.0011

van Leeuwen, J.K.W. 1980. "The decisive years for international science publishing in the Netherlands after the Second World War." In *The Development of Science Publishing in Europe*, edited by A. J. Meadows, 251–67. Amsterdam: Elsevier.

Wang, Zuoyue. 2002. "Saving China through science: The Science Society of China, scientific nationalism, and civil society in Republican China." *Osiris*, 17: 291–322.

Watts, I. 2014. "'We Want No Authors': William Nicholson and the contested role of the scientific journal in Britain, 1797–1813." *British Journal for the History of Science*, 47, No. 3: 397–419. doi: 10.1017/S0007087413000964

Wigelsworth, Jeffrey R. 2010. *Selling Science in the Age of Newton: Advertising and the Commoditization of Knowledge*. Aldershot: Ashgate.

Zuckerman, Harriet, and Robert K Merton. 1971. "Patterns of evaluation in science: Institutionalisation, structure and functions of the referee system." *Minerva*, 9, No. 1: 66–100.

CHAPTER TWENTY-EIGHT

Textbooks

JOSEP SIMON[1]

Were a visitor from a distant planet on a mission to understand the human art of science to land by chance in the *Zócalo* of Mexico City, she would be astonished in detecting, with her extraordinary cognition, the high concentration of textbooks by a Salvador Mosqueira R. in the homes, schools, libraries, and bookshops of that earthly city.

A brief inspection reveals that Salvador Mosqueira R. was an engineer who developed a prolific career as a science teacher and textbook author in the school network of the National Autonomous University of Mexico. His most successful textbooks, *Física General* (*General Physics*) (1944, 21 editions by 1976) and *Física Elemental* (*Elementary Physics*) (1947, 32 editions by 1980), were published by Patria, a company that still specializes in textbooks today. They were recommended by the Mexican ministry of education and circulated widely in Mexico.

Having prepared for her trip by reading the major human output in the genre of companions, encyclopedias, handbooks, articles, treatises and monographs, research reports, laboratory notebooks, conference proceedings, and grant proposals, our visitor's credulity in her authorities would be shaken in contrasting the extensive circulation of Mosqueira's writings with the absence of any mention to it in the aforementioned reference sources.

Mosqueira's *Física General* included a preface by the first president of the Mexican Physics Society and several appendixes by leading physicists working in Mexico on advanced research topics. He also translated several American textbooks that circulated extensively in Latin America, such as Resnick and Halliday's *Physics for Science and Engineering Students*.

The bewilderment of our visitor would not be lessened in realizing that the status of textbooks in the academic circles formed by humans has often been low, helped by the belief that research is clearly distinct to teaching and definitely more socially prestigious. Mosqueira's case might seem exceptional. In fact it is not. But it is illustrative of the hundreds (if not thousands) of case studies on science textbooks which still await their historian. Why should Mosqueira's textbooks be relevant? Should they

A Companion to the History of Science, First Edition. Edited by Bernard Lightman.
© 2016 John Wiley & Sons Ltd. Published 2020 by John Wiley & Sons Ltd.

concern historians of science or just historians of education or book historians? Are they worth researching for historians worldwide or just for historians of Mexico or for Latin American historians?

Because of his professional profile (teacher not researcher) and his nationality (Latin American not European or American), Mosqueira is unlikely to have a place in the history of science canonical narrative as it is today. Yet, his pedagogical production was likely to have a major role in the making of physics in a rapidly growing nation state. Furthermore, it displays a historiographical potential to deal with the problems associated with the production, circulation, and appropriation of scientific knowledge, locally, nationally, and internationally. This pattern can be extrapolated to other cases of textbook production and use in science.

Expectations

An advantage of textbooks and education as a focus for research in history of science is the truly international character of the problems they raise and their richness as sources. The study of scientific research is still commonly presented—explicitly or implicitly—as a matter of leading centers with national qualities radiating towards passive peripheries (Simon and Herran 2008). It seems more implausible to think that each country has not had its own science education and textbook cultures, regardless of its performance in international scientific research and its visibility in current history of science. While the history of science education displays a number of national contexts which were able to internationalize their textbook production better than others, this does not diminish the relevance of certain local or national textbook cultures over others (Simon 2012).

While nineteenth-century France, Germany, and Britain, and twentieth-century US feature prominently in the history of science, here I argue that the national hierarchies, periodizations, and criteria of relevance commonly applied by historians of science to scientific research do not necessarily match the study of science education and textbooks. Textbooks represent an opportunity to reconsider these national biases which should neither apply so forcefully to science education, nor to scientific research. This would require, though, raising the status of textbooks in the history of science, and fairly suggests the need for promoting further communication between historians of science, education, and book culture, respectively, and science education scholars (Simon 2015).

Historians of science have shown a longstanding interest in textbooks, but it has not resulted in more and better studies displaying significant historiographical and methodological contributions. Textbooks are commonly used as a resourceful tool for their collective multiplicity, but they are more rarely considered in their individuality: they are a vast repertoire of sources providing a sense of what knowledge was standard at a certain time, but not historical objects deserving to be considered on their own and able not only to reflect knowledge, but also to transform it. Textbook science has often been subordinated to research science, as historians have dealt preferably with those textbooks produced by prominent researchers, in order to complete the picture, but seldom to reshape it.

The status of textbooks in the history of science is for obvious reasons in contrast with the traditional emphasis on the scientific article and treatise, and the incorporation in the last decades of new sources such as laboratory notebooks. But, more paradoxically, the status of sources such as popular books and periodicals, which previously enjoyed a similar low standing but recently have become the center of some of the more dynamic developments in the history of science (Topham 2000). This should be an example to follow, furthered by the connections between *popularization* and *education*, but taking into account their major differences, too (Rudolph 2008; Simon 2009).

The study of science education and textbooks presents also an opportunity to break the traditional disciplinary divisions of training in science, and their projection into history of science. The transversal nature of pedagogy should be able to highlight historiographical problems relevant to all the scientific disciplines and their historical narratives (Kaiser 2005b). An example of such potential is illustrated by the generalized impact of Thomas Kuhn's *The Structure of Scientific Revolutions* across the history of science. Kuhn offered a wide range of case studies across the sciences to illustrate his epistemological proposal (Kuhn 1962). Over the years, his premises on textbooks and science education have efficiently permeated major works in history of science (Brush 1976; Warwick and Kaiser 2005).

Past Tenses

Kuhn's work is obviously a product of its time. It reflects particular developments in physics, education, history and philosophy of science, and politics, which need to be assessed in order to understand its virtues and limitations. It presents ideas on education and textbooks which are not particularly revolutionary, since they were and still are quite common, especially in academic circles. In 1951, in his first Lowell lecture, entitled "Textbook Science and Creative Science," Kuhn already contended that the "structure of knowledge in the textbook" masked "the nature of the creative process" by which knowledge is gained (Simon 2013a). Remarkably, this is not a special characteristic of textbook science, but arguably of most scientific writing (Holmes 1987; Myers 1990).

For Kuhn, a special characteristic of science from the early nineteenth century onwards was that education was conducted through textbooks to an extent unknown in other fields of knowledge. Textbooks presented a surprising uniformity in conceptual structure and only differed in subject matter or pedagogical detail according to their level. Textbook science was the driving agent in the transmission of scientific knowledge through education, and it involved indoctrination. Although this level of systematization was not present before the nineteenth century, works that are customarily characterized as "classics," such as the treatises by Aristotle, Ptolemy, Newton, Franklin, Lavoisier, or Lyell, could play a similar role to textbooks in representing "universally received paradigms" (Kuhn 1963; Kuhn 1962).

More than a decade earlier, George Sarton had issued a programmatic call for a history of science based on "The Study of Early Scientific Textbooks," by which he meant *treatises* published before the nineteenth century that constituted the canonical scientific literature in particular periods (Sarton 1948). His aim was to trace "scientific evolution" by examining the changes in textbook content over time (successive editions) and space (translations). He drew a clear distinction between early textbooks and those produced from the second half of the nineteenth century onwards—too abundant and incorporating new knowledge too quickly to be relevant for his aims.

Sarton emphasized that "a history of science dealing only with the leaders and scouts gives one a wrong view of the whole procession." But he focused on "classics" such as Lavoisier, Huygens, Newton, Franklin, and Euler. He contended that the success of the most popular textbooks "might be due in part to their mediocrity or to their syncretic and accommodating tendencies," instead of antagonizing their readers as the "more original and independent" authors would do (Sarton 1948, 138, 140).

A decade earlier, Gaston Bachelard had considered that contemporary physics textbooks "offer our children a very socialized and immobilized kind of science which [...] comes to be regarded as *natural*. But is not at all natural." In contrast, in the eighteenth century "science books started from nature" because "pre-scientific thought lives in the world. It is not *regular*, [...] it does not live under orders, like the scientific thought trained in official laboratories and codified in schoolbooks [post-eighteenth-century]" (Bachelard 2002, 34, 38).

Bachelard's, Sarton's, and Kuhn's views were similar in their rather basic distinction between *treatises* and *textbooks*, and their periodization of textbook writing around a nineteenth-century turning point. Bachelard and Kuhn endowed contemporary textbooks with a major role in the making of science. Their conception of textbooks as repositories of *normal science*, characterized by dogmatism and adulteration of research science, established a hierarchy analogous to that outlined by Sarton.

Two of Sarton's ideas are particularly striking today. First, there is his view that modern textbooks are too numerous and too quick in their updating of scientific information to be useful for historians of science. By contrast, in the last decades, historians of science have used large sets of nineteenth and twentieth-century textbooks as faithful indexes of change over long time periods. Furthermore, it is common to assume *à la* Kuhn that there is an important delay in incorporating new science in textbooks: *normal science* resists change, and teaching and textbooks are subsidiary to research and research articles. Second, Sarton's assertion that the most popular textbooks are not necessarily the most representative—also shared by Bachelard—is a rough statement in terms of historical sensitivity and current research on science popularization and reading. Nonetheless, it highlights an extant tension in our discipline concerning the criteria which make a source relevant. Is it its intrinsic qualities, or its historically and historiographically contextual values? Is it the status attained by the author and its work among an academic elite which defines the canon, or is it its quantitative number of readers and their qualitative readings?

Demarcation

The lexicographic definitions of *textbook* commonly include all these nuances. Between the mid-eighteenth and mid-nineteenth centuries, the term emerged in different languages to mean a book originally conceived specially for instructional purposes within formal education. The word had its roots in the previous centuries in which a *text*, whether extracted from the Scriptures or written by an author considered classic, was used by students, leaving spaces between lines to insert comments dictated by the teacher.

The practice of note taking in classrooms has a long and rich history which is still underexplored. These practices often led to the shaping of standard texts for teaching and the production of printed textbooks. Thus, they can illuminate the processes by which knowledge was standardized into a textbook. But they also offer an alternative to understand what actually happened in the classroom by unveiling evidence which is often masked in a printed text, offering data on textbook use in the classroom, or informing on pedagogical practices in contexts where textbooks were not central (Blair 2008; Warwick 2003a; Kaiser 2005a; García-Belmar 2006).

During the nineteenth century, in certain countries, the morphology of textbooks had variations in connection to the stratification of school systems targeting different age readerships and educational pathways, but its core definition was preserved. There were also a range of different terms used to refer to a textbook, which varied across languages, but in general indicated its connection with a *course*, its ad hoc teaching and learning purpose, its provision of the first *elements* of a subject, as an *elementary treatise*, or its *manual* quality as designed to be easily handled by young students. Twentieth-century uses of the word as an adjective have denoted both the quality of the standard or, conversely, of an ordinary stereotype.[2]

Beyond the simplicity of a strictly nominalist interpretation, what matters here is that the major features of a *textbook* are its use in teaching and learning, and its authority. Textbooks might be used preferably by the teacher or by the students, in the classroom, in the library, or at home. Books not originally designed for this purpose might acquire this quality by their use. The authority of textbooks is built upon their central role in teaching. Vice versa, the selection of certain books as textbooks is often based on their authority. This authority can come from various sources: the extensive use of a textbook, the approval given to it by a political, scientific, or educational authority, the marketing ability of its publisher, or the reputation of its author.

When Sarton and Kuhn used the word "classics," they were referring both to the scientific reputation of certain authors and to the place that their work had in the canon defined by their contemporaries and by modern historians of science. But the qualities of being a "classic" are also determined by a longstanding interest in a book among a wide range of readers and among publishers (Fyfe 2002; Olesko 2005; Simon 2009). The large readerships commanded by the school context and the cultural impact of schooling across generations, no doubt can endow certain textbooks with the aforementioned quality. Remarkably, this ambiguous term was used in France, after the Revolution (*livre classique*), to designate secondary school textbooks (Simon 2011). The source of authority for science textbooks is not only scientific or historico-scientific, but also pedagogical.

Scholars often use the term *treatise* for books endowed with scientific authority. These were sometimes used for teaching or even designed for that purpose, so that they can be sometimes designated as *textbooks* too. The distinction is relevant, since there are researchers who wrote textbooks and included their own research in them, while other textbook authors have been practitioners mostly or exclusively focused in teaching and writing. Thus, foundational chemistry textbooks such as Antoine Lavoisier's *Traité de chimie* (1789), Thomas Thomson's *A System of Chemistry* (1802), and Jacob Berzelius' *Lärbok i Kemien* (1808–1818), would be clearly different to Nicolas Deguin's *Cours élémentaire de chimie* (1845), Edward Turner's *Elements of Chemistry* (1827), and Nils Johan Berlin's *Elementar-lärobok i oorganisk kemi* (1857).

Nonetheless, all these authors produced textbooks influential because of their wide readerships, and they were all engaged in shaping their discipline through a combination of teaching and research.

Teaching and research have been commonly two connected activities in the working life of science practitioners, but there is a range of options in textbook writing for which contents and form are equally important. As an introductory, comprehensive, standard, or innovative presentation of a subject, a textbook can have a major impact in the shaping of a scientific discipline and this depends both on scientific and pedagogical factors. Hence, textbook writing has been a major activity in scientific practice, and an important source of prestige for its practitioners. But historians of science have often overlooked it in favor of a perception of status built exclusively upon *journal* or *frontier* science.

Summarizing, textbooks have special characteristics that distinguish them from other books, mainly their use in formal teaching and their pedagogical and scientific authority. Several authors have pinpointed the emergence of the textbook as a genre between the late eighteenth and the mid-nineteenth centuries (Choppin 1992; Bensaude-Vincent, García-Belmar, and Bertomeu 2003). In this period the expansion of national systems of secondary education (including science subjects), and publishers supplying that market, contributed to shape a well-defined product. However, a history of science textbooks should not be limited by this periodization. As long as there were texts having a central role in teaching the sciences (e.g., in medieval universities) research on textbooks could not only illuminate a wide range of disciplines across the sciences, but also contribute to a big picture across time. It is undoubtedly worthwhile to study the readers and readings of such classics as Copernicus' *De Revolutionibus Orbium Coelestium*, which has shown that such an abstruse treatise was used in classrooms (Gingerich 2004). But why not engage in similar terms with more humble but more widely read textbooks such as Sacrobosco's *Tractatus de Sphaera* (Gingerich 1988)? Extensive circulation over time and space are *per se* qualities that contribute to the shaping of a book into a *standard* or a *classic*, although historians of science have traditionally been reluctant to accept textbooks into their canon if their authors did not have credentials elsewhere.

Disciplines

In spite of their potential for historiographical transdisciplinarity, the history of science textbooks have been hitherto written in disciplinary niches. Historians of chemistry have led this endeavor, while historians of physics have produced some major case studies and introduced novel approaches to the study of science pedagogy, and textbook-oriented research is currently starting to surface in the history of biology. These are just a few major examples.

In 1919, the Nobel Prize winner in chemistry Wilhelm Ostwald considered that a history of chemistry textbooks would be valuable to solve contemporary methodological problems in science (Haupt 1987). A few years later, Helène Metzger's *Les doctrines chimiques en France du debut du XVIIe à la fin du XVIIIe siècle* used textbooks as main sources to write a history of (chemical) mentalities. In the 1970s, Owen Hannaway contended that the making of chemistry had taken place fundamentally in classrooms and through textbooks and didactic traditions (Hannaway 1975).

Subsequently, Frederic Holmes argued for the complementarity of teaching and research in chemistry (Holmes 1989). In the 1990s the development of a European research project led to the publication of *Communicating Chemistry* (Lundgren and Bensaude-Vincent 2000), a reference work providing a panoramic view on avenues for research in this field.

This edited collection included a general assessment of the historical value of textbooks (Brooke), and case studies which opened a spectrum of possibilities: surveys and preliminary analysis of sources in national context (García-Belmar and Bertomeu; Lundgren; Palló); attempts to problematize the intersections between textbooks and popular books (Orland; Dolan; Knight); exploratory studies of the relationship between lecture notes and textbooks (Bensaude-Vincent); research on textbooks as gendered literature (Pigeard), as sources of practical training (Nieto-Galan), as major agents in the making of disciplines and new theories (Gavroglu and Simoes; Lind; Kounelis), and the intermingling of pedagogical and investigative innovation (Brooks; Nye); textbook writing as an international enterprise (Blondel-Mégrelis); and textbooks at the intersection of historical and educational agendas (Izquierdo).

This collective effort has continued through a long list of publications on nineteenth-century chemistry textbooks led by García-Belmar, Bertomeu, and Bensaude-Vincent (e.g., 2005), a new international collection (Bertomeu, García-Belmar, Lundgren, and Patiniotis 2006), and *L'émergence d'une science des manuels*, a major work that synthesizes a decade of research (Bensaude-Vincent, García-Belmar, and Bertomeu 2003). This book provided a comprehensive picture of the birth of a genre (the general chemistry textbook) in France between the Revolution and the mid-nineteenth century. It analyzed the readerships of chemistry textbooks, the involvement of the state in promoting and controlling textbook production, the rise of a specialized publishing industry through the expansion of chemistry teaching, the pedagogical and scientific decisions that defined the structure and presentation of chemistry, and the place of experiment, theory, and history in textbooks. This comprehensive survey of chemistry textbook literature included reference to classic authors such as Lavoisier and Antoine-François Fourcroy, but also to unknown writers such as Nicolas Deguin and Alexandre Meissas. A major place in this account is given to Louis-Jacques Thénard's *Traité de chimie élémentaire, théorique et pratique* (1813–1816) as a model for textbook writing in French general chemistry.

Similar surveys have been produced for physics. Gunter Lind examined the physics textbook repertoire in German between the eighteenth and mid-nineteenth centuries, and provided case studies of the textbooks by Christian Wolff, Pieter van Musschenbroek, René-Just Haüy, and Karl Wilhelm Gottlob Kastner, around questions such as mathematization, theory, and experiment (Lind 1992). His account is mainly a history of the foundations of physics through the textbooks, which therefore is close in approach to Metzger's history of chemical mentalities, or John Heilbron's *Elements of Early Modern Physics* (1982). A virtue of these contributions is their broad survey of textbook literature and their conceptual analysis. A drawback is their lack of interest in textbooks *per se*, and their consideration of knowledge as an immaterial entity. A proper problematization of their main sources is lacking, justifying why they used textbooks and not other genres of scientific literature for their histories. A similar pattern is found in pioneering works such as those by Stephen Brush, which used large samples of textbook literature to build big pictures of the emergence of scientific ideas such as the kinetic theory of gases (Brush 1976). More recent additions, resuming a traditional interest in foundational textbooks in topical areas of contemporary physics, follow a similar path grounded in intellectual history, but influenced, too, by recent developments in the study of science pedagogy (Warwick 2003a; Kaiser 2005a) which have made a major impact in the history of science but little contribution to textbook studies.[3] Overlaying these perspectives is a focus which is mainly disciplinary and only secondarily pedagogical. Textbook research requires a more symmetrical balance between both (Olesko 2006; Simon 2013a).

A relevant case, which illustrates the necessity of a more balanced focus and interdisciplinary approach, is that of Adolphe Ganot's best-selling nineteenth-century physics textbooks. Ganot was a science teacher whose work has in general elicited little interest among historians. But his textbooks became standard during the second half of the nineteenth century as an introduction to physics across primary, secondary, university, and informal education worldwide. The study of the production, circulation, and appropriation of these textbooks highlights the blurry boundaries between medical and science education, and between teaching and research in nineteenth-century culture. A "humble" textbook helps us to historicize the making of physics, as a process driven by practices of school teaching and pedagogical writing, book production and distribution, and studying and reading, shaped by persistent international communication. The resulting picture contributes to challenge the standard disciplinary characterization, periodization, and geography of physics in the commonly held historiographical canon (Simon 2011). Textbooks can, therefore, be fruitfully placed at the center of analysis by emphasizing the creativity of textbook writing and its important role in the fashioning of scientific disciplines, the role of readers and the material culture of the book in the making of knowledge, and the changing boundaries between the sciences, teaching, and research. And historians can resort to approaches that transcend the disciplinary boundaries of fields such as history of science, technology and medicine, history of education, book history, and science education. There is life beyond Kuhn for textbook studies.

But the continuities with Kuhn's work are still important in research on biology textbooks. Here, textbook repertoires have been used as a collective source of *normal science* which allows characterizing standard knowledge, and tracking its changes across time, to determine when certain theories were accepted (Brush 2002; Skopek 2011), to identify key changes in the production and circulation of scientific tools (Hopwood 2015), and, more rarely, to characterize the shaping of specialized areas of research (Park 2008). The originality in this disciplinary field resides mainly in its tackling of the science–religion debates (Ladouceur 2008; Shapiro 2013). More generally, it is well known that textbooks can have a major political role as agents at the crossroad of governments, markets, and schools which shape pedagogical and scientific outlooks and cultural and national ideals (Nelkin 1977; Choppin 1992). This fundamental element, which has produced an enormous international scholarship on history and literature textbooks, for instance, has been rarely dealt with in the history of science.[4] Biology, and in particular, evolution, appears to be an area specially prepared to cover this deficit (Shapiro 2013).

Paradoxically, the study of evolution and natural history, which has been the site of chief developments in the study of book culture within history of science, has scarcely contributed to the study of textbooks and (formal) education. The work of James

Secord characterizes the enormous impact of the introduction of book history in our discipline and its framing in the field of *popularization* studies (Secord 2000). Several authors have exposed the interesting connections between popularization, popular and informal education, formal education, and communication at large (Fleck 1979; Whitley 1985; Topham 1992; Secord 2004; Simon 2009). But the specificities of formal education and textbooks against the several types of popularization and popular books are not always properly acknowledged. This is partly connected to national biases in historiographical traditions.

There are clear differences in how science education and its history are tackled in national context. Broadly, continental European and Latin American historians of science have in general shown interest in the study of formal education and textbooks (Roldán-Vera 2003; Bertomeu et al. 2006; Simon 2013b) coupled with an acknowledgement of the historical relevance of state-run educational initiatives. North American historians have increasingly produced relevant work in this field too (Rudolph 2002; Kohlstedt 2010). In contrast, historians of British science have been particularly reluctant to consider the relevance of formal structures of education for the teaching of science.[5]

The political action of the state in nineteenth-century Britain—especially England—was characterized by a reluctance to intervene in the organization of education which contrasts with other national experiences worldwide. But there were relevant actions in this direction, both through private initiatives and indirect, but substantial, government intervention. Historians of British science have rarely dealt with these historical events. Their primary focus on popularization and informal education has contributed to provide new historiographical tools and approaches, but also to overshadow other relevant historical evidence. By extension, the benefits of the international impact of British scholarship on popularization have contributed to reinforce a problem of disciplinary communication which in fact is international (Simon 2011; Simon 2015).

This state of affairs, historical, but historiographical, disciplinary, and cultural too, is one among several problems that have hindered a more vigorous integration of science textbook studies with those on popularization, and the promotion of a higher status for the former. The quest for a better integration of these two germane but distinct areas would have to deal with the convergence of distinct disciplinary agendas as well. Historians of science have in general focused on universities, as the site where research is produced. Historians of education have prioritized instead the study of primary schooling as the place in which society is built through universal access to education (Simon 2015).

Futures

One way to revitalize science textbook studies would involve addressing major targets at the intersection of history of science, history of education, and book history, and seeking fruitful connections with contemporary research in science education, science and technology studies, science communication, writing studies, discourse analysis, and visual studies. Another obvious avenue would consist in contributing alone to the questions which are central to current history of science as a whole. Although we live in a highly stratified academic world and the history of science has become an increasingly self-referential discipline, these two options might amount to the same thing.

If a new agenda for the history of science could be built on the premise of blurring the boundaries between the making of science and its communication (Secord 2004), education and textbooks should play a prominent role, as their quantitative and qualitative relevance in the communication of scientific knowledge cannot be ignored. The history of science has moved from a history of scientific thought and political elites, to a field which acknowledges that scientific knowledge is socially, politically, and geographically situated, that it has a material culture (in the laboratory but also in the printing workshop), that lay constituencies play a role in its production, and that its products are subject to market forces. At the crossroads of markets, government, and schooling, textbooks embody all these aspects.

The writing of a textbook involves the demarcation of the boundaries of that knowledge which is considered to constitute the core of a subject and the production of a particular form to communicate it, shaped by pedagogical, scientific, political, and economic aims. These choices are key factors in the formalization of scientific disciplines, which beyond their epistemological and institutional structure, in practice can be defined as "knowledge assembled to be taught" (Olesko 2014; Simon 2011). The production of a textbook involves the work of print technicians and the marketing strategies of publishers whose actions are often not limited to providing a physical container and a shop window, but have a direct impact in the shaping of knowledge (Secord 2000; Simon 2011). Local, regional, and national governments have traditionally been a major force in the shaping of textbook content and form, as textbooks have had for several centuries now a major role in the shaping of the cultural, political, and vocational constitution of the youth in schools (Choppin 1992).

As we move towards considering the role of politics and social judgments in the making of scientific expertise in contemporary societies (Collins and Evans 2007), it is too simple to consider that this process is performed by individuals floating free in a society without past, in which the role of education is restricted to (university)-*accredited expertise*, a bold difference between "doing and knowing," and tacitness as the driving force in knowledge acquisition (Olesko 1993; Olesko 2006). Schooling and learning through textbooks had and still have a major role not only in the making of *interactional expertise*, but largely in the building of the worldviews of citizens, what they know, what they do, what they are. The study of textbooks and their use could have a relevant role in our understanding of how expert knowledge is constituted in society.

Worldviews and the role of state governments in the production and control of textbooks are connected to another traditional agency of schooling: nation building. Although national histories are no longer fashionable in the history of science, research is commonly practiced—implicitly or explicitly—in the framework of the nation (Simon and Herran 2008). The study of textbooks can play a relevant part in problematizing the co-construction of science and the nation, the shaping of national identities in science, or conversely to understand the transformation of the nation by cross-national and transnational phenomena.

Research on science textbooks has shown their fitness to produce comparative and cross-national histories, as enterprises which have often transcended the nation through translation and international collaboration and circulation (Blondel-Mégrelis

2000; Simon 2011; Topham 2011; Gordin 2012). The rise of a capitalist economy of the book with the constitution of large publishing houses in the nineteenth century was driven by the production of textbooks for national and international mass education markets in which the expansion of science teaching had an important role (Mollier 1988; Simon 2011). Many of these publishing firms still exist today.

As history of science tries to move (not without difficulties) towards more global narratives, textbooks have an obvious potential for connecting research in local, regional, national, international, and transnational (but unfortunately not extraterrestrial) contexts, over large periods of time, and through research questions which are central to the nature of scientific practice.

Endnote

1   I am grateful to Pepe Pardo, José R. Bertomeu, and Adriana Minor for their generous reading.
2   There is no point in offering here a detailed historical thesaurus in different languages. I have surveyed, though, English, Spanish, French, and German online lexicographic repositories: *Oxford English Dictionary* (www.oed.com/), *Nuevo Tesoro Lexicográfico de la Lengua Española* (http://buscon.rae.es/ntlle/SrvltGUILoginNtlle), *Le Trésor de la Langue Française Informatisé* (http://atilf.atilf.fr/), CNRTL repository of old dictionaries (http://www.cnrtl.fr/dictionnaires/anciens/), and *Wörterbuchnetz* (http://woerterbuchnetz.de/). Further discussion in Bertomeu, García-Belmar, Lundgren, and Patiniotis (2006).
3   But see in contrast Warwick (2003) and Kaiser (2007).
4   An exception is Rudolph (2002).
5   An illustrative example is Secord (2007).

References

Bachelard, Gaston. 2002[1938]. *The Formation of the Scientific Mind: A Contribution to a Psychoanalysis of Objective Knowledge*. Manchester: Clinamen Press.

Bensaude-Vincent, Bernadette, Antonio García-Belmar, and José Ramón Bertomeu. 2003. *L'émergence d'une science des manuels: les livres de chimie en France (1789–1852)*. Paris: Éditions des archives contemporaines.

Bertomeu, José Ramón, Antonio García-Belmar, Anders Lundgren, and Manolis Patiniotis (eds.) 2006. "Textbooks in the scientific periphery." *Science and Education*, 15: 657–880.

Blair, Ann. 2008. "Student manuscripts and the textbook." In *Scholarly Knowledge: Textbooks in Early Modern Europe*, edited by Emidio Campi, Simone De Angelis, Anja-Sylvia Goeing, and Anthony Grafton, 39–73. Geneva: Librairie Droz.

Blondel-Mégrelis, Marika. 2000. "Berzelius's textbook: In translation and multiple editions, as seen through his correspondence." In *Communicating Chemistry: Textbooks and Their Audiences, 1789–1939*, edited by Anders Lundgren and Bernadette Bensaude-Vincent, 233–54. Canton: Science History Publications.

Brush, Stephen G. 1976. *The Kind of Motion We Call Heat: A History of the Kinetic Theory of Gases in the 19th Century*. Amsterdam: North-Holland Publishing.

Brush, Stephen G. 2002. "How theories became knowledge: Morgan's chromosome theory of heredity in America and Britain." *Journal of the History of Biology*, 35: 471–535.

Choppin, Alain. 1992. *Les manuels scolaires: histoire et actualité*. Paris: Hachette.

Collins, Harry and Robert Evans. 2007. *Rethinking Expertise*. Chicago: University of Chicago Press.

Fleck, Ludwik. 1979[1935]. *Genesis and Development of a Scientific Fact*. Chicago: The University of Chicago Press.

Fyfe, Aileen. 2002. "Publishing and the classics: Paley's *Natural Theology* and the nineteenth-century scientific canon." *Studies in History and Philosophy of Science*, 33: 729–51.

García-Belmar, Antonio. 2006. "The didactic uses of experiment: Louis Jacques Thenard's lectures at the Collège de France." In *Science, Medicine and Crime: Mateu Orfila (1787–1853) and His Times*, edited by José Ramón Bertomeu and Agustí Nieto-Galan, 25–53. Canton: Watson Publishing.

García-Belmar, Antonio, José Ramón Bertomeu, and Bernadette Bensaude-Vincent. 2005. "The power of didactic writings: French chemistry textbooks of the nineteenth century." In *Pedagogy and the Practice of Science. Historical and Contemporary Perspectives*, edited by David Kaiser, 219–51. Cambridge: MIT Press.

Gingerich, Owen. 1988. "Sacrobosco as a textbook." *Journal for the History of Astronomy*, 19: 269–73.

Gingerich, Owen. 2004. *The Book Nobody Read: Chasing the Revolutions of Nicolaus Copernicus*. New York: Walker.

Gordin, Michael. D. 2012. "Translating textbooks: Russian, German, and the language of chemistry." *Isis*, 103: 88–98.

Hannaway, Owen. 1975. *The Chemists and the Word: The Didactic Origins of Chemistry*. Baltimore: The John Hopkins University Press.

Haupt, Bettina. 1987. *Deutschsprachige Chemielehrbücher (1775–1850)*. Stuttgart: Deutscher Apotheker Verlag.

Holmes, Frederic L. 1987. "Scientific writing and scientific discovery." *Isis*, 78: 220–35.

Holmes, Frederic L. 1989. "The complementarity of teaching and research in Liebig's Laboratory." *Osiris*, 5: 121-64.

Hopwood, Nick. 2015. *Haeckel's Embryos: Images, Evolution, and Fraud*. Chicago: University of Chicago Press.

Kaiser, David. 2005a. *Drawing Theories Apart: The Dispersion of Feynman Diagrams in Postwar Physics*. Chicago: The University of Chicago Press.

Kaiser, David. 2005b. "Training and the generalist's vision in the history of science." *Isis*, 96: 244–51.

Kaiser, David. 2007. "Turning physicists into quantum mechanics." *Physics World*, 20(May): 28–33.

Kohlstedt, Sally Gregory. 2010. *Teaching Children Science: Hands-On Nature Study in North America, 1890–1930*. Chicago: University of Chicago Press.

Kuhn, Thomas S. 1962. *The Structure of Scientific Revolutions*. Chicago: University of Chicago Press.

Kuhn, Thomas S. 1963. "The function of dogma in scientific research." In *Scientific Change*, edited by A. C. Crombie, 347–69. New York: Basic Books.

Ladouceur, Ronald P. 2008. "Ella Thea Smith and the lost history of American high school biology textbooks." *Journal of the History of Biology*, 41: 435–71.

Lind, Gunter. 1992. *Physik im Lehrbuch, 1700–1850. Zur Geschichte der Physik und ihrer Didaktik in Deutschland*. Berlin: Springer-Verlag.

Lundgren, Anders and Bernadette Bensaude-Vincent (eds.) 2000. *Communicating Chemistry: Textbooks and Their Audiences, 1789–1939*. Canton: Science History Publications.

Mollier, Jean-Yves. 1988. *L'argent et les lettres: histoire du capitalisme d'édition, 1880–1920*. Paris: Fayard.

Myers, Greg. 1990. *Writing Biology: Texts in the Social Construction of Scientific Knowledge*. Madison: University of Wisconsin Press.

Nelkin, Dorothy. 1977. *Science Textbook Controversies and the Politics of Equal Time*. Cambridge, MA: MIT Press.

Olesko, Kathryn M. 1993. "Tacit knowledge and school formation." *Osiris*, 8: 16–29

Olesko, Kathryn M. 2005. "The foundations of a canon: Kohlrausch's Practical Physics." In *Pedagogy and the Practice of Science. Historical and Contemporary Perspectives*, edited by David Kaiser, 323–55. Cambridge, MA: MIT Press.

Olesko, Kathryn M. 2006. "Science pedagogy as a category of historical analysis: Past, present, and future." *Science and Education*, 15: 863–80.

Olesko, Kathryn M. 2014. "Science education in the historical study of the sciences." In *International Handbook of Research in History, Philosophy and Science Teaching*, edited by Michael R. Matthews, 1965–90. Amsterdam: Springer Verlag.

Park, Hyung Wook. 2008. "Edmund Vincent Cowdry and the making of gerontology as a multidisciplinary scientific field in the United States." *Journal of the History of Biology*, 41: 529–72.

Roldán Vera, Eugenia. 2003. *The British Book Trade and Spanish American Independence: Education and Knowledge Transmission in Transcontinental Perspective*. Aldershot: Ashgate.

Rudolph, John L. 2002. *Scientists in the Classroom: The Cold War Reconstruction of American Science Education*. New York: Palgrave.

Rudolph, John L. 2008. "Historical writing on science education: A view of the landscape." *Studies in Science Education*, 44: 63–82.

Sarton, George. 1948. "The study of early scientific textbooks." *Isis*, 38: 137–48.

Secord, James A. 2000. *Victorian Sensation: The Extraordinary Publication, Reception, and Secret Authorship of Vestiges of the Natural History of Creation*. Chicago: The University of Chicago Press.

Secord, James A. 2004. "Knowledge in transit." *Isis*, 95: 654–72.

Secord, James A. 2007. "Science." *Journal of Victorian Culture*, 12: 272–6.

Shapiro, Adam R. 2013. *Trying Biology: The Scopes Trial, Textbooks, and the Antievolution Movement in American Schools*. Chicago: University of Chicago Press.

Simon, Josep. 2009. "Circumventing the 'elusive quarries' of popular science: The communication and appropriation of Ganot's physics in nineteenth-century Britain." In *Popularizing Science and Technology in the European Periphery, 1800–2000*, edited by Faidra Papanelopoulou, Agustí Nieto-Galan, and Enrique Perdiguero, 89–114. Aldershot: Ashgate.

Simon, Josep. 2011. *Communicating Physics: The Production, Circulation and Appropriation of Ganot's Textbooks in France and England, 1851–1887*. London: Pickering & Chatto.

Simon, Josep. 2012. "Cross-national education and the making of science, technology and medicine." *History of Science*, 50: 251–6.

Simon, Josep. 2013a. "Physics textbooks and textbook physics in the nineteenth and twentieth century." In *The Oxford Handbook of the History of Physics*, edited by Jed Buchwald and Robert Fox, 651–78. Oxford: Oxford University Press.

Simon, Josep (ed.) 2013b. "Cross-national and comparative history of science education." *Science & Education*, 22: 763–866.

Simon, Josep. 2015. "History of science." In *Encyclopedia of Science Education*, edited by Richard Gunstone, 456–9. Dordrecht: Springer Verlag.

Simon, Josep and Néstor Herran. 2008. "Introduction." In *Beyond Borders: Fresh Perspectives in History of Science*, edited by Josep Simon and Néstor Herran, 1–23. Newcastle: Cambridge Scholars Publishing.

Skopek, Jeffrey M. 2011. "Principles, exemplars, and uses of history in early 20th century genetics." *Studies in History and Philosophy of Biological and Biomedical Sciences*, 42: 210–25.

Topham, Jonathan. 1992. "Science and popular education in the 1830s: The role of the Bridgewater treatises." *British Journal for the History of Science*, 25: 397–430.

Topham, Jonathan. 2000. "Scientific publishing and the reading of science in nineteenth-century Britain: A historiographical survey and guide to sources." *Studies in History and Philosophy of Science*, 31: 559–612.

Topham, Jonathan. 2011. "Science, print, and crossing borders: Importing French science books into Britain, 1789–1815." In *Geographies of Nineteenth Century Science*, edited by David N. Livingstone and Charles W. J. Withers, 311–44. Chicago: University of Chicago Press.

Warwick, Andrew. 2003a. *Masters of Theory: Cambridge and the Rise of Mathematical Physics*. Chicago: Chicago University Press.

Warwick, Andrew. 2003b. "'A very hard nut to crack' or making sense of Maxwell's treatise on electricity and magnetism in mid-Victorian Cambridge." In *Scientific Authorship: Credit and Intellectual Property in Science*, edited by Mario Biagioli and Peter Galison, 133–61. New York: Routledge.

Warwick, Andrew and David Kaiser. 2005. "Conclusion: Kuhn, Foucault, and the power of pedagogy." In *Pedagogy and the Practice of Science. Historical and Contemporary Perspectives*, edited by David Kaiser, 393–409. Cambridge, MA: MIT Press.

Whitley, Richard. 1985. "Knowledge producers and knowledge acquirers: Popularisation as a relation between scientific fields and their publics." In *Expository Science: Forms and Functions of Popularisation*, edited by Terry Shinn and Richard Whitley, 3–30. Dordrecht: D. Reidel.

CHAPTER TWENTY-NINE

Lectures

DIARMID A. FINNEGAN

As a form of talk, the lecture is hard to characterize. For the sociologist Erving Goffman (1981) a lecture is a mode of communication that creates a powerful impression of a direct and spontaneous apprehension of the truth. According to Goffman, this is achieved by the bodily presence and performance of the speaker, the use of rhetorical devices, and through ritual framing. By such means, public lectures animate and authorize ideas that would otherwise remain obscure or inaccessible. For others, public speech unsettles rather than facilitates "truth telling." According to the historian of communication, Walter Ong (1974, 230), an "oral mode of knowledge in public life" is essentially agonistic and polemical. Lectures and other forms of public speaking, rather than help forge consensus, tend to sharpen social tensions and disturb civic harmony. These two somewhat contrasting accounts suggest that lecturing on science presented opportunities but also brought significant risks. Certainly, introducing an audience to scientific subjects connected the speaker and their science to a wider set of expectations about the roles and influence of public speech.

The vexed question about how to speak effectively and authoritatively about science, given the potential and pitfalls thought to attach to public address, was one that occupied the founders of the Royal Society. Famously, Thomas Sprat (1667, 113) in his history of the Royal Society included an assault on the abuses of "fine speaking" and eloquent discourse and recommended a return to a "natural way of speaking" that approached as closely as possible a "mathematical plainness." At the same time, Sprat's anxiety about the abuses of language was predicated on the assumption that the "ornaments of speaking" could be used to powerful effect. As Tina Skouen (2011) has argued, rather than banishing it from scientific discourse Sprat aimed at the recovery of the right use of rhetoric. This same ambivalence towards the rhetorical arts continued to haunt those concerned with communicating science, whether through speech or writing in the centuries that followed.

The problem of clearly distinguishing themselves from other forms of public speech without losing the attention and interest of their audiences was a perennial issue for the

A Companion to the History of Science, First Edition. Edited by Bernard Lightman.

science lecturer. In the seventeenth century, for example, Fellows of the Royal Society of London worried over the extent to which lectures should include theatricality or provoke a sense of wonder. On the one hand, these elements lent interest to the subjects they wished to promote. On the other, they risked associating polite and serious science with vulgar entertainment (Golinski 2005, 92). In the eighteenth century, as science increasingly moved from the semi-private world of learned institutions into the public sphere, this challenge, if anything, intensified. Lecturers in Britain promoting Newtonian science outside the patronage of the Royal Society had to make strenuous efforts to overcome accusations of charlatanry and vulgarity (Stewart 1993). In the 1790s, the chemist and radical dissenter Joseph Priestley and his followers were condemned for promoting civil unrest by exciting wonder and promoting enthusiasm through their "sublime" chemical demonstrations (Golinski 2008).

The professionalization of science from the nineteenth century increasingly informed the question of who could speak authoritatively on scientific matters to a general public. This emerged, however, only gradually and in a piecemeal fashion. In Britain, those lobbying for greater state recognition of science and pushing for more carefully defined scientific standards could be ambivalent about making science a recognized profession. Until the late nineteenth century, professional science could still suggest knowledge pursued for pecuniary reward rather than for the sake of truth. The same was true in the United States (Lucier 2009). Space remained, then, for science lecturers who were not necessarily certified practitioners to retain a large share of the lecture market. Indeed, the creeping professionalization of science created new opportunities for popularizers to present themselves as mediators between the emerging professional and the general public (Lightman 2007).

However the task of defining what properly constituted expert public speech about science was worked out, it has been a largely masculinist enterprise. Until relatively recently, securing a reputation as a female lecturer was almost prohibitively difficult. It is not hard to find evidence to confirm this. None of the large cast of lecturers invited to speak at the famous Lowell Institute between 1839 and 1898 were women. Although women had lectured at the sectional meetings of the British Association for the Advancement of Science since its early years, it was not until 1968 that a female President addressed the Association. In 1994, Susan Greenfield became the first woman to deliver the celebrated Christmas lectures at the Royal Institution in London, a series that had run since 1825. Earlier exceptions, in many respects, prove the rule. One such was the Italian natural philosopher Laura Bassi, the first women to take up a teaching position at a European university in 1732. In her first public lecture she thanked the Senate of the University of Bologna for raising her, "beyond what I asked for and dreamed of, to the highest dignity of speaking in public" (as quoted in Findlen 1993, 450). Subsequently Bassi made many appearances at the annual anatomy carnival holding forth on a range of scientific subjects before a select learned audience and a frequently unruly crowd of revelers. These public discourses were not, however, as radical as they seemed. The carnival in particular, with its ritualistic reversals of social hierarchies, rendered Bassi's scientific performances less threatening.

It must also be acknowledged that non-Europeans are almost entirely absent from the cast of lecturers that frequently appear in English-language historical scholarship. Yet, the increasing amount of attention given to the significance of inter-cultural exchange in the production of scientific knowledge suggests that regarding the science lecture as a quintessentially Western mode of communication is, at the very least, open to question. Like other scientific practices, the science lecture was not an entirely stable tradition or one sealed off from wider trans-cultural influences. A recent set of essays on the natural philosopher and itinerant lecturer James Dinwiddie (1746–1815), for example, show in detail how a distinctly European tradition of experimental demonstration in the lecture hall could be re-worked through sustained encounters between European and Asian cultures (Lightman, McOuat, and Stewart 2013).

The science lecture, then, can be understood as a dynamic and at times unstable collection of communicative practices that were informed, and in turn helped to transform, a wider set of (inter-) cultural domains. The remaining sections of this essay expand on this definition by examining in turn how science lectures adopted and adjusted a range of visual and vocal techniques, exploited the communicative potential of the human body, and negotiated the challenges of creating venues and audiences for scientific address.

Visualizing Science

It is hard to imagine a science lecture without the dramatic use of visual illustrations or experimental demonstrations. The highly visual nature of science lectures should not, however, be taken for granted. At the beginning of the nineteenth century, the German philosopher Johann Gottlieb Fichte complained that professors at German universities continued to dictate their lectures to students from prepared scripts, a practice Fichte felt was obsolete given the ready availability of books (Phillips 2012, 132–3). Fichte's grievance came on the back of a long-established tradition of dictation and note taking. In the early modern period, this did not facilitate the creative use of images or experimentation. While it was a more complex experience than Fichte's complaint suggests, the early modern practice of a professor paraphrasing and interpreting textual material to note-taking students from a raised pulpit or *cathedra* (see Grafton 2003) was a communicative strategy that emphasized the verbal more than the visual and had more in common with the cathedral than the theatre.

There were, of course, striking exceptions. The anatomy theatre provides one early example. Anatomical dissections created a spectacle and the spaces in which demonstrations took place were aptly described as *theatrum* or a "place for seeing." The oldest surviving example, the permanent anatomy theatre at the University of Padua completed in 1595, could accommodate over 200 spectators. The lectures were delivered in winter partly to prevent the rapid decay of the cadaver and took place by candlelight (Klestinec 2011). The suboptimal viewing conditions underlined the visual nature of the event. That such anatomical lectures had an appeal and cultural significance beyond medical students was largely due to the creation of a dramatic ocular event. The dissected body became a staple part of lectures on anatomy and later ethnography in part because they provided a sensational and ghoulish attraction for members of the public with an appetite for the macabre. In Georgian London, for example, anatomy lecturers functioned not only as pedagogues but also as "showmen of the body" (Porter 1995, 91). Medical lectures exploited the blurred lines between anatomical instruction and freak-show entertainment.

By the eighteenth century it was not just dead bodies that attracted the public eye. The growing popularity of natural philosophy lectures was accompanied by the burgeoning use of experimental demonstration (Schaffer 1983). Building on the experimental philosophy that emerged in the seventeenth century, lecturers vied with one another to create spectacular events at which the visual threatened to engulf the verbal. The popularity of these events was also to bolster the authority of itinerant lecturers and erode the credibility of university professors. The remarkable expansion of this experimental culture, which helped to transform the science lecture into a public spectacle, can be explained in large part by what Bernadette Bensaude-Vincent and Christine Blondel (2008, 8) call the "epistemic power of sensationalism." The science lecture had truly become a transformative part of a wider visual and aesthetic culture.

If the experimental culture of eighteenth-century science lecturers relied in part on the general appeal of public spectacle, their own different accounts of nature informed how experimental demonstrations were performed and given meaning. This was true of two inheritors of this eighteenth-century tradition, Michael Faraday and William Sturgeon. As Iwan Morus has argued, while Faraday did not generally draw attention to the mechanical workings of his experimental apparatus, Sturgeon deliberately developed instruments that made visible the practical workings that produced the experimental effects. As Morus (1998, 50) puts it, for Sturgeon, "the universe itself operated like an item of electrical apparatus." Faraday, on the other hand, had a different understanding of the relationship between mechanical apparatus and the physical world. Nature was best understood as composed of "lines of force in space," something not visible in the operation of an electrical apparatus (Morus 1998, 52). It was thus necessary to point beyond the instruments themselves to more intangible physical realities. It is also worth noting that Faraday's public experiments employed techniques of concealment akin to those used by professional magicians (Watt-Smith 2013).

The emphasis on visual spectacle continued through the Victorian period. Science lecturers prioritized visual effects to win an audience in a quintessentially visual age. Among the best known was John Henry Pepper, a popular lecturer at London's Royal Polytechnic Institution (Morus 2007). Pepper traded on the popularity of optical illusions and borrowed not only from experimental physics but also, even more conspicuously than Faraday, from an older tradition of natural magic. His trademark illusion became known as Pepper's ghost, a phantasmagoria that drew thousands to hear Pepper lecture on optics. In Pepper's case, visual effects acted as medium that connected the worlds of cloistered scientific research to sensationalist theatre and parlor magic.

It is evident, then, that use of visual aids by science lectures had a powerful effect and huge appeal both for lecturers and for their audiences. Yet, whatever the advantages, an emphasis on visual instruction also brought significant risks. The abiding challenge of resisting the identification of science communication with other forms of public performances was redoubled by a shift towards the visual. The cost of the spectacular display could be a blurring of the boundary between science and other kinds of ways of seeing. In addition, at various times and places observers frequently expressed concern that an emphasis on wonderment and visual display could be readily co-opted by political forces to serve ends that threatened social stability. For these reasons, it is as important to investigate what was concealed by lecture illustrations as much as what was said to be revealed.

Voicing Science

The increasing use and variety of visual display did not necessarily undermine or demote the importance of the spoken word. On the contrary, the widespread use of visual instruction could make verbal performance more, not less, significant if only to avoid the impression that untutored looking was a sufficient basis for grasping scientific truth. Modern science lectures were and remained strongly bi-modal and had to negotiate the different demands of seeing and listening. There were also contexts in which scientific speech had to compete with other forms of public address to win a hearing. In those contexts, how science was spoken—the prosody and rhetoric of scientific speech—was a vital concern. Such debates sat alongside discussions about the distinctive character of science itself. Modern science widely came to be defined as a neutral discourse not subject to personal whimsy or ideological preference. It was important, therefore, to distinguish scientific speech from other forms of oral performance. The science lecture should not sound (exactly) like a sermon, a political speech, a theatrical performance, or a legal discourse. At the same time, it was frequently imperative to borrow from those different modes of speaking in order to connect with an audience and make space for science in a wider culture of public speaking.

Victorian Britain provides one particularly pertinent example of a context in which scientific lecturers not only tried to catch the public eye but also win a public hearing. As Martin Hewitt (2012) has observed, by the Victorian period lecture culture in Britain was no longer dominated by science. In an age of lecturing, talks on other subjects (literature, art, religion, travel, and much more) captured most of the market and led, in certain contexts, to less emphasis on visualization. Crafting a "spectacle of words" (Hewitt 2012, 95) became at least as important as displaying arresting illustrations. This was often true, for example, of addresses on scientific topics delivered to Victorian lecture societies dedicated to organizing annual series of talks on a wide range of subjects (Finnegan 2011). In this case, a well-established culture of hearing was arguably more important than a culture of looking for determining the relative success of a scientific discourse.

This helps to make sense of the importance placed on learning to speak effectively by a number of celebrated Victorian science lecturers. Michael Faraday, for example, took lessons and advice from the elocutionist Benjamin Smart at the start of his career, believing that "the most prominent requisite to a lecturer … is a good delivery" (as quoted in James 1991, 58). For Faraday the science of effective speech was essential for effective speech about science. Indeed, it was partly because of Faraday's conviction that the spoken word was an essential mode in which to communicate science to an interested audience that he was reluctant to publish his lectures. Most obviously, the printed word could not replicate the experiments performed during his lectures. But neither could it capture the "vivacity of speech" (as quoted in James 2008, 476) that Faraday believed breathed life and meaning into his scientific discourse.

Although Faraday may have been unusual in his careful and extended study of the arts of effectual speech, Thomas Henry Huxley came to grudgingly admit the value of learning the craft of public address after his much-discussed encounter with Samuel Wilberforce in 1860. At the very least, Huxley knew he had work to do on projection and vocal strength. Joseph Hooker had reported to Darwin that Huxley, while he had "turned the tables" failed to "throw his voice over so large an assembly."[1] Yet, if Huxley afterwards honed his diction and delivery, it was not only in the interests of being heard. Huxley's style of delivery—his studied stillness and his forensic precision—was in keeping with his views about the relations between the passions and science. Echoing Sprat, Huxley retained a deep suspicion of "mellifluous eloquence" throughout his life and stuck as far as he could to "the plainest of plain language" (Huxley 1893, 3). At the same time, however, he believed that exercising the power of "the living voice" was crucial in generating public sympathy for a form of knowledge that, "by the nature of its being cannot desire to stir the passions" (Huxley 1894, vi). This conviction, expressed late in life, had been consolidated by Huxley's experience during his lecture tour in North America in 1876. Although well enough received in some quarters, Huxley's oratorical persona was widely criticized. One reporter compared his address to that of "a not very well fed evangelical clergyman" noting his lack of vocal strength and inability to put himself in sympathy with his audience (Anon, 1876a). Another noted that he spoke "as a lawyer might talk to bench of judges on an abstruse point of law" (Anon, 1876b).

Of course, whatever its importance, it was not simply vocal performance that mattered when lecturing on science. Finding an effective rhetorical register has also been considered vital. How and where scientific knowledge was communicated mattered, particularly when dealing with subjects that had the potential to unsettle cherished cultural beliefs (Livingstone 2007). One tactic was to deliberately flout the etiquette of decorous and inoffensive speech in order to underline the revolutionary power of scientific thought and practice. Among the accusations directed at John Tyndall after his famous Belfast address in 1874 was his flagrant breaching of speech protocols to which any president of the British Association was expected to adhere (Livingstone 2014, 8, 70–1). For Tyndall, this insensitivity towards rhetorical propriety was pre-meditated. While lecturing at the Royal Institution he was, "as obedient to the laws of the institution as any son of the Church is to the Vatican" (as quoted in DeYoung 2011, 43). Beyond those walls he claimed intellectual freedom. Other famous lecturers used different rhetorical tactics to communicate science's dangerous ideas. William Buckland, for example, ensconced in Anglican Oxford, employed buffoonery and humor to blunt the radical potential of his geological theories (O'Connor 2007, 80–4). The rhetorical management of scientific speech thus varied from speaker to speaker and across different venues and settings.

Body Language

The coordination of verbal and visual rhetoric by science lecturers was, crucially, a corporeal affair. The human body, and its associated repertoire of communicative practices, has long been thought an important part of effective science communication. In basic terms, a lecturer's comportment and dress was often an integral aspect of their reputation and persona. Humphry Davy's affected demeanor and foppish dress, for example, was regarded as a threat to his scientific reputation. Davy invited the accusation that he was feminizing and thus marginalizing science (Golinski 2011, 23–4). A flamboyant sartorial style combined with dramatic delivery was also an important component of the appeal, and notoriety, of Robert Knox's lecture performances in Regency Edinburgh (Lonsdale 1870, 124–42).

Other successful lecturers in the same period maintained a close interest in the kinesics of science lecturing. For Michael Faraday, for example, the art of "natural" gestures was an important part of his efforts to communicate the wider meaning of scientific facts and induce sympathy for scientific discovery. Faraday's notes from the lectures of the elocutionist Benjamin Smart show a detailed familiarity with Gilbert Austin's *Chironomia*. It was imperative, Faraday noted (1818, f. 280) to ensure that "the tone, the gesture and the look of the [speaker] shall exactly correspond with the subject and with the occasion." As Iwan Morus (2010, 815) has suggested, such attention to bodily discipline while lecturing was "a key strategy in demonstrating mastery over nature." It was also a way to engender emotional response in an audience. It was common for Faraday to close his lectures with an appeal to the sublime operations of divine providence (Cantor 1991). These perorations were given extra force by following Smart's (1819, 145) instructions on how to induce a solemn and reverent mood: slow the pace of delivery, lower the tone of voice, and employ actions that suggest the lecturer's frame has been "overcome by the feelings that press upon it." Smart's techniques were thus an important element of Faraday's attempts to unite his scientific work to a religious sensibility and extend its cultural reach.

Bodily comportment mattered even for those who rejected the gestural systems associated with the so-called elocutionary revolution. Thomas Henry Huxley, for example, generally avoided bodily actions in a bid to give an impression of delivering unadorned scientific truth. According to one observer, his lecture performances were, "sparing of gesture, sparing of emphasis, careless of mere rhetorical or oratorical art … the force was in the thought and the diction, and he needed no other" (Huxley 1908, 384). That did not mean that the physicality of Huxley's performance was unimportant. One admirer noticed Huxley's

> square forehead, the square jaw, the tense lines of the mouth, the deep flashing dark eyes, the impression of something more than strength he gave you, an impression of sincerity, of solid force, of immovability, yet with the gentleness arising from the serene consciousness of his strength—all this belonged to Huxley and to him alone. The first glance magnetized his audience. (Huxley 1908, 383–4)

In this account, Huxley's message and his manner merged and his undisturbed body gestured toward a fixed and fervent commitment to scientific facts.

As well as helping to communicate concepts or feelings, the body of a lecturer or, indeed, auditor or assistant, could be used dramatically to illustrate scientific truth. The dangers involved in this form of public experimentation were deployed in part to underwrite the authority of the lecturer. The chemist Humphry Davy (1778–1829), for example, was injured several times while performing experiments before his fashionable audience at London's Royal Institution. Using the body as a site of experiment or as a visual aid made the science lecture a multi-modal experience for lecturer and auditor alike. William Hunter's lectures on anatomy, famously depicted by Johann Zoffany, included anatomical figures and live bodies (Figure 29.1). Early lectures on electricity commonly called on members of an audience to experience an electric shock. Chemical lecturers, too, connected with the bodies of their auditors through the creation of curious or repulsive smells. The body thus became not only a multi-faceted

Figure 29.1   Dr. William Hunter lecturing at the Royal Academy, c.1772, oil on canvas by Johann Zoffany. © Royal College of Physicians.

communicative device, but also a somatic and sensory instrument for public scientific experimentation.

Auditors and Auditoriums

As "trans-medial" events (Frieson 2011) and multi-modal performances, science lectures have long been associated with certain kinds of venues. Yet surprisingly, perhaps, custom-designed spaces for hosting lectures on scientific subjects were not particularly common beyond the anatomy theatres attached to medical establishments before the nineteenth century. It was only during the so-called second scientific revolution, beginning in the late eighteenth century, that lecture theatres were constructed on a wide scale to accommodate the growing popularity of science among the middle classes. These halls, as Sophie Forgan (1986) has shown, borrowed heavily from the design of the anatomy theatres and generally adopted a semi-circular shape with steeply raked seats or benches sloping upwards and outwards from the central floor area. This provided good lines of sight for the audience, indicating again the importance of visual demonstration and illustration in early nineteenth-century lectures. Attention to lighting confirms this. It was common to include a skylight above the lecture podium to enhance the visibility of experimental demonstrations. A table positioned directly below was also common—most famously in the lecture theatre of the Royal Institution of London.

The acoustic properties of science lecture theatres also mattered. By the nineteenth century, the science of acoustics was being applied to public buildings of all types including lecture theatres. When a new upstairs lecture room was opened in the Smithsonian Institution in 1855, with a seating capacity of 1,500, it was hailed as a "triumph of acoustical science applied to public buildings" (Anon, 1857, 17). The hall was in the form of "an immense trumpet" with the speaker positioned at its mouth (Henry 1886, 420). Designed by Joseph Henry, the secretary of Smithsonian and an expert in acoustic science, the hall's surfaces were carefully finished to prevent the kind of reverberations that confuse the clarity of a speaker's voice. In writing of the design of the hall Henry also noted that, "the arrangements for securing unobstructed sight did not interfere with those necessary for distinct hearing" (Henry 1886, 419). Here, then, was a scientifically designed space that doubly facilitated the growing authority of science and the science lecturer.

Paradoxically, however, such model venues for science communication may have placed certain limitations on where and how scientific knowledge could be articulated. In making specific conditions essential for a successful lecture, custom-built halls risked immobilizing science communication. This, of course, clashed with the idea of science as "universal knowledge." Scientific truth should, in principle, be communicable anywhere. The associated impetus to spread scientific truth through the spoken word was behind the rise of the peripatetic science lecturer. Yet the mobile speaker, precisely because they often had to make do with substandard venues, faced a number of practical challenges. For any travelling lecturer, public halls designed to prioritize certain kinds of looking could distort sonic performance. Venues that privileged acoustic properties, perhaps to accommodate a pipe organ or an orchestra, could disrupt the kind of visual field necessary for effective illustrations. Lecturing on science in these places could threaten the all-important cultural association between clarity—whether visual or vocal—and credibility. The same challenges faced organizers of the annual meetings of the peripatetic national science associations that became part of an increasingly globalized scientific culture in the nineteenth century. An early emphasis on the lecture as a crucial tool for enrolling a wider public in scientific advance meant that each year the question of whether or not a town or city had adequate auditoriums to host them was particularly pressing.

There were other reasons to resist—as did Michael Faraday, for example—lecturing beyond the carefully controlled space of a custom-designed lecture theatre. An emphasis on live experiments, for example, meant that effective science lectures required a substantial amount of equipment and technical support. Michael Faraday's lectures at the Royal Institution were, for that practical reason, not transportable to other venues. The same was true of John Tyndall's experimental performances in the same venue, performances that, as Simon Schaffer (2012) has shown, relied on the careful management of the visual field and visual projections created within the lecture hall. Many other lecturers did manage to find ways to travel, bringing with them the equipment necessary for their demonstrations. Even so, that brought its own demands and again threatened the credibility of science, particularly in the period before it was widely accepted as carrying by definition epistemic and cultural authority.

Deliberations over venues also involved anxieties about audiences. In many cases, the style and substance of science lectures was often related to assumptions about the intellectual capacities of different social or cultural groups. Humphry Davy, for example, grew increasingly suspicious of disseminating scientific knowledge as widely as possible (Secord 2014, 47–9). The polite and fashionable audiences that attended his Royal Institution lectures were unlikely to use science for seditious and radical ends. But to educate the laboring classes about the portentous results of science might risk fomenting political revolution. This view, however, was not widely shared in Davy's own day, or in later periods. The general trend was towards extending the influence of science beyond elite institutions and using it as a means to transform all levels of society.

Thomas Henry Huxley is the obvious case in point. His lectures to workingmen were among his most popular and Huxley himself expressed his own preference for addressing that particular group. He appreciated the need to adjust what and how he communicated. As Adrian Desmond (2007, 211) comments, Huxley "plebeian-ized" his language for the workingmen that came to hear him lecture at the School of Mines even while resisting the fiery speech of cloth-capped orators. While his lectures attracted good numbers, Huxley's success was not easily maintained. As Paul White (2003) has argued, this struggle intensified during the 1880s and 1890s when Huxley found himself at odds with powerful cultural and political movements within the working classes and his rhetoric—written and spoken—shifted somewhat from the projected ideals of plain speaking and common sense.

Gender, too, could be thoroughly implicated in the mutual relations between lecturer, venue, and hearer. As Rebekah Higgitt and Charles Withers (2008) have argued, the women who made up a sizeable portion of the audiences for public lectures organized for the annual meetings of the British Association for the Advancement of Science came to hear lecturers for a variety or reasons. While some attended in order to actively engage with the subject matter, many had other motives such as filial or marital duty, a desire to be entertained, or to engage in social conversation. This was used to advantage by leading men of science to define the lecture audience as passive consumers of expert knowledge. In turn, this representational strategy helped to create a clear demarcation between science (created and controlled by expert practitioners) and its public (pictured as a passive patron). As Higgitt and Withers point out, an emphasis on the active agency of listeners has to be set alongside the fact that audiences could collude with a powerful image of the auditor as a passive consumer of scientific expertise.

In all of this, it is worth remembering that audiences were often not restricted to those present at the lecture. Though individual lecturers, such as the anatomist William Hunter (1718–1783) for example, deliberately prevented their lectures from achieving a wider circulation in published form (see Porter 1995, 94–5), it was increasingly the case—particularly with the proliferation of cheap print—that science lectures reached a much larger audience than the one physically present during their delivery. Platform and print culture increasingly overlapped in significant ways and science lectures were often written with an eye on the final printed form. The full script of John Tyndall's famous Belfast address had been handed to the editor of *Nature* for publication before it was delivered. The complete text was also reproduced and endlessly debated in newspapers and periodicals within and far beyond Britain and was released in revised form in several standalone editions (Lightman 2004). The science lecture, like any other kind of public address, was more often than not inter-medial and lecturers had to negotiate the complexities of that reality not least with respect to the multiple audiences it created.

Legacies

The "remediation" (Bolter and Grusin 2000) of lectures in print has long been deployed both to extend their influence and secure a lasting legacy. As a result, to attempt to disaggregate the influence of lectures from printed texts is, for most periods, a futile endeavor. Yet it is also important to recognize the potential significance of the speech event itself. To be sure, before the advent of the phonograph and other recording devices, it is impossible to precisely recover the performative elements that made lectures significant (or, indeed, rendered them forgettable). It can nevertheless be suggested that science lectures could be particularly effective at stamping an indelible mark on popular consciousness because of their character as ritualized, formal but evanescent talk. Although the impact of Tyndall's Belfast address rippled across the world through the medium of print (Lightman 2004), its notoriety arguably owed much to it being, in its first airing, a speech event. The transgressing of rhetorical decorum, as much as the intellectual content, swept Tyndall into a storm of public controversy and created the conditions for an enduring, if contested and fragmented, cultural legacy (Livingstone 2007, 80–2). The ephemeral nature of the speech event may, paradoxically, be the reason why it becomes entrenched in popular imagination. This also may be why the lecture has, and continues to have, an important place in the repertoire of communicative practices used to disseminate and popularize scientific knowledge.

Whatever the potential of "fresh talk" to foster interest in scientific research and mark it out as an authoritative and distinct form of public knowledge, lecturing also implicated science in ways of knowing and communicating that could threaten its independence or public credibility. There is a need to be cognizant of varying perceptions about the potential and the perils of publicizing science in verbal form. How the act and art of lecturing was understood changed from one period to another and from one "speech space" (Livingstone 2007) to another. There is a need, in other words, to be attentive to temporal shifts and geographical variations. Making this point also reminds us that the dedicated science lecture is only one form of public address in which scientific concerns have been presented and discussed. More work is needed on how scientific talk has been performed in the courtroom, theatre, church or parliament and on the legacies of those rather different speech events. Much scope remains, too, for further work on the cultural history and geography of science lectures in the twentieth century. The emergence and proliferation of new media make this a demanding but important and interesting task. It might be assumed that in the twentieth century the science lecture became significantly less important as a pedagogic form and, more especially, as a mode of science popularization. Yet this would be to artificially extract the lecture from a wider multimedia environment and to leave unexplored its continued influence and pervasiveness. Science lectures have surely been transformed by the introduction of new media—not only in terms of form and content but also with respect to audiences and audience experiences. New media have also, in various ways, arguably facilitated their continued relevance and cultural prominence. There is surely room for detailed studies of, for example, Richard Feynman's recorded lectures on physics, Carl Sagan's lectures on cosmology (whether on television or from a podium), and the Royal Institution's popular Christmas lectures. Such work would doubtless enrich our understanding of a dynamic form of communication that has continually drawn science into wider worlds of public speech.

Endnote

1 Darwin Correspondence Database, http://www.darwinproject.ac.uk/entry-2852 accessed September 23, 2015.

References

Anon. 1857. *Scientific America*, 8: 17.

Anon. 1876a. "Professor Huxley's first lecture." *New York Herald*, September 19, p. 6.

Anon. 1876b. "Prof. Huxley's first lecture." *The Sun*, September 19, p. 2.

Bensaude-Vincent, Bernadette, and Christine Blondel (eds.) 2008. *Science and Spectacle in the European Enlightenment*. Aldershot: Ashgate.

Bolter, Jay David, and Richard Grusin. 2000. *Remediation: Understanding New Media*. Cambridge, MA: MIT Press.

Cantor, Geoffrey. 1991. "Educating the judgment: Faraday as a lecturer." *Bulletin for the History of Chemistry*, 11: 28–36.

Desmond, Adrian. 1997. *Huxley: From Devil's Disciple to Evolution's High Priest*. Reading, MA: Helix Books.

DeYoung, Ursula. 2011. *A Vision of Modern Science: John Tyndall and the Role of the Scientist in Victorian Culture*. New York: Palgrave Macmillan.

Faraday, Michael. 1818. *Common-place book*. Vol. 1. Institution of Engineering and Technology Archives.

Findlen, Paula. 1993. "Science as a career in Enlightenment Italy: The strategies of Laura Bassi." *Isis*, 84, No. 3: 441–69.

Finnegan, Diarmid A. 2011. "Placing science in an age of oratory: spaces of scientific speech in mid-Victorian Edinburgh." In *Geographies of Nineteenth-Century Science*, edited by David N. Livingstone and Charles W. J. Withers, 153–77. Chicago: University of Chicago Press.

Forgan, Sophie. 1986. "Context, image and function: A preliminary enquiry into the architecture of scientific societies." *British Journal for the History of Science*, 19: 89–113.

Frieson, Norm. 2011. "The lecture as a transmedial pedagogical form: A historical analysis." *Educational Researcher*, 40: 95–102.

Goffman, Erving. 1981. *Forms of Talk*. Oxford: Blackwell.

Golinski, Jan. 2005. *Making Natural Knowledge: Constructivism and the History of Science*, new edition. Chicago: University of Chicago Press.

Golinski, Jan. 2008. "Joseph Priestley and the chemical sublime." In *Science and Spectacle in the European Enlightenment*, edited by Bernadette Bensaude-Vincent and Christine Blondel, 117–27. Aldershot: Ashgate.

Golinski, Jan. 2011. "Humphry Davy: The experimental self." *Eighteenth-Century Studies*, 45: 15–28.

Grafton, Anthony. 2003. "Classrooms and libraries." In *The Cambridge History of Science. Volume 3: Early Modern Science*, edited by Katharine Park and Lorraine Daston, 238–50. Cambridge: Cambridge University Press.

Henry, Joseph. 1886. *Scientific Writings of Joseph Henry*, vol. 2. Washington: Smithsonian Institution.

Hewitt, Martin. 2012. "Beyond scientific spectacle." In *Popular Exhibitions, Science and Showmanship, 1840–1910* edited by Joe Kember, John Plunkett, and Jill A. Sullivan, 79–97. London: Pickering & Chatto.

Higgitt, Rebekah, and Charles W. J. Withers. 2008. "Science and sociability: Women as audience at the British Association for the Advancement of Science, 1831–1901." *Isis*, 99: 1–27.

Huxley, Leonard. 1908. *Life and Letters of Thomas Henry Huxley*, vol. 3. London: Macmillan.

Huxley, Thomas Henry. 1893. *Selected Works*, vol. 1. New York: Appleton.

Huxley, Thomas Henry. 1894. *Discourses Biological and Geological*. London: Macmillan.

James, Frank A. J. L. (ed.) 1991. *Correspondence of Michael Faraday*, vol. 1. London: Institution of Electrical Engineers.

James, Frank A. J. L. (ed.) 2008. *Correspondence of Michael Faraday*, vol. 5. London: Institution of Electrical Engineers.

Klestinec, Cynthia. 2011. *Theaters of Anatomy: Students, Teachers, and Traditions of Dissection in Renaissance Venice*. Baltimore: Johns Hopkins University Press.

Lightman, Bernard. 2004. "Scientists as materialists in the periodical press: Tyndall's Belfast address." In *Science Serialized*, edited by Geoffrey Cantor and Sally Shuttleworth, 199–237. Cambridge, MA: MIT Press.

Lightman, Bernard. 2007. *Victorian Popularizers of Science: Designing Nature for New Audiences*. Chicago: University of Chicago Press.

Lightman, Bernard, Gordon McOuat, and Larry Stewart (eds.) 2013. *The Circulation of Knowledge between Britain, India and China*. Leiden: Brill.

Livingstone, David N. 2007. "'Science, site and speech: Scientific knowledge and the spaces of rhetoric." *History of the Human Sciences*, 20: 71–98.

Livingstone, David N. 2014. *Dealing with Darwin: Place, Politics and Rhetoric in Religious Engagements with Evolution*. Baltimore: Johns Hopkins University Press.

Lonsdale, Henry. 1870. *A Sketch of the Life and Writings of Robert Knox*. London: Macmillan.

Lucier, Paul. 2009. "The professional and the scientist in nineteenth-century America." *Isis*, 100: 699–732.

Morus, Iwan R. 1998. *Frankenstein's Children: Exhibition, Electricity and Experiment in Early-Nineteenth Century London*. Princeton, NJ: Princeton University Press.

Morus, Iwan R. 2007. "'More the aspect of magic than anything natural': The philosophy of demonstration." In *Science in the Marketplace: Nineteenth-Century Sites and Experiences*, edited by Bernard Lightman and Aileen Fyfe, 336–70. Chicago: University of Chicago Press.

Morus, Iwan R. 2010. "Worlds of wonders: Sensation and the Victorian scientific performance." *Isis*, 101: 806–16.

O'Connor, Ralph. 2007. *The Earth on Show: Fossils and the Poetics of Popular Science, 1802–1856*. Chicago: University of Chicago Press.

Ong, Walter J. 1974. "Agonistic structures in academic life: past to present." *Daedulus*, 103: 229–38.

Phillips, Denise. 2012. *Acolytes of Nature: Defining Natural Science in Germany, 1770–1850*. Chicago: University of Chicago Press.

Porter, Roy. 1995. "Medical lecturing in Georgian London." *British Journal for the History of Science*, 28, No. 1: 91–9.

Schaffer, Simon. 1983. "Natural philosophy and public spectacle in the 18th century." *History of Science*, 21: 1–43.

Schaffer, Simon. 2012. "Transport phenomena: Space and visibility in Victorian physics." *Early Popular Visual Culture*, 10, No. 1: 71–91.

Secord, James A. 2014. *Visions of Science: Books and Readers at the Dawn of the Victorian Age*. Oxford: Oxford University Press.

Skouen, Tina. 2011. "Science versus rhetoric? Sprat's *History of the Royal Society* reconsidered." *Rhetorica: A Journal of the History of Rhetoric*, 29, No. 1: 23–52.

Smart, Benjamin H. 1819. *The Theory of Elocution*. London: John Richardson.

Sprat, Thomas. 1667. *The History of the Royal Society of London*. London.

Stewart, Larry. 1993. *The Rise of Public Science: Rhetoric, Technology and Natural Philosophy in Newtonian Britain, 1660–1750*. Cambridge: Cambridge University Press.

Watt-Smith, Tiffany. 2013. "Cardboard, conjuring and a 'very curious experiment'." *Interdisciplinary Science Reviews*, 38: 306–20.

White, Paul. 2003. *Thomas Huxley: Making the Man of Science*. Cambridge: Cambridge University Press.

CHAPTER THIRTY

Film, Radio, and Television

DAVID A. KIRBY

Print media were the first mass media technologies. Books, newspapers, and magazines allowed people in geographically distant locations to share information, but their reach was limited by literacy rates. Cinema, radio, and television spread ideas even further since audiences did not have to be able to read to understand their content. For scientists this inclusivity made them ideal tools for popularization. However, the history of science in cinema and on radio and television reveals a tension within the scientific community between those who saw in them the promise of universal education and those who were suspicious that their commercialized nature meant foregrounding entertainment over authenticity. This tension led to frequent clashes between the scientific community and media producers as scientists attempted to exert control over media content. Scientists also tried to establish clear boundaries for categories that were never stable, such as fiction vs. non-fiction, art vs. science, natural vs. artificial, science vs. sensationalism, and research vs. entertainment. It must be noted that scholarship on the history of science in cinema, radio, and television has overwhelmingly focused on the US and the UK. The cost of developing the infrastructure for production, dissemination, and reception of these technologies limited their initial development to countries willing to devote significant resources. This meant that the US and the UK became hubs for media production and the primary exporters of these entertainment products.

Film: Science on the Silver Screen

Science and movies have been intimately linked from cinema's beginnings in the late nineteenth century. Cinema, unlike radio and television, was not conceived of as a device for mass communication. Eadweard Muybridge and Etienne Jules Marey developed motion picture technologies in 1878 as scientific research tools to study animal movement (Tosi 2005). This technology could not only record phenomena that were inaccessible to human eyesight, it fostered a perception that the camera *objectively captured* time itself. Science took a "cinematographic turn" at the turn of the twentieth

A Companion to the History of Science, First Edition. Edited by Bernard Lightman.
© 2016 John Wiley & Sons Ltd. Published 2020 by John Wiley & Sons Ltd.

century as motion pictures began displacing established research techniques across a wide range of disciplines from astronomy to psychiatry (Canales 2002). Cinema's supposed "reality effect" allowed for truth claims about the objects and events fixed on the screen. Film theorist Andre Bazin provides a striking metaphor that captures what it was about cinema that led scientists to believe that movies represented "objectivity in time" (Bazin 1960, 8). Bazin claims that while photography "embalms time" leaving it as inert as the bodies of insects trapped in amber, cinematic time is more like an insect in a glass jar where it was confined but visibly alive.

More than any other discipline the life sciences embraced motion pictures. Medical scientists quickly adopted film as a standard teaching and training tool. Common themes for instructional films were medical disorders related to movement such as epileptic seizures. Cinema also represented an opportunity for studying "life" as opposed to learning from death. Paradoxically, biologists had previously gained knowledge about living creatures through the study of dead bodies. By using the movies he made in 1898 Ludwig Braun could repeatedly study a perpetually beating dog's heart rather than rely on observations gleaned through the lifeless dissection of an embalmed animal (Cartwright 1995). The epistemological interplay between life and death also came into play when researchers began employing cinema in conjunction with other scientific instruments that possessed a similar "penetrating vision," such as X-rays and the microscope. Skeletons are traditionally symbols of death. Yet, a skeleton came alive when the Scottish medical doctor John McIntyre combined cinema with newly discovered X-rays in 1897 to film a moving frog's leg (Tsivian 1996).

For most of the nineteenth century the study of microscopic structures and microbes relied on fixed histological slides. This static observational method did not allow for studying motion, so researchers did not ask questions about movement (Curtis 2013). With advances in tissue culturing and growing evidence for the germ theory of disease at the end of the nineteenth century biologists began asking questions about cellular functions and processes. These new questions required techniques for observing movement rather than examining static representations, so biologists began attaching movie cameras to microscopes. Temporality was central to the rapid development of micro-cinematography at the beginning of the twentieth century. Cinema permitted researchers to manipulate time through compression or extension. By using slow motion, scientists could explore phenomena that moved too quickly to be observed in real time, such as the physics of Brownian movement. Researchers could also employ time-lapse photography to observe phenomena that developed over days or months as if they were happening in real time, which was how Swiss biologist Julius Ries studied the fertilization and development of the sea urchin in 1909 (Landecker 2006).

Early scientific research films also proved to have significant entertainment value. Film scholar Tom Gunning (1990) describes early cinema's fascination with spectacular images as the "cinema of attractions." The technology's novelty was more important than narrative, so early filmmakers focused on what they could show instead of what they could tell. Pre-cinematic technologies employing projected images, such as Pepper's ghost, magic lantern shows, and *camera obscura*, had already shaped audience's emotional and intellectual engagement with scientific spectacle on a screen. In these demonstrations science *was* the special effect (Pierson 2002). For early cinema audiences scientific research films showing X-rays of a frog's leg or a virus attacking red corpuscles were no less magical than the trick cinematographic work of George Méliés. Micro-cinematography, in particular, was a major source of scientific spectacle for early entertainment films. Several movie companies in the early 1900s specialized in producing micro-cinematographic films for both scientists and for popular films. In the UK, zoologist Francis Martin Duncan produced scientific films for documentary film pioneer Charles Urban's company, while the French company Pathé supported physiologist Jean Comandon's film work and its rival Gaumont contributed to the research of Julius Ries. Films featuring microorganisms, such as Urban's *Typhoid Bacteria* (1903), were popular because they invoked simultaneous reactions of "attraction and repulsion." Audiences could marvel at the previously invisible wriggling, writhing aliens that actually lived around them, on them, and in them. Yet, these films were also frightening because of the role these microscopic monsters played in disease (Gaycken 2015).

We might refer to early popular science films as "proto-science documentaries" in that they presented scientific images in their "raw" form with little in the way of story or plot. Robert Flaherty's *Nanook of the North* (1922) not only changed the science documentary but also documentary films in general by adding a narrative structure with identifiable characters to what were nominally scientific recordings of the historical world. Flaherty did not rely on a linear reconstruction of the film footage he captured of the Inuk tribe living near Canada's Hudson Bay. Instead, he constructed *Nanook of the North*'s dramatic story primarily through staged re-creations and editing techniques. Despite the film's artificiality, Flaherty believed that his film provided authentic ethnographic insights into this culture because all of the staged events had happened at some point in the tribe's history whether or not his camera had captured the actual events at the time they had occurred. Flaherty's film was highly influential on subsequent science documentary filmmakers, who were no longer content to just document events and phenomena (Griffiths 2002).

For some early science documentarians, cinema's artificiality was not a flaw, it was what helped filmmakers expose the "truth." Avant-garde filmmaker Jean Painlevé believed that techniques like slow motion and close-ups allowed him to create art using scientific images during the interwar period. For Painlevé *art* was merely an alternative means for acquiring *truth*, so science documentaries could be both great art and legitimate science (Bellows and McDougall 2000). Russian filmmaker Vsevold Pudovkin did not consider editing to be a deceptive storytelling technique; instead he believed film editing created meaning out of randomness. Pudovkin's *Mechanics of the Brain* (1925) juxtaposed physiologist Ivan Pavlov's conditioning experiments alongside scenes from a film studio to create associations between the science of behavioral conditioning and cinema's psychological impacts (Sargeant 2000). British filmmaker Paul Rotha adopted Pudovkin's technique of dialectical montage for his films of the 1930s and 1940s celebrating modern technological innovations including aircraft, telephone networks, and electrical grids. Rotha was strongly influenced by his association with left leaning scientists like J. B. S. Haldane and Lancelot Hogben. He believed that documentary film could be a powerful persuasive tool in promoting their shared beliefs that science was a force for social progress (Boon 2008).

Public health officials, medical researchers, and progressive reformers also considered movies to be tools for social persuasion. By the 1920s there were already a massive number of educational films covering almost every aspect of public health, especially those dealing with what were referred to as "social diseases," such as tuberculosis, syphilis, and alcoholism, as well as other health-related social issues like eugenics. But the various groups making these films often held opposing opinions about the moral dimensions of disease, making cinema a battleground over competing visions about the best way to prevent these health crises (Pernick 1996). Medical scientists were also concerned that films aimed at lay audiences might undermine confidence in the medical professions by empowering the public to challenge medical authority. Despite this concern, movies continued to be an important component of health information campaigns throughout the twentieth century. By the early 1950s international health organizations such as the World Health Organization were largely responsible for coordinating the production of these films (Ostherr 2013).

Wildlife films quickly evolved from research films for studying animal behaviors into a popular film genre. Like travelogue films, wildlife films brought exotic locations to people who were not able to travel to places like the Masai Mara or the Arctic Circle. Wildlife films also shared similarities with the adventure film genre and they often mimicked the style of big game hunting expeditions as in Martin and Osa Johnson's *Simba* (1928). But, the commercial need for spectacle meant that wildlife films of the 1920s and 1930s frequently supplemented genuine footage with dramatic narrative structures and staged confrontations. This led to more sensationalized wildlife films including what were referred to as "nature faking" films such as *Ingagi* (1930). Scientists found the brazen phoniness of these pictures problematic because they called into question the authenticity of every wildlife film, even those used for scientific research. In 1931 wildlife filmmakers Ernest Schoedsack and Merian Cooper recognized that it would be easier, cheaper, and less controversial to create their own fictional "wildlife film" using special effects. The release of *King Kong* (1933) signaled the triumph of cinematic artifice over authentic nature at the box office (Mitman 1999).

An even more artificially constructed fictional "wildlife" film led to the brief re-emergence of wildlife documentaries as a commercially viable enterprise in the late 1940s. The success of *Bambi* (1942) convinced Walt Disney that the public still enjoyed movie narratives built around the lives of wild animals. Disney realized that by combining cheaply produced footage of real animals with playful voiceovers he could replicate *Bambi*'s dramatic storyline and its focus on an animal "star" for his "True Life Adventures" film series. The series was critically successful and highly influential, but the last theatrically released film in the series came only 11 years after the first. These films ultimately became staples on television in the 1960s and 1970s, which underscored the fact that while non-fiction science films were rarely shown in movie theaters after 1960 they found a welcome home on television (Bousé 2000).

Just as non-fiction films have adopted techniques from fictional storytelling, fictional movies have borrowed science's authority to legitimate their fantastical stories. The history of fictional scientific depictions reveals an overwhelming picture of a medium expressing deep-rooted fears of science in the twentieth century. Anxieties about recent discoveries from the end of the nineteenth century, including X-rays and developments with electricity, formed the basis for many films of the 1900s, while films in the 1920s featured chemistry's "dark side" after the use of chemical warfare in World War I. Numerous popular films in the 1930s revolved around killer microbes, but their narratives also featured heroic microbiologists saving humanity from the menace of infectious diseases. Films also expressed contradictory attitudes towards scientific developments, as was the case in the 1950s and 1960s where movies highlighted the destructive power of nuclear science alongside depictions of its progressive possibilities. By the 1970s, ecological disaster had replaced radiation as a movie threat, but these films also visualized how science could prevent these potential catastrophes (Frayling 2005).

Movies are constructed objects, which means that filmmakers have made very specific decisions to tell their cinematic stories about science in a particular way. Scientists attempted to influence how filmmakers made these decisions by serving as science consultants on movie productions going back to cinema's earliest days. Scientists' assistance allowed filmmakers to claim verisimilitude for films such as *A Blind Bargain* (1922) whether they followed scientists' advice or not. Many scientists willingly signed up to act as consultants because they believed that popular films were an effective way to raise public awareness for a scientific issue. This belief motivated scientists to work as consultants for the eco-disaster films of the 1970s. Scientists have also helped filmmakers create depictions of future technologies in the hopes that these films would convince audiences that these fictional technologies should become real technologies, as was the case for scientists working on space themed films of the 1950s and 1960s. Ultimately, studies of cinematic science have brought into focus the power of fictional narratives to shape our cultural meanings of science (Kirby 2011).

Radio: Broadcasting Science over the Airwaves

The first commercially licensed radio station, KDKA in Pittsburgh, began transmitting daily radio programs in 1920. Radio was a broadcast medium, which made it a radically different form of mass communication. Radio created shared experiences across geographic distances, which generated a sense of community amongst listeners that transcended race and class. At the same time, radio's status as a technology of the home provided it with a deceptive intimacy that made its presenters seem personable and accessible. Many science advocacy organizations immediately seized on radio as an opportunity to promote science and improve science literacy. However, there were a small number of gatekeepers who controlled the right to use the fixed number of radio frequencies available for broadcasting (LaFollette 2008).

In the UK, domestic broadcasting rights were limited to a single gatekeeper, the British Broadcasting Corporation (BBC) until the mid-1950s. The UK government considered broadcasting too important a public resource to be left to the free market, so in 1927 they restructured the BBC into a public service broadcasting monopoly. The BBC's first Director-General, John Reith, built the corporation's broadcasting policy around three values: education, information, and entertainment. Reith believed that radio programming could be a means of culturally improving the citizenry, so he favored the tenets of information and education and saw entertainment as an unfortunate necessity. Given Reith's preferences, science seemed to be an ideal topic and science broadcasting became a significant part of the BBC's Talks Department's output in the late 1920s and early 1930s (Farry and Kirby 2012).

BBC science programming concentrated so much on informing and educating listeners that entertainment was almost completely absent. The earliest science broadcasts consisted primarily of didactic lectures by eminent scientists, such as the 1930 series *The Stars in their Courses* presented by astronomer James Jeans. While there were no direct domestic broadcast competitors to the BBC, there were foreign commercial radio stations whose signal carried into the UK, such as Radio Luxembourg. Competition for listeners with these more entertainment-oriented stations forced the BBC to reconsider its policy of neglecting entertainment, but the corporation still wanted radio content that satisfied its public service values. A new breed of BBC radio producer emerged in the 1930s and 1940s who specialized in creating entertaining science broadcasts that were also informational. Most prominent amongst these new science producers were Mary Adams and Ian Cox, whose media expertise enabled them to develop engaging but highly educational shows like *The Night Sky* and *Inquiring into the Unknown* (Jones 2012).

Science producers like Adams defined success by broadcasting standards and audience numbers. But, their more entertaining approach to radio science did not always sit well with scientists. In fact, many scientists construed these producers' influence over the content of programs as a problem since these producers' expertise was in media production not science. The history of the BBC's science broadcasting is notable for the continuous attempts by scientific organizations and individual scientists to exert control over the BBC's practices regarding science coverage (Jones 2013). Four major scientific institutions—the Royal Society, the British Association for the Advancement of Science, the Department of Scientific and Industrial Research, and the Association of Scientific Workers—routinely brought pressure to bear on the BBC to establish a scientific advisory panel or an oversight committee for science programming. Occasionally this pressure forced the BBC to bring on board an official scientific advisor, as was the case in 1950 when physiologist Henry Dale was appointed senior scientific advisor for an experimental period of one year. The BBC finally established the Science Consultative Group (SCG) in 1964 after the influential Pilkington report in 1962 chastised the BBC for its lack of quality programming across its broadcast media. The SCG ultimately pacified these scientific institutions and it provided scientific advisors for BBC production staff into the 1990s (Jones 2014).

The US federal government also tightly regulated access to radio frequencies in the early 1920s. Organizations needed to demonstrate that they would be fulfilling a public service before they could obtain a broadcast license. As in the UK, this restriction proved to be a favorable environment for science on US radio since museums and universities were among the few types of organizations who could meet the public service requirement (Rinks 2002). Science departments within these organizations frequently produced their own radio programs, especially at agriculturally oriented land-grant universities. Oregon State University's radio station's official slogan "Science for Service" nicely captured one of the primary objectives of radio stations at these agricultural colleges (Slotten 2006, 255). While the UK government restricted radio access to a single public service corporation, the US Federal Radio Commission allowed commercial organizations to compete for radio frequencies in the late 1920s. This competition pushed out most of the museum and university based broadcasters in favor of a small number of commercial broadcasters, including CBS and NBC, who soon dominated the American broadcasting landscape (Streeter 1996, 98–101).

Even in this more commercialized environment there was a place for science programming on American radio. Advertisers on commercial radio avoided certain times, such as late at night; broadcasters considered science programming ideal content to fill these unsold time slots. Science programs were low cost and they allowed broadcasters to meet any public service mandates still demanded by their broadcast license. This is not to say that science was a prominent feature of US radio in the 1930s and 1940s. Science programs still had to compete with other types of educational programming for access to even the limited radio slots available through unpurchased airtime. Many scientists were also uneasy and suspicious of the new medium, which made them reluctant to take part in radio broadcasts. So, while science programs had some utility for commercial broadcasters, it was still a relatively small part of overall radio content on early US radio (LaFollete 2008).

The Smithsonian Institution was among the most prolific producers of American science broadcasts from the 1920s through to the 1950s. However, most scientists and scientific organizations did not have the kinds of resources that the Smithsonian Institution possessed. In order to provide every scientist with an outlet for popularization, newspaper publisher E. W. Scripps collaborated with national scientific organizations to found the news clearinghouse Science Service in 1921. The organization quickly abandoned their attempt to make science-based films in 1922. Compared to cinema, radio was cheaper, easier to produce, and Science Service could have total control over content. Although American scientific organizations embraced radio, it was clear that scientists themselves did not understand how to take advantage of the new medium's capabilities. Unlike the BBC with its specialized science broadcasters, scientists in the US had to produce their own radio shows. This meant that early science programs on radio differed little from formal scientific lectures. Most scientists failed to understand that reading notes into a radio microphone was not the same thing as speaking in front of a live audience. On radio scientists could not perform demonstrations, show illustrations, use hand gestures, or respond to audience feedback. The time sensitivity of radio also did not allow for spontaneous deviations from prepared remarks. Scientists measured radio success by the accuracy of the information and prestige of the speaker. This was a stance that ultimately clashed with radio broadcasters, who defined success by audience numbers and entertainment value (LaFollette 2002).

By the 1930s scientific organizations had to adapt their broadcasting approaches to compete with the comedians and detective dramas dominating the airwaves. This meant providing scientific personalities and dramatic science programming. CBS changed the name of Science Service's *Science News of the Week* program in 1938 to the more exciting sounding *Adventures in Science* (1938–1957), while also transforming it from a lecture-based show into a dramatized program. The new program became Science Service's most successful science radio program (Terzian 2008). Broadcasters' demand for entertaining science programs required tradeoffs. More engaging programs were far more expensive to produce than shows with simple lecture formats. Scientists were also concerned that any compromises to scientific integrity could potentially undermine public confidence in science. Even Smithsonian marine biologist Austin Hobart Clark, a scientist with extensive radio experience, saw his primary goal as protecting the reputation of science, not as producing entertaining programs. Most US scientific organizations did not see radio as enough of a priority to devote significant resources towards developing more exciting programs. So, unlike in the UK where the late 1940s and early 1950s represented a golden age for radio science, science broadcasting in the US became increasingly marginalized as broadcasters favored more commercially viable content that scientists were unable or unwilling to produce (LaFollette 2008).

There were two major exceptions to the dearth of science on postwar American radio: agriculture and the medical sciences. Strong governmental backing made agricultural sciences on radio possible. Medical-themed radio programs benefitted from the participation of the American Medical Association (AMA). The medical community initially had a number of misgivings about radio, including concerns that medical radio programs would create hypochondriacs, promote self-diagnosis, and arouse false hopes of miracle cures. But, the number of unlicensed "quack" doctors peddling phony medicines on the radio in the 1920s alarmed the AMA. So, they decided to offer advice to radio producers and to produce their own radio programs. The AMA conveyed legitimacy to any programming with its approval and the medical community's ability to speak with a single voice gave it significant influence over the quality of radio broadcasts (Turow 2010). On a global scale, agriculture and medicine remain the two biggest areas for science on the radio. The use of fictional radio dramas to disseminate agricultural or medical information, what practitioners call "entertainment education," began in the 1950s. It is still important today in countries with numerous remote regions where the public does not have ready access to televisions or computers and which have high illiteracy rates, particularly in Africa, South America, and Asia (Poindexter 2004).

Television: Science in Every Living Room

In 1936 the BBC launched the first television broadcasts in the UK. NBC broadcast the first regularly scheduled television programs in the US in 1939. The outbreak of World War II stunted television's development, as the BBC suspended all television activities and American broadcast companies like RCA turned their attention towards military production. The BBC resumed its broadcasts in 1946 and by 1948 four US networks (NBC, CBS, ABC, and DuMont) were broadcasting full prime-time schedules. Television receivers became more affordable after the war and the medium's popularity skyrocketed. Broadcasters' experiences with radio strongly influenced how they approached science on television. Many of the first science television programs, like DuMont's 1946 series *Serving through Science*, simply mimicked radio's lecture model but with added visuals (LaFollette 2013). Producers quickly learned, though, that television was not just "radio with pictures." Television shared its narrative structures and conventions with radio but its visuality allowed for more dramatic representations of science using illustrations, demonstrations, and reconstructions. Despite being a visual medium like cinema, though, science programs on television were not just movies transposed on to a smaller screen. Television's uniqueness forced producers to develop their own styles for science programming more suited to the new medium (Boon 2008).

Radio's wartime service had enhanced its reputation and the BBC's postwar structure was oriented towards furthering radio's development. In contrast, many BBC executives considered television to be a lowbrow medium and they isolated it as one of six "radio" divisions. The prospect of a commercialized television rival in the early 1950s, however, forced BBC executives to accord more importance to television. The establishment of the Independent Television Authority in 1954 and the subsequent creation of the ITV network in 1955 accelerated the BBC's shift towards television as competition for audiences intensified (Farry and Kirby 2012). Scientific programs were not a regular feature on UK television in the 1940s, with *Inventor's Club* (1948–1956) representing one of the few programs dedicated to science. But, in the early 1950s BBC television producers such as Aubrey Singer and Grace Wyndham Goldie convinced executives that science's prestige combined with its televisual appeal could make it an asset for the BBC in meeting the challenge of external rivals while also maintaining its public service ideals. The Television Talks division, founded in 1953, was particularly eager to exploit science's televisual potential. One of the division's heads was Mary Adams who brought with her significant experience as a radio producer specializing in science (Boon 2008).

The launch of the Soviet satellite Sputnik in October 1957 moved scientific developments to the center stage in world affairs, furthering a belief that scientific programs could help the BBC compete with ITV within the restrictions of educational programming. The popularity of Jodrell Bank director Bernard Lovell's 1958 Reith lectures *The Individual and the Universe* also showed them that stories about science could attract large audiences. However, the influential 1962 Pilkington Committee Report on the state of British broadcasting called into question the perception that scientific programs were inherently educational. The Pilkington Committee expressed a concern that British television had become trivialized and prone to American-style commercialism. Although supportive of science programming in general, the report was critical of the fact that much of the BBC's science programming exploited the dramatic nature of the "space race" rather than focusing on science's educational merits (Farry and Kirby 2012). The scientific community were already dismayed at the privileging of drama over scientific integrity in televised science, so they saw the report as another opportunity to seek influence over broadcasters. The Pilkington Report led to a number of changes within the BBC including the launch of BBC2 as a showcase for educational programming in 1964, the creation of a Science and Features Department in 1963, the launch of the influential science documentary series *Horizon* in 1965, the development of the science fiction drama *Dr Who* in 1963, and the establishment of the Science Consultative Group in 1964 (Boon 2008, 209–32).

Unlike the BBC, there was no conflict for American broadcasters about shifting their resources almost entirely towards television's development. As was the case with radio, many in the scientific community saw television as a golden opportunity for mass education. This meant that universities, museums, observatories, and zoos produced many of the earliest science programs. CBS broadcast *The Johns Hopkins Science Review* from 1948 to 1955, while the Franklin Institute's Fels Planetarium in Philadelphia hosted NBC's *The Nature of Things* (1948–1953). One of the longest running of these early science shows, *Science in Action* (1950–1966), was produced by the California Academy of Sciences. These programs were an improvement on early shows like *Serving through Science*, but their approaches were still relatively conservative. *Science in Action*'s format was standard, with a scientist in a white lab coat explaining concepts using props and demonstrations (see Figure 30.1). This somewhat dry, didactic style satisfied the scientific community's educational goals, but it did not generate the kinds of audience numbers that satisfied advertisers. Science did offer arresting visuals like solar eclipses, swarms of birds on the Kenyan plains, and images from cloud chambers,

Figure 30.1   Photo of presenter Earl Herald on the set of Science in Action. © Californian Academy of Sciences.

but, just as with early cinema, the appeal of visual novelty was limited. Television producers required drama and compelling characters if science on television was going to be competitive on commercial television in the 1950s (LaFollette 2008).

Television was a more expensive medium than radio, so it was difficult for scientific organizations to create their own programming without generous underwriting from sponsors like the Ford Foundation, Bell Laboratories, or AT&T. But, sponsors demanded some influence over the content of the programs they were funding. Negotiations between the Smithsonian Institution and television producers in 1954 collapsed, for example, when the producers suggested that the proposed series feature comedian Fred Allen lightly riffing on Smithsonian exhibits. Yet, some shows in the 1950s demonstrated how science programs could be both educational and entertaining, such as the game shows *The Big Idea* (1952–1953) and *What in the World?* (1951–1955). Children's science programming also successfully melded education and entertainment. Highly charismatic and enthusiastic hosts like *Watch Mr. Wizard*'s (1951–1965) Don Herbert made science appear fun and exciting. Despite these profitable fusions of science and entertainment, American commercial networks shied away from prime-time science programming in the 1960s. The situation for science on television changed in 1970 with the creation of the Public Broadcasting Service (PBS). As with the BBC, PBS's mandate to broadcast educational programming provided an avenue for scientific shows to strike a balance between education and entertainment that would satisfy the scientific community. Shows on PBS like *NOVA* (1974–present) and Carl Sagan's *Cosmos* (1980) demonstrated to television executives that it was possible to produce quality science programming that millions of people would tune in for (LaFollette 2013).

Television was an expensive undertaking for both audiences and broadcasters. This meant that most countries did not start producing their own programs until the 1960s and 1970s. In the 1950s network programming for most countries came primarily from US imports. Science programming was a low priority for national broadcasters, so they did not begin producing their own science shows until the 1970s and 1980s. South Korea's history of science on television is typical for non-European countries. The first television station in South Korea began broadcasting in 1961, but its programming drew from American imports until the late 1970s. When South Korean studios did begin producing their own programs, however, the focus was on entertainment shows like soap operas. South Korea did not develop its own science programming until the 1990s with shows like *Paradise for Curiosity* (1998–present; Lee 2003). Once science programs began to be produced outside North America and Europe they often took on a nationalistic flavor by promoting their country's scientific achievements. In Israel, for example, scientists worked with state-run Channel 1 to produce propagandistic science programs in both Hebrew and Arabic that were meant to convey the image of a scientifically elite nation in the aftermath of the Six Day War in 1967 (Katz-Kimchi 2012).

Wildlife films were one of the most successful genres of early science television. Television quickly adapted the narrative styles established in Disney's successful "True Life Adventure" films, but wildlife filmmakers also had to develop new codes and conventions for a medium that required programming on a weekly basis. In the 1950s the UK and US developed divergent approaches to wildlife television. The British style was informative and closer in style to a nature documentary with an emphasis on scientific inquiry. The UK's first wildlife program, *Zoo Quest* (1954–1963), exemplified this approach with its focus on the educational value of bringing animals back to the UK for study and display. The show turned its presenter David Attenborough into one of the UK's most recognizable voices of scientific authority (Davies 2000). The American tradition incorporated danger-oriented narratives with dramatic action, storytelling, and funny animal characters, while filming in controlled conditions. This style can be seen in the US's first wildlife television program, *Zoo Parade* (1950–1957), which was renamed *Wild Kingdom* (1963–1985). The series featured its presenter Marlin Perkins and his assistant Jim Fowler in a series of dangerous confrontations with animals. The difference between these two styles was evident in the re-editing of the 1990 BBC program *Trials of Life* for American television. The American version highlighted the show's violent scenes and was marketed with the tagline "find out why we call them animals." The American tradition's continual need for dramatic encounters led to a number of high profile scandals involving staged confrontations and animal mistreatment, many of which were covered in the 1986 television documentary *Cruel Camera* (Bousé 2000).

Just as with cinema and radio, medical dramas became a staple on television in the 1950s. The first medical shows on US television, *City Hospital* (1951–1953) and *The Doctor* (1952–1953), continued a dramatic formula that began with Dr. Kildare in the 1930s by portraying doctors as heroic figures using the weapons of biomedical research to fight death. The medical establishment was extensively involved in the production of television shows throughout the 1960s and 1970s. The AMA not only encouraged its members to act as consultants, they also provided assistance to producers who wanted to film scenes inside real hospitals. The AMA's cooperation helped television producers achieve an added level of realism, but it also benefitted the medical community by fostering popular representations of doctors as impassioned professionals who would do whatever it took to save a patient. Producers' desire to get the AMA's seal of approval also enabled the medical organization to veto storylines they felt could harm their image, such as stories promoting socialized medicine. Medical dramas proved to be among the most consistently popular genres on television across the decades, with shows like *Marcus Welby, M.D.* (1969–1976), M*A*S*H (1972–1983), and *St. Elsewhere* (1982–1988) proving to be ratings hits (Turow 2010).

As with cinema and radio, the history of televised science reveals that scientific content could be successful in competitive media markets. Science satisfied audiences' desire for spectacle, while it also offered media producers legitimacy for their dramatic stories. But media professionals' need to transform science into entertainment led to tensions with the scientific community who continually tried to exert control over productions. There are still unanswered questions about the history of science in cinema, radio, and television, in particular about dissemination practices and the reception of these entertainment texts amongst different cultural groups. We also need more studies into locally based productions as well as those not emerging from Hollywood or Western Europe. Ultimately, studying the history of science in mass media technologies can help us understand how the scientific community will respond to media technologies of the future like podcasts and YouTube.

References

Bazin, André. 1960. "The ontology of the photographic image." *Film Quarterly*, 13: 4–9.

Bellows, Andy M., and Marina McDougall (eds.) 2000. *Science Is Fiction: The Films of Jean Painlevé*. Cambridge, MA: MIT Press.

Boon, Tim. 2008. *Films of Fact: A History of Science in Documentary Films and Television*. London: Wallflower Press.

Bousé, Derek. 2000. *Wildlife Films*. Philadelphia: University of Pennsylvania Press.

Canales, Jimena. 2002. "Photogenic Venus: The "cinematographic turn" and its alternatives in nineteenth-century France." *Isis*, 93: 585–613.

Cartwright, Lisa. 1995. *Screening the Body: Tracing Medicine's Visual Culture*. Minneapolis: University of Minnesota Press.

Curtis, Scott. 2013. "Science lessons." *Film History*, 25: 5–54.

Davies, Gail. 2000. "Science, observation and entertainment: Competing visions of postwar British natural history television, 1946–1967." *Ecumene*, 7: 432–60.

Farry, James, and David A. Kirby. 2012. "The universe will be televised: Space, science, satellites and British television production, 1946–69." *History and Technology*, 28: 311–33.

Frayling, Christopher. 2005. *Mad, Bad and Dangerous? The Scientist and the Cinema*. London: Reaktion.

Gaycken, Oliver. 2015. *Devices of Curiosity: Early Cinema and Popular Science*. Oxford: Oxford University Press.

Griffiths, Alison. 2002. *Wondrous Difference: Cinema, Anthropology, and Turn-of-the-Century Visual Culture*. New York: Columbia University Press.

Gunning, Tom. 1990. "The cinema of attraction: Early film, its spectator and the avant-garde." In *Early Cinema*, edited by Thomas Elsaesser, 56–75. London: British Film Institute.

Jones, Allan. 2012. "Mary Adams and the producer's role in early BBC science broadcasts." *Public Understanding of Science*, 21: 968–83.

Jones, Allan. 2013. "Clogging the machinery: The BBC's experiment in science coordination, 1949–1953." *Media History*, 19: 436–49.

Jones, Allan. 2014. "Elite science and the BBC: A 1950s contest of ownership." *British Journal for the History of Science*, 47, No. 4: 701–23.

Katz-Kimchi, Merav. 2012. "Screening science, producing the nation: Popular science programs on Israeli television (1968–88)." *Media, Culture & Society*, 34: 519–36.

Kirby, David A. 2011. *Lab Coats in Hollywood: Science, Scientists, and Cinema*. Cambridge, MA: MIT Press.

LaFollette, Marcel C. 2002. "A survey of science content in US radio broadcasting, 1920s through 1940s: Scientists speak in their own voices." *Science Communication*, 24: 4–32.

LaFollette, Marcel C. 2008. *Science on the Air: Popularizers and Personalities on Radio and Early Television*. Chicago: University of Chicago Press.

LaFollette, Marcel C. 2013. *Science on American Television: A History*. Chicago: University of Chicago Press.

Landecker, Hannah. 2006. "Microcinematography and the history of science and film." *Isis*, 97: 121–32.

Lee, Dong-Ho. 2003. "A local mode of programme adaptation: South Korea in the global television format business." In *Television Across Asia: TV Industries, Programme Formats and Globalisation*, edited by Albert Moran and Michael Keane, 36–53. London: Routledge.

Mitman, Gregg. 1999. *Reel Nature: America's Romance with Wildlife on Film*. Cambridge, MA: Harvard University Press.

Ostherr, Kirsten. 2013. *Medical Visions: Producing the Patient Through Film, Television and Imaging Technologies*. Oxford: Oxford University Press.

Pernick, Martin. 1996. *The Black Stork: Eugenics and the Death of "Defective" Babies in American Medicine and Motion Pictures Since 1915.* Oxford: Oxford University Press.

Pierson, Michele. 2002. *Special Effects: Still in Search of Wonder.* New York: Columbia University Press.

Poindexter, David. 2004. "A history of entertainment education, 1958–2000." In *Entertainment-Education and Social Change: History, Research, and Practice*, edited by Arvind Singhal, Michael Cody, Everett Rogers, and Miguel Sabido, 21–37. Mahwah, NJ: Lawrence Erlbaum.

Rinks, J. Wayne. 2002. "Higher education in radio 1922–1934." *Journal of Radio Studies*, 9: 303–16.

Sargeant, Amy. 2000. *Vsevolod Pudovkin: Classic Films of the Soviet Avant-Garde.* London: I. B. Tauris.

Slotten, Hugh. 2006. "Universities, public service radio and the 'American system' of commercial broadcasting, 1921–40." *Media History*, 12: 253–72.

Streeter, Thomas. 1996. *Selling the Air: A Critique of the Policy of Commercial Broadcasting in the United States.* Chicago: University of Chicago Press.

Terzian, Sevan. 2008. "'Adventures in science': Casting scientifically talented youth as national resources on American radio, 1942–1958." *Paedagogica Historica*, 44: 309–25.

Tosi, Virgilio. 2005. *Cinema Before Cinema: The Origins of Scientific Cinematography.* London: British Universities Film & Video Council.

Tsivian, Yuri. 1996. "Media fantasies and penetrating vision: Some links between X-rays, the microscope, and film." In *Laboratory of Dreams*, edited by John Bowlt and Olga Matich, 81–99. Stanford, CA: Stanford University Press.

Turow, Joseph. 2010. *Playing Doctor: Television, Storytelling, & Medical Power.* Second edition. Ann Arbor, MI: University of Michigan Press.

PART IV

Tools of Science

CHAPTER THIRTY-ONE

Timing Devices

RORY MCEVOY

Our most basic concept of time is shaped by the apparent motions of stars and moon across our skies. The Earth's rotation, tilted axis, and annual orbit around the Sun give us the most basic intervals: the day, seasons, and equatorial year. For this reason, time finding was necessarily an astronomical endeavor and time measurement was standardized by the Earth's rotation, until 1967 when it was superseded by atomic timekeeping. The arrangement of many prehistoric monuments suggests that their founding purpose was connected to time measurement and perhaps that they served as ancient observatories. But in dealing with prehistory careful analysis of existing evidence can only offer tentative suggestions regarding function or practice. Of the many suspected archaeo-astronomical sites, there are some in particular that bear strong evidence that their function was partly for astronomical determination of time.

For example, Wurdi Youang, near Melbourne, Australia, and the Chankillo temple complex in Peru, offer very different levels of evidence suggesting likely function as a time-finding tool. The elliptical placing of stones at the Aboriginal site yielded slight indication of its possible purpose. When referenced from a particular point within the monument, evidence showed a likelihood that the site was, in part, used to observe the annual solar cycle with stones positioned to mark the setting sun at the equinoxes and solstices. In contrast to Wurdi Youang, the 13 stone towers at Chankillo offer more convincing evidence of their builders' astronomical knowledge and the site's horological function. They are placed along a north–south line between two viewpoints, from which the solstices can be precisely observed as the Sun rises and sets alongside the extremities of the outermost towers. The towers could have been used throughout the year to ascertain the date, but because the Sun's passage along the horizon slows around the solstices, this would have only been accurate to within a few days (Ghezzi and Ruggles 2007; UNESCO-IAU 2012; Norris, Norris, Hamacher, and Abrahams 2013).

The oldest known surviving tools used for subdividing the day into hours come from ancient Egypt. For the determination of time a T-shaped shadow clock dates to

A Companion to the History of Science, First Edition. Edited by Bernard Lightman.
© 2016 John Wiley & Sons Ltd. Published 2020 by John Wiley & Sons Ltd.

around 1500 BCE, and for keeping time or measuring short intervals the remains of a water clock, made around a century later. The water clock or clepsydra (translates as water thief) is a simple open vessel, which tapers towards the base where there is a small aperture to allow the water to leak out slowly; the time was indicated by the level of the water against a scale on the inner wall of the vessel, and the tapered shape helped to maintain an even water pressure as the jar emptied, allowing for an almost linear indication of elapsed time. The sundial was the most reliable time-finding device and provided the time standard to calibrate the water clocks so that they could be used at night or to measure short intervals of time (Mills 1982; Ackermann 1999).

The lasting influence of Egyptian astronomy was underpinned not only by favorable climatic conditions, but most importantly a secular civil calendar, which divided the year into 12 months, each of 30 days with an additional five days to correspond to the solar year. Its simplicity facilitated the comparison and study of older astronomical observations in much in the same way as the modified Julian date eases the work of modern-day astronomers. It is also through the Ancient Egyptians that we inherited our system for measuring time—that is counting in blocks of 12 hours, 60 minutes, and 60 seconds. Through their continuation of the use of the Sumerian and Babylonian sexagesimal system for mathematics and by dividing the day into ten parts with a further two parts for twilight at dawn and dusk today's 24-hour system gradually evolved (Neugebauer 1969).

Civil time, determined by the sundial, observed equal numbers of hours for both day and night and so the length of these hours varied according to the season. This unwieldy system of measurement was impractical for astronomy. Instead, sidereal time was found by observing the transits of bright stars, known as decans, which divided the ecliptic into ten-degree segments. These "clock stars" were observed using a notched staff, called a *merkhet*, and a plumb line aligned to the pole star. Whereas a sidereal day is defined by the time it takes for one full rotation of the Earth on its axis, a solar day is longer due to the additional factor of the Earth's orbit. The Earth rotates by an extra degree before the Sun is at its highest point again, which equates to just under four minutes time. Tables for conversion of sidereal time to civil time survive inside some sarcophagi (Neugebauer 1969; Whitrow 1988).

The water clocks that were used to measure time between star transits saw further improvement in Egypt. Vitruvius in *De Architectura* (c.20 BCE) described the new design and attributed its invention to Ktesibios around 250 BCE. The improved design overcame the problem of unequal water pressure by employing two cisterns. The first vessel was a simple outflow clepsydra kept full by a constant flow of water to provide constant pressure and regular flow into the second vessel, a straight sided tank. The rising water level measured the elapsed time, which could be indicated by a float and pointer against a linear scale or, if a rotating indication was required, the float could be attached to a rack and pinion[1] or a counter-weighted rope and pulley.

Vitruvius also described a clepsydra with a planispheric view of the stars marked onto a disc that rotated behind fixed reference wires to delineate the local horizon, meridian, ecliptic, and tropics, which gave a real-time view of the heavens. These are referred to as anaphoric clepsydra. Roman remains of such planispheric dials found across Europe indicate that their use was reasonably widespread. It is unlikely that these were consulted in the same manner as public clocks today, instead they were more likely to have been used as astrological tools for casting horoscopes (Noble and Price 1968; MacKay 1969).

In classical antiquity the study of the influences of celestial bodies on corporeal affairs was a branch of natural philosophy and was arguably one of the principal motivations for astronomical study. Claudius Ptolemaeus (Ptolemy), born towards the end of the first century CE, is best known for his compilation of mathematical astronomy, *Almagest*, and to a lesser extent for its accompanying work, *Tetrabiblos*, which covered astrological effects of astronomical motion. By the Hellenic era astrology had spread from the exclusive domain of the royal courts and military to the individual, with natal astrology having particular cultural importance. Astrological practitioners had to be proficient in time finding, have a good theoretical knowledge of astronomy, and be able to compile or even use an ephemeris (North 2005).

Description and illustration of astrologers at work in Islamic manuscripts generally show the three tools as being the ephemeris, dust board (similar to a chalk board or slate), and astrolabe. The astrolabe was the most widely used time-finding instrument of the medieval period and its existence owed much to Ptolemy's *Planisphaerium*, which preserved the method of stereographic projection, essential to project the celestial sphere onto its plate. Exactly when the astrolabe was invented is not known as the survival rate of early examples is very low, but a treatise on its construction and use by Theon of Alexandria (c.375 CE) gives us some indication. The oldest known surviving Islamic example, now part of the al-Sabah collection, dates to 927/8 CE (Neugebauer 1969; Saliba 1992; Ackermann 1999).

The astrolabe was a highly versatile instrument with various potential uses. As a portable map of the stars that could be adjusted to represent the sky view for any given time and date it was a useful learning tool for students of astronomy. The astrolabe can be used to find the time in both seasonal and equinoctial hours. Space on the reverse of the instrument was rarely wasted, often inscribed with astrological information or equipped with other features such as an alidade[2] and engraved shadow square for surveying heights (Ackermann 1999).

An early example of mathematical gearing is featured on the reverse of an astrolabe made around 1221/2 CE by Abi-Bakr of Isfahan (Museum of the History of Science, Oxford, inv. 48213). The engaging wheels have different tooth counts to reduce the motion of the rete[3] to indicate the moon's phase and age, with a zodiacal calendar showing the relative positions of the Sun and Moon. The positioning and design of the toothed wheels in Abi Bakr's astrolabe correspond closely with an illustrated design produced some 200 years earlier by the Muslim polymath, al-Biruni (973–1048). To facilitate precise manual division of the wheels, the tooth count in Abi Bakr's lunar train was modified to include more even numbers (Price 1974).

The equilateral triangle tooth form found in both the manuscript and on the wheels of the lunar model was identical to that found in the ancient mechanical astronomical model, known as the Antikythera mechanism. Thanks to improved methods of X-ray analysis employed in 2005, the Antikythera mechanism's original function is now well understood. Its mathematically complex geared train modeled the Greek geocentric view of the cosmos and the motions of the Sun, Moon, and the then known five planets. As well as calculating lunar and solar eclipses it also had a social calendric function with a subsidiary indication for the quadrennial pan-Hellenic games cycle (Price 1974; Freeth, Jones, Steele, and Bitsakis 2008).

The extraordinary survival, which is the Antikythera mechanism, testifies to the level of sophistication of mathematical gearing that the Greek mechanics had attained, but it also suggests an abrupt technological loss. Whilst the geared astrolabe shows tangible evidence of surviving knowledge of Greek gearing, it was more likely transmitted via mechanical texts, such as those written by Vitruvius and Heron, who described and illustrated the 60-degree tooth form in simpler machines such as clepsydrae with a rack and pinion. Islamic clocks of the middle ages, such as the example presented to Charlemagne by Harun ar-Rashid in 807, are unrelated as their complex automata operated on largely pneumatic principles with levers and pulleys rather than mathematical gearing (Price 1974; Kurz 1975).

A twentieth century translation and study of Chinese material revealed that during this apparent interregnum in the West, advanced monumental clepsydra were constructed, using gears to model the heavens for royal calendric and astrological needs. The tradition of building monumental water powered astronomical models in China begins in the second century CE with Chang Heng (born c.78 CE), who, in an incredible feat of engineering, managed to power the slow rotation of a bronze celestial sphere of around fourteen and a half feet in diameter using a low-powered clepsydra. The sphere was contained indoors to protect it from the weather so that the heavens could be consulted in comfort by the emperor and his officials whilst a platform on the roof was equipped with an observational armillary sphere[4] that enabled astronomers to check that the model was performing as it should. Essentially, Chang Heng had achieved the same end as the Western anaphoric clepsydrae, albeit on a grander scale.

China was far from insular and during the eighth century, as part of a general calendar reform, underwent a fundamental change in astronomical practice toward Ptolemaic methods thanks to Indian influence. Tantric Buddhist monk, I-Hsing (672–727), was commissioned to reorganize the calendar and, in order to improve the prediction of eclipses, modified the tubes on the armillary spheres to observe along the ecliptic rather than the celestial equator, which had been traditional in China beforehand. His influence went beyond astronomy into time measurement. His "Water-driven Spherical Bird's-eye-View Map of the Heavens" could equally be called a clock for, as well as its astronomical model, it featured two jacks, one of which struck a bell to indicate the passing of hour, the other a drum to sound the quarters. This feature is an indication of probable Western influence through diplomatic relations with Byzantium. Of particular importance to the history of the mechanical clock is that the rotation of the driving waterwheel was smoothed by a stop–start mechanism that regulated its rotation by means of a clepsydra arrangement that periodically released and locked the wheel (Needham, Ling, and Price 1960, 74–9).

Such a mechanism is known as an escapement and is the beating heart of any mechanical clock. Needham (1959) argued that the Chinese "Cosmic Engine," as it was referred to by Su Sung in the eleventh century, was the missing link in the development of the mechanical clock. Whilst there is some evidence in the last manuscript work of Ismail ibn al-Razzaz al-Jazari (1136–1206) of transmission of knowledge from China to the West of this form of escapement, it does not bear any similarity to the verge[5] escapement, prevalent in early European clocks, that was first documented in the mid-fourteenth century. This evidence suggests that the Chinese did invent the first mechanical clock, but that the European clock does not share the same mechanical heritage (Needham 1959).

In tracing the beginnings of the mechanical clock in Europe, a significant reference is found in Robertus Anglicus' commentary (1271 CE) on Sacrobosco's *Sphere* (about 1230 CE). Within the section on sidereal and apparent solar time, the author describes a clock that could make a full rotation "between sunrise and sunset except for as much time as it takes to cover one degree." This description alone is of little consequence as it makes no advance on Vitruvius's description of the clepsydra with anaphoric dial, but the fact that Anglicus was familiar with weight-driven clockwork is. Maybe he had seen a compartmented clepsydra, which slowed geared rotation by acting as an inertial brake on the pull of a driving weight. This device is described in *Libros del Saber de Astronomi,* a pivotal transfer of astronomical and mechanical knowledge from the Islamic world into Medieval Europe, commissioned by Alphonse X of Castile and produced within a decade after the commentary. However, it is just one part of the evolution of the mechanical clock in Europe and the critical component, the escapement, is not documented before the mid-fourteenth century (Thorndike 1941; Bedini 1962).

This pivotal period in horological history is obscured by the ambiguity of the term *horologium,* which at the time was applied to any time-related instrument including sun dials, sand glasses, clepsydrae, mechanical clocks, automated bells, and even manually rung bells. The mechanical escapement is first documented in manuscript descriptions of clockwork planetaria by Richard of Wallingford (1292–1336) and Giovanni de' Dondi (1318–1389). Both of these were driven by a simple clock with a mechanical escapement. Neither claimed invention of the escapement and it was, in both descriptions, of minor importance compared to the astronomical gearing. Indeed, the nascent mechanical clock on its own offered little use as a scientific tool, but when driving an astronomical model it fulfilled Dondi's initial intention of bringing a greater appreciation of scholarly astronomy, which he claimed had been muddled by careless astrology. Both Dondi and Wallingford's designs served as primitive computers that made astronomy more accessible by negating the need for arduous mathematics. Wallingford's clock was mathematically capable of representing the lunations with an error of "1.8 parts in one million" (Bedini and Madison 1966; Landes 1983, 84; North 2005).

In Europe, royal interest in astronomy and patronage of mathematicians and instrument makers provided the financial support needed for horological improvement and Dondi's masterpiece played an important, albeit passive, role in this progression. Holy Roman Emperor, Charles V (1500–1558), became interested in Dondi's astrarium at around the time of his coronation in Bologna in 1529. The emperor charged the ingenious mechanic, Gianello Torriano (Juanelo Turriano) (1501–1575), with the task of its restoration. Torriano apparently declined to restore the machine on account of its poor condition, offering instead to build a new version. Spanish historian, Ambrosio de Morales (1513–1591), recorded how Torriano spent almost a quarter of a century working on the machine and, as byproduct of the exercise, the invention of a gear-cutting machine enabled better quality gearing (Woodbury 1958; Reti and Turriano 1967; Moran 1977).

There is an inherent problem with the verge escapement that causes its timekeeping to accelerate when the driving force is increased and vice versa. Torriano's wheel-cutting device would have gone some way to reduce the variations in driving force caused by inequalities in wheel teeth, but when one considers that his astrarium used

1,800 engaging wheels, the clock drive had an inordinate amount of work to do and so timekeeping could only have been erratic. Dondi described how the clock could be roughly regulated by adding or removing weight from the crown-shaped foliot. Even simple mechanical clocks were affected. Tycho Brahe complained to Landgrave of Hesse-Kassel, Wilhelm IV (1532–1592), that the rate of his weight-driven clock gradually sped up over the course of its duration. He observed that this acceleration was caused by the mass of the rope, from which the driving weight was suspended, increasing the driving force as the weight descended (Lloyd 1958).

Wilhelm's clockmaker, Jost Bürgi (1552–1632), substantially reduced the problem of variable driving force by creating a device known as a remontoire,[6] which, coupled with his cross-beat[7] escapement, provided an unprecedented stability in timekeeping of around one minute per day. Whether or not Tycho's clocks were made by Bürgi is unknown, but given the correspondence between him and Wilhelm IV, it is more than likely that they followed Bürgi's design. Tycho outlined how he used two clocks side-by-side, so that any deviation in their rate could be noted and factored into the observations. Bürgi's skills were so highly valued, that when he was appointed to the royal court in Prague in May 1604, he received the third highest salary in the household, after the Emperor's physician and the treasurer. Bürgi's clocks provided the first practical time-standard for the observatory and Tycho's skilled usage of the clocks provided the data that enabled his successor, Johannes Kepler (1571–1630), to make the ground-breaking assertion that planetary orbits were elliptical (Staudacher 2014).

Without a doubt, one of the most important scientific developments of the seventeenth century was Galileo Galilei's (1564–1642) understanding and harnessing the timekeeping properties of the pendulum, "marking the commencement of the early modern era in physics" (Drake 1990, 6). Given the simplicity of a pendulum, its invention cannot be ascribed, but Galileo's important assertion, derived from experimentation with pendulum bobs made from materials of differing mass, was that a pendulum's period was entirely governed by the length of the suspension. After empirical study he stated: "the lengths are to each other as the squares of the times; so that if one wishes to make the vibration-time of one pendulum twice that of another, he must make its suspension four times as long" (Crew & de Salvio 1954, 96).

In practice, using a simple free-swinging pendulum as a tool of measurement was impractical and so Galileo used it as a form of time standard to calibrate existing tools of measurement: the clepsydra and weighing scales. By collecting water from a simple outflow clepsydra in a glass over a number of swings of the pendulum and then using the scales he was able to convert time to weight of water and thereby obviate the need for the pendulum. His scales were capable of weighing to the nearest sixtieth of a grain and so offered a very precise measurement of time in the laboratory. But, in order to provide a constant for natural acceleration, the pendulum's period had to be quantified. In a letter dated August 1, 1632 to a long-term correspondent, Giovani Battista Balliani (1582–1666), Galileo outlined his approach to timing the descent of a ball rolling along an inclined plane. Assisted by four "patient and curious friends," he maintained the swing of a pendulum and counted the number of vibrations between two successive transits of a bright star (a sidereal day) against a fixed marker (Drake, 1978, 399). Galileo and his assistants counted 234,567 vibrations between consecutive culminations of the same star equating to a period of around one third of a sidereal second. However, in his *Dialogo* (1638) natural acceleration of a ball rolling along an inclined plane was expressed in terms of weight, rather than time in seconds, over distance, which suggests that Galileo was not confident in his reckoning of the pendulum's period (Crew and de Salvio 1954, 179).

This study of the pendulum continued and was developed independently by French theologian, Marin Mersenne (1588–1648), and Italian astronomer and Jesuit Priest, Giambattista Riccioli (1598–1671). Riccioli's experiments are recorded in *Almagestum Novum* (1651) and include the use of hourglasses and sundials to attempt to ascertain the length of a seconds-beating pendulum. He, like Galileo, attempted a 24-hour counting exercise, but he recruited nine other assistants. It is worth noting that Riccioli and his assistants used song to count the higher-frequency vibrations of short pendulums, which supports Drake's (1978) speculation as to Galileo's use of music to time short intervals (Koyre 1953).

Mersenne investigated natural acceleration using the half swing of a pendulum and a free-falling ball. By suspending the pendulum from a hook on a wall, he was able to hold the pendulum bob and the ball in one hand. Then drawing the pendulum bob away from the wall to a prescribed angle, he could simultaneously release both the pendulum and the ball. The height of the suspension was adjusted until the sounds of the pendulum bob striking the wall and the free-falling ball hitting the floor coincided. Then, by changing the length of the pendulum he was able to extrapolate the results to provide the time of descent for any given elevation. Mersenne's method provided good results, but they were limited by the finite speed of human perception. He noted that he was unable to discern any difference between the impacts for drops from three to six feet (Koyre 1953; Yoder 2004, 13).

The Dutch diplomat, Constantijn Huygens (1596–1687), was one of the best-connected men in Europe at the time, and amongst his correspondents were both Galileo and Mersenne. In 1638 the Dutch court sent Galileo a valuable gold chain, which was declined for political reasons, in recognition of his work in determining longitude by observation of the moons of Jupiter. Galileo had also proposed to the Dutch court that the pendulum could be a longitude solution. He suggested a clock-work counter that could be driven by a pawl[8] attached to the pendulum rod, which would ratchet a saw-toothed wheel, causing it to turn one tooth spacing per swing. Curiously, Leonardo da Vinci sketched a very similar device in what is now known as the Madrid Codex. Da Vinci's annotation does not specify its purpose, but describes a modification of the then common verge escapement with sprung pallets fixed to a pendulum to provide a unidirectional drive to the crown wheel (da Vinci 1490–1496; Robertson 1931, 85; Drake 1978, 386).

It was Constantijn's son, Christiaan Huygens (1629–1695), who famously, though not without some controversy, established priority for invention of the mechanical pendulum clock by publishing and carefully distributing a brief illustrated description of his mechanical pendulum clock amongst influential courtiers, mathematicians, astronomers, and men of letters across Europe. In the pamphlet he bombastically likened his achievement to that of Marcus Phillipus, who provided Rome with its first correctly aligned sundial. A careful acknowledgement of Galileo's work followed:

> at last, from the original teaching of that most wise man, Galileo Galilei, the astronomer initiated this method: that they should impel manually a weight suspended by a light chain ... by this method they effected observations of the eclipses more accurately than

before; in like manner they measured— not unsuccessfully—the Sun's diameter and the distances of the stars (Robertson 1931; Edwardes 1970, 43; Howard 2008).

The pamphlet entitled *Horologium* (1658) could not have served to gain patents in other countries, but instead, as Huygens put it, it could prevent "the audacity of men of ill-spent leisure, lest—as customary with them—they should seize upon inventions and, most injuriously, sell them as their own." Huygens saw a greater reward, and therefore reason to establish his priority of invention, in the "science of Longitude, which … could have been obtained … by taking to sea the most exquisitely constructed timepieces free from all error" (as quoted in Edwardes 1970, 39–44).

Huygens collaborated with Alexander Bruce (1629–1681), a Scottish Royalist living in exile in Holland, to produce and trial sea-going pendulum clocks. After yielding encouraging results in early trials, he published *Kort Onderwijs* (1665) to instruct mariners in the use of his sea clocks. Not all were convinced of their reliability and Huygens hoped that a sea trial to the West Indies would assuage any doubters of his clocks. Opportunity for fair and controlled trial came with the foundation of the French Academy of Sciences in 1666, formed partly in response to the French need for improved marine navigation. The French astronomer, Jean Richer (1630–1696), was entrusted with the clocks and sailed westwards aboard the *Saint-Sebastien,* which sailed into a storm not far from the French coastline. The storm had a catastrophic effect on the clocks and Richer elected not to continue with their trial, much to Huygens's displeasure (Mahoney 1980).

Richer's decision to abandon the trial denied Huygens the corroborative data that he had hoped to publish in his seminal mathematical treatise *Horologium Oscillatorium* (1672). When the French Academy of Sciences organized what John Olmstead (1942) described as the original prototype modern scientific expedition in 1672 to Cayenne, principally to study the size of the solar system by observation of the parallax of Mars and the Sun, Huygens did not send his sea-going pendulum clock and this was most likely due to Richer's presence as astronomer on the voyage.

The pendulum clock may not have been useful at sea as a navigational instrument, but on land it proved to be a considerable boost to the science of astronomy and, as an unexpected consequence of the 1672–1673 voyage, geodesy. Richer found that his pendulum clock kept time in Paris, but ran slower at Cayenne and needed to have its pendulum shortened. To eliminate the possibility that this was caused by chance, Richer made a simple seconds pendulum calibrated by his clock in Cayenne to take back for comparison in Paris on his return. Isaac Newton (1642–1727) praised Richer's diligence in his third edition of *Principia* (1726). Newton collated numerous observations of the length of seconds pendulums made at different latitudes by others, including his friend Edmond Halley (1656–1742), who had observed the same phenomenon when cataloguing the stars of the Southern hemisphere from St. Helena in 1677. Newton's proposal, based on the figures, was that change in rate of the pendulums was caused by a reduction in gravity because the Earth's axis is shorter than its diameter at the equator—that the earth is not spherical but an oblate spheroid (Newton 1713 [2010], 341–9).

The pendulum clock saw further development in London with Sir Jonas Moore's (1627–1679) patronage of John Flamsteed (1646–1719). Moore commissioned Thomas Tompion (1639–1713) to make two year-going pendulum clocks that were installed in the newly built Royal Observatory. The clocks were unparalleled in terms of accuracy and provided the bedrock for the work that followed by proving that Earth's speed of rotation was constant, which up to that time had been assumed. These clocks can be considered the blueprint for the precision pendulum clocks that were to serve astronomers for the next 200 years. They had heavy pendulums swinging over a small arc and a dead-beating escapement. It is often suggested that the dead-beat escapement contributes to greater accuracy, but this is not necessarily the case as any other carefully constructed equivalent can be just as reliable. Its key advantage is the fact that the seconds hand on the dial remains static between beats of the pendulum and enables the astronomer to read the time to the nearest second at a glance (Howse 1970).

Shortly after the foundation of the Royal Observatory in late 1675 (old style) Christiaan Huygens published on his application of spiral spring to the balance of a pocket watch in the Royal Society's journal, *Philosophical Transactions*. The Society's curator of experiments, Robert Hooke (1635–1703), was incensed by this claim of invention. Protesting that he had demonstrated the same some 17 years earlier, he went further and accused the Dutchman of "magnifying himself in the plumes of others" (as quoted in Jardine 2009, 268). From a practical point of view, Huygens illustrated design worked and became the standard in subsequent watches (Jardine 2009).

It is true that watches made in the pre-balance spring era were unstable timekeepers and so most examples indicated time to within 15 minutes using a single hour hand. These watches suffered from the aforementioned weakness of the verge escapement, but added to that were manifold problems, most notably the non-linear output of power from the coiled mainspring and their portability meant that they were subject to unpredictable motions. Robert Moray (c.1609–1673), who later became a founder of the Royal Society, succinctly quantified the limits of a pre-balance spring watch in a letter to his friend and fellow exile, Alexander Bruce, by saying that no two watches could ever be made to keep two minutes the same. In the same letter he also states that he owns a watch that showed seconds for the timing pulses. Only a few examples of such watches are known to exist today (Jardine 2009; Thompson 2006).

It is not known when the first stopwatch was made, but Tompion's successor, George Graham (c.1673–1751), was certainly producing good examples from around the mid-1720s onwards. A contemporary description describes his design as a "very exact automaton" (Owen and Woodward 1981, 22), and that it could measure short-intervals to within half of a second. The outward appearance of these watches, which prevailed for most of the eighteenth century, was the same as a standard pocket watch with indication for hours, minutes, and seconds. Despite being widely used by astronomers, their use went unrecorded probably because it was deemed too basic to merit mention. Correspondence about sea trials of the lunar distance method for determining longitude in the 1760s acknowledge, at least, that the stopwatch was used in conjunction with a quadrant. Nevil Maskelyne (1732–1811) noted in his personal memorandum that he had purchased one that measured time to a tenth of a second (Figure 31.1), but never wrote about its use (Higgitt 2013, 65, 180).

It may well be that Maskelyne had used the watch to check his timing of star transits. He was confident that he could time the moment a star passed across wires, aligned vertically across the field of view of his telescope, to within one tenth of a second. He described glancing at the clock to note the number of seconds shown, then turning his

Figure 31.1 A short-interval timer, which was capable of measuring to within one tenth of a second, commissioned by the fifth Astronomer Royal, Nevil Maskelyne. Image © National Maritime Museum, Greenwich, London.

eye to the telescope and, whilst counting the seconds, mentally logging the two positions of the star either side of the wire, that coincided with the ticking of the clock. It was then a relatively simple spatial judgment to determine the timing to within a tenth of a second. The only reason that he wrote about the "eye-and-ear" method, devised by one of his predecessors, James Bradley (1693–1762), was to explain the dismissal of one of his assistants in 1796. Were it not for these unfortunate circumstances, we would be none the wiser to Maskelyne's method of timing observations (Maskelyne 1795).

In a similarly anonymous fashion, the stopwatch was used in other scientific disciplines. For example, it was used in conjunction with William Thomson's (1824–1907) quadrant electrometer to measure electrical current. The unquantified electrical source was passed through a light metal vane, on a torsional suspension within the quadrants. The quadrants were charged with a known and constant potential from a battery so that any potential difference caused a deflection of the vane. The stopwatch was used to determine the speed of deflection, which could then be used to calculate the current.

Improved electrometers facilitated the research of Jacques and Pierre Curie, who announced in 1880 their discovery of piezoelectricity. They measured the electrical potential generated by compressing quartz and other crystals and concluded that the electricity generated was directly proportional to the weight applied. This discovery led to an interesting parallel to Galileo's use of the weighing scale to measure time. When Marie Curie (1867–1934) was using an electrometer to measure radioactivity, she passed a known current between two metal plates separated in a condenser. Radioactive material spread on one of the plates caused ionization of the air within the chamber allowing some current to pass through. Rather than time the speed of the vane's deflection, she employed piezoelectricity to gain a more precise quantification. By charging the quadrants on her electrometer from a compressed quartz crystal she could fine tune the instrument by adding weight until both of the currents were equal and the vane returned to its resting positon (Curie 1898; Katzir 2003).

Despite its redundancy in the study of radioactivity, the stopwatch continued to see use alongside the electrometer throughout the first quarter of the twentieth century. Astronomers used photo-electric cells, connected to the electrometer, in their telescopes to observe the eclipses of binary stars, and by measuring the reduction of light, and therefore the change in current generated by the photoelectric cell, it was possible to measure and weigh stars using a stopwatch (Kron 1940).

It appears that the usage of the stopwatch as a subsidiary part of scientific investigation was deemed too insignificant, compared to the outcome of the experiments or observations, to deserve mention. Isaac Newton, for example, only gives two scant mentions of timing the short interval in one of his most important laboratory notebooks, both of which appear to be copied from Robert Boyle (1627–1691). The first describes a frigorific action that took about 15 seconds, timed by a pre-balance spring watch with a minute hand, and the second is a description of a remedy for hemorrhoids, which acted: "in the little time requisite to recite the Lord's Prayer" (Newton 1669, 39, 173).

Whilst the development of tools for measuring and determining time is not a constant linear development, their usage followed a lasting model. Clocks, be they clepsydra or mechanical, have for the most part been secondary to the clockwork of the heavens and a hierarchy of instruments is revealed by their visibility in historical scientific writing. The tools of the astronomer, for example, provided benchmarks for lesser instruments and are more frequently acknowledged and described. Though, excepting the descriptions of the early masterpieces and later milestones in precision, the most complete published accounts of the use of tools for time measurement tend to be catalyzed by a problem in the scientific process. Instruments at the lower end of this order, such as the stopwatch, were destined for relative obscurity due to the mundaneness of their use in comparison to the scientific outcomes that they contributed towards.

Endnotes

1   Simple gearing that converts linear motion into rotation. The rack is straight rod with teeth cut along one of its edges. The teeth engage with the pinion, which is a series of evenly spaced radial teeth attached to the shaft that carries the rotating component.

2   An alidade is a sighting device, often found on astrolabes as a pivoted bar spanning the diameter of the instrument with a sight on either end. Once the distant object has viewed through the two aligned sights, the bar is used to read the angle against an engraved scale.

3   The rete is a pierced plate that sits within the mater (the body of the astrolabe) and in simple terms describes the relative positions of bright stars. The rete is adjustable and can be rotated on a central pivot to align the star positions against the projection of the celestial sphere, according to a specific latitude, engraved onto a static plate known as a tympan.

4   A three-dimensional model of the celestial sphere.

5   The verge is a shaft with two integral flag-like pallets, placed at approximately 90° from each other and spaced so that their centers correspond with the circumference of the crown-shaped wheel. The clock's driving force attempts to push the pallet away from engagement from a tooth, when this is achieved the other pallet is placed to arrest a tooth on the opposite side of the crown wheel. The interaction causes the verge to oscillate and the weighted arms attached to the verge, the foliot, slow and regulate its oscillation.

6   By simplifying the clock train to one wheel (for example) frictional problems are largely eliminated, but this will reduce the clock's duration considerably. A remontoire is a clockwork auto-wind for such a simplified mechanism, which allows for a long duration without the problems associated with friction within the train.

7   This is a variation of the verge escapement, with two geared foliots connected to individual pallets.

8   The pivoted component in a ratchet mechanism that only allows the saw-toothed wheel to pass in one direction.

References

Ackermann, Silke. 1999. "Sun, moon and stars: Telling the time with astronomical instruments from the British Museum." *Antiquarian Horology*, 25: 31–46.

Bedini, Silvio. 1962. "The compartmented cylindrical clepsydra." *Technology and Culture*, 3: 115–41.

Bedini, Silvio, and Francis Madison. 1966. "Mechanical universe: The astrarium of Giovanni de' Dondi." *Transactions of the American Philosophical Society*, 56: 1–69.

Crew, Henry, and Alfonso de Salvio. 1954. [Galileo's] *Dialogues Concerning Two New Sciences*. New York: Dover Publications.

Curie, Marie. 1898. "Rayons émis par les composés de Vuranium et du Thorium." *Comptes Rendus Hebdomaires des Seances de l'Académie des Sciences*, 126: 1101.

Da Vinci, Leonardo. 1490–1496. *Madrid Codex* I: 61v: 111. http://ima.udg.edu/~dagush/Projects/Leonardo/Codex_Madrid_I.pdf Accessed September 27, 2015.

Drake, Stillman. 1978. *Galileo at Work: His Scientific Biography*. Chicago: University of Chicago Press.

Drake, Stillman. 1990. *Galileo: Pioneer Scientist*. Toronto: University of Toronto Press.

Edwardes, Ernest, L. 1970. "Horologium, Christiaan Huygens, 1658. Latin text with translation." *Antiquarian Horology*, 7: 35.

Freeth, Tony, Alexander Jones, John Steele, and Yanis Bitsakis. 2008. "Calendars with Olympiad and eclipse prediction on the Antikythera mechanism." *Nature*, 454: 614–17. DOI: 10.1038/nature07130.

Ghezzi, Ivan and Cliver Ruggles. 2007. "Chankillo: A 2300-year-old solar observatory in coastal Peru." *Science*, New Series, 315: 1239–43.

Higgitt, Rebekah (ed.) 2013. *Maskelyne: Astronomer Royal*. London: Robert Hale.

Howard, Nicole. 2008. "Marketing longitude: Clocks, kings, courtiers and Christiaan Huygens." *Book History*, 1: 59–88.

Howse, Derek. 1970. "The Tompion clocks at Greenwich and the dead-beat escapement part 1." *Antiquarian Horology*, 7, No. 1: 18–34.

Jardine, Lisa. 2009. *Going Dutch: How England Plundered Holland's Glory*. London: Harper Perennial.

Katzir, Shaul. 2003. "The discovery of the piezoelectric effect." *Archive for History of Exact Sciences*, 57: 61–91.

Koyre, Alexandre. 1953. "An experiment in measurement." *Proceedings of the American Philosophical Society*, 97: 222–37.

Kron, Gerald E. 1940. "Recent methods and technique of photoelectric photometry." *Publications of the Astronomical Society of the Pacific*, 52: 250–6.

Kurz, Otto. 1975. *European Clocks and Watches in the Near East*. London: The Warburg Institute.

Landes, David, S. 1983. *Revolution in Time*. Cambridge, MA: The Belknap Press of Harvard University.

Lloyd, H. Allan. 1958. *Some Outstanding Clocks over Seven Hundred Years 1250–1950*. London: Leonard Hill.

MacKay, Pierre. 1969. "A Turkish description of the tower of the winds." *American Journal of Archaeology*, 73, No. 4: 468–9.

Mahoney, Michael S. 1980. "Christiaan Huygens: The measurement of time and longitude at sea." In *Studies on Christiaan Huygens*, edited by H. J. M. Bos, M. J. S. Rudwick, H. A. M. Melders, and R. P. W. Visser, 234–70. Lisse: Swets. http://www.princeton.edu/~hos/Mahoney/articles/huygens/timelong/timelong.html Accessed September 27, 2015.

Maskelyne, Nevil. 1795. *Astronomical Observations made at the Royal Observatory at Greenwich*, Volume III. London: Royal Society.

Mills A. A. 1982. "Newton's water clocks and the fluid mechanics of clepsydrae." *Notes and Records of the Royal Society of London*, 37, No. 1: 35–61.

Moran, Bruce T. 1977. "Princes, machines and the valuation of precision in the 16th century." *Sudhoffs Archiv*, Bd. 61, H.3: 209–28.

Needham, Joseph. 1959. "The missing link in horological history: A Chinese contribution." *Proceedings of the Royal Society of London, Series A, Mathematical and Physical Sciences*, 250: 147–79.

Needham, Joseph, Wang Ling, and Derek de Solla Price. 1960. *Heavenly Clockwork: The Great Astronomical Clocks of Medieval China: A Missing Link in Horological History*. London: Cambridge University Press.

Neugebauer, Otto, 1969. *The Exact Sciences in Antiquity*. New York: Dover Publications.

Newton, Isaac. 1669. "Laboratory notebook." Portsmouth Collection Add. MS. 3975, Cambridge University Library, Cambridge University. http://webapp1.dlib.indiana.edu/newton/browse Accessed September 27, 2015.

Newton, Isaac. 1713 [2010]. *The Principia: Mathematical Principles of Natural Philosophy*. New York: Snowball Publishing.

Noble, Joseph, and Derek de Solla Price. 1968. "The water clock in the tower of the winds." *American Journal of Archaeology*, 74, No. 4: 345–55.

Norris, Ray P., Cilla Norris, Duane W. Hamacher, and Reg Abrahams. 2013. "Wurdi Youang: An Australian Aboriginal stone arrangement with possible solar indications." *Rock Art Research*, 30, No. 1: 55–65.

North, John. 2005. *God's Clockmaker: Richard of Wallingford and the invention of Time*. London and New York: Hambledon and London.

Olmsted, John W. 1942. "The scientific expedition of Jean Richer to Cayenne (1672–1673)." *Isis*, 34, No. 2: 117–28.

Owen, Dorothy M., and S. W. Woodward. 1981. "The minute-books of the Spalding Gentlemen's Society 1712–1755." *Lincoln Record Society*, 73: 22.

Price, Derek J. de Solla. 1974. *Gears from the Greeks, the Antikythera Mechanism – a Calendar Computer ca. 80 B.C.* Philadelphia: The American Philosophical Society. DOI: 10.2307/1006146.

Reti, Ladislao and Juanelo Turriano. 1967. "The Codex of Juanelo Turriano." *Technology and Culture*, 8, No. 1: 53–6.

Robertson, J. Drummond. 1931. *The Evolution of Clockwork*. London: Cassell & Company Ltd.

Saliba, George. 1992. "The role of the astrologer in medieval Islamic society." *Bulletin d'Études Orientales*, T. 44: 45–67.

Staudacher, Fritz. 2014. *Jost Buergi—Kepler und der Kaiser—Instrumentenbauer, Astronom, Mathematiker 1552–1632*. Zurich: Verlag Neue Zuercher Zeitung.

Thompson, David. 2006. "Jan Janssen Bockelts the younger 'puritan' style verge watch with centre seconds, Haarlem, c.1630." *Antiquarian Horology*, 29: 827–30.

Thorndike, Lynn. 1941. "Invention of the mechanical clock about 1271 A.D." *Speculum*, 16, No. 2: 242–3.

UNESCO-IAU, 2012. "Portal to the heritage of astronomy." http://www2.astronomicalheritage.net/ Accessed September 27, 2015.

Whitrow, G. J. 1988. *Time in History: Views of Time from Prehistory to the Modern Day*. Oxford: Oxford University Press.

Woodbury, Robert S. 1958. *History of the Gear-Cutting Machine*. Cambridge, MA: Technology Press.

Yoder, Joella G. 2004. *Unrolling Time, Huygens and the Mathematization of Nature*. Cambridge: Cambridge University Press.

CHAPTER THIRTY-TWO

Weights and Measures

HECTOR VERA

Historiography on Weights and Measures

The history of weights and measures involves a wide set of topics that make its study complex and perplexing. First of all, there are the instruments and techniques of measurement, along with the abstract systems that define the units of measurement and articulate how they are grouped or subdivided. Secondly, there are the measurers—the persons who do the measuring and the organizations they work for—and their relation with the measured—the people whose bodies and possessions are measured. Finally, there are the broader institutional orders and systems of belief that shape measurement practices: political authorities, economic structures, numeracy and shared understandings of quantification, ideas of justice and fairness, and so forth. The number of concrete activities in which these dimensions interplay is quite vast—for example, taxation, commerce transactions of all sizes, land surveying, census taking, architecture, navigation, mining, artillery, and cartography.

Using a broad categorization of the historiography on weights and measures one may talk of an "encyclopedic," a "social," and a "cultural" history of measurement. The encyclopedic history of measures can be described as the research focused primarily on the compilation and organization of basic information about the units, arithmetic, instruments, legislation, and systems of measurement used in past societies (e.g., Klein 1988; Connor, Simpson, and Morrison-Low 2004; Williams and Jorge 2008)—a painful job that includes specifying the approximate equivalences of old measures into metric units (e.g., Cardarelli 2003). This kind of work may sometimes lack analytical insight, but is of great utility. Zupko's dictionary of the weights and measures in France before the revolution (1968), for instance, shows the astounding diversity of measures prior to the invention of the metric system (there were in Europe 391 units called *pound* and 282 units named *foot*, all different in magnitude). Other scholars in this area work with archeological artifacts to determine which possible units of measurement may have been used in ancient civilizations (O'Brien and Christiansen 1986; Morley and Renfrew 2010).

A Companion to the History of Science, First Edition. Edited by Bernard Lightman.
© 2016 John Wiley & Sons Ltd. Published 2020 by John Wiley & Sons Ltd.

When limited to the unearthing and compilation of data, historical metrology has been considered as a mere "auxiliary" historical science. However, historians in the twentieth century advanced the idea that weights and measures can be an entrance point to ask broader questions. As Marc Bloch, one of the founders of the Annales School, put it: "Unthankful only in appearance, metrological studies, in the hands of an intelligent researcher, became an instrument capable of revealing the great streams of civilization" (1934, 280). Inspired by these ideas multiple historians developed a social history of measurement, closely related to topics like material life, economic relations of production, and class struggle (e.g., Hocquet 1985). The main bearer of this perspective was the Polish historian Witold Kula, author of presumably the single most influential book in historical metrology, *Measures and Men* (1986).

More recently, a series of studies, which could be labeled as "cultural history of measurement," have stressed other salient dimensions in the institutions and practices of measurement, underlining the role of previously forgotten actors, and showing the exchanges between scientific and local knowledges. Here the role of meanings, values, and ideas like trust, precision, quantification, agreement, identity, and objectivity are in the forefront of historical analysis (e.g., Schaffer 1992; Porter 1995; Olesko 1996; Alder 2002; Gooday 2004; Safier 2008).

Considering these trends, today it seems desirable to find some productive combinations among the three historiographical models. When dealing with the history of weights and measures, it is indispensable to know how and what is weighed and measured in any particular society, and that requires a detailed reconstruction. Likewise, it is essential to relate the local meanings, practices, and knowledges that are embodied in measuring behaviors and institutions with larger social, political, and economic processes that aid or deter the expansion of measurement systems.

Measurement: Politics and Economy

For the most part, scientific speculation has not provided the main impulse to systematize and standardize weights and measures; political and economic needs have been more determinant. Metrology is, above all, a science of the state. The right to regulate measures is an attribute of authority. Political authorities regulate who can measure, what can be measured, and how things must be measured. States are interested in controlling weights and measures and establishing what may be called a monopoly on the legitimate means of measurement. States crave to centralize metrological issues because stable systems of measurement help administrators to fulfill essential political functions: enhancing the extraction of taxes; consolidating internal markets; undermining the influence of local authorities; making the population and the economic resources legible; reducing commercial frauds and heightening, then, the administration of justice (Kula 1986; Curtis 1998; Scott 1998, 25–33; Carroll 2006). Concurrently, a considerable amount of state power is required to enact a measurement system in a given territory (Porter 1995, 26).

The implementation of a system of measurement in a territory is a work of distribution of knowledge (as present and future users of the system need to learn it); and it also requires the production of knowledge (i.e., state-sponsored knowledge) in the form of manuals, reports, translations of metric and customary units, catalogs of local measures, and tables of conversion. Modern nation states have thus recruited and trained legions of engineers and second tier scientists to work as metrologists, surveyors, and inspectors of weights and measures to undertake those duties.

States are the most effective institutions in helping, compelling, and, if necessary, forcing people to learn and employ measurement units. Scientists and economic agents by themselves cannot do this. Successful metrological standardization can only be achieved when two actions are combined: first, policing the employment of authorized units; second, providing populations with the intellectual and material means to learn the legal measurement system. Effective metrological standardization requires *compulsion*—and compulsion is essentially a political issue, an issue of the state. The futile attempts to attain a "voluntary transition" to the metric system in countries like the United States confirm this.

In addition to all this, economic interests have often been the main factor behind metrological regulation. Weighing and measuring are among the key cognitive processes involved in economic activities. Systems of measurement are economic institutions. Setting shared standards of weights and measures is of special importance for an economy, as economic exchange requires measurement. When different products and commodities are sold or bartered it is necessary to know how much of a product is being exchanged. Commodities are sold using measures of weight, length, and volume. Economic activities require accepted conventions about measurement. As North puts it, "Underlying all exchange is measurement. [...] The very terms price and quantity imply the ability to measure those two dimensions" (1987, 593–4).

The genesis and development of systems of measurement is related to particular economic relations. Measurement of quantity is an operational use of number that is defined mostly in economics terms (Crump 1992, 72). Systems of measurement are developed within specific sets of economic relations. A system of measurement used by peasants in a self-sustained village would vary greatly from one used by merchants in a big commercial city; their exactness, complexity, and level of standardization will diverge considerably in part because they satisfy different economic requirements (Kula 1986, 102–19).

Ancients units of measurement were closely related to the human body, its dimensions, extensions, and productive capacities. In pre- and post-Columbian Mesoamerica, for example, there were linear measures like the cubit, foot, and step. *Tlamamale* expressed the weight that could be transported by a carrier in one day. *Mano* (hand) referred to a unit of five and a *mano of milpa* (cornfield's hand) was an area that could be sown using the grains from five ears of maize. Among the Maya the *auat* (shout) was a unit equivalent to a quarter-league, due to the idea that it is the distance reached by a man's shout. Anthropometric measures identical or similar to these were present in the economic practices of virtually every civilization.

Another important set of ancient customs surrounding measurement was derived from the objectives and outcomes of human activities and labor—what Kula (1986, 3–8) described as the functional character of past measures. Techniques of production and transportation, the nature of the products measured, and the needs of consumption were crucial elements to define units of measurement. In different cultures stone's throw and arrow shot were used to measure distance. The *loads* that could be carried by a mule, a cart, or a ship were employed as units of weight.

For complex economic systems the standardization of measures fulfills the function of reducing transaction costs and ameliorating asymmetric information. A significant part in reducing transaction costs is the accurate specification of what and how much is being exchanged. When weights and measures are poorly standardized those costs are higher—diversity of weights and measures makes the search for information laborious, uncertain, and irregular. Conversely, the development of uniform weights and measures reduces the costs of measurement. Nevertheless, there are other dimensions in the practice of measurement that should be taken into account here, because even in conditions of deficient standardization economies could function and grow. Studies on the industrial revolution have shown that. Pollard (1983) proved how the intricacies of customary definitions and techniques of measurement actually provided some handy flexibility within the context of capitalist calculation; and Velkar (2012) emphasized how the shortcomings to centralize and standardize measures were solved by localized metrological practices in specific contexts that serve to build trust among economic actors.

These political and economic aspects ingrained in measurement practices can be seen in action in the history of the invention, rise, and global dissemination of the present most widely used measurement system: the decimal metric system.

Decimal Metric System

The creation of the decimal metric system during the French revolution was a watershed in the history of weights and measures; and the metric system itself was one of the defining achievements of the 1789 revolution. In the words of Gillispie, "in the regular doings of scientists and other people, the metric system of weights and measures remains the most pervasive legacy of the French revolution" (2004, 223–4). In a similar tone, Hobsbawm claimed that "the most lasting and universal consequence of the French revolution is the metric system" (1995, 57). It is not surprising then that this episode has been the object of ample inquiry.

The political and economic aspects of the metric reform (Kula 1986, 185–264; Heilbron 1989; Alder 1994), the work of savants and scientific academies (Gillispie 2004), and the intrepid journey of the astronomers Delambre and Mechain to measure the meridian arc that served to determine the length of the meter, the base unit of the metric system (Ten 1996), have all been studied in great detail. Guedj (2000) and Alder (2002) wrote captivating books with comprehensive interpretations of the creation of the metric system, displaying rich sociological insight to connect the scientific endeavor required to put in place the new measurement system with the social and cultural milieu of the revolution.

One of the central elements of the social life of measures in Europe and the Americas until the beginning of the nineteenth century was *multiple metrological sovereignty*— that is, the lack of a single unified political hierarchy to regulate weights and measures and the presence of competing claims by opposing parties over metrological authority. There usually was a proclaimed sovereign metrological authority, but in practice that authority was ignored or challenged. This situation ended in France on August 4, 1789 with the abolition of feudal privileges that deprived lords of the right to have final say in metrological matters in their own estates. The revolution unified all the French under a sole authority that set a single system and settled all disputes—of course it is not a coincidence that the birth of the metric system and the establishment of this sovereign metrological power occurred simultaneously.

The metric system was designed by some of the most eminent scientists of the eighteenth century—mathematicians, astronomers, and chemists like Lagrange, Laplace, and Lavoisier. For a few decades it was only employed by fellow scientists and intellectuals; but today the immense majority of the world's population uses the metric system in their daily dealings. The metric system made an incredibly successful transition from being a special knowledge possessed by a small group of specialists to becoming general knowledge; and not only general knowledge of a particular society, since it has become part of the social stock of knowledge of virtually all humanity.

This was an impressive achievement, especially considering that the metric reform was quite radical. Reforming a system of measurement involves changing one or all the nuclear elements of the system to be modified: 1) the magnitude, size, or amount of the units of measurement; 2) the names of the units; and 3) the system of grouping and division. The inventors of the metric system decided to create units with new magnitudes (meter, liter, and kilogram), provided a novel nomenclature for those units, and obliterated duodecimal and sexagesimal divisions in favor of a decimal system. They drastically and triumphantly broke with the past.

The magnitude of their achievement can be better appreciated if we contrast the metric system with the outcomes of the ill-fated Republican calendar and the decimal time, which were invented by the same group of thinkers during the revolution (Shaw 2011; Vera 2009). The time reform—guided by the similar intellectual principles of the metric system—was fruitless, while the weights and measures reform became a global institution.[1]

In 1791, when describing the aspirations behind the creation of the metric system, Condorcet underlined that it was made "For all time, for all people." That vision—so ambitious and optimistic—has come very close to reality. The metric system has become the only universal language to express quantities of length, mass, and volume in all realms of human activity. Today 95 percent of the world's population lives in countries where the metric system is the only legal system of measurement. How exactly the metric system went from a radical revolutionary intention in France to a global language of measurement is a question that remains insufficiently explored.

The efforts to design a rational system of measurement by the French revolutionaries culminated in 1799 with an international meeting of scientists held in Paris—the Congress on Definite Metric Standards (Crosland 1969)—where French savants collaborated with colleagues from Spain, Denmark, and some republics controlled by France, to finish the calculations and verifications to determine the length of the meter. The congress was a first step to achieve the internationalization of the system, which was one of the central aims of its creators. Even if the congress was seen by many as a cosmetic political move, it served as a diffusion point of metric expertise. For example, the Spanish delegate, the military engineer and mathematician Gabriel Ciscar, published in 1800 *Memoria elemental sobre los nuevos pesos y medidas decimals*, a brief treaty on the metric system. It served as essential reference for men of knowledge who sought technical literature in Spanish all across Latin America in the first half of the nineteenth century.

In the original proposal to create the metric system, the collaboration of British scientists was considered, but the hostility between France and England sank that part of the plan. And something similar happened with the United States, with Thomas Jefferson distancing himself from the proposal of the French National Assembly, despite having shown some initial interest in the idea of an international measurement system. The fact that neither England nor America participated in the 1799 Paris Conference had lasting consequences for the further development of the metric system, as those two countries became formidable obstacles in the metric global propagation.

Globalization of the Metric System

The historical accounts of the international circulation of the metric system have been centered on when, where, and how different groups of experts agreed to use the system: scientific conferences, international fairs and expositions, and diplomatic summits served as the points of contact to share information and reach agreements (Cox 1958). In this context the Meter Convention (signed by 17 countries) and the creation of the International Bureau of Weights and Measures in 1875 have been a focal point for research (e.g., Quinn 2012).

In the second half of the nineteenth century several international initiatives were developed, like the creation of the international standard time, the International Telegraph Union, the Universal Postal Union, and the International Meteorological Organization. The metric system was part of this series of projects that helped to build international and interlinguistic mechanisms of standardization and coordination in a context when the world became increasingly unified. Metric global diffusion was helped by a vision of internationalism that pushed forward many worldwide initiatives; and the metric system, in return, enhanced greatly the idea of internationalism by being a living proof of its benefits (Geyer 2001).

Assertive scientific diplomacy was involved in this period, which pitted France against England—where the imperial measurement system was promulgated in 1824 to counter the influence of the metric system (Zupko 1990). The 1884 International Meridian Conference in Washington, DC, for example, brought together delegates from 30 countries "for the purpose of fixing upon a meridian proper to be employed as a common zero of longitude and standard of time-reckoning throughout the globe" (International Conference 1884, 1). It was also a scientific and diplomatic confrontation between England and France. The French proposed that England would adopt the decimal metric system in exchange for France's acceptance of the British meridian. The adoption of systems of weights and measures, however, was not part of the topics established for the conference and the proposal was ruled out. From then on time has had its center in Greenwich, and weights and measures in Paris.

Social Forces of Metrication

The diplomatic vantage point is illustrative, but it does not take into account the decisive problem of how lay populations around the world came to learn and utilize the metric system. In other words, we know very well how thousands of experts embraced the metric system, but we know very little about how millions of non-experts were forced into metrication. To fill this gap it is necessary, at least partly, to pay more attention to the larger social dynamics behind metrication, like colonialism and state formation.

Revolutions—and other forms of rapid social and political change—have been the midwife of metrication. It is not an accident that the metric system was adopted in

China in 1912 (when the Qing Dynasty was abolished and the Republic of China established) and in Russia in 1918 (a year after the October Revolution ended tsarism). Radical metrological reforms follow radical social change (Vera 2015). In other cases, the metric system has been promoted by national unifications (like in Germany, Italy, and Bulgaria), with the aim of linking together previously autonomous cities and regions that had different administrative structures and diverse metrological arrangements. And in other cases the metric system was adopted after drastic changes in a political regime (like civil wars, the adoption of a new constitution, and aggressive policies of "modernization"), as happened in Colombia, Mexico, and Turkey.

Colonialism is one of the hidden stories behind global metrication. European control of overseas territories was one of the main motors propelling the dissemination of the metric system. Metrication by colonization helped to increase the number of metric territories all over the world, from Algeria and Senegal in 1840 to Macau and Timor-Leste in 1957. One out of every four currently existing countries outside Europe received the metric system as a colonial imposition. As of today, no former colony has rejected the metric system after their independence—a phenomenon that may be called *voluntary retention* of the metric system.

If colonization was an important factor in global metrication, decolonization was even more influential. Some newly formed nation states opted in favor of metrication to solve the problems produced by the heterogeneity of the colonial measurement systems; others, especially former British colonies, switched to the metric system as an act of emancipation, to underline their differences with the empire. A first large wave of postcolonial nations that voluntarily embraced the metric system was the former Spanish colonies in the Americas. The big majority of the Latin American countries were voluntary adopters. They were the first voluntary adopters outside France's military and geographical area of influence and they made the metric system a truly extra-European reality. Among the 20 countries that first adopted the metric system after France, 11 were from Latin American. They greatly helped the cause of international metrication by creating, alongside Western Europe, the critical mass of countries necessary for the system to be a convincingly multinational and multiregional metrological language in the second half of the nineteenth century.

A second wave came after the decline of the British Empire in the aftermath of World War II. Between 1950 and 2000 seventy countries adopted the metric system. Two interrelated historical developments account for this: the breakdown of the British control on its overseas dominions and the subsequent birth of a large number of newly formed nation states. Among these former British colonies, those using a system other than the metric system quickly decided after their independence to make the meter and kilogram their official standards. Once countries gained independence, people interested in the introduction of the metric system saw new opportunities to push their plans forward. In India, for instance, where a movement in favor of decimal coinage and metric weights and measures existed since the beginning of the twentieth century, all attempts in favor of decimalization were blocked because colonial authorities preferred to wait for England to adopt it first. After emancipation, groups like the Decimal Indian Society found more receptive ears in the government to attain the reform, which finally came about in 1954 (Verman and Kaul 1970).

In the few occasions when the metric system was adopted by multinational agreements it only happened among "peripheral" nations, like the Central American convention on regional unification of weights and measures in 1910, the joint government declaration to adopt the metric system signed by Kenya, Uganda, and Tanzania in 1967, and the metrological agreement among 15 members of the Caribbean Community in 1969.

Overall, by 1950 the globe was completely divided into two large metrological areas, with all the nations of the world using either the metric or the English system (the latter group was a shrinking minority). The rest of the hundreds of measuring systems that existed in the world had been displaced, partially tolerated in specific countries, or simply used underground in local communities. Seen from a larger perspective, the uncanny success of these two systems meant a great loss for humanity's stock of knowledge. As with the extinction of languages, the death of local and regional measures has meant the disappearance of collective experience accumulated by hundreds of generations. It is difficult to put in numbers how much has been lost. The few catalogs of units of measurement in the world are extremely incomplete. A survey made by the United Nations in the 1960s enumerated approximately 898 of those units (Statistical Office of the United Nations 1966). It is foreseeable that the majority of those measures will disappear rather soon. And this loss cannot be estimated only in terms of the number of units that have been swept out by the metric and imperial waves, but also by the antiquity and resiliency of some of the systems that are today forbidden from all official transactions and maybe doomed to disappear from everyday life in the not so distant future. Japan's shaku-kan system, for example, was outlawed in 1966 after being used in a relatively stable form for a millennium after its arrival from China in the tenth century.

The Metric System and Its Adversaries

To understand the historical origins of the present institutional order in world metrology, two questions ought to be asked. First, how was the metric system able to displace hundreds of traditional measurement systems around the world? Second, why was the metric system, and not any other system, the one that became a global language?

One of the most significant developments in the history of the metric system has been the marginalization and obliteration of thousands of units of measurement around the world. A lesser-known aspect of this process is the metric confrontation against a number of measurement systems created *ex professo* to challenge it. The overwhelming triumph of the metric system in scientific circles over those challengers contributed greatly to its ultimate global success, as experts in different countries were able to present plans for metrological reform centered only on the metric system (avoiding thus intestine disputes among experts to define what system should supplant customary measures). Since the second half of the nineteenth century the lack of national metrological uniformity and the absence of international coordination in weights and measures were afflictions that found in every nation the same prescription: metrication.

The inadequate understanding of this phenomenon and the lack of research on who challenged the metric system with newly invented systems has created the impression that those plans were not a factor in the history of metrication. The metric system actually faced plenty of competition (Zupko 1990, 209–25; Vera 2012, 364–9). Initially, at the end of the eighteenth century, there were several plans for metrological reform that came about shortly before or simultaneously to what the French savants were doing in the early 1790s when the metric was created. Cesare Beccar, James Watt, and Thomas Jefferson formulated some of these plans, just to mention a few of them (Jefferson 1953; Maestro 1980).

The metric system had multiple resemblances with the others—and it was itself quite similar to a plan sketched by Gabriel Mouton in the seventeenth century (Klein, 1988, 108–10; Zupko, 1990, 124–30). So it is hard to believe that the metric victory was due to its purely technical virtues. The main difference was that the metric system was *part of a social revolution.*

In the nineteenth and twentieth centuries, when nation states looked for the proper instruments to secure a metrological monopoly, the metric became a ready-made system, legitimized by its scientific aura and its proven record of success. Once the metric system was established and started its global dissemination many other proposals were advanced to challenge it (Vera 2011, 364–9). These projects included new units of measurement and some were based on non-decimal systems (they proposed instead base-8 and base-16 systems for the grouping and subdivision of units), like A. B. Taylor's *Octonary Numeration, and its Application to a System of Weights and Measures* (1887).

One of the few thinkers who gave some consideration to one of these publications was the philosopher Charles S. Peirce, who at the time was a metrologist at the US Coast and Geodetic Survey. Peirce reviewed E. Noel's *The Science of Metrology, or Natural Weights and Measures: A Challenge to the Metric System*, published in 1889, and he showed great skepticism towards his plan and the whole idea of challenging the metric system

> Mr. Noel's system is nearly as complicated and hard to learn as our present [English system], with which it would be fearfully confused, owing to its retaining the old names of measures while altering their ratios [...]; to "challenge" the metric system is like challenging the rising tide (1982, 378–9).

Some of the most visible movements of opposition to the metric system took surprising turns. During the second half of the nineteenth century a handful of scientists in Great Britain advocated for the so called "pyramid metrology," claiming that the Great Pyramid of Giza was a storehouse for a divinely-inspired metrological system, which had the "sacred Hebrew cubit" and the "Pyramid inch" as its main units. This idea was married with the notion that the modern British people and the "Anglo-Saxon race" inherited these standards through their own system of weights and measures. Pyramid metrology tried to assert an inherent superiority of the traditional system of weights and measures over the metric system (Schaffer 1997; Reisenauer 2003). The main advocate of these notions was Charles Piazzi Smyth, Astronomer Royal for Scotland, who made extensive fieldwork in Egypt to study the pyramids (Brück and Brück 1988, 95–134; Barany 2010). This unconventional theory was later continued in the United States by a group of engineers congregated in the 1880s in the International Institute for Preserving and Perfecting Weights and Measures, which was the first explicitly anti-metric association in America (Cox 1959; Crease 2011, 151–9).

The controversial nature of the pyramid metrology marginalized their advocates and even made them the object of ridicule by pro-metric intellectuals. However, they set the foundation (if not intellectually, at least organizationally) for future anti-metric groups in the United States that became more effective in halting metrication. The colorful pyramid metrology has received ample attention by scholars; sadly it is not the same for the debates that followed it. Two important episodes are still waiting for further research. First, scholars have neglected Hebert Spencer's defense of customary English weights and measures and his feud in the 1890s against Lord Kelvin, who was an enthusiast metric supporter (Vera 2011, 341–64). Second, there is the long-lasting confrontation of pro-metric scientists and educators in the US, who assembled the American Metric Association, against engineers and industrialists who were congregated in the American Institute of Weights and Measures (Treat 1971).

Moral Economy of Measurement and Opposition to Metrological Reform

The adoption of the metric system among the population at large is always an imposition from the top—in different degrees, but always from the top. Sometimes professional groups have adopted the metric system voluntarily, but no large segment of a general population has voluntarily abandoned its customary measures and adopted the metric ones before their government decided to make the meter, the liter, and the kilogram the only legal units. Metrication "from below" has been quite anemic and it can only be found in certain occupational groups (in scientific and technological areas), and that is one of the main reasons why non-mandatory adoptions of the metric system within countries have never achieved much.

Expressions of popular opposition to metrological change were recurrent whenever the metric system was introduced. Millions and millions of persons in all corners of the world were forced to abandon their customary instruments and units of measurement and to replace them with those of the metric system. Forced metrication triggered diverse forms of opposition, the majority of which were peaceful actions (though active and creative) aimed to continue using pre-metric measures while hiding from the authorities (Ros Galiana 2004; Vera 2011, 370–419). But sometimes resistance was also violent and explosive. Alder describes how in 1840—when the French state made its second and definitive try to impose the mandatory use of metric units—a group of dockworkers in Clemency "smashed decimal measures, and the government had to call in the cavalry." The riot was initiated due to suspicions that the change of measures would affect the workers and open the town to harmful competition (Alder 2002).

The substitution of a familiar and functional system of measurement with a new one was often problematic, and it could spark explosive reactions. One of the most famous and better-documented episodes of violent popular opposition to metrication occurred in Brazil, in the so-called "Quebra-Quilos revolt" of 1874 (Barman 1977; Richardson 2010). In numerous villages in northern Brazil rioters entered local markets and smashed the recently introduced kilogram-weights. Besides the destruction of metric implements, protestors prevented people from paying taxes, refused to answer census questions, and systematically destroyed records in tax offices. Tensions and confrontations were common and some episodes of violence occurred (including hundreds of armed rioters beating market sellers and killing a few). Due to their destruction of metric weights the movement was named *Quebra-Quilos* ("smash the kilos"). Luckily for researchers interested in historical metrology, it was labeled like that; with such an unmistakable name, researchers have paid attention to the metrological element in the rebels' actions. More attention is needed to recognize the moral economy of measurement and the metrological dimension in contentious politics.

Concluding Remark

Measurement is what social scientists call a total social fact. It links apparently distinct practices and institutions; it is a phenomenon at once scientific, economic, political, legal, and moral. None of these aspects ought to be studied in isolation. A full-fledged historical-metrological analysis should join these parts together. The cultural aspects of measurement have to be linked with the large economic and political forces; and what happens within laboratories and "centers of calculation" should be seen as inter-weaved with the broader social institutions and groups that affect and are affected by measurement practices.

Endnote

1   Those reforms were part of a larger project of decimalization, which included angles and currency (Tschoegl 2010).

References

Alder, Ken. 1994. "A revolution to measure: The political economy of the metric system in France." In *The Values of Precision*, edited by Norton Wise, 39–71. Princeton: Princeton University Press.

Alder, Ken. 2002. *The Measure of all Things: The Seven Years Odyssey and Hidden Error that Transformed the World*. New York: The Free Press.

Barany, Michael. 2010. "Pyramid metrology and the material politics of basalt." *Spontaneous Generations: A Journal for the History and Philosophy of Science*, 4: 45–60.

Barman, Roderick. 1977. "The Brazilian peasantry reexamined: The implications of the Quebra-Quilo Revolt, 1874–1875." *Hispanic American Historical Review*, 53: 401–24.

Bloch, Marc. 1934. "Le témoignage des mesures agraires." *Annales d'histoire économique et sociale*, 6: 280–2.

Brück, Hermann A., and M. T. Brück. 1988. *The Peripatetic Astronomer: The Life of Charles Piazzi Smyth*. Philadelphia, PA: Adam Hilger.

Cardarelli, François. 2003. *Encyclopaedia of Scientific Units, Weights and Measures: Their SI Equivalences and Origins*. New York: Springer.

Carroll, Patrick. 2006. *Science, Culture, and Modern State Formation*. Berkeley: University of California Press.

Connor, R. D., A. D. C. Simpson, and A. D. Morrison-Low. 2004. *Weights and Measures of Scotland: A European Perspective*. Edinburgh: National Museums of Scotland.

Cox, Edward Franklin. 1958. "The metric system: A quarter-century of acceptance (1851–1876)." *Isis*, 13: 358–79.

Cox, Edward Franklin. 1959. "The International Institute: First organized opposition to the metric system." *Ohio Historical Quarterly*, 58: 54–83.

Crease, Robert. 2011. *World in the Balance: The Historic Quest for an Absolute System of Measurement*. New York: W. W. Norton.

Crosland, Maurice. 1969. "The congress on definitive metric standards, 1798–1799: The first international scientific conference?" *Isis*, 60: 226–31.

Crump, Thomas. 1992. *The Anthropology of Numbers*. Cambridge: Cambridge University Press.

Curtis, Bruce. 1998. "From the moral thermometer to money: Metrological reform in pre-confederation Canada." *Social Studies of Science*, 28: 547–70.

Geyer, Martin. 2001. "One language for the world: The metric system, international coinage, gold standard, and the rise of internationalism, 1850–1900." In *The Mechanics of Internationalism*, edited by Martin Geyer and Johannes Paulmann, 55–92. New York: Oxford University Press.

Gillispie, Charles Coulston. 2004. *Science and Polity in France: The Revolutionary and Napoleonic Years*. Princeton, NJ: Princeton University Press.

Gooday, Graeme J. N. 2004. *The Morals of Measurement: Accuracy, Irony, and Trust in Late Victorian Electrical Practice*. New York: Cambridge University Press.

Guedj, Denis. 2000. *Le mètre du monde: Historie politique, scientifique et philosophique de l'invention du système métrique décimal*. Paris: Editions du Seuil.

Heilbron, John L. 1989. "The politics of the meter stick." *American Journal of Physics*, 57: 988–92.

Hobsbawm, Eric. 1995. *The Age of Extremes*. London: Abacus.

Hocquet, Jean-Claude. 1985. "Le pain, le vin et la juste mesure à la table des moines carolingiens." *Annales. Histoire, Sciences Sociales*, 40: 661–86.

International Conference Held at Washington for the Purpose of Fixing a Prime Meridian and a Universal Day. 1884. Washington, DC: Gibson Bros.

Jefferson, Thomas. 1953. "Notes on the establishment of a money unit, and of a coinage for the United States." In *The Papers of Thomas Jefferson*, edited by Julian Boyd, volume VII, 175–88. Princeton, NJ: Princeton University Press.

Klein, H. A. 1988. *The Science of Measurement: A Historical Survey*. New York: Dover.

Kula, Witold. 1986. *Measures and Men*. Princeton, NJ: Princeton University Press.

Maestro, Marcello. 1980. "Going metric: How it all started." *Journal of the History of Ideas*, 3: 479–86.

Morley, Iain and Colin Renfrew (eds.) 2010. *The Archaeology of Measurement: Comprehending Heaven, Earth and Time in Ancient Societies*. New York: Cambridge University Press.

North, Douglass. 1987. "Review of *Measures and Men*, by Witold Kula." *The Journal of Economic History*, 47: 593–5.

O'Brien, Patricia, and Hanne Christiansen. 1986. "An ancient Maya measurement system." *American Antiquity*, 51: 136–51.

Olesko, Kathryn. 1996. "Precision, tolerance and consensus: Local cultures in German and British resistance standards." *Archimedes*, 1: 117–56.

Peirce, C. S. 1982. "Review of Noel's *The Science of Metrology*." In *Writings of Charles S. Peirce*, 6, 377–379. Bloomington: Indiana University Press.

Pollard, Sidney. 1983. "Capitalism and rationality: A study of measurement in British coal mining, ca. 1750–1850." *Explorations in Economic History*, 20: 110–29.

Quinn, Terry. 2012. *From Artefacts to Atoms: The BIPM and the Search for Ultimate Measurement Standards*. New York: Oxford University Press.

Porter, Theodore. 1995. *Trust in Numbers: The Pursuit of Objectivity in Science and Public Life*. Princeton, NJ: Princeton University Press.

Reisenauer, Eric Michael. 2003. "'The battle of the standards': Great pyramid metrology and British identity, 1859–1890." *The Historian*, 65: 931–78.

Richardson, Kim. 2010. *Quebra-Quilos and Peasant Resistance: Peasants, Religion, and Politics in Nineteenth-Century Brazil*. Lanham, MD: University Press of America.

Ros Galiana, Fernando. 2004. *Así no se mide. Antropología de la medición en la España contemporánea*. Madrid: Ministerio de Cultura.

Safier, Neil. 2008. *Measuring the New World: Enlightenment Science and South America*. Chicago: University of Chicago Press.

Schaffer, Simon. 1992. "Late Victorian metrology and its instrumentation: A manufactory of ohms." In *Invisible Connections: Instruments, Institutions, and Science*, edited by Robert Bud and Susan Cozzens, 438–74. Bellingham, WA: SPIE Optical Engineering Press.

Schaffer, Simon. 1997. "Metrology, metrication, and Victorian values." In *Victorian Science in Context*, edited by Bernard Lightman, 438–74. Chicago: University of Chicago Press.

Scott, James C. 1998. *Seeing Like a State*. New Haven, CT: Yale University Press.

Shaw, Matthew. 2011. *Time and the French Revolution: The Republican Calendar, 1789–Year XIV*. Woodbridge: Royal Historical Society.

Statistical Office of the United Nations. 1966. *World Weights and Measures*. New York: United Nations.

Ten, Antonio. 1996. *Medir el metro: La historia de la prolongación del arco de meridiano Dunkerque-Barcelona, base del sistema métrico decimal*. Valencia: Universitat de Valencia.

Treat, Charles. 1971. *A History of the Metric System Controversy in the United States*. Washington, DC: NBS.

Tschoegl, Adrian. 2010. "The international diffusion of an innovation: The spread of decimal currency." *Journal of Socio-Economics*, 39: 100–9.

Velkar, Aashish. 2012. *Markets and Measurements in Nineteenth-Century Britain*. Cambridge: Cambridge University Press.

Vera, Hector. 2009. "Decimal time: Misadventures of a revolutionary idea, 1793–2008." *KronoScope*, 9: 29–48.

Vera, Hector. 2011. "The social life of measures: Metrication in the United States and Mexico, 1789–2004". PhD Dissertation, Sociology and Historical Studies, New School for Social Research.

Vera, Hector. 2015. "The social construction of units of measurement: Institutionalization, legitimation, and maintenance in metrology." In *Standardization in Measurement: Philosophical, Historical and Sociological Issues*, edited by Lara Huber and Oliver Schlaudt, 173–87. London: Pickering & Chatto.

Verman, Lal, and Jainath Kaul (eds.) 1970. *Metric Change in India*. New Delhi: Indian Standards Institution.

Williams, Barbara, and María del Carmen Jorge. 2008. "Aztec arithmetic revisited: Land-area algorithms and Acolhua congruence arithmetic." *Science*, 320: 72–7.

Zupko, Ronald. 1968. *French Weights and Measures before the Revolution: A Dictionary of Provincial and Local Units*. Bloomington: Indiana University Press.

Zupko, Ronald. 1990. *Revolution in Measurement: Western European Weights and Measures since the Age of Science*. Philadelphia, PA: American Philosophical Society.

CHAPTER THIRTY-THREE

Calculating Devices and Computers

MATTHEW L. JONES

Trumpeting the dramatic effects of terabytes of data on science, a breathless *Wired* article from 2008 described "a world where massive amounts of data and applied mathematics replace every other tool that might be brought to bear." No more theory-laden era:

> Out with every theory of human behavior, from linguistics to sociology. Forget taxonomy, ontology, and psychology. Who knows why people do what they do? The point is they do it, and we can track and measure it with unprecedented fidelity. With enough data, the numbers speak for themselves (Anderson 2008; Strasser 2012; Leonelli 2014).

A new empirical epoch has arrived.

Such big-data positivism is neither the first nor the last time that developments in information technology have been seen as primed to upset all the sciences simultaneously. Rarely have digital computers and claims of their revolutionary import been far apart. This chapter explores the machines, mathematical development, and infrastructures that make such claims thinkable, however historically and philosophically unsustainable. The chapter focuses upon computation, storage, and infrastructures from the early modern European period forward.[1] A remarkable self-reflexive approach to the very limits of computational tools has long been central to the productive quality of these technologies. Whatever the extent of computational hubris, much generative work within the computational sciences rests on creative responses to the limits of technologies of computation, storage, and communication. Scientific computation works within a clear eschatology: the promised land of adequate speed and storage is ever on the horizon, but, in the meanwhile, we pilgrims in this material state must contend with the materialities of the here and now.

However revolutionary in appearance, the introduction of electronic digital computers as processors, as storage tools, and a means of communication often rested initially upon existing practices of computation and routinization of data collection, processing, and analysis, in science and industry alike.[2] But just as computing

A Companion to the History of Science, First Edition. Edited by Bernard Lightman.
© 2016 John Wiley & Sons Ltd. Published 2020 by John Wiley & Sons Ltd.

altered the sciences, the demands of various sciences altered scientific computing. Transforming the evidence of existing scientific domains into data computable and storable in electronic form challenged ontology and practice alike; it likewise demanded different forms of hardware and software. If computers could tackle problems of such complexity that would otherwise prove infeasible, if not intractable, they did so through powerful techniques of simplification, through approximation, through probabilistic modeling, through means for discarding data and many features of the data.

This chapter focuses upon computational science and science using information technology, rather than the discipline of computer science. Centered on computation (arithmetical operations, integration) and data storage and retrieval, first in Europe, then primarily in the US, it omits the story of networks and the Internet, and the role of computational metaphors and ontologies within the sciences.[3] The approach here is episodic and historiographic, rather than comprehensive or narrative.

The constraints of computing technologies, and not just their possibilities, are essential for the path of computational sciences in recent years. To borrow a modish term, we need to give more epistemic attention to the *affordances* of different systems of calculation. The ideational history of computing must thus pay close attention to its materiality and social forms, and the materialist history of computing must pay attention to its algorithmic ingenuity in the face of material constraints.

Calculation "By Hand"

The first detailed publication concerning an electronic digital computer appeared in the prestigious journal *Mathematical Tables and other Aids to Computation*, sponsored by no less than the US National Academy of Sciences (Polachek 1995). The first issues of this revealingly named publication in 1943 included a detailed review of basic mathematical tables of logarithms, trigonometric functions, and so forth. The spread of mechanical calculating machines from the late nineteenth century had made the production of tables more, not less, important. "As calculating machines came into use, the need for seven-place tables of the natural values of the trigonometric functions stimulated a number of authors to prepare them" (C[omrie] 1943, 8). Beneath the surface, however, the calculations behind these tables were of surprising antiquity. The foremost major advocate for scientific computation using mechanical calculating machines in the early twentieth century, Leslie Comrie, argued that careful examination of the errors in tables strongly indicated that most of these new sets were simply taken, usually with no attribution, from sixteenth- and seventeenth-century tables.[4]

The upsurge of mathematical astronomy in early modern Europe, most associated with Nicolas Copernicus, Tycho Brahe, and Johannes Kepler, spurred the development of new methods for performing laborious calculations, particularly techniques for abridging multiplication by some sort of reduction to addition (Thoren 1988). While the first widespread techniques involved trigonometric identities, John Napier devised the more straightforward technique of logarithms early in the seventeenth century. With logarithms, multiplication and division were reduced to addition and subtraction. Put into a more elegant as well as base ten form by Napier's collaborator, the English mathematician Henry Briggs, logarithms soon became a dominant tool in astronomical and scientific calculation well into the twentieth century (Jagger

2003). "Briggs' industry," Comrie explained, "in tabulating logarithms of numbers and of trigonometrical functions made Napier's discovery immediately available to all computers"—that is, people performing calculations (C[omrie] 1946, 149). So basic was logarithmic computation that tables of functions by and large provided logarithms of those functions, rather than regular values, well into the first half of the twentieth century.

Far from replacing tables, mechanical calculating machines gained their currency within scientific applications largely by abridging the labor of the additions of results taken from the tables. One tool complimented the other. The first known mechanical digital calculating machine, that of Kepler's correspondent Wilhelm Schickard, was designed precisely to ameliorate the addition of results taken from Napier's Bones. In the mid-seventeenth century Blaise Pascal and Gottfried Wilhelm von Leibniz envisioned machines to aid financial and astronomical calculation. Despite decades of work Leibniz never brought to any sort of completion his envisioned machine for performing addition, subtraction, multiplication, and division directly. And despite a long process of invention and re-invention throughout the eighteenth century, mechanical calculating machines had not become robust enough for everyday financial or scientific use (Aspray 1990; Marguin 1994; Jones forthcoming). "In the present state of numerical science," a learned reviewer remarked in 1832, "the operations of arithmetic may all be performed with greater certainty and dispatch by the common method of computing by figures, than almost by any mechanical contrivance whatsoever." More manual devices were far more significant:

> we must except the scale and compasses, the sector, and the various modifications of the logarithmic line with sliders, all of which are valuable instruments. ... The chief excellence of these instruments consists in their simplicity, the smallness of their size, and the extreme facility with which they may be used.

In sharp contrast, these "qualities which do not belong to the more complicated arithmetical machines, ... render the latter totally unfit for common purposes" (Adam 1832, 400). Mechanical calculating machines entered into wide use, notably among actuaries and accountants, in Western Europe and the United States only in the 1870s at the earliest—and not without continuing skepticism (Warwick 1995; Cortada 2000; Nolan 2000; Yates 2000; Heide 2009). Simpler mechanical devices, above all the slide rule, a device using logarithms, remained important into the 1970s.

Charles Babbage envisioned his Difference Engine early in the nineteenth century just to produce mathematical tables in an automated way: the Engine was to be a machine for automating the production of the paper tools central to scientific and business calculation. Babbage sought to ameliorate two aspects of table making: the calculation of the values and, nearly as important, their typesetting (Schaffer 1994; Swade 2001). Although none of Babbage's machines was completed, others, notably the Swedish son and father Scheutz team, produced a working device that saw some use (Lindgren 1990). Securing the order of calculation meant securing the printing process. Problems with print greatly worried all those concerned with scientific computation well into the mid-twentieth century.

Mechanical contrivances were not limited to arithmetical operations. Initially developed for aiding census taking, tabulating equipment soon became central to the data intensive life insurance market. Life insurance firms pushed the corporations selling tabulators to develop and refine these machines, to encompass printing, automatic control, sorting, and the introduction of non-numerical and alphabetical data (Yates 1993). They offered a materialization of data processing at a large scale that had been brought to a very high level of reliability by the 1920s. At that time, they began to be used for scientific computation in larger numbers (Priestley 2011, chapter 3). Two major advocates, Comrie, and his American analogue, Wallace Eckert, preached the virtues of connecting two largely independent traditions of business machines: calculating machines and register machines, capable of arithmetical operations, and tabulating machines, capable of recording and reading large amounts of data.

Calculation "by hand" did not exclusively comprise manual arithmetic. Calculation by hand encompassed an array of techniques and tools aiding computation, from slide rules to mechanical calculators, and especially mathematical tables of important mathematical functions (Kidwell 1990). And it often involved teams of human calculators, in many cases groups of women (Light 1999; Grier 2005). Well after the advent of electronic digital machines following World War II, scientists in the US and UK weighed the costs and benefits of using teams of human computers and punch-card tabulators rather than expensive and hard to access electronic computers (Chadarevian 2002, 111–18).

Analog Computing

In 1946, Leslie Comrie remarked,

> I have sometimes felt that physicists and engineers are too prone to ask themselves "What physical, mechanical or electrical analogue can I find to the equation I have to solve?" and rush to the drawing board and lathe before enquiring whether any of the many machines that can be purchased over the counter will not do the job.

Comrie was decrying a rich tradition of building highly specialized devices that served as physical analogues allowing the solution to problems not otherwise tractable (Owens 1986; Mindell 2002; Care 2006). Such computers were "analog" in two senses: they measured continuous quantities directly and they were "analogous" to other physical phenomena. The best known of these machines, exemplified by Vannevar Bush's differential analyzer, allowed for mechanical integration, and thus were important in the solution of differential equations. Rather than an exact, analytical solution using highly simplified equations, mechanical integrators promised approximate solutions to problems in their much fuller complexity. The superiority of analog computation to numerical approximation for many purposes was still felt in 1946. Praising Bush's differential analyzer, Comrie noted,

> Although differential equations can be (and are) solved by finite difference methods on existing machines, the quantity of low-accuracy solutions required today is such that time and cost would be prohibitive. The use of machines for handling infinitesimals rather than finite quantities has fully justified itself … (C[omrie] 1946, 150).

Although digital electronic computers soon eclipsed analogue computers, they did so in many cases less by explicitly solving numerical problems, as by simulating them—a new form of analogical reasoning.

Electronic Computing, Numerical Analysis, and Simulation

The demands of war, first World War II and then the early Cold War, provided impetus and funding alike for the development of electronic digital computing in the United States, Britain, and the Soviet Union.[5] In 1946, John von Neumann and his collaborator H. H. Goldstine declared, "many branches of both pure and applied mathematics are in a great need of computing instruments to break the present stalemate created by the failure of the purely analytical approach to non-linear problems" (Von Neumann and Goldstine 1961, 4; Dahan Dalmenico 1996, 175). Working with electronic computers meant recognizing their affordances and limits. The "computing sheets of a long and complicated calculation in a human computing establishment" can store more than all the new electronic computers. They concluded,

> in an automatic computing establishment there will be a "lower price" on arithmetical operations, but a "higher price" on storage of data, intermediate results, etc. Consequentially, the "inner economy" of such an establishment will be very different from we are used to now, and what we were uniformly used to since the day Gauss. ... new criteria for "practicality" and "elegance" will have to be developed (Von Neumann and Goldstine 1961, 6; Aspray 1989, 307–8).

The new electronic digital computers produced just after World War II offered great possibility for speedy computation, while demanding their users rework older methods of numerical analysis. In the context of work around the atomic bomb, von Neumann altered numerical methods for solving partial differential equations in fluid dynamics better to allow them to be digitally calculated. Modifying existing approaches to numerical analysis to comport with this "inner economy," von Neumann and others spurred the development of new numerical analyses ever more tailored for the constraints and power of digital electronic machines, in particular the challenges of round-off error.

As the science of computerized numerical analysis developed, its limits became ever more clear, especially in the context of designing thermonuclear weapons (Galison 1997, chapter 8). Before the war, the physicist Enrico Fermi had worked on the idea of creating mathematical simulations of atomic phenomena. Stansilaw Ulam, along with von Neumann and Nicolas Metropolis, devised an approach dubbed "Monte Carlo" to tackle the challenging problems of studying the interactions with a nuclear weapon. The idea was to sample a large set of simulated outcomes of a process or situation, rather than attempting to solve analytically, or even numerically, the differential equations governing the process. Ulam began with the game of solitaire. One could generate a large number of different solitaire games, without enumerating them all, then analyze statistically the properties of that set of games. The same sort of analysis could be applied to the study of nuclear phenomena. Such simulations, remarkably, worked for many classes of problems without any stochastic content, such as the solution of an integral or the value of π. Something currently intractable theoretically became quasi-experimental. As Ulam and Metropolis noted, the potency of Monte Carlo came just because it could sidestep computationally intractable problems:

> The essential feature of the process is that we avoid dealing with multiple integrations or multiplications of the probability matrices, but instead sample single chains of events. We obtain a sample of the set of all such possible chains, and on it we can make a

statistical study of both the genealogical properties and various distributions at a given time (Metropolis and Ulam 1949, 339).

Monte Carlo and other such simulations rested then on a critique of human and artificial reasoning.[6]

Monte Carlo heralded the emergence of simulation as a central form of scientific knowledge in the years following the war (Galison 1997, 779; Seidel 1998; Lenhard, Shinn, and Küppers 2006). Computer simulation provided a novel sort of science sitting uncomfortably between experiment and theory and required a dramatic reconfiguration of adequate scientific knowledge. This reconfiguration was in many cases bitterly resisted, before becoming naturalized and now a central aspect of scientific practice. Originally used to sidestep the intractability of differential equations, simulations now come in many forms. Some generate simulations using underlying theoretical models; others eschew any claim to represented underlying theoretical structure and aim simply at behavioral reproduction. As so often in the history of science, the lack of closure about the philosophical issues around such a transformation has not precluded widespread adoption of the approach. Indeed, that lack of closure created the space for the creation of new—if often tendentious—approaches to the study of complex systems without the need for reduction to covering laws and highly simplified models.

Beyond Artillery, Bombs, and Particles

"How could a computer that only handles numbers be of fundamental importance to a subject that is qualitative in nature and deals in descriptive rather than analytic terms?" (as quoted in November 2012, 20). Such a concern, here about biology, was true of numerous domains of knowledge. The success of early electronic computers following World War II within traditionally heavily quantitative domains such as atomic physics and ballistics did not make it evident that computers had much to offer to rather different sciences. In field after field, pioneers nevertheless sought to transform the evidence and forms of reasoning of scientific subfields into new, more computationally tractable forms. The adoption of computing was neither natural nor easy (Yood 2013). In his recent history of biological computing, Hallam Stevens argues against the contention that superior computer power and storage capacity allowed biologists finally to adopt computerized tools in great number. Instead, he argues, biology "changed to become a computerized and computerizable discipline" (Stevens 2013, 13). Even in highly quantitative domains, the means for rendering problems appropriate to computation came from an array of disciplines, many initially created in wartime work: developments in fluid mechanics, statistics, signals processing, and operations research each provided distinctive ways of making problems computationally tractable (Dahan Dalmenico 1996). The plurality of approaches remains marked in the multiple names attached to many roughly similar computational techniques.[7]

For all the recent philosophical and historical work on models and simulations, we have no solid taxonomy of the varied forms of reflective simplification, reduction, and transformations of problem domains so that they become computationally tractable.[8] A great deal of the ingenuity of the application of computers to the sciences comes just in the creative transformation of problem domains conjoined to arguments about the scientific legitimacy of that transformation. As might be expected, reductions and simplifications that were initially bitterly contested became standard practice in subfields, and their contingency was lost. These reductions involved simplifications of data and of underlying possible models alike; they can also involve transformations in what suffices as scientific knowledge. We have a highly ramified set of different mixes of instrumentalism and realism still in need of good taxonomies.

In his commanding study of climate science, for example, Paul Edwards describes the emergence of a new ideal of "reproductionism" within computation of science that "seeks to stimulate a phenomenon, regardless of scale, using whatever combination of theory, data, and 'semi-empirical' parameters may be required." In this form of science, he argues, the "familiar logics of discovery and justification apply only piecemeal. No single, stable logic can justify the many approximations involved in reproductionist science" (Edwards 2010, 281). The line between the empirical and the theoretical has become productively blurred.

Herbert Simon, to take a second example, famously offered a contrast between approaches in operations research and then current artificial intelligence perspectives on decision problems. The algorithms of operations research, he noted,

> impose a strong mathematical structure on the decision problem. Their power is bought at the cost of shaping and squeezing the real-world problem to fit their computation: for example, replacing the real-world criterion function and constraint with linear approximation so that linear programming can be used.

In contrast, he explained, "AI methods generally find only satisfactory solutions, not optima ... we must trade off satisficing in a nearly-realistic model (AI) against optimizing in a greatly simplified model (OR)" (Simon 1996, 27–8; November 2012, 274).

These debates have continued into the era of data mining and big data. In 2001, the renegade statistician Leo Breiman polemically described the divide between two major statistical cultures:

> Statistics starts with data. Think of the data as being generated by a black box in which a vector of input variables **x** (independent variables) go in one side, and the other side the response variables **y** come out. Inside the black box, nature functions to associate the predictor variables with the response variables. ... These are two [distinct] goals in analyzing the data:

> *Prediction.* To be able to predict what the responses are going to be to future input variables;

> *Information.* To extract some information about how nature is associating the response variables to the input variables (Breiman 2001, 199).

Against the dominant statistical view, Breiman argued for an "algorithmic modeling culture" that is satisfied with the goal of prediction without making physical claims about the actual natural processes. Variants of such epistemic modesty are central to much recent work in machine learning, yet many scientists and statisticians find it far too instrumentalist.

Big Data Avant Big Data

In the late 1940s, Soviet cryptography abruptly became very strong and largely impervious to decryption by the US and its allies. The signals intelligence agencies of the West, notably the newly established National Security Agency (NSA), found themselves early in the Cold War needing the capacity to process large amounts of data far more than the capacity to perform arithmetic quickly. Under the sponsorship of the US national laboratories concerned with nuclear weapons, computer developments had been focused to a great extent upon improving the processing speed needed for simulations using floating-point arithmetic (MacKenzie 1991, 197). In contrast, the NSA needed to be able to sort through large amounts of traffic quickly: "the Agency became as much or more a data processing center than a 'cryptanalytic center'." As a result NSA sought "high speed substitutes for the best data processors of the era, tabulating equipment" (Burke 2002, 264). In focusing "on the manipulation of large volumes of data and great flexibility and variety in non-numerical logical processes," NSA had needs more akin to most businesses than to physicists running simulations. Just as substantial federal funds promoted the creation of ever faster arithmetical machines, substantial federal funds for cryptography sponsored intense work on larger storage mechanisms. The two came together, with great friction, in funding IBM's attempts to create a jump in capability in the mid 1950s. "AEC's computer requirement emphasized high-speed multiplication, whereas NSA's emphasis was on manipulation of large volumes of data and great flexibility and variety in non-numerical logical processes" (Snyder 1980, 66).

The sciences followed suit. In 1950, Mina Rees of the Naval Research Office noted the "great emphasis" in early machines "that would accept a small amount of information, perform very rapidly extensive operations on this information, and turn out a small amount of information as its answer." Now, she wrote, the interest "seems to lie in a further exploration of the use of machines to accept large amounts of data, perform very simple operation upon them, and print out, possibly, very large numbers of results" (Rees 1950, 735). The experimental data produced in high-energy physics quickly challenged storage and processing abilities alike (Seidel 1998, 54). In science as in snooping, the data potentially to be analyzed and stored has ever outstripped processing power, memory, and storage capacity. "Over the past 40 years or more," a piece in *Science* noted in 2009, "Moore's Law has enabled transistors on silicon chips to get smaller and processors to get faster. At the same time, technology improvements for disks for storage cannot keep up with the ever increasing flood of scientific data generated by the faster computers" (Bell, Hey, and Szalay 2009).

These material constraints challenged scientists and applied mathematicians to develop ways of abridging and reducing data and to account for the legitimacy of such reductions. In an early effort at computing Fourier syntheses for use in X-ray crystallography, J. M. Bennett and J. C. Kendrew explained, "In a machine such as the EDSAC [Electronic Delay Storage Automatic Calculator], ... it is impossible to accommodate all the terms of a typical two-, and more, especially of a three-, dimensional synthesis..." The authors devised and defended numerous techniques for representing the data more compactly without losing too much significant information, including "smoothing": "in many cases the synthesis obtained from such smoothed-off data is not significantly different from that compounded of unsmoothed data" (Bennett and Kendrew 1952, 112). Such reflective reductions of the data—with potentially dangerous loss of information—remain integral to nearly all data intensive scientific work, and have only become more central with the petabytes of "big data." Algorithms drawn from statistics, initially developed for smaller data sets, often require dramatic transformation; one 1996 paper, for example, explains that approaches from artificial intelligence and statistics

> do not adequately consider the case that the dataset can be too large to fit in main memory. In particular, they do not recognize that the problem must be viewed in terms of how to work with a [*sic*] limited resources (e.g., memory that is typically, much smaller than the size of the dataset) to do the clustering as accurately as possible while keeping I/O [input/output] costs low (Zhang, Ramakrishnan, and Livny 1996, 104).

Large amounts of data thus often become the grounds for a technological deterministic account of the necessity of computational choices. Not just the number of observations, but also the dimensionality of those observations requires transformative techniques for reducing and choosing among the data.

Data Infrastructures

Computerized storage of data is not neutral, obvious, or natural.[9] Obtaining data from the world is hard work; standardizing it often more so. Standardizing data is intensive and non-trivial. Contemporary data scientists often quip that some 95 percent of analytical time involves "data munging." This commonplace is borne out within science studies. In his study of twentieth-century climate science, Paul Edwards discusses the distinct process of "making global data"—collecting weather data from around the world and of "making data global"—"building complete, coherent, and consistent global data sets from incomplete, inconsistent, and heterogeneous data sources" (Edwards 2010, 251).

Standardizing data is challenging for individual scientific research groups, but even more contested while crossing institutional, disciplinary, and national lines. In their study of model organism databases, Sabine Leonelli and Rachel Ankeny describe the development of formalized data curators responsible for "(1) the choice of terminology to classify data and (2) the selection and provision of information about experimental settings in which data are produced, including information about specimens and protocols" (Leonelli and Ankeny 2012, 31). To be useful, datasets must include metadata: information about the production of the data, essential for evaluating it in further analysis. Dispute over the approach can produce what one group calls "science friction" (Edwards et al. 2011).

Each database depends on a set of decisions—and compromises—about how to represent which aspects of data collection and how to store them on actual computer systems.[10] These choices at once limit and make possible particular forms of knowledge production. The cataloging of genomes saw the "evolution of GenBank from flat-file to relational to federated data-based," which Hallam Stevens has argued "paralleled biologists' moves from gene-centric to alignment-centric to multielement views of biological action" (Stevens 2013, 168). Leonelli and Ankeny likewise note,

> Through classification systems such as the Gene Ontology, databases foster implicit terminological consensus within model organism communities, thus strengthening

communication across disciplines but also imposing epistemic agreement on how to understand and represent biological entities and processes (Leonelli and Ankeny 2012, 32).

The point is not that databases allow only one sort of theory: different databases lend themselves to particular types of investigation and make others more challenging. Different ways of storing data have different investigative affordances. Like models, databases can be performative (Bowker 2000, 675–6).

Advocates of the introduction of computation into various scientific fields draw heavily upon technological determinist narratives to justify the necessity of new epistemic practices and differently skilled practitioners. To justify the intrusion of computational statistical methods into taxonomy, for example, the biologist George Gaylor Simpson explained that they "become quite necessary as we gather observations on increasing large numbers of variables in large numbers of individuals" (Simpson 1962, 504).

The Social Organization of Expertise

In 1962, Simpson envisioned new forms of computational taxonomy in zoology:

> the day is upon us when for many of our problems, taxonomic and otherwise, freehand observation and rattling off elementary statistics on desk calculators will no longer suffice. The zoologist of the future, including the taxonomist, often is going to have to work with a mathematical statistician, a programmer, and a large computer. Some of you may welcome this prospect, but others may find it dreadful (Simpson 1962, 504–5; see Hagen 2001).

Practices of computation rest on social organizations of expertise. Debates about the propriety of using calculating tools often hinge on the distribution of skill and boundaries of expertise. Having just advocated the necessity of statistical computing, Simpson defended the continuing necessity of the trained human biologist against "extremists" who "hold that comparison of numerical data on samples by means of a computer automatically indicates the most natural classification of the corresponding populations." While "computer manipulation has become not only extremely useful and indispensible," he explained, it is false that "it can automatically produce a biologically significant taxonomic result" (Simpson 1962, 505).

Such demarcation battles figure prominently in the many sciences computerized in the second half of the twentieth century. Peter Galison documented the conflict within postwar microphysics concerning the necessity of human interpretation of high-energy events. Committed to the discovery of novel, startling events, the physicist Luis Alvarez stressed the distinctiveness of human cognitive capacities. Insisting on a "strong positive feeling that human beings have remarkable inherent scanning abilities," Alvarez declared, "these feelings should be used because they are better than anything that can be built into a computer" (as quoted in Galison 1997, 406). Attendant upon this epistemic claim was the need for an industrial organization of human scanners possessing such feelings.

Programming—or teaching—computers to perform acts of judgment and inference motivated major work in artificial intelligence. Notable successes included attempts to formalize the judgment of scientists concerning organic chemical structures, as in the case of the expert system DENDRAL (November 2012, 259–68). By the early 1970s, many practitioners worried greatly about the challenge of converting human expertise into "knowledge-bases" and formal inference rule. In a move akin to Harry Collins' reinvigoration of "tacit knowledge" in the sociology of science, artificial intelligence researchers became worried about the "knowledge acquisition bottleneck" (Forsythe 1993; Feigenbaum 2007, 62–3). J. Ross Quinlan noted that part "of the bottleneck is perhaps due to the fact that the expert is called upon to perform tasks that he does not ordinarily do, such as setting down a comprehensive roadmap of a subject" (Quinlan 1979, 168). Rather than attempting to simulate some aspect of the cognitive process of judgment, new forms of pattern recognition and machine learning attempted to predict the expert judgments based on the behavior of experts in some task of classification.

> the machine learning technique takes advantage of the data and avoids the knowledge acquisition bottleneck by extracting classification rules directly from data. Rather than asking an expert for domain knowledge, a machine learning algorithm observes expert tasks and induces rule emulating expert decisions (Irani, Cheng, Fayyad, and Qian 1993, 41).

Just such a positivist dream about the possibilities of such instrumentalist learning algorithms ultimately inspired the breathless *Wired* article with which I began.

While attempts to automate aspects of human cognition inspired machine learning, another strand of research sought to optimize computer output best to draw upon human potential. A National Science Foundation sponsored report in 1987 noted, the "gigabit bandwidth of the eye/visual cortex system permits much faster perception of geometric and spatial relationship than any other mode, making the power of supercomputers more accessible." The goal was to harness the brain, not sidestep it.

> The most exiting potential of wide-spread availability of visualization tool is … the insight gained and the mistakes understood by spotting visual anomalies while computing. Visualization will put the scientist into the computing loop and change the way science is done (McCormick, DeFanti, and Brown 1987, vii, 6).

A celebration of embodied minds, scientific visualization brought together the affordances and limits of human beings and machines alike.[11]

Hubris and Materiality

A 2006 piece in *Science* located the coming of a new data-focused science within a classical narrative of the history of science:

> Since at least Newton's laws of motion in the 17th century, scientists have recognized experimental and theoretical science as the basic research paradigms for understanding nature. In recent decades, computer simulations have become an essential third paradigm: a standard tool for scientists to explore domains that are inaccessible to theory and experiment, such as the evolution of the universe, car passenger crash testing, and predicting climate change.

Information systems, the authors claim, have now moved beyond simulation: "As simulations and experiments yield ever more data, a fourth paradigm is emerging, consisting of the techniques and technologies needed to perform data-intensive science …"

And yet this prophecy of a coming age lacks eschatological vim; its concerns are infrastructural and material. The vast data now available outstrips storage, processing, and communications resources.

> In almost every laboratory, "born digital" data proliferate in files, spreadsheets, or databases stored on hard drives, digital notebooks, Web sites, blogs, and wikis. The management, curation, and archiving of these digital data are becoming increasingly burdensome for research scientists.

The problem rests on a lack of understanding of the material conditions for data-intensive science: "data-intensive science has been slow to develop due to the subtleties of databases, schemas, and ontologies, and a general lack of understanding of these topics by the scientific community." Too ideational a conception of computational science, in other words, has slowed the development of a data-driven computation science:

> In the future, the rapidity with which any given discipline advances is likely to depend on how well the community acquires the necessary expertise in database, workflow management, visualization, and cloud computing technologies (Bell, Hey, and Szalay 2009, 1297–8).

Devices for computing and information storage have long challenged their users: far from leading users into a virtual world without the challenges of the material one, they require their users to contend with their affordances and material limits. These limits—in processing power, in storage size and speed, in bandwidth—demand much of users, and users have done much with them.

Endnotes

1 For an overview of the historiography, which has taken a decided turn toward business history, see Haigh 2011; for sharp historiographical insight on the histories of computing, Mahoney 2011; for "computing" before the digital computer, with good reference to engineering traditions, see Akera 2007, chapter 1. The classic study of the early development of the digital computer for scientific applications is Goldstine 1972. For the spread of information technologies internationally, see Cortada 2012.

2 A crucial corrective to simple narratives of computerization is Agar 2006, 873; compare Mahoney 2005; Hashagen 2013.

3 Among many studies, see, e.g., Kay 2000.

4 For broader concerns about tables, see Warwick 1995, 317–27.

5 For the UK, see the revisionist account Agar 2003; for the Soviet Union, see Crowe and Goodman 1994; Goodman 2003.

6 For the ENIAC and Monte Carlo, see Haigh, Priestley, and Rope 2014.

7 For an international survey, see Brezinski and Wuytack 2001.

8 See, however, the fine Winsburg 2010. For models in the history of science, see Morgan and Morrison 1999; Creager, Lunbeck, and Wise 2007.

9 For histories of data, see, for example, Edwards 2010; Strasser 2012; Sepkoski 2013; Leonelli 2014.

10 The main academic histories of database systems are Bergin and Haigh 2009 and Haigh 2009; more generally, see Nolan 2000.

11 See Burri and Dumit 2008 for visualization studies in STS.

References

Adam, Anderson. 1832. "Arithmetic." In *The Edinburgh Encyclopaedia Conducted by David Brewster, with the Assistance of Gentlemen Eminent in Science and Literature*, edited by David Brewster, 2: 345–400. Philadelphia, PA: J. and E. Parker.

Agar, Jon. 2003. *The Government Machine: A Revolutionary History of the Computer*. Cambridge, MA: MIT Press.

Agar, Jon. 2006. "What difference did computers make?" *Social Studies of Science*, 36, No. 6: 869–907. DOI: 10.1177/0306312706073450.

Akera, Atsushi. 2007. *Calculating a Natural World: Scientists, Engineers, and Computers During the Rise of U.S. Cold War Research*. Cambridge, MA: MIT Press.

Anderson, Chris. 2008. "The end of theory: The data deluge makes the scientific method obsolete." *Wired Magazine On-Line*. http://archive.wired.com/science/discoveries/magazine/16-07/pb_theory [Accessed September 28, 2015].

Aspray, William. 1989. "The transformation of numerical analysis by the computer: An example from the work of John von Neumann." In *History of Modern Mathematics*, edited by David E. Rowe, and John McCleary, 2: 307–22. Boston, MA: Academic Press.

Aspray, William (ed.) 1990. *Computing before Computers*. Ames: Iowa State University Press.

Bell, Gordon, Tony Hey, and Alex Szalay. 2009. "Beyond the data deluge." *Science*, 323: 1297–8.

Bennett, John M., and John C. Kendrew. 1952. "The computation of Fourier synthesis with a digital electronic calculating machine." *Acta Crystallographica*, 5, No. 1: 109–16.

Bergin, Thomas J., and Thomas Haigh. 2009. "The commercialization of database management systems, 1969–1983." *Annals of the History of Computing*, 31, No. 4: 26–41.

Bowker, Geoffrey C. 2000. "Biodiversity datadiversity." *Social Studies of Science*, 30, No. 5: 643–83.

Breiman, Leo. 2001. "Statistical modeling: The two cultures." *Statistical Science*, 16: 199–215.

Brezinski, C., and L. Wuytack. 2001. "Numerical analysis in the twentieth century." In *Numerical Analysis: Historical Developments in the 20th Century*, edited by L. Wuytack and C. Brezinski, 1–40. Amsterdam: Elsevier.

Burke, Colin B. 2002. *It Wasn't All Magic: The Early Struggle to Automate Cryptanalysis, 1930s–1960s*. Fort Meade, MD: Center for Cryptological History, NSA. http://archive.org/details/NSA-WasntAllMagic_2002 [Accessed September 28, 2015].

Burri, Regula, and Joe Dumit. 2008. "Social studies of scientific imaging and visualization." In *The Handbook of Science and Technology Studies*, edited by Edward J. Hackett, 3rd edition, 297–317. Cambridge, MA: MIT Press.

Care, Charles. 2006. "A chronology of analogue computing." *The Rutherford Journal: The New Zealand Journal for the History and Philosophy of Science and Technology*, 2: July. http://www.rutherfordjournal.org/article020106.html [Accessed September 28, 2015].

Chadarevian, Soraya de. 2002. *Designs for Life: Molecular Biology After World War II*. Cambridge: Cambridge University Press.

C[omrie], L. J. 1943. "Recent mathematical tables." *Mathematical Tables and Other Aids to Computation*, 1, No. 1: 3–23. DOI: 10.2307/2002683.

C[omrie], L. J. 1946. "The application of commercial calculating machines to scientific computing." *Mathematical Tables and Other Aids to Computation*, 2, No. 16: 149–59. DOI: 10.2307/2002577.

Cortada, James W. 2000. *Before the Computer: IBM, NCR, Burroughs, and Remington Rand and the Industry They Created, 1865–1956*. Princeton, NJ: Princeton University Press.

Cortada, James W. 2012. *The Digital Flood: The Diffusion of Information Technology across the U.S., Europe, and Asia*. New York: Oxford University Press.

Creager, Angela N. H., Elizabeth Lunbeck, and M. Norton Wise (eds.) 2007. *Science Without Laws: Model Systems, Cases, Exemplary Narratives*. Durham, NC: Duke University Press.

Crowe, G. D., and S. E. Goodman. 1994. "S. A. Lebedev and the birth of Soviet computing." *Annals of the History of Computing, IEEE*, 16, No. 1: 4–24. DOI: 10.1109/85.251852.

Dahan Dalmenico, Amy. 1996. "L'essor des mathématiques appliquées aux États-Unis: l'impact de la seconde guerre mondiale." *Revue d'histoire des mathématiques*, 2, No. 2: 149–213.

Edwards, Paul. 2010. *A Vast Machine: Computer Models, Climate Data, and the Politics of Global Warming*. Cambridge, MA: MIT Press.

Edwards, Paul, Matthew S. Mayernik, Archer Batcheller, Geoffrey Bowker, and Christine Borgman. 2011. "Science friction: Data, metadata, and collaboration." *Social Studies of Science*, 41, No. 5: 667–90.

Feigenbaum, Edward. 2007. *Oral History of Edward Feigenbaum*. Interview by Nils Nilsson. http://archive.computerhistory.org/resources/access/text/2013/05/102702002-05-01-acc.pdf [Accessed September 28, 2015].

Forsythe, D. E. 1993. "Engineering knowledge: The construction of knowledge in artificial intelligence." *Social Studies of Science*, 23, No. 3: 445–77. DOI: 10.1177/0306312793023003002.

Galison, P. 1997. *Image and Logic: A Material Culture of Microphysics*. Chicago: University of Chicago Press.

Goldstine, Herman H. 1972. *The Computer from Pascal to von Neumann*. Princeton, NJ: Princeton University Press.

Goodman, Seymour. 2003. "The origins of digital computing in Europe." *Communications of the ACM*, 46, No. 9: 21–5.

Grier, David Alan. 2005. *When Computers Were Human*. Princeton, NJ: Princeton University Press.

Hagen, Joel B. 2001. "The introduction of computers into systematic research in the United States during the 1960s." *Studies in History and Philosophy of Science Part C: Studies in History and Philosophy of Biological and Biomedical Sciences*, 32, No. 2: 291–314. DOI: 10.1016/S1369-8486(01)00005-X.

Haigh, Thomas. 2009. "How data got its base: Information storage software in the 1950s and 1960s." *Annals of the History of Computing, IEEE*, 31, No. 4: 6–25.

Haigh, Thomas. 2011. "The history of information technology." *Annual Review of Information Science and Technology*, 45, No. 1: 431–87.

Haigh, Thomas, Mark Priestley, and Crispin Rope. 2014. "Los Alamos bets on ENIAC: Nuclear Monte Carlo simulations, 1947–1948." *Annals of the History of Computing, IEEE*, 36, No. 3: 42–63.

Hashagen, Ulf. 2013. "The computation of nature, or: Does the computer drive science and technology?" In *The Nature of Computation. Logic, Algorithms, Applications*, edited by Paola Bonizzoni, Vasco Brattka, and Benedikt Löwe, 263–70. Springer Berlin: Springer.

Heide, Lars. 2009. *Punched-Card Systems and the Early Information Explosion, 1880–1945*. Baltimore: Johns Hopkins University Press.

Irani, Keki B., Jie Cheng, Usama M. Fayyad, and Zhaogang Qian. 1993. "Applying machine learning to semiconductor manufacturing." *IEEE Expert*, 8, No. 1: 41–7.

Jagger, Graham. 2003. "The making of logarithm tables." In *The History of Mathematical Tables: From Sumer to Spreadsheets*, edited by Martin Campbell-Kelly, Mary Croarken, Raymond Flood, and Eleanor Robson, 49–78. Oxford: Oxford University Press.

Jones, Matthew L. forthcoming. *Reckoning with Matter: Calculating Machines, Innovation, and Thinking about Thinking from Pascal to Babbage*. Chicago: University of Chicago Press.

Kay, Lily E. 2000. *Who Wrote the Book of Life?: A History of the Genetic Code*. Stanford, CA: Stanford University Press.

Kidwell, Peggy A. 1990. "American scientists and calculating machines: From novelty to commonplace." *IEEE Annals of the History of Computing*, 12, No. 1: 31–40.

Lenhard, Johannes, Terry Shinn, and Günter Küppers. 2006. "Computer simulation: Practice, epistemology, and social dynamics." In *Simulation*, 25: 3–22. Sociology of the Sciences Yearbook. Dordrecht: Springer.

Leonelli, Sabina. 2014. "What difference does quantity make? On the epistemology of big data in biology." *Big Data & Society*, 1, No. 1. DOI: 10.1177/2053951714534395.

Leonelli, Sabina, and Rachel A. Ankeny. 2012. "Re-thinking organisms: The impact of databases on model organism biology." *Studies in History and Philosophy of Science Part C: Studies in History and Philosophy of Biological and Biomedical Sciences*, 43, No. 1: 29–36. DOI: 10.1016/j.shpsc.2011.10.003.

Light, Jennifer S. 1999. "When computers were women." *Technology and Culture*, 40, No. 3: 455–83.

Lindgren, Michael. 1990. *Glory and Failure: The Difference Engines of Johann Müller, Charles Babbage and Georg and Edvard Scheutz*. Cambridge, MA: MIT Press.

MacKenzie, Donald. 1991. "The influence of the Los Alamos and Livermore national laboratories on the development of supercomputing." *Annals of the History of Computing*, 13, No. 2: 179–201.

Mahoney, Michael S. 2005. "The histories of computing(s)." *Interdisciplinary Science Reviews*, 30, No. 2: 119–35. DOI: 10.1179/030801805X25927.

Mahoney, Michael S. 2011. *Histories of Computing*. Edited by Thomas Haigh. Cambridge, MA: Harvard University Press.

Marguin, Jean. 1994. *Histoire des instruments et machines à calculer: Trois siècles de mécanique pensante, 1642–1942*. Paris: Hermann.

McCormick, Bruce H., Thomas A. DeFanti, and Maxine D. Brown. 1987. *Visualization in Scientific Computing*. Vol. 21. Computer Graphics. New York: ACM Press. http://www.sci.utah.edu/vrc2005/McCormick-1987-VSC.pdf [Accessed September 28, 2015].

Metropolis, Nicholas, and Stanislaw Ulam. 1949. "The Monte Carlo method." *Journal of the American Statistical Association*, 44, No. 247: 335–41.

Mindell, David A. 2002. *Between Human and Machine: Feedback, Control, and Computing before Cybernetics*. Baltimore: The Johns Hopkins University Press.

Morgan, Mary S., and Margaret Morrison (eds.) 1999. *Models as Mediators: Perspectives on Natural and Social Science*. Cambridge: Cambridge University Press.

Nolan, Richard L. 2000. "Information technology management since 1960." In *Nation Transformed by Information: How Information Has Shaped the United States from Colonial Times to the Present*, edited by Alfred Dupont Chandler and James W. Cortada, 217–56. New York: Oxford University Press.

November, Joseph Adam. 2012. *Biomedical Computing: Digitizing Life in the United States*. Baltimore: Johns Hopkins University Press.

Owens, Larry. 1986. "Vannevar Bush and the differential analyzer: The text and context of an early computer." *Technology and Culture*, 27, No. 1: 63–95.

Polachek, Harry. 1995. "History of the Journal *Mathematical Tables and Other Aids to Computation*, 1959–1965." *Annals of the History of Computing*, 17, No. 3: 67–74.

Priestley, Mark. 2011. *A Science of Operations.* London: Springer.

Quinlan, J. R. 1979. "Discovering rules by induction from large collections of examples." In *Expert Systems in the Micro-Electronic Age,* edited by Donald Michie, 168–201. Edinburgh: Edinburgh University Press.

Rees, Mina. 1950. "The federal computing machine program." *Science,* 112: 731–6.

Schaffer, Simon. 1994. "Babbage's intelligence: Calculating engines and the factory system." *Critical Inquiry,* 21: 203–27.

Seidel, Robert W. 1998. "'Crunching numbers': Computers and physical research in the AEC laboratories." *History and Technology,* 15, No. 1–2: 31–68. DOI: 10.1080/07341519808581940.

Sepkoski, David. 2013. "Towards 'a natural history of data': Evolving practices and epistemologies of data in paleontology, 1800–2000." *Journal of the History of Biology,* 46, No. 3: 401–44. DOI: 10.1007/s10739-012-9336-6.

Simon, Herbert A. 1996. *The Sciences of the Artificial.* 3rd edition. Cambridge, MA: MIT Press.

Simpson, George Gaylord. 1962. "Primate taxonomy and recent studies of nonhuman primates." *Annals of the New York Academy of Sciences,* 102, No. 2: 497–514. DOI: 10.1111/j.1749-6632.1962.tb13656.x.

Snyder, Samuel S. 1980. "Computer advances pioneered by cryptologic organizations." *Annals of the History of Computing,* 2, No. 1: 60–70.

Stevens, Hallam. 2013. *Life Out of Sequence: A Data-Driven History of Bioinformatics.* Chicago: University of Chicago Press.

Strasser, Bruno J. 2012. "Data-driven sciences: From wonder cabinets to electronic databases." *Studies in History and Philosophy of Science Part C: Studies in History and Philosophy of Biological and Biomedical Sciences,* 43, No. 1: 85–7. DOI: 10.1016/j.shpsc.2011.10.009.

Swade, Doron. 2001. *The Difference Engine: Charles Babbage and the Quest to Build the First Computer.* New York: Viking.

Thoren, Victor E. 1988. "Prosthaphaeresis revisited." *Historia Mathematica,* 15, No. 1: 32–9. DOI: 10.1016/0315-0860(88)90047-X.

Von Neumann, John, and Herman H. Goldstine. 1961. "On the principles of large scale computing machines (1946)." In *Collected Works,* by John Von Neumann, 1–33. New York: Pergamon Press.

Warwick, Andrew. 1995. "The laboratory of theory, or, what's exact about the exact sciences." In *Values of Precision,* edited by M. Norton Wise, 135–72. Princeton: Princeton University Press.

Winsberg, Eric B. 2010. *Science in the Age of Computer Simulation.* Chicago: University of Chicago Press.

Yates, JoAnne. 1993. "Co-evolution of information-processing technology and use: Interaction between the life insurance and tabulating industries." *Business History Review,* 67, No. 1: 1–51. DOI: 10.2307/3117467.

Yates, JoAnne. 2000. "Business use of information and technology during the industrial age." In *Nation Transformed by Information: How Information Has Shaped the United States from Colonial Times to the Present,* edited by Alfred Dupont Chandler and James W. Cortada, 107–36. New York: Oxford University Press.

Yood, Charles N. 2013. *Hybrid Zone: Computers and Science At Argonne National Laboratory, 1946–1992.* Chestnut Hill, MA: Docent Press.

Zhang, Tian, Raghu Ramakrishnan, and Miron Livny. 1996. "BIRCH: An efficient data clustering method for very large databases." In *Proceedings of the 1996 ACM SIGMOD International Conference on Management of Data, Montreal, Quebec, Canada, June 4–6, 1996,* edited by H. V. Jagadish and Inderpal Singh Mumick, 103–14. New York: ACM Press.

CHAPTER THIRTY-FOUR

Specimens and Collections

MARY E. SUNDERLAND

Defending Collections

Why should scientific institutions spend valuable research money and allocate highly coveted space to house extensive collections of preserved specimens? Natural history museums have long been poised to answer this question and are ready to justify their actions to seemingly uninformed audiences. Scientists who work with collections are concerned about more than just public understanding, they also lament their colleagues' apparent ignorance and disdain regarding their research—a worry that was recently substantiated by a high-profile *Science* publication, which concluded that collecting was "no longer required to describe a species or to document its rediscovery" (Minteer, Collins, Love, and Puschendorf 2014). One of the co-authors, herpetologist Robert Puschendorf, professed that his contribution to the paper was motivated by a fit of rage surrounding his colleagues' decision to collect the only detected specimen of a species that was presumed to be extinct (Conniff 2014). Not surprisingly, a flood of angry responses erupted.

Just days after the article's publication, one of its critics declared victory on Twitter, writing that the "post-publication peer-review" that took place through social media had effectively "smacked down" the article (Jackson 2014). Blog posts listed the many ways in which the article was unwarranted, poorly researched, and just plain false (e.g., Chakrabarty 2014; University of Alaska 2014; Wheeler 2014). Decrying Minteer et al.'s (2014) suggestion that photographs, tissue samples, and sound recordings made the collecting of actual specimens unnecessary, Terry Wheeler from the Lyman Entomological Museum at McGill University pointed out the obvious: "a fruit fly is not a mammal" (Wheeler 2014). Ornithologists from the University of Alaska explained why collecting continues to be the scientific "gold standard" despite recent technologies that might make it seem antiquated and unnecessary. By emphasizing the role that voucher specimens play in the identification of a new species, Minteer et al. (2014) mask the real value of collections, which is realized in the long-term; a collection is only as valuable as the comparisons that it enables. Each specimen is

A Companion to the History of Science, First Edition. Edited by Bernard Lightman.
© 2016 John Wiley & Sons Ltd. Published 2020 by John Wiley & Sons Ltd.

scientifically valuable with respect to the specific hypothesis that motivated its collection, not merely its ability to facilitate an accurate identification (University of Alaska 2014). Surveying the well-referenced responses to Minteer et al. (2014), it is evident that many of the authors had previously marshaled evidence in support of collections (e.g., Lane 1996; Suarez and Tsutsui 2004; Pyke and Ehrlich 2010; Tewksbury et al. 2014). Indeed, collectors have been defending their work as scientific for centuries (Sunderland 2012; Sunderland, Klitz, and Yoshihara 2012).

This chapter shows how specimens and collections have acquired scientific utility throughout history and, in doing so, draws attention to their role in shaping the life sciences (Star and Griesemer 1989; Griesemer 1990; Griesemer 1991; Winsor 1991; Griesemer and Gerson 1993; Pickstone 1994; Gerson 1998; Nyhart 2009; Shavit and Griesemer 2009; Strasser 2012; Sunderland 2012). Viewing a particular collection in different contexts and at different times reveals how technological and theoretical changes have the capacity to remake collections by imparting new scientific and social value. Much of the scholarship on natural history museums and collections emphasizes their cultural content and ability to reveal social insights while fewer studies have examined the science of collecting and collections (Kohlstedt 1995; Asma 2001; Kohler 2006; Kohler 2007; Strasser 2012). Rather than exploring the public-facing side of collections, this chapter focuses on two case studies to highlight the dynamic production and functioning of collections as scientific instruments in different contexts and time periods.

Beginning with Carl Linnaeus (1707–1778), the first section explores the famous Swedish naturalist's pioneering work with collections. Although he is best known for his theoretical contributions to classification nomenclature, he also made important technological innovations that formed the basis of collections work into the nineteenth century (Müller-Wille 2006; Müller-Wille 2007). The next section moves into the twentieth century with an examination of the collections at the University of California, Berkeley's Museum of Vertebrate Zoology (MVZ). While the MVZ is indeed somewhat of an outlier in the museum world because it was and is strictly a research museum, this unique emphasis allowed it to play a lead role in reinventing collections and defending their scientific utility during a period when many university museums shed their collections and public museums shifted their focus away from natural history (Rader and Cain 2008; Sunderland 2013b).

Making Collections Scientific

Although collections-based researchers continue to be on the defensive, ready to respond to accusations that their work is outdated and unnecessary, they collectively lowered their guard in 2009 to celebrate the 150-year anniversary of Charles Darwin's *On the Origin of Species*. Arguably the most highly esteemed and famous scientific collector, Darwin colors collections in a positive light. Historian of science Mary P. Winsor recognized that the glow was sure to fade and seized the occasion to consider why so many non-taxonomists devalue collections and taxonomic work. This image problem, Winsor argues, stems from widespread and popular misinformation about the history of taxonomy and collections (Winsor 2009, 43). Many biology students today, for example, continue to learn that systematics had little scientific value until the mid-twentieth-century modern synthesis: a view that has been championed and perpetuated by one of its key architects, Ernst Mayr (1904–2005) (Smocovitis 1992;

1996; Winsor 2006b). Analyzing the shortcomings and aftermath of the modern synthesis are beyond the scope of this paper, but it is important to flag that some of biology's most vocal popularizers, such as Mayr, helped to perpetuate a very divisive picture of the life sciences in which natural history was old fashioned, descriptive, and essentialist, while molecular biology was modern, empirical, and experimental (Wilson 1994; Winsor 2006b). This portrayal detracts attention from the important contributions of pre-Darwinian taxonomy. Darwin owes a great debt to taxonomy, especially to its methodologies and technologies that enabled him to study the relationships between species (Winsor 2009, 45).

These taxonomical practices on which Darwin relied are the great legacy of Linnaeus. While Linnaeus has become a symbol of the sort of old-fashioned natural history that the new modern synthesis aimed to replace, recent historical scholarship challenges this depiction with evidence that it is not only uncharitable, but also unfounded (Winsor 1991; Winsor 2006a). This section looks at how Linnaeus assembled his herbarium to bring into focus his methodological efforts that imparted scientific value onto collections, effectively making them into scientific instruments that were capable of identifying taxonomic relationships (Müller-Wille 2006; Müller-Wille 2007).

Although Linnaeus is most famous for laying the foundations of modern taxonomy with his invention of binomial nomenclature, this was not his most significant contribution. At first glance, his methodological work might appear ordinary, but a close examination of Linnaeus' daily practices reveals the transformative nature of his approach. Recognizing that his collection of plants was far from static, Linnaeus was one of the first botanists to not bind his collection and instead separately store sheets of mounted specimens in a custom-built cabinet. This innovation turned his collections from a repository of specimens into a scientific instrument with the capacity to discover the natural world's true order (Müller-Wille 2006; Müller-Wille 2007).

Linnaeus, perhaps, would be pleased with an assessment of his work that emphasized the scientific relevance of his methods. Indeed, he and his Enlightenment contemporaries demarcated their collecting and collections from the earlier cabinets of curiosities by emphasizing the scientific relevance of their approach. By separating their work from the early modernists' contributions, the Enlightenment naturalists rewrote their historiography to exclude the collecting practices that emerged in the Italian Renaissance. Discounting this earlier work, however, is misleading because it played a key role in transforming natural history from a predominantly textual, erudite endeavor into a tactile, empirical study. This shift rearranged the natural history community and its audiences in a way that changed who was permitted to build collections and make scientific contributions (Findlen 1994). In the sixteenth century, naturalists argued that the establishment of museums distinguished the new natural history from the old, but by the eighteenth century, naturalists argued that their collecting methods paired with strategies for ordering collected objects made their approach scientific and capable of discovering the natural world's order. The now older cabinets of curiosity were viewed as forms of entertainment. Yet these earlier collections provided more than just entertainment; they had the capacity to transform status and personal identity, particularly of individuals who were living in Britain's expanding empire. French nationals living in India, for example, came to see themselves as British gentlemen, in part, because of their participation in amassing collections. Furthermore, just as European nobility built cabinets of curiosity to strengthen their social status, Indian rulers collected rarities to show their global reach and connections to distant foreign elites (Jasanoff 2004).

Enlightenment collectors distanced their work from collections' earlier associations with entertainment and power by introducing new practices, capable of transforming collections into scientific tools. Looking at Linnaeus' botany textbook offers a window into these new practices for building scientific collections. Notably, Linnaeus discouraged the binding of specimen sheets into volumes. At the time, the common practice for assembling a collection involved gluing several specimens to a single sheet of paper and then binding them into volumes. Rather than creating a static tome of specimens, Linnaeus advised his students to mount only one specimen per sheet and described how to construct a special storage cabinet with movable shelves that would allow easy access to each individual sheet. Once accessed, the sheet could then be reinserted almost anywhere in the cabinet. This was important because the collection was not fixed or permanent, but dynamic (Müller-Wille 2006, 61–2). Designing the cabinet in this way allowed it to function as a tool that enabled Linnaeus to do his work.

Before Linnaeus, plant specimens had been identified by assigning a species to its genus and then establishing a character of that particular specimen that allowed it be distinguished from its congeners. Linnaeus worried that this method led to artificial species definitions because it was impossible to establish that the same character would work to distinguish different congeners. To address this concern, Linnaeus developed a method that emphasized the importance of re-examining archived specimens in comparison to new material as it entered the collection—a process of collation. First, the botanist who located a potentially new specimen compiled a detailed descriptive account of the plant's morphology. Next, as additional representative specimens were acquired, each would be compared to the first species and its description. If one of these newer specimens had a character that deviated from the first specimen, the character would be canceled from the description of the original. As a result, the remaining characters were assumed to be constant and natural. Within Linnaeus' herbarium, each specimen was defined by its relations to the other specimens in the collection. It was, therefore, the collection as a whole, rather than a specific specimen, that served as the scientific instrument that enabled the determination of plant species and genera (Müller-Wille 2006). From this perspective, we see how Linnaeus' methodology and curatorial technologies introduced new standard practices of collection, description, and collation, which formed the foundation of natural history throughout the eighteenth and nineteenth centuries (Müller-Wille 2007).

Thanks, in part, to Linnaean methodology and evidenced by the success of Darwin's famous Beagle journey (1831–1836), scientific natural history collecting and collections flourished during the eighteenth and nineteenth centuries. Expeditions of discovery often carried a token naturalist, who was charged with collecting and documenting new specimens. Throughout Europe and the United States, institutions were established to house and facilitate research on the growing specimen collections (Kohlstedt 1988; Winsor 1991; Farber 2000; Endersby 2008). While the infrastructure of these diverse institutions varied, they were all built to enable the production of scientific knowledge through the systematic comparison and juxtaposition of specimens. Ranging from the famous Muséum National d'Histoire Naturelle in Paris (refounded in 1793), where George Cuvier (1769–1832) and Jean-Baptiste Lamarck (1744–1892) made their mark, to the British Museum in London where Robert Owen dreamed of amassing all of the world's species, the curators of these collections shared scientific aspirations founded in the value of studying physical specimens (Outram 1996; Secord 1996; Strasser 2012).

Linnaean practices traveled overseas, where they gave non-scientifically trained collectors the ability to transform specimens they acquired while working for places such as the British Consular Services in China into "objective" and "useful" scientific objects. When China opened its treaty ports to Westerners in the mid-nineteenth century, natural history collecting became a way for British government employees to contribute "factual" information about China. Indeed, many government employees were excited by the opportunity to collect the foreign flora and fauna of areas that were largely unknown to Westerners. Although very few of the collectors had any kind of scientific reputation, collecting was recognized as a respectable hobby that was particularly appealing to many of the young diplomats who loved to be out in nature. In order for these specimens to be useful to scientific institutions like Kew Gardens, they needed relevant accompanying information, rooted in Linnaean standards. Discovering a new specimen was dependent on proper Linnaean classification, which required collectors, for example, to gather information about each specimen's geographic locality (Fan 2003). This way of acquiring specimens privileged Western practices and ways of knowing that shaped how collections were built.

Modernizing Collections in the Twentieth Century

The gilded age was a golden age for naturalists in the United States, whose work was funded and supported at unprecedented levels (Pauly 2000, 47), but natural historians found themselves defending the scientific value of collections by the early twentieth century. Addressing his colleagues in a 1910 publication, Joseph Grinnell, the founding director of the MVZ, described the functions of a natural history museum (Grinnell 1910). Venerated by both the ornithology and mammalogy communities, Grinnell is probably most famous for his role as the MVZ's founding director (Miller 1964). His directorial position was made possible by the patronage of Annie M. Alexander, heir to the C & H sugar fortune, who selected Grinnell because of his scientific vision for the institution (Stein 2001). Grinnell envisioned the MVZ as a center for evolutionary research that would house collections that could enable long-term studies of evolution. As a result, the MVZ's collections were intentionally developed as a database and research instrument to inform studies of evolutionary biology (Sunderland 2013c).

Grinnell stressed that the scientific value of each specimen was entirely dependent on its attached information. The MVZ's infrastructure was designed to keep track of each specimen's detailed locality information as well as behavioral and ecological information. Locality mattered to Grinnell because this information was central to his own research interest: biogeography, speciation, and subspeciation. Grinnell aimed to understand how species formation was related to geographical and ecological boundaries, such as mountains or lakes. The information was critical for Grinnell's work on developing his speciation theory, which is expressed in his publications but also evident in his field notes, correspondence, and lecture notes (Grinnell 1943; Sunderland 2013a). According to Grinnell, habitats were dynamic in two important ways: they were being altered by human use (which was particularly apparent

Figure 34.1 This image was taken in Berkeley's Museum of Vertebrate Zoology to depict representative storage of mammal specimens. The animals shown here are chipmunks, genus Tamias. The skin and skull are housed in large metal cases for long-term storage and arranged alphabetically by family, genus, species, and subspecies, then by geographical area. A tag is attached to each specimen to show the specimen's locality, genus, and species name, as well as information about the collector, including the collector's name and the date of the collection. The skull and jaw each have a catalog number written on the bone and are stored in a labeled box or glass vial. Photo supplied courtesy of the Museum of Vertebrate Zoology, University of California, Berkeley.

in California) and they varied spatially and temporally (Grinnell 1919). Looking to the rapid urban and agricultural developments that were changing the Californian environment, Grinnell saw an unprecedented opportunity to amass a collection that could facilitate long-term studies of evolutionary mechanisms. Grinnell set out to document what happened when species were forced to move into new environments. California was also appealing because of its extraordinarily diverse landscapes that included valleys, mountains, rivers, oceans, deserts, and plains (Grinnell 1914a; Grinnell 1914b). Grinnell was interested in documenting both the inter- and intraspecies diversity within, and on the edges of, these diverse places. It mattered, therefore, if a mouse specimen was collected on black lava rock in the desert or on the adjacent white sand.

The MVZ collections were organized to keep track of all of these details and to enable comparative analysis and data retrieval. For example, a detailed tag was tied to each specimen in the field that complemented the affiliated field notes of the excursion when it was collected. Like Linnaeus, Grinnell implemented record-keeping procedures that can be viewed as technological innovations, which imparted scientific value on the collections. In his instructions for taking field notes, Grinnell stressed the importance of keeping track of maps, weather conditions, and general ecology (Figure 34.1). And, to ensure that these notes would last the test of time, Grinnell insisted on using permanent materials such as India ink and archival (100 percent rag bond) paper. With an eye to the future, almost everything was deemed relevant for recording. Grinnell even stressed recording seemingly redundant observations because the utility of each specimen was bolstered by the quality of its affiliated field notes (Sunderland 2013c).

Grinnell intended for the MVZ's collections to grow into a center for evolutionary research, but their growth demanded more resources, which brought them to the attention of university administrators who required increasingly sophisticated justifications of their worth (Sunderland 2012). During the tumultuous 1960s, it wasn't just collections that were scrutinized. Widely recognized as a period of broad social change, the 1960s environment encroached on scientists' protected bubbles within academia's ivory towers. Growing skepticism of the epistemic and moral value of science fueled an anti-science sentiment that emerged in the context of social and environmental concerns. For the first time, scientists had to justify their funding and their activities with reference to their ability to solve complex problems, such as environmental degradation, loss of ecological diversity, use of pesticides, and overpopulation (Appel 2000, 236). To cope with these changes, the National Science Foundation (NSF), for example, designed new initiatives that were oriented toward both scientifically significant and politically relevant goals; one of these initiatives was systematic biology (Appel 2000, 236).

In an effort to generate support from the NSF, systematic biologists joined forces to compile the report, *Systematic Biology Collections of the United States: An Essential Resource* (1971), which explained why systematists were ready to play a central role in responding to the environmental movement. The report identified the importance of maintaining a diverse ecosystem and that knowledge of the "kinds of creatures" in the ecosystem was fundamental to learning how to manage the world's burgeoning environmental problems (Appel 2000, 257). Thanks to the vocal support of a few advocates who argued that collections were an invaluable source of comparative environmental data from a less polluted time, the NSF promised additional funding for collections (Appel 2000, 258). The MVZ was one of the institutions that was consistently awarded funding from the NSF to build collections toward desirable epistemic and political ends. Looking at how the MVZ modernized their collection during the latter part of twentieth century provides a window into the forces that shaped collections.

One of the MVZ's initiatives involved augmenting their existing collections with new kinds of collections, including karyotype preparations and frozen tissue. In the 1960s, while he was a graduate student at the University of Arizona, James Patton, now emeritus curator of mammals at the MVZ, developed a technique to observe chromosomes in rodents. When he applied this technique to study the chromosomes of pocket gophers, he noticed that pocket gophers in the same population had extremely diverse karyotypes. Because different species have different looking chromosomes, Patton was able to carefully study hybrids and ask questions about population dynamics and species formation (Figure 34.1). In the 1970s starch gel electrophoresis made it possible to study protein differences in natural populations. When Patton's group applied these techniques to pocket gophers they discovered extreme variability both within and between populations of pocket gophers, indeed much more than the average rodent. Incorporating technologies that allowed protein analysis of the specimens in the collections revealed substantial variability within populations, but it also showed that some supposed subspecies were actually the same species (Sunderland, Klitz, and Yoshihara 2012).

Along with a handful of institutions, the MVZ also received funding in the 1970s to begin computerizing their collections, which was seen as the way of the future. Computers promised to modernize natural history, improve curatorial procedures, and make new kinds of research possible (Strasser 2012; Sepkoski 2013). Looking at the MVZ's early digitization efforts during the late 1970s and 1980s, before the era of widespread personal computing and the Internet, provides a window into how electronic databases shaped collections (Sunderland 2013c). Into the 1960s, most institutions relied on ledger catalogs and index cards to organize their collections. All written by hand, these notes, tags, and catalogs were time consuming to produce and far from error free. Automation technologies appeared in the 1960s, but were only introduced sparingly. Natural history museums were not quick to adopt electronic data processing because of scarce resources and concerns about added value.

After some effort, adjustment, and time, it became clear that computerizing collections resulted in more efficient and effective collections management. Writing about the roles of natural history collections in 1996, Meredith Lane from the University of Kansas Natural History Museum, emphasized the wealth of data that was stored in collections, including the genetic and phylogenetic information contained within samples, and the biogeographic, ecological, and biographical information that was stored on the labels of specimens. She notes that computerization efforts had made it easier to retrieve the data, but she foresaw a transformative change ahead; the Internet had the potential to enable unprecedented interconnectivity of physically dispersed specimens and collections. Lane called for an end to competition and rallied naturalists around the shared goal of discovering and describing the world's biota (Lane 1996). With the aim of facilitating collaborative work, the MVZ was among the first institutions to migrate its database to a relational online, open-access system in the early 2000s. Today, the MVZ's database includes multiple institutions and continues to expand through the development of VertNet (vertnet.org), which is an international effort to consolidate vertebrate databases and develop new data analysis applications to transform collections-based research (Constable et al. 2010). Although beyond the scope of this paper, the computerization of collections is a rich area of future research for both historians and philosophers of science (Shavit and Griesemer 2009).

Conclusions

The recent collecting controversy sparked by Minteer et al. (2014) shows that natural history museums must be ready to explain the utility of their collections to a variety of audiences, including the scientific community. Efforts to do so are demonstrated by recent public displays in prominent natural history institutions that allow visitors to peer through windows and see scientists at work, actively studying specimens. Putting scientists on display highlights ongoing efforts to justify the scientific value of specimen collections (Sunderland, Klitz, and Yoshihara 2012).

As the core tool and research subject of natural history, collections and specimens have been closely scrutinized throughout the twentieth century as the life sciences underwent a transformation. While the steady development of new technologies made it possible to study the molecular and genetic features and processes of organisms in an increasing amount of detail, the scientific utility of specimens and whole-organism collections came under scrutiny. Why bother collecting animals at all? Can't scientists learn everything they need to know from a blood or tissue sample? Shouldn't the focus be on establishing genetic libraries instead of curated collections of preserved organisms? Questions like these were the source of heated debates throughout the second half of the twentieth century, and continue to inspire controversy about the allocation of research funding today. Examining practitioners and administrators ongoing defense of collections brings into focus the shifting roles and identities of specimen collections against a backdrop of enormous scientific and technological change.

Specimens, samples, and collections are both objects of inquiry and research tools. This dual identity is made through careful collecting processes that move individual objects, or specimens, from their home, or "natural" environments into organized collections (Shavit and Griesemer 2009). During this movement, specimens become research tools with the capacity to test questions about evolutionary processes, but simultaneously function as research objects, which become the focus of close analysis. As tools, both specimens and collections act as boundary objects, around which different practitioners gather; the diverse cultural and professional identities of these practitioners reflects important changes in this history of the life sciences, especially evolutionary biology (Star and Griesemer 1989).

Focusing on collections as an analytical category challenges the common narrative of the history of the life sciences that dichotomizes experimentalism and natural history by revealing previously hidden connections and opening up new questions (Strasser 2012; Sunderland 2012). These new questions have the potential to initiate fruitful dialogs with today's practitioners who are worried about the extinction of natural history collections (Gropp 2003). Is the extinction of collections a real risk? Who and what would be comprised by the removal of scientific collections? How might answers to these questions be shaped by deeply embedded and widespread misinformation regarding the history of collections-based research? The *Science* controversy provides a window into what, from a distance, looks like a long-standing, icy divide between natural historians and the rest of the scientific community. Studying the history of collections as scientific instruments has the power to erase this divide. Since the early modern period, collections have brought diverse practitioners together to produce new scientific knowledge.

Guide to Further Readings

An excellent starting point for further readings is the edited volume, *Cultures of Natural History* (Jardine, Secord, and Spary 1996), as it provides a comprehensive and accessible overview of the different contexts that shaped collections from the sixteenth to the nineteenth century. For a window into the intricate networks and practices that made collections-based research in Victorian England possible, Jim Endersby's *Imperial Nature: Joseph Hooker and the Practices of Victorian Science* (2008) shows us how Hooker's time was consumed with a diverse range of activities and relationships. Although not covered in depth in this chapter, collecting and collections played an important role in early-modern Europe (Findlen 1994; Ogilvie 2006). In the United States, Mary P. Winsor's *Reading the Shape of Nature: Comparative Zoology at the Agassiz Museum* (1991) provides a picture of how collections were built to realize a specific research vision. From the scientific community, there are a number of recent edited collections that were assembled in celebration of Linnaean and Darwinian milestones (e.g., Winsor's 2009 article is part of a special issue of *Taxon*).

References

Appel, Toby. 2000. *Shaping Biology: The National Science Foundation and American Biological Research, 1945–1975*. Baltimore: Johns Hopkins University Press.

Asma, Stephen, T. 2001. *Stuffed Animals and Pickled Heads: The Culture and Evolution of Natural History Museums*. Oxford: Oxford University Press.

Chakrabarty, Prosanta. 2014. "Collecting organisms to save their species." http://www.southernfriedscience.com/?p=16957 [Accessed September 29, 2015].

Conniff, Richard. 2014. "Should scientists be killing species they thought were extinct?" http://www.takepart.com/article/2014/04/16/when-should-scientists-kill-extinct-species [Accessed September 29, 2015].

Constable, Heather, Robert Guralnick, John Wieczorek, Carol Spencer, and A. Townsend Peterson. 2010. "VertNet: A new model for biodiversity data sharing." *PLOS Biology*, 8, No. 2: e1000309. DOI: 10.1371/journal.pbio.1000309.

Endersby, Jim. 2008. *Imperial Nature: Joseph Hooker and the Practices of Victorian Science*. Chicago: University of Chicago Press.

Fan, Fa-ti. 2003. "Victorian naturalists in China: Science and informal empire." *The British Journal for the History of Science*, 36: 1–26.

Farber, Paul L. 2000 *Finding Order in Nature: The Naturalist Tradition from Linnaeus to E. O. Wilson*. Baltimore: John Hopkins University Press.

Findlen, Paula. 1994. *Possessing Nature: Museums, Collecting, and Scientific Culture in Early Modern Italy*. Berkeley: University of California Press.

Gerson, Elihu M. 1998. "The American system of research: Evolutionary biology, 1890–1950." PhD thesis, University of Chicago.

Griesemer, James R. 1990. "Modeling in the museum: On the role of remnant models in the work of Joseph Grinnell." *Biology and Philosophy*, 5: 3–36.

Griesemer, James R. 1991. "Material models in biology." In *PSA 1990*, v. 2, edited by A. Fine, M. Forbes, and L. Wessels, 79–93. East Lansing, MI: Philosophy of Science Association.

Griesemer, James R., and Elihu M. Gerson. 1993. "Collaboration in the Museum of Vertebrate Zoology." *Journal of the History of Biology*, 26: 185–204.

Grinnell, Joseph. 1910. "The methods and uses of a research museum." *Popular Science Monthly*, 77: 163–9.

Grinnell, Joseph. 1914a. "The Colorado River as a hindrance to the dispersal of species." Selection from "An account of the mammals and birds of the Lower Colorado Valley with especial reference to the distributional problems presented." *University of California Publications in Zoology*, 12: 51–294, on 100–7. As reprinted in Grinnell 1943, 1968, 47–55.

Grinnell, Joseph. 1914b. "Barriers to distribution as regards birds and mammals." *The American Naturalist*, 48: 248–54. As reprinted in Grinnell 1943, 1968, 57–63.

Grinnell, Joseph. 1919. "The English sparrow has arrived in Death Valley: An experiment in nature." *The American Naturalist*, 53: 468–72. As reprinted in Grinnell 1943, 1968, 89–94.

Grinnell, Joseph. 1943, 1968. *Joseph Grinnell's Philosophy of Nature: Selected Writings of a Western Naturalist*. Berkeley: University of California Press.

Gropp, Robert E. 2003. "Are university natural science collections going extinct?" *BioScience*, 53: 550.

Jackson, Morgan. 2014. Tweet. April 21. https://twitter.com/BioInFocus.

Jardine, Nicholas, James A. Secord, and Emma C. Spary (eds.) 1996. *Cultures of Natural History*. Cambridge: Cambridge University Press.

Jasanoff, Maya. 2004. "Collectors of empire: Objects, conquests and imperial self-fashioning."
    Past & Present, 184: 109–35.
Kohler, Robert E. 2006. *All Creatures: Naturalists, Collectors, and Biodiversity, 1850–1950*.
    Princeton, NJ: Princeton University Press.
Kohler, Robert. 2007. "Finders, keepers: Collecting sciences and collecting practice." *History
    of Science*, 45: 428–54.
Kohlstedt, Sally Gregory. 1988. "Curiosities and cabinets: Natural history museums and edu-
    cation on the antebellum campus." *Isis*, 79: 405–26.
Kohlstedt, Sally Gregory. 1995. "Museums: Revisiting sites in the history of the natural sci-
    ences." *Journal of the History of Biology*, 28: 151–66.
Lane, Meredith A. 1996. "Roles of natural history collections." *Annals of the Missouri Botanical
    Gardens*, 83: 536–45.
Miller, Alden H. 1964. "Joseph Grinnell." *Systematic Zoology*, 13: 235–42.
Minteer, Ben, A., James P. Collins, Karen E. Love, and Robert Puschendorf. 2014. "Avoiding
    (Re)extinction." *Science*, 344: 260–1.
Müller-Wille, Staffan. 2006. "Linnaeus's herbarium cabinet: A piece of furniture and its func-
    tion." *Endeavour*, 30: 60–4.
Müller-Wille, Staffan. 2007. "Collection and collation: Theory and practice of Linnaean
    botany." *Studies in the History and Philosophy of Biological and Biomedical Sciences*, 38: 541–
    62.
Nyhart, Lynne K. 2009. *Modern Nature: The Rise of the Biological Perspective in Germany*.
    Chicago: Chicago University Press.
Ogilvie, Brian W. 2006. *The Science of Describing: Natural History in Renaissance Europe*.
    Chicago: University of Chicago Press.
Outram, Dorinda. 1996. "New spaces in natural history." In *Cultures of Natural History*, edited
    by Nicholas Jardine, James A. Secord, and Emma C. Spary, 249–65. Cambridge: Cambridge
    University Press.
Pauly, Philip J. 2000. *Biologists and the Promise of American Life: From Meriwether Lewis to
    Alfred Kinsey*. Princeton, NJ: Princeton University Press.
Pickstone, John V. 1994. "Museological science? The place of the analytical/comparative in
    nineteenth-century science, technology and medicine." *History of Science*, 32: 111–38.
Pyke, Graham H., and Paul R. Ehrlich. 2010. "Biological collections and ecologi-
    cal/environmental research: A review, some observations and a look to the future." *Bio-
    logical Reviews*, 85: 247–66.
Rader, Karen A., and Victoria E. M. Cain. 2008. "From natural history to science: Display and
    the transformation of American museums of science and nature." *Museum and Society*, 6:
    152–71.
Secord, James A. 1996. "The crisis of nature." In *Cultures of Natural History*, edited by
    Nicholas Jardine, Jim A. Secord, and Emma C. Spary, 447–59. Cambridge: Cambridge
    University Press.
Sepkoski, David. 2013. "Towards 'a natural history of data': Evolving practices and epistemolo-
    gies of data in paleontology, 1900–2000." *Journal of the History of Biology*, 46: 401–44. DOI:
    10.1007/s10739-012-9336-6.
Shavit, Ayelet, and James Griesemer. 2009. "Transforming objects into data: How minute tech-
    nicalities of recording 'species location' entrench a basic challenge for biodiversity." In *Sci-
    ence in the Context of Application*, edited by M. Carrier and A. Nordmann, 169–93. Boston
    Studies in the Philosophy of Science 274. DOI: 10.1007/978-90-481-9051-5_12
Smocovitis, Vassiliki B. 1992. "Unifying biology: The evolutionary synthesis and evolutionary
    biology." *Journal of the History of Biology*, 25: 1–65.
Smocovitis, Vassiliki B. 1996. *Unifying Biology: The Evolutionary Synthesis and Evolutionary
    Biology*. Princeton, NJ: Princeton University Press.

Star, Susan Leigh, and James R. Griesemer. 1989. "Institutional ecology, 'translations' and boundary objects: Amateurs and professionals in Berkeley's Museum of Vertebrate Zoology, 1907–1939." *Social Studies of Science*, 19: 387–420.

Stein, Barbara R. 2001. *On Her Own Terms: Annie Montague Alexander and the Rise of Science in the American West.* Berkeley: University of California Press.

Strasser, Bruno J. 2012. "Collecting nature: Practices, styles, and narratives." *Osiris*, 27: 303–40.

Suarez, Andrew V., and Neil D. Tsutsui. 2004. "The value of museum collections for research and society." *BioScience*, 54: 66–74.

Sunderland, Mary E. 2012. "Collections-based research at Berkeley's Museum of Vertebrate Zoology." *Historical Studies in the Natural Sciences*, 42: 83–113.

Sunderland, Mary E. 2013a. "Teaching natural history at the Museum of Vertebrate Zoology." *British Journal for the History of Science*, 46: 97–121. DOI: 10.1017/S0007087411000872.

Sunderland, Mary E. 2013b. "Modernizing natural history: Berkeley's Museum of Vertebrate Zoology in transition." *Journal of the History of Biology*, 46: 369–400.

Sunderland, Mary E. 2013c. "Computerizing natural history collections." *Endeavour*, 37: 150–61.

Sunderland, Mary E., Karen Klitz, and Kristine Yoshihara. 2012. "Doing natural history." *BioScience*, 62: 824–9.

Tewksbury, Joshua J., John G. T. Anderson, Jonathan D. Bakker, Timothy J. Billo, Peter W. Dunwiddie, Martha J. Groom, et al. 2014. "Natural history's place in science and society." *BioScience*, 64: 300–10.

University of Alaska Museum Department of Ornithology. 2014. "(Re)affirming the specimen gold standard." http://www.universityofalaskamuseumbirds.org/reaffirming-the-specimen-gold-standard/ [Accessed September 29, 2015].

Wheeler, Terry A. 2014. "A fruit fly is not a mammal, and other revelations from the museum." http://lymanmuseum.wordpress.com/2014/04/18/a-fruit-fly-is-not-a-mammal-and-other-revelations-from-the-museum/ [Accessed September 29, 2015].

Wilson, Edward O. 1994. *Naturalist.* Washington, DC: Island Press.

Winsor, Mary P. 1991. *Reading the Shape of Nature: Comparative Zoology at the Agassiz Museum.* Chicago: University of Chicago Press.

Winsor, Mary P. 2006a. "Linnaeus's biology was not essentialist." *Annals of the Missouri Botanical Garden*, 93: 2–7.

Winsor, Mary P. 2006b. "The creation of the essentialism story: An exercise in metahistory." *History and Philosophy of the Life Sciences*, 28: 149–74.

Winsor, Mary P. 2009. "Taxonomy was the foundation of Darwin's evolution." *Taxon*, 58: 43–9.

CHAPTER THIRTY-FIVE

Recording Devices

JIMENA CANALES

We can hardly imagine a world without recording devices. We have come to know the world by recording it. From photography to cinema, from phonographs to MP3 technologies, from bulky apparatuses to miniature ones, recording devices have indelibly changed our history as much as our selves. What are the essential, defining characteristics of these devices? Can the surface of a cave, used for some of the earliest human inscriptions, be considered a "recording device"? Under what circumstances can a retina be considered in these terms? Since their inception, recording instruments have changed how we have thought of the most intimate aspects of our private identities and our communities to the most public ones. They have led us to reconsider some of the central foundations of the law and ethics through the role they have played in changing standards of evidence, privacy, and reproduction rights. Our current understanding of the Earth is tied to the proliferation of recording devices used to portray it as a place that could exist without humans, and, according to various factions of the environmental movements from the 1960s onwards, might be better off without them.

Why did a growing number of researchers, starting in the middle of the nineteenth century, develop and use instruments to automatically record natural phenomena? Motivations were varied and the adoption of recording devices was not uncontested. Different scientists highlighted different benefits for each instrument, depending on what disciplines they came from and what particular research practices they supported. By the twentieth century, recording devices were so widely used in scientific laboratories that Bruno Latour and Steve Woolgar estimated that researchers "spend two-thirds of their time working with large inscription devices" (Latour and Woolgar 1986, 69).

The widespread use of recording devices has vast repercussions beyond science, affecting modern culture and society more generally. Photography and cinematography became increasingly important in societies characterized by surveillance and spectacle. The drive to develop and use recording instruments often relied on the possibility and promise of storing, calculating, and sending information across expanding networks of communication, starting with traditional transportation technologies

A Companion to the History of Science, First Edition. Edited by Bernard Lightman.
© 2016 John Wiley & Sons Ltd. Published 2020 by John Wiley & Sons Ltd.

followed by telegraph, telephone, and radio, and culminating with the World Wide Web. Computing and scanning tasks, performed first by men (computing) and women (scanning), and later by machines, became increasingly necessary for analyzing automatically produced records. The history of recording devices in science thus intersects with the Information Revolution, the development of communications media, the establishment of modern bureaucracies, and with the rise of post-Fordist economies. The increasing miniaturization of electronic technologies, from the Cold War to the Personal Computer, aided their appearance in satellites, drones, hand-held devices (primarily cell phones), and later wearable technology. They have profoundly altered our beliefs about automation and non-human agency.

By the first decades of the twentieth century imaging and auditory recording devices had improved so much that it was possible to take still and moving pictures as well as to record and reproduce sound. Kodak led the photographic industry with the "you press the button, we do the rest" concept. Commercial gramophone and phonograph industries flourished during these years, with technologies first based on tinfoil and wax and later vinyl. By the 1920s, recording instruments for sound and image became widespread. Magnetic (mostly analog) and electronic (mostly digital) technologies became increasingly important after WWII, culminating with the audiocassette, the video cartridge, and the computer "floppy" disk. By the end of the twentieth century, sound and video were recorded on digital laser CDs and DVDs, semiconductor chips, and silicon-based memory cards.

Spontaneous Reproduction

How did the modern world end up populated by these new registering instruments? How were they different from the machines that characterized the Industrial Revolution, such as clocks, pumps, and engines? The meaning of "recording device" varies widely, although the term today usually refers to "new" instruments or machines that differed from older recording and preservation media, such as casts, monuments, scrolls, and books. In contrast to other media, recording devices are largely considered to produce records that came *directly from nature*. They were often described as producing "exact" and "precise" records "automatically," "spontaneously" or "instantaneously." Nicéphore Niépce, one of the inventors of photography, considered it a technique of "automatic reproduction" (as quoted in Trachtenberg 1980, 5). In one of the first presentations of photography, Louis Daguerre described it as a means for "spontaneous reproduction." "The daguerreotype is not merely an instrument which serves to draw nature," he claimed, but rather one that "gives her the power to reproduce herself" (as quoted in Trachtenberg 1980, 11–13). The astronomer François Arago in France and Henry Fox Talbot in England ascribed a similar role to photography by referring to it as the "pencil of nature." The physiologist Étienne-Jules Marey described graphic recording techniques as revealing the "language of the phenomena themselves" (Marey 1878, iii). For these reasons, they were largely considered to be privileged instruments for studying the natural world and for preserving and communicating knowledge.

A recorded image, explained the film theorist Rudolf Arnheim, "is not only supposed to resemble the object, but it is also supposed to guarantee this resemblance by being the product of the object itself, i.e. by being mechanically produced by it" (as quoted in Kittler 1999, 11–12). Recording devices produced a "double of the world." While the "principle of sufficient reason" characterized the Enlightenment age, a "principle of sufficient photography," reigned in the age of mechanical reproduction, where proof, evidence, and even existence were considered almost exclusively in terms of what could potentially be recorded (Laruelle 2011, viii).

Expanding and Replacing

How does the history of recording instruments intersect with that of other kinds of scientific instruments? Since the scientific revolution, sensorial abilities were often understood and described by reference to instruments. Early recording devices were also considered in these terms. Niépce, in his early attempts to fix the image of a camera obscura, considered it as "a kind of artificial eye" (Fouque 1973, 29).

Defenders of the graphic method, such as Helmholtz and Marey, often described the benefits of recording instruments in terms of their ability to produce records of effects that lay well beyond the threshold of our limited sensorial capacities. Phenomena that were too feeble or too fast to be perceived could be recorded instead. In addition to the benefits of imitating and expanding the senses, the impetus for adopting recording devices was often described in terms of the capacity of these instruments for *replacing* the senses.

From the late nineteenth century on, a growing number of commercial companies became deeply invested in portraying recording instruments in terms of their potential to sense phenomena in exactly the same way than humans sensed the world. For example, a 1926 advertisement for the Photo-Miniature camera boasted: "What you can see you can Photograph" (Figure 35.1).

The criterion defining "sense-data" used by certain analytical philosophers of science often depended on its potential to be recorded. Thus, the mathematician and philosopher Bertrand Russell in the "The Relation of Sense-Data to Physics," claimed

Figure 35.1   What You Can See You Can Photograph. Advertisement for Ermanox miniature camera from 1926. Source: Beaumont Newhall. 1949. *The History of Photography.* New York: The Museum of Modern Art, 188.

that sense-data could be grasped equally by machines as they could by living observers, making any philosophical distinction between the two irrelevant (Russell 1918). In Russell's later work on Einstein's theory of relativity, he insisted on this same point: "It is natural to suppose that the observer is a human being, or at least a mind; but it is just as likely to be a photographic plate or a clock" (Russell 1925, 138).

By the second half of the twentieth century, Philipp Frank, one of the main representatives of Logical Positivism, could "say that man is himself a self-registering instrument and what we call 'sense observation' is not different from the registering by an instrument" (Frank 1948/1949, 460). The assumption that machines could "see" was by then prevalent.

In Philosophy of Science

How did our understanding of the world change with the proliferation of machines designed to take over tasks traditionally belonging to our senses? Scientific materialism, the intellectual framework for much of the scientific thought of modern times, was first developed by reference to bodies and machines which were considered to be "mechanical." New philosophical systems were later conceived to understand a host of different instruments designed to replace our senses rather than our bodies. By the end of the nineteenth-century, modern industry was no longer characterized primarily by mechanical instruments, but by chemical, electromagnetic and later, electronic ones. While, traditionally, scholars and philosophers would rarely refer to these new instruments, a new generation of thinkers, some of them inspired by the philosopher Henri Bergson's references to photography, phonography and cinematography, slowly started to incorporate them into their work, beginning in the 1920s and into the 1930s (Canales 2015). From astronomical observatories to laboratories of experimental psychology, observers were increasingly compared to machines that were no longer primarily conceived as "automata" (as in Descartes) or as "engines" (as in Marx), but rather as "self-registering instruments."

Certain kinds of recording instruments, especially those used to obtain numerical data, were often referred to as "inscription" devices. In the first decades of the twentieth century, a growing number of scientific laboratories started to be equipped with inscription devices and direct-reading instruments that replaced older measurement taking practices. "The whole subject-matter of exact science consists of pointer readings and similar indications," explained the astronomer Arthur Eddington (Eddington 1935 [1928], 252). These changes came with profound philosophical repercussion in terms of how scientists understood these instruments.

How should we think of recording devices? In 1909, inspired in large part by Bergson, the philosopher Édouard Le Roy urged scientists to admit that scientific instruments, many of them recording devices, were really "materialized theories" whose ostensibly objective results were already compromised by what researchers were interested in finding (Anon. 1909, 183). Philosophers of science, from Pierre Duhem and Gaston Bachelard to Thomas Kuhn and other anti-positivists, followed Le Roy in calling attention to the theoretical presuppositions that affected the facts obtained by instruments of this kind. While philosophers' of science understanding of recording devices has varied greatly, many thinkers noted, by direct reference to them, the need to think of the progress of science as much more than the result of applying analytical methods on observations obtained by ever more perfect instruments. The philosopher Susanne K. Langer, one of the earliest thinkers to focus on the proliferation of inscription devices in science, explained how they had "begotten a new philosophical issue" which made debates about *meaning* and *symbolic* interpretation central to philosophy once again: "The problem of observation is all but eclipsed by the problem of meaning. And the triumph of empiricism in science is jeopardized by the surprising truth that our sense-data are primarily symbols" (Langer [1942] 1951, 21).

The Disappearing Device

Recording devices—except for the printing press—were nearly absent in the work of Karl Marx. Later thinkers inspired by Marx would develop new methods for thinking about them. Theodor Adorno, working closely with Walter Benjamin, started to analyze recording devices by reworking Marxist ideas of commodity fetishism. He focused on "the occultation of production by means of the outward appearance of the product" across fields, from theater and film to consumer culture (Adorno 1981, 90).

Starting in the 1930s, numerous scholars noticed that recording devices were successful at producing "automatic" documents only because they were considered as things that could be set aside from the rest of the world. Walter Benjamin, in his work on photography and film, remarked on their uncanny ability to hide themselves: "The equipment-free aspect of reality [*apparatfreie Aspekt der Realität*] here has become the height of artifice," he explained. It was "precisely because of the thoroughgoing permeation of reality with mechanical equipment," that spectators in the age of film, ironically, perceived "an aspect of reality that is free of equipment" (Benjamin 1968, 233–4).

Inspired by the Frankfurt School's attempt to think about recording devices in relation to capitalism and mass movements, numerous other thinkers strove to show how the uncontested adoption of recording devices stemmed from particular social, political and commercial forces. In the 1970s the film theorist Jean-Louis Baudry inaugurated "apparatus theory" as a way to understand how recording devices functioned in broader socio-political terms, and how they depended on a particular view of human subjectivity. "Does the technical nature of optical instruments, directly attached to scientific practice, serve to conceal not only their use in ideological products but also the ideological effects which they may themselves provoke?" he asked (Baudry [1970] 1999, 354).

The influential work of Jacques Derrida, critiquing common assumptions of "presence" and "logocentrism" underpinning Western thought, drew some of its most important conclusions by reference to recording devices. Derrida turned to them to illustrate how they influenced how we commonly thought of human subjectivity. The common understanding of a sense of self as a coherent entity with a "voice" could shift with the appearance of new technologies. What was a "tape-recorder" he asked? In one sense, it was a technology for recording our voice. But in another sense, it was also a writing machine. "Tape recordings are writings in some sense," he insisted. The idea that voice, writing and other representational media always referred back to an essential original source was suddenly imperiled (Derrida 1983, 42). Poststructuralist philosophers (led by Derrida) devised new ways of understanding the relation between man and machines, idealism and materialism, in a manner far from Cartesian binaries (body and soul) and where the unitary subject of the Enlightenment tradition gave way to a *decentered* one.

Recent philosophers and sociologists of science have continued to think about the wider repercussions that follow from the widespread use of recording and inscription devices in science. They have sought to overcome the evident lacunae of most explanations for the success of science where inscription and recording devices are largely absent and that rely too heavily on cognitive abilities. The sociologist and philosopher Bruno Latour coined the term "immutable mobile" to explain the power of recorded inscriptions in shaping the modern world. For him, their success showed why these and other material entities should be accorded a rightful place in accounts of the development and success of Western science. The philosopher and historian of science Hans-Jörg Rheinberger, in his influential study of molecular biology, has similarly focused on these instruments, considering scientific work as a kind of "tracing game" where scientists "move around graphic objects or numbers" and thus "produce new entities by surpassing others, concatenate them into ever changing chains" (Rheinberger 1997, 225).

From Media Studies to Poststructuralism

With the postwar expansion of the media industry, a new generation of thinkers focused on recording devices from the perspective of the new discipline of "media studies," where they were largely considered alongside communications technologies. Marshall McLuhan, for example, described the motivations for building a phonograph primarily in terms of its benefits for interpersonal communication. They "consider[ed] it as a 'talking repeater': that is, a storehouse of data from the telephone, enabling the telephone to 'provide invaluable records, instead of being the recipient of momentary and fleeting communication'" (McLuhan 1964, 276). Similarly, Friedrich Kittler focused on recording devices only as a small subset of a larger chain of storage, transmission, and processing technologies. Media theorists (such as McLuhan and Kittler) consider recording devices in terms of wide-ranging cultural transformations from the introduction of the alphabet to the electronic age.

In recent years, the common commercial and philosophical claim that human and machine sensation could be equally considered in terms of recording devices has been contested. Alternative philosophical movements, drawing from phenomenology and poststructuralism, have been concerned with offering a new explanation for the proliferation of recording devices in the modern world. Has their success been due to how they "sense" the world in a manner analogous to how humans perceive it? In *What is Philosophy?*, Félix Guattari and Gilles Deleuze (1991, 131) provided a different answer. They answered Russell's call to consider observers and recording machines as simply interchangeable by developing an alternative explanation for their success. Russell, they argued, had "assimilated [observers] to apparatuses and instruments like Michelson's interferometer or, more simply the photographic plate, camera or mirror that captures what no one is there to see." By reference to these devices, analytical philosophers had led us to believe in the existence of "sense-data without sensation." But "sense-data without sensation" argued Deleuze and Guattari, were—ironically— always "waiting for a real observer to come and see." How to deal with the paradoxical relationship of real observers to recorded data? Deleuze and Guattari offered to understand both by proposing the new term "sensiblia" as an intermediary term connecting "sensation" to our belief in "sense-data without sensation." Recording instruments, they argued, have only functioned because they have "presupposed" an "ideal partial observer" (Deleuze and Guattari 1994, 131) who is only temporarily *in absentia*. It is now paramount to investigate these presuppositions and paradoxes in order to understand how recording instruments work in the first place (Canales 2014).

Desires

Historians Lorraine Daston and Peter Galison have connected the history of early recording devices, particularly photography, to the ideal of "mechanical objectivity" that arose in the second half of the nineteenth century. A growing number of researchers during this period advocated new techniques for representing natural phenomena that were "non-interventionist," aiming for results "uncontaminated by interpretation," and where the "human hand" associated with skilled artistic work was largely absent. Daston and Galison have found that these ideals were driven by "a morality of self-restraint" that underpinned much of the rhetoric defending their adoption. Scientists' drive to obtain "mechanical objectivity" fueled the development of recording technologies in two ways: first, by creating instruments to produce images "without the intervention of an artist," and second, by improving on previous technologies for the "'automatic' multiplication of images" such as lithography (Daston and Galison 2007, 135).

Advocates of recording instruments were split between those who favored technologies that produced records that could be mechanically reproduced, versus those who preferred more *precise* records. Most French astronomers up to the end of the nineteenth century often preferred the possibility of taking precise measurements on glass daguerreotypes instead of using Talbot's negative-based methods, from which copies could be more easily made on paper. Arago considered that the possibility of easily scratching off a daguerreotype image from a glass plate was a distinct advantage. Others, however, advocated more permanent records that could be stored for long periods of time. The astronomer Jules Janssen, inventor of the pre-cinematographic photographic revolver, worked to develop methods that permitted the conservation and multiplication of standardized records that could be publicly witnessed (Janssen 1888).

Desires driving research into new recording technologies went from the modest (such as obtaining mechanical objectivity) to the outright fantastical (such as the dream of reaching immortality through a perfect reproduction). Numerous researchers mused about the possibility of having complete, perfect records of historical events or of people. A Victorian writer who marveled at the possibility of recording sound, optimistically claimed that "death has lost some of its sting since we are able to forever retain the voices of the dead" (Sterne 2003, 308). At a presentation of Gaumont's Chronophone, a machine that synchronized recorded sound with film, presenters argued that with further improvements this machine could make us immortal: "That day, we will no longer need to communicate ourselves; we could do it even after death. It is then that we would become truly immortal" (Carpentier 1910, 1325).

In some disciplines, such as meteorology and particular branches of astronomy, recording devices were adopted because they permitted researchers to obtain continuous, uninterrupted observations. In other cases, such as in the precision sciences and for the growing field of biometrics used by anthropologists and as much as by the police, they were defended because they could be used to produce measurable numerical data. Recording devices used by Alphonse Bertillon, the inventor of the mugshot, were employed primarily because they permitted him to determine numerically the precise anthropometric measurements of his subjects in ways that could be sent telegraphically.

Recording and Quantification

Not all recording devices were used for producing pictures or perspectival images. Many of them left traces drawn by inscription pens and styluses in lensless instruments that recorded on smoked paper rolls on rotating drums driven by clockwork. These types of instruments were central to the "graphic method" developed primarily by Carl Ludwig and Hermann Helmholtz in Germany and Étienne-Jules Marey in France (Brain 1996).

Some scientists showed a preference for recording devices that produced quantitative data over those that produced hard-to-interpret graphs or images. Since the early years of the nineteenth century, scientists increasingly understood their work as one that involved reading and calculating numbers instead of writing, smelling, or feeling (Roberts 2005). Engraved scales on rulers, bars and balances, needles and dials on gauges, and grated leveling instruments permitted scientists to use "direct reading" methods to take measurements in ways that could be easily recorded and transmitted (Gooday 2004). These kinds of measurements could either be recorded automatically or by assistants, permitting scientists to calculate and transmit them at a later time. Recording instruments were often used in combination with these new number-based measuring technologies.

Servant instruments

Machines that produced records continuously and were powered by electric motors or clockwork mechanisms were often described as "self-registering instruments." They contrasted with others that worked by triggering (photography) or by cranking (early film cameras). One of the main motivations for developing self-registering apparatuses arose from the desire to automatically track the rotation of the Earth against the stars in order to record time continuously. Instruments for recording temperature and other meteorological data were later employed at weather stations and observatories for the same purpose.

Throughout a wide array of disciplines, scientists considered the benefits of recording devices in terms of labor efficiency and new possibilities for delegation. When using the term "automatic," Niépce did not mean that it was not laborious or fatiguing. On the contrary, the process required careful, time-consuming preparations for the plates and their development. Exposure times usually took several hours. What most early researchers meant when they described early recording devices as "automatic" "spontaneous" or "instantaneous" was that the required labor was simple and repetitive, rather than skilled, and that preparations took place before and after the machine was employed. Daguerre, who collaborated with Niépce and was able to reduce exposure times from several hours to three to thirty minutes, claimed that "the little work it entails will greatly please ladies." Arago, in his famous presentation of Daguerre's invention to the Chambre des députés, boasted that all manual labor involved in it could be performed "step by step" by "anyone" with no "special knowledge" and requiring "no special dexterity" (as quoted in Trachtenberg 1980, 19).

Most scientists extolled automatically recording devices for permitting them to delegate tasks, and described them as efficient servants (Krajewski 2010). The astronomer Edward S. Holden titled his book on the topic as *Photography, The Servant of Astronomy* (1886). Astronomers could have records permanent enough so that they could examine them leisurely at a later time, or, even better, so that someone else could comb through them.

The early use and adoption of automatic recording instruments by scientists took place alongside a reorganization of labor practices in laboratories and astronomical observatories. As laboratories and observatories grew in size and complexity, they increasingly employed a growing staff of assistants and amanuenses. In some cases, the organization of scientific labor started to resemble that of factories where repetitive tasks, such as simple observations and measurements, were increasingly relegated to machines (Schaffer 1988). The new division of labor was highly gendered and hierarchical, privileging mental over manual work and furthering their separation. Additionally, the adoption of recording instruments entailed large-scale changes in the sensori-motor organization of labor where the fingers and eyes of a growing number of users increasingly landed and focused on keys and screens.

The Personal Equation

Although motivations for adopting recording devices in science varied widely, the possibility of producing records that eliminated the individual idiosyncrasies of observers, sometimes referred to as their personal equation, was frequently noted across many fields. An instrument's ability to produce quantitative data free from the personal equation of the observer was frequently noted as a distinct benefit. Chronographs, for example, were often adopted because they helped translate a temporal measurement into a spatial one, where individual variations in the measured quantity were minimal.

Recording devices were frequently combined with other instruments used to combat differences in observers' subjective assessments. An inventor of one of the most widely-used recording devices (the chronograph) identified the rise of modern science with the elimination of individual subjectivities, portraying the entire progress of science as a gradual mastery over the personal equation: "The purely sensorial information [collected during diagnosis] was affected by that which the *astronomers call the personal equation* of the observer, which varied according to different sensibilities and momentary judgments" (Chauvois 1937, 425–6). Arago also highlighted photography's potential for eliminating individual differences. Complaining about "the great discrepancies between the determinations of the comparative light intensities ... as given by equally able scientists" (as quoted in Trachtenberg 1980, 21), he considered photography as the best available means for eliminating these discrepancies. Similarly, the need to eliminate individual differences in the perception of moving phenomena was one of the main motivations for developing pre-cinematographic cameras (Canales 2009).

Advocates of recording devices in science often listed the possibility of publicly and collectively witnessing records as a distinct benefit. Scientists immediately investigated ways to couple them with projection devices that could be used in large lecture halls and theaters. The history of these technologies thus intersects with the larger history of spectacle, from the classroom to the movie theater.

Due to the growing awareness of the effects of individual differences on scientific observations, various philosophers of science, most prominently amongst them Karl Popper, started to understand notions of scientific "objectivity" in terms of intersubjectivity. While Popper largely overlooked the role played by recording instruments in the production of intersubjective knowledge, recent sociologists of science consider them as essential: "This 'objectivity' is not simply an intersubjectivity between rational human beings who decide to agree on observed results (a solution proposed by Popper [...]): it is an intersubjectivity inscribed in instruments, protocols, procedures; which *in fine*, is anchored in these indisputable bodily experiences defining our common humanity" (Callon 1999, 274).

The Modern Subject and the Discipline of Psychology

The central place of recording devices in the history of psychology shows how our ideas of the modern "subject" emerged in parallel to these instruments. Arguments for adopting recording devices were often based on a particular conception of human subjectivity. The astronomer Hervé Faye urged scientists to adopt them because of how they could "eliminate nervousness" in cases when it was impossible to observe calmly. The discipline of experimental psychology, from Wilhelm Wundt to B. F. Skinner, was one of the first areas of science to rely centrally on recording devices, starting with simple chronographs that measured reaction and moving on to much more complicated tests and technologies.

By the end of the nineteenth century, recording devices were such an integral component of laboratories that scientific work was often characterized by a postindustrial labor organization. This new organization not only depended on the separation between "thinking" versus "doing," but the act of "feeling" increasingly appeared as a third, distinct component. Thus, Edward Scripture, one of the strongest advocates of furnishing scientific laboratories with new recording devices, felt justified in naming "the first book on the *new*, or experimental psychology" written in English as *Thinking, Feeling, and Doing* (Scripture 1895). The use of recording technologies by psychologists was part of a growing interest across fields in the role of advertisement, marketing and mass culture. The work of Hugo Münsterberg is a paradigmatic example showing the overlapping relation between recording devices, labor organization, experimental psychology and film theory.

Art and Hand-drawings

Since their inception, the importance of recording devices for modern art was evident. One of the first statements about the effects of photography on the arts, written by the artist Paul Delaroche, shows that artists were excited about photography for exactly the same reasons that excited scientists, namely "precision," "accuracy," its unique ability to capture "details," and how it could be used to save "time and labor" (as quoted in Trachtenberg 1980, 18). Recent work in art history and in the history of science has shown that the relation between recording devices and art is much more complex than some early models suggested, which were usually based on the one-way influence from science to art. In *Before Photography: Painting and the Invention of Photography*, art historian Peter Galassi eloquently countered this narrative by showing how "photography was not a bastard left by science on the doorstep of art, but a legitimate child of the Western pictorial tradition" (Galassi 1981, 12). The development of recording devices was driven as much by scientific as by artistic concerns. They continue to affect science as much as art. The association of recording devices with science rather than art was contested on many fronts, by scientists (who shunned them) and by artists (who adopted them).

The adoption of graphic and recording methods in science was not uncontested. In the early years of their adoption, an opponent of Marey argued that automatically recorded physiological phenomena, including the horse's gallop, could be grasped in a better way without complicated "self-registering" apparatus. Marey was accused of ignoring the fact that graphic traces themselves had to be "read," and that this introduced some of the same challenges of direct observations: "The registering apparatus does nothing but to inscribe undulating lines that fall on our senses; but once it comes to interpreting the traces, the graphic method has no more certitude than direct observation." His contradictor contested the supposed "universality" of the "language" of recording devices, urging for caution: "In the sciences of observation, all instruments, no matter who simple or complicated, are aids... that speak a special language. Before using them, one must strive to learn their language" (Gavarret 1878, 622).

The adoption of photography in science was slow and contested throughout nearly half a century. In 1875 the astronomer Charles Wolf found claims that it "eliminated the observer" absurd. Observers, he claimed, would always be needed for obtaining "absolute and authentic knowledge." Wolf considered direct observations as superior, because they were more coherent and stable across time. While different cameras and photographic methods produced different results (for example, collodion versus gelatin and bromide), "the human eye, on the contrary, is an organ which remains the same, and the observations of the eye are, at all times, comparable amongst themselves" (Wolfe 1875, 19). Another argument against photography was based on the advantages of using one's memory to obtain a more complete image. Experienced observers could use their memory to their advantage, learning from accumulated impressions obtained through years of practice and distilling, in this manner, occurrences with short periods of time.

Criticisms of the use of recording devices in science were so powerful that, by 1882, most astronomers recommended drawings over photography (Canales 2009). Étienne Léopold Trouvelot, widely celebrated for his hand-drawn images of celestial bodies, expressed the benefits of drawings as stemming from their "accuracy" "fidelity" and "detail" (qualities often associated with photography) as well as for their ability for "preserving the natural elegance" to "reproduce upon paper the majestic beauty and radiance of the celestial objects" (Trouvelot 1882).

Altered Distances

Since their appearance, automatic recording devices were considered in terms of their potential to alter distances between people and things, not only in spatial terms, but also temporally. The possibility of producing records where one could condense or expand time (using speed photography or slow-motion and fast-motion cinematography) excited biologists as much as physicists. One of the strongest statements underlining photography's ability to alter our sense of space was written in *The Age of Photography* by Oliver Wendell Holmes (1859): "Form is henceforth divorced from matter… Matter in large masses must always be fixed and dear; form is cheap and transportable" (as quoted in Trachtenberg 1980, 80). Archeologists stressed the benefits of having permanent records that could be transferred to Europe where they could be kept safely and studied. Arago tied the benefits of photography to French colonial interests when he mused how photographic records of Egyptian monuments and hieroglyphics would protect them from "the greed of the Arabs and the vandalism of certain travelers" (as quoted in Trachtenberg 1980, 17). As automatically produced records were used in combination with communications networks reaching farther and farther away from European and North American centers, an increasing number of scholars noted how our general conception of distances was changing with them. One of the main thinkers to investigate the effects of recording devices on our sense of distances was Walter Benjamin. He drew from the work of the poet Paul Valéry, who had articulated the effects of recording devices on spatial distances in "La conquête de l'ubiquité" (1928). In *The Work of Art in the Age of Mechanical Reproduction,* Benjamin explained how thanks to the photograph and the phonograph "the cathedral leaves its locale to be received in the studio of a lover of art; the choral production, performed in an auditorium or in the open air, resounds in the drawing room" (Benjamin 1968, 221).

Automatically produced records could affect our sense of the past and future by virtue of their permanency, datability or lack thereof. Martin Heidegger, in his widely cited lecture "The Thing" (1950), considered the effects that records had on our understanding of temporal as much as spatial distances: "Distant sites of the most ancient cultures are shown on film as if they stood this very moment amidst today's traffic." For Heidegger, distance-altering records were at the center of some of the most vexing problems of modern culture, constraining our possibility of acting on the future and augmenting our sense of helplessness and anxiety in the face of modern technology. "Despite all conquest of distances the nearness of things remains absent," he pessimistically concluded (Heidegger [1950] 1992, 405).

Recorded History

Recording devices, particularly in connection to photo-journalism, have played essential roles in world historical events, from their earliest use depicting the Crimean War and the California Gold Rush to our current events. How have recording devices changed our view of history and historical documents? Walter Benjamin investigated how particular kinds of records were tied to particular conceptions of history. Before the advent of mechanical reproducible technologies, "historical testimony" rested "on authenticity," he explained. But because mechanically reproduced records offered no distinction between one copy and the next, the traditional concept of authenticity was imperiled. In consequence, he argued, as he wrote during the rise of Nazism in the 1930s, a traditional understanding of history was being by replaced by a new Fascistic aesthetics (Benjamin 1968, 221).

Historians—along with scientists and philosophers—thought about perfect, complete records and their potential. Although most historians and philosophers of history did not refer directly to new technologies (with the rare exception of film), they have often discussed the repercussions to their discipline arising from the possibility of obtaining ever more precise, complete and perfect records. The philosopher Arthur Danto coined the term "Ideal Chronicler" to describe a being that not only could know everything (as Laplace's famous observer), but who would also have the gift of instantaneous transcription. An Ideal Chronicler "knows whatever happens the moment it happens, even in other minds. He is also to have the gift of instantaneous transcription: everything that happens across the whole forward rim of the Past is set down by him, as it happens the *way* it happens"(Danto 1965, 149). The question of what constitutes an adequate or perfect record, often by reference to actual recording devices, is central to the philosophy of history.

Across science, philosophy, history and the arts, recording devices did much more than just record—they appeared alongside broader transformations that altered our understanding of the world (in personal, historical and scientific terms) as much as of ourselves.

References

Anon. 1878. "Communications 1. Sur l'importance au point de vue medical des signes extérieurs des fonctions de la vie." *Bulletin de l'Académie de Médecine*, 7: 610–26.

Anon. 1909. "La Théorie de la physique chez les physiciens contemporains." *Bulletin de la Société française de philosophie*, 9: 161–91.

Adorno, Theodor. 1981. *In Search of Wagner*. New York: Schocken Books.

Baudry, Jean-Louis. [1970] 1999. "Ideological effects of the basic cinematographic apparatus." In *Film Theory and Criticism*, edited by L. Braudy and M. Cohen, 345–55. Oxford: Oxford University Press.

Benjamin, Walter. 1968. *The Work of Art in the Age of Mechanical Reproduction. Illuminations*, edited by H. Arendt, 217–51. New York: Schocken.

Brain, Robert M. 1996. "The Graphic Method: Inscription, visualization and measurement in nineteenth-century science and culture." Ph.D. Thesis, University of California Los Angeles.

Callon, Michel. 1999. "Whose imposture? Physicists at war with the third person." *Social Studies of Science*, 29, No. 2: 261–86.

Canales, Jimena. 2009. *A Tenth of a Second: A History*. Chicago: Chicago University Press.

Canales, Jimena. 2014. "Living color: Jimena Canales on the art of Eliot Porter." *Artforum*, 53, No. 1: 334–41.

Canales, Jimena. 2015. *The Physicist and the Philosopher: Einstein, Bergson and the Debate That Changed Our Understanding of Time*. Princeton, NJ: Princeton Press.

Carpentier, J. 1910. "M. J. Carpentier présente à l'Académie le Chronophone Gaumont." *Comptes rendus de l'Académie des sciences*, 151: 1324–5.

Chauvois, Louis. 1937. *D'Arsonval Soixante-cinq ans à travers la Science*. Paris: J. Oliven.

Danto, Arthur C. 1965. *Analytical Philosophy of History*. Cambridge: Cambridge University Press.

Daston, Lorraine, and Peter Galison. 2007. *Objectivity*. New York: Zone Books.

Deleuze, Gilles, and Félix Guattari. 1994. *What Is Philosophy?* New York: Columbia University Press.

Derrida, Jacques. 1983. "Excuse me, but I never said exactly so: Yet another Derridean interview." *On The Beach*, 1 (August): 42–3. http://www.egs.edu/faculty/jacques-derrida/articles/excuse-me-but-i-never-said-exactly-so/ Accessed September 28, 2015.

Eddington, Arthur Stanley. [1928] 1935. *The Nature of the Physical World.* London: J. M. Dent.

Fouque, Victor. 1973. *The Truth Concerning the Invention of Photography: Nicéphore Niépce, His Life, Letters, and Works.* New York: Arno Press.

Frank, Philipp. 1948/1949. "Logical empiricism I: The Problem of physical reality." *Synthese*, 7, No. 6-B: 458–65.

Galassi, Peter. 1981. *Before Photography: Painting and the Invention of Photography.* New York: The Museum of Modern Art.

Gavarret. 1878. "Observations à l'ocassion du procès-verbal. III. Méthode Graphique." *Bulletin de l'Académie de Médecine*, 7: 759–69.

Gooday, Graeme J.N. 2004. *The Morals of Measurement: Accuracy, Irony and Trust in Late Victorian Electrical Practice.* Cambridge: Cambridge University Press.

Heidegger, Martin. [1950] 2009. "The Thing." In *The Craft Reader*, edited by Glenn Adamson, 404–8. London: Bloomsbury Publishing.

Janssen, Jules. 1930. "En l'honneur de la photographie: Discours prononcé au Banquet annuel de la Société Française de la photographie, juin 1888." In *Oeuvres scientifiques recueillies et publiées par Henri Dehérain*, edited by Henri Dehérain, 86–90. Paris: Société d'Éditions Géographiques, Maritimes et Coloniales.

Kittler, Friedrich A. 1999. *Gramophone, Film, Typewriter*, Stanford, CA: Stanford University Press.

Krajewski, Markus. 2010. *Der Diener: Mediengeschichte einer Figur zwischen König und Klient.* Frankfurt am Main: S. Fischer.

Langer, Susanne K. [1942] 1951. *Philosophy in a New Key.* Cambridge, MA: Harvard University Press.

Laruelle, François. 2011. *Le Concept de non-photographie.* Falmouth: Urbanomic; Sequence Press.

Latour, Bruno, and Steve Woolgar. 1986. *Laboratory Life: The Construction of Scientific Facts.* Princeton, NJ: Princeton University Press.

Marey, Etienne-Jules. 1878. *La méthode graphique dans les sciences expérimentales et principalement en physiologie et en médecine.* Paris: G. Masson.

McLuhan, Marshall. 1964. *Understanding Media: The Extensions of Man.* New York: McGraw-Hill.

Rheinberger, Hans-Jörg. 1997. *Toward a History of Epistemic Things: Synthesizing Proteins in the Test Tube.* Stanford, CA: Stanford University Press.

Roberts, Lissa. 2005. "The Death of the Sensuous Chemist: The 'New' Chemistry and the Transformation of Sensuous Technology." In *The Empire of the Senses*, edited by D. Howes, 106–127. Oxford: Berg.

Russell, Bertrand. 1918. *Mysticism and Logic.* New York, Longmans, Green and Co.

Russell, Bertrand. 1925. *The ABC of Relativity.* New York: Harper.

Schaffer, Simon. 1988. "Astronomers mark time: Discipline and the Personal Equation." *Science in Context*, 2: 115–45.

Scripture, Edward Wheeler. 1895. *Thinking, Feeling, Doing.* Meadville, PA: Chautauqua-Century Press.

Sterne, Jonathan. 2003. *The Audible Past: Cultural Origins of Sound Reproduction.* Durham, NC: Duke University Press.

Trachtenberg, Alan (ed.) 1980. *Classic Essays on Photography*. New Haven, CT: Leete's Island
    Books.
Trouvelot, Étienne Léopold. 1882. *The Trouvelot Astronomical Drawings Manual*. New York:
    C. Scribner's Sons.
Wolf, Charles. 1875. "Conférence sur les applications de la photographie à l'astronomie et en
    particulier à l'observation du passage de Vénus." *Bulletin de la Société Française de Photogra-
    phie*, 21 (8 January): 16–28.

Chapter Thirty-Six

Microscopes

Boris Jardine

The microscope is one of the truly emblematic tools of science. The familiar image of a "scientist at work" will inevitably feature a white lab coat (conveying disinterested authority), and a microscope on the lab bench: a sign of diligence, of control over nature, and access to hidden worlds. Yet, unlike its close relative the telescope, the microscope's role in the history of science is uncertain. The question of its origins is vexed, and its early use is typically seen as desultory. After initial fumblings (so the standard accounts relate) came a period of great achievement—that of Robert Hooke, Antoni van Leeuwenhoek, Marcello Malpighi, and Jan Swammerdam. But once this golden age was over we supposedly return to futility: microscopy in the eighteenth-century has been labeled "fallow, scientifically" (Turner 1980, 161). The modern age of microscopy then appears *ex nihilo* with the invention of lenses fully corrected for the various aberrations that hinder clear vision.

My main purpose in this chapter is to provide an alternative to this narrative, emphasizing continuity where others have seen startling change; interweaving stories—of popularization, discovery, trade, invention and theory—that others have seen as separate. This is not done to downplay the achievements of Hooke, Lister *et al.*, rather to set those in their proper context and give better shape to the history of what remains a vitally important scientific instrument. The chapter is laid out chronologically—however I make only preliminary suggestions for a history of the microscope in the twentieth century, a project for which the foundations have not even been laid at present.

Origins

The microscope, as one historian has wryly commented, was "never invented" (Lüthy 1996, 2). Objects appear larger through water or uneven pieces of glass—surely we cannot talk of the origins of magnification. The development of lens-grinding, meanwhile, had nothing to do with the systematic study of the very small; rather, its concern was providing aids to defective vision. If we think of a microscope as an instrument for

A Companion to the History of Science, First Edition. Edited by Bernard Lightman.
© 2016 John Wiley & Sons Ltd. Published 2020 by John Wiley & Sons Ltd.

studying minutiae, we could start with Giovanni Rucellai, who is said to have made drawings of a bee in an enlarging mirror in 1523 (Ronchi 1970, 107). But his work was dismissed as either illusion or magic, and has escaped the notice of most historians. Instead, accounts of the microscope's early years have tended to begin with the recognizably modern *instrument*, that is, the particular arrangement of lenses connected by a cylindrical tube that was first contrived in the Low Countries in the years around 1600—perhaps the work of the same craftsmen who invented the telescope (Turner 1985).

For the first half-century of the microscope's use we have a rather fragmented picture. We know of a handful of early pioneers, who elaborated on the basic two-lens system and invented new stands, and we know that these instruments circulated around Europe along routes of courtly exchange (Fournier 1996). The main use in these early years was entomological: Galileo examined "a certain insect in which each eye was covered with a rather thick membrane"; French *savant* Nicolas-Claude Fabri de Peiresc set the tone for early modern microscopic interpretation when his investigations into fleas and lice led him to celebrate the "effects of divine providence, which was far more incomprehensible to us when that aid to our eyes was wanting" (as quoted in Fournier 1996, 25). In Rome, the first microscopic illustrations were published, showing the exterior parts of the bee and titled *Melissographia* (1625).

Soon, dissection was added to observation, and internal structures were described. For example, Giovanni Battista Odierna, a Sicilian priest, made a major advance when he dissected the eye of the fly and speculated on the mechanism of perception. Here we have one of the first instances of microscopic observation playing a role in natural explanation. In England, under the banner of Baconianism, this was taken a step further, and made into a more programmatic (if idealized) project. Not only was there Bacon's general claim that it would be "by instruments and helps that the work is done," but, further, Solomon's House would be replete with "Glasses and Meanes to see the small and Minute Bodies, perfectly and distinctly: As shapes and Colours of Small Flies and Wormes, Graines and Flawes in Gemmes which cannot otherwise be seen, Observations in Urine and Bloud not otherwise to be seen" (as quoted in Wilson 1995, 50).

Gradually, the microscope was deployed in new disciplines: botany, mineralogy, and the anatomy of higher organisms. In mid-century England, for example, two important works advocated the use of microscopes in embryological research. These were Nathaniel Highmore's *History of Generation* (1651), which makes casual mention of the microscope in answering the question of whether the embryo chick develops first in the yolk or the white of the egg, and William Harvey's *Exercitationes de generatione animalium* (1651), which encourages the use of "Perspectives" in the same connection.

In a century of high rhetoric, the microscope was quickly and eloquently brought into line with the dominant physico-theological justification of natural inquiry. This project, of "reading in the features of the world the existence, presence, and characteristics of a supernatural being" (Wilson 1995, 176), functioned not only as an apologia for natural philosophy, but also as a guarantor of objectivity—albeit one quite distant from later and more familiar paradigms that prize procedure and verisimilitude over baroque verbiage. This justification for microscopy developed in tandem with an involuted notion of the link between the eye of the observer, the nature of the microscope itself, and the eye as an object of inquiry. This was just the combination that Odierna had begun research into in the 1640s, and it was to become one of the enduring subjects of microscopical research, with developments in microscopy and optical theory alike refining the geometrical and anatomical understanding of vision (Schickore 2007).

Practically, too, microscopy was on the rise across Europe. The expert skill required to make small highly convex lenses was now relatively widespread, with centers of production in all the major capitals.

Hooke and After

It was in this context—of high speculation and growing craft knowhow—that Robert Hooke first embarked on his famous microscopical researches. These were begun in 1663 as a series of demonstrations to the Royal Society, for whom Hooke was Curator of Experiments.

Preparing his samples of moss, cork, mold, vinegar and flint fully occupied even Hooke's great experimental talents. In spite of the excitement the microscope was beginning to generate, there was almost no protocol for the preparation and mounting of specimens, method of illumination, configuration of lenses or sharing of results. Indeed, the nature of the microscope itself was up for grabs, with Hooke writing that he had experimented with "a Microscope with one piece of Glass [...] another only with a plano-concave [...] others of Waters, Gums, Resins, Salts, Arsenick, Oyls, and with divers other mixtures of watery and oyly Liquors" (Hooke 1665, unpaginated preface). But in the main Hooke worked with an instrument purchased from the London tradesman Christopher Cock: a large microscope that could be used with two or three lenses, which he had adapted with various devices for mounting and viewing the specimen (Clay and Court 1932, 20–31).

For Hooke, as for so many who followed, illumination was the main problem. Sitting at his south-facing window, there was often insufficient light—but then suddenly there could be too much, putting the specimen at risk of incineration. In this case Hooke either inserted a sheet of greased paper between the condensing lens and the specimen or reflected the light via a sanded mirror. Artificial illumination was also possible, and here Hooke constructed an elaborate system using an oil lamp and two condensing lenses.

This serves to remind us how complicated and strenuous observation could be—after all, Hooke had constructed this set-up in order to avoid using single lenses, which, though powerful, were difficult to grind and had to be pressed almost onto the surface of the eye. In addition, the practice of microscopic manipulation itself was a key part of Hooke's natural philosophy, not merely as a precursor to observation but as an active part of observation itself. As Hooke explained, his "usual manner" of examination involved "varying the degrees of light and altering [the specimen's] position to each kinde of light" in order to reveal details of physical structure. In his description of the eye of a fly, for example, Hooke noted the different structures (apparently) visible under different kinds of light. This is Hooke as the true mechanical philosopher, drawing no distinction between manipulation, mechanism, and "Philosophical Inquiry," and indeed critiquing his contemporaries on the grounds of their limited instrumental procedures (Bennett 1986).

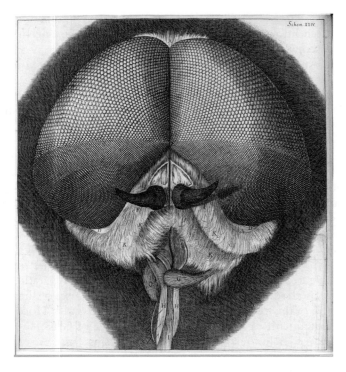

Figure 36.1   "Eyes and head of a grey drone fly," plate XXIV from Hooke's *Micrographia*
(1665, facing p. 175). Wellcome Library, London.

The labor was worth it. On 6 July 1663 Hooke was ordered to prepare his observations for publication. In only eighteen months he had overseen the production
of *Micrographia* (1665), the undoubted masterpiece of early microscopical research,
containing large folding plates, stunningly engraved with fleas, fungi, plants, needles
and spiders (Figure 36.1). These, of course, are not simply an embellishment: one
of Hooke's main difficulties was how best to share his observations—a question of
great importance throughout the history of microscopy (Taub 1999, 732). Even if the
preparation and lighting could be managed so that each of his colleagues at the Royal
Society could see through the instrument (no mean feat), the multitude of observations that went into the analysis of each specimen—through different arrangements of
lenses, for example—could not be repeated in full. So the engravings, which captured
this multitude of views, *were* the observations. And what they revealed was remarkable: amongst the welter of insights and speculations Hooke gave the first description
of micro-fungi and adopted the word "cell" to describe constituents of cork. He had
demonstrated, for the first time, that the microscopic world was various and strange
enough to require examination in its own right, rather than as a mere adjunct to the
established disciplines.

Inspired by Hooke's work, others soon added to the list of microscopic discoveries.
By far the most dramatic of these came from an unknown Dutchman named Antoni
van Leeuwenhoek, who truly set out to map the "new discovered world" that Hooke

had begun to describe. Using lenses that could magnify more than 250 diameters, Leeuwenhoek was the first to see spermatozoa, bacteria, blood cells and a veritable zoo of microorganisms, which he called "animalcules."

The case of Leeuwenhoek brings out perhaps the most important aspect of early microscopy: credibility. Leeuwenhoek's discoveries were remarkable, to be sure, but could his reports be trusted? He was a respectable member of Delft society, and in gaining the respect of the members of the Royal Society he took a huge step towards more general acceptance. But the latter was hard-won, even in his own country. Leeuwenhoek's technique for making lenses was a closely guarded secret and his skill in preparing specimens was formidable, so he went to the extraordinary length of opening his house and making special demonstrations for visitors. Two anecdotes give a sense of the drama and labor of Leeuwenhoek's practice: on one occasion he reported spending days killing over one hundred mosquitoes in order to prepare an observation of the insect's mouth-parts; on another the artisan-scholar Nicholas Hartsoeker accosted Leeuwenhoek on the latter's doorstep in order to grill him over the impossibility of dissecting a flea without obscuring the light with the knife (Ruestow 1996, 151–3). Long-range communication was perhaps easier for the harried Leeuwenhoek. In order to keep the Royal Society onside he even prepared specimens, sending them attached to his letters; eventually he donated over 200 microscopes, a necessary step given that Leeuwenhoek's manner of working was effectively to build an entire system—specimen and microscope fixed together—for each separate observation.

For Leeuwenhoek's contemporaries on the continent it seems that, like the pre-*Micrographia* researchers in England, discussion of optics and instrumental set-ups was not deemed relevant when it came to publishing results. Here, clearly, other measures of authenticity were being used. For example the Bologna-based Royal Society correspondent Marcello Malpighi, in developing an extremely ornate explanation of living processes based on his observations of microscopic tubes and pores (most notably the capillaries), provided almost no information on the instruments he used—the quality of his observations seemingly transcending the need for the kind of extensive instrumental digressions given by Hooke. Far less urbane, but just as prolific, the Dutch entomologist Jan Swammerdam began a remarkable series of studies in insect physiology; while in England Nehemiah Grew pioneered the microanatomy of plants.

These projects, remarkable in their scope as they may seem, were not at the time considered uniformly successful. Grew and Malpighi, for example, failed to find organs and systems in plants analogous to those in animals; Swammerdam struggled throughout his life against his father's wishes that he return to medicine, and against his own crises of faith in experimental philosophy as a means of spiritual fulfilment. Most dramatically of all, the virtuosi—and in particular their researches into the very small—were pilloried on the stage and in satirical poems (Nicolson, 1956). Towards the end of the century, harassed by the playwrights and witnessing a decline in microscopical research, Hooke lamented that he knew "of none that make any other Use of that Instrument, but for Diversion and Pastime" (as quoted in Wilson 1995, 226).

The Development of the Instrument Trade

It is perhaps not surprising, therefore, that if we follow traditional accounts we find the microscope in poor health *circa* 1700. Historians have typically taken Hooke's pronouncement at face value. Hooke himself had set the microscope aside in favor of other researches; Leeuwenhoek's immense productivity was soon to dwindle, and with Swammerdam and Malpighi both dead, the instrument's champions would seem to be few. And yet, looked at from another, perhaps less lofty vantage point, this supposedly "fallow" period in the microscope's history is transformed—for it is at precisely the turn of the eighteenth century that the popular use of the instrument began.

To understand this transformation it is necessary to look in some detail at the optical instrument trade *circa* 1700. In England, artisans were struggling free of the guild system, which in principle defined who could trade and what they could offer. Meanwhile, the clientele for those selling microscopes was diverse, ranging from fellow artisans all the way to royal appointment. Also important, in instrument manufacture as elsewhere, was the increasing division of labor. Where the first instrument makers—just over a century prior—were highly specialized, working each piece from beaten brass to divided scale, the new breed would put their name to an instrument assembled from parts made elsewhere. A microscope, for example, required expert lens-grinding, tube manufacture and stand construction, and these three functions were by no means all carried out by the "maker" whose name appeared on the finished product.

So just as the scholar Hooke could lament that the microscope was becoming otiose, an entrepreneur like James Wilson could boast that "The Use of Microscopes is so well known, that it's as needless to attempt their Recommendation to the Inquisitive, as it would be tedious to numerate their particular advantages in Natural Inquiries" (Wilson 1702, 1241). With these words Wilson introduced the new portable "screw-barrel" microscope to the English market—and though the patter of the marketplace may be in evidence, Wilson was not merely aiming for popular appeal. The context of his essay—the Royal Society's *Philosophical Transactions*—speaks of the complex interrelation of scholarship and commerce in the period. Indeed, as he continues Wilson deftly adds a touch of drama to Hooke's physico-theology, writing of the "entertainment" of confirming the superiority of natural objects over "the most Celebrated pieces of Art" (Wilson 1702, 1241).

Here Wilson evokes a point Hooke had made in *Micrographia*, doing so for an audience who would understand the reference, in order to justify the development and marketing of a cheap microscope that would allow repetition of experiments first published over a quarter of a century earlier. And the design advocated by Wilson—an elegant single-lens microscope that packed neatly into a small case—proved immensely successful.

While microscopes of Hooke's design had been used by a handful of specialists across Europe, now portable microscopes were sold in their hundreds, at prices that made them accessible to the large number of interested amateurs. We might usefully compare this situation with the near-contemporary rise in the popularity of pocket globes—just like the portable source of microscopic knowledge, a portable source of geographic knowledge catered to a rapidly growing market for polite learning. Wilson's pocket microscope was soon offered by a large number of opticians and instrument makers, almost always sold with a pre-prepared selection of slides and apparatus for viewing the very paradigm of microscopic achievement, the circulation of the blood (normally in the tail of a small fish encased in a narrow vessel) (Figure 36.2).

So Hooke's words were highly prescient—indeed one of his charges against the microscope was that it was merely "a portable Instrument, and easy to be carried in one's pocket." Far from being a fall from grace, as Hooke would have it, the

Figure 36.2   A screw-barrel microscope made to James Wilson's design, with a variety of lenses and slides. Wellcome Library, London.

development and marketing of cheaper instruments across Europe constituted a remarkable success for the new experimental philosophy.

Each country—even each region—saw subtle differences in the economy of knowledge production and circulation. For example, Wilson's microscope was depicted fully assembled in the *Philosophical Transactions*, so that a potential user would have to go direct to the instrument maker. In France, Louis Joblot proposed a microscope of very similar design, whose illustration was a technical one showing the separate parts of the microscope, the intention being to allow any interested researcher to construct their own instrument (Ratcliff 2009, 22). In the German lands, meanwhile, highly ornate instruments were being made by artisans like G. F. Brander alongside so-called "toy" wooden microscopes; in Italy optical know-how was often in the hands of amateurs, sometimes learned priests.

Amongst those who benefited directly from the new popular microscopy was Edmund Culpeper (c. 1670–1737), the instrument-maker who most successfully exploited the design Wilson had introduced. Trading under the sign of the Cross Daggers at the "Old Mathematical Shop" in Moorfields, London, Culpeper was the son of an Oxford-educated clergyman and was apprenticed to the mathematical instrument-maker Walter Hayes. Through his father's college, Merton, he may have had contact with some of the savants who resided there. Operating quite literally on the borders of the city of London's guild system and clearly a master craftsman, Culpeper took what would have seemed to his contemporaries a drastic step, away from a strictly mathematical range of instruments and into the marketing and manufacture of optical and "philosophical" instruments (Turner 1980, 9–11; Bennett 2011, 703).

Culpeper first experimented with the incorporation of the simple microscope into an all-purpose compendium, but he quickly saw the possibilities of Wilson's smart design. With his background primarily in the engraving of scales and delineation of sundials, it is highly likely that Culpeper collaborated with a lens-maker, perhaps Wilson himself. He had competition in the manufacture of single-lens microscopes from Edward Scarlett, and elsewhere in the capital John Marshall and a few others offering compound microscopes largely derived from Hooke's design. So Culpeper was opportunistic, versatile and dismissive of the trade restrictions that ought to have limited his output in both size and range—in all of these respects he was a pioneer, both creating and catering to a new market for mathematical *and* philosophical education amongst a growing bourgeoisie.

The scientific enterprise of the Enlightenment was as much a matter of commerce, spectacle, and politics as it was one of discovery and invention—and this in no way diminishes its importance with regard to what went before and what came after. The new breed of popular lecturers also made extensive use of the microscope, adding sublime rhetoric to the physico-theological patter and developing instruments that facilitated group viewing.

Microscopy in this period was not a self-contained and discipline-bound practice, but rather part of a Europe-wide culture of collecting, lecturing and display. Both Leeuwenhoek's and Swammerdam's collections had been on the itinerary of learned tours around Europe, and in England the cultivation of a complete system of philosophical apparatus was an important part of gentlemanly culture. The wide range of types of specimen prepared by microscopists of the time and the marketing of the microscope with other mathematical and philosophical instruments were of a piece, and only make sense in the context of collections of *naturalia* (Taub 1999, 738). Microscopes themselves could constitute, as well as reveal, objects of curiosity. Hence, where historians have tended to emphasize the undoubted decline in the number of microscopical *discoveries* between c.1650 and c. 1750, we ought instead to acknowledge continuity in the rhetoric and display associated with microscopical practice.

By mid-century, the repertoire of microscopical observations was firmly established, not least due to the tireless work of the popular writer Henry Baker, a regular contributor to the *Philosophical Transactions*, and by trade a tutor to the deaf. His most successful book, *The Microscope Made Easy* (1742), drew on the growing market for popular natural philosophy, and the 1000 copies of the first edition sold out in a mere five months. Soon new impressions and then editions were prepared, and the work was translated into numerous languages (Turner 1980, 195). Baker's books are as much about microscope design and manufacture as they are about the natural world. *The Microscope Made Easy* opens with yet another discussion of Wilson's simple microscope, before showing how it can be adapted for tabletop use and turned into a compound instrument. The merits of various other designs are assessed, all of them available from Baker's favored artisan and collaborator John Cuff. Indeed, in spite of Cuff's perpetual financial difficulties, the pair's next collaboration was to be momentous. In preparation for a sequel to *The Microscope Made Easy*, Baker was performing a series of experiments on crystals. Finding that the crude focus mechanism and obstructed stage of the current "Culpeper-type" hindered his research, he worked with Cuff to design a new instrument, in which the body—supported on a side-pillar with a fine-screw focus mechanism—was made entirely from brass, supported on a sturdy wooden base. Though other designs persisted—for example Culpeper's, Wilson's and a new portable "drum" microscope by Benjamin Martin—Cuff's design was highly popular, eventually to become standard; it is recognizably the ancestor of the modern instrument.

Nor were Baker's successes purely in marketing and design: seizing quickly on a report from the Swiss naturalist Abraham Trembley, Baker published a report of his own experiments on the behavior of the "polypes"—waterborne creatures with remarkable powers of regeneration. If these objects of inquiry were new in the eighteenth century, there was much that was continuous with earlier work. For example, echoing Hooke and the earlier microscopists, popular writers of the eighteenth century gave introductions to their works in which the optical system of the eye and the microscope were explained—the wonderment of the earlier writers giving way to an "improving" discourse that at once explained the basics of modern optics and justified the use of the microscope as a mere extension of the senses, no different in kind from our own optical apparatus (Schickore 2007, 21–3). And the nature of the insect eye allowed writers like Benjamin Martin to advance novel theories while maintaining the older metaphysical notion of the perfection of natural optics, or "divine geometry" as he called it. This theorizing, in addition to contemporary debates over spontaneous generation and preformation, and about the division of the plant and animal kingdoms and microanatomy show that microscopical research in this period was in fact a lively enterprise—perhaps undervalued by historians because of an interpenetration of trade, popular exposition and discovery unknown before or since.

In the eighteenth century, then, we see discontinuities with what went before, but they are not of the sort typically identified by historians of the microscope. The microscope was no longer the province of a handful of scholars keen to extol the instrument's virtues. Popular lecturers had taken over; the instrument trade was booming; new designs took the microscope into the field, into the public hall and into the drawing room. The emphasis was now on having a microscope by one of the major makers—the names of Dollond, Adams, Martin, and Cuff signaled quality, though not necessarily any improvement from the previous century in the optics. Sure enough, the initial burst of great discovery was over, but the physico-theology that was developed in the seventeenth century was consolidated and the number of manufacturers grew. As we shall see, these were vital for the rapid expansion of microscopy that was to come.

The Microscope in the Nineteenth Century

At the turn of the nineteenth century, a naturalist wishing to investigate insects, polyps or the circulation of the blood could choose from a vast range of instruments. Nor were these only the large new stands designed by the various makers. Because the compound microscope was known to exacerbate the problems of aberration, the simple microscope remained extremely popular. In addition, the growing enthusiasm for natural history in the second half of the eighteenth century led to new designs of single-lens instrument, notably the folding portable design associated with the botanist William Withering and John Ellis's "aquatic" microscope, in which a small vessel replaced the stage and water-borne microorganisms could be observed in their natural environment.

These more private pursuits were matched in the public sphere by the microscopy on offer in the "shows of London" (Altick 1978). The solar and lucernal microscopes, popularized in the second half of the eighteenth-century, projected an enlarged image onto a wall and therefore allowed collective viewing of microscopic phenomena. Drawing room demonstrations had given way to spectacular shows, first conducted by itinerant lecturers like Benjamin Martin, and then installed into the capital's "exhibitions", for example that hosted by Gustavus Katterfelto, whose "wonders" included insects "as big as birds". Even as late as 1839 the proprietor of the Colosseum at Marylebone, John Braham, could breathlessly (and of course impossibly) boast of magnifications of over 4.6 million times (Altick 1978, 83–5, 152).

But at the same time, other sorts of demonstrations were taking place. Amateur enthusiasts were gathering informally to discuss all manner of "philosophical" subjects, and microscopy was a major part of the trend. Gentlemanly meetings often used the same techniques as the larger public shows—giant diagrams and projecting microscopes—but the tone was of learned pleasure rather than raucous spectacle. In many ways the microscope was a model instrument of self-improvement, combining fieldwork, dexterous preparation, artistic skill and patience, and the knowledge of classificatory schemes. Instruments, drawings and slide arrangements were compared, and perplexing discoveries debated. The more refined enthusiasts could choose from a wide range of table-top instruments, most of them set up to work with both simple and compound configurations, some with multiple eyepiece tubes to facilitate collective viewing.

It is important to note that this amateur, semi-private microscopy was the cutting edge of practice in Britain—but elsewhere the microscope was a tool of the nascent laboratory. Experimental physiologists like Johannes Müller made extensive use of the microscope, in particular in his researches into blood globules and the development of the embryo. As Jutta Schickore has shown, microscopy in the German lands was, in this period, fully compatible with the *Naturphilosophie* tradition of self-experiment, and with the view that the experimental set-up bridged the epistemological gaps opened by Immanuel Kant's transcendental idealism (Schickore 2007). In spite (or perhaps because) of the excellent lenses being developed on the continent at this time, assessment of the microscope's performance was secondary to the careful cultivation of experimental skill.

National differences in experimental style were marked throughout the nineteenth century, and yet instruments and texts could move freely. Henry Baker's work, mentioned above, was cited as authoritative well into the nineteenth century in France and Germany, and competition was intense between the main firms of Chevalier (France), Amici (Italy), Fraunhofer (Germany) and the numerous manufacturers working in London. As rivalries between instrument-making firms deepened and extended to matters of international competition, designs proliferated. Nor were practitioners unaware of the problems that such a fractured market could cause. How were observations to be verified when two quite different kinds of microscope might be in use? How were the merits of simple *versus* compound instruments to be decided? There were, of course, lofty imperatives, largely unchanged from the preceding century: having a good instrument from an established maker was a prerequisite, moral standing remained important, and speculation was to be avoided. And yet standards of comparison—of lenses, specimen preparation and observational practice—were few and far between. In this context, new theories generated lively controversy, which eventually turned to matters of technique.

The "globular" theory, for example, was an early nineteenth-century attempt to describe the formation of living tissues (Pickstone 1973; Schickore 2009a). The questions that have vexed historians have concerned the relation of the globular theory to cell theory, and the reasons for its demise—generally taken to be linked to improvements in microscopy. Yet what is important here is not so much the success or failure of the theory itself, rather the attention it drew to the importance of comparative standards in microscopical work. How could theories about the importance of microscopic objects be debated when estimates of their size varied so wildly? It was in this context, of collaborative witnessing and debate over the ultimate constituents of living tissue, that a key technology was developed: the test object.

This innovation in microtechnique was primarily the work of Charles Goring, an Edinburgh-trained physician who had—in an age increasingly favoring reliable single-lenses—a "lurking fondness" for the compound microscope. This encouraged Goring to devise methods of comparison of what he termed "penetrating power" (Schickore 2009b). While Goring struggled to establish this terminology, his test objects themselves proved wildly successful. The earliest were the parts of insects that displayed regular patterns, for example the striations on the scales of butterfly wings. Soon the remarkable forms of diatoms were being used instead, and eventually mechanically-ruled gratings were developed by Friedrich Adolf Nobert, these being so fine that attempts to resolve them resulted in dramatic improvements in lens design (Turner 1980, 141–58).

Goring's innovation can be taken as representative of a general move in the early nineteenth century towards a concern with the quality and combinations of different kinds of glass used in microscope lenses. Some instrument-makers—for example Giovan Battista Amici in Modena and John Cuthbert and Andrew Pritchard in London—rejected glass lenses altogether, working with diamond and sapphire, or attempting to make reflecting microscopes on the same principle as the long-established reflecting telescope. But these alternatives were eventually trumped by the work of Joseph Jackson Lister, who collaborated with the instrument-maker William Tulley to make, in 1827, the first lenses corrected for both chromatic and spherical aberration—that is, the distortion of the image owing to variant refraction of different colors, and to the range of focal points across the lens (Turner 1980, 13–15).

Appropriately enough, the first use to which Lister's instrument was put was the observation of blood, carried out with Thomas Hodgkin at Guy's Hospital. In an 1827 paper Lister and Hodgkin critically assessed the globular theory, finding no evidence of the formative particles within blood cells (Bracegirdle 1977). They also provided a comparison of the widely divergent measurements of blood cells with their own estimation, which closely matches modern figures. The authors stated that the instrument was superior even to Amici's reflecting microscope, and that, although faithful to what they had seen, advocates of the globular theory had been misled by defective lenses and preparation techniques.

If the globular theory had been a valiant attempt to make microscopy central to studies of living tissue, it was work done with the new microscopes that proved decisive in the establishment of experimental histology. The paper by Hodgkin and Lister served to reiterate one of the key conclusions of earlier microscopy: namely that there would always be a layer or stage in natural processes inaccessible to experiment. It is not surprising, therefore, that the microscopic paradigm that emerged in this period—the cell theory of Schleiden and Schwann—involved much larger constituent parts than the globular theory. Indeed the most notable work on the nature of the cell in this early phase was done by Robert Brown, who identified and named the nucleus using a single-lens microscope.

Other disciplines, too, were undergoing microrevolutions. For example, research into the question of spontaneous generation took a remarkable turn with the announcement, by Christian Gottfried Ehrenberg in Berlin, that fossilized "infusoria" made up vast swathes of the earth's strata. Where the new *Naturphilosophie* was welcomed, for example at Edinburgh, there was a particular clamor for infusorial research. Elsewhere, amongst the most influential of Ehrenberg's interlocutors was a young Charles Darwin, just back from his voyage on board the Beagle and keen to engage with the new discoveries about infusoria in order to establish himself as an expert microscopist. Ultimately Darwin and others were to reject Ehrenberg's more salacious claims about the complexity of these microorganisms—but the boost to the technique of microscopy and the spur to reconsider the relationship between speculation and empiricism had lasting effects (Jardine 2009).

Ehrenberg's was just one of a host of new theories that seemed to indicate that the microscopic world was a universal explanatory resource. Fossilized infusoria made up many of the strata; cells were the organizing principle of life, and germs the carriers of disease. As the century drew on the microscope played an ever greater part in public health, especially following revelations about the role of polluted water in typhoid and cholera outbreaks in London (Figure 36.3).

With the microscope now established as an instrument of the classroom and teaching laboratory, it became the primary means by which heavily mediated "Nature" was brought under control in the new academic laboratories of the second half of the nineteenth century (Gooday 1991). Here the careful stage management of observations allowed laboratory science to attain the kind of authentic validity previously reserved for field investigation. Experimental rigor now served to separate the communities of microscopy, with reformers like T. H. Huxley claiming that the instrument could only be correctly used with expert training and in the context of the new professional science.

At the close of the nineteenth century Lister's and Goring's work of making microscope optics a matter of systematic inquiry was effectively completed by Ernst Abbe, who provided the theoretical limit of microscope optics and then worked with Carl Zeiss to produce apochromatic lenses that approached optical perfection. As microscopists began to understand the ways in which the nature of light and the human eye limited the potential power of microscopes, and as the cell theory gave a programmatic direction to research, methods of sample preparation became increasingly important.

Figure 36.3 "Monster Soup, commonly called Thames Water," an 1828 engraving by William Heath. Wellcome Library, London.

Techniques in staining specimens, for example—pioneered by Ehrenberg and brought to a wider audience in the work of Robert Koch—allowed ever greater resolution of the parts of the cell and the nature of bacteria. With the new techniques of microphotography, scientists could trade on the supposed objectivity of the camera. And the establishment of the microscope as the preeminent instrument of laboratory research meant that it entered into countless syllabi, and standardized student microscopes were made in their tens of thousands.

Modern Microscopy

It is beyond the scope of the present essay to deal even cursorily with twentieth-century microscopy. Indeed there are good reasons for denying that a history of the instrument in this period would make sense. The success in the nineteenth century of the new and self-contained discipline of microscopy swiftly led to its dissolution: the microscope became a tool of professionalization, which in turn sent the instrument into the separate domains of zoology, anatomy, entomology, and so on.

Perhaps the most promising approach to recent microscopy—and one which would maintain the interconnection between theorizing, trade and manufacture that I have pursued here—is the tracing of microscopes as they circulated between laboratories. For example Pedro Ruiz-Castell (2006) has shown, with a case study of a Soviet microscope in use in Franco's Spain, how the movement of instruments could overcome the most formidable of political obstacles. The movement of microscopes amongst colonial networks was of course a much more routine and long-established business. With the development in the nineteenth century of botanical and microbiological research on a global scale, and through the growth of research in tropical medicine in the early twentieth century, microscopy came to be deployed far from its earlier European locales.

Thinking of the most dramatic change in microscopy in the twentieth century—the invention in the 1930s of the electron microscope—it is clear that instrument studies has much to contribute to the history and philosophy of science more generally. Following the techniques and materials involved in electron microscopy, we can see that—although it was often cast as merely an extension of light microscopy—in fact the new instrument had much to do with X-ray crystallography, and was therefore bound up with the nascent discipline of biophysics. At the more mundane level, the persistence of light microscopy offers an excellent opportunity for analyzing "the shock of the old" in the laboratory setting (Edgerton 2007).

For some advocates of the study of the microscope, its story has offered an intriguingly materialist and technologically determinist alternative to mainstream historiography (Turner 1980). Yet this has had its pitfalls. For example, commonly made arguments about the "failure" of a field to develop owing to primitive optics are obviously and damagingly Whiggish. Here I have endeavored to show that continuities and important developments in the history of science need not solely be theoretical—they can be found in altered relations of trade and scholarship and in the techniques of showmanship, rhetoric, and illustration. The microscope is perhaps the most recognizable tool of science and yet, especially for its recent history, there is much still to understand about its use and significance.

References

Altick, Richard D. 1978. *The Shows of London*. Cambridge, MA: Belknap Press.

Bennett, Jim A. 1986. "The mechanics' philosophy and the mechanical philosophy." *History of Science*, 24, No. 1: 1–28. DOI: 10.1177/007327538602400101

Bennett, Jim A. 2011. "Early modern mathematical instruments." *Isis*, 102, No. 4: 697–705. DOI: 10.1086/663607

Bracegirdle, Brian. 1977. "J. J. Lister and the establishment of histology." *Medical History*, 21: 187–91. DOI: 10.1017/S0025727300037716

Bradbury, Savile. 1967. *The Evolution of the Microscope*. Oxford: Pergamon Press.

Butler, Stella, R. H. Nuttall, and Olivia Brown. 1991. *The Social History of the Microscope*. Cambridge: Whipple Museum of the History of Science.

Clay, Reginald S., and Thomas S. Court. 1932. *The History of the Microscope: Compiled from Original Instruments and Documents, up to the Introduction of the Achromatic Microscope*. London: Charles Griffin.

Edgerton, David. 2007. *The Shock of the Old: Technology and Global History since 1900*. Oxford: Oxford University Press.

Fournier, Marian. 1996. *The Fabric of Life: Microscopy in the Seventeenth Century*. Baltimore: Johns Hopkins University Press.

Gooday, Graeme. 1991. "'Nature' in the laboratory: Domestication and discipline with the microscope in Victorian life science." *British Journal for the History of Science*, 24, No. 3: 307–41. DOI: 10.1017/S0007087400027382

Hooke, Robert. 1665. *Micrographia: Tabled & Illustrated*. London.

Jardine, Boris. 2009. "Between the *Beagle* and the barnacle: Darwin's microscopy, 1837–1854." *Studies in History and Philosophy of Science*, 40, No. 4: 382–95. DOI: 10.1016/j.shpsa.2009.10.007

Lüthy, Christoph H. 1996. "Atomism, Lynceus, and the fate of seventeenth-century microscopy." *Early Science and Medicine*, 1, No. 1: 1–27. DOI: 10.1163/157338296X00097

Nicolson, Marjorie. 1956. *Science and Imagination*. Ithaca, NY: Great Seal Books.

Pickstone, John V. 1973. "Globules and coagula: Concepts of tissue formation in the early nineteenth century." *Journal of the History of Medicine*, 28, No. 4: 336–56. DOI: 10.1093/jhmas/XXVIII.4.336

Ratcliff, Marc. 2009. *The Quest for the Invisible: Microscopy in the Enlightenment*. Farnham: Ashgate.

Ronchi, Vasco. 1970. *The Nature of Light: An Historical Survey*. London: Heinemann.

Ruestow, Edward. 1996. *The Microscope in the Dutch Republic: The Shaping of Discovery*. Cambridge: Cambridge University Press.

Ruiz-Castell, Pedro. 2006. "An instrument from communist Europe in Franco's Spain." In *East and West: The Common European Heritage, Book of Abstracts of the XXV Scientific Instrument Symposium*, 59–64. Krakow: Jagiellonian University Museum.

Schickore, Jutta. 2007. *The Microscope and the Eye: A History of Reflections*. Chicago: University of Chicago Press.

Schickore, Jutta. 2009a. "Error as historiographical challenge: The infamous Globule Hypothesis." In *Going Amiss in Experimental Research*, edited by Giora Hon, Jutta Schickore, and Friedrich Steinle, 27–45. Dordrecht: Springer. DOI: 10.1007/978-1-4020-8893-3_3

Schickore, Jutta. 2009b. "Test objects for microscopes." *History of Science*, 47: 117–45. DOI: 10.1177/007327530904700201

Taub, Liba. 1999. "Heroes of microscopy and museology." *Studies in History and Philosophy of Science*, 30, No. 4: 729–44. DOI: 10.1016/S0039-3681(99)00024-2

Turner, Gerard L'E. 1980. *Essays on the History of the Microscope*. Oxford: Senecio.

Turner, Gerard L'E. 1985. "Animadversions on the origins of the microscope." In *The Light of Nature: Essays in the History and Philosophy of Science presented to A. C. Crombie*, edited by J. D. North and J. J. Roche, 193–207. Dordrecht: Nijhoff.

Wilson, Catherine. 1995. *The Invisible World: Early Modern Philosophy and the Invention of the Microscope*. Princeton, NJ: Princeton University Press.

Wilson, James. 1702. "The description and manner of using a late invented set of small pocket-microscopes[…]." *Philosophical Transactions*, 23, No. 281: 1241–7.

Chapter Thirty-Seven

Telescopes

Jim Bennett

Of all the "tools" of science, the telescope has probably the most conspicuous popular standing; it will surely be the most common response to an invitation to name a scientific instrument. It is also the instrument that most identifies a scientific discipline: in a crowded field of candidates that includes computers, cameras, rockets, and spacecraft, the public will still think most readily of astronomers using telescopes. The survival over centuries of a popular distinguishing link, however inadequate and misleading, between a discipline and a particular instrument is remarkable (King 1979; Brandl, Stuik, and Katgert-Merkelijn 2010; Morrison-Low, Dupré, Johnston, and Strano 2012).

The notion of a telescope was a powerful one. It revealed things that were otherwise invisible. Instruments had long been fundamental to the practice of astronomy but their role was to introduce, manage, or regulate the mathematical work of astronomers through a mechanical routine or algorithm. Armillary spheres or astrolabes did not discover anything and they told their users nothing about the physical nature of the heavens. The name for this new instrument, adopted from among several early alternatives, combined the Greek for "afar off" and "to look," and captured a compelling idea that could accommodate the many disparate forms of "telescope" that emerged in subsequent centuries. The name is generally attributed to a scholarly poet and theologian from Greece, one John Demisiani, who is otherwise scarcely remembered. His single but hugely influential creative intervention came when he was in Rome in 1611 and attended a banquet given in Galileo's honor, where the telescope was successfully and memorably demonstrated (Rosen 1947).

Such were the novel pretensions of the telescope that in 1611 it would not have been regarded as an instrument of astronomy, as the subject was conventionally understood. Astronomy was a mathematical science dealing with the motions of the heavenly bodies and the geometry that could calculate their measured positions; the revelations of the telescope offered no direct contribution to this discipline but seemed much more relevant to the separate concerns of natural philosophy, albeit the natural

A Companion to the History of Science, First Edition. Edited by Bernard Lightman.
© 2016 John Wiley & Sons Ltd. Published 2020 by John Wiley & Sons Ltd.

philosophy of the heavens. Galileo, while anxious to adopt the mantle of a natural philosopher, worked to make his telescopic observations relevant to the arrangement and motions of the heavens. In the unfolding of his dramatic career he would use them to assemble a compelling discourse, constructed boldly around what he called "sistemi del mondo." The outcome would be only the first of many examples of how the science of astronomy—not just its technical content but its methods and boundaries—would be changed by developments in the telescope.

The Early Refractor

The earliest, unambiguous, documentary record of a telescope is a letter of recommendation written in September 1608 from the authorities of the Dutch province of Zeeland to the States General in the Hague, introducing Hans Lipperhey, a spectacle-maker of Middelburg, who claimed to have a new invention whereby "one can see all things very far away as if they were nearby" (Van Helden, Dupré, van Gent, and Zuidervaart 2010, 11). Soon, others came forward with similar claims, notably Jacob Metius of Alkmaar and Zacharias Jansen, also from Middelburg. Discussion, argument, and research around the identity of the true inventor has a long and continuing history among scholars and other protagonists (Van Helden et al. 2010). It would be wrong to call this debate "fruitless," as it has generated much valuable information and understanding, but it seems unlikely to yield a single, definitive attribution and the very idea that such an outcome would be appropriate to the historical context is now largely repudiated. Similar passions characterize the telescope's prehistory, where claimed feats of artificial observation, using such phrases as "by a certain instrument," sound as though they might indicate the use of a telescope, but there is a similar absence of a sustained consensus among historians.

News of the telescope spread through diplomatic and scholarly channels, as well as by travellers, one of whom was offering to sell this "secret" to the Venetian Senate in July 1609. Galileo, according to his own account, managed to deduce the arrangement of the instrument and formally offered his own version, with a power of 9× and anticipated military applications, on August 24. By now, however, this "invention" was becoming too well known to be a "secret" that could be sold. Galileo continued to improve his telescopes, eventually achieving a power of 30×, and took the profound step of turning them away from buildings and ships and towards the heavens. He was not the first such observer. In England, Thomas Harriot had made a telescopic drawing of the moon on August 5 1609 (dated July 26 in the Julian Calendar), but it was Galileo who published his observations, in his *Sidereus Nuncius*, which appeared in March 1610, and applied them to the natural philosophy of the heavens.

The Galilean telescope was, like its Flemish ancestors, a combination of a convex objective lens and a concave eyepiece. There were different accounts of how the telescopic effect of such a combination might occur, but certainly the resultant image was small and the best position for the eye not easily found. Such a telescope was difficult to use and required patience and practice, which could make it difficult to convince new users of its efficacy in natural philosophy or in practical applications. An early systematic attempt to provide a geometrical account of telescopic optics was made by Johannes Kepler and resulted in his book *Dioptrice* of 1611, which would have a profound effect on the development of astronomy and its instrumentation. Kepler proposed a different construction, introducing a convex eyepiece. In Galileo's instrument the concave eye lens had intervened before the rays refracted by the object glass had come to a focus, diverging them so that they entered the eye as though originating from a larger object than the source, that is, creating a magnified, virtual image. Kepler moved his eyepiece further from the objective, allowing the rays to come together in a real image, that the convex eye lens would then magnify in the manner of a reading glass (King 1979).

The result was a larger field of view and the potential for much greater magnification. The image was inverted, but it was quickly realized that, whatever the drawbacks for terrestrial use, this mattered little for studying the heavens. Greater magnification, however, brought its own frustrations. Kepler's analysis also demonstrated that lenses with spherical curvature—the only form that could be manufactured in practice—did not bring parallel rays of light to a single focus: incident rays at different distances from the center came together at different points on the optical axis, so the image was blurred and excessive magnification simply increased its fuzziness, revealing nothing. Lack of distinction came also from images having a spread of focal planes for different colors, a form of aberration ("chromatic") not then understood or distinguished from the "spherical" aberration Kepler had explained. He determined that spherical aberration could be removed by using lenses with hyperbolical instead of spherical surfaces, but such lenses could not be manufactured with the available techniques.

It is not unusual for familiar reputation to flout historical record. The strong popular link between the telescope and the figure of Galileo, for example, completely overshadows any awareness of its association with Kepler. But Kepler's invention of the basic astronomical version of the instrument is only one of his major contributions. Even more fundamental was his approach to the analysis and understanding of the telescope, making this a branch of geometry and accommodating it in his *Dioptrice* to the discipline he had set out in his *Astronomiae Pars Optica* of 1604. Further, Kepler had insisted in his *Astronomia Nova* of 1609 that traditional mathematical astronomy, in particular the geometrical treatment of planetary motion, must be consonant with its physical explanation. He had argued passionately against the separation of the natural philosophy of the heavens and the mathematical calculation of positional records and predictions. Appropriately enough, it was his form of telescope that brought the instrument into the discipline of positional astronomy, that is to say, into the traditional understanding of astronomy. Galileo, by contrast, had used the telescope to escape from the restrictions of that tradition by becoming a natural philosopher.

Because in Kepler's "astronomical" telescope the converging rays from the object glass are allowed to come to a focus within the tube, anything placed alongside the real image so formed will be magnified with it by the positive eyepiece and also be presented in focus to the observer. A reference point, marked for example by the point of a pin or the intersection of wires or threads placed at the focal point, would turn a telescope used only for observing into a telescopic sight to be used for measurement. It could replace the traditional sighting rule of an astronomical instrument, allowing much more precise positioning. Alternatively, if the visual angle to be measured were small enough to be contained within the field of view, it could be measured by a graduated rule, or by a sharp edge or wire moved by a micrometer screw. A pair of such edges or wires might be moved in opposite directions by the single action of a screw cut with two opposed threads, and the wires set to span the required angle—whether the distance between the components of a double star or features in the topography of the Moon. This may not be the popular view of the telescope but most instruments from the seventeenth to the nineteenth century, other than those for terrestrial use on land or at sea, were designed for measurement.

In fact, the earliest such application of the astronomical telescope came not from one of the major historical figures at all but from William Gascoigne, a country gentleman in Yorkshire, England. He and two associates, Jeremiah Horrocks and William Crabtree, had astronomical interests that were notably Keplerian, both in telescopic optics and in planetary theory. It happened that a spider, whose action Gascoigne interpreted as providential, spun a line in the shared focus of the lenses of his Keplerian telescope, whereupon the line stood out in sharp relief when the instrument was next used. Two crossing lines would provide an ideal reference point for aligning the telescope, then applied as a sight on a 4-foot astronomical sextant. Gascoigne also devised a micrometer, where a screw gave opposing motions to two knife edges and an index registered their displacement on a graduated disc. In the 1650s, such techniques were developed by Christopher Wren and independently by Christiaan Huygens, and later applied further by Robert Hooke, John Flamsteed, and Johannes Hevelius and (again independently from the early English work) by the French astronomers Adrien Auzout and Jean Picard. With the establishment of permanent "national" observatories in Paris and Greenwich, such techniques could become incorporated into a regime of astronomical measurement whose development would outlive the commitment of an individual astronomer. They extended the role of the telescope from being a tool only of observation to being essential for measurement, the traditional *sine qua non* of astronomy, or as Wren put it simply in 1656, "we make the Tube an Astronomical Instruēnt" (British Library MS Add. 25,071, ff.92–3). In facilitating the convergence of the tools and techniques of the different students of the heavens, the mathematician and the natural philosopher, the material resource of the telescope played an influential part in reforming the discipline of astronomy.

There were important developments also in the telescope as an instrument for observing, and again Kepler's work was pivotal. If his recommendation of hyperbolical lens surfaces was not practicable, an *ad hoc* mitigation of the problem Kepler had identified of spherical aberration (and, as it happened, of chromatic aberration, that was yet to be explained) was to use lenses of slighter curvature. This meant that, although it was often found necessary to "stop down" to the more central area of the object glass, larger apertures could be used to collect more light and form images of sufficient coherence to withstand increased magnification. The material result, of course, was that telescopes became longer—eventually very much longer. It meant also that their lenses became the province of specialist makers. Formerly spectacle makers had dabbled in the promotion of telescopes and microscopes as a promising sideline. Now a different business model evolved: in the later seventeenth century, optical instrument makers would make spectacles as a routine source of income to underpin their more exotic and challenging work. In Italy Eustachio Divini and Giuseppe Campani excelled in telescope making, urged on by their fierce and public rivalry; in Augsburg the talented and innovative Johann Wiesel gained an international reputation, while in London the leading makers were Christopher Cock and Richard Reeve. At the same time, astronomers might themselves be engaged with lens making, perhaps in collaboration with a maker. Huygens was active in grinding glass, while in England Paul

Neile and Wren, and then Hooke, worked closely with specialist makers. Both Divini and Campani were observers as well as makers.

By the late 1650s a serious refractor for astronomical work could be some 30 feet in length and mounting and controlling it was a challenging business. The usual approach was to have a pulley suspension carried by a substantial pole or mast, with the telescope suspended at about its mid-point, though with increasing length a variety of bracing arrangements, included the use of guy ropes, were added to maintain a straight tube. Both Hooke and Hevelius had telescopes with conventional tubes of 60 or 70 feet in length, but when Hevelius aspired to something even longer—150 feet—he reduced the weight by having only two sides of the "tube," braced along its length by a series of squares with circular apertures. Huygens took a different approach for his instrument of 123 feet: in his "aerial" telescope the object glass was set in a short tube, carried by a ball-and-socket mounting whose elevation could be adjusted on the mast, while the eyepiece was in a similarly short tube linked to the upper tube only by a line. Neither of these expedients produced observational results and the progress of telescopic astronomy may have seemed to be thwarted by the mechanical limits of mounting long instruments.

One feature of these extravagant creations, however, was to stay with the culture of the telescope. There was now a "heroic" dimension to this branch of astronomy. The illustration of the long telescope in the sumptuous *Machina Coelestis* published by Hevelius in 1673 shows it attended by a crowd of assistants and onlookers in the countryside, while a party of well-dressed visitors is received with some formality, and the city of Danzig forms an impressive backdrop. The mounting of the telescope is presented as a significant public occasion. The telescopes of Divini and Campani were pitted in public competitions, or "paragoni" as they were called, sponsored by Italian aristocrats and cardinals, or by the Accademia del Cimento in Florence. In London, Charles II accommodated a 36-foot refractor by Reeve in the garden of Whitehall Palace in the early 1660s.

A more productive development was the introduction of additional lenses. The inverted image of the Keplerian telescope could be "corrected" by means of a third lens, but this had little advantage for astronomers and the poor quality of early lenses would mean additional distortion. One significant improvement, however, was the inclusion of a "field" lens, first by Wiesel in a compound eyepiece having two plano-convex lenses in a small tube. Huygens followed his lead in developing the improved compound eyepiece that bears his name. It had a superior field of view and, incidentally, some correction for chromatic aberration, a problem we have mentioned several times as being present without being clearly identified (as was still the case with Huygens) and to which we should now turn.

The Beginning of the Reflector

It was Isaac Newton who determined that the problem of chromatic aberration arose from the differently colored rays of light having different refractive properties, so that their paths inevitably deviated when they were refracted. Chromatic aberration was the result of these sets of rays forming differently colored images in different planes. Because he held that the degree of dispersion for a given refraction was the same for all types of glass, he announced that there was no possibility of recombining the rays to form a single image. Newton's early interest in telescopes had included attempts to grind aspherical lenses—an example of the practical engagement we have noted from other mathematicians—but he now concluded that, even if such work were to succeed, the chromatic aberration, which had a greater effect than spherical, would remain. He resolved to seek a design based principally on reflection and produced a telescope whose concave primary mirror he ground and polished himself in an alloy of copper and tin.

Newton's was not the first design for a reflecting telescope. One proposal that would prove influential had come from the Scottish mathematician James Gregory, who published a design in 1663, where the incident light reflected from a parabolic concave objective or primary mirror was intercepted by a much smaller elliptical secondary concave and passed through a central hole in the primary into the eyepiece. Gregory was disappointed with attempts to have his telescope made by commercial opticians. Newton succeeded by making his own. His secondary mirror was flat and set at 45 degrees to the optical axis of the primary, so the eyepiece received the rays to one side of the main tube.

News of Newton's telescope reached the Royal Society, and in 1671 he sent them the second example he had made. This was followed by the theory of colors that had led him to concentrate on the reflector, which appeared in the *Philosophical Transactions* in 1672. Hooke also had a number of telescope designs combining mirrors and lenses, though he was motivated not by a theory of light but by the need to shorten telescopes and make them more manageable. In spite of this interest, it would be some time before reflectors became commonly manufactured and used.

Telescopes in Observatories

By the late seventeenth century, the refractor, used as a telescopic sight, was firmly and irrevocably part of the design of instruments for astronomical measurement. Objections from Hevelius to the application of such sights, despite his leading position in the development of the refractor for observing and even for measurement by micrometer, had created a heated public argument with Hooke but did not delay the otherwise general adaption of the telescope to fundamental measurement. The enhanced precision it enabled was a stimulus to the development of improved mountings for meridian instruments in particular, as was the establishment of permanent observatories, linked in turn to the urgent needs of navigation at sea.

The observatories of Paris and Greenwich adopted telescopic sights on their fixed instruments for measurement from the beginning. At Paris there were two 6-foot mural quadrants, while Greenwich acquired a 10-foot mural quadrant designed by Hooke and made by the clockmaker Thomas Tompion, a 7-foot mural arc and a 7-foot equatorial sector. The apparent versatility of an instrument on an equatorial mount, such as the Greenwich sector with its two telescopes, able to work in a wide range of orientations, was effectively eclipsed by the disadvantage of its relative instability in comparison with a meridian instrument, such as a mural arc (typically a quadrant). For fundamental measurement, movement needed to be minimal and confining motion to the meridian, with the telescope pivoted at the center of curvature and moving over the graduated arc, produced the most consistent results. The quadrant, measuring declination, came to be complemented by a second telescope mounted as a "transit instrument" on a horizontal axis that could be corrected for meridian alignment; right ascension would be measured by timing the passage of a body across the central reference wire as the earth rotated.

The first transit instrument of this sort was introduced not in Paris or Greenwich but in Copenhagen. The Danish astronomer Ole Rømer had been an associate of Picard and spent time at the Paris Observatory before becoming a Professor of Astronomy, in charge of the Copenhagen Observatory. Here he installed some remarkably innovative instruments, which included not only a transit instrument but a transit circle. This had a full altitude circle and could take both coordinates: declination, and right ascension. Rømer's instruments were made in the royal armory: this local arrangement worked well for his own needs but his designs would not have the wider impact that could come from their reproduction and sale by a more commercial enterprise that might take orders based on a successful precedent. This would be a feature of the London workshops that would have a profound influence on astronomy in the eighteenth century.

Several factors were important to London's success. The Royal Observatory was focused on an astronomical solution to the longitude problem and this foregrounded all the qualities that were vital to the development of precision astronomy. The Royal Society was happy to admit the leading mathematical instrument makers to fellowship, and to encourage their contributions to the *Philosophical Transactions*. Finally, the Board of Longitude encouraged and rewarded productive innovation. In 1725 the second Astronomer Royal, Edmond Halley, installed an 8-foot mural quadrant he had commissioned from George Graham, an outstanding watch and clockmaker, who had formerly been in partnership with Tompion. Graham followed Tompion's example by also making observatory instruments, but his influence in that respect went far beyond that of his former partner. The instruments he provided for Greenwich set a standard and offered what might be called a "technical vocabulary" that was widely adopted. In addition to Graham's quadrant, Greenwich acquired a transit instrument, an equatorial sector and a zenith sector, intended for the detection of stellar parallax. Each instrument might be regarded as a telescopic sight mounted to facilitate a particular class of measurements, whether right ascension, declination, or zenith distance. They were complemented by a precision clock, Graham's innovative design of astronomical regulator, which was essential to the complete program of work, but not directly part of the story of the telescope.

The whole suite of instruments, the regulator included, was enormously influential in the eighteenth century and this was possible because London had a vigorous, thriving community of mechanical makers. The instruments had been created in this commercial setting, rather than, as in Rømer's case, for example, in an exclusive official workshop, and the London makers were keen to build more such instruments for the growing number of observatories. In 1750 John Bird completed a second quadrant for Greenwich, based on Graham's original, and his design was published by the Board of Longitude. Other London makers who achieved international reputations in precision astronomical instrumentation and orders from abroad were Jonathan Sisson, Jesse Ramsden, and Edward Troughton, taking London's pre-eminence through the century and beyond.

New Telescopes and Audiences in the Eighteenth Century

There were important developments also in telescopes themselves, even if they were not applied to fundamental measurement. In part this was because London had an expanding popular market for instruments as well as a professional one. The earliest occurrence of the word "popular" in the title of a scientific text came in a book on optics that was very much concerned with the telescope. Robert Smith's *Compleat System of Opticks* of 1738, probably the most widely read work on the telescope in the eighteenth century, describes itself as *A Popular, a Mathematical, a Mechanical, and a Philosophical Treatise*. The reflecting telescope had not had much impact on telescope making but Smith played an important part in its change of fortune and it is no coincidence that he saw himself generally working in a Newtonian intellectual tradition. The same was true of others who had tried to forward the reflector and Smith reports the mirror-making techniques of Samuel Molyneux, James Bradley, and John Hadley in the "mechanical" section of his treatise. It is notable that here was an area of tedious mechanical work—the grinding and polishing of metal mirrors—undertaken by prominent mathematicians and natural philosophers. This too was in keeping with the Newtonian precedent.

Gradually, the commercial makers joined in the project, encouraged by the growth in experimental philosophy as a virtuous recreation. Popular lectures and books were prepared by makers and accommodated in fashionable shopping streets, such as Haymarket and the Strand in London, but also by itinerant lecturers in the provinces. The reflecting telescope, especially the Gregorian, was in many ways the ideal instrument for this area of astronomical interest. Its followers were not concerned with the exact measurement of positions but they needed as much light as possible to view the planets and nebulae, as well as the detail of the lunar surface. A division began to grow between the refractor as used by the academic or professional astronomers in observatories, principally as a telescopic sight for measuring positions, and the reflector, used more for viewing the wonders of the heavens.

Yet the most successful commercial maker of reflectors in the mid-century, James Short, who moved his business from his native Edinburgh to London in 1738, managed to straddle both these worlds. He traded only in reflecting telescopes and his reputation for making the finest mirrors was second to none. Yet he was a Fellow of the Royal Society and a serious candidate for Astronomer Royal in 1764. The Newtonian credentials of his very specialized work no doubt helped, as did an endorsement by the mathematician and Newton acolyte Colin Maclaurin, printed in Smith's *Opticks*, but he benefited also from a development that allowed the Gregorian reflector to be used for making measurements. This was an invention of a particular micrometer by the optician John Dollond. An object lens would be cut across a diameter and the two halves mounted in separate cells, so that they could be moved laterally by a micrometer screw and their displacement measured. Two images were presented to the viewer, identical but fainter than the original one, and a range of measurements was possible: for example, the components of a double star might be brought into coincidence, the displacement required to achieve this being a measure of their original separation. The whole lens could be rotated to any orientation to facilitate whatever angle was required. While Dollond saw his invention as a feature of the refractor, in which he specialized, by adding a long focal length divided lens to the mouth of a Gregorian tube, it could be adapted to the reflector.

Dollond's more celebrated contribution to the history of the telescope is the introduction of the achromatic lens as a serious manufacturing and commercial possibility. To what extent his work amounted to an "invention" is still a matter of discussion among historians (Sorrenson 2013; Gee 2014). Dollond began as a London silk weaver but, as optics became a field of more general interest and the subject of a "popular" book, he both began his own experimental investigations and encouraged his son Peter to enter this more promising trade. John was surprised and at first disconcerted to find that Newton's belief that the dispersion of light for a given degree of refraction was the same for all types of glass was not true. He showed that the high-lead "flint" glass had a significantly greater dispersive power than the more common optical "crown" glass, and this opened the possibility for an "achromatic" combination of two lenses that could produce overall deviation without dispersion. Simply put, a negative flint element in a lens combination could undo the dispersive effect of a stronger positive crown element without undoing all the refractive effect on which its focusing properties depended.

Dollond explained his technique in a paper to the Royal Society in 1758 and he was awarded the Copley Medal. In the same year he also took out a patent on his new lenses. He was elected a fellow of the Society in May 1761, but died suddenly in November, and his share in the patent passed to Peter. Having bought out his father's partner in this enterprise, the maker Francis Watkins, Peter's attempts to enforce the patent against other makers (including Watkins) were largely successful. He held that, whatever previous attempts there had been to make achromatic lenses, his father had been successful independently and on the basis of his theoretical command of Newtonian optics. He had not simply been a jobbing optician, but a natural philosopher. The challengers to the patent put forward the case of an amateur optician, the lawyer Chester Moor Hall, whose idea for a combination of crown and flint lenses had become known through the common practice of subcontracting between optical instrument makers and lens grinders. Hall had tried to keep his orders for lens elements separate but had been thwarted by this practice, of which he was, presumably, unaware. In upholding Dollond's patent, the courts were inclined to the view that, even if he was not the first to devise an achromat, he deserved credit for bringing it to public attention and to use.

The Large Reflector: William Herschel and His Successors

The achromatic object glass had a profound effect. It quickly became an essential feature of astronomical refractors, whether as telescopic sights on fixed observatory instruments or as telescopes for more general observing. The reflector was still characteristic of amateur observing, where attempts to discover the best methods of working mirrors remained a challenge, not least because Short never revealed the techniques that had brought him such success, and the measuring instruments of the observatory astronomers, with their telescopic sights, remained a largely separate astronomical culture. The observer who illustrates that division perfectly is William Herschel: making his own reflectors, having first learnt the basic principles from Smith's *Opticks*, and observing in his spare time, he was a classic amateur of the eighteenth century. This made it all the more uncomfortable when he had to be accommodated by the world of academic and professional astronomers.

Herschel was a professional musician from Hanover, who had settled in England in 1762 to pursue a musical career. He became organist of the Octagon Chapel in Bath, which he combined with a range of other musical work, while pursuing his real passion for telescopes and observing. In his relative isolation from other makers and observers, he evolved unique designs for mountings and an individual approach to making mirrors. It should be added, however, that every approach to grinding and polishing had a strong individual character, as it was largely a manual skill that had to be learned through practice and all attempts to codify and communicate the skill in words had failed. Herschel's own attempt to do the same would fail also.

Herschel's ambition for large reflectors was already apparent by early 1781 but not the occasion that would bring his very individual, if not bizarre, program for astronomy to public attention. In March, however, though engaged on his usual priority of stellar observation, he spotted an object that he could see immediately was not stellar: he thought it was a comet but it turned out to be the first planet (Uranus) ever discovered in historical times. Herschel was using a Newtonian with a focal length of 7 foot and an aperture of 6.2 inches—large by contemporary standards, but he also had a 10-foot of 9-inch aperture and a 20-foot of 12-inch aperture (Bennett 1976). Two years later he would introduce a 20-foot with an aperture of 18.7 inches (see Figure 37.1), and eventually his famous, but not very successful, 40-foot, where the diameter of the primary mirror was 4 feet. This sequence is worth rehearsing because it reflects Herschel's cosmological ambition. Ever larger telescopes were the tools he needed to discover the arrangement of the entire cosmos and, because he realized that by looking far into space he was looking into the past, to discover also how the cosmos and its components had evolved over time. This may sound today like the expected tools and intellectual agenda of astronomy, but in the late eighteenth century it could sound deranged and its protagonist, it was said, "fit for Bedlam" (Royal Astronomical Society MSS Herschel W.1/13.W.12). Herschel would doubtless have been dismissed as a crank where it not for the fact that he could not be ignored, having become a Fellow of the Royal Society, been awarded the Copley Medal, been given a royal pension, and become internationally famous, thanks to his extraordinary discovery. Astronomers did not alter their priorities overnight, of course, but once again their discipline would change in parallel with developments in the nature of the telescope.

Herschel's successors in the building of large reflectors were at first amateurs like himself. As the instruments became larger, this was a field for men of means, sufficient leisure and a deep interest in mechanical engineering. Given the importance of engineering ambition and mechanical flair and confidence, it is not surprising that Britain was where the program developed in the nineteenth century. James Nasmyth was a very successful engineer whose early retirement allowed him to concentrate on his astronomy; his friend William Lassell was a wealthy brewer with a strong mechanical bent, who achieved a 4-foot aperture equatorial, with a steam-driven polishing machine; William Parsons, third Earl of Rosse, was a wealthy landowner with a passion for engineering and a large estate in Ireland, where his largest telescope had a primary mirror of 6-foot diameter (Mollan 2014). All of these telescopes had some success, while the first state-sponsored large reflector (and the last to have a speculum metal objective mirror), a 48-inch built by Thomas and Howard Grubb of Dublin for

Figure 37.1   William Herschel's large 20th telescope, from an engraving he published in February 1794. RAS MSS Herschel W 5/15.2. Permission to reproduce courtesy of the Royal Astronomical Society.

the Melbourne Observatory in Australia and delivered in 1869, was a disappointing failure.

The Essential Tool of Modern Astronomy

The development of the telescope was predominantly a Western project and its introduction elsewhere generally came through missionaries or colonial institutions. Even China, so often the cradle of invention, first encountered the telescope through the Jesuits. However, the history of production in the West is a dynamic one of changing commercial as well as technical success.

The distribution of the manufacture of astronomical refractors shifted in the ninetenth century, with Germany, France, and later the US becoming important players.

The "Munich School" of precision instrument making would have a profound influence on the history of the telescope, following the establishment of a "Mathematical-Mechanical Institute" in the monastery of Benediktbeuern, after its secularization in 1803. Georg von Reichenbach, Joseph Liebherr, and Joseph Utzschneider were joined by the apprentice optician Joseph von Fraunhofer in 1806 and the Swiss glassmaker Pierre Louis Guinand in 1809, to create a formidable cluster of talent in both the mechanical and the optical aspects of instrument making, which were optimally combined in the building of observatory instruments. On the basis of painstaking research and practical trials, Fraunhofer achieved unprecedented control over the properties and quality of his glass and the figures of his lenses. It was in the course of this research, in seeking a source of monochromatic rays to determine accurately the refractive properties of his glass, that he invented the spectrometer and discovered the dark lines in the solar spectrum. The heliometer Fraunhofer built for the Königsberg Observatory—a clock-driven equatorial with a divided 6.2-inch achromatic object glass—was an outstanding combination of optical and mechanical skill. It was this instrument, in the hands of Friedrich Bessel in 1838, that achieved the long sought for measurement of a stellar parallax. The German makers improved and rationalized instrument design in a range of ways—finally, for example, making the transit circle the standard meridian instrument—and observatories rushed to install their instruments, just as they had done with the London makers in the previous century.

Later in the century other makers—French and American—were making important instruments, while the major "British" maker was the firm of Grubb in Dublin, founded by Thomas, who had been encouraged in his early work by Lord Rosse, joined and later succeeded by his son Howard. The availability of optical glass improved in Britain when Chance Brothers of Birmingham managed to import Guinand technology and personnel in the mid-century. In France the leading optician of the second half of the century was Paul Gautier, who had the misfortune to end a successful career by building what remains the world's largest-ever refractor—a 1.25-meter lens of nearly 60 meters focal length (alternative lenses were intended for observing and for photography), mounted horizontally and fed by a giant siderostat whose 2-meter mirror weighed seven tons. This extraordinary instrument was part of the "Palais de l'Optique" at the Paris international exhibition of 1900, and its failure as a productive instrument was not the fault of Gautier but of the impossible setting for serious astronomical work. In America Alvan Clark founded a telescope manufactory with his sons in 1846, working glass blanks from, for example, Chance Brothers, and had outstanding and sustained success with large equatorial refractors. It is remarkable that they exported a 30-inch refractor to Pulkova in Russia, following this success with a 36-inch for the Lick Observatory and the famous 40-inch for Yerkes in 1897. This is still the largest working refractor in the world and is generally seen as the apogee of this class of instrument.

Once again it was developments in telescopes that would change the nature of astronomy. The application of cameras and spectrographs to equatorial refractors in the later nineteenth century opened up previously unimagined areas of research and paved the way for an "astrophysical" approach to the heavens. Spectroscopy in particular offered an intellectual liberation for observers. The light from the stars had been understood in the traditional observatory to indicate the position of its source; the large reflector had been used to speculate about structure and evolution; now it seemed that encoded within the light was information on chemical composition. The change was not unlike Galileo's use of the telescope to draw out a new agenda of the natural philosophy of the heavens, and the change in the discipline was just as profound.

The twentieth century was the era of giant equatorial reflectors and the extension of the telescope to new areas of the electromagnetic spectrum. Replacing speculum metal by silver-on-glass, a technique pioneered by Léon Foucault in Paris in the mid-nineteenth century, was making it possible to construct lighter, larger mirrors and the Mount Wilson Observatory had a 60-inch equatorial by George Willis Richie as early as 1908, followed by a 100-inch, also by Richie, in 1917. The driving force behind both these telescopes, the astronomer George Ellery Hale, then led a success-ful project to build a 200-inch telescope at Mount Palomar, with an aluminized mir-ror of Pyrex glass, completed in 1948. These and the many subsequent instruments in the great expansion of design and building in the century were major technical, financial, engineering, and even political projects, and it would be absurd to pretend to deal with them at the close of an article already beyond its permitted size. Instru-ments were combined in clusters and configurations, and the concept of the telescope was extended also to include radio astronomy and instruments sent into space. The meaning constructed for those little tubes at the beginning of the seventeenth cen-tury proved remarkably flexible and resilient: "to look afar off" captured a compelling ambition that has helped give shape, coherence, and continuity to the changing prac-tice of astronomy ever since.

References

Bennett, J. A. 1976. "'On the power of penetrating into space': The telescopes of William Herschel." *Journal for the History of Astronomy*, 7: 75–108.

Brandl, Bernhard R., Remko Stuik, and J. K. Katgert-Merkelijn (eds.) 2010. *400 Years of Astro-nomical Telescopes*. Berlin: Springer.

Gee, Brian. 2014. *Francis Watkins and the Dollond Telescope Patent Controversy*. Aldershot: Ashgate.

King, Henry C. 1979. *The History of the Telescope*. New York: Dover.

Mollan, Charles (ed.) 2014. *William Parsons, 3rd Earl of Rosse*. Manchester: Manchester Uni-versity Press.

Morrison-Low, Alison D., Sven Dupré, Stephen Johnston, and Giorgio Strano (eds.) 2012. *From Earth-Bound to Satellite: Telescopes, Skills and Networks*. Leiden: Brill.

Rosen, Edward. 1947. *The Naming of the Telescope*. New York: H. Schuman

Sorrenson, Richard. 2013. *Perfect Mechanics*. Boston, MA: Docent Press.

Van Helden, Albert, Sven Dupré, Rob van Gent, and Huib Zuidervaart (eds.) 2010. *The Origins of the Telescope*. Amsterdam: KNAW Press.

CHAPTER THIRTY-EIGHT

Prisms, Spectroscopes, Spectrographs, and Gratings

Klaus Hentschel

The Prism—From a Toy to a Tool

Colorful effects induced by light diffraction, such as rainbows, were observed by all cultures, but intense experimentation with this phenomenon had to await the early modern period, when recipes for clear and striation-free glass, the *cristallo* in Murano near Venice, spread through the continent. Giovanni Battista Della Porta in *De iride et colore* (1593), John Peacham in *Gentlemanly Exercises* (1612), René Descartes in his *Dioptice* (1637), and Marcus Marci in his *Thaumantias* (1648) mention "three square cristal prisme," "iris trigonia," or "dreyecktes Glas." The popular term for prisms at the time, "fool's paradise," reflects a still widespread skepticism concerning the reality of these strange colored phenomena, also prevalent in the oldest definitions of "spectre" in the sixteenth century (Oxford English Dictionary, 2nd edition 1989): "a) an apparition, phantom, or ghost, esp. one of a terrifying nature or aspect, b) an unreal object of thought; a phantasm of the brain."

The mental model in all theories of light prior to Newton was that color was an artificial change of natural (white) light, induced by transition through glass or other media. Thus Descartes, for instance, assumed color to be a rotation of aether particles, induced by their hitting the surface of glass, water, or other diffracting media non-vertically, whereas Marci hypothesized some kind of tiring effect of light within its path through the glass.

Isaac Newton's (1642–1727) first prism experiments on the "celebrated Phaenomena of Colours" are recorded in his *Questiones quædam philosophicæ* (1664–1665). He had obtained his first prism at a market fair at Sturbridge in 1664 where it had been sold as a toy.[1] His "new theory of light and colours," published in the *Philosophical Transactions of* 1672, deviated from all former modification theories of light by assuming that "light consists of rays differently refrangible." Colored light is thus in its natural state, whereas white light is nothing but a suitable mixture of different colors. Prisms for him thus became analytic instruments with which these mixtures could be resolved into their ultimate components. Newton's "crucial experiment" showed that

A Companion to the History of Science, First Edition. Edited by Bernard Lightman.
© 2016 John Wiley & Sons Ltd. Published 2020 by John Wiley & Sons Ltd.

once analyzed into their basic color components (of which he assumed exactly seven to serve an analogy with full tones in musical scales), these spectral colors cannot further be decomposed by a second prism but stay unaltered.[2] After initial controversies with Jesuits, his arch-rival Robert Hooke, and a few others, Newton's corpuscular theory of color and light became the standard theory, dominating throughout the eighteenth century until the rise of the wave theory of light in the early nineteenth century.

Dark Lines in the Solar Spectrum: Wollaston and Fraunhofer

The discovery of discontinuous line spectra was one of the most important in early nineteenth-century optics. The novel addition of a narrow slit to the prism apparatus as well as improved prism-glass quality made it possible. Disregarding Thomas Melvill's (1726–1753) observation of a striking yellow line in flame spectra it was the English physician chemist and mineralogist William Hyde Wollaston (1766–1828) who noticed in 1802—while observing a beam of sunlight led through an 1.5-millimeter wide slit and a high-quality flint prism placed at a 3-meter distance from it—that four color ranges of the spectrum (red, yellow-green, blue, and violet) were separated by dark zones or "divisions" which he also called "lines." Interestingly Wollaston interpreted these dark zones as noticeable divisions between the variously colored regions of the spectrum. But it was only with much wishful thinking that these dark zones could be identified with the color boundaries. The rediscovery of the dark lines by Josef Fraunhofer (1787–1826), a young optician in the Munich Mechanical Institute founded in 1802, initiated a more directed investigation of this problem. In response to Napoleon's continental embargo of 1806 against the British Isles, Fraunhofer's firm had taken up the production of scientific instruments to help meet the growing demand mainly of astronomical and optical instruments (Jackson 2000). Its best-selling geodetic instruments for angular measurements known as theodolites were designed by Georg Friedrich von Reichenbach (1772–1826), who had lifted the knowhow from Ramsden's workshops in England. The Munich theodolites had a superb circular graduation into units of 10 feet and which, with the aid of a microscopic vernier, could be read to an accuracy of within 10 inches. Such precision instruments were indispensable for the survey of Bavaria ordered by the Elector Maximilian IV in 1801—initially in collaboration with the French *Bureau topographique*.[3]

After the secularization of the Bavarian monasteries in 1803, the Munich Institute's optics division was moved to the former Benediktbeuern cloister where from 1806 Fraunhofer worked principally for Joseph von Utzschneider (1763–1840) on improvements in glass and lens production. He was so successful in this that within three years he became head of the optical workshop and in 1814 became a partner. Fraunhofer was particularly interested in the dark lines of the solar spectrum because they always appeared at the same places (i.e., at a specific shade in the color scale of the continuous spectrum). Hence he could use them as markers for a specific wavelength (i.e., for monochromatic light) in his study of the "refractive power"—in modern terminology the refractive index n—of various types of glass, which was necessary for the production of achromatic lenses and highly dispersive prisms. In measuring the angles of incidence and reflection for individual demarcation lines, Fraunhofer obtained a clear picture of the change in "refractive power" n with the color of the light (that is its dispersion in relation to the wavelength λ: $dn/d\lambda$), which he measured to five

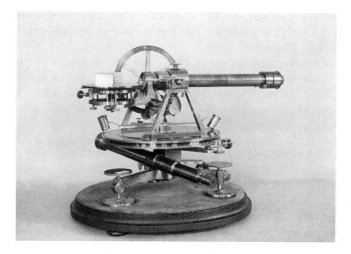

Figure 38.1 The first prism spectrometer by Fraunhofer: a modified theodolite with which the Fraunhofer lines were measured in 1814/1815. The lower disk attached to the viewing telescope has an angular scale of 10 feet which can be read off to an accuracy of 10 inches using the two magnifying lenses; above it is the rotatable fixture for the 60° single prism (top left) and a theodolite telescope with a small 25mm aperture (top right) whose position can be adjusted by means of the micrometer screw (back) to precisely defined degrees relative to the incident beam of light. Photograph of the original apparatus at the Deutsches Museum, Munich.

decimal places. In the course of these investigations he also developed the prototype spectroscope. Its prisms disperse the incident light and the telescope magnifies the resulting spectrum for subsequent observation (see Figure 38.1). Both were mounted on a theodolite, a rotatable metal disk with a precision angular scale.

The practical importance of the dark lines as markers for the optical industry motivated Fraunhofer to record 574 of them now known as "Fraunhofer lines" and in 1814 to publish the first accurate drawing true to scale of the positions of a selection of about 350 distributed over the entire solar spectrum. He labeled them alphabetically from the red to the violet. Above this spectral map he added a curve indicating the estimated intensity as a function of color, which showed that to the human eye the highest intensity is in the yellow-green region in the middle of the spectrum decreasing steadily towards the red and violet ends. For this compilation using a semi-transparent plane mirror set at a 45-degree angle he superimposed onto the eyepiece the comparison light from a flame whose distance from the theodolite telescope could be varied. To gauge the relative intensity he determined the flame's distance at which the spectra from both the flame and the Sun appeared of approximately equal intensity and took the inverse square of this distance. A closer examination of the relative distances between these Fraunhofer lines showed that they were dependent on the type of glass used. A flint prism, for example, produces less red but more blue and violet than one made out of crown glass and the spectrum it generates is about twice as wide—even broader spectra could be obtained from hollow prisms filled with liquid carbon disulphide or cassia oil. Thus the dark lines were no longer merely a convenient means of defining a particular color of light. They themselves came increasingly into the focus of systematic study to which Fraunhofer also contributed, especially after his workshop was removed back to Munich in 1819.[4]

In Fraunhofer's pioneering article of 1814, and later in his short "addendum" to a paper of 1823, we find comments on his prismatic analyses of light emitted from other objects. These included Venus, some fixed stars of the first magnitude, as well as the electric arc. He noticed, for instance, that unlike the solar spectrum with its dark lines the spectrum produced from the flame of a candle contained bright lines and specifically that "in the orange a bright line distinguishing itself from the rest of the spectrum is duplicated and is situated at the spot where the double line D lies in the spectrum of sunlight" (Fraunhofer 1888, 140). He also dispersed light generated from an electricity engine and moonlight which revealed "in the lighter colors the same fixed lines as in sunlight likewise in exactly the same place" (Fraunhofer 1888, 141). But this finding was not pursued further until 1849 when the French physicist and astronomer Jean Bernard Léon Foucault (1819–1868) made the same observation while spectroscopically analyzing the bright yellowish light of an electric arc. To check this coincidence Foucault superimposed the solar spectrum onto that of the electric arc and to his astonishment found that the Fraunhofer line D became even darker than without the addition of the arc spectrum. This was the first suggestion of a close connection between the emission and absorptive properties of a luminous gas, albeit demonstrated with a single spectral line and appearing in an obscure publication. William Allen Miller (1817–1870), a chemistry professor in London, made a similar discovery in 1845 when he noticed the intensification of various Fraunhofer lines by an oil flame spectrum; his observations of the flame spectra of calcium copper barium chloride and strontium nitrate were published in the same year.[5]

The Emergence of Spectrum Analysis: Swan, Bunsen, and Kirchhoff

Miller's and Foucault's isolated references to a possible link between spectral lines and the chemical composition of luminous bodies in the 1840s did not go completely unheeded. The chemist Robert Bunsen's (1811–1899) improvement of the gas burner, which he developed together with the university mechanic Peter Desaga (1812–c.1879), and his former student then collaborator in his photochemical investigations Henry Enfield Roscoe (1833–1915) in the years 1853/1854, was decisive. Unlike the coal-gas and oil flames then in common use, the gas–air mixture of what became known as the Bunsen burner could generate much higher temperatures of over 1800°C without producing any unwanted carbon deposits (i.e., it was smokeless). Furthermore it had a much fainter color of its own so that the specific color characteristics of flames and spectra could be much better analyzed with this burner. The Scottish chemist William Swan (1818–1894) then undertook to examine the "prismatic spectra of flames" combusting a particular element or class of substance going from the premise that the occurrence of specific lines or line groups was causally related to the presence of a specific substance or substance class. Just six years later this physical connection postulated by Swan was found.

Since the mid-1850s Bunsen himself also had been trying to conclude the chemical composition of vaporized salts from their color characteristics using his new type of burner. Together with his British student Rowlandson Cartmell (1824–1888),

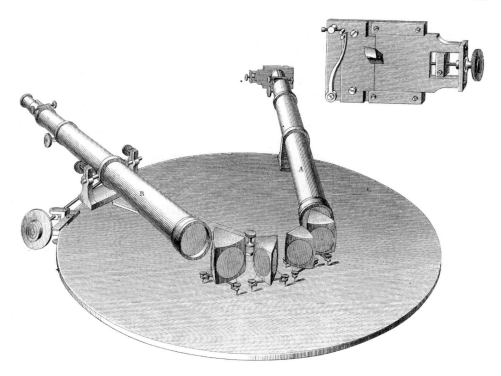

Figure 38.2 A multiprism spectroscope constructed by Steinheil for Kirchhoff which with the aid of a deviating prism at the slit (top right) permits the simultaneous observation of spectra from two flames to which is added a superimposed scale illuminated by a separate flame to permit an exact determination of the line intervals. Fig. from Gustav Robert Kirchhoff: *Untersuchungen über das Sonnenspectrum und die Spectren der chemischen Elemente*, *Abhandlungen der kgl. Preußischen Akademie der Wissenschaften zu Berlin* 1861, pl. I.

Bunsen studied the effect of various filters on the flame colors in the hope that they would help distinguish between the often similar flames of some elements. But his results were far from satisfactory not least because, as we now know, owing to the production methods of colored glass the range of transmittance of the filters was too large to be sufficiently selective. His colleague and friend from Heidelberg, the physicist Gustav Robert Kirchhoff (1824–1887), suggested using prisms to decompose the light into its spectral components. Kirchhoff was acquainted with the spectroscope from his training in physics; for the chemist it was an unfamiliar instrument. As a result of this advice Bunsen's original color study was transformed into a search for characteristic spectral lines of specific chemical elements. Being more than commonly well versed in inorganic analysis using experimental arrangements (as in Figure 38.2), Bunsen soon identified unique emission lines for a whole series of the chemical elements and their compounds in his Bunsen-burner flame.[6]

The clever way in which the light from two different flames was guided along the same beam path along with a superimposed line scale made it possible to register the positions of individual spectral lines very accurately. Their first publication on "chemical analysis by spectral observations" included a plate depicting the characteristic spectral lines of a whole range of alkaline metals along with some other elements. This plate was later reproduced in countless books, articles, and lectures by chemists and physicists alike and disseminated in laboratories, observatories, and even schools (Figure 38.2). In order to record the spectra of different elements more precisely in 1860/1861 Bunsen and Kirchhoff designed a more sophisticated multi-prism spectroscope which Karl August Steinheil (1801–1870) then manufactured in a small series (Hentschel 2002a, 124–8; Hennig 2003, 27–30).

Multifacetted Applications of Spectroscopy

Bunsen revealed the great potential of qualitative spectral analysis when he demonstrated that even microscopic amounts of sodium (0.0000003 milligram) were sufficient to detect the characteristic yellow D lines in its burner-flame spectrum. In March 1860 Bunsen noticed conspicuous blue and red emission lines in certain spectra. He distilled a more concentrated sample from 44,000 liters of Dürkheim mineral water and discovered a new chemical element: caesium after the Latin word *caesia* for the blue of the sky. In the following year, using the same method which required not only the sharp eye of an experienced spectroscopist but also the analytical knowledge of a good chemist, he isolated from about 150 kilograms of lepidolite which had been mined in Saxony, the new element rubidium, Latin *rubidus* for dark red. Further discoveries by similar methods soon followed, such as the elements thallium by William Crookes already in 1861, indium in 1863, and gallium in 1875; but this deceptively simple method also produced a number of premature reports based upon the erroneous reassignment of spectral lines associated with already known elements.

The success of spectrum analysis in 1859/1860 was so dramatic that it was not long before the analytic technique appeared in the chemist's and pharmacist's laboratory, the astronomer's observatory, the physician's hospital, the judge's courtroom, the engineer's bessemer steel works, and within a few years also in the teacher's classroom lab. Apart from its widespread use in qualitative analysis the new technique was certainly furthered by the great fascination expressed in the many introductory texts written after 1860 to all kinds of audiences (see, e.g., Roscoe 1869, a lecture series for the London *Society of Apothecaries*; Schellen 1870; or Secchi 1870).

The myriad new applications were astonishing, indeed. In metallurgy, for instance, the superviser could now easily identify the exact moment of decarbonization of molten steel just before it loses its fluidity using a simple direct vision pocket spectroscope. Previously considerable skill had been required to "see" this from the subtle change in color of the red-hot metal inside the vessel. It was now sufficient to be able to discriminate between the many iron lines and CO-bands in the spectrum—the latter would vanish just at the completion of the decarbonization process. Similarly, chemical analysis of various gases and absorbing fluids, such as blood, for instance, now simply required the ability to distinguish the various patterns of line and band spectra. The mapping of all kinds of spectra under controlled laboratory conditions and their systematic comparison with spectra of unknown substances or substances under unknown physical conditions became a major preoccupation for experimentalists at various science departments for the remainder of the nineteenth century (Kayser 1900; Baly 1929; Hentschel 2002a). Various improvements in spectroscope design such as the introduction of a collimator (1839), of multi-prism chains (1860s) automatically set at the angle of minimum deviation (Browning 1870), or the so-called Littrow-mounting in which the same prism chain was used twice, redirecting the light back through it with the aid of a mirror (1863), increased the spectral resolution obtainable for precision spectroscopy and for the analysis of spectral fine structure.

Solar Physics and Chemistry: The Origins of Astrophysics

From its role in the discovery of the new elements and from its many chemical applications spectral analysis won canonical status. But it also gave Bunsen and Kirchhoff insights of quite a different nature. While vaporizing chemicals in the burner flame Kirchhoff noticed bright spectral lines in a dark background. How did these lines relate to the dark lines dividing the otherwise continuous solar spectrum? Kirchhoff gives the following description of the yellow sodium D doublet:

> I formed a solar spectrum by projection and allowed the solar rays concerned before they fell on the slit to pass through a powerful salt-flame. If the sunlight were sufficiently reduced there appeared in place of the two dark lines D two bright lines; if on the other hand its intensity surpassed a certain limit the two dark lines D showed themselves in much greater distinctness than without the employment of the salt-flame (Stokes 1860, 195).

This experiment showed the convertability of the bright lines of the sodium vapor into dark ones similar to those found in the Fraunhofer spectrum of the Sun. Kirchhoff interpreted the dark bands as absorption lines and the bright ones as the emission lines of luminous gases. So both spectral types were seen as the reversible negative image of each other which he praised as an admirably simple key to finding out the chemical composition of the Sun. In a letter to Roscoe written before February 1860 he states:

> The sun possesses an incandescent gaseous atmosphere which surrounds a solid nucleus having a still higher temperature. If we could see the spectrum of the solar atmosphere we should see in it the bright bands characteristic of the metals contained in the atmosphere and from the presence of these lines should infer that of these various metals. The more intense luminosity of the sun's solid body however does not permit the spectrum of its atmosphere to appear; it *reverses* it according to the proposition I have announced; so that instead of the bright lines which the spectrum of the atmosphere by itself would show dark lines are produced. Thus we do not see the spectrum of the solar atmosphere but we see a negative image of it. This however serves equally well to determine with certainty the presence of those metals which occur in the sun's atmosphere. For this purpose we only require to possess an accurate knowledge of the solar spectrum and of the spectra of the various metals. (Kirchhoff to Roscoe, published in Roscoe 1869, 185–6, original emphasis).

The coincidence of the dark and bright lines was verified in a growing number of chemical elements such as iron with over 60 coincident lines. Kirchhoff calculated the probability for all these coincidences to occur simultaneously purely by chance to be one to a quintillion! Thus it provided at once not only a plausible explanation for the formation of the hitherto completely puzzling Fraunhofer spectrum but also a new model of the Sun which left open, though, why the solar core should remain solid and luminous despite the evidently extreme high temperatures. Upon the discovery that under certain conditions gases also can emit continuous spectra this latter assumption of Kirchhoff's could later be dropped.

The prospect of finding out the chemical composition of the Sun from the presence or absence of specific lines in the solar spectrum by comparison against emission spectra generated in the laboratory inspired Kirchhoff in 1860 to compile a detailed series of plates of the solar spectrum. In observing the spectra with a three-prism spectrometer built by Steinheil according to his own specifications Kirchhoff strained his eyes so much that the second half of the series had to be completed by one of students in 1861. Amateurs took up the chemical analysis of the composition of astronomical objects by the physical means of prismatic dispersion while the traditional astronomers kept conspicuously aloof in the early stages of the new discipline which Johann Karl Friedrich Zöllner (1834–1882) from Leipzig had coined "astrophysics" in 1865. Unaffiliated hobby scientists soon realized that they could quickly arrive at important results with the relatively modest means available to them through the innovative use of prism spectroscope and electric arc (cf. Hufbauer 1991 and Lankford 1997). After the invention of photography, amateur astronomers, daguerreotypists, and photographers such as John William Draper (1811–1882) and his son Henry Draper (1837–1882) were the first to obtain high-quality photographs of solar (1843), planetary, and stellar spectra (1872)—spectro*scopes* were thus transformed into spectro*graphs* that also allowed them to examine wavelength ranges beyond human vision, such as the infrared or ultraviolet (Hentschel 2002a; Hearnshaw 2009, chapter 6).

Joseph Norman Lockyer (1836–1920) was one of the first to examine the spectrum of solar spots and prominences; and, parallel with Pierre Jules César Janssen (1824–1907), he developed the procedure to observe the solar prominences in daytime thus avoiding the necessity to wait for a solar eclipse. These observations in particular made Lockyer aware of the possibility that some lines were shifted. The tremendous gaseous clouds expelled from the solar surface at high speeds of a few hundred kilometers per second were found to produce spectroscopically significant Doppler shifts. In the solar chromosphere—that is, according to Lockyer in the narrow region in the solar atmosphere between the hot interior and the absorbing outer region—he found line shifts that interpreted as Doppler shifts indicated vertical currents of up to 1400 km/sec and horizontal or spiraling currents of up to 300 km/sec.[7]

The first application of the Doppler principle to extraterrestrial light sources was made in 1868 with the observation of red and violet shifts in the spectra of some bright stars for instance the Fraunhofer F (= H_β) line of Sirius relative to the hydrogen spectrum from a Geissler tube (on the history of stellar spectroscopy see, for example, Hearnshaw 1986, chapter 4ff.). The measurements of the gentleman astronomer William Huggins (1824–1910) and his wife Margaret at their private observatory built in 1856 on Tulse Hill, then on the outskirts of London, yielded relative velocities of individual astronomical objects such as Sirius of approximately 40 km/s. Huggins' findings contradicted the expectations of the time about the magnitude of astronomical velocities; furthermore the Jesuit priest Pietro Angelo Secchi (1818–1878) had just recently announced at the Paris *Académie des Sciences* the negative result of his search for Doppler shifts in stellar spectra (Figure 38.3).[8]

Figure 38.3 Huggins' stellar spectroscope with two flint prisms and cylindrical lens (A) which can be shifted up or down along the tube TT in dashed outline. Heinrich Schellen. 1871 2nd edition. *Die Spektralanalyse in ihrer Anwendung auf die Stoffe der Erde und die Natur der Himmelskörper*. Braunschweig: Westermann, 446.

The radial velocities initially determined for these stellar objects were later corrected, sometimes quite drastically, with a change in sign—above all by Huggins and Secchi but soon also by other astrophysicists specializing in stellar spectroscopy such as the director of the *Potsdam Astrophysikalisches Observatorium*, Hermann Carl Vogel (1841–1907), and the director of the *Harvard College Observatory*, Edward Charles Pickering (1846–1919).

A reliable determination of the velocities of stars relative to the Earth to an accuracy of about 3 km/s became possible in the 1870s when Henry Draper in New York, William Huggins in England, and, from 1887, Hermann Carl Vogel in Potsdam could photograph the faint stellar spectra with the aid of specially designed spectrographs and improved dry-plate emulsions. In 1889 E. C. Pickering, and independently from him H. C. Vogel, first discovered the existence of binary stars spectroscopically via the Doppler effect. These objects either exhibited temporarily duplicated lines or periodically changing line shifts which could be attributed to traveling velocities of a few 100 km/s for the binary-star systems whose components intermittently approach or recede from view (cf., e.g., Hearnshaw 1986, 87ff. for commentary and references). Further technological improvements along with skillful fundraising efforts, particularly by the American astrophysicist George Ellery Hale (1868–1938), who inaugurated as many as three astrophysical observatories, then triggered the institutionalization of astrophysics at the close of the nineteenth century, which continued to flourish, especially in the United States.[9]

E. C. Pickering became a pioneer in the technique of objective prism spectrographs, which photographed stellar fields with a thin prism of 5° to 15° in front of the objective. All stellar images were thus transformed into thin stripes of spectra—one photograph could thus register hundreds of stellar spectra. More than 225,000 of them were catalogued and classified in the so-called *Henry Draper Catalogue* (published 1918–1924, with extensions of another 133,000 stellar spectra published until 1941) by a large group of lowly paid, but highly qualified female specialists, called, tongue in cheek, "Pickering's harem" (Lankford 1997; Hearnshaw 2009, chapter 5).

The emphasis of nineteenth-century spectroscopical research lay on the empirical comprehension of the multiplicity of atomic and molecular spectra. With the systematic use of the correspondence between emission and absorption spectra in spectrum analysis the innumerable and diverse applications soon followed. The decoding of the strange spectral patterns in terms of atomic and molecular physics had to await the twentieth century.[10]

Diffraction Gratings and Precision Spectroscopy

The refraction of prisms very much depends upon the chemical composition and homogeneity of the material of which they are composed: glass has high absorption coefficients especially in the violet and ultraviolet; furthermore, prisms also have irregularities in their index of refraction and dispersion; finally, they are sensitive to temperature. Gratings, on the other hand, are based upon the diffraction of light, stemming from a constructive interference of light waves with wavelengths λ, emitting from the parallel lines of the grating separated by a distance ε and at the angle θ, thus having a constant phase difference of $\sin \theta = n \cdot \lambda / \varepsilon$, with n a low integer number called the order of the spectrum. So, in contrast to prisms, the quality of a diffraction grating does not so much depend upon its material constitution but rather upon the regularity of its ruled surface, which was usually produced by the scratching of a metal coated glass plate or a metallic surface with a carefully chosen diamond point, because a stable interference can only occur if the formula above is valid for many wave fronts. In case of irregularities in the distances between the lines of the grating, the angles θ vary along the surface of the grating, and its optical image becomes diffuse, or worse, so-called ghosts appear. Progress in spectroscopy in the second half of the nineteenth century can mostly be attributed to the improvement in the manufacture of these diffraction gratings.[11]

As contemporary comparisons of prismatic and diffraction spectra revealed, diffraction gratings were clearly superior as regards their overall dispersive power, but this advantage was somehow counterbalanced by the greater loss of light and high astigmatism, making them particularly unsuitable for stellar spectroscopy for which other types such as coudé and echelle spectrographs were developed in the second half of the twentieth century (Hearnshaw 2009, chapters 3, 7–8).

Table 38.1 Development of diffraction gratings in the nineteenth century. Bracketed names stand for observers who did not manufacture their own spectroscopes but obtained them from other instrument makers. Units for the size of gratings and line distances, ε, are converted into centimeters.

Name	Year	Instrument (type of grating)	Size (cm)	Number of grooves	ε (cm)
Fraunhofer	1821	First diffraction grating with grooves in a gold-coated glass plate		3,600	0.0048
Nobert	1851	Diamond grooves in glass	c.2.5	c.1000	0.0025
(Ditscheiner)	1864	Grating by Nobert	1.38	3000	0.00046
(Ångström)	1864	Grating by Nobert	c.2	c.4500	0.00046
Rutherfurd	1868	Grating by Nobert in glass / metal	c.5	c.20,000	0.00025
Rutherfurd	1881	Grating by Nobert in metal (for Mendenhall)	4.4	c.30,000	0.00015
Rowland	1882	Concave grating	c.7.2	c.45,000	0.00017
Rowland	1896	Concave grating	c.14.5	c.110,000	0.00015
Michelson	1907	Concave grating	22 × 11	c.110,000	0.00010
Jacomini		Plane grating	10 × 12		
Jacomini		Plane grating	25 × 25		

David Rittenhouse's (1732–1796) primitive grating of 1785 as well as Fraunhofer's first grating of 1821 were simply made of thin wires, stretched parallel to each other in the notches of a long screw.[12] Later, Fraunhofer in Munich, by that time head of the best optical instrument shop in the world, was the first to employ the technique of scratching parallel lines with a diamond point into a glass surface coated with a very thin layer of gold or silver. This technique was then considerably improved during the nineteenth century and extended to the scratching of grooves into metal surfaces, where the diffraction effect was, of course, not used in transmission, but in the reflected part of the light. The disadvantage of the lower intensity of the reflected light produced by the metal surfaces as compared with coated glass surfaces was more than compensated for by the much higher regularity obtained in their manufacture. In so-called blazed gratings, the grooves are very regularly shaped and higher angles of incidence are chosen so that only one or two orders of diffraction get most of the light intensity. Thomas Young (1773–1829) was the first to use a simple diffraction grating with around 500 lines per inch to measure solar wavelengths and thereby also to demonstrate the wave nature of light in 1801.

Great progress in the ruling of gratings was achieved by Friedrich Adolf Nobert (1806–1907) in Pomerania, who had ruled small plates mainly for the purpose of testing the resolution of microscopes ranging from nearly 450 lines per millimeter in 1851 to a line distance, ε, of 0.11 μm in 1873.

Thus, the early development of ruling engines is closely related to the history of microscopy: the finest test plates made by Nobert in the 1870s were no more resolvable by any light microscope. They have been analyzed with electron microscopes revealing their quality as well as their remaining ruling defects (Turner and Bradbury 1966).

Although Nobert's main motivation was thus not so much to produce diffraction gratings but rather test plates for the examination of the ultimate resolution of microscopes, he was the main source of well-ruled diffraction gratings for two decades from 1850 on, until Rutherfurd and finally Rowland superseded him as producers of gratings for spectroscopic precision measurements. In comparison to these defects in Nobert's gratings mostly ruled in glass surfaces, Rowland's gratings ruled on speculum metal.

Because the highest resolution of a grating $\lambda/\Delta\lambda$ is equal to the product of total number of lines # and the order of interference n, and since Rowland's gratings were as accurate as to allow observation in the third and even fourth order of the spectrum, his gratings allowed resolutions of up to 400,000, which was then considered the practical limit of resolution for precision spectroscopy. While prism spectrographs and the earlier gratings could resolve maximally two spectral lines that were as close as 1/40th the distance of the two sodium D-lines to each other, Rowland's gratings could resolve lines of 1/100th this distance. With one of Rowland's best gratings, his research assistance Lewis E. Jewell discovered the redshift of spectral lines in the solar spectrum around 1890 (Hentschel 1993). The instrument revolution in spectroscopy achieved by these precision instruments was so immense that in his handbook Edward Charles Cyrill Baly (1871–1948) counted it as "one of the greatest inventions ever made in spectroscopy."[13]

After Rowland's death in 1901, Baltimore remained the center for diffraction grating production, but in the 1930s, George Harrison (1898–1978) at MIT, and in the 1940s, commercial producers of diffraction gratings such as Bausch and Lomb took over the market. Today, industrially produced transmission gratings on plastic foils or glass copied from ultraprecise masters are available very cheaply.[14] Spectroscopy with prism, spectrographs and gratings is still a crucial research technology in many branches of modern science, from astrophysics to analytic chemistry and materials science.

Endnotes

1   See Gjertsen 1986, 507–8 for a compilation of all prisms traceable from Newton's manuscripts, papers, and correspondence.

2   See Newton 1672; [1979]; and 1984 on Newton's theory of light, furthermore Schaffer 1989 and Shapiro 1996 on the controversies around Newton's crucial experiment which several continental experimenters initially could not replicate, partly for lack of high-quality prisms, partly for lack of skill in experimentation.

3   On this context cf. Jackson (2000).

4   cf. Rohr 1929 and Jackson 2000 on Fraunhofer's context, and Hentschel 2002a, 33–6 on Fraunhofer's mapping of the solar spectrum.

5   On the study of bright lines in the spectra of flames, arcs and sparks before 1859 see Hentschel 2002a, 36–47 and references given there.

6   On the (pre)history of spectrum analysis see McGucken 1969 and Hentschel 2002a, chapter 1; cf. Bunsen 1904, vol. 3 and Hennig 2003 for a detailed description of the hollow-prism spectroscope used by Bunsen and Kirchhoff in 1859.

7   On Lockyer's life and work in spectroscopy see Meadows 1972.

8 On the work of Sir and Lady Huggins see Huggins 1909; Hearnshaw 1986; Becker 2011.

9 See Wright 1966 on Hale, and Lankford 1997 for social historical data on astrophysics in the US, Hufbauer 1991 and Hentschel 1998 on Europe, Hearnshaw 2009, 113ff. on the design and function of Hale's tower telescopes, and Hentschel 1997 on the Einstein tower in Potsdam.

10 On the atomic theory of series and band spectra in the twentieth century that also range into the infrared and ultraviolet, cf. Baly 1929 and Brand 1995.

11 For an overview of the development of technology for the ruling of gratings as well as scratching into surfaces for other purposes before Rowland, see Warner 1986, 125ff.

12 Up to 325 wires per inch in Fraunhofer's case; cf. Fraunhofer 1888 and Rohr 1929.

13 Baly 1929, vol. 1, p. 28; cf. also Sweetnam 2000 on Rowland and his collaborators.

14 On the progress in diffraction grating production after Rowland see Harrison 1973; Palmer 2005; Hearnshaw 2009, chapter 2.

References

Ames, J. S. 1900. *Prismatic and Diffraction Spectra*. Edited and translated by J. S. Ames. New York: Harper. Reprinted in J.S. Ames, 1981. *The Wave Theory, Light and Spectra*. New York: Arno.

Baly, E. C. C. 1st edition 1905; 3rd edition 1929 in 4 vols. *Spectroscopy*, London: Longmans, Green & Co.

Becker, Barbara Jean. 2011. *Unravelling Starlight: William and Margaret Huggins and the Rise of the New Astronomy*. Cambridge: Cambridge University Press.

Brand, John C. D. 1995. *Lines of Light. The Sources of Dispersive Spectroscopy 1800–1930*. Luxembourg: Gordon & Breach.

Bud, Robert, and Deborah Jean Warner (eds.) 1998. *Instruments of Science: An Historical Encyclopedia*. New York: Garland.

Bunsen, Robert Wilhelm. 1904. *Gesammelte Abhandlungen*. Edited by Wilhelm Ostwald and Max Bodenstein, 3 vols. (esp. vol. 3 on spectroscopy). Leipzig: Engelmann.

Fraunhofer, Joseph von. 1888. *Gesammelte Schriften*. Munich: Akademie in Kommission Franz.

Gjertsen, Derek. 1986. *The Newton Handbook*. London: Routledge & Kegan Paul.

Harrison, George R. 1973. "The diffraction grating: An opinionated appraisal." *Applied Optics*, 12: 2039–49. DOI: 10.1364/AO.12.0020392.

Hearnshaw, John B. 1986. *The Analysis of Starlight: 150 Years of Astronomical Spectroscopy*. Cambridge: Cambridge University Press.

Hearnshaw, John B. 2009. *Astronomical Spectrographs and their History*. Cambridge: Cambridge University Press.

Hennig, Jochen. 2003. *Der Spektralapparat Kirchhoffs und Bunsens*. Berlin: GNT-Verlag.

Henry, Richard C., and David H. DeVorkin. 1986. "Henry Rowland and astronomical spectroscopy: Celebration of the 100th anniversary of Henry Rowland's introduction of the concave diffraction grating." *Vistas in Astronomy*, 29, No. 2: 119–236.

Hentschel, Klaus. 1993. "The discovery of the redshift of solar spectral lines by Rowland and Jewell in Baltimore around 1890." *Historical Studies in the Physical and Biological Sciences*, 23, No. 2: 219–77.

Hentschel, Klaus. 1997. *The Einstein Tower: An Intertexture of Dynamic Construction, Relativity Theory, and Astronomy*. Stanford, CA: Stanford University Press.

Hentschel, Klaus. 1998. *Zum Zusammenspiel von Instrument, Experiment und Theorie*. Hamburg: Verlag Dr. Kovač.

Hentschel, Klaus. 2002a. *Mapping the Spectrum: Techniques of Representation in Research and Teaching*. Oxford: Oxford University Press.

Hentschel, Klaus. 2002b. "Why not one more imponderable? John William Draper and his 'Tithonic' rays". *Foundations of Chemistry*, 4, No. 1: 5–59.

Hufbauer, Karl. 1991. *Exploring the Sun: Solar Science since Galileo*. Baltimore: Johns Hopkins University Press.

Huggins, William. 1909. *The Scientific Papers*, edited by Lady Margaret Huggins. London: Wesley.

Jackson, Myles. 2000. *Spectrum of Belief: Joseph von Fraunhofer and the Craft of Precision Optics*. Cambridge, MA: MIT Press.

Kayser, Heinrich. 1900f. *Handbuch der Spektroscopie*, esp. vol. 1. Leipzig: Hirzel.

Lankford, John. 1997. *American Astronomy: Community, Careers, and Power, 1859–1940*. Chicago: University of Chicago Press.

McGucken, William. 1969. *Nineteenth-Century Spectroscopy: Development of the Understanding of Spectra 1802–1897*. Baltimore: Johns Hopkins University Press.

Meadows, Arthur Jack. 1972. *Science and Controversy: A Biography of Sir Norman Lockyer*. London: Macmillan.

Newton, Isaac. 1672. "New theory of light and colours." *Philosophical Transactions of the Royal Society*, 7: 5004–7. Reprinted with commentary in I. B. Cohen and Robert Schofield (eds.) 1958. *Isaac Newton's Papers and Letters on Natural Philosophy and Related Documents*. Cambridge, MA: Harvard University Press.

Newton, Isaac. 1707 [1979]. *Opticks* (1st edition. 1704), reprint of the 4th edition. London 1730. New York: Dover.

Newton, Isaac. 1984. *Optical Papers*. Edited by Alan Shapiro. Cambridge: Cambridge University Press.

Palmer, Christopher. 2005. *Diffraction Grating Handbook*. 6th edition. Rochester, NY: Newport.

Rohr, Moritz von. 1929. *Fraunhofers Leben, Leistung und Wirksamkeit*. Leipzig: Akademische Verlagsgesellschaft.

Roscoe, Henry E. 1869. *Spectrum Analysis. Six Lectures Delivered in 1868*. London: Macmillan.

Rowland, H. A. 1902. *The Physical Papers of Henry August Rowland*. Baltimore: Johns Hopkins University Press.

Schaffer, Simon. 1989. "Glass works." In *The Uses of Experiment*, edited by David Gooding, Trevor Pinch, and Simon Schaffer, 64–104. Cambridge: Cambridge University Press.

Schellen, Heinrich. 1870 1st edition; 1871 2nd edition. *Die Spektralanalyse in ihrer Anwendung auf die Stoffe der Erde und die Natur der Himmelskörper*. Braunschweig: Westermann. (English translation by C. Lasell, 1872. *Spectrum Analysis in its Applications to Terrestrial Substances, and the Physical Constitution of the Heavenly Bodies*. London: Longman).

Secchi, Angelo. 1870. *Le Soleil*. Paris: Gauthier-Villars.

Shapiro, Alan. 1996. "The gradual acceptance of Newton's theory of light and color, 1672–1722." *Perspectives on Science*, 4: 59–140.

Stokes, George Gabriel. 1860. "On the simultaneous emission and absorption of rays of the same definite refrangibility." *Philosophical Magazine (4th ser.)*, 19: 193–7.

Sweetnam, George Kean. 2000. *The Command of Light: Rowland's School of Physics and the Spectrum*. Philadelphia, PA: American Philosophical Society.

Turner, Gérard L'Estrange, and S. Bradbury. 1966. "An electron microscopical examination of Nobert's finest test-plate of twenty bands." *Journal of the Royal Microscopical Society*, 85: 435–47.

Warner, Deborah. 1986. "Rowland's gratings: Contemporary technology." *Vistas in Astronomy*, 29: 125–30.

Wright, Helen. 1966. *Explorer of the Universe: A Biography of George Ellery Hale*. New York: Dutton.

CHAPTER THIRTY-NINE

Diagrams

CHARLOTTE BIGG[1]

Archeologists' sketches, botanical drawings, photographs of particle tracks or brain scans belong to a family of scientific images that purport to represent reality or some aspect of it. These images most obviously raise questions about their evidentiary status, the kind of representation of nature they offer, and consequently the mediations operated by technologies of visualization such as those discussed in Chapters 35–38. They have given rise to a rich literature in the sociology and history of science, which has profited from fertile interactions with other fields such as the history of art or the history of photography.

An equally important though perhaps less obvious, or less obviously visual, type of scientific iconography includes tables, graphs, diagrams, chemical formulae, maps and other abstract two-dimensional representations. These are not usually thought of as images since they do not follow realistic conventions of representation but rather aim to portray relations, to manipulate numbers and other entities, or to organize, display and communicate information. They have proven very powerful and effective in many scientific enterprises and are routinely used today in all fields of science.

In their most basic definition they may be described as spatial arrangements on a two-dimensional surface (e.g. paper, chalkboard, computer screen) of letters, numbers and/or words in combination with lines, shapes or images. This corresponds to the most inclusive definition of the term *diagram*, closely in line with its originally Greek meaning of "that which is marked out by lines, a geometrical figure, written list, register, the gamut or scale in music" (*OED Online*, accessed December 2014). This term will be used here as the overarching category to speak of the different types of two-dimensional abstract visual representations.

Theoretical Approaches

Diagrams have been studied by a select number of authors from Aristotle to Deleuze (for an anthology see Ernst, Wöpking, and Schneider 2015). Their writings have recently seen a surge in interest, no doubt related to the recent development of new

A Companion to the History of Science, First Edition. Edited by Bernard Lightman.
© 2016 John Wiley & Sons Ltd. Published 2020 by John Wiley & Sons Ltd.

techniques of data visualization and information design that have crucially contributed to the contemporary proliferation of computerized visualizations and simulations in scientific and generalist publications (Tufte 1997; Börner 2015).

Among the theoretical resources used for the study of diagrams, semiological approaches feature prominently. Charles Sanders Peirce is frequently invoked, who proposed to distinguish between three kinds of signs: the icon, the index, and the symbol. In a nutshell, Peirce defined these in the following way:

- the icon "represents its object by resembling it", e.g. a portrait
- the index represents objects "by being actually connected with them;" e.g. a weathercock
- the symbol refers to the object "because it will be so interpreted" by some convention, habit or rule, e.g. words (Peirce 1991a, 270).

Peirce briefly discussed the diagram: "though it will have Symbolide features, as well as features approaching the nature of Indices, it is nevertheless in the main an Icon of the forms of relations in the constitution of its Object" (Peirce 1991b, 252). Peirce's remarks underline both the difficulty of pinpointing the exact nature of the relationship of the diagram to the object represented, and one of its essential features: the portrayal of relations. For Peirce, the diagram is "a kind of icon, and an icon of intelligible relations" (Peirce 1991b, 252).

Whereas whole academic disciplines are devoted to the study of images on the one hand (such as history of art) and of texts and language on the other (such as linguistics or literary studies), none has paid much attention to the peculiar form of the diagram, whose properties exceed that of a simple addition or combination of text and image. In their plea for a specific theory of the diagram, Steffen Bogen and Felix Thürlemann argued in 2003 that it is its very ability to be read as a web of dynamic relations that best characterizes it: the topological and geometrical arrangement of textual and pictorial elements creates relationships between them. Diagrams thus do not simply recapitulate information available in other formats but possess a specific "pragmatic power" that confers to them "an immense significance as a tool for the constitution of meaning in Western culture" (Bogen and Thürlemann 2003, 21–2).

Semiotics are used here less to broach the issue of representation than to insist on the practical importance of diagrams as a *tool for thinking*, both for those creating and for those using diagrams, including in Peirce's own practice of thinking through "existential graphs" (Bogen and Thürlemann 2003, 17). Bogen and Thürlemann focused on medieval diagrams as used in legal, religious, and scientific settings. But their insistence on the operative and performative dimensions of diagrams echoes developments in other areas of scholarship.

By "attempting to reconstruct the diagrammatical structure of some instances of philosophical thought" in Plato, Descartes, and Wittgenstein's writings, philosopher Sybille Krämer has thus recently proposed an "epistemology of the line," arguing that:

> In the tension between the hand that *does* and the eye that *sees*, the stroke constitutes the elementary action of operative iconicity. The plane of inscription creates a space for the movement of thought in which theoretical entities are made visible and thus manageable. The perceptual nature of this act may be described as suddenly being able to see a non-empirical state of affairs within empirical arrangements. With regard to tactility, it

means that operations of configuring and reconfiguring graphical markings simultaneously carry out ideal/intellectual operations. The interstitial world of planar inscriptions mediates between intuition and thought; it intellectualises intuition and sensualises thinking. In brief, *we think on paper* (Krämer 2010a, 278–9).

Krämer analyses in detail the properties of diagrams and their functions, concluding with Peirce that the epistemic function of diagrams lies in "the interplay of visualization, demonstration and the production of new insights using diagrams" (Krämer 2009, 106, my translation). As Krämer points out, studies of graphism and of the line exist within a range of fields beyond philosophy and art history, including cultural history, palaoanthropology, cognitive semantics, and literature, all of which, she suggests, could profitably be harnessed towards the elaboration of a "diagrammatology" (Krämer 2010b).

The work of Bogen, Thürlemann and Krämer is exemplary of the lively field of "diagram studies" that has developed in the German-speaking academic world at the intersection of visual studies, media studies, philosophy, history of art, and history of science. This field is characterized by the ambition to elaborate general theories of the diagram, drawing upon a wide range of texts in different languages and from a range of traditions, though it has remained relatively little known elsewhere.[2] It is also worth noting that alongside this body of work other philosophical traditions exist that have focused on particular types of diagrams, notably in the philosophy of logic and mathematics (e.g. Shin 1995; Mancosu 2005).

Diagrams in the History of Science

In the history of science, there has also been a growing interest in diagrams, though the investigations have usually been less theoretical in nature. Many studies of individual diagrammatic forms exist, of their invention and uses in specific settings, but these studies and their authors often do not usually interact with each other. Diagrams indeed pertain to a wide variety of historical contexts, forms, and functions. Some trace their origins to ancient times, with recorded evidence of tables and diagrams in the ancient world and throughout the middle ages. Others are more recent creations: the graph or the chemical formula did not appear until the nineteenth century. Some kinds of diagrams such as tables or maps are by no means limited to learned environments. Others, such as chemical formulae, are closely connected to specific sciences. Moreover, the historiography on each of these forms is very unevenly developed: while only few studies focus specifically on graphs or tables, the history of cartography is a well-established field with a distinguished tradition of research into the history of geographical maps and the philosophical issues relating to the conceptualization and representation of space.

Still, it can generally be said that within the history of science, our knowledge of specific diagrams, their origins, the forms they take, and of their uses has very much been shaped by the approaches developed to study scientific *practices and material cultures*. Initially focusing on the laboratory sciences, historians of science started insisting three decades ago on the necessity of looking beyond theories and studying the day-to-day work, the instruments, the actors, the tacit knowledge, and the objects that characterized scientific activity. They showed that even the most abstract of sciences, such as mathematics or theoretical physics, also involved skills, know-hows, ways of thinking, of writing and of calculating, or "theoretical technologies" (Warwick 1992; Warwick 1993).

Historians soon began looking at the most mundane aspects of scientists' activities, for instance how they organize, classify, and store information; and this not only in fields obviously concerned with such practices such as natural history, but also, for instance, in linguistics or economics (Jacob 2011). Joining forces with historians of the book and media historians, some set out to investigate scientific work as part of the history of scholarly and non-scholarly techniques of information management. Studies of the archive (Blair 2010), the document (Gitelman 2014), the bibliography (Csiszar 2010), the commonplace book (e.g. Moss 1996), but also notebooks (e.g. Blair and Yeo 2010), newspaper clippings (Te Heesen 2014), or atlases (Akerman 1995; Didi-Huberman 2011) have thrown new light on the nature of academic work.

In a similar vein, diagrams have attracted attention as a family of devices, paper techniques developed for presenting, condensing, or representing information for a range of purposes. This approach to diagrams has been shaped by Bruno Latour's conceptualization of inscriptions and the role they play in scientific communication and knowledge production. These he defines as following:

> INSCRIPTION: A general term that refers to all the types of transformations through which an entity becomes materialized into a sign, an archive, a document, a piece of paper, a trace. Usually but not always inscriptions are two-dimensional, superimposable, and combinable. They are always mobile, that is, they allow new translations and articulations while keeping some types of relations intact. Hence they are also called "immutable mobiles," a term that focuses on the movement of displacement and the contradictory requirements of the task. When immutable mobiles are cleverly aligned they produce the circulating reference (Latour 1999, 307–8).

For Latour, diagrams are one kind of inscription, whether made by humans or machines: one important step in the production, circulation, and stabilization of knowledge. Through inscriptions, natural phenomena and their observation are managed, systematized, and made communicable. The specificity of science and technology as human activities and its "vast effects," Latour suggests, can be identified not by studying some specific kind of "rationality," but rather in "the transformation of rats and chemicals into paper" (Latour 1990, 22). Philosophers' debates about representation are thereby sidelined, or rather recast: here again, the focus is less on the relationship between the image and its referent than on what inscriptions *do* and how effective they are at being "*mobile* but also *immutable, presentable, readable* and *combinable* with one another" (Latour 1990, 26).

Latour's work has been important in providing an impetus and an epistemological justification to the study of scientific practices and its non-textual productions. Simultaneously, it has contributed to shape a widely-held view in the history of science (as distinct from some histories of art) that images cannot be studied independently of their context of production, perception, and use. Approaching scientific iconography, including diagrams, from the perspective of practice and material culture means that their interactions with instruments and texts are taken into account, together with the material cultures in which they are embedded, for instance the culture of print in Renaissance Europe (Kusukawa and Maclean 2006). This characteristic is important to bear in mind: it is common to most studies of diagrams in the history of science. In the remainder of this chapter, different types of diagrams are discussed that are of particular importance for the history of science.

Paper Tools

The notion of "paper tool" was introduced by historian of chemistry Ursula Klein in her study of the emergence of modern laboratory chemistry in the early nineteenth century, and more specifically of the elaboration of Berzelian chemical formulas, with the aim of accounting for "the striking duality of a world of signs and a world of laboratory manipulations and instruments, and the more or less obvious interaction of both worlds [that] has become emblematic of the experimental science of chemistry" (Klein 2003, 2). Defined by analogy with instruments and other laboratory tools, the paper tool is for Klein "an analytical category that serves to focus the historical analysis and reconstruction on the performative, cultural and material aspects of the development of inscriptions, models, concepts and theories" (Klein 2003, 3). Thereby, Klein brings together studies of the practice of theory with semiological insights, most notably the work of philosopher Nelson Goodman and of art theorist Rudolf Arnheim.

She suggests that Berzelian formulas were "'iconic symbols' that were particularly suited for the representation of 'chemical portions' rather than 'atoms'," and helped to reify the former, turning formulas into a "material resource for building chemical models." This was a dynamic process: "Chemists applied Berzelian formulas not as a medium for merely expressing and illustrating already existing knowledge, but as paper tools for developing the theory of chemical portions, or 'chemical atomism' as Alan Rocke has called it, and for producing chemical models and classificatory systems in organic chemistry" (Klein 2001, 28–9). She insists that paper tools "generate new goals, objects, inscriptions, and concepts linked to them" (Klein 2003, 3).

Klein argues that the "'graphic suggestiveness' and 'maneuverability' of Berzelian formulas was an important material precondition for their application as paper tools" (Klein 2003, 6). Pursuing the study of paper tools in the field of postwar physics in the US, David Kaiser picked up on this particular characteristic, emphasizing the malleability of Feynman diagrams as they were taught, appropriated and adapted for many uses in a growing number of communities. Referring to anthropologist Claude Levi-Strauss's conception of tool use as bricolage, Kaiser underscored against Latour how mutable, malleable and multivalent these kinds of inscriptions could be (Kaiser 2005, 18):

> To some, the diagrams functioned as pictures of physical processes—they seemed to capture something essential about the mechanisms of the microworld. To others, the diagrams were no more than helpful mnemonic aids for wading through long strings of complicated mathematical expressions—they were not to be confused with the stuff of the real world. Still others developed the diagrams as tools for a new kind of diagrammatic reasoning—the diagrams' structural or topological features prompted and enabled the investigation of various symmetries that their associated mathematical expressions should obey. Most often these distinct roles blurred together in practice (Kaiser 2005, 21).

Kaiser's in-depth study of the elaboration, teaching, and communication of Feynman diagrams revealed a myriad of local uses and evolutions as they dispersed over time and space, thus providing support to the view that scientific work is a situated set of intellectual and material practices. It also underlines the importance of diagrams in pedagogical settings in the physical sciences, insights which have been extended to other contexts, such as eighteenth-century Scottish chemistry (Eddy 2014) or 1960s General Relativity (Wright 2014).

The work on paper tools demonstrates that, just like the three-dimensional models discussed in Chapter 40, and sometimes alongside them, diagrams fulfill a range of functions: heuristic, expository, and didactic. Accordingly, they have been studied as part of the growing body of work focusing on the visual forms by which knowledge is presented, communicated and taught. Attention is paid not only to the question of what these visual forms represent and by what means but also to the contexts in which such forms appear, circulate, evolve, and are apprehended by different publics. Since ancient times, the mnemonic and pedagogical value of diagrams has been recognized (e.g. Murdoch 1984; North 2004; Salonius and Worm 2014). In ancient Greece and Rome, and up to the early modern period and beyond, diagrams also have linked practical and learned knowledge and help uncover interactions between a range of actors including mathematicians, surveyors, navigators, instrument makers, draftsmen, and printers (e.g. Lefèvre, Renn, and Schöpflin 2003; Kusukawa and Maclean 2006; Jardine and Fay 2014; Valleriani 2014). This continues to be true: diagrams have, for instance, been a method of choice in the appropriation of mathematical modelling by communications and control engineers (Bissell 2004).

Maps

Whereas the paper tool is an analytical category recently introduced by historians, other kinds of diagrams have a much longer history and have acquired the status of historical entities. Historians of cartography consider maps as "undoubtedly one of the oldest form of human communication" that "have impinged upon the life, thought, and imagination of most civilizations that are known through either archeological or written records." (Harley 1987, 1). Evidence of maps has been found in all regions of the world and going back to prehistoric times.

Notwithstanding, contemporary Western conceptions about maps should not be projected onto different or earlier cultures. Before the emergence of modern maps, and alongside them, there existed a wide variety of maps whose functions could be cosmological, ideological, religious, or symbolic, and which may or may not have had geographical spaces as their object. The ideas that a map should be an accurate representation, or that it primarily represents territory, are both artefacts of the emergence of science and of the discipline of geography in the past few centuries and of a retrospective take on the development of mapping (Harley 1987, 1–5).

Maps—and by extension diagrams—are thus relevant to the history of science in two major ways. They are objects of study for the practitioners of archeology, anthropology, history, and more recently the cognitive sciences. If we ask about the historical emergence of *scientific* maps, however, the answer is tautological: it is inseparable from the emergence of scientific and technical ideals, practices, and institutions. Harley points out that "maps did not become everyday objects in many areas of the world until the European Renaissance," and that "by the seventeenth century maps… were becoming more narrowly associated with geography and terrestrial survey and charting." During the Enlightenment, mapping practices reflected the growing emphasis put on measurement and precision by investigators of nature. From the mid-nineteenth century, mapping entered a new era with the emergence and institutionalization of geography as a discipline (Harley 1987, 1–12). This is true of other scientific maps: in his classic study of the emergence of a visual language in geology, Martin Rudwick pointed out that the geological map "paralleled—and indeed was an essential part of—the development of a self-conscious new science" at about the same time (Rudwick 1976, 177).

But mapping was also crucially affected by other factors, such as the transformation of print cultures and navigation techniques and, especially in the case of geographical maps, of the worldwide exploratory, commercial, and colonial enterprises undertaken by Europeans from the Renaissance onwards. This points again to the performative function of diagrams: not only do they put on display and communicate a particular form of knowledge, they also have practical and ideological purposes. Historians of cartography have long been aware of the political dimensions of mapping. Even in their apparently most technical or mundane aspects, such as the choice of the type of projection or of the order of maps within atlases, the history of mapping is unavoidably marked by the history of European nationalism and imperialism (Wood 1992; Harley 2001). This is true more generally: the abstract or technical appearance of diagrams does not preclude them from (and sometimes makes them particularly suitable for) carrying particular values and motives and being put to work in enterprises to transform both science and society such as in economics (Charles and Giraud 2013), not to speak of cybernetics' flagship diagrams that stood for the new science's totalizing approach to systems mechanical and organic.

Tables

Tables take the form of spatial arrangements, typically but not necessarily in rows and columns, of numbers, words or other symbols that are thereby put in relation with each other. Like maps, they are both ancient and ubiquitous. The earliest traces of tables were found on Sumerian tablets, and they appear in most known cultures. Tables have been used for a range of purposes, from ordering and presenting information (quantitative or qualitative, verbal or numerical) to representing mathematical functions, and as practical aids for calculation, for instance logarithm tables. Astronomers, mathematicians, and actuaries were long recognized as experts in the fabrication of tables, but their users encompassed a wide constituency that included shopkeepers and businessmen, navigators, farmers, and the general population—for centuries, ephemerides could be found in every household (Figures 39.1a and 39.1b). Since the nineteenth century, tables have become important tools in industrial, engineering, and military contexts, as for instance in ballistics (Aubin 2014).

Though tables occur in a great number of publications pertaining to the history of the different sciences and technologies, they have only recently become a focal point of investigation as such. Comparative studies or overviews of the history of scientific tables are still very rare, making any generalization somewhat tentative. A recent volume made a first attempt, dividing the history of tables into four broad periods:

INTERVALLE semi-diurne.	CORRECT. POUR LES LEVERS ET COUCHERS DE LA LUNE.						
	LATITUDE : 42°			43°			
	30'	40'	50'	0'	10'	20'	30'
3h30m	36m1	35m3	34m5	33m7	32m9	32m0	31m2
40	33,4	32,6	31,9	31,1	30,3	29,6	28,8
50	30,8	30,1	29,3	28,6	27,9	27,2	26,5
4. 0	28,2	27,6	26,9	26,3	25,6	24,9	24,3
10	25,8	25,2	24,6	24,0	23,4	22,8	22,2
20	23,4	22,9	22,3	21,8	21,2	20,7	20,1
30	21,1	20,6	20,1	19,6	19,1	18,6	18,1
40	18,9	18,4	18,0	17,5	17,1	16,6	16,2
50	16,7	16,3	15,9	15,5	15,1	14,7	14,3
5. 0	14,5	14,2	13,8	13,5	13,2	12,8	12,5
10	12,4	12,1	11,8	11,6	11,2	10,9	10,7
20	10,4	10,1	9,9	9,6	9,4	9,1	8,9
30	8,3	8,1	7,9	7,7	7,5	7,3	7,1
40	6,3	6,1	6,0	5,8	5,7	5,5	5,4
50	4,3	4,2	4,1	4,0	3,9	3,8	3,7
6. 0	2,3	2,2	2,2	2,1	2,1	2,0	2,0
10	0,3	0,3	0,3	0,3	0,3	0,2	0,3
	+	+	+	+	+	+	+
20	1,7	1,7	1,6	1,6	1,5	1,5	1,4
30	3,7	3,6	3,5	3,4	3,3	3,2	3,2
40	5,7	5,6	5,4	5,3	5,1	5,0	4,9
50	7,7	7,5	7,3	7,2	7,0	6,8	6,6
7. 0	9,7	9,5	9,3	9,1	8,8	8,6	8,3
10	11,8	11,5	11,3	11,0	10,7	10,4	10,1
20	13,9	13,6	13,3	12,9	12,6	12,2	11,9
30	16,0	15,7	15,3	14,9	14,5	14,1	13,7
40	18,2	17,8	17,4	16,9	16,5	16,1	15,6
50	20,4	20,0	19,5	19,0	18,5	18,0	17,5
8. 0	22,7	22,2	21,7	21,1	20,6	20,0	19,5
10	25,1	24,5	23,9	23,3	22,7	22,1	21,5
20	27,5	26,9	26,2	25,6	24,9	24,3	23,6
30	30,0	29,3	28,6	27,9	27,2	26,5	25,8
40	32,6	31,8	31,1	30,3	29,6	28,8	28,1
50	35,3	34,5	33,7	32,9	32,1	31,2	30,4

Correction + : ajoutez au lever, retranchez du coucher.
Correction − : retranchez du lever, ajoutez au coucher.

a

TABLE I.

Loi de la mortalité en France suivant la Table de Deparcieux, complétée dans les premières années.

AGES.	VIVANTS à chaque âge.	SOMME des vivants.	DURÉE DE LA VIE	
			Moyenne.	Probable.
			Ans. Mois.	Ans. Mois.
0	1286	51467	39 8	42 0
1	1071	50181	46 4	53 2
2	1006	49110	48 4	54 11
3	970	48104	49 1	55 4
4	947	47134	49 4	55 2
5	930	46187	49 2	54 10
6	917	45257	48 10	54 4
7	906	44340	48 5	53 9
8	896	43434	48 0	53 2
9	887	42538	47 5	52 6
10	879	41651	46 11	51 10
11	872	40772	46 3	51 1
12	866	39900	45 7	50 3
13	860	39034	44 11	49 6
14	854	38174	44 2	48 9
15	848	37320	43 6	47 11
16	842	36472	42 10	47 2
17	835	35630	42 2	46 5
18	828	34795	41 6	45 8
19	821	33967	40 10	44 11
20	814	33146	40 3	44 2
21	806	32332	39 7	43 5
22	798	31526	39 0	42 9
23	790	30728	38 5	42 0
24	782	29938	37 9	41 3
25	774	29156	37 2	40 6
26	766	28382	36 7	39 10
27	758	27616	35 11	39 1
28	750	26858	35 4	38 4
29	742	26108	34 8	37 7
30	734	25366	34 1	36 10

b

Figure 39.1a and 39.1b A table for correcting the times of the rise and setting of the moon given for Paris and a table giving mortality in France as a function of age. Such tables were made by astronomers combining observation, extrapolation, and computations and were designed as bases for further computation (Bureau des Longitudes 1869, 48 and 265).

From around 2500 BC to 150 AD, the story is one of the invention of the table as a concept and its realization in a number of forms for different purposes. Over the next millenium and a half the second period saw some of the great achievements of the human mind, in the astronomical and trigonometric tables which lay at the heart of progress in the hard sciences leading up to the scientific revolution. The third period, from the early seventeenth century to the mid-nineteenth centuries, was the heyday of work on logarithm tables which formed the basis of calculation needs for the industrial revolution. The fourth period, from the mid-nineteenth century up to the present, has seen a number of developments in the production of a range of ever more sophisticated tables for physical, mathematical, industrial and economic purposes, as well as the development of technology to help in their calculation. The story is by no means over, as the development of spreadsheets, dynamic tables in computers, shows that there is still a lot of life in the deep idea of presenting information on a two-dimensional tabular screen (Campbell-Kelly, Croaken, Flood, and Robson 2003, 2).

This chronology will no doubt have to be reassessed and refined in light of the investigations currently being carried out on the use of tables in a much wider range of cultures including non-Western ones and across the whole period. Noteworthy is the growing interest not only in the technical dimensions of table-making and use but also in its socioeconomic contexts: how tables participate in the reorganization of labor, their interaction with machines, and how they reflect social relations. Tables and other graphic methods were, for instance, at the heart of eighteenth-century debates among astronomers and naval officers on the kind of mathematical training required of sailors to determine the longitude of their ships (Boistel 2010). They were also a focus for the English accountants and scientists who founded the Astronomical Society in 1820 with the intention of improving "astronomical book-keeping" (Ashworth 1994, 409). Such studies reveal the interactions between different groups of actors in scientific and non-scientific pursuits and the moral and social values that the promotion of particular diagrammatic practices often entailed.

Graphs

Like chemical formulae, graphs are a recent invention—and indeed may be related to chemical visualizations since the term "graph" was coined in the English language by the mathematician J. J. Sylvester in 1878 to describe mathematical diagrams that looked like the figures representing chemical bonds in molecules (Hankins 1999, 52). If a graph is taken in its modern meaning of a curve displaying a relationship between coordinates, as distinguished from the older practice of graph drawing (Kruja Marks, Blair and Waters 2002), then its origins can be traced to the second half of the eighteenth century, in particular with the graphic innovations made in James Playfair's statistical atlases, James Watt's indicator diagrams, and Johann Heinrich Lambert's attempts to display measurements graphically, for instance of temperature variations over time (Tilling 1976). Later, French military and civil engineers played an important role in developing graphical techniques (Hankins 1999). In the course of the nineteenth century, graphs came into widespread use, partly as a means of coping with the "avalanche of numbers" produced in many sciences (Hacking 1990, 2; Hacking 2006). The history of graphic methods of representation is thus closely related to the history of statistics (Funkhouser 1937, Klein 1997). Together with chemical formulae, modern tables or scientific maps, the spread of graphs went hand in and with the institutionalization of science and engineering in the nineteenth century, and these modern diagrams were often seen to embody "the image of precision" and a new conception of rigor and accuracy (Holmes and Olesko 1995).

The history of graphical methods is revealing of one fundamental characteristic of diagrams, and no doubt one reason for their success: their versatility. Playfair was probably inspired by the visual representations developed by his brother, the mathematician and geologist John Playfair, as well as by Watt's diagrams. Alexander von Humboldt's invention of isotherms was heavily indebted both to Playfair and to the French engineering tradition (Hankins 1999). Graphs were early on grasped as tools applicable to a wide variety of objects, a conceptualization that materialized in the commercialization of graph paper in the late nineteenth century.

The plasticity of graphs is also made evident by their different functions: from the beginning they were used to represent mathematical functions, to facilitate calculation, to display series of measurements, and to compensate for errors in observation series, but they could also be traces produced by self-recording instruments. While

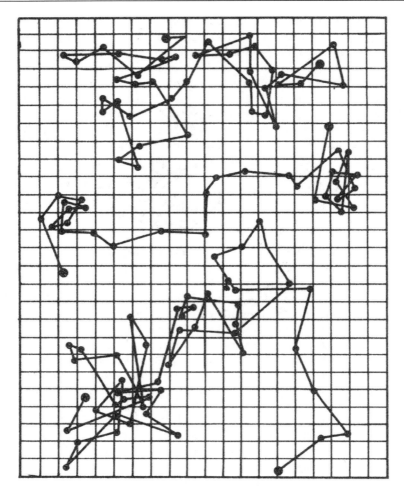

Figure 39.2 Measuring the displacement of three particles: using a camera lucida, Perrin marked the successive positions of each particle at regular intervals of time, before drawing straight lines to join the dots (Perrin 1909, 81).

in many cases these functions remained distinct, sometimes they merged or overlapped. Jean Perrin's measurements of the Brownian motion of microscopic particles were explained using a drawing (Figure 39.2) that recapitulated his observation technique (pointings of the positions of individual particles taken at regular intervals) while it proposed a visual rendering of the novel method of measurement proposed by Einstein (average square displacement). This mode of visualization was itself inspired by chronophotographic conventions, though it was made with pencil and paper using a microscope. But it could also be read, Perrin suggested, as a visualization of "continuous, nowhere differentiable functions that were wrongly seen as mathematical curiosities, since nature can suggest them as well as differentiable functions" (Perrin 1909, 31; Bigg 2011).

Conclusion

Like Feynman diagrams "that had been developed as convenient ways to talk about the world came to be treated as pictures of how the world really was," diagrams often evolve from being "mnemonic to mimetic" and back again (Kaiser 2005, 369). Experimental traces can become mathematical curves; maps or chemical formulae can display data visually or extrapolate it, turning them effectively into models (Besse 2008). They can be evidence or simulation. Conversely, the same data can be represented in different diagrammatic ways: it was only when graphs stopped being seen as capturing reality in a "photographic" manner and were perceived as engines of discovery that they were adopted in economics, displacing tables (Maas 2012). Because of their slippery nature and varying perceived degree of abstraction, diagrams can be at the heart of disputes such as the nineteenth-century debates among chemists and physicists on the existence of atoms or the postwar discussions in electronics about the nature of transistors (Jones Imhotep 2008). They can also escape the local conditions of their making and gain wider currency, and become standardized tools that are widely applicable.

Diagrams unsettle any simple distinctions between realistic and abstract representations. Their multiplicity and flexibility of forms and functions complicate theories of the scientific image (Lüthy and Smets 2009). Among the many directions in which the history of diagrams could be developed and broadened in scope, one of the more ambitious would be to focus on this plasticity and to replace diagrams in a wider history of lines as outlined by Tim Ingold (2007), that includes threads, curves, traces, and songs and emphasizes their ability to turn into each other, to transcend their material medium and change meaning; and thereby historicizes the physical experience of space and time and its rendering in two dimensions.

Another important dimension which has only been alluded to is computerization and the multilevel ways in which it is affecting diagrammatic practices—and the history of science. For these ongoing developments are surely one impetus for the new interest in "analog" scholarly and non-scholarly techniques of managing and displaying information (e.g. Rosenberg and Grafton 2010). As digital data storage and imaging techniques transform practices and social organization in many scientific and technological fields, they also recast our understanding of the status and functions of images, and of diagrams in particular (Coopmans, Vertesi, Lynch, and Woolgar 2014). And finally, as we investigate these developments we are caught up in them, as the calls grow louder for the development of "visual thinking" and more effective means of presenting research using data visualization and information graphics (Bender and Marrinan 2010; Börner 2015).

Endnotes

1 The author wishes to express her thanks to David Aubin, Annemieke Verboon, and Ronan Le Roux for their expert advice on different parts of this chapter.

2 See also, e.g., Heßler and Mersch 2009; Bauer and Ernst 2010; Schmidt-Burkhardt 2012. See Batt 2004 for a point of comparison in the, largely, French philosophical context.

References

Akerman, James R. 1995. "From books with maps to books as maps: The editor in the creation of the Atlas idea." In *Editing Early and Historical Atlases*, edited by Joan Winearls, 3–48. Toronto: University of Toronto Press.

Ashworth, William J. 1994. "The calculating eye: Baily, Herschel, Babbage and the business of astronomy." *British Journal for the History of Science*, 27: 409–41.

Aubin, David. 2014. "'I'm just a mathematician:' Why and how mathematicians collaborated with military ballisticians at Gâvres." In *The War of Guns and Mathematics: Mathematical Practices and Communities in France and its Western Allies around World War I*, edited by David Aubin and Catherine Goldstein, 307–50. Providence, RI: American Mathematical Society.

Batt, Noëlle (ed.) 2005. *Penser par le Diagramme de Gilles Deleuze à Gilles Châtelet. (Théorie. Littérature. Enseignement 22)*. Saint-Denis: Presses Universitaires de Vincennes.

Bauer, Matthias, and Christoph Ernst. 2010. *Diagrammatik: Einführung in ein kultur- und medienwissenschaftliches Forschungsfeld*. Bielefeld: Transcript Verlag.

Bender, John, and Michael Marrinan. 2010. *The Culture of Diagram*. Stanford, CA: Stanford University Press.

Besse, Jean-Marc. 2008. "Cartographie et pensée visuelle. Réflexions sur la schématisation graphique." In *Les usages des cartes (XVIIe–XIXe siècle). Pour une approche pragmatique des productions cartographiques*, edited by Isabelle Laboulais, 19–32. Strasbourg: Presses Universitaires de Strasbourg.

Bigg, Charlotte. 2011. "A visual history of Jean Perrin's Brownian motion curves." In *Histories of Scientific Observation*, edited by Lorraine Daston and Elisabeth Lunbeck, 156–79. Chicago: Chicago University Press.

Bissell, Chris C. 2004. "Models and 'black boxes': Mathematics as an enabling technology in the history of communications and control engineering." *Revue d'Histoire des Sciences*, 57, No. 2: 305–38.

Blair, Ann. 2010. *Too Much to Know: Managing Scholarly Information before the Modern Age*. New Haven, CT: Yale University Press.

Blair, Ann, and Richard Yeo (eds.) 2010. *Note-taking in Early Modern Europe*. Special issue of *Intellectual History Review* 20, No. 3: 1–177.

Bogen, Steffen, and Felix Thürlemann. 2003. "Jenseits der Opposition von Text und Bild: Überlegungen zu einer Theorie des Diagramms und des Diagrammatischen." In *Die Bildwelt der Diagramme Joachims von Fiore. Zur Medialität religiös-politischer Programme im Mittelalter*, edited by Alexander Patschovsky, 1–22. Ostfildern: Jan Thorbecke Verlag.

Boistel, Guy. 2010. "Training seafarers in astronomy: Methods, naval schools and naval observatories in eighteenth- and nineteenth-century France." In *The Heavens on Earth: Observatories and Astronomy in Nineteenth-Century Science and Culture*, edited by David Aubin, Charlotte Bigg, and Otto Sibum, 148–73. Durham, NC: Duke University Press.

Börner, Katy. 2015. *Atlas of Knowledge: Anyone Can Map*. Cambridge, MA: MIT Press.

Bureau des Longitudes. 1869. *Annuaire pour l'an 1869*. Paris: Gauthier-Villars.

Campbell-Kelly, Martin, Mary Croaken, Raymond Flood, and Eleanor Robson. 2003. *The History of Mathematical Tables from Sumer to Spreadsheets*. Oxford: Oxford University Press.

Charles, Loïc, and Yann Giraud. 2013. "Economics for the masses: The visual display of economic knowledge in the United States (1910–45)." *History of Political Economy*, 45, No. 4: 567–612.

Coopmans, Catelijne, Janet Vertesi, Michael Lynch, and Steve Woolgar. 2014. *Representation in Scientific Practice Revisited*. Cambridge, MA: MIT Press.

Csiszar, Alex. 2010. "Seriality and the search for order: Scientific print and its problems during the late nineteenth century." *History of Science*, 48, Nos. 3–4: 399–434.

Didi-Huberman, Georges. 2011. *Atlas ou le gai savoir inquiet*. Paris: Editions de Minuit.

Eddy, Matthew Daniel. 2014. "How to see a diagram: A visual anthropology of visual affinity." *Osiris*, 29: 178–96.

Ernst, Christoph, Jan Wöpking, and Birgit Schneider (eds.) 2015. *Diagrammatik: Ein interdisziplinärer Reader*. Berlin: Akademie Verlag.

Funkhouser, H. Gray. 1937. "Historical development of the graphical representation of statistical data." *Osiris*, 3: 269–404.

Gitelman, Lisa. 2014. *Paper Knowledge: Toward a Media History of Documents*. Durham, NC: Duke University Press.

Hacking, Ian. 1990. *The Taming of Chance*. Cambridge: Cambridge University Press.

Hankins, Thomas L. 1999. "Blood, dirt, and nomograms: A particular history of graphs." *Isis*, 90: 50–80.

Hankins, Thomas L. 2006."A 'large and graceful sinuosity': John Herschel's graphical method." *Isis*, 97: 605–33.

Harley, John Brian. 1987. "The map and the development of the history of cartography." In *The History of Cartography. Volume 1: Cartography in Prehistoric Ancient, and Medieval Europe and the Mediterranean*, edited by John Brian Harley and David Woodward, 1–42. Chicago: Chicago University Press.

Harley, John Brian. 2001. *The New Nature of Maps. Essays in the History of Cartography*. Baltimore: Johns Hopkins University Press.

Heßler, Martina, and Dieter Mersch. 2009. *Logik des Bildlichen: Zur Kritik der ikonischen Vernunft*. Bielefeld: Transcript Verlag.

Holmes Frederic L., and Kathryn M. Olesko. 1995. "The images of precision: Helmholtz and the graphical method in physiology." In *The Values of Precision*, edited by M. Norton Wise, 198–221. Princeton, NJ: Princeton University Press.

Ingold, Tim. 2007. *Lines: A Brief History*. London: Routledge.

Jacob, Christian (ed.) 2011. *Lieux de Savoir 2: Les Mains de l'Intellect*. Paris: Albin Michel.

Jardine, Nicholas, and Isla Fay (eds.) 2014. *Observing the World through Images: Diagrams and Figures in the early modern Arts and Sciences*. Leiden: Brill.

Jones-Imhotep, Edward. 2008. "Icons and electronics." *Historical Studies in the Natural Sciences*, 38, No. 3: 405–50.

Kaiser, David. 2005. *Drawing Theories Apart: The Dispersion of Feynman Diagrams in Postwar Physics*. Chicago: University of Chicago Press.

Klein, Judy L. 1997. *Statistical Visions in Time: A History of Time Series Analysis, 1662–1938*. Cambridge: Cambridge University Press.

Klein, Ursula. 2001. "The creative power of paper tools in early nineteenth-century chemistry." In *Tools and Modes of Representation in the Laboratory Sciences*, edited by Ursula Klein, 13–34. Dordrecht: Kluwer.

Klein, Ursula. 2003. *Experiments, Models, Paper Tools: Cultures of Organic Chemistry in the Nineteenth Century*. Stanford, CA: Stanford University Press.

Krämer, Sybille. 2009. "Operative Bildlichkeit: Von der 'Grammatologie' zu einer 'Diagrammatologie'? Reflexionen über erkennendes 'Sehen'." In *Logik des Bildlichen: Zur Kritik der ikonischen Vernunft*, edited by Martina Heßler and Dieter Mersch, 94–122. Bielefeld: Transcript Verlag.

Krämer Sybille. 2010a. "'The mind's eye': Visualizing the non-visual and the 'Epistemology of the Line'." In: *Image and Imaging in Philosophy, Science, and the Arts. Preproceedings of the 33rd International Wittgenstein Symposium 2*, edited by Elisabeth Nemeth, Richard Heinrich, and Wolfram Pichler, 275–95. Kirchberg am Wechsel: Austrian Ludwig Wittgenstein Society.

Krämer Sybille. 2010b. "'Epistemology of the Line': Reflections on the diagrammatical mind." In *Studies in Diagrammatology and Diagram Praxis. (= Logic and Cognitive Systems,*

Studies in Logic 24), edited by Olga Pombo and Alexander Gerner, 13–39. London: College Publications.

Kruja, Eriola, Joe Marks, Ann Blair, and Richard Waters. 2002. "A short note on the history of graph drawing." *Proceedings of the 9th International Symposium on Graph Drawing, GD 2001 (Springer Lecture Notes in Computer Science*, Vol. 2265): 272–86.

Kusukawa, Sachiko, and Ian Maclean (eds.) 2006. *Transmitting Knowledge: Words, Images, and Instruments in Early Modern Europe.* Oxford: Oxford University Press.

Latour, Bruno. 1990. "Drawing things together." In *Representation in Scientific Practice*, edited by Michael Lynch and Steve Woolgar, 19–68. Cambridge, MA: MIT Press.

Latour, Bruno. 1999. *Pandora's Hope: Essays on the Reality of Science Studies.* Cambridge, MA: Harvard University Press.

Lefèvre, Wolfgang, Jürgen Renn, and Urs Schöpflin. 2003. *The Power of Images in Early Modern Science.* Basel: Birkhäuser.

Lüthy, Christoph, and Alexis Smets. 2009. "Words, lines, diagrams, images: Towards a History of scientific imagery." *Early Science and Medicine*, 14: 398–439.

Maas, Harro. 2012. "The photographic lens: Graphs and the changing practices of Victorian economists." In *The Victorian World*, edited by Martin Hewitt, 500–518. London: Routledge.

Mancosu, Paolo. 2005. "Visualization in Logic and Mathematics." In *Visualization, Explanation and Reasoning Styles in Mathematics*, edited by Paolo Mancosu, Klaus Frovin Jørgensen, and Stig Andur Pedersen, 13–30. Dordrecht: Springer.

Moss, Ann. 1996. *Printed Common-Place Books and the Structuring of Renaissance Thought.* Oxford: Oxford University Press.

Murdoch, John E. 1984. *Album of Science: Antiquity and the Middle Ages.* New York: Scribner's Sons.

North, John. 2004. "Diagram and thought in medieval science." In *Villard's Legacy: Studies in Medieval Technology, Science and Art, in Memory of Jean Gimpel*, edited by Marie-Thérèse Zenner, 265–87. Aldershot: Ashgate.

Peirce, Charles Sanders. 1991a [1908]. "A neglected argument for the reality of God." Reprinted in *Peirce on Signs: Writings in Semiotic by Charles Sanders Peirce*, edited by James Hoopes, 260–78. Chapel Hill, NC: University of North Carolina Press.

Peirce, Charles Sanders. 1991b [1906]. "Prolegomena to an apology for pragmatism." Reprinted in *Peirce on Signs: Writings in Semiotic by Charles Sanders Peirce*, edited by James Hoopes, 249–52. Chapel Hill: University of North Carolina Press.

Perrin, Jean. 1909. "Mouvement brownien et realité moléculaire." *Annales de chimie et de physique*, 8, No. 18: 5–114.

Rosenberg, Daniel, and Anthony Grafton. 2010. *Cartographies of Time: A History of the Timeline.* Princeton, NJ: Princeton Architectural Press.

Rudwick, Martin. 1976. "The emergence of a visual language for geological science." *History of Science*, 14: 149–96.

Salonius, Pippa, and Andrea Worm. 2014. *The Tree: Symbol, Allegory, and Mnemonic Device in Medieval Art and Thought (*International Medieval Research 20). Turnhout: Brepols.

Schmidt-Burkhardt, Astrit. 2012. *Die Kunst der Diagrammatik: Perspektiven eines neuen bildwissenschaftlichen Paradigmas.* Bielefeld: Transcript Verlag.

Shin, Sun-Joo. 1995. *The Logical Status of Diagrams.* Cambridge: Cambridge University Press.

Te Heesen, Anke. 2014. *The Newspaper Clipping: A Modern Paper Object.* Trans. Lori Lantz. Manchester: Manchester University Press.

Tilling, Laura. 1975. "Early experimental graphs." *British Journal for the History of Science*, 8, No. 3: 193–213.

Tufte, Edward R. 1997. *Visual Explanations. Images and Quantities, Evidence and Narrative.* Cheshire, CT: Graphics Press.

Valleriani, Matteo (ed.) 2014. *Appropriation and Transformation of Ancient Science*. Special issue of *Nuncius*, 29: 301–432.

Warwick, Andrew. 1992. "Cambridge mathematics and Cavendish physics: Cunningham, Campbell and Einstein's relativity, 1905–1991." Part 1, "The uses of theory." *Studies in the History and Philosophy of Science*, 23: 625–56.

Warwick, Andrew. 1993. "Cambridge mathematics and Cavendish physics: Cunningham, Campbell and Einstein's relativity, 1905–1991." Part 2. "Comparing traditions in Cambridge physics." *Studies in the History and Philosophy of Science*, 24: 1–25.

Wood, David. 1992. *The Power of Maps*. New York: Guilford Press.

Wright, Aaron Sidney. 2014. "The advantages of bringing infinity to a finite place: Penrose diagrams as objects of intuition." *Historical Studies in the Natural Sciences*, 44, No. 2, 99–139.

CHAPTER FORTY

Three-Dimensional Models

JOSHUA NALL AND LIBA TAUB

I am very busy with Saturn … He is all remodelled and recast, but I have more to do to him yet, for I wish to redeem the character of mathematicians, and make it intelligible.
James Clerk Maxwell to H. R. Droop, 14 Nov. 1857 (as quoted in Harman 1990, 557)

Introduction

The physics of the rings of Saturn seems like an archetypically abstract, mathematical problem. The first solution to the questions of the rings' structure and stability, presented by James Clerk Maxwell in 1856, is typically held up as an exemplary achievement of mathematical analysis (Harman 1990, 20–24; Brush, Everitt, and Garber 1983, ix–31). But what Maxwell realized was that for his work to be intelligible—for it to be understood, accepted, and disseminated amongst a wide audience—he needed more than mathematical formulae: he needed a physical model (Figure 40.1).

Three-dimensional (3-D) models have been used to help understand and explain natural phenomena since antiquity. In the fourth century BCE eponymous Platonic dialogue (40c), Timaeus claimed that it is impossible to describe the motions of the planets without visible models; the object referred to may have been some type of armillary sphere. Here, we discuss the variety of 3-D scientific and mathematical models that have been developed and used for a range of purposes including research, teaching, demonstration, and entertainment, particularly post-1700. The term "model" is used by historians and philosophers of science in various ways; we focus specifically on 3-D models as sources for history of science (cf. Hopwood and de Chadarevian 2004, 1–2). We consider them in research environments, as representational objects mediated to various audiences, as devices mass-produced for teaching, and as modeling nature.

3-D models are an under-studied historical resource. We are fortunate to be able to write this chapter as historians of science curating an internationally renowned collection of scientific instruments and models in the Whipple Museum of the History

Figure 40.1 Maxwell's model for illustrating the wave motions generated in a ring of satellites around Saturn, made by the Aberdeen instrument maker Charles Ramage. Maxwell wrote to a friend that it was designed "for the edification of sensible image worshippers" (as quoted in Harman 1990, 576). © Cavendish Laboratory, Department of Physics, University of Cambridge.

of Science, part of the University of Cambridge, itself a place with long traditions of making, using, and studying scientific and mathematical models. Choosing as examples some particular favorites to illustrate more general points, our chapter has benefited from access to a great variety of 3-D models, and to the original work done on these by generations of research students in the Department of History and Philosophy of Science. To fully understand models it is important to examine and work with actual examples, and not only read about them.[1]

Models in Research

Models are used at every step of scientific practice and mediation, from laboratory bench to public exposition. Not only end-products of the scientific enterprise—optional representations of work already accomplished—models are often a central facet of research work. Providing material space in which scientists can think through problems, models have myriad uses as tools for inquiry, experimentation, and speculative "play". Crucially, models' third dimension enables scientists to achieve results not attainable on paper, or with analytical manipulations alone (Baird 2004, 21–40; Griesemer 2004).

This kind of research work with models is particularly clear in mathematics. For topology, in particular, practitioners from the mid-nineteenth century onwards faced challenges contemplating the shape and manipulation of complex 3-D surfaces. Striking artifacts attest to mathematicians' construction of physical models using wool, string, wood, and brass as part of their everyday practices (Kidwell 1996; Mehrtens 2004). Exemplary are the knitted mathematical models of the Scottish chemist and mathematician Alexander Crum Brown, surviving in collections of several British historic science collections.[2] Crum Brown is best known for introducing now-familiar graphical representations of organic molecules as strings of atoms conjoined by lines in positions corresponding to the hypothesized structure of molecules. This practice in abstract chemical modeling is closely related to another of his interests, the mathematics of "knot theory." In thinking through topographic problems, Crum Brown shifted to physical modeling, knitting a variety of models of interpenetrating surfaces. Much more than mere visual aids, Crum Brown stressed how working in wool enabled him to think through surfaces that he then published as abstract mathematical formulations. Physical models were "helps to the imagination"; though they "cannot [be] perfectly accurate representations of the objects about which we reason; they serve their purpose if they enable us to see these objects accurately with the mind's eye, and so reason correctly about them" (Crum Brown 1914). Whilst historians of science are familiar with the kinds of formulaic manipulations organic chemists perform using diagrams on paper, Crum Brown's knitted surfaces demonstrate how modeling in three dimensions allows the manipulation and exploration of immaterial objects of enquiry too complex to represent in words, diagrams or formulae alone (Dunning 2015).

As physical representations bridging the immaterial and material, models can play a central role in the interplay between theory and experiment. In his investigations of the efficiency of waterwheels, the engineer John Smeaton found that theoretical analysis of the complex dynamic interactions of water on wheel failed to determine the best design. Building and experimenting with a working scale model, he developed a practical test-bed for design trials. This model work demonstrated Smeaton's mastery over complex mechanics, as well as warranting the costly engineering works for which he sought patronage (Schaffer 1994, 172–8; Baird 2004, 29–32). Such use of models is prevalent in the history of engineering, where the complexity of physical parameters can far outstrip even modern computational techniques.

Repeated trials with models produce more than data. By introducing a tactile element to scientific work they can also help train scientists in manual skills, often an essential part of research practice. Models therefore find places in laboratories and observatories as vital components in training regimes. When the Royal Observatory at Greenwich was put in charge of the British effort to observe the 1874 transit of Venus, considerable national expense and honor rode on the ability of staff to perform precise measurements of the transit under exacting—and unrepeatable—conditions. As the London *Graphic* explained, "the Astronomer Royal, acting on the principle which induces navy and military commanders to organize sham fights as a preparation for real battle, constructed an artificial model of Venus," from metal, wood, clockwork, and mirrors, to help train the many observers that would be sent to far-flung observation posts across the globe. Using this model on the lawns below Greenwich Observatory, astronomers sharpened their skills whilst gathering valuable information about the vicissitudes of observing what would remain, until the moment of transit itself, an otherwise inaccessible phenomenon (Ratcliff 2008, 77–88, on 78).

This use of models within a research environment as tactile embodiments of inaccessible phenomena has been particularly scrutinized in the case of chemistry. In chemists' day-to-day work models can be built and 'played with' as aids to thinking through problems that depend upon an understanding of molecular structure and shape. As Eric Francoeur has noted, a remarkable array of types of atomic model can be found in chemistry labs, reflecting "the wide variety of needs and purposes chemists face in their exploration of chemical structures" (Francoeur 1997, 11–12; also de Chadarevian 2004, 343–9; Meinel 2004). As with their mathematical counterparts, molecular models embody spatial relationships, providing material analogies that can be freely manipulated and 'thought with.' As interactive objects that can be made and remade, such models expand our conception of 'thinking', constituting one aspect of what Davis Baird has called "thing knowledge" (Baird 2004, 21–40). This material work has had a profound impact on chemistry's practice; even as chemical modeling shifts to the digital realm, new techniques of the small screen draw on earlier practices of physical modeling (Francoeur and Segal 2004).

Examples of using chemical models as research tools can be traced back to the earliest attempts by chemists to develop a theory of 3-D chemical structure. As Christoph Meinel has argued, for stereochemists such as August Kekulé and Jacobus Henricus van 't Hoff, physical modeling played a central role in the wider acceptance from the 1870s of a new steric conception of matter. Cardboard and 'ball-and-stick' models were neither generative of, nor derivative from, abstract chemical theory; rather, they were constitutive parts of an emergent research school, helping create a "symbolic and gestic space into which theoretical notions, bodily actions, cultural values, and even professional claims could be convincingly inscribed" (Meinel 2004, 270; van der Spek 2006).

In the twentieth century, wooden, metal, and then plastic model kits became a ubiquitous feature of the laboratory, with their most striking impact evident in the emergence of structural thinking in biochemistry and molecular biology. Faced with the challenge of interpreting complex crystallographic data, and spurred by a growing emphasis by many practitioners on molecules as 3-D atomic arrangements, molecular chemists and biologists developed new, more complex ways of physically modeling the structures they studied. At Caltech from the 1930s Linus Pauling and his colleagues' pioneering work on the structure of proteins proceeded hand-in-hand with the design and production of a new 'space-filling' modeling system. Based on bond angle and atomic radii structural data, what came to be dubbed CPK models (after Robert Corey, Pauling, and Walter Koltun) gave a topology of molecules that allowed testing of spatial relationships and steric hindrance in large, complex structures (Francoeur 1997, 26–31; Nye 2001). As Soraya de Chadarevian has noted, this architectural approach to chemistry "shaped the way crystallographers talked and 'thought' about molecules"; when Max Perutz produced his structural model of hemoglobin, he explained "[t]hat is what it really is" (de Chadarevian 2004, 348 and 362, n.15). This way of working is encapsulated in the canonical account of James Watson and Francis Crick's determination of the double helix structure of DNA. In Watson's own discovery story he makes a great deal of Pauling's influential model work: "We could thus see no reason why we should not solve DNA in the same way," not with pencil and paper, but rather with "a set of molecular models superficially resembling the toys of preschool children" (Watson 1968, 50). Watson's own account has come to exemplify the significance of 3-D models in research, as physical tools providing material space for scientists to solve complex problems that extend beyond the capacities of traditional analytic techniques on paper and, now, screen (see, for example, Baird 2004, 32–6).

Models as Representations

However, models are rarely just research tools. Models also function as representational media, utilized in a variety of settings to publicize, promote, familiarize, celebrate, and critique. Models thus act as objects capable of mediating between the laboratory space and wider scientific and cultural settings. Through this maker–audience dialectic the meanings and authority of models are continually renegotiated. Whilst canonical stories of model use in research almost always present such meanings and authority as self-evident, careful analysis of the representational role of models challenges these seemingly straightforward accounts.

The case of DNA is exemplary. Watson's account is recognized as a rational reconstruction, with the simple trajectory from Pauling's models to Watson and Crick's "play" with metal "backbones" and cardboard bases drawing attention to certain aspects of their work whilst simultaneously obscuring much else (Gross 1990, 54–65; de Chadarevian 2002, 164–98). As Robert Bud has argued, the metal model familiar through Anthony Barrington Brown's iconic photographs was constructed *after* the determination of DNA's structure, and was built to an extravagant scale to be photographed alongside the scientists, as a rhetorical device to represent their materialist solution to "the Secret of Life." Even at the moment of its creation "the model was seen as having a public as well as a strictly scientific value" (Bud 2013, 328).

As Maxwell knew, the representational power of models derives in part from their ability to show aspects of complex physical phenomena that otherwise could only be understood through mathematical formulations or diagrams. Models like that of Saturn's rings functioned as media for promoting the validity and comprehension of novel scientific claims by rendering them in a tangible form. Maxwell was a proponent of what he called "experiments of illustration," and an advocate of the translation of abstruse dynamics into terms comprehensible to "persons not trained in high mathematics" (Schaffer 2002, 135–7). His work therefore exemplifies important links between Victorian physics, instrument makers, and physical modeling as a technique of representation. A lifelong fan of physics lessons embodied in toys and games, Maxwell tackled the question of the motion of the Earth by turning to dynamic tops as models: "No illustration of astronomical precession can be devised more perfect than that presented by a properly balanced top" (Maxwell 1857, 559). This argument was substantiated by the production of a number of precision tops—as with the rings-of-Saturn model manufactured by Charles Ramage—for use in his own lectures, for gifting to "various parties who teach rigid dynamics," and for distributing to "various seats of learning" (as quoted in Harman 1990, 557 and 576).

Distribution of models to various audiences is therefore an integral part of their success as representational media. However, understanding audiences' responses to models can be challenging. Claims embodied in models are not easily conveyed, especially in comparison with the conventional communicative medium of science, print matter (Hopwood and de Chadarevian 2004). Direct sensory experience of models is obviously the ideal mechanism for their mediation, but this can be difficult to achieve. Models may be deployed as part of lecturing or laboratory teaching, sometimes as mass-produced objects for wide distribution, but rarely in quantities that rival print publication.

Many model-makers therefore rely on museums to secure a wide audience for their work, but even then the objects' physicality can impose limitations on the scale and kinds of audience reached. Given these constraints, a great deal of what others may learn from models, and what we may be able to recover about historical models, comes from their depiction on the printed page, as textual description, diagrammatic portrayal, or photographic portrait. More recently, these forms of mediation have been augmented by televisual and computational technologies (Francoeur and Segal 2004). But these forms of representation, too, come with their own range of challenges and problems. Photographs may only display one angle of a complex model, whilst illustrations and complex verbal descriptions convey an abstraction of an abstraction. The crystallographer Lawrence Bragg neatly summed up the problem of 'publishing' a complex molecular structure when he declared that "[t] he 'paper' sent to colleagues should be a model" (as quoted in de Chadarevian 2004, 349). As this is rarely possible, scientists must commonly employ the less satisfactory solution of describing a model and how it works. Maxwell, in his essay "On the Stability of the Motion of Saturn's Rings," committed three pages to a detailed description of the ring model, as well as a series of diagrams that dominate the essay's only plate. Maxwell recognized the superiority of witnessing the working model versus the protrayal in print: "By considering these figures [on the plate], and still more by watching the actual motion of the ivory balls in the model, we may form a distinct notion of the motions of the particles of a discontinuous ring" (Maxwell 1859, 59–62).

Furthermore, by extending the range of a model's representation beyond the laboratory or museum environment, print mediation of models opens them up to a range of interpretations that can be harder to control than in the case of the physical model itself. Images of Watson and Crick's double-helix model proliferated, and interpretations of what it represented and the meanings that audiences ascribed to it have been complex and protean ever since (de Chadarevian 2002, 236–59; de Chadarevian 2003a; de Chadarevian 2003b; de Chadarevian and Kamminga 2002). Whilst peer groups of professionals such as the International Union of Biochemists can establish formal rules for how complex structures are represented in print (de Chadarevian 2004, 352), distribution of models through various communications media often precludes the imposition of controls over representation and interpretation. As one example, Seikei Sekiya's intricate model of the motion of an idealized particle of earth during an 1887 earthquake received markedly different publicity in Britain than it did in Japan. Sekiya's model deliberately shifted the representation of seismological data away from its usual two-dimensional format (seismograms), using instead a complex array of bent wires to demonstrate "the complexities and irregularities of the earthquake motion" (Sekiya 1887, 360). Part of a campaign to raise the status of Japanese academic seismology, Sekiya promoted his model through an English-language article in the Western-oriented *Journal of the College of Science, Imperial University, Japan*. However, when the prestigious British periodical *Nature* reported Sekiya's article, it described his model as a "pretty and instructive Japanese 'curio'". *Nature* repurposed the model for its Western audience, effacing the accuracy of Sekiya's data, instead representing the model as a suitable acquisition for "private persons, not to speak of curators of museums and others officially interested in scientific novelties" (Anon. 1888; Sprevak 2000). *Nature*'s implication that Sekiya's model was not suitable for academic study, but was appropriate for museum display, points to tensions in the public role of scientific models, between entertainment and edification, aesthetic appeal and practical utility. When a model moves beyond the limits of the laboratory and enters the public sphere, there is wide scope for interpretation, a fact that offers practitioners challenges and opportunities.

Models and their Audiences

In the West, this has been the case since at least the Enlightenment, when models were amongst an array of demonstration devices representing the natural philosophical project to various audiences. As Simon Schaffer has noted, within the emergent entrepreneurial culture of the eighteenth century, carefully managed demonstrations of well-behaved scientific models before paying crowds, royal audiences, and learned societies cultivated the authority of men of science, bolstering their claims to "the right to govern and represent the macroscopic systems these models represented" (Schaffer 2004, 72).[3] Lightning rods are exemplary, for their correct design and use generated fierce debate throughout the 1770s. At stake were both the practical defense of national infrastructure, such as military powder magazines, and the reputations of rival natural philosophical claims regarding the nature and management of atmospheric electrical phenomena. Benjamin Franklin famously advocated the use of tall, pointed, grounded lightning rods, a design that his supporters promoted through the use of ingenious "thunder houses": model buildings that collapsed—or exploded if primed with gunpowder—when struck with an electrical charge, but stood unscathed if defended with a grounded rod (Warner 1997). But Franklin's design was contested by the English painter and electrician Benjamin Wilson on the grounds that it would dangerously draw electricity out of passing clouds. Wilson highlighted the disjuncture between small-scale electrical experiments and real-world meteorological phenomena, and then deployed his own models to advocate spherical balls as the safest design for defending the English Purfleet arsenal. With Royal backing, Wilson demonstrated a massive 1:36 scale model of the magazine at the trendy London Pantheon in 1777, replete with giant model clouds discharged daily before paying audiences. His opponents in the Royal Society responded by denigrating these performances as artful shows that lacked the appropriate neutrality of enlightened philosophy (Schaffer 2004, 76–87; Schaffer 1983, 15–31).

Such demonstrations highlight a problem central to natural philosophy and subsequent scientific enterprises. By appealing to wide audiences, model demonstrations made the natural philosopher's practice an easy target for accusations of being mere spectacle. Displays of otherwise invisible nature did not sit easily with the functions of public performance, and the politics of trade and industry fed tensions between natural philosophical claims and commerce (Schaffer 1983; Schaffer 1994; Schaffer 2004). This is clear in the case of astronomical models, which were amongst the most important commercial outputs of eighteenth- and nineteenth-century scientific instrument makers. As demonstrations of Sun-centered cosmology, devices like the 'Copernican' armillary sphere and orrery were utilized as teaching aids in diverse spaces, from the private drawing room to the public lecture hall. 'Entrepreneurs of science' such as Benjamin Martin sold orreries, lectured with them, and wrote about them as well. Yet the utility and value of such devices, beyond their rough demonstration of the helio-centric system, was contested (King and Millburn 1978; Taub 2006). Joseph Priestley contrasted the "philosophical instruments" that "exhibit the operations of nature" (the air pump, the pyrometer), with "the globes, the orrery, and others" used by "ingenious men … to explain their own conception of things to others" (as quoted in Schaffer 1994, 158). Nearly a century later, the Great Exhibition juror James Glaisher lamented that "the time, ingenuity, and expense" devoted to orreries, planetariums, and astronomical machines "are wasted; they are of no use to the student of astronomy, and the erroneous impressions which they give are always displeasing to the eye of the astronomer" (Glaisher 1852, 306–7). Yet these models' success with some audiences, their continued importance to the instrument-making profession, and their uses at different times and places serves as a reminder that the use, meaning, and value of a model and what it represents varies for different actors in different spaces across peri-ods of time. Objects themselves therefore serve as signposts to specific details of the context in which they were made and used (Taylor 2009).

Again, Maxwell's passion for modeling is telling. Traditional narratives of the rise of experimental physics have tended to portray it as part of a significant shift for the sciences away from demonstrational, museum- and theater-based practices towards private, laboratory- and journal-based research. Yet as we have seen, for Maxwell—the first Professor of Experimental Physics at Cambridge—models and their demon-stration formed a fundamental part of his scientific methodology (Hunt 1991; Haley 2002, 92–129; Brenni 2004). Knowing that he co-founded the Cambridge Mod-elling Club in 1873, a group that brought together likeminded workers "to promote the making of models, machines, and drawings illustrative of geometry," helps us bet-ter understand Maxwell's transformation of Gibbs' 'thermodynamic surface' from a mathematical formulation into a physical plaster model, which he cast and distributed to colleagues (Anon. 1874; Garber, Brush, and Everitt 1995, 49–50). When Maxwell learnt that William Thomson had constructed a similar model he wrote requesting a copy for the Cavendish Laboratory, as "[w]e have now got an excellent case with glass front … and we shall have a special place for models such as yours" (as quoted in Harman 2002, 231). Whilst this kind of model work has been analyzed in the context of theoretical/philosophical thought styles (e.g. "British" mechanism versus "French" dynamism), the significance of its *materiality*—and thus its power to mediate and rep-resent to diverse audiences—is often overlooked (cf., for example, Purrington 1997, 19–25; Harman 1998, 91–112).

Models Mass-produced for Teaching

If there is still much to be learnt about what models and modeling meant to a prac-titioner like Maxwell and his colleagues, one fruitful area of exploration is likely to be the role of models in teaching. Just as the foundation of the Cavendish Labora-tory signaled a shift towards experimental science at the University of Cambridge, so too did it signal a shift towards practical demonstration as a core component of science teaching. As Maxwell told his Cambridge audience in 1871, his new laboratory's program would augment the mathematicians' "familiar apparatus of pen, ink, and paper" with material practices designed to "exercise our senses in observation, and our hands in manipulation" (as quoted in Garber, Brush, and Everitt 1986, 241). Models were at the center of this physical engagement, not least because seeing and using them helped fix—and therefore teach—abstract ideas, illustrating them in a comprehensible way without loss of rigor (Hunt 1991, 75). Thus the Cavendish, in copying a style of teaching already well established in Scottish and European universities, stocked itself with models illustrating electrical phenomena, physical structures, and mathematical concepts. Whilst some of these were custom-made, others came from the stock of instrument manufacturers. These firms' trade catalogues are a valuable resource for the history of model use, attesting to the rising demands of scientific education for mass-produced pedagogical devices.

Physics was by no means the first subject to undergo this transition to commonplace model use in classroom and laboratory pedagogy. Mathematics itself, despite Maxwell's assumptions, used models for Euclidian education from at least the eighteenth century, though the kind of ornate boxed sets that the instrument maker George Adams produced were unlikely to have been manufactured in large numbers. Adams's Euclidian models were elaborate and exquisite, targeting an audience of Georgian gentlemanly society and polite drawing room pedagogy (Rich 2006). Likewise, early models mass-produced for astronomical education, such as Edward Mogg's cardboard dissected celestial globe kit (1813), were designed as preparation for polite discourse on astronomy. As scientific education became a requisite of middle- and upper-class childhood, assembly and reassembly of such models familiarized students with astronomical knowledge appropriate for genteel conversation. Designed for use alongside explanatory treatises, such models are a reminder that scientific pedagogy often must be understood as text-object-user interaction, with models forming one important component of a wider world of books, instruments, teachers, and students (Taylor 2009).

This world of pedagogical science expanded as the sites of scientific education multiplied. Nineteenth-century political reform across Europe and North America extended school education, and the foundation of mechanics' institutes, public libraries, technical colleges and new universities opened up possibilities for scientific study to new publics. Modern specialist disciplines such as geology developed alongside these pedagogical reforms, making the design and distribution of teaching models a constituent part of their work. The engineer, surveyor, and geologist Thomas Sopwith was both an agitator for the expansion of miners' education and a maker of geological models. In 1841 he began marketing sets of wooden models of geological structures "intended to afford a familiar explanation of various phenomena, a knowledge of which is essential to the study of Geology as connected with practical mining ... which cannot be so well explained by ordinary drawings" (as quoted in Turner and Dearman 1979, 333). Sopwith promoted the models through mechanics' institutes, the Geological Society of London, and an instructional book, securing the patronage of Charles Lyell and William Buckland. Lyell illustrated some of the models in his *Elements of Geology*, cautioning that "the beginner may find it by no means easy to understand such copies, although if he were to examine and handle the originals, turning them about in different ways, he would at once comprehend their meaning" (Lyell 1865, 57). That numerous sets of Sopwith's models survive in teaching collections and museums attests to their popularity, reminding us that 3-D models need to be incorporated into historical understandings of new "visual languages" for the sciences after 1800 (Rudwick 1976, Turner 2011).

In chemistry the development of such a visual language was intimately connected, as we have seen, with the emergence of 3-D structural thinking and its associated molecular modeling. But chemical models were also vital pedagogical devices. Stereochemical theories were disseminated and inculcated through inexpensive model kits, their guided use alongside textbooks and lab work teaching students the norms, values, and problems of the discipline. Their manufacture became big business, with numerous firms producing diverse arrays of model types and sets, targeting markets from the child's playroom to university teaching labs. In 1940 the American company Science Service began a monthly "Things of Science" postal subscription, mailing kits and models to customers accompanied by "museum-type labels" (Moody 2009). In the post-war era what one chemical educator called the subject's "new and exciting three-dimensional look" came to pervade popular culture, with crystallographic molecular structures chosen as the design motif for the 1951 Festival of Britain (Fieser 1963; Forgan 1998). Seven years later molecular design reached its apotheosis at the Brussels World's Fair, where visitors could climb inside the Atomium pavilion, a 102-metre tall model of a unit cell of an iron crystal, magnified 165 billion times. At these varied sites of pedagogy and display the public learnt through models, inculcating physical conceptions of matter whilst at the same time normalizing 3-D representations. These effects were especially evident in the medical and biological sciences.

Modeling Nature

Modeling as a core scientific practice has one of its longest and richest histories in the medical and biological sciences. The model-maker's ability to produce stable, robust, 3-D representations of the natural world, often on a scale vastly larger or smaller than the object of study, has been of vital utility to scholars since at least the Renaissance. Wax anatomical models could be used and reused in pedagogical settings where the procurement and storage of cadavers was problematic; in moulages they could fix 3-D representations of complex pathological changes in the human body. Plant and fruit models in wax, silk, glass, and plaster could serve as aids for the study and improvement of agricultural technologies, for the display of botanical classifications, and as representations of the exotic commercial produce of empire. Seemingly capturing ephemeral nature, models such as Harvard's famous Blaschka glass flowers, enabled the scientist and artist to demonstrate mastery of the natural world. Representing processes of development over time, such as in the embryo models of the Ziegler studio, models also opened up new visions of animal growth and evolution. And as a form of public spectacle, models like the dinosaurs at the Crystal Palace merged commerce and education to open up new views of past life on Earth. As Nick Hopwood has argued, research and pedagogy in the biological sciences was mediated by not only the printed word, but also through "plastic publishing," the manufacture and distribution of models (Hopwood 2002; Daston 2004; Secord 2004).

The skill and artistry of makers was crucial to the practices of these sciences. Such skills were hard-won, both through artistic labor and carefully managed relations with research scientists, and were often jealously guarded. Family dynasties like those of the Blaschkas and the Zieglers could nearly monopolize the market and, in the most striking example, that of the modeling empire of Louis Auzoux, a unique recipe for durable papier-mâché helped secure a considerable grip over the nineteenth- and early twentieth-century demand for models of nature. Lifelike, detailed, affordable, yet extremely robust, Auzoux's models exemplify the growth of a mass market for complex pedagogical models capable of replicating the experience of anatomical dissection or botanical study. Responding to developments in scientific research and education, Auzoux's firm branched out into an extraordinary array of subjects, from life-size mannequins composed of hundreds of demountable parts, to sets of replica teeth used to determine the age of horses, to models of plants, fruit, and fungi (Grob 2000; Olszewski 2009).

Such models encapsulate many of the categories considered here. Models of nature are made and used in every stage of biological and medical practice, from research to teaching (including patient education) to display. In the Whipple Museum of the History of Science, visitors see a remarkable example of this in the glass models of microfungi made by the Cambridge mycologist W. A. R. Dillon Weston (Figure 40.2). Modeled as seen through the microscope, and intricately manufactured at a scale of × 400 using pliers, Bunsen burner, and imported Czechoslovakian glass, Dillon Weston's models derive from his interwar research into fungal diseases of commercial crops. Made as demonstrational objects for use in lectures, and as quick reference guides for identifying microfungi that are hard to represent in pictures, the models had a diverse and peripatetic life in and beyond the academy. Displayed at the agricultural Royal Show, featured on BBC TV, illustrated in academic books and the national press, and exhibited before learned societies, his models ultimately became museum pieces representing applied mycology and its history (Tribe 1998; Horry 2008). Dillon Weston's work demonstrates the difficulty and subjectivity inherent in any attempt at separating the types and uses of scientific models into strict categories. Though such categories have been employed in this essay, we have done so only to focus attention on specific aspects of the diverse history of 3-D model use in the sciences. More work is needed before these strands can be comprehensively bound together in a way that would enable a clear understanding of the changing roles and uses of models over time. Rather, our accounts have focused on various makers and users of models, and their places of use, providing windows onto the practices and mediation of past science, and the tenuous boundaries between these topics.

Guide to Further Reading and Further Viewing

Whilst there is an extensive philosophical literature on models in the sciences, historical accounts remain limited in number. De Chadarevian and Hopwood (2004) is the best general introduction, and both authors have contributed other important works on models in medicine and biology (see bibliography). Klein (2001) is another useful edited volume. Model use in chemistry is a rich area of research, with the work of Francoeur and Meinel good starting points. Works cited by Schaffer encapsulate the public and private roles of models in the eighteenth century, whilst Hunt (1991), Haley (2002), and Brenni (2004) are good sources on models in classical physics. King and Millburn (1978) remains the reference text for mechanical astronomical models.

Beyond works cited here, the most vibrant sources available are the many collections of models in museums and other institutions. The University of Göttingen's Collection of Mathematical Models and Instruments is particularly rich, for example.[4] Details

Figure 40.2 A 25cm-tall glass model of *Bremia lactucae* (downy mildew), a microfungus that attacks lettuce crops. One of over 90 glass models made in the 1930s and 1940s by the Cambridge mycologist W. A. R. Dillon Weston. © The Whipple Museum (Wh.5826.24).

of the Whipple Museum's collection can be found on its 'Explore' website.[5] Many history of science collections in museums contain a wide variety of models, a salient reminder of their central importance to the practice and pedagogy of the sciences.

Endnotes

1 The authors are indebted to Boris Jardine and David Rowe for sharing their insights on models.

2 Two such models are held in the collection of the Whipple Museum in Cambridge, with others in the University of Edinburgh School of Chemistry, the National Museum of Scotland, and the Science Museum, London.

3 Numerous examples of Georgian demonstration models can be found in Morton and Wess 1993.
4 http://www.math.uni-goettingen.de/historisches/modelcollection.html.   Accessed September 30, 2015.
5 http://www.hps.cam.ac.uk/whipple/explore/models/. Accessed September 30, 2015.

References

Anon. 1874. "The Modelling Club." *University of Cambridge Reporter*, 55: 236.

Anon. 1888. "A Model of an Earthquake." *Nature*, 37, No. 952: 297.

Baird, Davis. 2004. *Thing Knowledge: A Philosophy of Scientific Instruments.* Berkeley: University of California Press.

Brenni, Paolo. 2004. "Mechanical and hydraulic models for illustrating electromagnetic phenomena." *Nuncius*, 19, No. 2: 629–58.

Brush, Stephen G., C. W. F. Everitt, and Elizabeth Garber (eds.) 1983. *Maxwell on Saturn's Rings.* Cambridge, MA: MIT Press.

Bud, Robert. 2013. "Life, DNA and the model." *British Journal for the History of Science*, 46, No. 2: 311–34.

Crum Brown, Alexander. 1914. "Mathematical models." In *Modern Instruments and Methods of Calculation: A Handbook of the Napier Tercentenary Exhibition*, edited by E. M. Horsburgh, 302–13. London: G. Bell and Sons.

Daston, Lorraine. 2004. "The glass flowers." In *Things That Talk: Object Lessons from Art and Science*, edited by Lorraine Daston, 223–54. New York: Zone Books.

de Chadarevian, Soraya. 2002. *Designs for Life: Molecular Biology after World War II.* Cambridge: Cambridge University Press.

de Chadarevian, Soraya. 2003a. "Portrait of a discovery: Watson, Crick, and the double helix." *Isis*, 94, No. 1: 90–105.

de Chadarevian, Soraya. 2003b. "Relics, replicas and commemorations." *Endeavour*, 27: 75–9.

de Chadarevian, Soraya. 2004. "Models and the making of molecular biology." In *Models: The Third Dimension of Science*, edited by Soraya de Chadarevian and Nick Hopwood, 339–68. Stanford, CA: Stanford University Press.

de Chadarevian, Soraya, and Harmke Kamminga. 2002. *Representations of the Double Helix.* Cambridge: Whipple Museum of the History of Science.

de Chadarevian, Soraya, and Nick Hopwood (eds.) 2004. *Models: The Third Dimension of Science.* Stanford, CA: Stanford University Press.

Dunning, David E. 2015. "What are models for? Alexander Crum Brown's knitted mathematical surfaces." *Mathematical Intelligencer*, 37, No. 2: 62–70.

Fieser, Louis F. 1963. "Plastic dreiding models." *Journal of Chemical Education*, 40, No. 9: 457.

Forgan, Sophie. 1998. "Festivals of science and the two cultures: Science, design and display in the Festival of Britain, 1951." *British Journal for the History of Science*, 31: 217–40.

Francoeur, Eric. 1997. "The forgotten tool: The design and use of molecular models." *Social Studies of Science*, 27, No. 1: 7–40.

Francoeur, Eric, and Jerome Segal. 2004. "From model kits to interactive computer graphics." In *Models: The Third Dimension of Science*, edited by Soraya de Chadarevian and Nick Hopwood, 401–29. Stanford, CA: Stanford University Press.

Garber, Elizabeth, Stephen G. Brush, and C. W. F. Everitt (eds.) 1986. *Maxwell on Molecules and Gases.* Cambridge, MA: MIT Press.

Garber, Elizabeth, Stephen G. Brush, and C. W. F. Everitt (eds.) 1995. *Maxwell on Heat and Statistical Mechanics.* Cranbury: Associated University Presses.

Glaisher, James. 1852. "Orreries, planetariums, and astronomical machines." In *Exhibition of the Works of Industry of All Nations, 1851: Reports by the Juries*, 306–7. London: William Clowes.

Griesemer, James. 2004. "Three-dimensional models in philosophical perspective." In *Models: The Third Dimension of Science*, edited by Soraya de Chadarevian and Nick Hopwood, 433–42. Stanford, CA: Stanford University Press.

Grob, Bart. 2000. *The World of Auzoux: Models of Man and Beast in Papier-Maché*. Leiden: Museum Boerhaave.

Gross, Alan G. 1990. *The Rhetoric of Science*. Cambridge, MA: Harvard University Press.

Haley, Christopher. 2002. "Envisioning the unseen universe: Models of the ether in the nineteenth century." Unpublished PhD dissertation, University of Cambridge.

Harman, Peter M. (ed.) 1990. *The Scientific Letters and Papers of James Clerk Maxwell, Vol. 1, 1846–1862*. Cambridge: Cambridge University Press.

Harman, Peter M. 1998. *The Natural Philosophy of James Clerk Maxwell*. Cambridge: Cambridge University Press.

Harman, Peter M. (ed.) 2002. *The Scientific Letters and Papers of James Clerk Maxwell, Vol. 3, 1874–1879*. Cambridge: Cambridge University Press.

Hopwood, Nick. 2002. *Embryos in Wax: Models from the Ziegler Studio*. Cambridge: Whipple Museum of the History of Science.

Hopwood, Nick, and Soraya de Chadarevian. 2004. "Dimensions of modelling." In *Models: The Third Dimension of Science*, edited by Soraya de Chadarevian and Nick Hopwood, 1–15. Stanford, CA: Stanford University Press.

Horry, Ruth. 2008. "Glass models of fungi." *Explore Whipple Collections*, Whipple Museum of the History of Science, University of Cambridge. http://www.hps.cam.ac.uk/whipple/explore/models/glassfungi/ Accessed September 30, 2015.

Hunt, Bruce J. 1991. *The Maxwellians*. Ithaca, NY: Cornell University Press.

Kidwell, Peggy. 1996. "American mathematics viewed objectively: The case of geometric models." In *Vita Mathematica: Historical Research and Integration with Teaching*, edited by Ronald Calinger, 197–207. Washington, DC: Mathematical Association of America.

King, Henry, with John Millburn. 1978. *Geared to the Stars: The Evolution of Planetariums, Orreries, and Astronomical Clocks*. Toronto: University of Toronto Press.

Klein, Ursula (ed.) 2001. *Tools and Modes of Representation in the Laboratory Sciences*. Dordrecht: Kluwer.

Lyell, Charles. 1865. *Elements of Geology*. London: John Murray.

Maxwell, James Clerk. 1857. "On a dynamical top, for exhibiting the phenomena of the motion of a system of invariable form about a fixed point, with some suggestions as to the Earth's motion." *Transactions of the Royal Society of Edinburgh*, 21, No. 4: 559–70.

Maxwell, James Clerk. 1859. *On the Stability of the Motion of Saturn's Rings*. Cambridge: Macmillan.

Mehrtens, Herbert. 2004. "Mathematical models." In *Models: The Third Dimension of Science*, edited by Soraya de Chadarevian and Nick Hopwood, 276–306. Stanford, CA: Stanford University Press.

Meinel, Christoph. 2004. "Molecules and croquet balls." In *Models: The Third Dimension of Science*, edited by Soraya de Chadarevian and Nick Hopwood, 242–275. Stanford, CA: Stanford University Press.

Moody, George B. 2009. "Rediscovering Things of Science." Online: http://ecg.mit.edu/george/tos/ Accessed September 30, 2015.

Morton, Alan, and Jane Wess. 1993. *Public and Private Science: The King George III Collection*. Oxford: Oxford University Press.

Nye, Mary Jo. 2001. "Paper tools and molecular architecture in the chemistry of Linus Pauling." In *Tools and Modes of Representation in the Laboratory Sciences*, edited by Ursula Klein, 117–32. Dordrecht: Kluwer.

Olszewski, Margaret. 2009. "Designer nature: The papier-mâché botanical teaching models of Dr. Auzoux in nineteenth-century France, Great Britain and America." Unpublished PhD dissertation, University of Cambridge.

Purrington, Robert D. 1997. *Physics in the Nineteenth Century.* New Brunswick, NJ: Rutgers University Press.

Ratcliff, Jessica. 2008. *The Transit of Venus Enterprise in Victorian Britain.* London: Pickering and Chatto.

Rich, Michael. 2006. "Representing Euclid in the eighteenth century." In *The Whipple Museum of the History of Science: Instruments and Interpretations, to Celebrate the 60th Anniversary of R. S. Whipple's Gift to the University of Cambridge*, edited by Liba Taub and Frances Willmoth, 319–44. Cambridge: Whipple Museum of the History of Science.

Rudwick, Martin. 1976. "The emergence of a visual language for geological science, 1760–1840." *History of Science*, 14: 149–95.

Schaffer, Simon. 1983. "Natural philosophy and public spectacle in the eighteenth century." *History of Science*, 21: 1–43.

Schaffer, Simon. 1994. "Machine philosophy: Demonstration devices in Georgian mechanics." *Osiris*, 9: 157–82.

Schaffer, Simon. 2002. "James Clerk Maxwell." In *Cambridge Scientific Minds*, edited by Peter Harman and Simon Mitton, 123–40. Cambridge: Cambridge University Press.

Schaffer, Simon. 2004. "Fish and ships: Models in the Age of Reason." In *Models: The Third Dimension of Science*, edited by Soraya de Chadarevian and Nick Hopwood, 71–105. Stanford, CA: Stanford University Press.

Secord, James A. 2004. "Monsters at the Crystal Palace." In *Models: The Third Dimension of Science*, edited by Soraya de Chadarevian and Nick Hopwood, 138–69. Stanford, CA: Stanford University Press.

Sekiya, Seikei. 1887. "A model showing the motion of an Earth-particle during an earthquake." *Journal of the College of Science, Imperial University, Japan*, 1: 359–62.

Sprevak, Mark. 2000. "Representing time and motion: Sekiya, the Gilbreths, and chronophotographic art." Unpublished MPhil Essay, University of Cambridge Department of History and Philosophy of Science.

Taub, Liba. 2006. "Are orreries Newtonian?" In *The Whipple Museum of the History of Science: Instruments and Interpretations, to Celebrate the 60th Anniversary of R. S. Whipple's Gift to the University of Cambridge*, edited by Liba Taub and Frances Willmoth, 403–25. Cambridge: Whipple Museum of the History of Science.

Taylor, Katie. 2009. "Mogg's Celestial Sphere (1813): The construction of polite astronomy." *Studies in History and Philosophy of Science*, 40, No. 4: 360–71.

Tribe, Henry. 1998. "The Dillon Weston glass models of microfungi." *Mycologist*, 12, No. 4: 169–73.

Turner, Susan. 2011. "Thomas Sopwith, miners' friend: His contributions to the geological model-making tradition." In *History of Research in Mineral Resources*, edited by J.E. Ortiz, O. Puche, I. Rábano, and L.F. Mazadiego, 177–92. Madrid: Instituto Geológico y Minero de España.

Turner, Susan, and W.R. Dearman. 1979. "Sopwith's geological models." *Bulletin of the International Association of Engineering Geology*, 19: 331–45.

Van der Spek, T. M. 2006. "Selling a theory: The role of molecular models in J. H. Van't Hoff"s stereochemistry theory." *Annals of Science*, 63, No. 2: 157–77.

Warner, Deborah J. 1997. "Lightning-rods and thunder houses." *Rittenhouse*, 11: 124–7.

Watson, James D. 1968. *The Double Helix: A Personal Account of the Discovery of the Structure of DNA.* London: Weidenfeld and Nicolson.

Index

A Companion to the History of Science, First Edition. Edited by Bernard Lightman.
© 2016 John Wiley & Sons Ltd. Published 2020 by John Wiley & Sons Ltd.